T0224554

Energy Beam-Solid Interactions
and Transient Thermal Processing/1984

MATERIALS RESEARCH SOCIETY SYMPOSIA PROCEEDINGS

ISSN 0272 - 9172

MATERIALS RESEARCH SOCIETY SYMPOSIA PROCEEDINGS

Energy Beam-Solid Interactions and Transient Thermal Processing/1984

Symposium held November 26-30, 1984, Boston, Massachusetts, U. S. A.

EDITORS:

D. K. Biegelsen
Xerox Palo Alto Research Center, Palo Alto, California, U. S. A.

G. A. Rozgonyi
Microelectronics Center of North Carolina, Research Triangle Park, North Carolina, U. S. A.

C. V. Shank
AT&T Bell Laboratories, Holmdel, New Jersey, U. S. A.

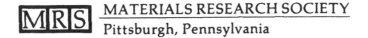 **MATERIALS RESEARCH SOCIETY**
Pittsburgh, Pennsylvania

CAMBRIDGE UNIVERSITY PRESS
Cambridge, New York, Melbourne, Madrid, Cape Town,
Singapore, São Paulo, Delhi, Mexico City

Cambridge University Press
32 Avenue of the Americas, New York NY 10013-2473, USA

Published in the United States of America by Cambridge University Press, New York

www.cambridge.org
Information on this title: www.cambridge.org/9781107405776

Materials Research Society
506 Keystone Drive, Warrendale, PA 15086
http://www.mrs.org

First published 1985
First paperback edition 2012

Single article reprints from this publication are available through
University Microfilms Inc., 300 North Zeeb Road, Ann Arbor, MI 48106

CODEN: MRSPDH

ISBN 978-0-931-83700-5 Hardback
ISBN 978-1-107-40577-6 Paperback

Contents

PART VII: SILICON CRYSTAL GROWTH FROM
THE MELT OVER INSULATORS

PART VIII: SILICON SOLID-PHASE CRYSTAL
GROWTH ON INSULATORS

Preface

This volume contains papers presented at the symposium on "Energy Beam Solid Interactions and Transient Thermal Processing." held in Massachusetts, November 26-30, 1984. The symposium was the seventh in a series of Materials Research Society symposia devoted to those topical areas evolving from the subject of "laser annealing." The conference opened with a plenary session highlighting representative areas of the symposium. A special concluding talk was presented by Walter Brown (AT&T Bell Labs), summarizing the meeting and pointing to areas requiring further study. Symposium sessions were coordinated with those in other symposia (e.g. impurity diffusion) to enhance the coupling in areas of common interest.

The material covered in this symposium has three distinct foci. First is the field of sub-microsecond (to femtosecond) interactions of energy beams with solids and the resultant material changes on similar time scales. Issues of energy transfer, internal equilibration, solid phase regrowth of amorphized layers, ultrarapid quenching of amorphous semiconductors dominate this topic. A second focus is the maturing area of rapid thermal processing (10^{-3} to 10^{2} seconds). The materials science issues of defect and impurity kinetics, and silicide and alloy development underlie a strong technological thrust. The third focus is the area of semiconductor growth over insulators by lateral epitaxy. Studies concerning low angle grain boundaries, the characteristic microscopic defects in zone melting recrystallization of thin films, and increasing emphasis on solid phase lateral epitaxy are showing strong progress towards technologically suitable preparation techniques.

Symposium Co-Chairmen

D.K. Biegelsen G.A. Rozgonyi C.V. Shank

January 1985

Acknowledgments

We wish to thank all the contributors and participants who made the symposium so successful. We particularly would like to acknowledge the invited speakers, who provided excellent summaries of specific areas and set the tone of the meeting. They are:

W.L. Brown	G. Mourou
S.U. Campisano	J. Narayan
J. -P. Colinge	G.L. Olson
A. Compaan	P.S. Peercy
A.G. Cullis	T.E. Seidel
R.B. Fair	M. Tabe
J.C.C. Fan	C.V. Thompson
H. Kurz	B. -Y. Tsaur
E. -H. Lee	J.S. Williams
N.M. Johnson	R.F. Wood

We are deeply indepted to the session chairs, who directed the sessions, guided the discussions, and gave invaluable help in having the papers refereed. They are:

B.R. Appleton	D.H. Lowndes
H. Baumgart	L. Pfeiffer
N. Bloembergen	R.F. Pinizzotto
W.L. Brown	J.M. Poate
G.K. Celler	D.K. Sadana
R.T. Hodgson	T.O. Sedgwick
B.C. Larson	H.I. Smith
R.A. Lemons	M.O. Thompson

It is our great pleasure to acknowledge, with gratitude, the financial support provided by the Army Research Office, Electronics Division (Dr. J. Hurt and Dr. R. W. Griffith); the Office of Naval Research (Dr. L. R. Cooper); the Air Force Office of Scientific Research (Dr. K. J. Malloy); and the Xerox Corporation.

Finally, we are grateful to F. Soto (Xerox) and A. Walker (AT&T) for excellent secretarial support.

PART I

Introductory Overviews

PICOSECOND PHOTON-SOLID INTERACTION

H. KURZ AND N. BLOEMBERGEN
Division of Applied Sciences, Harvard University, Cambridge, MA 02138

ABSTRACT

   Some of the fundamental aspects of pulsed laser interactions with semi-conductors are reviewed within the frame of recent experimental work with picosecond laser pulses. Main emphasis is placed on data obtained in space-time resolved measurements of transmission and reflectivity of picosecond pulse irradiated silicon at different probing wavelengths. Four distinct interaction regimes can be defined, depending on the fluence of the laser pulse. Special attention is devoted to the phase transition from solid to liquid state at the surface. The optical heating model is found to be valid even on a time scale of picoseconds. Firm data of the electron-hole plasma density and lattice temperature are derived. The thermal nature of the phase transition is confirmed.

1. INTRODUCTION

   Numerous investigations of the pulsed laser-solid interactions have been carried out in the past with high power ns-laser pulses [1-7]. They provide ample evidence for the validity of the optical heating model, which assumes an instantaneous energy transfer from the electronic carrier system to the lattice phonons [8-13]. Time-resolved optical reflectivity and transmission measurements have traditionally been unique tools to study the dynamics at nonequilibrium carriers in semiconductors [14]. Since the optical properties at laser irradiated surfaces are noticeably changed by the generation of free carriers, the time evolution of reflectivity and transmission provides quan-titative information about the energy relaxation and carrier recombination. While recent optical investigations have been restricted to the investigation of nonequilibrium carrier dynamics, the optical analysis presented in this paper is extended to the time-resolved study of lattice temperature. Under certain spectroscopic conditions the electron-hole plasma contributions to the optical dielectric function are minimized. The optically detected trans-mission and reflectivity changes are mainly caused by thermally induced modi-fication of the indirect energy gap and the occupation statistics of the phonons involved. Thus, the evolution of lattice temperature can be monitored with a signal-to-noise ratio superior to that of recent Raman scattering experiments, where several thousand laser pulses have to be superposed [15].
   Basically, the picosecond laser interaction with semiconductors can be divided into four regimes, depending on the absorbed fluence of the incident pulses. In regime I we deal with nonequilibrium carrier distributions at a constant lattice temperature T_L. The maximum density of carriers out of equilibrium does not exceed 10^{19} cm^{-3}. The lattice temperature is not elevated significantly by the energy loss to the phonons. In the case of semiconductors, nondegenerate carrier distributions are involved.
   In regime II the nonequilibrium carrier density is increased to the highest levels possible in the solid phase of the media. Significant lattice heating occurs, governed by the energy relaxation of hot carriers. Most of the semiconductor systems are driven into degeneracy. The temperature dependence of material parameters starts to play a decisive role for the modeling of the interaction processes.
   The transition from regime II to III is marked by the appearance of structural changes. Most prominent examples are the transition from solid to liquid phase as soon as the absorbed laser fluence exceeds a critical

value. It is one of the key points of this review to discuss the nature of this phase transition, based on time-resolved optical investigation on silicon.

In semiconductors the transition into the liquid state is manifested by drastic changes of the optical properties. As soon as the surface starts to melt, metallic properties dominate. The excess pulse energy heats up further the highly absorbing liquid metallic layer far above the melting point. The boiling point is easily passed. At higher fluences even the critical point of the phase diagram is reached. The transition to the liquid state corresponds to an abrupt increase of the free carrier concentration from 10^{21} to $\sim 10^{23} cm^{-3}$. The band gaps collapse and sudden changes of the band structure are encountered. Regime III is characterized by ultrafast melting and thermal evaporation processes. Nonequilibrium thermodynamics phenomena such as superheating are expected, due to limitations of interface velocities.

Regime IV is reached if the incident fluences are increased roughly by one order of magnitude above the fluence level required to melt the surface. The incident radiation is strongly absorbed by evaporated material. Nonthermal processes, such as laser-induced plasma formation and emission of highly energetic ions, become dominant.

The main experimental work at Harvard has been concerned with the physical processes encountered in regimes II and III. The aim of this paper is to summarize the data obtained in time-resolved optical measurements with picosecond pulses and to clarify whether the optical heating [10] and thermal melting model is still valid on a time scale of picoseconds.

2. BASIC DESCRIPTION

In this brief section we shall summarize the dominant physical mechanism of pulsed laser interaction. For reasons of brevity we shall concentrate on crystalline silicon. If the photon energy E_1 exceeds the fundamental energy gap E_g, electron-hole pairs are formed with an initial excess energy of E_1-E_g. The space-time evolution of electron-hole pair density $N(x,t)$ is given by the one-dimensional continuity equations:

$$\frac{\delta j(x,t)}{\delta x} + \frac{\delta N(x,t)}{\delta t} = G(x,t) - \gamma N^3(x,t) \tag{1}$$

with the generation rate $G(x,t)$, particle current $j(x,t)$ and Auger recombination rate γN^3 defined in ref. [16]. The optically coupled states cover a narrow energy range and a small volume of momentum space. Rapid intercarrier collisions ensure thermal equilibrium between electrons and between holes ($\tau_{e-e}, \tau_{h-h} \leq 10^{-14}$ s), but not necessarily between electrons and holes ($T_e \neq T_h$). The thermalized subsystems are described by generalized Fermi-Dirac distributions with separate Fermi levels E_e^F and E_h^F [10].

For parabolic bands the local carrier temperatures (T_e, T_h) and quasi-Fermi levels are related to the carrier density N through [17]:

$$N = 2\left\{\frac{2\pi m^e kT_e}{h^2}\right\}^{3/2} F_{1/2}(\eta_e) = 2\left\{\frac{2\pi m^h kT_h}{h^2}\right\}^{3/2} F_{1/2}(\eta_h) \tag{2}$$

where $m^{e,h}$ are the density of state effective masses, F_n the Fermi integral of order n and $\eta_m = E_m^F/kT_m$ the reduced quasi-Fermi energy for carrier of type m. The formation of excess carriers causes a splitting of the quasi-Fermi levels ($E_e^F \neq E_h^F$). While in regime I the carrier distribution is nondegenerate, the carrier density in regime II reaches levels at which the electron-hole plasma is degenerate ($\eta \gg 0$) even at high temperatures. At these high densities the independent particle picture of band structure becomes questionable. Exchange and correlation effects narrow the band gap, modify the near band gap transition by band filling effects and possibly influence the electron-phonon interaction by screening [18].

In degenerate carrier distributions the electron-hole scattering is less efficient due to the lack of available final states of lower energy. Thus the coulomb thermalization between electrons and holes is slowed down. Thermal equilibrium between electrons and holes will be established in times significantly larger than τ_{e-e} and τ_{h-h}, but still in times short compared to the recombination time $1/\gamma N^2$. Without going into details, one may safely assume that on a time scale of picoseconds the $T_e = T_h = T_c$ equilibrium is established, but the quasi Fermi levels remain separated.

The build-up of carriers is opposed by recombination and diffusion processes. Band-to-band Auger recombination dominates the temporal behavior of high-density carrier distributions in regime II, while the much slower carrier dynamics in regime I are determined by diffusion, radiative and surface recombination. For carriers accumulated in the lowest conduction band valleys, this recombination is mediated by phonons. For silicon weakly temperature dependent Auger recombination coefficients are reported [19,20]. The recombination and diffusion reduces the splitting of the quasi-Fermi levels and increases the carrier temperature. Diffusion by density gradients is determined by the electron-hole pair current $j(x,t) = -D_a(\delta N/\delta x)$, where the ambipolar diffusion coefficient D_a is given by the generalized Einstein relation [17].

The initial temperature of carriers T^i is determined by the number of electron-hole pairs created and the excess energy shared among them:

$$T^i = (E_1 - E_g)/kH, \quad H = F_{3/2}(n_e)/F_{1/2}(n_e) + F_{3/2}(n_h)/F_{1/2}(n_h) \qquad (3)$$

Due to the Fermi exclusion principle the initial carrier temperature is lowered by the degeneracy factor H.

The actual carrier temperature is given by the balance of energy fed into the electron-hole plasma and the energy loss by phonon emission. Interaction with energetic optic and large wave vector "intervalley" phonons rapidly transfers energy to the lattice. The energy loss rate by deformation potential scattering in silicon is calculated as 1 eV/ps [21].

The optical heating model assumes thermal equilibrium between electron-hole plasma and phonons and between phonons themselves ($T_e = T_L = T$). Thus energy relaxation and phonon thermalization has to be completed in times short compared to the duration of the excitation pulse. Then the lattice heating can be described by a simple energy balance equation [10]:

$$\rho C_p \, dT/dt = \frac{\delta(K_L \delta T/\delta x)}{\delta x} + S \qquad (4)$$

where ρC_p is the specific heat of the lattice and K_L the thermal conductivity. The heating source term S is given by:

$$S = G(E_1 - E_g - kT H) + \gamma N^3 (E_g + kT H) \qquad (5)$$

Each electron-hole pair transfers immediately the difference between initial (E_1-E_g) and average energy kT H to the phonons. If a large amount of excess carrier is generated, the split quasi-Fermi levels approach the initial excess energy. As a consequence, the second term in Eq. (5) grows at the expense of the first one. Direct heating is reduced, and delayed heating by Augur recombination becomes dominant. There the recombination energy $E_g + kT H$ is transferred to a third carrier, which again shares its energy immediately with the other carriers and phonons present. The second heating source scales as γN^3, leading to a highly nonlinear character of Eq. (7).

The solution of the coupled Eqs. (1-5) is shown in Fig. 1, where the calculated temporal evolution of lattice temperature $T_s(0,t)$ and plasma density $N_s(0,t)$ at the surface are presented. For an incident fluence at 532 nm, $F(2\omega) = 100$ mJ/cm^2 and a pulse duration of $t_p = 20$ ps, the high density regime is closely controlled by γN^3. The plasma density rises rapidly and reaches,

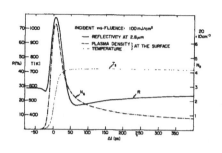

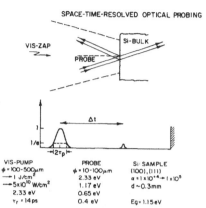

SPACE-TIME-RESOLVED OPTICAL PROBING

Fig.1. Calculation of surface temperature T_s, electron-hole pair density
at the surface N_s and reflectivity at
2.8 μm versus time. Fluence at 532 nm:
100 mJ/cm².

VIS-PUMP	PROBE	Si-SAMPLE
φ = 100-500μm	φ = 10-100μm	(100),(111)
→ 1 J/cm²	2.33 eV	α = 1 x 10⁺⁴ → 1 x 10⁵
→ 5x10¹⁰ W/cm²	1.17 eV	d ~ 0.3mm
2.33 eV	0.65 eV	
τᵣ = 14 ps	0.4 eV	Eg = 1.15 eV

Fig.2. Details of set-up for space-
time resolved optical probing.

at ten picoseconds after the excitation peak, a maximum of 7×10^{20} cm^{-3}. The
temporal decay profile of the density is mainly governed by Auger recombination, while diffusion plays a minor role. The temperature rise is delayed
with respect to plasma evolution. A final value of 700 K is reached at
Δt = 30 ps because of delayed Auger heating.

3. TIME-RESOLVED OPTICAL MEASUREMENTS

    Picosecond optical pulses derived from mode-locked Nd:YAG laser systems
offer a unique possibility to study the dynamics of high density nonequilibrium carrier distributions. Free carriers modify the optical properties of
the irradiated semiconductors mainly through intraband and interband transitions [14-22]. Since the optical properties depend on the detailed nature of
the carrier distribution, the time evolution of reflectivity and transmission
provides information about dynamics and energy loss of a dense electron-hole
plasma. A variable time delay Δt between excitation pulse and interrogating
pulse enables time-resolved reflectivity and transmission measurements of thin
silicon films on sapphire (SOS) or bulk samples. As listed in Fig. 2, the
photon energy of the highly focussed probe beam is varied between 2.33 eV and
0.4 eV. Further details of the space-time resolved optical probing set-up
are noted in Fig. 2.
    The real part of the dielectric function is given by:

$$\varepsilon' = n^2 - k^2 = n_L^2 - k_L^2 - AN + \delta\varepsilon' \qquad (6a)$$

where n_L is the refractive index and k_L is the absorption coefficient of the
material with no carriers present (N = 0). Their dependence on lattice temperature is reported in ref. [23].
    The Drude coefficient

$$A \approx 4\pi e^2 \left[\frac{1}{m_e} \left\langle \frac{\tau_e^2}{1 + \omega^2 \tau_e^2} \right\rangle + \frac{1}{m_h} \left\langle \frac{\tau_h^2}{1 + \omega^2 \tau_h^2} \right\rangle \right] \qquad (7)$$

describes the optical polarizability of an electron-hole plasma, characterized
by the optical effective masses m_e and m_h and energy averaged expressions containing the carrier-momentum relaxation times $\tau_{e,h}$. The dielectric function

is further affected by interconduction or intervalence band transitions described by $\delta\varepsilon'$. As long as spatial density gradients are below the critical value for inhomogeneous broadening effects [24], reflectivity changes can be directly related to ΔN. Thus carrier recombination kinetics and diffusion phenomena are favorably probed at a frequency small compared to the band gap frequency.

Under highly degenerate conditions, where the Fermi levels are driven deeply into the conduction (valence) band, nonparabolicities of the conduction and valence band lead to a density dependence of the optical effective masses. For the case of spherical constant energy surfaces, the masses are expressed by [22]:

$$\frac{1}{m_{e,h}} = \frac{1}{\hbar^2 k_{e,h}^F} \left| \frac{\delta E}{\delta k} \right|_{E_{e,h}^F} \tag{8}$$

where $k_{e,h}^F$ is the Fermi momentum and $|\delta E/\delta k|_{E_{e,h}^F}$ the curvature of conduction and valence bands at the position of the quasi-Fermi level $E_{e,h}^F$. Thus, nonparabolicities of the band structure imply additional modifications of A, if the quasi-Fermi levels are changed according to Eq. (2).

The incremental change $\Delta\varepsilon' = -\Delta N$, associated with intraband transitions, is increased considerably if the probing frequency approaches a direct transition marked by the frequency ω_g [14]:

$$\Delta\varepsilon' = -\Delta N \, \omega_g^2/(\omega_g^2 - \omega^2) \tag{9}$$

when $\omega_g \gg \omega$ the classical Drude formula is valid with $\Delta\varepsilon' < 0$. For $\omega_g < \omega$ the sign of $\Delta\varepsilon'$ reverses, resulting in an increase of reflectivity. The correction of the Drude term becomes important for probing wavelengths shorter than 1 μm.

The imaginary part of the dielectric function:

$$\varepsilon'' = 2nk = 2n_L k_L + BN + \delta\varepsilon'' \tag{6b}$$

is composed by contributions associated with band gap transitions ($2n_L k_L$), intraband (BN) and weak intervalence and interconduction band transitions ($\delta\varepsilon''$).

The free carrier coefficient B is given by:

$$B = \frac{4\pi e^2}{\omega} \left[\frac{1}{m_e} \left\langle \frac{\tau_e}{1 + \omega^2 \tau_e^2} \right\rangle + \frac{1}{m_h} \left\langle \frac{\tau_h}{1 + \omega^2 \tau_h^2} \right\rangle \right] \tag{10}$$

4. EXPERIMENTAL RESULTS

At optical frequencies far below the indirect band gap of silicon, the reflectivity and transmission changes can be directly related to density changes via the Drude coefficients A and B. Interconduction and intervalence transitions can be neglected in a first approximation, because of the low oscillator strength. The solid line in Fig. 1 illustrates the reflectivity of 2.8 μm radiation expected from the N_s and T_s signatures. The reflectivity is calculated assuming a constant optical reduced mass $1/m^* = 1/m_e + 1/m_h = 0.164\, m_0$.

The actual reflectivity transients are convoluted with the temporal gaussian profile of the probing laser pulse $I(t) = I_0 \exp(-t/t_p)^2$. Shortly before the intensity reaches a maximum, the reflectivity passes through a first minimum. At this time the density $N_s(0,t)$ is raised to $N_{min} = (\varepsilon_\infty - 1) m^* (c/e\lambda)^2$, where ε_∞ is the high frequency dielectric constant and λ the wavelength of the probing pulse. The time needed to pass through the minimum is shorter than the pulse duration t_p. Thus the measured minimum appears less pronounced than

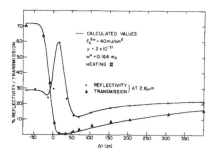

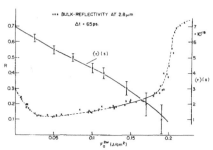

Fig.3. Comparison between calculated and measured data at 2.8 μm.

Fig.4. Bulk reflectivity minimum at 2.8 μm versus incident fluence at 532 nm. Solid line: momentum relaxation times evaluated from the reflectivity minimum.

the actual (unconvoluted) one. At density slightly higher than N_{min} the $\varepsilon' = 0$ condition is met. The transverse optical field is in resonance with the collective (longitudinal) plasma oscillations.

The conditions for plasma resonances last only for 30-40 picoseconds. As a consequence, the reflectivity rises rapidly to a maximum of 80% reached at $\Delta t = 10$ ps. As soon as the density drops again to N_{min}, a second reflectivity minimum appears, at $\Delta t = 65$ ps. The recovery to the original value of the reflectivity corresponds to the relatively slow drop of the plasma density, as expected from the Auger recombination rate γN^3.

In Fig. 3 the measured reflectivity and transmission data for a low fluence level of 40 mJ/cm^2 are compared with calculated values. Again a constant optical reduced mass $m^* = 0.164\ m_0$ is used. The momentum relaxation times τ_e and τ_h are assumed to be equal and inversely proportional to the lattice temperature ($\tau_e = \tau_h = \tau_0\ (T_L/300)^{-1}$, $\tau_0 = 7 \times 10^{-15}$s).

The agreement between measured and calculated data is excellent. The application of the optical heating model in the picosecond range seems to be justified. Using $\gamma = 4 \times 10^{-31}$cm^6/s, much faster recovery of reflectivity and transmission than is measured are calculated. The value of $\gamma = 2 \times 10^{-31}$cm^6/s is in agreement with data already reported in early optical transmission experiments with ns pulses [19].

By probing with 2.8 μm pulses, physical information on the maximum carrier density can only be obtained for fluence level $F = 40$ mJ/cm^2, far below the threshold value $F_{th} = 200$ mJ/cm^2 required to transform the surface into the liquid state. Above the critical densities for plasma resonance at 2.8 μm, the reflectivity does not provide any further information, and the transmission drops to undetectable small values. To analyze the carrier dynamics at fluences close to F_{th}, shorter probe wavelengths have to be used [25-28].

The second reflectivity minimum, however, offers a unique possibility for studying some specific details of the optical properties of a medium-dense ($\sim 2 \times 10^{20}$ cm^{-3}) electron-hole plasma in silicon. An approximate ($D_a = 0$) solution of Eq. (1) for times Δt longer than the pulse duration is:

$$N_s = N_0 (1 + 2\gamma \Delta t\ N_0^2)^{-1/2} \qquad (11)$$

where N_0 is the maximum density generated during the pulse. For $\Delta t = 65$ ps, where the reflectivity exhibits the second minimum, the density becomes independent of the fluence $F(2\omega)$: ($N_s \sim (2\gamma\Delta t)^{-1}$). The nonlinear recombination reduces the density to $N_s \sim 2 \times 10^{20}$ cm^{-3} for $N_0 > 3 \times 10^{20}$ cm^{-3}. The lattice temperature, however, increases strongly with $F(2\omega)$.

The time at which the minimum appears is quite sensitive to changes in the optical reduced mass or in the recombination coefficient γ. Due to the finite width of the probe laser pulse, the measured reflectivity minimum appears later than defined by N_{min}. By numerical modeling of the reflectivity minimum, taking into account the convolution with the probe pulse, a value of $m^* = 0.16\ m_0$ is deduced for $\gamma = 2 \times 10^{31}\ cm^6/s$.

The position of the second minimum does not change with increasing pump fluence. The critical parameter $(\gamma m^{*2})^{-1}$ appears to be insensitive towards variations of the lattice temperature. Both γ and the optical reduced mass are supposed to increase with temperature, moving the reflectivity minimum to shorter times. The stability of m^* is clear evidence for degenerate carrier distributions and for strict parabolic curvatures of the bands close to the band edges through the whole range of incident fluence. The carrier temperature remains close to the lattice temperature.

The value of R_{min} of the second reflectivity minimum shown in Fig. 3 ($R_{min} = 10\%$) indicates strong damping of the plasma resonance by carrier scattering. In Fig. 4 the values of reflectivity minima at $\Delta t = 65$ ps are plotted as a function of pump fluence $F(2\omega)$. For $F = 30$ mJ/cm^3, the density remains constant, but the lattice temperature is increased. The continuous increase of R_{min} with increasing fluence indicates a strong decrease of carrier momentum relaxation times τ_e and τ_h, as already noticed in Fig. 3. Contrary to the optical effective mass, the momentum relaxation times appear temperature dependent. To estimate these times, the measured R_{min} values have been fitted by numerical modeling of the reflectivity around $\Delta t = 65$ ps. In this fitting procedure the average over the carrier energy distribution, indicated by the $\langle\ \rangle$ brackets in Eq. (7) and Eq. (10), has been omitted and $\langle\tau\rangle = \tau_e = \tau_h$ has been assumed. The momentum relaxation times, required to fit the measured R_{min} values, drop from 7×10^{-15} to 1×10^{-15} s, as F increases up to the melting threshold $F_{th} = 200$ mJ/cm^2.

Similar results have been obtained in time-resolved transmission experiments, where the free carrier absorption has been studied at 1064 nm. The free carrier absorption cross section σ has been found to increase linearly with the lattice temperature ($\sigma = \sigma_0(T_L/300)$, $\sigma_0 = 5 \times 10^{-18}$ cm^2), indicating again extremely short relaxation times [29].

For optical phonon and intervalley scattering at low carrier densities, the momentum relaxation times are measured on the order of $\tau = 2.2 \times 10^{-13}$ sec, exhibiting a strong temperature dependence ($(T/300)^{-x}$, $2.4 \lesssim x \lesssim 2.6$). For large excess carrier densities, the phonon scattering mechanism is accompanied by electron-hole collisions. They are assumed to strongly damp plasma resonances and to significantly reduce the momentum relaxation times. Despite the fact that the role of electron-hole collisions requires further investigations, the results in Fig. 4 can be taken principally as evidence for extremely short collision times. They invalidate the k-selection rules for phonon emission. Consequently, the energy relaxation to the phonons and the decay of phonons is strongly enhanced in the presence of a highly dense electron-hole plasma.

At a probing wavelength of 532 nm ($E_1 = 2.33$ eV), the optical properties are mainly changed by lattice heating. The linear optical absorption associated with the indirect band gap transition exceeds by far the free carrier absorption $\alpha(FCA) = \sigma N$, even at the highest carrier densities. The bulk reflectivity is barely affected by the largely reduced (ω^{-2}) Drude term.

In silicon on sapphire samples (SOS), however, multiple reflections from air-silicon and silicon-sapphire interfaces enhance the optical detectivity of $\Delta\varepsilon'$ changes considerably and allow the quantitative analysis of plasma contributions and thermal effects. In addition, thermally induced absorption changes can be monitored by time-resolved transmission experiments.

The reflectivity and transmission are oscillatory functions of $4\pi nd/\lambda$ (see Fig. 5). The modulation depth is determined by the absorbance of the thin silicon film ($\alpha = 0.5$ μm). For a certain wavelength the reflectivity of the unexposed sample can be tuned to a minimum (A) by varying the angle of incidence. As soon as the plasma is formed, the negative Drude term raises

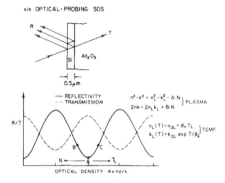

vis OPTICAL-PROBING SOS

Fig.5. Sketch of thin film response to changes in the dielectric functions.

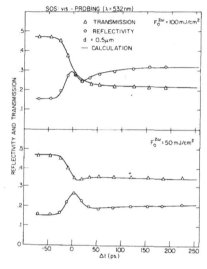

SOS: vis - PROBING (λ = 532 nm)

Fig.6. Time-resolved reflectivity and transmission of SOS samples, measured at 532 nm.

the reflectivity towards point B. The subsequent lattice heating opposes the plasma effect by the thermally induced enhancement of n_L. When the two contributions cancel each other, the reflectivity drops again to a minimum. As time proceeds ($\Delta t > 50$ ps), the electron-hole plasma vanishes, while the lattice temperature remains fairly at the same level. Thus the reflectivity moves toward point C.

The time-resolved reflectivity and transmission signatures of picosecond irradiated SOS samples exhibit exactly this behavior. In Fig. 6 the data obtained with 532 nm probing pulses are compared with calculated values. Because of the short probe wavelengths the correction in Eq. (9) is used. The thermo-optic data concerning the temperature dependence of n_L and k_L are taken into consideration [23]. Again, a constant optical reduced mass ($m = 0.16$ m_0) is used.

At low fluences ($F = 50$ mJ/cm^2) the rise of reflectivity towards point B is mainly terminated by reaching the recombination limited density, while the thermal compensation is less important. The interplay between plasma formation and lattice heating is changed at $F = 100$ mJ/cm^2. The reflectivity maximum is reached nearly at the same time. The following minimum, however, occurs earlier, indicating a more dominant role of thermal effects.

The transmission drops suddenly due to thermal band gap shrinkage, masking the copious absorption by free carriers at 532 nm. The salient features of the data in Fig. 6 are entirely consistent with the optical heating model, outlined in Eqs. (1-5). A nearly instantaneous energy transfer from the plasma to the phonons leads to lattice heating within the duration of picosecond pump pulses.

The conclusive answer to the question of whether the phase transition at $F_{th} = 200$ mJ/cm^2 is caused by thermal or plasma effects requires the direct measurement of lattice temperature, or at least the energy redistributed among the different vibrational modes. The results in Fig. 6 show that at $\Delta t = 200$ ps the changes of reflectivity and transmission are solely caused by thermal effects. The Harvard group has used the reflectivity and transmission changes observed at $\Delta t = 200$ ps to determine both the spatially averaged temperature and the surface temperature as a function of incident laser fluences [26,28].

Contrary to earlier Raman results [15], lattice temperatures close to the melting point are derived at the threshold fluence F_{th}. Consistent with

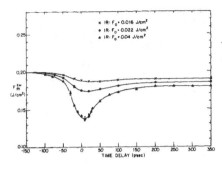

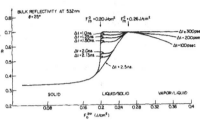

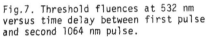

Fig.7. Threshold fluences at 532 nm
versus time delay between first pulse
and second 1064 nm pulse.

Fig.8. Fluence dependence of the bulk
reflectivity at 532 nm monitored at
different time delays.

the thermal melting model, the phase transition from solid to liquid state is
caused by an excess of vibrational energy, destabilizing the lattice structure.
Using a novel three-pulse technique, L.A. Lompré et al. could clarify the
role of impact ionization in regime II [27]. In this experiment the first
laser pulse at 532 nm creates a certain amount of carriers and heats the
lattice to a certain level. A second pump pulse at 1064 nm couples energy
into the existing electron-hole plasma via free carrier absorption, without
creating new carriers. A third pulse at 532 nm probes the optical behavior
of the plasma and the lattice under the combined influence of the two pump
beams. By varying the time delay Δt_{12} between the two heating pulses, the
plasma kinetics can be observed via the amount of additional lattice heating,
monitored at a fixed time delay $\Delta t = 400$ ps. Beside this thermometric determination of plasma densities, the double pulse heating technique can be
applied to study the phase transition itself. By monitoring the fluence
necessary to induce this phase transition, $F_{th}(2\omega)$, as a function of fluence
$F(\omega)$ of the second pulse and time delay Δt_{12}, the density of electron-hole
pairs formed can be ruled out as a critical parameter. As shown in Fig. 7,
the threshold $F_{th}(2\omega)$ is lowered by the 1064 nm pulse, depending on its
fluence. Maximum reduction occurs at $\Delta t_{12} = 10$ ps, where the plasma density,
generated by the first pulse, reaches a maximum.

From experiments below the critical fluence it is known that the second
pulse does not change the N/m* ratio but increases further the lattice temperature [27]. There is no observable increase of the plasma density by additional electron-hole pairs creation, neither by indirect band transition nor
by impact ionization. The phase transition threshold $F_{th}(2\omega)$ can be lowered
obviously, even if the plasma density is not changed. In conclusion, the
only essential process encountered in the double pulse irradiation experiments is an additional increase of vibronic energy. The excess energy of the
carriers deposited by the second pulse is again rapidly transferred to the
phonons.

The maximum density of electron-hole pairs in regime II, close to the
phase transition, is limited to less than 1.4×10^{21} cm^{-3}, as calculations and
transmission experiments at 1064 nm show [29]. Below 0.08 J/cm^2 the bulk
reflectivity of silicon is unaffected by the formation of electron-hole pairs
and the lattice heating. As demonstrated in Fig. 8, the negative Drude contribution, as well as the thermal enhancement of n_L, are below the detection
limit for 532 nm radiation. Above 0.16 J/cm^2 the reflectivity rises considerably by thermal effects, until a sudden jump occurs to high values at
0.2 J/cm^2. The high value level of 70% corresponds exactly to the optical
refraction index of liquid silicon at the melting point ($T_m = 1680$ K). In
regime III, at higher fluence levels, the reflectivity drops again. The slope

to lower values depends on the time delay between pump and probe pulse. After 300 ps the reflectivity remains constant at the T_m = 1680 K level of liquid silicon. The reflectivity drop above 0.28 J/cm^2 is consistent with an ultra-fast heating of the liquid layer far above the melting point. After the pulse rapid cooling occurs within several hundred picoseconds. The large cooling rate is plausibly explained by thermal diffusion processes, driven by large temperature gradients developed in the liquid film. The evaporation of material prevents a quantitative determination of the surface temperature by optical techniques.

An alternative drop of reflectivity is observed between 0.2 J/cm^2 and 0.26 J/cm^2 after 1 nsec. Within this narrow fluence range the surface is transformed from crystalline into amorphous states. The drop in the surface reflectivity in Fig. 8 indicates the resolidification of the liquid film at the surface into an amorphous layer. The data in Fig. 8 illustrate when the resolidification process is completed. A time-resolved analysis of these reflectivity drops yields an estimate of the lower limit of the resolidification speed v_m^-. Just above the threshold fluence values of $v_m^- \geq 25$ m/s are found. Closer inspections by model calculations, where the reflectivity of a liquid silicon film bonded to a hot solid substrate is calculated as a function of film thickness, reveal that the resolidification speed slows down when the solid-liquid interface approaches the surface.

5. CONCLUSIONS

Space-time-resolved optical techniques, applied in experiments with high intensity picosecond pulses, provide ample evidence for the dominance of thermal effects. The maximum plasma density in regime II is limited by Auger recombination to 1.4×10^{21} cm^{-3}. This value is one order of magnitude below the densities envisaged in nonthermal models. The phase transition from the solid to the liquid state is entirely explained by vibrational destabilization of the lattice. The validity of the optical heating and thermal melting model is based on an ultrafast energy transfer from the electron plasma to the phonons. Rapid intercarrier collisions are observed in optical reflectivity and transmission data. The ultrashort momentum relaxation times support the picture of an intense electron-phonon coupling mechanism at high carrier densities. The stability of the optical reduced masses allows a precise determination of densities in degenerate plasmas by optical techniques.

ACKNOWLEDGMENTS

We should like to thank L. A. Lompre, J. M. Liu and A. M. Malvezzi for their valuable contributions to this paper. This work was supported in part by the U.S. Office of Naval Research under contract no. N00014-83K-0030, and in part by the Alexander von Humboldt Foundation, Bad Godesberg, F.R. Germany.

REFERENCES

1. *Laser Solid Interactions and Laser Processing*, edited by S.D. Ferris, H.J. Lemay and J.M. Poate (American Institute of Physics, New York, 1979).
2. *Laser and Electron Beam Processing of Materials*, edited by C.W. White and P.S. Peercy (Academic Press, New York, 1980).
3. *Laser and Electron Beam Solid Interactions and Materials Processing*, edited by J.F. Gibbons, L.D. Hess and T.W. Sigmon (Elsevier North-Holland, New York, 1981), Mat. Res. Soc. Symp. Proc. 1 (1981).

4. *Laser and Electron Beam Interactions with Solids*, edited by B.R. Appleton and G.K. Celler (Elsevier North-Holland, New York, 1982), Mat. Res. Soc. Symp. Proc. 4, 49 (1982).
5. *Laser-Solid Interactions and Transient Thermal Processing of Materials*, edited by J. Narayan, W.L. Brown and R.A. Lemons (Elsevier North-Holland, New York, 1983), Mat. Res. Soc. Symp. Proc. 13 (1983).
6. *Laser Annealing of Semiconductors*, edited by J.M Poate and J.W. Mayer (Academic Press, New York, 1982).
7. *Energy Beam-Solid Interactions and Transient Thermal Processing*, edited by J.C.C. Fan and N.M. Johnson (Elsevier North-Holland, New York, 1984) Mat. Res. Soc. Symp. Proc. 23 (1984).
8. C.M. Surko, A.L. Simons, D.H. Auston, J.A. Golovshenko and R.E. Slusher, Appl. Phys. Lett. 34, 655 (1978).
9. P. Baeri, U. Campisano, G. Foti and E. Rimini, J. Appl. Phys. 50, 788 (1979).
10. J.R. Meyer, F.J. Bartoli, M.R. Kruer, Phys. Rev. B 21, 1559 (1980).
11. R.F. Wood and G.E. Giles, Phys. Rev. B 23, 2925 (1981).
12. D.H. Lowndes, Phys. Rev. Lett. 48, 267 (1982).
13. M.I. Gallant and H.M. van Driel, Phys. Rev. B 26, 2133 (1982).
14. D.H. Auston, S. McAfee, C.V. Shank, E.P. Ippen and O. Jeschke, Solid State Electronics 21, 147 (1978).
15. D. von der Linde, G. Wartmann and A. Ozols, ref. 6, p. 7.
16. H. Kurz, L.A. Lompré and J.M. Liu, J. de Physique 44, C5-23 (1983).
17. R. Smith, *Semiconductors* (Cambridge University Press, Cambridge, U.K., 1968), pp. 234-257.
18. E. Yoffa, Phys. Rev. B 21, 2415 (1980).
19. Woerdmann, Thesis, University Amsterdam, 1971.
20. J. Dzievior and W. Schmid, Appl. Phys. Lett. 31, 346 (1977).
21. W.P. Dumke, Phys. Lett. 78 A, 477 (1980).
22. W.G. Spitzer and H.K. Fan, Phys. Rev. (5), 406, 882 (1957).
23. G.E. Jellison, F.A. Modine, Appl. Phys. Lett. 41, 180 (1982).
24. I.F. Vakmneno and V.L. Stritzmevskii, Sov. Phys. Semicond. 3, 1562 (1970).
25. J.M. Liu, H. Kurz and N. Bloembergen, Appl. Phys. Lett. 41, 643 (1982).
26. L.A. Lompré, J.M. Liu, H. Kurz, N. Bloembergen, Appl. Phys. Lett. 43, 168 (1983).
27. L.A. Lompré, J.M. Liu, H. Kurz, N. Bloembergen, Appl. Phys. Lett. 44, 3, (1984).
28. L.A. Lompré, J.M. Liu and H. Kurz, Mat. Res. Soc. Symp. Proc. 23, 57 (1984).
29. L.A. Lompré, J.M. Liu, H. Kurz and N. Bloembergen, in *Ultrafast Phenomena IV*, edited by D.M. Auston and K.B. Eisenthal (Springer, New York, 1984)

FUNDAMENTAL ASPECTS OF HIGH SPEED CRYSTAL GROWTH
FROM THE MELT

A.G. CULLIS
Royal Signals and Radar Establishment, St. Andrews Road, Malvern,
Worcs. WR14 3PS, England

ABSTRACT

   Advances in the study of high speed crystal growth from the melt are
reviewed, with special emphasis on the fast melting and solidification of
silicon achieved by use of Q-switched laser radiation pulses. Rapid melting
of amorphous Si is confirmed to yield a liquid undercooled by several hundred
Kelvins and, under suitable conditions, explosive crystal growth processes
can occur. The latter involve the self-sustaining propagation of melt bands
buried within the initially amorphous material. When the highest quench-
rate conditions are established melting of even crystalline Si can yield
a final amorphous solid phase. This breakdown in crystal growth is orientation dependent and can give regimes of crystal defect formation when amorphization does not take place. The processes which characterize this limiting growth behaviour are discussed.

INTRODUCTION

   During the past few years, the use of transient annealing techniques
has led to a substantial improvement in our understanding of high speed
melt-growth phenomena. Transient melting over a wide range of timescales
has allowed the direct study of solidification at rates of up to tens of
metres per second. This paper will review some of the important advances
which have taken place, with particular reference to the limiting growth
behaviour of elemental Si.

   The fastest solidification regime can be attained by initial melting
of Si with Q-switched and mode-locked laser pulses. If a Si sample has an
amorphous surface layer, rapid melting may induce recrystallization events
which have recently been shown [1] to be explosive in character. The formation of an undercooled liquid is a key feature of this process and the
amount by which the melting temperature of amorphous Si (T_{al}) is depressed
below that of the crystal (T_{cl}) is an important parameter. After direct
melting of crystalline Si with sufficiently short laser pulses the transient
liquid can also become extremely undercooled. Indeed, the subsequent recrystallization can be driven so rapidly that atomic ordering processes are
overwhelmed and a disordered, amorphous final solid phase is produced. The
factors influencing this growth breakdown phenomenon will be outlined and
correlated with the observed [2] substrate orientation dependence.

EXPLOSIVE RECRYSTALLIZATION OF AMORPHOUS SILICON

   The phenomenon of self-propagating, explosive recrystallization of an
amorphous material was first observed [3] in the last century. Just a few
years ago, at a previous meeting of this society, its occurrence during the
scanning continous-wave laser annealing of amorphous Ge films was reported
[4,5]. This explosive film growth often results in the formation of characteristic disc- or arc-shaped crystallization patterns. The mechanism by which
the process takes place was soon identified [6,7] and relies upon the formation of a narrow, travelling band of undercooled liquid. The motion of

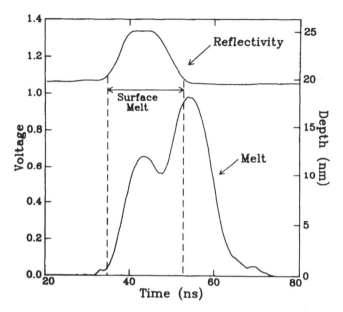

Fig. 1. Conductivity and surface reflectivity transients for a 0.2J/cm² laser pulse incident on a 320nm amorphous Si film. The left axis shows the observed voltage during the conductivity transient and the right axis gives the corresponding molten layer thickness [1].

this liquid band is self-sustaining within certain limits, since heat liberated locally by crystallization is greater than that needed to melt a corresponding amount of amorphous semiconductor. Still more recently, related explosive recrystallization events have also been observed [8,9] during the scanning laser annealing of amorphous Si films. The crystal growth behaviour can be quite complex but the underlying mechanism is the same as that described above.

When a Q-switched laser pulse of tens of nanoseconds duration is incident upon an amorphous Si layer formed, for example, by ion implantation in a single crystal substrate it is well known that, if the transient melt penetrates to the underlying crystal, high quality epitaxial recrystallization can take place. In contrast, with low radiation energy densities, the amorphous layer is only partially consumed and final solidification leads to the formation of misoriented polycrystallites [10,11]. Particularly in this subthreshold regime, by analogy with the scanning laser annealing results it might be expected that explosive growth processes should occur under appropriate conditions. Nevertheless, evidence supporting this conjecture has proved difficult to obtain. However, during the past year, detailed work in several laboratories has shown that explosive solidification does occur in nanosecond regime experiments, although quite elaborate measurements are required for verification.

In order to establish the occurrence of unusual crystal growth processes, direct determination of the location of molten regions within a sample as a function of time is especially important. Figure 1 shows transient reflectivity and conductivity data [1] obtained from a 320nm amorphous Si layer during irradiation with a 28ns ruby laser pulse (694nm) at 0.2J/cm².

Fig. 2. Cross-sectional transmission electron images of 215nm
initial amorphous layers in Si on
sapphire after laser irradiation
at: a) 0.2J/cm^2 and b) 0.4J/cm^2.
Note the polycrystalline Si formed
by the laser melting. (Inclined
microtwins in underlying crystalline Si were formed during the
original layer growth.) [1]

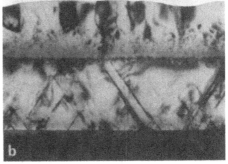

Both measurements indicate that surface melting begins after 36ns. However,
the surface reflectivity returns to the solid value after just 54ns, while
the transient conductivity yields a large signal (due to the presence of
liquid) for an additional 10ns. The conductivity curve exhibits a double
peak shape and, since the reflectivity is only sensitive to the first 20nm
near the sample surface, the liquid indicated by the second conductivity
peak must exist within the interior of the sample. This buried band of melt
propagates through the initial amorphous layer in an explosive manner [1],
similar to that described earlier for other annealing conditions.

The explosive crystal growth occurring at low pulse energy densities
results in the formation of fine Si polycrystallites throughout the melted
portion of the layer. This is illustrated in Fig. 2a for a 0.2J/cm^2 laser
pulse incident on a 215nm amorphous Si film (the film is in Si on sapphire
used for transient conductivity measurement work). The explosive melting
and recrystallization leads to the production of fine Si polycrystals with
a mean size of ~5nm and extending to a depth of ~150nm. This type of annealed film morphology was known previously [11] although the formation mechanism
was not understood. At somewhat higher laser pulse energy densities two
strata of polycrystallites (large and fine) are produced [12]. This is shown
in Fig. 2b for a pulse energy density of ~0.4J/cm^2. The upper coarse-grained
polycrystalline Si is formed at somewhat higher temperatures and extends
to a depth corresponding to slightly less than that of the primary melt [1].
The underlying fine-grained material results from explosive growth in more
greatly undercooled liquid Si. The final polycrystalline Si is considered
to nucleate heterogeneously at appropriate boundaries of the liquid layers.
An alternative growth model based upon conjectured bulk liquid nucleation
[13] appears to overestimate bulk nucleation rates and is not in accord with
the results of recent impurity segregation studies [14]. The latter work
strongly favours the explosive growth model presented earlier.

18

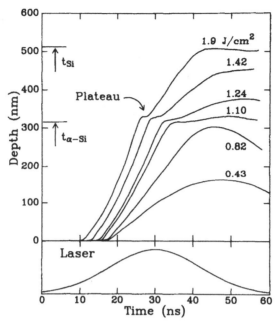

Fig. 3. Melt depth transients at various incident laser pulse energy densities. The 1.9J/cm² pulse was adequate to melt the Si layer fully to the sapphire interface (510nm). The lower trace indicates the timing of the laser pulse [1].

MELTING TEMPERATURE OF AMORPHOUS SILICON

The direct formation of an undercooled liquid by the fast melting of amorphous Si is an important first step in the explosive crystal growth sequence described in the last section. Accordingly, it is important to have a good estimate of T_{al} and previous determinations [15-17] have placed this parameter in the approximate range 1185-1435K (compare T_{cl} at 1685K). However, it is also possible to deduce the value of T_{al} from transient conductivity (melt depth) measurements carried out for an amorphous Si layer in contact with crystal [1]. In Fig. 3, the melt depth (conductivity) measurements of Fig. 1 are extended to greater radiation energy densities. In particular, where melting occurs beyond the amorphous/crystal interface, a transient plateau is observed at this depth with a duration between 2 and 6ns. A plateau of this type represents a delay in melt penetration since $T_{al} < T_{cl}$ and the plateau width can be used to calculate T_{al}. The result obtained is [1] $T_{al} \approx 1480 \pm 50K$.

Another method of determining T_{al} relies upon computer modelling the melt penetration through an amorphous Si layer as a function of Q-switched laser pulse energy density [2]. However, in order to render the computations tractable, it is important to reduce complexities in the heat flow by experimentally suppressing explosive recrystallization events. The latter can be achieved by reducing the undercooled melt dwell time below that required for random polycrystal nucleation to occur. The use of both short laser pulse lengths and relatively thin amorphous Si layers can give suitable solidification conditions. This is illustrated in Fig. 4 for 2.5ns ruby

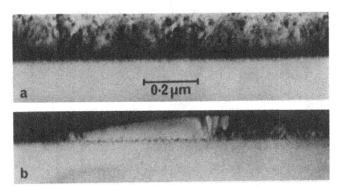

Fig. 4. Cross-sectional transmission electron images of amorphous layers in (001) Si after irradiation with 2.5ns ruby laser pulses at 0.3J/cm² : a) initial 180nm amorphous layer and b) initial 120nm amorphous layer.

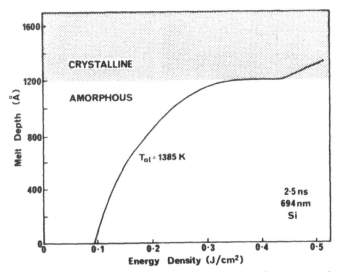

Fig. 5. Computed melt depth as a function of pulse energy density for 2.5ns, 694nm laser pulses incident on a 120nm amorphous Si layer (T_{al} = 1385K) with an underlying crystalline Si matrix [2].

laser pulses incident at 0.3J/cm² on amorphous Si layers produced in (001) Si by As⁺ ion implantation. In Fig. 4a the relatively thick 180nm amorphous layer has experienced polycrystal formation, probably by explosive growth processes, since the low thermal conductivity of the amorphous material has given an extended melt dwell time. In contrast, Fig. 4b shows that, for the thinner 120nm amorphous layer, single crystal growth off the matrix interface has occurred in many areas but, where it is incomplete, amorphous Si has reformed without any polycrystal nucleation. The melting processes in this latter system can be modelled by computation and, by use of appropriate materials parameters including a reduced thermal conductivity for amorphous Si [16], Fig. 5 shows the predicted melt depth/energy density characteristic

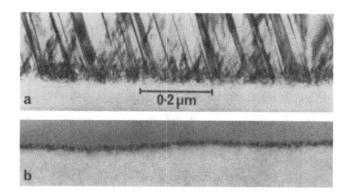

Fig. 6. Cross-sectional transmission electron images of (111) Si layers irradiated with 2.5ns, 347nm laser pulses: a) 0.9J/cm^2 and b) 0.5J/cm^2 [2].

for T_{al} = 1385K. A plateau in the curve at a depth corresponding to the amorphous/crystal interface is again produced since $T_{al} < T_{cl}$ and the plateau length is dependent upon the difference between these two parameters. The curve and plateau shown in Fig. 5 are in good agreement with experimental observations [2] so that the chosen value of T_{al} is given support. Nevertheless, the expected error range overlaps that of the previously described determination so that we may conclude that a best estimate for the depression of T_{al} below T_{cl} is ∿200-300K.

ORIENTATION DEPENDENCE OF AMORPHIZATION PHENOMENA

When crystalline Si is melted with laser pulses of especially short duration (≳5ns) and short wavelength (preferably ultra-violet) the thermal gradients induced in the material can be so steep that solidification is fast enough to defeat the atomic ordering processes at the recrystallization interface. Under the most extreme conditions, with large liquid undercooling, the surface layer reforms as a disordered, amorphous phase [18-20]. For the Si crystal (001) surface orientation, the solidification transition between ordered and totally disordered growth is quite sharp and, based on work with 2.5ns, ultra-violet laser pulses, the transition occurs at a measured [21] interface velocity of ∿15m/s. In contrast, for the Si (111) orientation, crystal growth breakdown first occurs at a very much lower velocity of ∿5-6m/s to give highly defective material containing inclined stacking faults and microtwin lamellae [20]. A typical defective layer produced in this manner is shown in Fig. 6a, with growth twins formed on the inclined (111) plane at 19.5° to the surface normal. However, there is still a higher limiting crystal growth velocity beyond which an amorphous solid is formed, as illustrated in Fig. 6b.

Recent work [2] on high speed growth breakdown phenomena exhibited by Si has extended the observations to the (011) and (112) growth surfaces. With increasing growth velocity, both of these surfaces exhibit limited but well-defined growth regimes of defect formation just prior to amorphization. A summary of results is presented in Fig. 7 where interface velocities are estimated from radiation energy densities with the help of both experiment and computation. From the curve, it is clear that the amorphization threshold velocities lie in the order (011)>(001)>(112)>(111). However, due to the finite extent of the defect formation regime for (011) Si, the maximum velocities at which defect-free single crystal can grow lie in the different

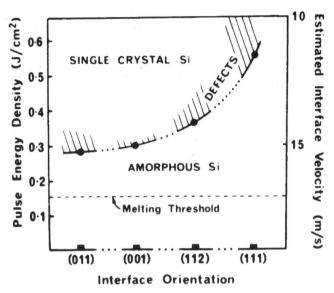

Fig. 7. Variation in amorphization threshold (measured energy density
(2.5ns, 347nm) and estimated solidification interface velocity) with
change in low index interface orientation. Defect formation regimes
are shown hatched [2].

order (001)>(011)>(112)>(111). This result should be compared with the
predictions of Monte Carlo simulations [22] of high speed Si crystal growth
based on the kinetic Ising model of the crystal-melt interface. For a wide
range of undercoolings the computations indicate that the growth velocities
of the various surface orientations lie in the order (001)>(011)>(111), in
excellent agreement with experiment.

It is important to gain a detailed understanding of the way in which
crystal growth breakdown occurs under high quench rate conditions. The
atomic-scale processes are conveniently studied by direct high resolution
imaging of breakdown interfaces in the transmission electron microscope
[2]. Examples are shown of (001) and (111) interfaces in Fig. 8, where each
cross-sectional image presents the complete transition from single crystal
growth (bottom) to amorphous layer formation (top). In the crystalline
regions a crossed (111) lattice fringe array is visible, with an internal
spacing of 0.314nm. In Fig. 8a, as the breakdown interface is approached
from the crystal matrix there is a rapid increase in defect density, with
profuse fault and twin generation on (111) planes both vertically aligned
and inclined to the electron beam direction. This structure is generated
at a stage in the solidification sequence when the temperature of the already
highly undercooled melt is falling still further and the gradually decreasing
melt diffusivity promotes the occurrence of additional errors in atomic
stacking. Finally, rapid error multiplication leads to the nucleation and
growth of an amorphous overlayer.

It is also important to note that, in Fig. 8a, islands of crystal which
protrude furthest into the amorphous region generally have a twinned component. Indeed, the presence of closely spaced twin boundaries (for example
at X and Y) leads to a corrugation of the growth interface with the provision of many growth nucleation sites (compare corresponding twinning in

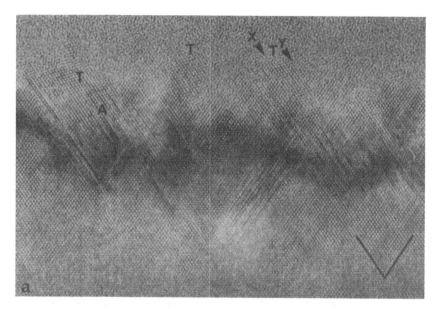

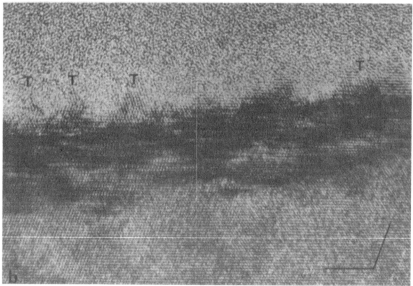

Fig. 8. High resolution, cross-sectional transmission electron images of Si layers irradiated with 2.5ns, 347nm laser pulses: a) (001) Si: 0.25J/cm^2; b) (111) Si: 0.4J/cm^2. Twinned regions at interfaces are marked T. In (a) twin/matrix overlap fringes are visible at A and edge-on twin boundaries are arrowed X and Y [2].

dendrite growth models [23,24]). Thus, in principle, inclined twin formation can yield a growth speed advantage. However, this is negligible for the (001) surface since it exhibits a relatively high intrinsic growth velocity for any particular melt undercooling, each atomic growth site being doubly bonded. Accordingly, the defect formation regime for (001) Si in Fig. 7 is also negligible. In contrast, crystal growth on the (111) Si surface is kinetically blocked by the continual need for plane nucleation. Under conditions of forced growth breakdown, Fig. 8b shows that atomically-spaced faulting and twinning at first occurs along the growth plane, although this does not enhance the crystal growth rate since the barrier to (111) plane nucleation still exists and amorphization can follow when the melt undercooling is sufficiently large. However, under slightly less extreme conditions, there is opportunity for twins to form on inclined {111} planes and, in this case, their presence does increase substantially the crystal growth rate (by provision of nucleation sites as outlined above). This accounts for the broad defect formation regime shown in Fig. 7 for (111) Si and the inclined twins which are involved are clearly evident in Fig. 6a. Thus, the orientation dependence of high speed Si crystal growth involves an interplay between a number of different atomistic processes. The reader is referred to an extended discussion elsewhere [2] for more detailed consideration of the various points.

REFERENCES

1. M.O. Thompson, G.J. Galvin, J.W. Mayer, P.S. Peercy, J.M. Poate, D.C. Jacobson, A.G. Cullis and N.G. Chew, Phys. Rev. Lett. 52, 2360 (1984).
2. A.G. Cullis, N.G. Chew, H.C. Webber and D.J. Smith, J. Crystal Growth 68, 624 (1984).
3. G. Gore, Phil. Mag. 9, 73 (1855).
4. R.B. Gold, J.F. Gibbons, T.J. Magee, J. Peng, R. Ormond, V.R. Deline and C.A. Evans Jr., in Laser and Electron Beam Processing of Materials, edited by C.W. White and P.S. Peercy (Academic Press, New York, 1980) pp. 221-226.
5. H.J. Zeiger, J.C.C. Fan, B.J. Palm, R.P. Gale and R.L. Chapman, in Laser and Electron Beam Processing of Materials, edited by C.W. White and P.S. Peercy (Academic Press, New York, 1980) pp. 234-240.
6. G.H. Gilmer and H.J. Leamy, in Laser and Electron Beam Processing of Materials, edited by C.W. White and P.S. Peercy (Academic Press, New York, 1980) pp. 227-233.
7. H.J. Leamy, W.L. Brown. G.K. Celler, G. Foti, G.H. Gilmer and J.C.C.Fan, Appl. Phys. Lett. 38, 137 (1981).
8. R.A. Lemons and M.A. Bösch, Appl. Phys. Lett. 39, 343 (1981).
9. G. Auvert, D. Bensahel, A. Perio, V.T. Nguyen and G.A. Rozgonyi, Appl. Phys. Lett. 39, 724 (1981).
10. M. Bertolotti, G. Vitali, E. Rimini and G. Foti, J. Appl. Phys. 50, 259 (1979).
11. A.G. Cullis, H.C. Webber and N.G. Chew, Appl. Phys. Lett. 36, 547 (1980).
12. J. Narayan and C.W. White, Appl. Phys. Lett. 44, 35 (1984).
13. R.F. Wood, D.H. Lowndes and J. Narayan, Appl. Phys. Lett. 44, 770 (1984).
14. W. Sinke and F.W. Saris, this proceedings volume.
15. P. Baeri, G. Foti, J.M. Poate and A.G. Cullis, Phys. Rev. Lett. 45, 2036 (1980).
16. H.C. Webber, A.G. Cullis and N.G. Chew, Appl. Phys. Lett. 43, 669 (1983).
17. E.P. Donovan, F. Spaepen, D. Turnbull, J.M. Poate and D.C. Jacobson, Appl. Phys. Lett. 42, 698 (1983).
18. P.L. Liu, R. Yen, N. Bloembergen and R.T. Hodgson, Appl. Phys. Lett. 34, 864 (1979).

19. R. Tsu, R.T. Hodgson, T.Y. Tan and J.E. Baglin, Phys. Rev. Lett. 42, 1356 (1979).
20. A.G. Cullis, H.C. Webber, N.G. Chew, J.M. Poate and P. Baeri, Phys. Rev. Lett. 49, 219 (1982).
21. M.O. Thompson, J.W. Mayer, A.G. Cullis, H.C. Webber, N.G. Chew, J.M. Poate and D.C. Jacobson, Phys. Rev. Lett. 50, 896 (1983).
22. G.H. Gilmer, in Laser-Solid Interactions and Transient Thermal Processing of Materials, edited by J. Narayan, W.L. Brown and R.A. Lemons (North-Holland, New York, 1983) pp. 249-261.
23. R.S. Wagner, Acta. Met. 8, 57 (1960).
24. H.F. John and J.W. Faust, Jr., in Metallurgy of Elemental and Compound Semiconductors, edited by R.O. Grubel (Interscience, New York, 1961) pp. 127-148.

KINETICS AND MECHANISMS OF SOLID PHASE EPITAXY AND COMPETITIVE PROCESSES IN SILICON

G.L. OLSON
Hughes Research Laboratories, 3011 Malibu Canyon Road, Malibu CA 90265

ABSTRACT

Recent progress in studies of temperature dependent kinetic competition during solid phase crystallization of silicon is reviewed. Specific areas which are emphasized include: the enhancement of solid phase epitaxial growth rates by impurity-induced changes in electronic properties at the crystal/amorphous interface, the influence of impurity diffusion and precipitation in amorphous silicon on the kinetics of epitaxial growth, the effects of impurities on the kinetic competition between solid phase epitaxy and random crystallization, and the kinetics of solid phase crystallization at very high temperatures in silicon.

INTRODUCTION

A comprehensive description of the kinetics and mechanisms of solid phase crystallization and competitive processes in silicon is important both for the development of a fundamental understanding of transitions between metastable and stable phases and for the practical application of laser and incoherent beam processing methods. With the advent of transient beam heating and laser-based diagnostic techniques [1] it has become possible to extend studies of solid phase crystallization and the interactions between kinetically competitive solid phase processes into regimes of time and temperature that were previously inaccessible. These techniques have proven to be especially useful for investigating the kinetics of intrinsic crystallization in ion-implanted and deposited amorphous films over a wide temperature range; the influence of impurity atoms on the rate of solid phase epitaxy; the concentration and temperature dependent competition between SPE, impurity diffusion, and precipitation in amorphous silicon; and the interplay between SPE and random crystallization in amorphous films at high temperatures.

In this paper we review recent progress in studies of solid phase crystallization dynamics and temperature dependent kinetic competition during epitaxial growth at high temperatures. Particular emphasis is placed on the temperature dependence of impurity effects on solid phase epitaxial growth kinetics in ion-implanted layers. We examine: (1) the enhancement of SPE rates by impurity-induced changes in electronic properties at the crystal/amorphous (c/a) interface; (2) the influence of impurity diffusion and precipitation in As$^+$-implanted amorphous layers on the epitaxial crystallization kinetics; (3) the temperature dependence of enhancement and retardation of the SPE rate in films implanted with boron and fluorine; and (4) effects of impurities on the competition between SPE and random crystallization in ion-implanted layers. In addition, we discuss results obtained on solid phase transformations at very high temperatures in amorphous silicon films.

EXPERIMENTAL

In order to investigate solid phase crystallization kinetics and the intervention of competitive processes during crystallization at high temperatures, it is essential to rapidly heat the sample to the desired temperature at a sufficient rate so that minimal transformation occurs at lower temperatures and to simultaneously monitor the rapid crystallization rate during the heating process. These requirements are satisfied by the use of cw laser heating together with measurements of time-resolved reflectivity (TRR) during crystallization. The experimental techniques used for these measurements have been described in detail elsewhere [1] and therefore will only briefly be reviewed here.

A simplified schematic of the experimental configuration which is used for measurement of crystallization dynamics during cw laser heating is shown in Fig. 1. Samples are held on a temperature controlled, resistively heated vacuum stage and irradiated with a focused cw argon laser beam. A HeNe laser beam which is spatially filtered and focused to a spot coincident with the center of the laser-heated spot is used for the TRR measurements. An important element of the TRR system is a rapid opto-mechanical shutter which can be used to controllably gate the heating beam on and off the sample, thereby allowing the heating duration to be precisely controlled.

During solid phase crystallization of an amorphous thin film, the reflectivity changes with time due to constructive and destructive interference that occurs between light reflected from the surface and from the buried, advancing growth interface. The temporal change of the net reflected light intensity is directly related to the spatial position of the buried c/a interface; hence, measurements of the time dependence of the reflectivity can be used to determine the rate of SPE, as well as any possible variation of SPE rate with depth. As discussed previously [1], TRR techniques can also be used for determining temperature during rapid crystallization.

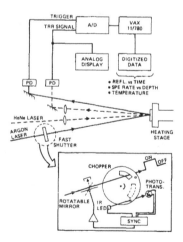

Fig. 1. Simplified schematic of experimental configuration employed for TRR data acquisition and analysis; fast shutter used to controllably gate heating beam on/off sample.

KINETIC COMPETITION DURING SOLID PHASE EPITAXY

SPE Rate Enhancement and Retardation

The rate of SPE can be increased significantly relative to the
intrinsic value by the addition of dopants such as B, As, and P to the
amorphous layer. Although rate enhancement by ion-implanted dopants has
been known for some time [2], until recently there has been a scarcity of
data on the temperature and concentration dependence of this phenomenon. By
comparing TRR measurements of the SPE rate vs. interface position with
measurements of the impurity concentration profile by secondary ion mass
spectrometry (SIMS) it is now possible to obtain detailed information on the
SPE rate as a function of concentration at different temperatures. As an
example, in Fig. 2 we show TRR data obtained during crystallization of an
amorphous Si film implanted with boron. In this example the Si(100)
substrate was first implanted with Si^+ (120 keV, $2 \times 10^{15} cm^{-2}$; 80 keV,
$1 \times 10^{15} cm^{-2}$; 40 keV, $5 \times 10^{14} cm^{-2}$, 77°K) to produce an amorphous layer approximately 2500 Å thick. Then boron was implanted at 30 keV to a dose of
$1 \times 10^{15} cm^{-2}$. The resulting boron concentration profile was determined by
SIMS.

The effect of the implanted boron on the SPE rate is manifested in the
TRR data of Fig. 2a by the nonuniform temporal separation of the interference features. The variation of SPE rate with depth determined from these
data is compared with the boron concentration profile in Fig. 2b. Comparison of the rate variations with the boron concentration profile allows SPE
rate vs. boron concentration to be directly determined. This analysis has
been performed for crystallization at three different temperatures in sam-
ples containing a range of doses ($5 \times 10^{14} cm^{-2}$ to $1 \times 10^{16} cm^{-2}$). The resulting

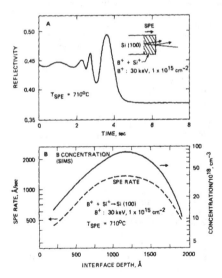

Fig. 2. TRR measurement during SPE in
B^+-implanted Si. Si(100)
pre-amorphized by Si^+
implantation. (A) Reflec-
tivity vs. time for growth
at 710°C. (B) SPE rate vs.
depth determined from data
in (A); SIMS measurement of
boron concentration profile
given for comparison.

28

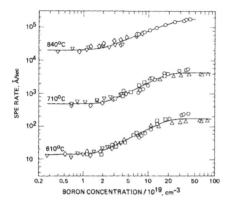

Fig. 3. SPE rate vs. boron concentration at three different temperatures. Preamorphized Si(100) samples implanted with 30 keV B^+, different doses: $\nabla 5\times10^{14}cm^{-2}$, $\Diamond 1\times10^{15}cm^{-2}$, $\square 2.5\times10^{15}cm^{-2}$, $\bigcirc 5\times10^{15}cm^{-2}$, $\triangle 1\times10^{16}cm^{-2}$.

rate vs. concentration data are plotted in Fig. 3. At concentrations $\lesssim 2\times10^{19}cm^{-3}$ boron has essentially no effect on the SPE rate. As the concentration increases to approximately $2\times10^{20}cm^{-3}$ there is a corresponding increase in the SPE rate with a roughly linear dependence of rate on concentration. Then, at higher concentrations the rate saturates, with the saturation occurring at slightly higher concentrations when the crystallization is performed at higher temperatures. These saturation effects are believed to be a consequence of the limited solid solubility of boron in Si and the increase of solubility with temperature, since it is presumably only the dissolved boron which enhances the rate of SPE.

The origin of the SPE rate enhancement continues to be a topic of active research. Considerable insight was provided by experiments conducted by Suni et al. [3,4] and Lietoila et al. [5] in which it was shown that although n- and p-type dopants when acting separately enhance the SPE rate, when both are present at comparable concentrations the SPE rate tends to the intrinsic value characteristic of undoped material. Subsequently we showed [1] that this dopant compensation effect holds over a wide temperature range (550°C to 900°C) in samples implanted with phosphorus and boron. In explaining this effect, Suni and co-workers hypothesized that the SPE rate is proportional to the concentration of vacancies at the c/a interface, as first suggested by Csepregi et al. [6]; they also noted that the total vacancy concentration is dominated by charged vacancies at the doping levels for which SPE rate alteration effects are observed [7]. Since the concentration of charged vacancies responds to the Fermi level position which in turn is established by the dopant concentration, enhancement of the SPE rate will occur whenever the Fermi level moves toward either band edge, i.e. in heavily doped samples of either conductivity type. When the Fermi level is driven back to mid-gap by the coexistence of compensating impurities, the vacancy concentration, and hence the SPE rate, will return to their intrinsic values.

These effects have been interpreted in an alternate way by Mosley and Paesler [8] who suggest that the operative mechanism involves migration of charged dangling bonds and subsequent capture of these defects at the c/a interface. Likewise, Williams and Elliman [9] relate the rate enhancement process to impurity-enhanced formation of kinklike growth sites at the interface. Although the details of the rate enhancement process are yet to be fully elucidated, we emphasize that in all of the descriptions a mechanism based on impurity-induced changes in the electronic properties at the interface appears to be necessary to account for the observed phenomena.

In the case of boron-implanted Si films we saw that changes in the SPE rate were related directly to changes in the dopant concentration, and that the results could be described by changes in the electronic structure at the c/a interface. In contrast, arsenic-implanted silicon is a system in which a substantially more complicated dependence of SPE rate on concentration is observed [10], and we use it as an example to illustrate the complex temperature and concentration-dependent kinetic competition that can occur during epitaxial crystallization of an amorphous layer. As is the case with boron, arsenic is a dopant in silicon which is known to enhance the rate of SPE when present at concentrations between ~5×10^{19}cm^{-3} and ~5×10^{20}cm^{-3} [2]; however, at higher concentrations arsenic can also retard the SPE rate [10-12]. The interplay between the rate enhancement and retardation mechanisms is a strong function of temperature as is illustrated in Fig. 4 by the data for SPE rate vs. depth obtained at five different temperatures in a sample produced by high dose arsenic implantation (2×10^{16}cm^{-2}, 220 keV; peak concentration ~1.4×10^{21}cm^{-3}). In this example, the peak of the arsenic concentration profile is located about 1300 Å from the surface, and the total amorphous layer thickness is approximately 2600 Å. In the lowest temperature case, the rate vs. depth behavior is characterized by an initial enhancement of the SPE growth rate as the interface moves to about 2300 Å from the surface (the intrinsic rate at this temperature is 50 Å/sec). However, as the interface approaches the region containing the highest arsenic concentration, a dramatic decrease in the SPE rate is evident. This large reduction in the SPE rate near the peak concentration becomes progressively less pronounced with increasing temperature. Microstructural analysis of these films by cross-sectional TEM has shown [10] that in the region where a large rate reduction is observed, there is also a high density of precipitates. A substantial decrease in the density of precipitates is observed in samples crystallized at the highest temperatures.

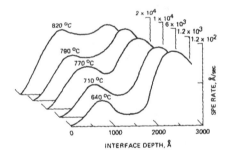

Fig. 4. Effect of temperature on SPE rate vs. interface depth in samples produced by high dose As$^+$ implantation (2×10^{16}cm^{-2}, 220 keV) into Si(100).

These results can be explained in terms of the temperature dependent competition between solid phase epitaxy and impurity diffusion and precipitation in the amorphous phase. At low temperatures the time required for the impurities to diffuse and form stable precipitates in the amorphous phase ahead of the advancing interface is shorter than the time for the epitaxial growth front to propagate through the region containing the high arsenic concentration. Rate retardation occurs when the interface propagates through the region in which precipitation has taken place. However, with increasing temperature the rate of epitaxial growth becomes greater than the rates for diffusion and precipitation in the amorphous phase. Consequently, there is progressively less SPE rate retardation as the temperature is increased. We have examined the concentration and temperature dependence of these competitive interactions in detail in samples implanted at single energies at different doses [10] and in samples containing a uniform concentration profile produced by multiple energy implantation. We are currently using these data to quantify the temperature dependence of arsenic diffusion and precipitation in amorphous silicon films.

An important example in which temperature-dependent competition between rate enhancing and rate retarding processes can alter the net crystallization kinetics is illustrated by the case of BF_2^+-implanted silicon. This is of particular practical importance because BF_2^+ implantation is commonly used to introduce boron into thin surface layers for semiconductor device applications. When BF_2^+ is implanted into silicon the molecular ions break up into their atomic constituents which can each act to influence the crystallization process in different ways. For example, in contrast to the rate enhancement produced by boron, fluorine atoms act to retard the SPE rate [13], and as shown in Fig. 5 the amount of retardation depends strongly on temperature. The data shown in Fig. 5 were

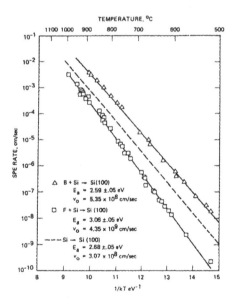

Fig. 5.   SPE regrowth kinetics in boron and fluorine-implanted Si; B^+: 10 keV, $3 \times 10^{15} cm^{-2}$; F^+: 18 keV, $6 \times 10^{15} cm^{-2}$; Si(100) sample previously amorphized by Si^+ implantation.

obtained using samples prepared by implantation of B and F into Si(100) samples previously amorphized by Si$^+$ implantation. In both cases the thickness of the amorphous layer (2500 Å) was greater than the thickness of the region containing the implanted impurities.

The SPE rates presented in Fig. 5 were determined at the peak of the impurity concentration profile. Temperature in these experiments was obtained from the rate of SPE in the undoped portion of the film, using the well-established temperature dependence of the intrinsic SPE rate in Si [1]. The growth kinetics for both fluorine and boron-implanted layers are described by an Arrhenius expression. Although the activation energies for the undoped and boron doped films are similar (2.7 and 2.6 eV, respectively), the activation energy for the fluorine implanted layer is considerably greater (3.1 eV). Moreover, it is apparent that the activation energy for SPE is greater than that for impurity migration in the amorphous phase and segregation at the a/c interface as is evidenced by decreased rate retardation at higher temperatures.

Since the temperature dependence of the crystallization rates in B$^+$ and F$^+$-implanted layers is different, there is a significant temperature dependent variation in the net crystallization kinetics when these impurities are simultaneously present in the film (as in the case of BF$_2^+$ implantation). For example, at low temperatures ($\leq$650°C) fluorine retards the SPE rate to a greater extent than boron enhances the rate; consequently there is a net reduction in the rate as the interface propagates through the region containing the impurities. Since the amount of rate retardation due to the fluorine decreases with increasing temperature (see Fig. 5), the rate reduction in the BF$_2^+$-implanted samples becomes less pronounced as the temperature is increased.

Effects of Impurities on the Competition Between SPE and Random Crystallization

In addition to solid phase epitaxy, random crystallization is a process which can occur during heating of an amorphous film. In previous studies [1] we showed that in intrinsic films or in films containing low impurity concentrations, SPE is the dominant process over the entire range of temperatures (475°C to 1350°C) investigated. That conclusion is also supported by results of studies [14] on the temperature dependence of random crystallization rates in ultra-high vacuum (UHV) deposited films and the kinetics of SPE in ion-implanted and deposited layers. However, we have recently shown [15], that certain impurity atoms (As, F) at concentrations $\geq 4 \times 10^{20}$ cm^{-3} can increase the rate of random crystallization to the extent that at temperatures $\geq$1000°C nucleation and growth of polycrystalline material dominate the crystallization kinetics. The details of that work are presented in a companion paper in these Proceedings [15]; therefore only the salient features will be presented here.

Time-resolved reflectivity data obtained during cw laser heating of As$^+$, F$^+$, and BF$_2^+$-implanted layers (2500 Å thick) indicate the disruption of the motion of a planar epitaxial interface by a competitive process at temperatures $\geq$1000°C. This is illustrated in Fig. 6 by TRR data obtained during crystallization of a F$^+$-implanted (18 keV, 4×10^{15} cm^{-2}), pre-amorphized sample at three different temperatures. As in the case of the As$^+$-implanted samples (see Fig. 1 in Ref. [15]), progressive diminution of the last interference peak is observed as the temperature is increased above approximately 960°C. Transmission electron microscopic analysis of regions crystallized at temperatures greater than 1000°C revealed the presence of polycrystalline material in the near-surface region of the sample. This is

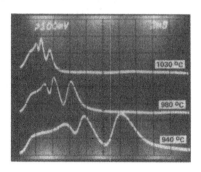

Fig. 6.   Time-resolved reflectivity data
obtained during crystallization
of F^+-implanted Si at three
different temperatures.

corroborated by results of selective chemical etching and electron
channeling analysis of the laser-irradiated regions. We emphasize that this
degradation in the epitaxial growth interface was not observed in samples
containing reduced As^+, F^+, and BF_2^+ doses; likewise, in extensive studies
of SPE of intrinsic layers and B^+-implanted films, disruption of epitaxy by
random crystallization was not observed.

Computer simulation of the TRR experiment which incorporates
reflectivity changes due to motion of an epitaxial interface, temperature
dependent changes in the volume fraction of polycrystalline material, and
the temperature dependence of the optical constants of silicon has been
performed, and by comparison with the experimental data, has allowed
quantitative information about the kinetics of random crystallization to be
obtained. In Fig. 7 we plot the time to form a volume fraction of
polycrystalline material equal to $1-1/e$ in the As^+ and F^+-implanted layers
as a function of $1/kT$. The significance of this particular volume fraction
is that if the nucleation rate is constant and growth occurs by a single
mechanism, then the volume fraction of randomly crystallized material can be
expressed as [16]: $X=1-\exp[-(t/\tau)^n]$. Thus, at $t=\tau$, $X=1-1/e$, independent of
the growth mechanism. Also shown in Fig. 7 is the temperature dependence of
the intrinsic random crystallization time for UHV-deposited amorphous
layers. The data show that over the temperature range from ~980°C to
~1060°C the random crystallization time in the As^+ and F^+-implanted layers
is approximately 10 times shorter than the intrinsic crystallization time.
To give an indication of the temperature dependent competition between
random crystallization and SPE we also plot the time required to crystallize
a 1600 A thick film by SPE (this is the thickness over which crystallization
occurs at constant temperature in these experiments). It is apparent that
at temperatures greater than 980°C the random crystallization time becomes
progressively shorter than the time to crystallize the layer by SPE;
consequently as the temperature increases, the random crystallization
process progressively dominates the net crystallization kinetics.

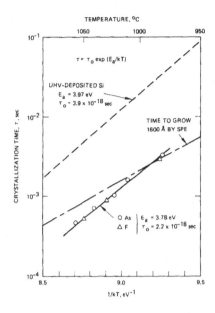

Fig. 7. Characteristic random crystallization time (time to form
polycrystalline volume fraction equal to 1-1/e) vs. 1/kT
for intrinsic film and layers
implanted with As+ and F+.
Shown for comparison is the
time to grow 1600 Å film by
SPE (1600 Å is growth distance
over which temperature was
constant).

It is important to note that the impurity-enhanced random
crystallization process is observed in films containing impurities that
increase (As) as well as retard (F) the rate of solid phase epitaxy. We
therefore suggest that the enhancement of random crystallization is not
controlled by the growth process, but instead is the result of impurity-
catalyzed nucleation. The energetics of the nucleation process is usually
described [16] in terms of the opposition between the negative free energy
change resulting in the formation of a particular volume fraction of
crystalline material, and the positive free energy change due to the
creation of an interface between the amorphous and crystalline phases. We
speculate that when arsenic or fluorine is present at a sufficiently high
concentration, there is a temperature dependent reduction in the free energy
required to form a stable interface. This results in an enhancement of the
nucleation rate and a concomitant increase in the rate of random
crystallization. We are currently investigating the influence of other
impurities on the competition between SPE and random crystallization and
will use this information together with microstructural analysis of
partially and totally crystallized films to develop a comprehensive
description of the interplay between these crystallization mechanisms.

KINETICS OF SOLID PHASE CRYSTALLIZATION AT HIGH TEMPERATURES

As discussed previously, cw laser heating and TRR diagnostic techniques
can be used to obtain crystallization rate data over a wide range of rates
and temperatures. This is illustrated in Fig. 8 by the temperature
dependence of the SPE rate in As+-implanted (150 keV, $5 \times 10^{14} cm^{-2}$) Si(100).
We note that the arsenic concentration in this sample is well below the
values for which impurity-catalyzed nucleation is observed. The data show
that an Arrhenius-type behavior characterizes the SPE process over a range
of rates from $\sim 10^{-2}$ Å/sec to $> 10^8$ Å/sec and a temperature range from $\sim 475°C$
to $\sim 1360°C$ (see Ref. [1] for description of the methods used for
measurements of rate and temperature over this range).

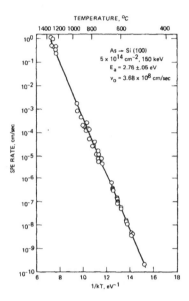

Fig. 8. Dependence of SPE rate on
temperature in As$^+$-implanted
(150 keV, 5x10^{14}cm^{-2}) Si(100).

To examine the behavior of the solid phase transformation process at
high temperatures in greater detail we performed experiments in which the
rapid epitaxial growth process was interrupted before the amorphous film was
totally crystallized (this was accomplished using the fast opto-mechanical
shutter shown in Fig. 1). It was then possible to evaluate whether
epitaxial crystallization could be resumed at a lower temperature and if so,
whether the epitaxial growth rate at that temperature was the same as the
SPE rate in a sample that had not been subjected to the high temperature
exposure. These experiments were initiated to determine whether the
possible effects due to formation of a thermally "relaxed" phase of
amorphous Si [17] would influence the kinetics of SPE, and to evaluate the
competition between SPE and random crystallization at very high
temperatures.

In Fig. 9 we show the TRR result obtained when SPE growth at ~1100°C
and a peak rate of 4.4x10^6 Å/sec is interrupted (the dashed line shows the
result that would have been obtained if crystallization had been allowed to
continue at that temperature). The sample used in this experiment was Si$^+$-
implanted Si(100) (100 keV, 2x10^{15}cm^{-2}, 77°K). The data show that motion of
the epitaxial interface stops abruptly when the shutter is closed. When the
resistively heated stage on which the sample was placed was then raised to
640°C, we found that epitaxial growth continued at a rate equal to that
obtained in companion samples which had not been exposed to the high
temperature pre-treatment. Based on the results discussed in the previous
section we note that if a significant amount of polycrystalline material had
formed at the c/a interface or in the amorphous region ahead of the

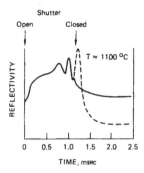

Fig. 9. Time-resolved reflectivity result obtained when rapid SPE growth is interrupted using opto-mechanical shutter shown in Fig. 1. Dashed line gives result that would be obtained if heating beam had been maintained on sample.

interface during the high temperature growth phase, we would expect epitaxy to be degraded during subsequent growth at reduced temperatures. However, this was not observed. Furthermore, since the rate of epitaxy during the low temperature growth portion was unaffected by the high temperature exposure we conclude that insofar as variations in the SPE rate are sensitive to structural changes, structural relaxation of the amorphous phase is not occurring in these samples at high temperatures. A more detailed discussion of these results as well as a comparison of the kinetics data with results of microstructural analysis will be presented in a future publication.

The observation of solid phase epitaxial crystallization at temperatures approaching the crystalline melting point raises interesting questions about the nature of amorphous silicon at high temperatures and the response of the material to high intensity cw laser irradiation. For example, in cw laser heating experiments we observe [1,18] a solid phase transformation at temperatures in excess of the melting temperature of amorphous Si (~1220°C) deduced from analysis of pulsed laser heating experiments (for details of these measurements and analyses see reviews by A.G. Cullis [19] and P.S. Peercy [20] in these Proceedings), as well as from differential scanning calorimetry measurements of the molar enthalpy of crystallization [21]. We have performed measurements on thick (4000 Å) ion-implanted layers as well as on UHV-deposited films on clean and on oxidized substrates and in all cases have observed solid phase transformations at temperatures greater than the proposed amorphous Si melting temperature, $T_m(a)$. As discussed in Refs. [1] and [22] a resolution of the discrepancy between results obtained in cw laser heating and pulsed laser heating experiments must involve a consideration of both the equilibrium thermodynamic parameters of the crystalline, amorphous and liquid phases and the kinetics of the transition between phases.

To consider first the static thermodynamic parameters, based on predictions from transition state theory [23] we would expect that if the melting temperatures of the crystalline and amorphous phases were equal, the epitaxial growth rate would "roll-over" at temperatures significantly lower than that value (for a more detailed discussion of this point see Refs. [1] and [22]). However, referring to Fig. 8 the data show no tendency to deviate from a purely Arrhenius behavior at high temperatures; therefore, assuming the validity of the transition state formulation in this temperature regime, the supposition of equal melting temperatures for the amorphous and crystalline material is ruled out. This conclusion together with the results of the pulsed laser heating experiments [19,20] implies that our highest temperature SPE measurements were conducted at temperatures greater than the point at which the amorphous and liquid silicon free energy curves cross. To examine why these films can be heated above $T_m(a)$ without melting we turn to a discussion of the amorphous-to-crystal and amorphous-to-liquid phase transition kinetics.

Referring to Fig. 10 it can be seen that if an amorphous Si film is heated to temperatures above the point at which the amorphous and liquid free energy curves cross, there are two states of lower free energy available: namely, the liquid and the crystalline solid. Consequently, two pathways exist for relaxation of the metastable amorphous phase: the layer can melt first and then freeze into the crystalline phase (path 1) or it can convert directly to the crystalline state (path 2). It is apparent that in cw laser heating experiments of the kind reported here, direct crystallization via path 2 is always observed. We therefore conclude that under cw laser heating conditions crystallization is a much faster process than melt nucleation. This is not unreasonable since epitaxial crystallization is an interface-controlled process which can occur without nucleation, whereas the process of melting and resolidification consists of two separate nucleation and growth steps each of which involves a significant change in physical and electronic structure (i.e. 4-fold coordinated solid to metallic liquid to 4-fold coordinated solid).

Although this description is consistent with the results obtained during cw laser heating it does not explain why superheating of the amorphous phase apparently does not occur in pulsed laser heating experiments even though the heating rates are much greater. To address this issue it may be very important to evaluate the competitive kinetics of solid phase crystallization and melt nucleation in a time regime intermediate between that of cw and Q-switched laser heating. In addition, important future work in this area should include the development of a nonequilibrium description of the thermodynamics and kinetics of transitions between metastable and stable phases in silicon.

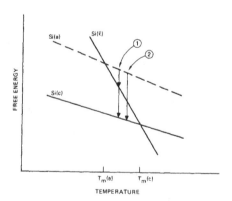

Fig. 10. Schematic illustration of free energy vs. temperature for amorphous, crystalline, and liquid silicon. Path 1: two step, amorphous-liquid-crystal transition; path 2: one step, amorphous-crystal transition.

SUMMARY AND CONCLUSIONS

We have examined the temperature and concentration dependent kinetic competition that can occur during solid phase crystallization in ion-implanted amorphous Si films. Enhancement of the SPE growth rate by implanted dopant atoms was related to impurity-induced changes in the electronic properties at the c/a interface. In boron-implanted films it was shown that the SPE rate follows the concentration profile over a range of concentrations from $\sim 2 \times 10^{19} cm^{-3}$ to $\sim 3 \times 10^{20} cm^{-3}$. In contrast, in arsenic-implanted silicon, impurity diffusion and precipitation in the amorphous phase ahead of the advancing interface can cause large concentration and temperature dependent rate-altering effects which can compete with the rate enhancement produced by the electronic changes. At high temperatures SPE is kinetically favored over arsenic diffusion in the amorphous material and rate alterations due to the latter effect become less important. In BF_2^+-implanted Si(100) there is significant competition between the rate retarding influence of the fluorine and the rate enhancement due to the boron. We show that the activation energy for SPE in the F^+-implanted samples (3.1 eV) is significantly greater than the intrinsic value (2.7 eV). Consequently, the retardation due to the presence of fluorine atoms becomes less pronounced with increasing temperature and leads to temperature dependent changes in the SPE process when boron and fluorine are simultaneously present in the film.

We also examined the effects of impurities on the kinetic competition between SPE and random crystallization in silicon. We showed that in films containing As^+, F^+, and BF_2^+ at concentrations $\geq 4 \times 10^{20} cm^{-3}$ epitaxial growth is disrupted at temperatures $\geq 1000°C$. Correlation with results of microstructural analysis and computer simulation of the TRR experiment indicates that the effect is due to enhancement of the random crystallization rate via impurity-catalyzed nucleation. The kinetics of solid phase crystallization in silicon at very high temperatures is a final topic which was addressed. We showed that epitaxy can be controllably interrupted at high temperatures and resumed at reduced temperatures and that the low temperature growth rate is indistinguishable from that obtained in samples which were not exposed to the high temperature. The observation of solid phase transitions at temperatures in excess of the proposed melting temperature of amorphous silicon was discussed in terms of competitive kinetics between direct crystallization and melt nucleation. We conclude that under cw laser heating conditions the transition directly to the crystalline state is kinetically favored over the two-step amorphous-to-liquid-to-crystal transition.

ACKNOWLEDGEMENT

The author gratefully acknowledges L.D. Hess, S.A. Kokorowski, J. Narayan, J.A. Roth, Y. Rytz-Froidevaux, L.W. Tutt, and P.K. Vasudev for their many contributions to this work.

REFERENCES

1. G.L. Olson, S.A. Kokorowski, J.A. Roth and L.D. Hess, Mat. Res. Soc. Symp. Proc. 13, 141 (1983) and references therein; G.L. Olson, J.A. Roth, L.D. Hess and J. Narayan, "Kinetics of Solid Phase Crystallization in Ion-Implanted and Amorphous Films", in Layered Structures and Interface Kinetics: Their Technology and Applications (RTK Scientific Publishers, D. Reidel Publishing Co.) in press.

2. L. Csepregi, E.F. Kennedy, T.J. Gallagher, J.W. Mayer and T.W Sigmon, J. Appl. Phys. 48, 4234 (1977).

3. I. Suni, G. Göltz, M-A. Nicolet and S.S. Lau, Thin Solid Films 93, 171 (1982).

4. I. Suni, G. Göltz, M.G. Grimaldi, M-A. Nicolet and S.S. Lau, Appl. Phys. Lett. 40, 269 (1982).

5. A. Lietoila, A. Wakita, T.W. Sigmon and J.F. Gibbons, J. Appl. Phys. 53, 4399 (1982).

6. L. Csepregi, R.P. Küllen, J.W. Mayer and T.W. Sigmon, Sol. St. Commun. 21, 1019 (1977).

7. J.A. Van Vechten and L.D. Thurmond, Phys. Rev. B 14, 3539 (1976).

8. L.E. Mosley and M.A. Paesler, Appl. Phys. Lett. 45, 86 (1984).

9. J.S. Williams and R.G. Elliman, Phys. Rev. Lett. 51, 1069 (1983).

10. G.L. Olson, J.A. Roth, L.D. Hess and J. Narayan, Mat. Res. Soc. Symp. Proc. 23, 375 (1984).

11. J.S. Williams and R.G. Elliman, Appl. Phys. Lett. 37, 829 (1980).

12. J. Narayan and O.W. Holland, Phys. Stat. Solidi(a) 73, 225 (1982).

13. I. Suni, U. Shreter, M-A. Nicolet and J.E. Baker, J. Appl. Phys. 56, 273 (1984).

14. J.A. Roth, S.A. Kokorowski, G.L. Olson and L.D. Hess, Mat. Res. Soc. Symp. Proc. 4, 169 (1982); J.A. Roth, G.L. Olson and L.D. Hess, Mat. Res. Soc. Symp. Proc. 23, 431 (1984).

15. G.L. Olson, J.A. Roth, Y. Rytz-Froidevaux and J. Narayan, these Proceedings.

16. J.W. Christian, The Theory of Transformations in Metals and Alloys, (Pergamon Press, Oxford, 1965).

17. G.K. Hubler, C.N. Waddell, W.G. Spitzer, J.E. Fredrickson and T.A. Kennedy, Mat. Res. Soc. Symp. Proc. 27, 217 (1984).

18. S.A. Kokorowski, G.L. Olson, J.A. Roth and L.D. Hess, Phys. Rev. Lett. 48, 498 (1982); Mat. Res. Soc. Symp. Proc. 4, 195 (1982).

19. A.G. Cullis, these Proceedings, and references therein.

20. P.S. Peercy, these Proceedings, and references therein.

21. E.P. Donovan, F. Spaepen, D. Turnbull, J.M. Poate and D.C. Jacobson, Appl. Phys. Lett. 42, 698 (1983).

22. D. Turnbull, in Metastable Materials Formation by Ion Implantation, S.T. Picraux and W.J. Choyke, Eds. (North-Holland, New York, 1982) p. 103.

23. For a review see Ref. [16], chapters 3 and 11.

APPLICATIONS OF BEAM-SOLID INTERACTIONS IN SEMICONDUCTOR MATERIAL
AND DEVICE PROCESSING

JOHN C. C. FAN
Lincoln Laboratory, Massachusetts Institute of Technology,
Lexington, Massachusetts 02173-0073

ABSTRACT

Pulsed or scanned energy beams have been shown to be very useful for
many semiconductor applications, ranging from annealing of ion-implanted
damage to preparation of materials. An overview of this important field is
given and its prospects assessed.

I. INTRODUCTION

In the decade since the initial work of Laff and Hutchins (1) and Fan,
et al. (2-4) on laser recrystallization of Si films (1-4) and GaAs films
(3,4), the field of energy-beam interactions with solids has grown
extensively. Many of the technological advances in this area are beginning
to find applications in semiconductor material and device processing. In
this paper, some of these advances and applications will be reviewed.
Emphasis will be placed on transient annealing of ion-implanted materials
and on semiconductor-on-insulator technologies. Some aspects of material
synthesis, but not laser-assisted photochemical processing, will also be
discussed.

II. TRANSIENT ANNEALING OF ION-IMPLANTED MATERIALS

In transient annealing of ion-implanted materials, an energy beam
induces a thermal cycle to remove ion-implanted damage and to activate the
ion-implanted species. The principal application is in the fabrication of
solid state circuits, where the competing established technology is
conventional furnace annealing.

Furnace Annealing

In furnace annealing, an implanted wafer is usually held between 15 and
30 minutes at elevated temperatures. Figure 1 shows the wafer temperature
as a function of time (t), lateral distance (x, y) and depth (z). The graph
on the left shows the surface temperature at the center of the wafer, which
rises to a maximum and remains there for about 15-30 minutes before
returning to room temperature. The maximum temperature, which is usually
designated as the annealing temperature, is about 900-1200°C for Si and
800-1000°C for GaAs. The annealing temperature is the same for the whole
wafer, as indicated by the middle and right plots of Fig. 1. The other key
features of furnace annealing are: annealing is a solid-phase process;
wafers remain flat after annealing; because of the uniform temperature
distribution, efficient and uniform activation of dopants is achieved; many
wafers can be annealed in a single batch. Despite the many advantages of
furnace annealing, the large dopant redistribution (1-2 μm) resulting from
the long annealing time makes it difficult to fabricate shallow junctions
and small devices for very-large-scale integration (VLSI) circuits. In
order to reduce the annealing time, three different transient-annealing
techniques are being developed.

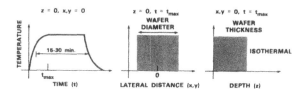

Fig. 1.   Wafer temperature as a function of time (t), lateral distance (x,y) and depth (z) for conventional furnace annealing.

Pulsed Laser or Electron-Beam Annealing

In pulsed laser or electron-beam annealing, the annealing time is only 10 to 100 ns, as shown at the left side of Fig. 2, where the wafer surface temperature is plotted against time. To obtain effective annealing in such a short time, the annealing temperature must exceed the melting temperature, resulting in marked dopant redistribution. The extent of redistribution depends on the depth of the molten zone, which is usually 0.3 to 0.5 µm (5).   Since it is difficult to produce a laser or electron beam with sufficient power over a broad area, the beam is generally incident on a region only 5 to 10 mm in diameter. Therefore beam or sample scanning is required for large-area activation. As a result, uniformity becomes an issue. With pulsed-beam annealing, the temperature gradient with depth is very sharp, as shown in Fig. 2. Because of the gradient, thermal stresses are induced in the wafer and thermal slip becomes a serious problem. In addition, line and point defects are often formed, particularly the latter. In spite of all these drawbacks, there may be some applications where pulsed-beam annealing is useful because controlled dopant redistribution is needed   (6).   Furthermore, pulsed-beam annealing can achieve the incorporation of dopants at concentrations above their solubility limits (7), and thus may be used in applications requiring supersaturated dopant solid solutions.   Nevertheless, pulsed-beam annealing is not expected to play a major role in the removal of ion-implanted damage.

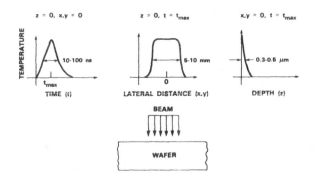

Fig. 2.   Wafer temperature as a function of time (t), lateral distance (x,y) and depth (z) for pulsed laser or electron-beam annealing.

Scanned Laser or Electron-Beam Annealing

In scanned laser or electron-beam annealing, the annealing time is longer, about 1-10 ms (see Fig. 3). In this time activation can be accomplished without melting. Generally the laser or electron beam is focused to a spot about 5-500 µm in diameter, and rapid beam rastering is used for large-area activation. Scanned annealing is achieved via surface heating (see Fig. 3), and thermal stresses are induced in the wafer, although the thermal gradients are not as high as in pulsed annealing.

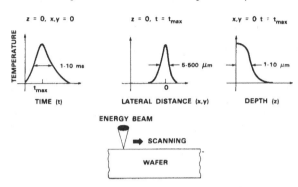

Fig. 3. Wafer temperature as a function of time (t), lateral distance (x,y) and depth (z) for scanned laser or electron-beam annealing.

Because annealing is of such short duration and takes place in the solid phase, line and point defects are major problems, as shown by Fig. 4. This

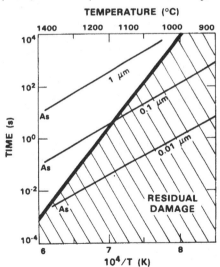

Fig. 4. Time versus temperature plots for arsenic-implanted Si wafers with 0.01 µm, 0.1 µm, and 1 µm impurity redistribution.

figure, which was presented by Gibbons, et al. (8), shows the combinations of time and temperature required for satisfactory annealing of arsenic-implanted Si wafers. If the temperature is too low and/or the time too short, the implanted damage will not be removed (as shown by the shaded area in the figure). Figure 4 also includes time vs temperature plots for 0.01 µm, 0.1 µm, and 1 µm impurity redistribution. These plots show that it is difficult for scanned annealing to achieve good activation without residual damage if impurity redistribution must be limited to only 0.01 µm. The optimal conditions are an annealing time of about 1 ms and an annealing temperature close to the melting point of Si (~ 1410°C). In spite of these drawbacks, scanned annealing is very appropriate for selected-area activation, and also for applications where minimal dopant redistribution (less than 100 Å) is required.

Rapid Thermal Annealing (RTA) with Incoherent Radiation

In the RTA process, the entire area of the sample is heated simultaneously by incoherent radiation, usually from a bank of heat lamps or graphite strip heaters. Like furnace annealing, RTA is a solid-state process. As shown in Fig. 5, the annealing time is between 1 and 10 s. This time, which is much longer than the times for laser or electron-beam annealing but still much shorter than furnace annealing times, is sufficient to remove most crystal defects but limits dopant redistribution to about 1000 Å (see Fig. 4). An important feature of RTA is that heating is essentially isothermal in the z direction (see Fig. 5). Temperature uniformity in the plane of the wafer is not as easily achieved as in furnace annealing, but effective, uniform dopant activation is routinely obtained (9). Since RTA can provide much higher throughput, it can be anticipated that this process will be readily accepted by the semiconductor industry, particularly for annealing shallow implants, if defect formation (10) and enhanced dopant redistribution (11) are not serious.

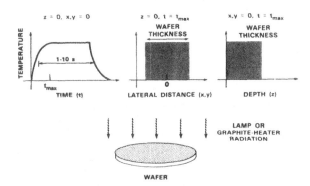

Fig. 5. Wafer temperature as a function of time (t), lateral distance (x,y) and depth (z) for rapid thermal annealing (RTA).

Another important feature of RTA, in contrast with most furnace annealing, is that compound semiconductors can be annealed without encapsulation. Kuzuhara, et al. (12) have described a lamp-annealing system (Fig. 6) in

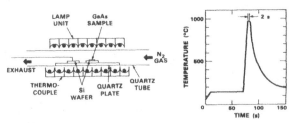

Fig. 6. Schematic diagram of RTA apparatus using lamps (from Ref. 12).

which GaAs samples are placed with their implanted surface down on either a
Si wafer or another GaAs wafer. For an annealing temperature of about
1000°C, it is necessary to limit the annealing time to about 2 s when the
GaAs sample is placed on a Si wafer, but the time can be somewhat longer if
a GaAs wafer is used. By employing the capless technique Kuzuhara, et al.
have obtained excellent GaAs transistors with sharp implanted-dopant
profiles (12).

Assessment

Of the available transient annealing techniques, RTA is the one most
likely to be adopted by the semiconductor industry. It has many of the
desirable features of furnace annealing, but provides much higher
throughput and produces less dopant redistribution. It also permits
convenient, effective capless annealing of compound semiconductors.

III. SEMICONDUCTOR-ON-INSULATOR (SOI) TECHNOLOGIES

A variety of SOI technologies, in which a high-quality semiconductor
layer (usually silicon) is formed on an insulating substrate (usually
silicon dioxide), are currently being developed. Their potential
applications, which are reviewed in detail by Tsaur elsewhere in this
volume, are VLSI circuits, high-voltage integrated circuits (ICs),
large-area ICs, and vertical ICs. Three of the SOI technologies under
development are based on beam-solid interactions: zone-melting
recrystallization, oxygen or nitrogen high-dose implantation, and
implantation-assisted solid-phase epitaxy. The competing conventional
technology is silicon-on-sapphire.

Silicon-on-Sapphire (SOS)

In SOS technology, silicon films 0.5-2 μm thick are formed by chemical
vapor deposition on single-crystal sapphire substrates. This technology has
been known for over a decade. Uniform SOS wafers 3 inches in diameter can
be obtained commercially, and even larger wafers are becoming available.
However, sapphire substrates are expensive, and currently SOS wafers cost
about three times as much as conventional Si wafers of the same size. In
addition, SOS films have defect densities as high as 10^6 cm^{-2}, and SOS
processing procedures are quite different from conventional Si procedures.
Therefore the principal applications for SOS are limited to
radiation-hardened circuits and some custom circuits.

Zone-Melting Recrystallization (ZMR)

A. Scanned Laser or Electron Beam. Figure 7 shows schematic diagrams
of scanned-beam recrystallization using either seeded or unseeded samples.

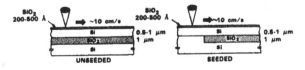

Fig. 7. Schematic diagram of scanned-beam recrystallization
of either a seeded or unseeded sample.

The Si film, which is usually about 0.5 µm to 1 µm thick, is deposited on a SiO$_2$-coated Si wafer. The energy beam is generally focused to a spot 5-50 µm in diameter, and a scanning rate of about 10 cm/s is often used. The key feature of this process is that the Si film is melted while the substrate remains at much lower temperatures. The process is ideal for selected-area melting. For large-area recrystallization, rastering is required, and this usually affects the quality of the recrystallized films (13,14). These films often contain defects, including grain boundaries, and it is not yet known whether they can be used for VLSI with sufficient yields. However, the fact that the substrates can be kept much cooler than the films makes this process very attractive for materials to be used in vertical ICs (or 3-D circuits).

    B.   Scanned Incoherent Radiation.   Large-area films can easily be recrystallized by using scanned incoherent radiation.   Figure 8 shows a schematic diagram of a ZMR process that employs two graphite heaters (15). The inset of Fig. 8 shows a cross section of a typical sample, which consists of a fine-grained Si film on an insulator-coated substrate, together with an Si$_3$N$_4$/SiO$_2$ encapsulating layer. The sample is placed on a stationary lower strip heater that is used to heat the sample to a base temperature of 1100-1300°C, generally in a flowing Ar gas ambient. Additional radiant energy, provided by the movable upper strip heater, is used to produce a narrow molten zone across the full width of the poly-Si layer. The molten zone is then translated across the sample by scanning the upper heater, typically at 1 mm/s, leaving a recrystallized Si film. Lamp heating has been shown to give similar results (16).

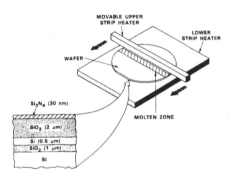

Fig. 8.   Schematic diagram of experimental configuration used for zone-melting recrystallization of encapsulated Si films. The inset shows a cross section through a typical sample.

In this ZMR process the motion of the recrystallization front is unidirectional. The process provides high throughput, is easily scaled up for large-diameter wafers, and is potentally inexpensive. Furthermore, the process is not limited to recrystallization of Si films on SiO$_2$-coated Si substrates, but can be applied to other semiconductors, such as GaAs or Ge, and other substrates, such as fused silica or sapphire. There are, however, several drawbacks. The substrates are often heated to temperatures just below the melting point of the materials, so that the process is not very suitable for 3-D circuits. Because of the very high temperatures, thermal stresses are induced, which can cause detrimental effects such as warping and slip. In addition, ZMR samples frequently contain grain boundaries (although these can be eliminated by seeding) and in most cases contain sub-boundaries.

    However, there have been many recent advances in this process (17). Warpage is now less than 4 µm over 3-inch wafers (comparable to the warpage of bare Si wafers), and both slip and the density of sub-boundaries have been much reduced. Furthermore, the process has been used to prepare recrystallized Ge films on SiO$_2$-coated Si wafers. Such films may provide excellent substrates for epitaxial growth of GaAs, which would be very important for the monolithic integration of GaAs and Si (18). Nevertheless, elimination of sub-boundaries may be necessary before this process can be widely adopted for the production of large-area SOI wafers.

Oxygen or Nitrogen High-Dose Implantation

By high-dose implantation of oxygen or nitrogen, followed by furnace annealing, insulating layers of SiO_2 (19,20) or Si_3N_4 (21) can be produced inside Si wafers. Figure 9 shows a Si wafer that has been implanted with over 10^{18} cm^{-2} O$^+$ ions at 150 keV. The wafer temperature is kept at about 500°C during implantation. An amorphous Si layer is buried below a monocrystalline Si layer that remains at the surface. The dotted area in Fig. 9 is the region where Si and O atoms are intermixed. After furnace annealing at 900-1100°C for 2 hours in N_2, this region forms a SiO_2 layer with a sharp Si/SiO_2 interface and the amorphous Si layer is converted to monocrystalline by solid-phase growth seeded by the surface layer. The

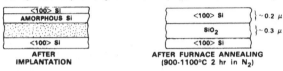

Fig. 9. Schematic diagram of the formation of buried oxide in a Si wafer by high-dosage oxygen implantation, followed by furnace annealing.

regrown layer has been found to contain threading defect densities exceeding 10^6 cm^{-2}. The quality of the Si next to the SiO_2 layer is especially poor. Because of these high defect densities, and because the Si top layer is very thin (~0.2 µm), an epitaxial Si layer is often needed. However, SOI wafers up to 4 inches in diameter have been prepared with excellent uniformity and flatness. With an epitaxial Si layer, LSI circuits have been obtained (22).

Another advantage of this process is that under certain conditions electric-field-shielding layers can be formed, with various device applications (23). Figure 10 shows the oxygen depth profiles obtained after furnace annealing for various implanted oxygen doses. For a dose of 1.2 x 10^{18} cm^{-2} the oxygen profile is roughly Gaussian, while at 2.4 x 10^{18} cm^{-2} the profile shows an abrupt Si/SiO_2 interface. At an intermediate dose of 1.8 x 10^{18} cm^{-2}, humps in the oxygen profile are observed on either side of

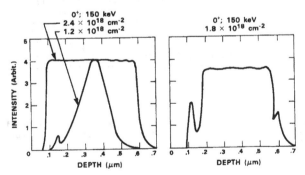

Fig. 10. Depth profiles of the implanted oxygen (in arbitrary units) in Si wafers after furnace annealing. The different profiles were obtained with different implanted doses.

the SiO_2 layer. The regions corresponding to the humps have been shown to be polycrystalline Si, which produces the electric-field shielding.

Some work has also been done with nitrogen implantation. The interface is not as satisfactory for devices as the $Si-SiO_2$ interface. The defects in layers after high dose oxygen or nitrogen implantation do not appear to cause serious circuit-yield problems, probably because of their threading nature, although it is still important to lower the defect densities. The key requirement for implementing the buried oxide or nitride technology is the development of a high-beam-current, high-voltage ion-implantation machine. A machine that can deliver at least 100 mA at an ion energy greater than 200 keV (in order to avoid the need for growing epitaxial layers) is necessary for the production of SOI wafers for commercial applications. Even if such a machine becomes available, its high cost, and the resultant wafer cost, may be a problem.

Implantation-Assisted Solid-Phase Epitaxy

The third SOI technology uses solid-phase epitaxy (SPE) to prepare epitaxial films on insulators at low temperatures. The major potential application is in 3-D circuits. Figure 11 shows a schematic diagram of this process. A polycrystalline Si film deposited on SiO_2 is amorphized by implanting selected impurity species and then crystallized by SPE. This procedure has been shown to greatly enhance lateral SPE--20 μm have been achieved (24). This process is still in an early stage of development, with relatively poor material quality and no device results.

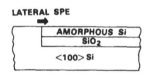

Fig. 11. Schematic diagram illustrating ion-implanta-tion-enhanced lateral solid-phase epitaxy over an insulator.

Assessment

For 3-D circuits, both ZMR using a scanned laser or electron beam and implantation-assisted SPE are being developed. Currently, the ZMR process is much more advanced, yielding SOI films that permit the fabrication of good-quality devices.

For VLSI applications, ZMR using incoherent radiation and high-dose implantation are the most promising technologies. Each process faces a critical problem: for ZMR, elimination of sub-boundaries; for implantation, development of a reliable, high-current, high-voltage implanter. The technology that solves its critical problem first may well be the one that is ultimately adopted.

For high-voltage or large-area ICs, and for preparing high-quality SOI films of materials other than Si, the ZMR techniques using a scanned laser or electron beam or incoherent radiation are currently the most appropriate.

IV. MATERIAL SYNTHESIS

In material synthesis, energy beams essentially provide a thermal source to facilitate the desired chemical reactions. The work so far reported in this area is limited.

Scanned Energy Beams

A technique has been developed (25) for using a scanned laser beam to produce compound semiconductor films by direct synthesis from layers of the constituent elements deposited on insulators such as SiO_2 (see Fig. 12).

Synthesis is performed in the liquid phase and the stoichiometry can be varied by adjusting the number and thickness of the deposited layers. Such binary compounds as AlSb, AlAs, CdTe, CdSe, ZnTe, and ZnSe have been prepared by this technique (25), which is well suited to selected-area synthesis.

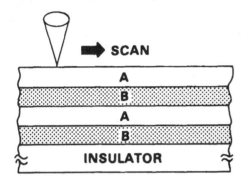

Fig. 12. Schematic diagram of synthesis of a compound on an insulator from a multilayer film by using scanned energy beam.

Pulsed Energy Beams

Larger area films are readily synthesized by the use of pulsed beams, especially when incoherent radiation is employed. The synthesis reaction usually occurs in the liquid phase, and the process is generally quite insensitive to interface contamination, so that epitaxial growth is obtained.

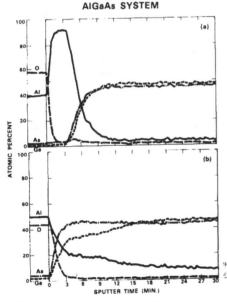

Fig. 13. Auger depth profiles of an Al-coated GaAs before and after RTA. After RTA, AlGaAs alloy was formed.

Excellent Ni and Co silicides have been prepared by depositing metallic Ni or Co films on Si wafers, then irradiating with a pulsed laser (26). Graded AlGaAs layers have been obtained (27) by depositing an Al film about 300 Å thick on a <100> GaAs wafer, then using a stationary graphite heater to heat the sample to 550-600°C for 10 s. The reaction appears to occur in the liquid phase. Figures 13(a) and (b) show the depth profiles of Al, Ga, As, and O obtained by Auger analyses before and after heating. There is a surface Al-O layer that apparently serves as an encapsulant, since the sample aggregates into beads upon heating, if the same experiment is carried out with a Si_3N_4 cap on the Al to eliminate the Al-O layer. From Fig. 13(b) it is apparent that Al has partially replaced Ga, with the As composition remaining constant. By using different Al thicknesses and time-temperature profiles, variously graded AlGaAs layers have been obtained.

Ion-Beam Mixing

Synthesis performed by the pulsed energy-beam techniques is relatively insensitive to interface contamination because it occurs in the liquid phase. For solid-phase synthesis, however, the interface properties are critical. Figure 14 shows the basic concept of ion-beam mixing. If a metal film is deposited on Si, native oxide is often found at the interface (as shown by the shaded areas in Fig. 14). If thermal annealing is performed without any other treatment, the reacted regions are not continuous. If bombardment with an energetic ion beam is used to modify the interface by ion-beam mixing (28), uniform reacted regions can be obtained. Since implanted As is an effective dopant in Si, by using As-ion bombardment for ion-beam mixing of W/Si samples it has been possible to obtain both excellent WSi_2 layers and shallow n-p junctions (29) after RTA with a graphite heater.

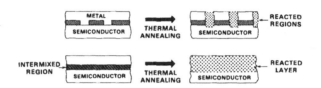

Fig. 14. Schematic diagram showing the effect of ion-beam mixing in eliminating non-uniform reacted regions.

Assessment

Material synthesis using energy beams is still in an early stage of development. The potential applications are in selected-area synthesis (using scanned beams) and large-area growth (using incoherent radiation). If the reactions take place in the liquid phase, they are relatively insensitive to interface properties. For solid-phase growth, ion-beam mixing is very useful in overcoming the effects of interface contamination.

V. SUMMARY

Beam-solid interactions have many potential applications. Three major areas have been reviewed: transient annealing of ion-implanted samples, semiconductor-on-insulator technology and material synthesis. Some of the techniques under development in these areas have a good potential for being adopted by the semiconductor industry.

VI. ACKNOWLEDGEMENT

The author acknowledges A. J. Strauss for helpful discussions. The Lincoln Laboratory portion of this work, which was performed in collaboration with many colleagues, was sponsored by the Department of the Air Force and the Defense Advanced Research Projects Agency.

REFERENCES

1. R. A. Laff and G. L. Hutchins, IEEE Trans. Electron Devices ED-21, 743 (1974).

2. J. C. C. Fan and H. J. Zeiger, Appl. Phys. Lett. 27, 224 (1975).

3. J. C. C. Fan, H. J. Zeiger, and P. M. Zavracky, in Proc. National Workshop on Low Cost Polycrystalline Silicon Solar Cells, Dallas, TX, 1976, Eds. T. L. Chu and S. S. Chu (Southern Methodist University, Dallas, TX, 1976) p. 89.

4. J. C. C. Fan and H. J. Zeiger, US Patent No. 4,059,461 (1977).

5. J. Narayan, C. W. White, and O. W. Holland, in Energy Beam-Solid Interactions and Transient Thermal Processing, Eds. J. C. C. Fan and N. M. Johnson (Elsevier North Holland, New York, 1984), p. 179.

6. C. Hill, in Laser-Solid Interactions and Transient Thermal Processing of Materials, Eds. J. Narayan, W. L. Brown, and R. A. Lemons (Elsevier North Holland, New York, 1983), p. 381.

7. M. J. Aziz, in Energy Beam-Solid Interactions and Transient Thermal Processing, Eds. J. C. C. Fan and N. M. Johnson (Elsevier North Holland, New York, 1984), p. 369.

8. J. F. Gibbons, D. M. Dobkin, M. E. Greiner, J. L. Hoyt, and W. G. Opyd, in Energy Beam-Solid Interactions and Transient Thermal Processing, Eds. J. C. C. Fan and N. M. Johnson (Elsevier North Holland, New York, 1984), p. 37.

9. B-Y. Tsaur, J. P. Donnelly, J. C. C. Fan, and M. W. Geis, Appl. Phys. Lett. 39, 93 (1981).

10. J. Narayan, in Energy Beam-Solid Interactions and Transient Thermal Processing, Eds. J. C. C. Fan and N. M. Johnson (Elsevier North Holland, New York, 1984), p. 335.

11. T. O. Sedgwick, R. Kalish, S. R. Mader, and S. C. Shatas, in Energy Beam-Solid Interactions and Transient Thermal Processing, Eds. J. C. C. Fan and N. M. Johnson (Elsevier North Holland, New York, 1984), p. 293.

12. M. Kuzuhara, H. Kohzu, and Y. Takayama, in Energy Beam-Solid Interactions and Transient Thermal Processing, Eds. J. C. C. Fan and N. M. Johnson (Elsevier North Holland, New York, 1984), p. 651.

13. D. K. Biegelsen, N. M. Johnson, W. G. Hawkins, L. E. Fennell, and M. D. Moyer, in Laser-Solid Interactions and Transient Thermal Porcessing of Materials, Eds. J. Narayan, W. L. Brown, and R. A. Lemons (Elsevier North Holland, New York, 1983), p. 537.

14. H. W. Lam, R. F. Pinizzotto, and A. F. Tasch, Jr., J. Electrochem. Soc. 128, 1981 (1981).

15. J. C. C. Fan, J. Cryst. Growth 63, 453 (1983).

16. A. Kamgar, G. A. Rozgonyi, and R. Knoell, in Energy Beam-Solid Interactions and Transient Thermal Processing, Eds. J. C. C. Fan and N. M. Johnson (Elsevier North Holland, New York, 1984), p. 569.

17. C. K. Chen, M. W. Geis, H. K. Choi, B-Y. Tsaur, and J. C. C. Fan, presented in this symposium.

50

18. J. C. C. Fan in Extended Abstracts of the 16th (1984 International) Conference on Solid State Devices and Materials, August 1984, Kobe, Japan, p. 115.

19. K. Izumi, M. Doken and H. Ariyoshi, Electron. Lett. $\underline{14}$, 593 (1978).

20. H. W. Lam and R. F. Pinizzotto, J. Cryst. Growth $\underline{63}$, 554 (1983).

21. R. J. Dexter, S. B. Watelski, and S. T. Picraux, Appl. Phys. Lett. $\underline{23}$, 455 (1973).

22. C. E. Chen, T. G. W. Blake, L. R. Hite, S. D. S. Malhi, B. Y. Mao, and H. W. Lam, IEEE SOS/SOI Technology Workshop, Hilton Head Island, South Carolina, October 1984.

23. K. Izumi, Y. Omura, and S. Nakashima, in Energy Beam-Solid Interactions and Transient Thermal Processing, Eds. J. C. C. Fan and N. M. Johnson (Elsevier North Holland, New York, 1984), p. 443.

24. H. Yamamoto, H. Ishiwara, and S. Furukawa, in Extended Abstracts of the 16th (1984 International) Conference on Solid State Devices and Materials, August 1984, Kobe, Japan, p. 507.

25. L. D. Laude, in Energy Beam-Solid Interactions and Transient Thermal Processing, Eds. J. C. C. Fan and N. M.. Johnson (Elsevier North Holland, New York, 1984), p. 611.

26. R. T. Tung, J. M. Gibson, D. C. Jacobson, and J. M. Poate, in Energy Beam-Solid Interactions and Transient Thermal Processing, Eds. J. C. C. Fan and N. M. Johnson (Elsevier North Holland, New York, 1984), p. 721.

27. J. C. C. Fan, B-Y. Tsaur, and G. H. Foley, unpublished.

28. G. J. Galvin, L. S. Hung, J. W. Mayer, and M. Nastasi in Energy Beam-Solid Interactions and Transient Thermal Processing, Eds. J. C. C. Fan and N. M. Johnson (Elsevier North Holland, New York, 1984), p. 25.

29. B-Y. Tsaur, C. K. Chen, C. H. Anderson, Jr., and D. L. Kwong, J. Appl. Phys. (to be published).

PART II

Ultrafast Processes in
Highly Excited Materials

THERMODYNAMIC AND KINETIC STUDIES OF PULSED-LASER ANNEALING FROM TRANSIENT CONDUCTIVITY MEASUREMENTS*

P.S. PEERCY and MICHAEL O. THOMPSON+
Sandia National Laboratory
Albuquerque, New Mexico 87185

ABSTRACT

Simultaneous measurements of the transient conductance and time-dependent surface reflectance of the melt and solidification dynamics produced by pulsed laser irradiation of Si are reviewed. These measurements demonstrate that the melting temperature of amorphous Si is reduced 200 ± 50 K from that of crystalline Si and that explosive crystallization in amorphous Si is mediated by a thin ($\lesssim$ 20 nm) molten layer that propagates at $\sim$ 15 m/sec. Studies with 3.5 nsec pulses permit an estimate of the dependence of the solidification velocity on undercooling. Measurements of the effect of As impurities on the solidification velocity demonstrate that high As concentrations decrease the melting temperature of Si ($\sim$ 150 K for 7 at.%), which can result in surface nucleation to produce buried melts. Finally, the silicon-germanium alloy system is shown to be an ideal model system for the study of superheating and undercooling. The $Si_{50}Ge_{50}$ alloy closely models amorphous Si, and measurements of layered Si-Ge alloy structures indicate superheating up to 120 K without nucleation of internal melts. The change in melt velocity with superheating yields a velocity versus superheating of 17 ± 3 k/m/sec.

I. INTRODUCTION

It has been recognized for some time from work in metals that rapid solidification following surface melting with pulsed laser or electron beam irradiation can be used to create novel metastable and amorphous structures. With the advent of "laser annealing," there have been numerous studies of crystallization in semiconductors at high solidification velocities. Such beam-induced rapid solidification processing is important for two reasons. First, it permits formation of metastable structures, and the unique materials that can be obtained by this new area of processing have significantly improved mechanical and electrochemical properties for selected applications. Second, the well-defined geometry available in beam-induced surface melting experiments permit detailed, quantitative measurements and theoretical modeling of solidification phenomena over an unprecedented range in solidification velocities. Such measurements of nucleation and crystallization are providing unique insights into the thermodynamic and kinetic processes that control rapid solidification.

Recently, a new technique based on transient electrical conductance was developed to measure melt and solidification dynamics during pulsed laser annealing [1]. This technique permits measurement of the solidification velocity, which is of central importance to rapid solidification processing. The measurements demonstrated that annealing in pure crystalline Si proceeds by surface melting followed by rapid solidification [2-4], and that in the nsec time regime the annealing phenomenon is well-described by thermal processes governed by the energy absorption and heat

*This work was performed at Sandia National Laboratories and supported by the U.S. Department of Energy under contract number DE-AC04-76DP00789.
+Present Address: Cornell University, Department of Materials Sciences, Ithaca, NY.

transport [5,6]. Applications of these techniques, in conjunction with transient reflectance [7] measurements of the melt duration at the irradiated surface, have been reported for Si [1-4], Si on sapphire (SOS) [8] and compound semiconductors [9,10].

Measurements of the melt dynamics to determine the melting temperature of the amorphous phase [11] and the mechanism by which explosive crystallization occurs will be reviewed. These measurements reveal that the origin of the unique microstructure, consisting of a layer of coarse-grained polycrystalline Si above a layer of fine-grained polycrystalline Si in annealed a-Si [12-13], is the result of explosive crystallization [11]. Bulk nucleation, as was proposed from some studies [14-16] where the melt dynamics were not measured, is inconsistent with the present measurements.

The interpretation of the melt dynamics in a-Si was verified by measurements in Si-Ge alloys [17]. The $Si_{50}Ge_{50}$ alloy yields a melting temperature and thermal conductivity comparable to a-Si. In addition, the effect of impurities on the solidification velocity of molten Si [18,19] was studied. These measurements can be used to deduce the metastable phase diagram for these alloys and to demonstrate that buried molten layers and internal melts [19] can be obtained in Si alloys when the melting temperature is depressed by impurities.

II. EXPERIMENTAL DETAILS

The melt and resolidification dynamics were obtained using the transient conductance technique [3]. Simultaneously, the surface melt duration was measured using transient optical reflectance [7] with the 488 nm wavelength of an argon laser. Data were acquired in digital form using a Tektronix 7912 AD transient digitizer.

Samples from either 0.5 or 1.0 μm Si on sapphire (SOS) were patterned photolithographically to give a length-to-width ratio (L/W) of 55.5. Because the electrical conductivity of Si increases by a factor of ~30 upon melting, the melt depth can be inferred from the change in conductance of the sample. For the charge-line configuration used in the present experiments, the transient conductance $\sigma(t)$, for an observed voltage $V(t)$, is given by

$$\sigma(t) = \frac{1}{R_L} [V_b/V(t) - 2]^{-1} \qquad (1)$$

where V_b is a bias voltage and R_L (50 Ω) is the cable impedance. The melt depth $d(t)$, neglecting the electrical conductivity in the solid phase, is given by

$$d(t) = \rho \frac{L}{W} \sigma(t) = \frac{L}{W} \frac{\rho}{R_L} [V_b/V(t) - 2]^{-1} \qquad (2)$$

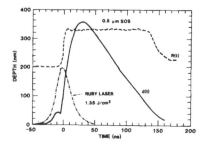

Fig. 1. Typical surface reflectance R(t) and melt depth d(t) versus time measured from the transient conductance for SOS.

where ρ is the resistivity of molten Si (80 $\mu\Omega$cm). A typical melt transient converted to d(t) via eq. 2 is shown for pure SOS in Fig. 1. Also shown in this figure is the simultaneous transient reflectance R(t).

For studies of the thermodynamic properties, amorphous Si layers were produced by Si implantation at various energies. Impurities were also introduced by ion implantation into 0.5 μm SOS to yield the desired depth distribution. The Ge-Si alloys were prepared by electron beam evaporation from Ge-Si mixtures and the resulting alloy composition was measured by Rutherford backscattering. Irradiations were performed with single pulses from a ruby laser at either the fundamental wavelength of 694 nm or at the frequency-doubled wavelength of 347 nm. Pulse lengths between 3 and 70 nsec were used in the experiments. During irradiation, the transient conductance and transient surface reflectance were measured simultaneously.

III. RESULTS AND DISCUSSION

A. Amorphous Silicon, a-Si

Estimates of the free energy of a-Si [20] indicate that this phase should melt at a temperature T_a substantially below the melting temperature T_c of crystalline silicon (c-Si). Because a-Si transforms rapidly to c-Si by a solid phase process at temperatures in excess of 600 C, this melting temperature reduction cannot be readily measured by conventional techniques. Prior attempts have yielded estimates that range from $T_a = T_c \pm 10$ [21] to $T_a = T_c - 500$ [22]. In addition to uncertainties in the values of such fundamental thermodynamic parameters as the melting temperature, pulsed laser melting of a-Si has raised questions concerning the kinetics which control the melt and resolidification process. For example, at incident laser energy densities insufficient to melt the amorphous layer completely to yield epitaxial regrowth, part, or all, of the a-Si film can be transformed to polycrystalline Si. The thickness for this polycrystalline layer, however, does not increase linearly with incident laser energy density [11,12]. Furthermore, detailed examination of the microstructure of the transformed material reveals two distinct microstructures: a fine-grained (<10 nm) polycrystalline layer beneath a coarse-grained surface layer. Prior to these measurements of the melt and resolidification

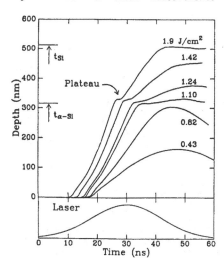

Fig. 2. Melt depth versus time for ~ 210 nm a-Si films on SOS irradiated with various laser energy densities.

dynamics, it was assumed that, as in the case of melting on single crystals, the melt front propagated to the full extent of the resulting polycrystalline layer. This assumption led to various interpretations of the kinetics of the process, ranging from low thermal conductivity in the a-phase [12] to homogeneous bulk nucleation and 'slush zones' [14,15]. As will be discussed below, however, direct measurements of the melt dynamics indicate that the melting temperature of a-Si is ~200 K below the melting temperature of c-Si and that the extent of the fine-grained polycrystalline Si does not represent the laser-induced melt depth. Rather, the fine-grained polycrystalline Si is produced by an explosive crystallization. No evidence has been obtained for homogeneous bulk nucleation from 'slush zones'.

Consider first the melting temperature of a-Si. Melt depths as a function of time, determined from the transient conductance measurements, are shown in Fig. 2 for a series of incident laser energy densities on 150 keV Si implanted SOS. Ion beam analysis techniques showed the samples consisted of a 330 nm amorphous layer on a total 510 nm SOS film. Transient reflectance measurements demonstrated that melt originated at the free surface before the peak of the laser pulse. As the melt reached the amorphous-crystalline interface, a plateau with a duration between 2 and 6 nsec, depending on the incident energy density, was observed. To confirm that this plateau is indeed produced by differences in the thermodynamic properties of a-Si and c-Si, similar measurements were performed on samples with 215 and 400 nm a-Si thicknesses. The results, in Fig. 3, show similar plateaus as the melt front reaches the amorphous-crystalline interface.

The plateaus observed in the traces in Figs. 2 and 3 are consistent with a first-order transition of a-Si to metallic liquid at a melting temperature reduced from the melting temperature of crystalline Si. The data are intepreted as follows. Melt initiates at a the free surface when the temperature exceeds the melting temperature T_a of a-Si. This melt propagates into the a-Si film at T_a until reaching the a-Si/c-Si interface. Because the temperature of the molten liquid is less than the melting temperature T_c of c-Si, the melt front pauses at this interface until enough energy is absorbed from the laser to raise the liquid temperature above T_c. At that time, the melt front propagates into the underlying c-Si. It should also be noted that the melt velocity (slope of the melt depth versus time) differs in a- and c-Si; in part, this difference reflects the change in thermal conductivity for the two phases and the different latent heats of melting.

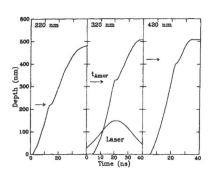
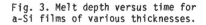

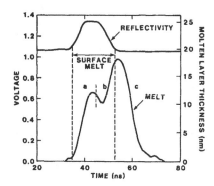

Fig. 3. Melt depth versus time for a-Si films of various thicknesses.

Fig. 4. Transient conductance and surface reflectance for a 0.20 J/cm² pulse incident on a 320 nm a-Si film.

The absolute temperature difference ΔT between the a-Si and c-Si melting temperature can be determined from the plateau width (Δt) and the incident laser intensity $I(t)$ by

$$C_p d(\Delta T) + \kappa \frac{\partial T}{\partial z} \Delta t = \int_{\Delta t} (1 - R_\ell) I(t) dt \qquad (3)$$

where C_p is the specific heat of the liquid, d is the thickness of the molten layer, κ is the thermal conductivity, $\partial T/\partial z$ is the temperature gradient at the interface and R_ℓ is the reflectivity of liquid Si at the laser wavelength. As noted above, the thermal properties of a-Si cannot be readily measured at high temperature. Lack of knowledge of the high temperature thermal conductivity of a-Si (κ_a) introduces an uncertainty in $\partial T/\partial z$. Upper and lower limits can be determined for ΔT, however, by assuming $\kappa_a = 0$ and $\kappa_a = \kappa_c$, the thermal conductivity of c-Si. This procedure yields a melting temperature reduction for a-Si of 200 ± 50 K.

If a-Si is irradiated at laser energy densities only slightly above the threshold for surface melting, markedly different behavior is observed from that for high energy irradiation. This behavior, for an incident energy density of 0.20 J/cm^2, is shown in Fig. 4 for both the transient conductance and the transient reflectance. Of particular interest are the double-peaked structure in the transient conductance and the fact that the surface reflectivity returns to its solid phase value before the molten layer (transient conductance) disappears. Similar behavior is observed for all laser energy densities <0.4 J/cm^2, and is interpreted as follows. Melt initiates at the free surface, and propagates into a depth of ~ 12 nm, whereupon this primary melt begins to solidify. With the solidification of the primary melt, latent heat $\Delta H_{\ell c}$ is released which heats the underlying a-Si above its the reduced melting temperature T_a to produce a thin molten layer that mediates the explosive crystallization. This self-sustaining explosive crystallization process transforms the a-Si to crystalline Si. Because the melt is severely undercooled and resolidification occurs from a disordered interface, fine-grained polycrystalline Si is observed after the explosive crystallization.

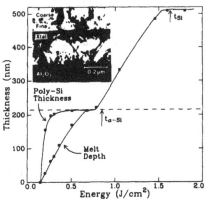

Fig. 5. Thickness of the primary melt and total poly-Si versus incident energy density on a 215 nm a-Si film. Inset: TEM cross section after 0.5 J/cm^2 illustrating coarse- and fine-grained poly-Si.

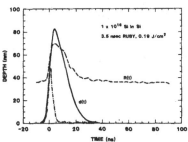

Fig. 6. Simultaneous reflectance and conductance measurements on a-Si irradiated with a 3.5 nsec laser pulse.

58

The explosively transformed depth can be determined from the micro-structure measured by cross-section TEM. An example of the microstructure, which shows coarse-grained polycrystalline Si on fine-grained polycrystalline Si on the underlying crystalline Si, is given in the inset of Fig. 5. Also shown in Fig. 5 is the primary melt depth and the total transformed polycrystalline thickness versus incident laser energy density. An estimate of the explosive crystallization velocity, obtained by dividing the transformed thickness by the duration of the molten layer, is ~ 15 m/sec. This velocity is comparable to the velocity at which liquid Si transforms to a-Si [23]. Slightly above the melt threshold energy density, essentially the entire amorphous thickness is transformed to crystalline Si for a 30 nsec laser pulse at 694 nm. The plateau in the transformed thickness versus incident energy density is another manifestation of the reduced melting temperature a-Si compared to c-Si.

The phenomenon known as explosive crystallization was originally observed in thin films of a-Ge on insulating substrates. In those experiments, it was found that self-sustaining crystallization occurred only for substrate temperatures above a critical temperature. For lower temperature, crystallization stopped because there was insufficient energy to raise the nontransformed material to the melting temperture. The transformation depths at the lowest laser energy are assumed to be limited by the same process. If the interpretation is correct, one would expect only limited explosive crystallization in a-Si films irradiated with ultrashort laser pulses at room temperatures; i.e., explosive crystallization would terminate rapidly because the large temperature gradients at the interface of the primary melt would prevent the neighboring a-Si from reaching T_a. To determine if such quenching occurred, measurements similar to those illustrated in Fig. 4 were performed with 3.5 nsec laser pulses at 694 nm. The results are shown in Fig. 6. The resolidification velocity is greater with the 3.5 nsec than for the 30 nsec irradiation, as expected; however, no evidence is observed for a buried molten layer for this pulse length. As noted above, the explosive crystallization is quenched because of the large amounts of energy required to heat the underlying a-Si to T_a in the presence of the large $\partial T/\partial z$. These results are therefore consistent with explosive crystallization mediated by a thin molten layer.

Further confirmation of explosive crystallization mediated by a molten layer has recently been obtained by Narayan and coworkers [24] and by Sinke and Saris [25] in measurements of the impurity motion for a-Si samples implanted with Cu and irradiated with 30 nsec pulses. Since copper has a low solubility in Si, surface segregation of implanted Cu is observed for samples irradiated with an energy density sufficient to melt through the a-Si film. However, when samples were irradiated at energy densities slightly above the melt threshold, Cu was observed to segregate to the

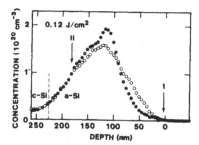 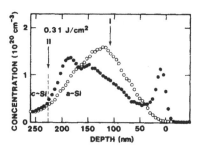

Fig. 7. Cu depth distribution for Cu in a-Si after explosive crystallization, showing Cu motion produced the buried molten layer (after Sinke and Saris, Ref. 25).

a-Si/c-Si interface. The effect is illustrated in Fig. 7 where depth profiles at selected energy densities are shown. Segregation of Cu to the surface is attributed to normal melt segregation, whereas the peak that appears near the a-Si/c-Si interface is attributed to Cu that was carried by the thin molten layer which mediates explosive crystallization. The primary melt depth, directly produced by the energy absorbed from the incident laser beam, and the secondary melt depth, which is the depth of the explosive crystallization, were estimated from the Cu migration. The results are very similar to those shown in Fig. 5 from the transient conductance measurements. It should be emphasized that the observed Cu transport is inconsistent with homogeneous bulk nucleation.

B. Silicon-Germanium Alloys

Silicon-Germanium alloys $Si_{100-x}Ge_x$ have well-understood thermodynamic properties and offer a model system for studies of the effects of melting temperature changes during pulsed-laser melting. Because these alloys form a continuous solid solution [26], the melting temperature can be varied from the 1210 K melting temperature of Ge to the 1685 K melting temperature of Si, which permits studies of the effects of superheating over a wide range of temperatures. In addition, because the thermal conductivity of the alloys are significantly lower [27] than that for either Si or Ge, the $Si_{50}Ge_{50}$ alloy, which has a melting temperture reduced by 240 K relative to that of a-Si, is a close analog of a-Si. Recent studies [17] of these alloys have provided support for the interpretation of the plateau in the melting depth versus time of a-Si as the effect of a reduced melting temperature in a-Si compared to that of c-Si and have permitted an experimental estimate of the relationship between the melt velocity and superheating.

The effects of a reduced melting temperature are illustrated by the data in Fig. 8. These data show the melt depth versus time measured for a 165 nm $Si_{50}Ge_{50}$ alloy layer on Si melted with a 30 nsec ruby laser pulse. Comparison of the melt depth versus time in the Si-Ge alloys with that shown in Figs. 2 and 3 for a-Si reveal that the behavior in these two systems is qualitatively similar. The behavior is also found to be quantitatively similar [17]. Detailed analysis of data on the $Si_{50}Ge_{50}$ alloys, which have the melting temperature reduced 240 K compared to that for c-Si, are consistent with the intepretation of the measurements of the reduced melting tempertaure of 200 ± 50 K for a-Si.

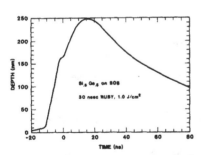

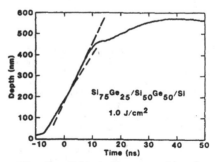

Fig. 8. Melt depth versus time for a $Si_{50}Ge_{50}$ alloy on Si illustrating the similarity of the behavior to that of a-Si.

Fig. 9. Melt depth versus time for a $Si_{75}Ge_{25}/Si_{50}Ge_{50}/Si$ alloy. Note the change in slope at the $Si_{75}Ge_{25}/Si_{50}Ge_{50}$ interface, emphasized by the dotted lines.

A second important point that should be noted for the data in Fig. 8 is the change in melt velocity at the Si-Ge/Si interface. As noted above, the melt dynamics of these alloys have been used to study the dependence of the velocity on the superheating [17]. For these studies, it is convenient to use samples with buried alloy layers which have lower melting temperature than the surface layer. Therefore, samples with the structure Si on $Si_{80}Ge_{20}$ on Si or $Si_{75}Ge_{25}$ on $Si_{50}Ge_{50}$ on Si were formed. Melt depth versus time data for a $Si_{75}Ge_{25}$ on $Si_{50}Ge_{50}$ on Si sample are shown in Fig. 9. There are several points to note in these data. First, comparison of the reflectance and conductance data demonstrates that melt initiated at the free surface. Second, the temperature of the buried $Si_{50}Ge_{50}$ alloy is superheated by at least 120 K, the difference between the melting temperature of the overlying $Si_{75}Ge_{25}$ surface layer and the $Si_{50}Ge_{50}$ alloy; however, no bulk nucleation of melt was observed. Third, the melt velocity increased 7 m/s when the melt front penetrated this superheated alloy. Since the change in the melting temperature of the interface is 120 K, this measurement yields a measurement of the slope of the superheating versus velocity of 17 ± 3 K/m/s. This result thus provides an experimental estimate of this superheating-velocity relationship, and extension of these measurements throughout the Si-Ge alloy system should permit detailed evaluation of theoretical models for the fundamental relationships that control melt and solidification dynamics under rapid solidification conditions.

C. Measurements of Interfacial Undercooling

Solidification occurs when the temperature of a liquid falls below the melting temperature of the solid. The free energy difference between the liquid and the solid phase provide the driving force for solidification. Several models have been developed to relate this undercooling driving force to the solidification velocity [28,29]. However, prior to the transient conductance technique, there were no reliable techniques for measuring the solidification velocity under the rapid solidification conditions following pulsed surface melting to test these theories. As will be discussed below, detailed measurements of the transition between melt and solidification permit estimates of this velocity-undercooling relationship.

Solidification of a SOS sample melted with 3.5 nsec laser pulse is shown in Fig. 10. The solidification velocity obtained from the numerical derivative of the melt depth is also shown. The important feature for our consideration is the time t_m required for the velocity to increase from $v = 0$ to its maximum value v_m. The heat transport across the liquid-solid interface obeys the relationship

$$J_s = J_\ell + v\Delta H \qquad (4)$$

where J_s is the flux of heat transported into the solid which is balanced by the heat $v\Delta H$ released by solidification and the heat flux J_ℓ that flows from the liquid to the solid. In general, J_ℓ will be comprised of two terms: the energy deposited by the laser and the energy released by the liquid as it undercools. These contributions can be evaluated numerically using computer-based heat transport codes for precise determination of the undercooling [30]. However, in the case of very short laser pulses which deposit negligible energy after the melt peak, approximate analytical solutions, which yield physical insight into the important processes, can be used to estimate the undercooling [31].

To illustrate these solutions, consider the total heat transported into the solid between the time when $v = 0$ and when $v = v_m$, denoted by Δt_m

$$\int_{\Delta t_m} J_s dt = \int_{\Delta t_m} J_\ell dt + \int_{\Delta t_m} v\Delta H dt \ . \qquad (5)$$

If no laser energy is deposited after v = 0 and changes in the temperature gradient in the molten Si can be neglected, the integral over J_ℓ is simply given by $[C_p \Delta T(d+d_m)/2]$, where C_p is the heat capacity of molten Si, d is the maximum melt depth, d_m is the melt depth at t_m, and ΔT is the under-cooling. Similarly, the integral of the latent heat released is $\Delta H \Delta d$, where $\Delta d = d - d_m$. The most difficult integral to evaluate is the one containing J_s. However, for the short times ($\lesssim$ 8 nsec) under consideration, we will assume that $J_s = \kappa \, \partial T/\partial z$ is constant and equal to in $v_m \Delta H$. In that case, this integral can be approximated by $[\Delta H v_m \Delta t_m]$. With these approximations, the relationship between the velocity and the undercooling bcomes

$$\Delta T = \frac{2}{C_p} \frac{\Delta H}{(d+d_m)} (v_m \Delta t_m - \Delta d) . \qquad (6)$$

Application of Eq. 6 to the data of Fig. 10 yields an undercooling $\Delta T \cong$ 90 K for an interface velocity of 6 m/sec. Assuming that the velocity is lin-early proportional to the undercooling in this velocity regime yields an undercooling-velocity slope of $\approx$ 15 $\pm$ 5 K/m/sec. It should be emphasized that the treatment given here is only approximate; however the estimates are in reasonable agreement with detailed numerical simulations [30].

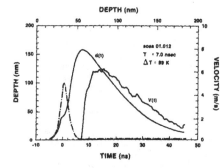

Fig. 10. Melt depth versus time, and the numerical derivative of this curve to yield velocity, after melting with a 3.5 nsec pulse.

D. Effects of As Impurities on the Solidification Velocity of Si During Pulsed Laser Annealing

It is well established that impurity atoms can be incorporated at con-centrations substantially in excess of the equilibrium solid solubility during the rapid solidification that follows surface melting with a pulsed laser [32-33]. Several theoretical treatments have been given for the kine-tics that control the formation of such supersaturated solutions [34-35]. In general these treatments are given in terms of a velocity-dependent interfacial segregation coefficient; however, because of the lack of direct measurements of the solidification velocity, theoretical treatments to date have assumed that the solidification velocity is independent of the impurity content of the solid.

To explore the effects of impurities on solidification dynamics under rapid solidification conditions, we have measured the melt and resolidifi-cation velocity of As in Si during pulsed laser annealing. Preliminary studies [36] suggested that impurities decrease the solidification velocity. Detailed measurements of As in Si demonstrate that the velocity is reduced at high As concentrations, which reduce the melting temperature. Buried molten layers have been observed for high As concentrations because of this reduction in T_c [17].

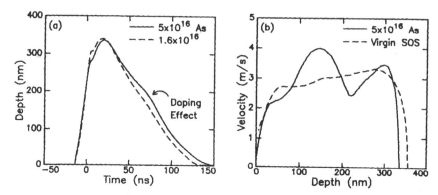

Fig. 11. (a) Melt depth versus time for Si implanted with 1.6 and
5 x 10^16 As/cm^2. (b) Solidification velocity versus depth for the
5 x 10^16 As/cm^2 data in (a) compared to unimplanted SOS.

Melt depths, determined from the transient conductance, for samples implanted with 1.6 and 5 x 10^{16} As/cm^2 are shown in Fig. 11. To obtain the melt depth the contribution of the activated As dopants to the electrical conductivity of the solid phase was deconvoluted from the transient conductance. This deconvolution was performed by assuming the depth dependence of the electrical conductance produced by the As doping followed the As depth distribution measured after solidification. The interface velocity versus depth, obtained from the numerical time derivative of the melt depth, is shown in Fig. 11b. For 1.6 x 10^{16} As/cm^2, the velocity is comparable to that for pure Si; however, the velocity decreases by ~ 20% for 5 x 10^{16} As/cm^2. The velocity reduction was interpreted [17] in terms of a reduced melting temperature for the Si-As alloy compared to that of pure Si. Assuming that the velocity reduction is dominated by this effect, we estimate a decrease of ~ 150 K in the melting temperature for 7 at.% As. A more detailed analysis of the full depth dependence indicates there may also be some interfacial kinetic effects.

Effects due to this decreased melting temperature have been observed in other experiments. Because of the high thermal conductivity of molten Si, the temperature of the overlying Si layer can be depressed to temperatures such that surface nucleation occurs. This temperature depression was achieved two ways: increasing the solidification velocity by decreasing the pulse width, and increasing the As concentration to decrease the melting temperature. In the latter case, surface nucleation occurred for 30 nsec pulses.

To illustrate surface nucleation, measurements for 3.5 nsec FWHM pulses at 694 nm samples implanted with 5 x 10^{16} As/cm^2 are shown in Fig. 12. Within the experimental uncertainty of ± 1 nsec, the reflectivity increases at the same time as the transient conductance, indicating that melt initiated at the free surface. Comparison of the transient reflectance and conductance demonstrates that the surface solidifies well before the solidification front from the bulk reaches the surface. Surface solidification begins when the solidification front is still ~ 150 nm below the surface -- i.e., near the maximum in the As distribution. Since surface solidification produces a buried molten layer, transient conductance measurements no longer yield a simple melt depth but rather, give the thickness of this buried molten layer.

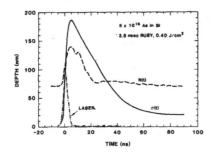

Fig. 12. Transient conductance $\sigma(t)$ and reflectance $R(t)$ showing surface solidification for Si implanted with 5×10^{16} As/cm^2 after melting with a 3.5 nsec pulse.

IV. CONCLUSION

Studies of melt and resolidification dynamics in a wide variety of Si-based systems were reviewed. These measurements confirm the existence of a substantial decrease in the melting temperature of a-Si compared to that of c-Si and permit direct observation of the mechanism by which explosive crystallization transforms a-Si to c-Si. Extension of the measurements to short (~ 3.5 nsec laser pulses) permit the undercooling versus velocity to be estimated. Extension to Si that contains As impurities permits one to determine the change of the melting temperature for metastable Si-As alloy systems. In addition, because As decreases the melting temperature of Si, surface nucleation and buried melts are observed for very short laser pulses or high As concentrations.

Finally, studies of Si-Ge alloys demonstrate that these alloys provide a model system for the study of superheating and undercooling. The Si$_{50}$Ge$_{50}$ alloy closely models a-Si and confirms the interpretation of the transient conductance data in a-Si in terms of a reduced melting temperature. Studies in layered samples provide direct observations of superheating of buried layers and measurements of the interface temperature-velocity relationship under rapid melting and solidification conditions.

REFERENCES

1. G. J. Galvin, Michael O. Thompson, J. W. Mayer, R. B. Hammond, N. Paulter and P. S. Peercy, Phys. Rev. Lett. 48, 33 (1982).
2. G. J. Galvin, Michael O. Thompson, J. W. Mayer, P. S. Peercy, R. B. Hammond and N. Paulter, Phys. Rev. B27, 1079 (1983).
3. Michael O. Thompson, G. J. Galvin, J. W. Mayer, R. B. Hammond, N. Paulter and P. S. Peercy, in Laser and Electron Beam Interactions with Solid, ed. by B. R. Appleton and C. R. Celler (North Holland, NY, 1983), p. 209.
4. Michael O. Thompson and G. J. Galvin, in Laser Solid Interactions and Transient Thermal Processing of Materials, ed. by J. Narayan, W. L. Brown and R. A. Lemons (North Holland, NY, 1983), p. 57.
5. P. Baeri, S. U. Campisano, G. Foti and E. Rimini, J. Appl. Phys. 50, 788 (1979).
6. R. F. Wood and G. E. Giles, Phys. Rev. B23, 2923 (1981).
7. D. H. Auston, C. M. Surko, T.N.C. Venkatesan, R. E. Slusher and J. A. Golovchenko, Appl. Phys. Lett. 33, 437 (1978).
8. Michael O. Thompson, G. J. Galvin, J. W. Mayer, P. S. Peercy and R. B. Hammond, Appl. Phys. Letts. 42, 445 (1983).
9. P. S. Peercy, G. J. Galvin, M. O. Thompson, J. W. Mayer and R. B. Hammond, Physica, 116B, 558 (1983).

10. G. J. Galvin, Michael O. Thompson, J. W. Mayer, P. S. Peercy and R. B. Hammond in Laser Processing of Semiconductor Devices, ed. by C. C. Tang, Proc. SPIE, Vol. 385, 38 (1983).
11. Michael O. Thompson, G. J. Galvin, J. W. Mayer, P. S. Peercy, J. M. Poate, D. C. Jacobson, A. G. Cullis and N. G. Chew, Phys. Rev. Letts. 52, 2360 (1984).
12. H. C. Weber, A. G. Cullis, and N. G. Chew, Appl. Phys. Letts. 43, 669 (1983).
13. J. Narayan and C. W. White, Appl. Phys. Letts. 44, 35 (1984).
14. R. F. Wood, D. H. Lowndes and J. Narayan, Appl. Phys. Letts. 44, 770 (1984).
15. D. H. Lowndes, R. F. Wood, C. W. White and J. Narayan, in Energy Beam-Solid Interactions and Transient Thermal Processing, ed. J.C.C. Fan and N. M. Johnson (North Holland, NY, 1984), p. 99.
16. D. H. Lowndes, R. F. Wood, and J. Narayan, Phys. Rev. Letts. 52, 5621 (1984).
17. Michael O. Thompson, P. S. Peercy, J. Y. Tsao, M. J. Aziz, to be published.
18. P. S. Peercy, Michael O. Thompson and J. Y. Tsao, to be published.
19. Michael O. Thompson, P. S. Peercy and J. Y. Tsao, this proceedings AP.6.
20. E. P. Donovan, F. Spaepan, D. Turnbull, J. M. Poate and D. C. Jacobson, Appl. Phys. Letts. 42, 698 (1983).
21. S. A. Kokorowski, G. L. Olson, J. A. Roth and L. D. Hess, Phys. Rev. Letts. 48, 498 (982).
22. P. Baeri, G. Foti, J. M. Poate and A. G. Cullis, Phys. Rev. Letts. 45, 2036 (1980).
23. Michael O. Thompson, J. W. Mayer, A. G. Cullis, H. G. Webber, N. G. Chew, J. M. Poate and D. C. Jacobson, Phys. Rev. Letts. 50, 896 (1983).
24. J. Narayan, C. W. White, O. W. Holland and M. J. Aziz, J. Appl. Phys. 56, 1821 (1984).
25. W. Sinki and F. W. Saris, Phys. Rev. Letts. 53, 2121 (1984).
26. J. P. Dismukes and L. Ekstrom, Trans. Metall. Soc., 233, 672 (1965).
27. B. Abeles, D. S. Beers, G. O. Cody and J. P. Dismukes, Phys. Rev. 125, 44 (1962).
28. J. W. Cahn, W. B. Hillig and G. W. Sears, Acta. Metall. 12, 1421 (1964).
29. F. Spaepen and D. Turnbull, in Laser Annealing of Semiconductors, ed. J. M. Poate and J. W. Mayer (Academic Press, NY, 1982) Ch. 2.
30. Michael O. Thompson and P. H. Bucksbaum, this proceedings, A3.5.
31. G. J. Galvin, J. W. Mayer and P. S. Peercy, to be published.
32. J. M. Poate in Ref. 3, p. 121.
33. C. W. White, B. R. Appleton and S. R. Wilson in Ref. 28, p. 111.
34. J. A. Jackson in Surface Modification and Alloying by Laser, Ion and Electron Beams, ed. J. M. Poate, G. Foti and D. C. Jacobson (Plenum, NY, 1983), Ch. 3.
35. M. J. Aziz, J. of Appl. Phys. 53, 1558 (1982).
36. G. J. Galvin, J. W. Mayer and P. S. Peercy, in Ref. 15, p. 111.

ELECTRON-PHONON AND PHONON-PHONON INTERACTIONS UNDER LASER ANNEALING CONDITIONS

ALVIN COMPAAN
Kansas State University, Dept. of Physics, Cardwell Hall, Manhattan, KS 66506

ABSTRACT

Since the first time-resolved Raman studies of pulsed laser annealing (PLA) effects in Si, a number of cw Raman studies have been performed which provide a much improved basis for understanding the consequences on Raman spectra of temperature-dependent resonance effects, high carrier density effects, phonon anharmonicity, and strain effects. Here we briefly review these effects and then analyze the latest pulsed Raman studies of PLA including Stokes/anti-Stokes ratios, the shift and shape of the first order line, and time-resolved second-order spectra. The Raman data indicate the existence of a Raman-silent phase followed by a rapidly cooling solid which begins within 300 K of the normal melting temperature of Si. The Raman data also give evidence of carrier densities in the recrystallizing solid of $\sim 1-2 \times 10^{19}$/cm^3.

INTRODUCTION

When visible lasers are used for pulsed laser annealing (PLA) the optical energy in most semiconductors is initially absorbed into the electronic system. As the carriers cool and electron-hole recombination occurs, energy is fed into the phonon system as energy of atomic motion in the crystal lattice. Better understanding of the rate of energy transfer from electrons to phonons and of the rate of thermalization within the many branches of the phonon system have been the stimuli for a large number of experiments, many of which are discussed in these proceedings.

A great deal of attention has focussed on whether the melting process may be assisted by the high electron and hole densities created by the laser [1,2] and whether perhaps a new plasma-induced phase may result [3]. At least two issues are central to this discussion: 1) the electron density and temperature and 2) the amount of energy and the degree of thermalization in the phonon system. Much useful information on carrier densities has been obtained by transient reflectivity studies from the nanosecond [4] to the subpicosecond [5] regimes and is reviewed in other papers in these proceedings.

The measurements of lattice expansion by time-resolved x-ray diffraction [6] and measurements of thermionic emission [7] indicate average properties of the lattice but cannot give phonon occupation factors directly. The question of whether melting occurs below the normal melting temperature and whether the phonon system is in thermal equilibrium is nicely addressed in a time-resolved Raman experiment. Such measurements have been performed by two independent groups using Raman scattering with pulsed lasers [8-12]. Early results of Stokes/anti-Stokes ratios indicated that the phonon populations (at least in the optic branch) remain well below the normal melting point of Si. However, incomplete knowledge of the high temperature behavior of Raman scattering in Si left some uncertainty in the interpretation of these experiments.

In the period of time since the first report of Raman data under PLA conditions [8], a number of cw laser Raman studies have been performed which have greatly improved the base of knowledge of the Raman effect in Si upon which the pulsed Raman measurements depend. The cw measurements include studies of phonon frequency shifts and broadening as a function of crystal temperature using oven heating [13,14] and cw CO_2 laser heating

[15]. Other studies have been performed on very heavily doped Si to elucidate the effect on phonons of high carrier densities [16,17,18]. Finally
the high temperature behavior of electronic resonance effects in phonon
Raman scattering is now much more completely understood [19].

In this paper we shall review these recent cw Raman measurements and
then apply this information to the most recent pulsed Raman data obtained
under PLA conditions. We shall examine not only Stokes/anti-Stokes ratios
but also the Raman intensity and the shift, broadening and asymmetry of the
observed Raman peaks. A second objective is to examine energy flow between
electron and phonon systems and within the phonon system. Raman data from
doped semiconductors and second order pulsed Raman data will help elucidate
this behavior.

ENERGY FLOW INTO THE PHONON SYSTEM: ELECTRON-PHONON INTERACTIONS

Energy and Momentum Conservation

Energy and crystal momentum conservation restricts the allowed electron-
phonon and hole-phonon scattering in Si. For parabolic bands and a disper-
sionless optic phonon, maximum and minimum phonon momenta, q, are given by
[20]

$$|q(^{min}_{max})| = |k| [1 \mp (1 - \frac{\hbar\omega_0}{E})^{1/2}] \tag{1}$$

where $\hbar\omega_0$ is the optic phonon energy and E and k are the initial excess
electron energy and momentum. For frequency-doubled Nd-YAG laser excitation
($h\nu$=2.33 eV) band-to-band absorption is phonon-mediated. About 1.2 eV of
excess energy is shared by the electron and hole. For a 1.0 eV electron
scattering in the [100] direction (m^*= 0.91 m_0), the minimum momentum change
would be $\Delta q_{min} \approx 1.6 \times 10^6$ cm^{-1}. Similarly in the transverse direction (perpendicular to [100]), m^*= .19 m_0 and $\Delta q_{min} \approx 7.1 \times 10^5$ cm^{-1}. For light holes
and heavy holes with ~1.0 eV a similar range of values is obtained [21].
Maximum momentum changes are of the order of the distance to the zone boundary for eV electrons and decrease in proportion to the square root of the
excess energy. Clearly energy may be deposited over a large fraction of the
optic phonon branch, but as the electrons and holes cool Δq_{min} increases and
Δq_{max} decreases. These electron-phonon scattering processes are illustrated
in Fig.1 for scattering near the conduction band minimum at 0.85 X, for phonon momentum parallel to [100] and perpendicular to [100].

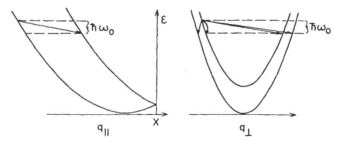

Figure 1: Portions of the band structure of Si near the band minimum at
$(0.85 ,0,0) \frac{\pi}{a}$ showing momentum-conserving phonon scattering events along the
X direction ($q_\parallel$) and the perpendicular direction ($q_\perp$).

For a backscattering Raman experiment in Si the crystal momentum transfer is $\Delta q = 4\pi n/\lambda = 1 \times 10^6$ cm^{-1} at $\lambda = 532$ nm and $\Delta q = 1.7 \times 10^6$ cm^{-1} at $\lambda = 405$ nm. Thus Raman experiments at these wavelengths do overlap with the range of phonon momenta initially produced in electron-phonon scattering. However as the electrons and holes cool, phonons are no longer directly generated at wave vectors probed by Raman scattering.

Effects of High Carrier Densities

It is now quite well established that Auger recombination limits the maximum carrier densities during nanosecond and picosecond pulse excitation in Si. For a 1 J/cm^2, 10 nsec pulse one obtains

$$n_{max} = (\alpha I/Ch\nu)^{1/3} \simeq 2 \times 10^{20} \text{cm}^{-3} \tag{2}$$

where α is the absorption coefficient (1×10^4/cm at 532 nm), C is the Auger rate ($C \simeq 4 \times 10^{-31}$ cm^6/s) [22], I is the laser intensity and $h\nu$ the photon energy. Recent ir reflectivity measurements give a similar number [4]. Carrier densities of $\sim 10^{20}$ cm^{-3} produce significant changes in electron-phonon coupling in Si and can be studied in phonon Raman spectra from heavily doped Si.

cw Raman Scattering in Doped Si

As free electrons or holes are added to Si by doping, a spectrum of possible electronic excitations is added with excitation energies which overlap the phonon energies. The optic phonon Raman spectrum displays this coupling in the form of an asymmetric line (Breit-Wigner-Fano shape) which is broadened and shifted in energy [16].

Figure 2a illustrates the effect observed in the spectrum of heavily B-doped Si [17]. This sample was prepared by ion-implantation and pulsed excimer laser annealing (λ=308nm) with a heavily doped surface layer $\sim$0.2μm thick. Some penetration of laser light into the unimplanted substrate leads to the Raman peak which occurs at the unshifted, 520 cm^{-1} position for pure Si. Both n- and p-type doping produce a phonon softening due to interaction with electronic excitations. The interaction also leads to increased broadening of the line which is also apparent in Fig. 2a.

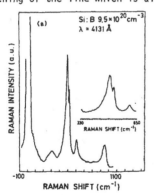
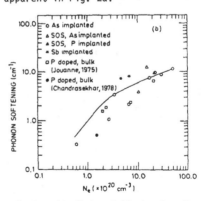

Figure 2: (a) Raman spectrum of extremely heavily B-doped Si showing the unshifted substrate peak at 520 cm^{-1} and the broadened and shifted peak from the heavily doped region. (b) Phonon softening as a function of carrier concentration for n-type dopants. From ref. 17.

The density dependence of the real part of the phonon self energy (phonon softening) is shown in Fig. 2b for n-type doping [17]. The imaginary contribution to the self energy, which appears as broadening of the Raman line, has a similar dependence on carrier density. For a hole density of $p=1\times10^{21}$ cm^{-3} we find a width of $\Delta\bar{\nu}\approx15$cm^{-1} compared with the normal room temperature optic phonon width of $\Delta\bar{\nu}\approx2.5$cm^{-1}. Thus the broadening is increased by nearly an order of magnitude due to interaction with single particle electronic excitations. In undoped Si the optic phonon line broadening is due only to anharmonic interactions with other phonon modes. In doped Si, however, the optic phonon can decay into electronic excitations, thus introducing an important new consideration into the phonon dynamics.

The origin of these single particle electronic excitations is shown in Fig. 3. Here the essential parts of the electronic band structure are qualitatively sketched with arrows indicating the type of transitions which may be produced by phonon decays. These excitations will affect not only near-zone-center optic phonons, cf. Fig. 3a and 3b, but via intervalley scattering, zone boundary optic and acoustic modes will also be affected. Finally low frequency acoustic modes [18] may be involved in the type of excitation sketched in Fig. 3c.

These carrier-phonon interactions may have important influence on the phonon thermalization. When the carrier concentration is high, phonons can decay into either hole excitations or electron excitations which can then decay into other phonons again. The net result is to speed thermalization of energy among the optic and acoustic modes reducing the possibility for occurrence of phonon hot spots at isolated regions of the Brillouin zone.

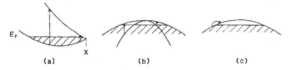

(a) (b) (c)

Figure 3: Portions of the Si band structure showing the type of single-particle excitations which produce phonon softening and broadening.

Anharmonic Phonon Interactions

In pure Si the lifetime of optic phonons is determined by decay into two acoustic branch phonons. This decay rate has recently been studied as a function of lattice temperature by Tsu and Hernandez [13] and by Balkanski, et al. [14] using cw Raman scattering. They find that the optic phonon lifetime decreases from 2 psec (2.5cm^{-1}) at room temperature to 0.4 psec at 1140K. Extrapolating the fit of ref. 12 using third and fourth order anharmonic terms yields an optic phonon lifetime of 0.2 psec at the Si melting point. Nemanich, et al. [15] have recently measured the Raman line in Si heated to melting with a cw CO$_2$ laser. The width of the Raman line which they measure is consistent with the extrapolation discussed above.

The phonon-phonon interactions also renormalize the phonon frequencies signficantly at high temperature. The cw Raman experiments described above show a general softening of frequencies due to these interactions and to the lattice expansion, which of course is also a consequence of the anharmonicity. The optic phonon shift is extrapolated to be -41 cm^{-1} at 1700 K [14].

The above measurements of phonon-phonon interactions have been performed essentially under thermal equilibrium conditions. But the existence of these data now allow the possibility of interpreting similar shifts in pulsed Raman data as a further check for consistency with the temperature inferred from Stokes/anti-Stokes ratios.

Resonant Raman Effects

Raman scattering with above-band-gap photon energies in a semiconductor is sensitive to details of the electronic band structure. The position and width of critical points are temperature-dependent and therefore resonance effects in Raman scattering are also temperature-dependent. Compaan and Trodahl have recently examined these effects in a cw Raman study of oven-heated Si [19]. We briefly review the results.

The rate of Stokes photon scattering, R'_s, is related to the Raman susceptibility χ''_s, by [19]

$$R'_s = [\frac{T_s T_L P'_L \omega_s^3}{(\alpha_s + \alpha_L) n_L n_s}] [\frac{\Delta\Omega'[n(\omega_0)+1]}{32\pi^2 c^4 \omega_0}] |\varepsilon_L \varepsilon_s \chi''_s|^2, \qquad (3)$$

where T_s and T_L are transmission coefficients of the Si surface, P'_L is the incident laser power, ω_s the scattered photon frequency, ω_0 the phonon frequency, $\alpha_s(\alpha_L)$ the absorption coefficient and $n_s(n_L)$ the index of refraction at the scattered (incident) frequency, $\Delta\Omega'$ the external collection solid angle, n the phonon occupation factor, and $\varepsilon_L(\varepsilon_s)$ the polarization unit vectors. Note that the cartesian indices have been suppressed on $\varepsilon_L, \varepsilon_s$ and χ'_s. Temperature dependence occurs in the factors $T_s, T_L, \alpha_s, \alpha_L, n_L, n_s$ and χ''_s in addition to the phonon population factor $n(\omega_0)$.

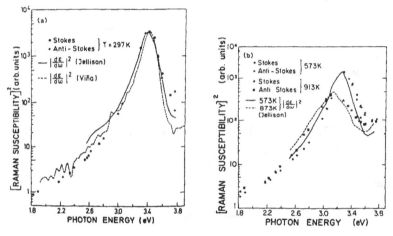

Figure 4: Raman susceptibility, χ''_s, as a function of laser photon energy for room temperature (a) and for two elevated temperatures (b), from ref. 19.

By far the strongest temperature dependence other than for $n(\omega_0)$ occurs in the absorption coefficients [23], α_s and α_L, and in the Raman susceptibility, χ''_s. (We have found [19], however, that over much of the spectrum the temperature dependence of $(\alpha_s + \alpha_L)^{-1}$ nearly cancels the temperature dependence of $|\chi''|^2$.) The temperature dependence of χ''_s is shown in Fig. 4. The shift and broadening of the resonance is clearly apparent in the curves of Fig. 4b. Note that we have compared our Raman measurements of $|\chi''_s|^2$ with the ellipsometric data of Jellison, et al. [23] and of Vina [24] after shifting upward in frequency by $\omega_0/2$ to account for both incoming and outgoing photon resonances. The Raman and ellipsometric data extend only to ~900 K. However the shift and broadening of the resonance are linear in temperature to 900 K (see Fig. 5) and we believe a linear extrapolation to the melting point is reasonable. This extrapolation is used in the data fit described below.

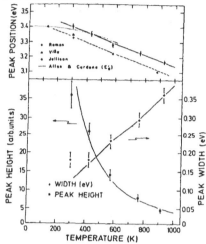

Figure 5: Shift and broadening of the Raman susceptibility as a function of temperature; from ref. 19.

The temperature-induced shift and broadening of the Raman resonance cause a gradual decrease in the strength of the optic phonon Raman signal. The Stokes Raman strength follows the empirical form $A=A_o \exp[-(T-297)/T_o]$ where $T_o=650K$ (1050K) at $\lambda=407nm$ (532nm). This decrease in Raman strength is included in the detailed line shape fits described below.

TIME-RESOLVED RAMAN EXPERIMENTS

Raman Scattering with 10 ns Pulses

The experimental configuration for the pulsed Raman measurements has been described elsewhere [8,10]. The main part of the amplified, frequency-doubled Nd:YAG pulse is used to pump a dye laser with two amplifier stages. The dye laser pulse at 565 nm has excellent homogeneity with a half-power width of 800 μm on the sample. A small fraction of the 532 nm beam is split off and delayed in an optical delay line to be used as the Raman probe. It is focussed to a diameter of ∿120μm on the sample. Pulse durations are 12ns.

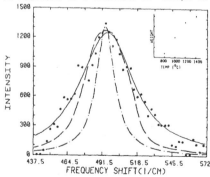

Figure 6: Stokes Raman spectrum (points) as described in text. Dot-dash curve is expected shape for uniform temperature of 1400 K, dashed curve adds thermal broadening from a 900-1700 K temperature distribution, solid curve includes broadening from carrier interactions and temperature weighting distribution shown in the inset.

Figure 6 shows the Stokes Raman signals obtained with 0.18 J/cm^2 probe energy at 532 nm with a probe delay of 90 nsec. The heat pulse power at 565 nm was adjusted to give a full reflectivity rise lasting 80 nsec. The probe power is about 3 times greater than used previously and was chosen so that the probe produces significant heating which keeps the sample temperature high during the probe pulse. We find, under these conditions, Stokes/anti-Stokes ratios of 1400±300K.

The dashed curve of Fig. 6 illustrates the peak position and width expected for a uniform temperature of 1400K. The shift observed in refs. 13 and 14 has been corrected for an assumed two-dimensional compressive strain for the pulsed case arising from one dimensional thermal expansion normal to the surface. This correction lowers the shift by about 5% [25]. Much of the additional broadening can be attributed to temporal and spatial temperature variations which arise from the large probe beam power. Thus, material near the center of the probe spot is held near the melting point but near the edges significant cooling occurs. However, even a temperature variation from 800 to 1400 K (dotted curve in Fig. 6) is insufficient to account for the wings of the lines. It is necessary to include significant additional broadening especially to account for the data above 520 cm^{-1}, the room temperature position of the Raman line. This additional broadening arises from the interaction of the optic phonons with photoexcited carriers in the same way as seen above in the doped samples. The solid curve is the fit obtained by allowing both carrier concentration n and temperature distribution to be adjustable. An excellent fit occurs for a carrier density of $2\pm1 \times 10^{19} cm^{-3}$ and the temperature weighting factors shown in the inset. The weighting includes the observed decrease in Raman strength with temperature discussed above.

Raman Scattering with 150 nsec Pulses

First order Raman scattering gives a direct sample of phonon populations at only one point on the phonon dispersion curves. Therefore we obtain no information on the degree of thermalization among the phonon branches. Second order scattering, however, leads to peaks in the Raman spectrum which generally arise from points of high phonon density of states. In fact most second order peaks in Si arise from overtones $(\omega_1+\omega_1)$ rather than combinations $(\omega_1+\omega_2)$ making identification simple [26]. Second order scattering, unfortunately, has only ~1% or less of the first order intensity in Si at 300K. Thus the second order spectrum is only marginally observable with low repetition rate (~ 10 Hz) lasers. The 10 kHz frequency of a repetitively Q-switched "cw" Nd:YAG, however, permits second order studies.

With intracavity frequency doubling, we obtained ~5 watts of average power in ~150 nsec pulses at 532 nm. Raman scattering was obtained from the same beam by using a 20 nsec gate on the photon counting electronics. This gate was adjusted to begin about 30 nsec after the peak of the laser pulse. A He-Ne reflectivity probe indicated this as the time when recovery from the high reflectivity phase occurred for the central region of the laser spot. A 300 µm aperture was used at the entrance to the spectrometer so that only the central 100µm region of the sample was imaged into the spectrometer. The sample was rotated at ~20 Hz to avoid damage from the laser.

Spectra obtained with this system are shown in Fig. 7. The low power spectrum, Fig. 7a, was obtained with the laser beam attenuated by a factor of 10 and the gate switched off. The high power spectrum Fig. 7b, shows the large enhancement of anti-Stokes peaks relative to the Stokes peaks (both first and second order) indicative of high phonon populations. Stokes/anti-Stokes ratios for the first order line at ~500 cm^{-1}, and for the second order features at ~950 cm^{-1} and 280 cm^{-1} indicate phonon "temperatures" of 1500±300 K. Thus phonon populations in the optic branch at zone center and zone boundary (X & L points) appear, within experimental error,

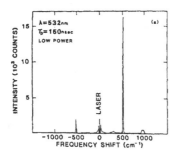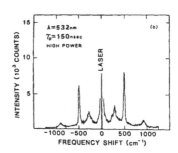

Figure 7: Raman spectra showing both first- and second-order Stokes and anti-Stokes peaks for (a) low power and (b) high power. Nd:YAG laser excitation (λ=532 nm $\tau_p \approx$150 ns) as described in the text.

to be in equilibrium with each other and with the zone boundary transverse acoustic mode (X and L points).

For 150 nsec pulses the data were acquired with a single beam serving for excitation and probing. Thus temperature gradients temporally and spatially are of some concern. The 20 nsec gate late in the laser pulse was chosen to allow most of the laser energy to be deposited before signal acquisition. Furthermore, 50-100 nsec should be sufficient time for temperature gradients normal to the surface to be reduced. (The Raman probe depth for 532 nm in Si above T=1000 K is less than $\sim$0.2μm.) Therefore the temperature estimates given above should apply to a reasonably homogeneous situation. The results indicate equilibration of the phonon modes after $\sim$100 nsec at temperatures near the normal melting point of Si.

CONCLUSIONS

Using the excite/probe technique, with $\sim$10 nsec pulses at 565 nm and 532 nm respectively, we have found the shape and shift of the Raman line as well as the Stokes/anti-Stokes ratio to be indicative of peak phonon temperatures of 1400±300K. This is much higher than previous studies with either a 532 nm/405 nm excite/probe combination or a 485 nm/405 nm combination where temperatures of 300-450°C were found. A variety of suggestions had been proposed earlier to account for the low temperatures inferred from the Raman measurements. These included possible laser pulse-to-pulse fluctuations [27], steep temperature gradients [28], and temperature-dependent correction factors to the Stokes/anti-Stokes ratio [28, 29]. We have addressed these issues in detail elsewhere [10,30] and have shown that they cannot account for the low inferred temperatures. For example, the correction factors to the Stokes/anti-Stokes ratio approach unity at high temperatures as a consequence of the broadening of the resonance seen in Fig. 4 [10,12,19].

Two factors account for the higher inferred temperatures. First, it has now been possible to include explicitly the fact that the Raman strength at the melting point decreases by factors extrapolated to be 0.13 for 532nm excitation and 0.26 for 405nm excitation [31]. The second and major factor is that we have substantially increased the probe power partly to compensate for this drop in signal strength. Thus we have sacrificed the ideal of using a weak probe which does not itself strongly disturb the system. The data of Fig. 6 were obtained with excite and probe powers of 1.0 J/cm^2 and 0.18 J/cm^2 respectively. The probe influence will be proportionately higher, however, because the absorption length is smaller (due to the higher temperature and shorter wavelength) and because the thermal conductivity of Si is much lower at high temperatures. Thus in the present work the probe significantly affects the Si by slowing the cooling after the high reflectivity phase. In fact we can observe a lengthening of the transient reflectivity rise when the probe is on.

Nonuniform probe heating across the probe spot accounts for the inhomogeneous broadening of the Raman peak seen in Fig. 6 since data is acquired over the entire probe spot. Because of this heating the Si cooling rate cannot readily be studied by varying the probe delay. However the 10 nsec data--Stokes/anti-Stokes ratio, shift, and broadening of the Raman line--consistently indicate optic phonon populations characteristic of temperatures of 1400±300K. These data also indicate carrier densities of $1-2 \times 10^{19}$ cm^{-3} consistent with recent transient ir reflectivity measurements [4].

Raman experiments with 150 nsec pulses have provided firm evidence indicating good equilibration among several regions of the phonon dispersion curves. We have found populations characteristic of T=1500±300°K in the optic mode at the zone center and zone boundaries and for the TA mode at the X and L points. Second order spectra have not yet been obtained with 10 nsec pulses, but the fact that the phonon frequency shifts are consistent with Stokes/anti-Stokes ratios suggests that thermalization is largely complete on this time scale as well.

ACKNOWLEDGEMENTS

The author is indebted to H. W. Lo, A. Aydinli, M. C. Lee and G. J. Trott for their collaboration on the experiments with 10 nsec pulses, and to M. Cardona, H. Vogt and the Max Planck Institut für Festkörperforschung - Stuttgart for making possible the work with 150 nsec pulses. The support of the U.S. Office of Naval Research and A. von Humboldt Foundation is gratefully acknowledged.

REFERENCES

1. V. Heine and J. A. Van Vechten, Phys. Rev. B 13, 1622 (1976); J. Bok, Phys. Lett. 84A, 448 (1981).
2. J. A. Van Vechten in Semiconductor Processes Probed by Ultrafast Spectroscopy, edited by R. R. Alfano (Academic, NY, 1983).
3. J. A. Van Vechten, J. Phys. (Paris) 44, C5-11 (1983).
4. J. S. Preston and H. M. Van Driel, Phys. Rev. B 30, 1950 (1984).
5. C. V. Shank, R. Yen, C. Hirliman, Phys. Rev. Lett. 50, 454 (1983); D. Hulin, M. Combescot, J. Bok, A. Migus, J. Y. Vinet, and A. Antonetti, Phys. Rev. Lett. 52, 1998 (1984).
6. B. C. Larsen, C. W. White, T. S. Noggle, J. F. Barhurst and D. M. Mills, Mat. Res. Soc. Symp. Proc. 13, 43 (1983).
7. J. M. Liu, R. Yen, H. Kurz and N. Bloembergen, Appl. Phys. Lett. 39, 7551 (1981); M. Bensoussan and J. M. Moisson, Physica 117B and 118B, 404 (1983); A. Pospieszczyk, M. Abdel Harith and B. Stritzker, J. Appl. Phys. 54, 3176 (1983).
8. H. W. Lo and A. Compaan, Phys. Rev. Lett. 44, 1604 (1980).
9. D. von der Linde and G. Wartman, Appl. Phys. Lett. 41, 700 (1982); D. von der Linde, G. Wartman, and A. Ozols, Mat. Res. Soc. Symp. Proc. 13, (1983); G. Wartman, D. von der Linde, and A. Compaan, Appl. Phys. Lett. 43, 613 (1983).

10. A. Compaan, Mat. Res. Soc. Symp. Proc. 13, 23 (1983); A. Compaan, H. W. Lo, M. C. Lee and A. Aydinli, Phys. Rev. B 26, 1079 (1982).
11. G. Wartmann, M. Kemmler and D. von der Linde, Phys. Rev. B 30, 4850 (1984).
12. A. Compaan, in Proc. 3rd Trieste IUPAP Semiconductor Symposium "High Excitation and Short Pulse Phenomena" July 2-4, 1984, [J. Luminescence (in print)].
13. R. Tsu and J. G. Hernandez, Appl. Phys. Lett. 41, 1016 (1982)
14. M. Balkanski, R. F. Wallis and E. Haro, Phys. Rev. B 28, 1928 (1983).
15. R. J. Nemanich, D. K. Biegelson, R. A. Street and L. E. Fennell, Phys. Rev. B 29, 6005 (1984).
16. See for example, M. Balkanski, K. P. Jain, R. Beserman and M. Jouanne, Phys. Rev. B 12, 4328 (1975) and M. Chandrasekhar, J. B. Renucci and M. Cardona, Phys. Rev. B 4, 1623 (1978).
17. A. Compaan, G. Contreras, M. Cardona and A. Axmann, J. Phys. (Paris) 44, C5-197 (1983); A. Compaan, G. Contreras and M. Cardona, Mat. Res. Soc. Symp. Proc. 23, 117 (1984); G. Contreras, A. K. Sood, M. Cardona A. Compaan, Sol. St. Commun. 49, 303 (1984).
18. A. K. Sood and M. Cardona, Sol. St. Commun. 49, 306 (1984).
19. A. Compaan and H. J. Trodahl, Phys. Rev. B 29, 793 (1984).
20. A. Battacharya, B. G. Streetman and K. Hess, J. Appl. Phys. 53, 1261 (1982); M. C. Lee, A. Aydinli, H. W. Lo and A. Compaan, J. Appl. Phys. 53, 1262 (1982).
21. Landoldt-Bornstein, New Series Gp. III, Vol. 7a ed. by O. Madelung, M. Schulz and H. Weiss (Springer, Berlin, 1982).
22. J. Dziewior and W. Schmid, Appl. Phys. Lett. 31, 346 (1977); K. G. Svantessen and N. G. Nilsson, Sol. St. Electron, 21, 1603 (1978).
23. G. E. Jellison and F. A. Modine, Phys. Rev. B 27, 7466 (1983); D. E. Aspnes and A. A. Studna, Phys. Rev. B 27, 985 (1983).
24. L. Vina and M. Cardona, Phys. Rev. B 29, 6739 (1984).
25. The effects on Raman spectra of compressive stress resulting from laser heating have been discussed by H. W. Lo and A. Compaan [J. Appl. Phys. 51, 1565 (1980)] and analyzed in detail by G. E. Jellison, Jr. and R. F. Wood [Mat. Res. Symp. Proc. 23, 153 (1984)]. We find the stress-induced phonon frequency shifts to be approximately 1/4 as large as calculated by Jellison and Wood. The difference probably arises from a numerical error in the original stress experiments of Anastassakis, et al. [E. Anastassakis, A. Pinczuk, E. Burstein, F. H. Pollak, and M. Cardona, Sol. St. Commun. 8, 133 (1970)].
26. P. A. Temple and C. E. Hathaway, Phys. Rev. B 7, 3685 (1973).
27. J. Narayan, J. Fletcher, C. W. White and H. Christie, J. Appl. Phys. 52, 7121 (1981); G. E. Jellison, Jr., D. H. Lowndes, and R. F. Wood, Mat. Res. Soc. Symp. Proc. 13, 35 (1983).
28. R. F. Wood, M. Rasolt and G. E. Jellison, Jr., Mat. Res. Soc. Symp. Proc. 4, 61 (1982).
29. R. F. Wood, D. H. Lowndes, G. E. Jellison, Jr. and F. A. Modine, Appl. Phys. Lett. 41, 287 (1982).
30. A. Compaan, M. C. Lee, H. W. Lo, G. J. Trott and A. Aydinli, J. Appl. Phys. 54, 5950 (1983).
31. M. C. Lee, G. J. Trott and A. Compaan (unpublished).

PICOSECOND PHOTOEMISSION STUDIES OF THE LASER-INDUCED
PHASE TRANSITION IN SILICON

A.M.MALVEZZI, H.KURZ and N.BLOEMBERGEN
Division of Applied Sciences, Harvard University,
Cambridge, Massachusetts 02138, USA

ABSTRACT

Four different regimes of photoelectric emission are observed over a wide fluence range of UV-laser pulses irradiating single-crystal silicon samples. The role of the electron-hole plasma in the nonlinear photoemission is demonstrated by temporal correlation measurements. A regime where ion thermal evaporation processes take place is observed above the critical fluence for melting. At higher laser fluences nonlinear ion acceleration is demonstrated by direct time-of-flight measurements.

INTRODUCTION

Pulsed laser-induced modifications of semiconductor surfaces find growing technical interest for the development of novel material processing technologies. In recent times much experimental effort has been devoted to studying laser-induced morphology changes by optical techniques [1]. The formation and relaxation of electron-hole pairs, the kinetics of the ambipolar solid state plasma and the energy transfer to the lattice phonons has been investigated by pump and probe experiments with ps and fs pulses [1,2,3,4].

Charged particle emission measurements provide an alternative approach to the study of the fundamental process at the surface of laser-irradiated material. Contrary to optical measurements, where bulk processes are dominant and the interpretation is complicated by spatial gradients of the parameters to be studied, here one probes the outermost layer of the surface. Nonlinear photoemission of silicon has been observed under nanosecond laser excitation [5]. The underlying mechanisms, however, could not be clarified. In recent picosecond photoemission experiments the Richardson-Dushman equation has been used to determine the upper limit of the electron temperature of picosecond-irradiated silicon at laser fluences much below [6] and above [7] the threshold values for melting.

In this contribution we present the results of extensive measurements on silicon irradiated by 4.66 eV picosecond pulses. By using correlation photoelectric techniques, we have been able to reveal the dominant role of the electron-hole plasma in a two-step, single-photon sequence. Above the critical fluence F_{th} at which amorphous spot formation on the silicon surface first occur, a regime of thermal ion velocities has been observed. This suggests that a Saha-type of equilibrium between the emitted ions and neutrals holds in this fluence range. Moreover, by further extending the measurements above F_{th}, direct evidence of nonlinear ion acceleration is established at approximately $6 F_{th}$.

EXPERIMENT

In our experiment the fourth harmonic ($h\nu = 4.66$ eV) of a 30 ps Nd:YAG laser pulse is focused on a crystalline silicon target placed in 10^{-9} torr vacuum. The emitted charges are collected by a diode configuration as described in [6,7]. A bias voltage of ± 4 kV is applied to ensure the collection of the total amount of charges emitted. The dependance of the collected charge density (Coulombs/cm^2) on the incident laser fluence is shown in Fig. 1. Four different photoelectric regimes are observed versus UV-laser fluence. In regime I, below 1 mJ/cm^2, a superposition of linear and quadratic response is observed independent of the applied collector voltage. The dashed line of Fig. 1 shows the calculated charge density using the yield data obtained in regime I. In regime II the photoemission deviates from the calculated behavior and is progressively reduced by space charge effects. The collected charge density is now dependent upon the applied voltage. By the rapid accumulation of

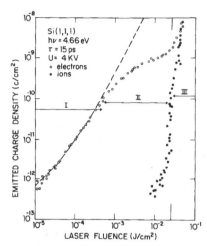 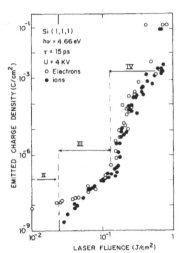

Fig. 1. Emitted charge density versus laser fluence. The dashed line refers to the calculated linear and quadratic photoelectric effect.

Fig. 2. Emitted charge density versus laser fluence in the high excitation regimes. The scales are different from those in Fig.1.

electrons in front of the surface, the applied field is screened and the collector current drops. The electron emission then exhibits a strong increase again as soon the critical laser fluence for melting, $F_{th} \sim 25 \, mJ/cm^2$ is reached (regime III). The sharp rise in the electron emission is accompanied by the emission of positive ions, which are collected when the bias voltage is reversed. The space charge field, which inhibits the electron emission in regime II, is neutralized by the injection of positively charged particles. Thus, the collection of electrons above the threshold for melting is governed by the availability of positive ions to remove the space charge field. In the low fluence part of the regime there is no charge neutrality. At higher fluences, charge neutrality is observed, i.e. the same amounts of electrons and ions are collected independently of the magnitude of the applied field. The electron and ion emission in regime IV is illustrated in Fig.2. At a laser fluence $\neq 6 \, F_{th}$, a sudden increase of both electron and ion emission is observed. The collected charges increase by about four orders of magnitude and reach a new space charge regime where the applied voltage is no longer able to collect the total amount of charges emitted by the sample. Only at the highest fluences of the present experiment is occasional structural damage of the surface observed.

PHOTOELECTRIC CORRELATION MEASUREMENTS IN REGIMES I AND II

In order to explore the temporal evolution of the photoemission process, we used a time correlation technique in which two consecutive picosecond pulses are focused on the same area of the sample. The synergetic effects on the photoemission are measured as a function of time delay τ between the two pulses in regimes I and II. In Fig. 3 the ratio of total collected charge Q to the sum of laser energy $E_t = E_1 + E_2$ is plotted versus the sum of laser fluences $F_t = F_1 + F_2$ (J/cm^2). The $\tau = \infty$ condition, shown by a solid line, is simulated by spatially separating the two laser spots. Below $F \sim 3 \, mJ/cm^2$ the quantum efficiency increases linearly, as expected from a quadratic process. At $7 \, mJ/cm^2$ the maximum efficiency is reached and at higher fluences (regime II) the value of Q/E_t drops again. As soon as the temporal separation of the two laser pulses is less than 100 ps, noticeable changes in the

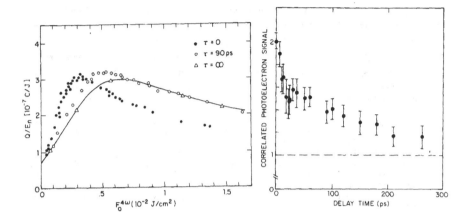

Fig 3. Ratio Q/E_n of the emitted negative charges over total laser energy for two overlapping pulses temporally separated by the delays τ shown, in regimes I and II.

Fig. 4. Correlated electron signal in region I versus time delay.

fluence response occur. The slope of the linear increase below 3 mJ/cm² increases significantly. The collected charge maximum moves towards smaller fluences and the drop in the space charge regime is more pronounced. The largest reduction of collection efficiency occurs in this regime if the two pulses strike the surface simultaneously (τ - 0). In this case the total charge emitted by the two pulses is less than the sum of the charges of two separated pulses, because of increased suppression by the space charge cloud in front of the surface. If the two pulses hit the same area, but are separated by a time interval sufficiently long to remove the space charge generated by the first pulse, then the combined charge emission should be equal to that of two spatially separated pulses. It is estimated from our sample and anode field geometry that the electron cloud is removed from the illuminated spot in about 10^{-10} s. The data in Fig. 3 appear to confirm these space charge effects. Longer lasting space charges resulting from thermionic emission are negligible.

The main part of our interest, however, is focused on the increase of quantum efficiency at fluences at which space charge effects can be neglected. To explore the origin of the quadratic photoemission process in regime I, we performed a complete time- resolved study of the correlation of the collected charges. In Fig. 4 the linear slopes of Fig. 3 are plotted versus the time delay between the two UV pulses. The data are normalized to the τ - ∞ condition. Despite the large experimental error, the temporal correlation behavior clearly indicates that the first pulse activates the surface for the second. This optical activation persists for times much longer than the pulse duration and is closely related to the dynamics of the elctron-hole plasma formed. At a fluence level of 1 mJ/cm² no significant lattice heating occurs, and the maximum carrier density generated at the surface is too low for Auger recombination processes to become dominant in the density equation. The penetration depth of 4.66 eV photons (α' - 5 x 10 $^{-7}$ cm) is much smaller than the diffusion length L_a - $(D_a \tau_p)^{1/2}$ on a picosecond time scale. Under these conditions an analytical solution of the density at the surface $N(0,t)$ can be derived using solutions of the surface heating problem [8]. Under irradiation with a laser pulse with gaussian temporal profile, the density time-dependance is given by:

$$N(0,t) = \frac{I_m(1-R)}{(D_a)^{1/2}} \, \tau_p^{1/2} \, \eta(t/\tau_p) \tag{1}$$

where $I_m(1-R)$ is the reflection corrected incidence irradiance, τ_p the $1/e$ half width of the laser pulse, D_a the ambipolar diffusion coefficient and $\eta(t/\tau_p)$ a strictly time-dependent form factor, which is developed in [8]. With $D_a = 20$ cm^2/s, a maximum density of 5×10^{19} cm^{-3} is reached after $0.55 \, \tau_p$, under the excitation levels used in this correlation experiment. The temporal decay is solely determined by $\eta(t/\tau_p)$. The striking resemblance between the temporal behavior of the photoelectric correlation signal and temporal shape of η leads us to the conclusion that the nonlinearity of the photoemission regime I is caused by the presence of the electron-hole pairs. It is basically a two-step, single-photon excitation process, where in the first step electrons are excited via direct transitions into high-lying conduction bands. Rapid intercarrier collisions ensure thermalization of the carrier distribution, and phonon assisted inter- and intravalley relaxation processes lead to the accumulation of a large carrier density in the vicinity of conduction band minima. Because of the lack of sufficient electron-hole recombination channels, their accumulation is mainly determined by the relaxation rate from the higher-lying states and diffusion into low-density regions. From the intermediate excited states, as well as from the final reservoir in the conduction band minimum, the electrons are reexcited by a second single-photon absorption process to final states above the vacuum level.

POSITIVE PARTICLE EMISSION IN REGIMES III AND IV

The transition from regime II to regime III is marked by the appearance of amorphous surface structures in the center of the irradiated spot. Consistently with reflectivity measurements, this transition is attributed to melting of the surface and subsequent ultrafast cooling and resolidification into a disordered phase. The critical laser fluence required to trigger this transition depends sensitively on the diffusion properties of the hot electronic carriers. The distance L^* over which these hot carriers have been diffused into the bulk before they release their excess energy to the lattice phonons can be treated in a first approximation as a diffusion length $L^* = (D_a^* \tau)^{1/2}$, with D_a^* the ambipolar diffusion coefficient of the hot plasma and τ the energy relaxation time [9]. As a consequence, the heating rate is reduced by $\alpha^{-1}/(\alpha^{-1}+L^*)$ where α^{-1} is the penetration depth of the heating laser pulse. Because of the small penetration depth, the reduction in heating rate or the increase of the critical laser fluence F_{th} by hot carrier diffusion is specifically pronounced in the case of UV irradiation ($\alpha^{-1} = 5 \times 10^{-7}$ cm). In numerical calculations the influence of L^* on the threshold fluence F_{th} has been studied. An experimental value of $F_{th} = 25$ mJ/cm^2 indicates an hot carrier diffusion of less than $L^* = 1 \times 10^{-6}$ cm. Since from time-resolved optical measurements an upper limit of the energy relaxation time $\tau < 1 \times 10^{-12}$ s has been evaluated [10], the diffusion coefficient of UV-generated carriers has to be on the order of 1 cm^2/s, far below the room temperature value at low carrier densities ($D_a = 20$ cm^2/s)[11]. Obviously, the momentum of the hot carriers is rapidly relaxed by intense electron-phonon and electron-hole scattering during UV excitation, overcompensating the linear increase of D_a with carrier temperature T_c.

Slight changes of the optical properties of silicon at 266 nm occur in the transition from regime II to III. The optical absorption constant drops from the solid state value $\alpha_s = 2 \times 10^{+6}$ cm^{-1} to the liquid state value of $\alpha_l = 1 \times 10^{+6}$ cm^{-1} and the reflectivity does not change significantly. The optical heating of the liquid surface layer, however, is marked by a sudden jump of the thermal diffusivity from the high temperature solid state value of $D^s_{th} = 0.16$ cm^2/s to $D^l_{th} = 0.273$ cm^2/s as soon as the process of fusion is completed. Thus, the heating rate of the liquid is roughly a quarter of the solid phase heating rate. The thermal conductivity K_m of liquid metal increases linearly with temperature, while the product of density and specific heat ρ_{cp} remains fairly constant. The thermal diffusivity D^l_{th} of the liquid layer scales with $(T^l)^{-1}$ therefore lowering the heating rate above the melting point. Close to the critical point ($T_{crit} \approx 5000°K$), cooling of the surface by evaporation losses has to be taken into account. Below this temperature, the surface heating model can be used to estimate the temperature of the liquid surface in regime III. According to this model the temporal dependence of the surface temperature is independent of the maximum temperature reached. After supplying the heat required to raise the temperature to the melting point and deliver the latent heat of fusion, a process which requires 25 mJ/cm^2, the liquid is rapidly heated far above the melting point. Near to the transition between region III and IV a surface temperature close to the critical temperature $T_{crit} = 5000°K$ can be estimated. At this point, the density of the vapour approaches the density of the liquid phase. The excessive optical heating of the highly dense vapour leads to completely different charged particle emission characteristics.

The emission of positive particles in regime II and III displays clearly an Arrhenius type of behavior. The charge collected scales with $Q = Q_0\exp(-E_a/kT)$, where Q_0 is a material constant and $E_a = (\Phi - \Delta H - I)$ the activation energy of the thermal evaporation and ionization [12].With the published data of work function of liquid silicon, $\Phi = 4.3$ eV, the enthalpy of evaporation $\Delta H = 4.68$ eV and the ionization potential $I = 8.15$ eV, the Saha-Langmuir equation describes semiquantitatively the signal increase in regime II and the subsequent decrease in the nonlinear response in regime III.

ION VELOCITY MEASUREMENTS

The picture of thermal emission and ionization is confirmed by ion velocity measurement. By measuring the time elapsed from the arrival of the laser pulse on the sample and the instant in which ions arrive on the collecting electrode, a direct estimate of the ion velocity can be obtained. Using zero accelerating voltage on the collector, the ion signals were observed at the oscilloscope with a suitable time resolution. The time of arrival of the peak of the ion signal at the collecting electrode, positioned 3 mm away from the silicon surface, ranges from several microseconds to a few tens of nanoseconds.At low fluences, close to the threshold value for melting, a broad positive peak, extending for a few microseconds, is observed. By increasing the laser fluence, the peak of the signal becomes higher, it shifts towards shorter times and its leading edge becomes steeper than the trailing edge. In Fig. 5 the ion velocities deduced from the temporal position of the peak of the signal are plotted versus the incident laser fluence.

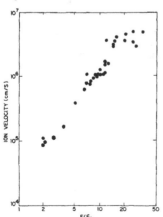

Fig. 5. Ion velocity versus laser fluence, normalized to the critical value $F_{th} = 25$ mJ/cm^2 in regimes III and IV.

Close to the melting threshold, the measured ion velocities are intrinsically thermal in magnitude and suggest that a Saha type of equilibrium at temperatures around the melting point of silicon is established between ions and neutrals. This picture is therefore consistent with the notion of thermal evaporation of silicon from a liquid layer. At higher laser fluences in regime IV, however, the measured ion velocities become exceedingly high, approaching values typical in laser produced plasmas. At about six times the critical fluence F_{th}, therefore, a new, highly nonlinear mechanism for ion generation and acceleration sets in, presumably related to the formation of an optically dense plasma in front of the surface. Details of this regime clearly need further investigation.

CONCLUSIONS

Four distinct regimes in the charged particle emission are observed in picosecond experiments using 4.66 eV photons. The quadratic photoelectric response is closely related to the formation of an electron-hole plasma during the pulse providing the intermediate states from which electrons are more easily excited into the vacuum level. A fluence range above the threshold fluence for melting exists, in which the number of ions collected and their velocities are consistent with thermal evaporation and ionization. At greater values of the laser fluence, highly nonlinear ion emission and acceleration indicates the onset of a new emission regime, where the laser energy seems directly coupled to the dense plasma in front of the surface.

ACKNOWLEDGEMENTS

This work was supported by the U.S. Office of Naval Research under contract N00014-83K-0030 and by the Joint Services Electronics Program of the U.S. Department of Defense under contract N00014-84-K-0465

REFERENCES

[1]   J.M.Liu, H.Kurz and N.Bloembergen, Appl. Phys. Letters 41, 693 (1982)
[2]   L.A.Lompre`, J.M.Liu, H.Kurz and N.Bloembergen, Appl. Phys. Letters 44, 3, (1984)
[3]   C.V.Shank, R.Yen, and C. Hirlimann, Phys. Rev. Letters, 50, 454, (1983)
[4]   C.V.Shank,R.Yen and C.Hirlimann, Mat. Res. Soc. Symp. Proc. 23, 53 (1984)
[5]   M.Bensoussan, J.M.Moison, B.Stoesz and C.Sebenne, Phys. Rev.B 23, 992 (1981)
[6]   A.M.Malvezzi, J.M.Liu and N.Bloembergen, Mat. Res. Soc. Sym Proc. 23, 135 (1984)
[7]   J.M.Liu, H.Kurz and N.Bloembergen, Mat. Res. Soc. Symp. Proc. 4, 23 (1982)
[8]   J.M.Bechtel, J.Appl. Phys. 46, 7585 (1975)
[9]   E.Yoffa, Phys. Rev. B 21, 2415 (1980)
[10]   H.Kurz, A.M.Malvezzi, L.A.Lompre`, Proc. of the 17th International Conference on the Physics of Semiconductors, S.Francisco, Ca, August 1984
[11]   A.M.Malvezzi, H.Kurz and N.Bloembergen, to be published in Appl. Phys. A
[12]   M.J.Dresser, J.Appl. Phys. 39, 338 (1968)

PHOTOELECTRON SPECTROSCOPY OF CONDUCTION-ELECTRON ENERGY DISTRIBUTIONS IN LASER-EXCITED SILICON

J.P. LONG, R.T. WILLIAMS, J.C. RIFE, and M.N. KABLER
Naval Research Laboratory, Washington, D.C. 20375-5000

ABSTRACT

Electron energy distributions, extending from equilibrium surface states through transiently populated conduction band states, have been observed in laser excited silicon by means of transient photoelectron spectroscopy. Atomically clean (111) wafers in UHV were excited by 90 nsec, 2.33 eV or 65 nsec, 4.66 eV pulses. Photoelectron spectra were obtained with the 4.66 eV pulses. Photovoltage shifts have been measured and found to saturate. We have identified a feature in the excited state portion of the spectra due to electrons which have equilibrated in the conduction band minima (CBM). Electron temperatures as a function of 2.33 eV pump fluence have been extracted from the spectra and are shown to be in equilibrium with the lattice above 1100 K. Very hot non-equilibrium electrons are also observed to ~ 4 eV above the CBM. A method of determining the diffusion depth of CBM electrons has been developed and applied to our data.

INTRODUCTION

The energy distribution of photoexcited electrons is determined by detailed balance between the rate of generation and the rate of energy relaxation through electron-phonon, electron-plasmon, electron-electron, impact ionization, diffusion, and recombination processes. The electron energy distribution is therefore a valuable indicator of the fundamental processes underlying the photoresponse of a solid. The ground state distribution of electrons has for many years been studied by photoelectron spectroscopy, but only recently have the transient energy distributions of photoexcited electrons been probed by photoelectron spectroscopy techniques [1,2]. In this paper we describe laser induced photoemission measurements of the distribution of electrons excited in crystalline silicon by intense ultraviolet (UV) and UV plus green laser pulses of ~ 100 nsec duration.

Principles underlying our measurements are given in Fig. 1. An incident UV photon (energy $h\nu = 4.66$ eV) is absorbed within 50 Å of the surface [3] by promoting an electron from an occupied to an unoccupied level. An electron excited above vacuum may escape the sample if it is within several tens of angstroms of the surface. A measurement of the escaping electron current versus electron kinetic energy yields an energy distribution curve (EDC) which reflects the initial distribution of occupied states. The states may be occupied by virtue of lying near or beneath the Fermi level (E_F) or by virtue of a previous photoexcitation (perhaps followed by a relaxation process) to a usually empty conduction band state. The former states will yield a photocurrent linear in the UV intensity by which they can be distinguished from the latter which yield a super-linear dependence on intensity. The sample can also be simultaneously excited by an intense green pump pulse (energy 2.33 eV, absorption depth .8μ [3]) thereby populating levels at or near the conduction band minimum in excess of the UV population. The green pump intensity may be increased to allow studies in the laser annealing regime.

EXPERIMENTAL

Experiments were performed on samples maintained in a stainless steel ultra high vacuum (UHV) chamber (base pressure 5 x 10^-11 torr) depicted in Fig. 2. Silicon surfaces were prepared in an integral side chamber as described below. An ion pumped gas handling system and leak valve permitted controlled H₂ dosing. Tungsten ribbons in the main chamber were heated to ~ 2000 K during dosing, which has been shown to be important for hydrogen adsorption [4].

Before being loaded into the vacuum system, samples were cleaved into suitable shapes from (111) and (100) oriented polished wafers of p-type (2 x 10^15 cm^-3 B) silicon. We have discovered that particulate contamination of the wafer surfaces is the dominant source of emission of electrons and ions at green fluence levels above the annealing threshold and affects electron emission studies below. Samples for this work were prepared in a clean room. Diamond scribing was limited to back surfaces only. Samples were examined with dark field optical microscopy after mounting and retained only if fewer than .03 particles per 100μ diameter area were counted.

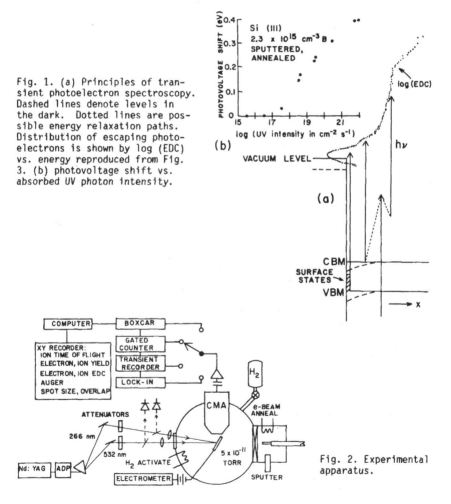

Fig. 1. (a) Principles of transient photoelectron spectroscopy. Dashed lines denote levels in the dark. Dotted lines are possible energy relaxation paths. Distribution of escaping photoelectrons is shown by log (EDC) vs. energy reproduced from Fig. 3. (b) photovoltage shift vs. absorbed UV photon intensity.

Fig. 2. Experimental apparatus.

Silicon surfaces were prepared in UHV by repeated cycles of argon ion bombardment and thermal annealing induced by electron bombardment heating. After transfer to the experimental chamber, surface cleanliness was verified with Auger electron spectroscopy.

Photoemission was induced by 65 nsec (FWHM) pulses of 266 nm radiation derived by doubling 532 nm pulses in ADP. The 90 nsec 532 nm pulses were obtained by intracavity doubling the Q-switched Nd:YAG fundamental with lithium iodate. The laser was operated in the TEM_{00} mode at a repetition rate of 1 kHz.

The green and UV wavelengths were separated in a quartz prism, directed through variable attenuators, and independently focused through quartz lenses onto the sample. A small portion of each beam was diverted to calibrated photodiodes for absolute power measurements. Spot sizes were determined in situ at the sample plane for both wavelengths by means of a dummy sample equipped with a 25μ pin hole and a 25μ tungsten wire which had been treated in NaCl solution to lower its work function. The pinhole was scanned in two dimensions through the green spot and transmitted light measured as a function of displacement. In a similar manner, the tungsten wire was scanned through the UV spot as the photoemission current was recorded. Green spot diameters were checked by microscopic examination of ripple pattern diameters after the method of Liu [5] which also determined the fluence threshold, F_r, for ripple formation. Spot diameters were typically 100μ for the green and 30μ for the UV.

As described below, portions of the EDC indicated increased electron emission when green illumination was superimposed on the UV. The emission increase was exploited to achieve overlap of UV and green spots on the sample by scanning the green spot in two dimensions through the UV spot until the enhanced emission was optimized.

Samples were biased - 15.6 eV to insure collection of photoelectrons ejected with small kinetic energy. Photoelectron kinetic energy was measured with a resolution of .09 eV by a Physical Electronics double pass cylindrical mirror analyzer (CMA) equipped with a high current channeltron electron multiplier (CEM). For large signals (greater than ~ .3 electrons per laser pulse), a boxcar integrator detected the CEM output pulses. Smaller signals were detected with a gated counter. An EDC was recorded on magnetic disk by repeatedly scanning the CMA pass energy under computer control.

RESULTS

An important result of our work is summarized in the EDC obtained on Si (111) under both UV and UV plus green illumination shown in Fig. 3. The zero of energy, E, has been taken as the conduction band minimum (CBM). This EDC represents linear photoemission for $E \leq 0$ and photoemission from transiently populated intermediate states for $E \geq 0$. The intensity dependence at various E is displayed in Fig. 4. In establishing the zero of energy it is important to account for surface photovoltage shifts. Fig. 1 shows the photovoltage shift versus UV intensity obtained by defocussing the UV beam and measuring the EDC position versus UV intensity. At high intensity the shift saturates indicating flat band conditions have been achieved. The assumption of flat bands together with the saturated shift of .37 eV and E_F for our doping level implies E_F is pinned at the surface .60 eV above the valence band in excellent agreement with a recent determination [6]. All data in this work were obtained under saturated conditions.

The prominent peak at -0.5 eV in Fig. 3 is due to linear emission from surface states lying in the silicon band gap near E_F. The assignment to surface states is supported by the energy of the peak relative to a tantalum reference, the peak's linear dependence on UV intensity, and its 3-to 5-fold reduction in strength after hydrogen dosing. The shoulder near 0.0 eV is caused by emission from electrons which have accumulated at the CBM. The conduction band signal falls off nearly exponentially with energy before

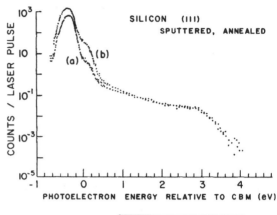

Fig. 3. EDCs for UV and UV plus green excitation. (a) 1.9×10^{23} UV photons absorbed $cm^{-2}s^{-1}$, (b) 1.7×10^{24} green photons absorbed $cm^{-2}s^{-1}$ plus UV intensity in (a).

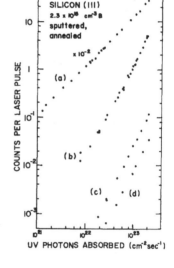

Fig. 4. Emission vs. absorbed UV flux, I_u, at various energies E in eV. Note emission $\propto I_u^m$. (a) E = -0.4, m = 1.0, (b) E = .08, m = 1.9 (c) E = 1.33, m = 2.0, (d) E = 2.73, m = 2.

merging with an extended distribution of very hot electrons. The hot electron cut-off near 3.6 eV is consistent with the energy of an electron at the valence band maximum undergoing two UV transitions. When the green pump is added, the EDC changes reversibly if the pump is kept below a threshold fluence to be discussed. Most significant is the increase at E = .05 eV which indicates a substantial increase in the CBM photo-population.

The electron and lattice temperatures during pulsed laser excitation are of great interest. Here we extend previous thermionic emission measurements of electron temperature [7] to well below the annealing threshold. The measurement is accomplished by computer fitting, over a suitably restricted range, the surface state signal S obtained with UV + green to the theory of Fowler-Dubridge [8]:

$$S \propto \frac{E_k \cdot D}{\exp((E_k - U)/kT_e) + 1} \tag{1}$$

E_k is the electron kinetic energy (i.e. energy above the vacuum level), U = $h\upsilon - \Phi$ where Φ is the work function, and D is the density of states from which the photoelectrons originate. The CBM signal, defined by the subtraction of the surface state fit from the raw data, was subsequently fit using eqn. (1) over a range excluding the very hot electrons. The parameter U was readjusted to·allow for a quasi-Fermi level. Largely because T_e is determined by the Maxwellian high energy tail of eqn. (1), T_e is found to be insensitive (±100 K) to other fitting parameters or the assumed energy dependence of D. For the data of Fig. 5, a constant D was used for the surface state fit and $D \propto (E_k + \chi - h\upsilon)^{1/2}$ for the CBM fit. χ is the electron affinity. T_e obtained in this way are <u>surface</u> temperatures because escaping photoelectrons originate within ~ $\overline{50 \text{ A of}}$ the surface.

The measured surface state and conduction band electron temperature is plotted versus <u>absorbed</u> green fluence in Fig. 5. Below ~ 900 K the temperatures obtained from the surface state and the conduction band are the same within experimental error. Above ~ 900 K the conduction band became unobservable as a distinct feature. T_e rises with fluence before leveling off at the melt temperature T_m. The fluence at T_m corresponds to the threshold F_r for ripple formation.

The fluence denoted F_T in Fig. 5 corresponds to a threshold fluence above which an irreversible increase in χ of 0.3 eV was observed. The temperature measured at F_T corresponds to the 1140 K transition temperature, T_S, of a well known surface phase transition wherein Si (111) 7 x 7 reconstructs to 1 x 1 [9]. (Our surface preparation procedures were designed to produce a 7 x 7 reconstruction although we presently lack LEED for verification). The coincidence of F_r to T_m and F_T to T_s, both equilibrium lattice temperatures, shows T_e to be the lattice temperature ± 100 K, at least above ~ 1100 K. Unexpectedly large T_e values are observed below ~ .25 J/cm^2 and are presently under investigation.

Once the transiently populated CBM is observed the laser induced photoemission technique provides a means of measuring the carrier diffusion depth $d = \sqrt{D\tau}$. Here D is the ambipolar diffusion constant and τ the recombination time. By way of illustration we adopt the simplest possible situation wherein the excitation flux is low enough to render nonlinearity in absorption and

Fig. 5. Temperature vs. absorbed green pump fluence.

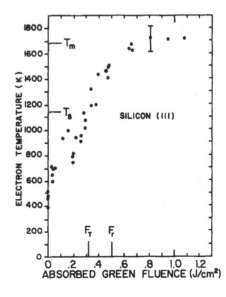

recombination unimportant. *Bands are assumed to be flat. If two wavelengths λ_1 and λ_2 are separately employed at intensities I_1 and I_2 to popu-late the conduction band, d may be determined in terms of the CBM signal strengths S_1, S_2 measured for each λ_i and the absorption depths a_i [10]:*

$$d = \frac{Ba_1 - a_2}{1 - B} \qquad (2)$$

where $B = S_1I_2/S_2I_1$. Eqn. 2 is obtained from the steady state solu-tion to the diffusion equation for n, the electron-hole pair density just beneath the surface. Steady state is not obtained for the low excitation in-tensities described above but detailed calculations show eqn. 2 to be accurate within ~ 10% for transient conditions if the recombination rate τ^{-1} is replaced by $\tau^{-1} + \tau_L^{-1}$ where τ_L is the laser pulse length. Unfortunately, our small absorption depths relative to d and uncertainties in B conspire to make measurements of d imprecise. We have however deter-mined d > 5μ which indicates that no significant confinement of the electron-hole plasma is taking place for n ≤ 10^{19} cm^{-3}.

Our experiments establish spectroscopy of laser induced photoemission as a valuable tool for assessing the photoexcited electron energy distri-bution in silicon. Electrons below E_F, through the CBM, and well into higher lying conduction band states are observed. The temperature of the electron distribution may be measured and is found to be equal to the lat-tice temperature for our 90 nsec pulses. The importance of diffusion in re-ducing the number density of photoexcited conduction band electrons has been demonstrated. Further publications [10] will present additional data and examine the relaxation processes of very hot electrons which have not been addressed in this paper.

REFERENCES

1. R.T. Williams, T.R. Royt, J.C. Rife, J.P. Long, M.N. Kabler, J. Vac. Sci. Technol., 21, 509 (82).

2. M. Bensoussan, J.M. Moison, Phys. Rev., B27, 5192 (83).

3. D.E. Aspnes, A.A. Studna, Phys. Rev., B27, 985 (83).

4. J.E. Rowe, S.B. Christman, H. Ibach, Phys. Rev. Lett., 34, 874 (75).

5. J.M. Liu, Optics Lett., 7, 196 (82).

6. F.J. Himpsel, G. Hollinger, R.A. Pollak, Phys. Rev., B28, 7014 (83).

7. J.P. Long, R.T. Williams, T.R. Royt, J.C. Rife, M.N. Kabler, in Laser-Solid Interactions and Transient Thermal Processing of Materials, ed. by J. Narayan, W.L. Brown, R.A. Lemons (North-Holland, N.Y., 1983).

8. Lee A. Dubridge, Phys. Rev., 43, 727 (33).

9. P.A. Bennett, M.W. Webb, Surf. Sci., 104, 74 (81).

10. J.P. Long, R.T. Williams, J.C. Rife, M.N. Kabler, to be published.

TIME-RESOLVED, LASER-INDUCED PHASE TRANSFORMATION IN ALUMINUM

S. WILLIAMSON[*], G. MOUROU[*], and J.C.M. LI[**]
* Laboratory for Laser Energetics, University of Rochester, 250 East ¿ River
River Road, Rochester,NY 14623-1299
**Department of Mechnical Engineering, University of Rochester,
250 East River Road, Rochester,NY 14623-1299   ;0 East

ABSTRACT

The technique of picosecond electron diffraction is used to time resolve the laser-induced melting of thin aluminum films. It is observed that under rapid heating conditions, the long range order of the lattice subsists for lattice temperatures well above the equilibrium point, indicative of superheating. This superheating can be verified by directly measuring the lattice temperature. The collapse time of the long range order is measured and found to vary from 20 ps to several nanoseconds according to the degree of superheating. Two interpretations of the delayed melting are offered, based on the conventional nucleation and point defect theories. While the nucleation theory provides an initial nucleus size and concentration for melting to occur, the point defect theory offers a possible explanation for how the nuclei are originally formed.

Phase transformation in condensed matter is an important area of study in solid state physics since it relates to the genesis and evolution of new microstructure. The mechanisms responsible in such critical phenomena are still not fully understood. Previous experimental information has left unmeasured such important parameters as the minimum number of nuclei and their common critical radius for a transformation to occur as well as the interphase velocity. The short-pulse laser has become a valuable tool serving many functions in this field. Energies large enough to melt (or even vaporize) condensed substances can now be applied in a time interval much shorter than the time for the transformation to take place. Several probe techniques have now been developed to time-resolve phase transformations in semiconductors during laser annealing. However, most of these probes (eg. electrical conductivity, [1,2] optical reflection, [3,4] optical transmission, [5] Raman scattering, [6] and time of flight mass spectrometry [7]) supply no direct information about the atomic structure nor the temperature of the material. Probing the structure can reveal when and to what degree a system melts as it is defined by degradation in the long range order of the lattice. True structural probes based on x-ray [8] and low energy electron diffraction, [9] and EXAFS [10] with nanosecond time resolution have been developed offering fresh insight into both the bulk and surface dynamics of material structure. Also, a subpicosecond probe based on structural dependent second harmonic generation [11] has been demonstrated. But at present, only the technique of picosecond electron diffraction [12] can produce on the picosecond timescale, an unambiguous picture of the structure of a given material. In addition to revealing the structure of the material, the diffraction pattern holds information about the temperature of the lattice. In this letter we report on the results of using this probe to directly observe the laser-induced melting of aluminum.

The technique takes advantage of the strong scattering efficiency of 25 keV electrons in transmission mode to produce and record a diffraction pattern with as few as 10^4 electrons in a pulse of 20 ps duration. The burst of electrons is generated from a modified streak camera that, via the photoelectric effect, converts an optical pulse to an electron pulse of equal duration. [13] Equal in importance is the fact that the electron pulse can be synchronized with picosecond resolution to the laser pulse. [14] The experimental arrangement is illustrated in Fig. 1. A single pulse from an active-passive modelocked Nd^{+3}:YAG laser is spatially filtered and amplified to yield energies up to 10 mJ. The streak tube (deflection plates removed), specimen, and phosphor screen are placed in vacuum at 10^{-6} mm Hg. The electron tube is comprised of the photocathode, extraction grid, focusing cone, and anode. A gold photocathode is used to permit the vacuum chamber to be opened to air. The photocathode is held at the maximum voltage (-25 kV) so that space charge, which can cause significant temporal broadening, is minimized. The portion of the laser irradiating the photocathode is first up-converted to the fourth harmonic of the funda-mental wavelength in order to produce the electrons efficiently. The duration of the UV pulse and thus the electron pulse is ~ 20 ps. Once the electron pulse is generated it accelerates through the tube past the anode and then remains at a constant velocity. The specimen is located in this drift region. The electrons pass through the specimen with a beam diameter of 500 μm and come to a 200 μm focus at their diffracted positions on the phosphor screen. A gated microchannel plate image intensifier in contact with the phosphor screen amplifies the electron signal ~ 10^4 times.

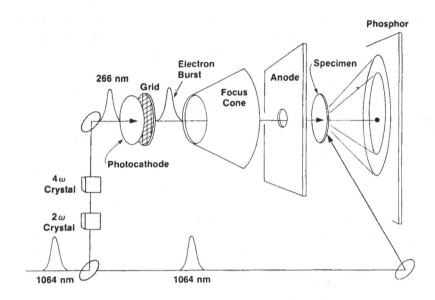

Fig. 1   Schematic of Picosecond electron diffraction apparatus. A streak
          camera tube (deflection plates removed) is used to produce the
          electron pulse. The 25 keV electron pulse passes through the Al
          specimen and produces a diffraction pattern of the structure with
          a 20 ps exposure.

A metal was chosen over a semiconductor as the specimen because of the
ease with which metals can be fabricated in ultra-thin polycrystalline
films. The specimens were fabricated by first depositing Al onto formvar
substrates and then vapor-dissolving away the formvar. Free standing films
250 ± 20 Å in thickness were required so that the electrons sustain, on the
average, one elastic collision while passing through the specimen. This
thickness of Al corresponds to twice the 1/e penetration depth at 1060 nm.
It is worth noting that the penetration depth for a metal does not vary
significantly from solid to liquid as is the case with a semiconductor
where a change in absorption of one to two orders of magnitude is possible.
Since the diffusion length $(D\tau)^{\frac{1}{2}}$, where D is the thermal diffusivity
coefficient (0.86 cm^2/s) and τ is time, is limited to 250 Å, the temperature
in the Al is uniformly established in less than 10 ps. The absorption of
the laser by aluminum is 13 ± 1 percent.

Since the diffracted electrons lie in concentric rings at discrete
radii from the zero order we can circular average to enhance the signal.
This is accomplished by rapidly spinning the photograph of the signal about

its center. The process acts to accentuate the real signal occurring at fixed radii while smoothing out the randomly generated background noise.

The diameter of the laser stimulus is ~ 4 mm ($1/e^2$) and is centered over the 2 mm specimen. A 1 mm pinhole positioned in place of the specimen facilitates accurate alignment of the laser beam profile to the electron beam. Using a 1 mm pinhole assures accurate measurement of the fluence within the probed region. Synchronization between the electron pulse and the laser stimulus is then accomplished by means of a laser-activated deflection plate assembly. [15]

The experimental procedure is then to stimulate the aluminum sample with the laser while monitoring the lattice structure at a given delay. The films are used only once even though for low fluence levels ($\lesssim 8$ mJ/cm^2) the films could survive repeated shots. Figure 2 shows the laser-induced time-resolved phase transformation of aluminum at a constant fluence of ~13 mJ/cm^2. The abrupt disappearance of rings in the diffraction pattern occurs with a delay of 20 ps. As is evident, the breakdown of lattice order can be induced in a time shorter than the resolution of our probe. However, the fluence required for this rapid transition exceeds F_{melt}, the calculated fluence required to completely melt the Al specimen under equilibrium conditions (~ 5 mJ/cm^2). At a constant fluence of 11 mJ/cm^2 the phase transition was again observed but only after a probe delay of 60 ps. Figure 3 shows the melt metamorphosis of Al where the points represent the delay time before the complete phase transition is observed for various fluence levels. We see that the elapsed time increases exponentially with decreased fluence and at 7 mJ/cm^2 the delay is ~ 1 ns. Because the fluence level that is applied is always in excess of F_{melt}, the observed delay time suggests that the Al is first driven to a superheated solid state before melting. It must be pointed out that as the fluence was decreased the abruptness with which the rings disappeared became less dramatic. Consequently, determination of the precise delay increases in difficulty with decreasing fluence. The temperature scale represents the temperature of the superheated Al assuming a linear increase in the specific heat with temperature. We see that temperatures in excess of 2000°K are expected with fluence levels near 13 mJ/cm^2. This temperature can be found by measuring the variation in the ring diameter as shown in Fig. 4. The change in lattice parameter is directly related to the thermal expansion coefficient. [16] The change in the lattice parameter observed in Fig. 4 corresponds to a large change in the specimen temperature of the order of 1000° K above the melting point. Note also the ~ 50% decrease in ring intensity for the 111 plane, which is in good agreement with the calculated change in the Debye-Waller factor of 60% that

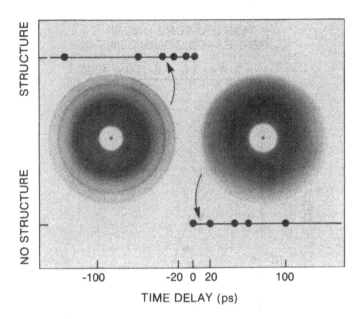

Fig. 2   Time-resolved laser-induced phase transition in aluminum. The pattern on the left is the diffraction pattern for Al and represents the points along the top line - where the electron pulse arrives before the laser stimulus. The pattern on the right shows the loss of structure in the Al 20 ps (or more) after applying the laser stimulus at a fluence of 13 mJ/cm^2. The fine line background structure occurring in both pictures is an artifact of the circular averaging technique.

corresponds to this increase in temperature.

Two interpretations for the observed delayed melting are offered. The first, is based on the traditional nucleation theory and gives information, on the initial nucleus size and density. The second uses the point defect theory to help in understanding the kind of defect responsible for initiating the melt and its activation energy.

1.   Nucleation Theory of Melting

Let us consider the melt to originate from a spherical cluster of atoms that begin to vibrate incoherently to the point where structural order is lost. The nucleation theory then tells us that beyond a critical radius r*, the sphere will rapidly expand throughout the volume until the

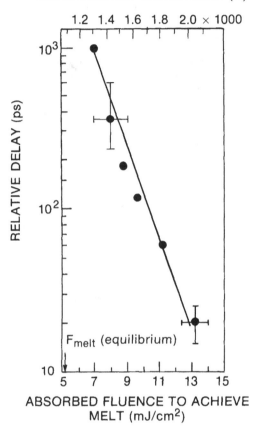

Fig. 3   Laser-induced melt metamorphosis for aluminum. The points mark
the elapse time for the diffraction rings to completely
dissappear. The vertical error bar represents the degree of
uncertainty in defining the moment of complete melt. The region
beneath the curve represents the conditions under which the Al is
left in a superheated solid state.

entire Al specimen is transformed into the liquid phase. The critical radius r* is given by

$$r^* = \frac{2\gamma_{\ell s}}{F_s - F_\ell} \qquad (1)$$

where $\gamma_{\ell s}$ is the interfacial free energy between the solid and liquid phases, and F_s and F_ℓ the free energy of each phase. An estimate of the critical radius is a few Angstroms at 1500° K and a few tens of Angstroms at 1000°K. According to the theory, the observed transition time (τ_{ob}) between the driving pulse and the complete disappearance of the diffraction

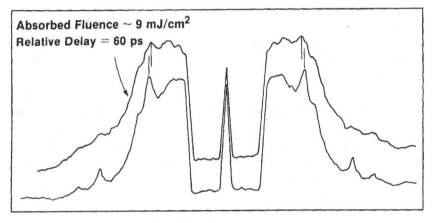

$$\frac{\Delta \text{ ring diameter}}{\text{ring diameter}} = \alpha \times \Delta T$$

$$0.03 = (24 \times 10^{-6} \text{ k}^{-1}) \times \Delta T$$

$$\Delta T = 1250^\circ \text{ K} \pm 20\%$$

Fig. 4   Time-resolved temperature measurement for aluminum. Upper trace depicts the diffraction pattern 60 ps after laser irradiation. The shift inwards of the 111 order relative to the room temperature pattern (lower trace) represents an expansion to the lattice which corresponds to a temperature of ~1500 °k.

rings should correspond to the time necessary for the expanding liquid spheres to fill the volume between nuclei. The average distance between nuclei is $2v\tau_{ob}$, where v is the radial rate of expansion.

Taking $v = 5\times10^5$ cm/s and $\tau_{ob} = 20$ ps, this distance is 2000 Å, a value much larger than the film thickness. Hence, the expansion is primarily cylindrical rather than spherical. The nuclei density is thus given by

$$N = \frac{1}{\pi(2v\tau_{ob})^2} \tag{2}$$

Eq. (1) is valid regardless of whether the nuclei originate on the surface or in the bulk. If we assume v to be constant with lattice temperature, we find that the density of nuclei increases from 10^5 cm^{-2} to 10^9 cm^{-2} when the fluence is increased from 7 mJ/cm^2 to 13 mJ/cm^2. From this interpretation, it is clear to see why determining the transition time becomes increasingly more difficult with decreasing fluence. The point in time where the diffraction rings have degraded to half normal intensity is reached just a few picoseconds prior to the full melt when the fluence is 13 mJ/cm^2 but a few hundred picoseconds prior to the full melt when the fluence is 7 mJ/cm^2.

2.   Point Defect Theory and Melting

The following scenario of melting or decay of long-range order is proposed. First under the action of laser irradition the material temperature is quickly raised above the melting point and is left in a superheated metastable state where long range order still predominates. Because of the high temperature, vacancies are formed at a rate dictated by a jump frequency γ_J defined as

$$\gamma_J = \gamma.e^{-E/kT} \tag{3}$$

where γ is the atomic vibrational frequency and E is the activation energy for vacancy formation. If we assume a purely adiabatic regime, where the energy is deposited instantaneously, the vacancy density n varies like

$$n = \gamma^\bullet\exp(-E/kT)^\bullet\tau_{ob} \tag{4}$$

where τ_{ob} is time. Let us consider the Schottky defect, where an atom is transferred from a lattice site in the crystal to a site on the surface, to be the predominant process forming defect sites. This we can assume since the activation energy for formation of a Schottky defect is ~ 1 eV compared to 4-5 eV for a Frenkel defect which is the process of transferring an atom from a lattice site to an interstitial site. The large number of defects resulting from the large excess temperature leads to a large vacancy gradient over the first monolayer of the specimen. This steep gradient is responsible for the subsequent rapid diffusion of atoms and vacancies through the specimen with an activation energy in the range of 1-2 eV, which culminates into a loosening of the lattice in the vicinity of the vacancy and a complete randomization of atomic position that is characteristic of the liquid state. If we assume that a critical density of vacancies is necessary to obtain the liquid state, the vacancy activation energy can be determined by measuring the slope of the delayed melting as a function of the reciprocal temperature

$$E = k \cdot \frac{\partial(\ln \tau_{ob})}{\partial(\frac{1}{T})} \qquad .$$

From Fig. 3, an activation energy $E = 1 \pm 0.2$ eV can be determined, in good agreement with the known value for the activation energy of Schottky defects in Al of 0.76 eV.

This model supposes only one type of vacancy and diffusion processes as well as a constant atomic vibrational frequency regardless of the atom position with respect to vacancy sities and as such can only offer a very qualitative view of this initial melting phase. To improve our understanding of solid-liquid phase transformation, time-resolved electron diffraction in reflection mode will be required to monitor the evolution of the lattice order from the first monolayer to the bulk as a function of time and termperature.

In conclusion we have demonstrated that before melting has occurred, a specimen can be driven to a superheated metastable solid-state. The time required for the melt to be completed varies from picoseconds to nanoseconds according to its degree of superheating. This overshoot in temperature has been monitored by looking directly at the variation in ring diameter. Two interpretations of the delayed melting are offered. The first is based on the nucleation theory of melting and provides information on the size and density of initial nuclei. The delayed melting can also be interpreted in terms of point defect theory where the structural order decays as a result of a large number of Schottky defects that are produced and rapidly diffuse away from the surface. In the near future timeresolved electron diffraction in a reflection mode will be implemented to differentiate the two theories.

96

ACKNOWLEDGMENT

This work was partially supported by the Laser Fusion Feasibility Project at the Laboratory for Laser Energetics which has the following sponsors: Empire State Electric Energy Research Corporation, General Electric Company, New York State Energy Research and Development Authority, Northeast Utilities Service Company, Southern California Edison Company, The Standard Oil Company (Ohio), the University of Rochester. Such support does not imply endorsement of the content by any of the above parties.

We would like to acknowledge the support of Jerry Drumheller who assisted in the fabrication of the Al films as well as to Hsiu-Cheng Chen for her help during the experiment.

REFERENCES

[1] M. Yamada, H. Kotani, K. Yamazaki, K. Yamamoto, and K. Abe, **J. Phys. Soc. Japan**, 49, 1299 (1980).

[2] G.J. Galvin, M.O. Thompson, J.W. Mayer, R.B. Hamond, N. Paulter and P.S. Peercy, **Phys. Rev. Lett.**, 48, 33 (1982).

[3] D.H. Auston, C.M. Surko, T.N.C. Venkatesan, R.E. Slusher, and J.A. Golovchenko, **Appl. Phys. Lett.**, 33, 437 (1978).

[4] C.V. Shank, R. Yen, and C. Hirlimann, **Phys. Rev. Lett.**, 50, 454 (1983).

[5] J.M. Liu, H. Kurz, and N. Bloembergen, **Appl. Phys. Lett.**, 41, 643 (1982).

[6] H.W. Lo and A. Compaan, **Phys. Rev. Lett.**, 44, 1604 (1980).

[7] A. Pospieszczyk, M.A. Harith, and B. Stritzker, **J. Appl. Phys.**, 54, 3176 (1983).

[8] B.C. Larson, C.W. White, T.S. Noggle, and D. Mills, **Phys. Rev. Lett.**, 48, 337 (1982).

[9] R.S. Becker, G.S. Higashi, and J.A. Golovchenko, **Phys. Rev. Lett.**, 52, 307 (1984).

[10] H.M. Epstein, R.E. Schwerzel, P.J. Mallozzi, and B.E. Campbell, **J. Am. Chem. Soc.**, 105, 1466 (1983).

[11] C.V. Shank, R. Yen, and C. Hirliman, **Phys. Rev. Lett.**, 51, 900 (1983).

[12] G. Mourou and S. Williamson, **Appl. Phys. Lett.**, 41, 44 (1982).

[13] D.J. Bradley and W. Sibbett, **Appl. Phys. Lett.**, 27, 382 (1975).

[14] G. Mourou and W. Knox, **Appl. Phys. Lett.**, 36, 623 (1980).

[15] S. Williamson, G. Mourou, and S. Letzring, **Proc. 15th Int. Cong. on High Speed Photography**, Vol. 348 (I), 197 (1983).

[16] R.O. Simmons and R.W. Balluffi, **Phys. Rev.**, 117, 52 (1960).

PULSE-DURATION-DEPENDENT—FREE-CARRIER-
ABSORPTION IN SEMICONDUCTORS

P. MUKHERJEE, M. SHEIK-BAHAEI AND H.S. KWOK
Department of Electrical and Computer Engineering, State University of New York at Buffalo, Bell Hall, Amherst, NY 14260

ABSTRACT

The free carrier absorption in InSb was found to depend on the pulse duration of the picosecond CO_2 laser pulse. This is interpreted as the effect of incomplete carrier relaxation within the laser pulse duration. An electron energy relaxation time of 5.6 ps was derived from the observation. The present experiment presents a new method of measuring relaxation times in semiconductors and metals.

INTRODUCTION

The availability of ultrashort duration pulses has enabled the measurement and characterization of ultrafast dynamics in a wide range of applications from biology to chemistry and electronics. In particular, the dynamics of nonequilibrium high density electron-hole plasmas in semiconductors have received much attention recently because of fundamental physical interests and their importance in ultrafast optical devices [1]. Most experimental investigations, however, involve the same laser pulse to create and to probe these nonequilibrium plasmas. We propose and demonstrate a technique where only one laser pulse is used and no interband transition is involved.

The basic idea is to use far infrared laser pulses to probe the free carriers which are created by doping the material. Because the photon energy is smaller than the bandgap, interband transitions are impossible. The predominant interaction will therefore be intraband interactions, which reveals itself as free carrier absorption. In our experiments, the free carrier absorption cross section is measured as a function of the laser duration. Hence relaxation effect can be detected if the laser pulse is short enough. To demonstrate this idea, the high mobility semiconductor InSb was studied. The results indicate that the electrons have an energy relaxation time of ~ 5.6 ps.

An important feature of this new method in measuring relaxation times is that instead of the usual "pump and probe" approach, a single pulse of variable pulse duration is employed. The variable pulse duration takes the place of the variable time delay in the two beam method. This eliminates the potentially troublesome coherent transient effect [2] and other alignment problems.

EXPERIMENTAL

The necessary variable duration far infrared laser pulses to accomplish these experiments are provided by the picosecond optical free induction decay CO_2 laser system [3]. This system delivers 2 MW 10.6 μm laser pulses with a duration continuously and conveniently changeable between 20 ps and 300 ps. A thorough characterization of this system has been completed recently giving details about the pulse shape, peak power, duration as a function of the gas pressure in the pulse former of this system [4].

The experimental arrangement and procedure will now be described. The unfocussed laser beam was transmitted through a wafer of InSb at near normal incidence. The Gaussian beam waist of the laser was 1.7 mm. A stack of

CaF_2 attenuators were used to change the intensity of the laser beam. The transmitted beam was filtered by a wide band 10 μm filter, attenuated if necessary, and detected by a 77°K Ge:Au detector.

The absolute transmission of the laser beam was measured by moving the InSb sample in and out of the laser beam. The reproducibility of the position of the InSb crystal was quite good as the signal did not change by moving the sample several times. The sample used had a thickness of 0.37 mm and was n-type doped to 1.5×10^{18} cm^{-3}.

The small signal absorption coefficient of the sample was measured. This was obtained by attenuating the laser beam by at least a factor of 100. Additionally, by varying the intensity of the laser beam it was confirmed that the transmission was linear in laser intensity. Because InSb was a well-known nonlinear material, care must be taken so that nonlinear transmission and reflection did not occur[5]. The intensities used in the experiment were below 1 MW/cm^2. Moreover to ascertain that no etalon effect was present, the sample was mounted on a rotation stage. The transmission was found to be independent of the small angle of rotation which would have corresponded to several Fabry-Perot fringes.

The above procedures were repeated carefully for several laser pulse durations. Figure 1 shows the measured transmission as a function of the laser pulse duration. There is one additional data point which corresponds to a pulse duration of 100 ns (the single axial mode TEA CO$_2$ laser). The very long pulse transmission is indicated by the horizontal dash line and should be regarded as the asymtotic value. It can be seen that the transmission increases as the pulse duration is decreased. This is indicative of pulse duration approaching the carrier relaxation time.

The same experiment was repeated with samples of n and p type silicon. No change in the small signal transmission value was observed for the same pulse duration range. This implies that silicon has a faster relaxation time and that the experimental observation with InSb is not an artifact.

DISCUSSIONS

Figure 2 presents the calculated FCA cross section as a function of the pulse duration. The quantitative interpretation of the result can be given by the following simplified model. Consider the FCA process as shown in Fig. 3. The electron is excited to a nonequilibrium state in the conduction band. It relaxes in a time τ_e by the emission of several phonons. Subsequent decays will eventually lead to an increase of energy in the lattice by $h\nu$, the photon energy. Without loss of generality, we can assume that the

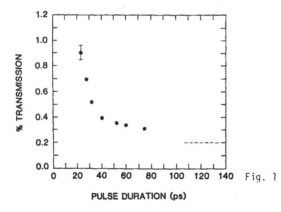

Fig. 1 Small signal transmission of n-InSb as a function of pulse duration.

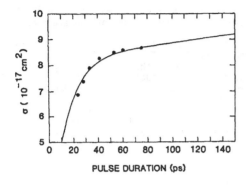

Fig. 2 Free carrier absorption cross section as a function of pulse durations. Solid curve is a theoretical fit with a τ_e of 5.6 ps.

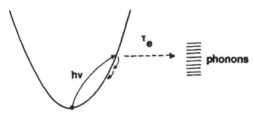

Fig. 3 Model for calculation of σ_{FCA}.

degree of excitation is small so number of excited electrons is given by

$$dn/dt = BNI - n/\tau_e \tag{1}$$

where B is the stimulated absorption coefficient, N is the total number of free electrons. Stimulated emission is ignored because n<<N. The energy absorbed by the lattice is given by

$$d\varepsilon/dt = nh\nu/\tau_e \tag{2}$$

Assuming a square pulse of duration t_p for the laser intensity I (t), (1) and (2) can be solved to give

$$\varepsilon = h\nu BN \, I \, [\, t_p - \tau_p(1-2^{-t_p/\tau_e})] \tag{3}$$

By definition $\varepsilon = N\sigma I t_p$ where σ is the FCA cross section, hence

$$\sigma = Bh\nu[1-\tau_e/t_p(1-e^{-t_p/\tau_e})] \tag{4}$$

It can be seen that for long pulses $\tau_e<<t_p$, $\sigma = Bh\nu$ which is a constant. For t_p approaching τ_e, σ decreases to zero as expected.

The solid line in Fig. 2 is a theoretical curve using a value of 5.6 ps for τ_e in (4). The fit to the data points is quite satisfactory. An improved calculation should take into account for the change in the Fermi-Dirac distribution function for the degenerate electron plasma in the conduction band.

The value of τ_e obtained should be compared with the momentum relaxation time τ_0 predicted by the carrier mobility $\mu_e = e\tau_0/m^*$. τ_0 has been measured to be ~ 1 ps using microwave dispersion technique[6]. τ_e in our model should be the energy relaxation time. Therefore the results implies that it takes several momentum reversing collisions before the electron energy is relaxed. This is not surprising because of the small effective mass of the electrons in InSb.

In conclusion, we have measured the energy relaxation time in InSb using a new technique of pulse duration dependent FCA. This technique can be applied to any type of free carrier which includes those in semiconductors and metals. However, for metals and other semiconductors such as silicon and GaAs, shorter IR pulses are needed. The subpicosecond pulses developed by Corkum [7] should be appropriate for these applications.

Further studies should also include the measurement of the temperature and doping dependence of τ_e. The CO_2 laser can be used to generate other IR frequencies by nonlinear techniques so that the value of τ_e at various levels of excitations can be measured. This will enable theories of FCA to be verified [8].

This work was supported partially by the National Science Foundation Grant No. ECS 8351264.

References

[1] Z.Y. Yu and C.L. Tang, Appl. Phys. Lett. 44, 1235(1984).
[2] A. Wiener, presented at Third Topical Meeting on Ultrafast Phenomena, Monterey, CA 1984.
[3] H.S. Kwok and E. Yablonovitch, Appl. Phys. Lett. 30, 158(1977)
[4] M. Sheik, H.S. Kwok, to be published.
[5] M. Hasselbeck and H.S. Kwok, Appl. Phys. Lett. 41, 1204(1982)
[6] T. Kobayashi, J. Appl. Phys. 48, 3154(1977).
[7] P. Corkum, Opt. Lett. 8, 514(1983).
[8] B. Jensen, IEEE J. Quant. Elect. QE-18, 136(1982).

TIME-RESOLVED STUDIES OF RAPID SOLIDIFICATION
IN HIGHLY UNDERCOOLED MOLTEN SILICON*

D. H. LOWNDES,[1] G. E. JELLISON, JR.,[1] R. F. WOOD,[1] S. J. PENNYCOOK,[1]
AND R. W. CARPENTER[2]
[1] Solid State Division, Oak Ridge National Laboratory, P. O. Box X, Oak
Ridge, TN 37831
[2] Center for Solid State Science and School of Engineering and Applied
Science, Arizona State University, Tempe, AZ 85287

ABSTRACT

A KrF (248 nm) pulsed laser was used to melt 90-, 190-, and 440-nm
thick amorphous silicon layers produced by Si ion implantation into (100)
crystalline Si substrates. Time-resolved reflectivity measurements at two
different probe wavelengths (633 nm and 1.15 μm) and post-irradiation TEM
measurements were used to study the formation of an undercooled liquid Si
phase and the subsequent solidification processes. The time-resolved
measurements provide new experimental information about the nucleation of
fine-grained Si crystallites in undercooled liquid Si, at low laser energy
densities (E_ℓ), and about the growth of large-grained Si in the near-surface
region at higher E_ℓ. Measurements with the infrared probe beam reveal the
presence of a buried, propagating liquid layer at low E_ℓ. Model calculations indicate that this liquid layer is generated in part by the release of
latent heat associated with the nucleation and growth process.

INTRODUCTION

Recent studies have shown that the melting temperature (T_a) of amorphous (a) silicon is approximately 200 (±50) K below the melting temperature
(T_c) of the crystalline (c) phase [1]. As a result, liquid (l) silicon that
is undercooled by several hundred Kelvins, relative to its normal (crystalline) freezing point, can be experimentally prepared by pulsed laser melting
of a-Si layers, using laser pulses of low energy density E_ℓ. (Pulse durations of 10—100 ns are used in order to avoid the a→c transformation that
can occur in the solid phase at lower heating rates.) Relatively well-
characterized a-Si can be prepared by Si ion implantation into a c-Si substrate at 77 K. By melting only partially through such an a-Si layer, the
undercooled l-Si that is produced is thermally and spatially separated from
the c-Si beneath by the remaining, unmelted low thermal conductivity ($K_a \approx$
0.02 W/cm-K) [2] a-Si barrier layer. Therefore, it is possible to study the
subsequent process of solidification from a highly undercooled melt, under
conditions such that epitaxial regrowth from c-Si cannot occur.

We recently reported preliminary results of time-resolved reflectivity
(R) measurements (at 633-nm wavelength) during pulsed Nd (532 nm) laser
irradiation [2], together with model calculations [2,3] and TEM measurements
[4,5], which were used to study the transformation of a-Si to the highly
undercooled l-phase and the subsequent solidification process. Typical TEM
micrographs following low E_ℓ laser irradiation of a 500-nm thick a-Si layer
on a c-Si substrate showed a region of fine-grained (FG) polycrystalline Si
(p-Si) at the surface, followed at increasing depth by the remaining a-Si,
the a-c interface, and the c-Si substrate. At higher E_ℓ a layer of large-
grained (LG) p-Si was found to lie at the surface, followed at greater depth
by FG p-Si, a-Si, and the c-Si substrate. Further increases in E_ℓ resulted

*Research sponsored by the Division of Materials Sciences, U. S. Department
of Energy under contract DE-AC05-840R21400 with Martin Marietta Energy
Systems, Inc.

in further growth of the LG region, reduction in the thickness of the FG region, and displacement of the FG region to greater depth, until melting entirely through the a-Si layer was accomplished at high E_ℓ. Individual grains in the FG region were found to be quite small (~10 nm) on average and were randomly oriented and nearly equiaxed, indicating that there was no preferred direction of growth during formation of this region. These obser-vations suggested that the FG p-Si marks a region in which bulk (volume) nucleation and crystallite growth (rather than growth from a planar inter-face) occurred in the undercooled 1-Si [2,3].

Model calculations were "calibrated" by comparison with R measurements of surface melt duration and were then used to determine temperature pro-files in the undercooled 1-Si [3]. The nucleation and growth behavior of undercooled 1-Si was inferred by comparison of these temperature profiles with the solidification morphology profiles observed by TEM. By using the theory of homogeneous nucleation and an estimated bulk nucleation tempera-ture of $T_n \simeq T_a + 50$ K a nucleation rate was estimated [3]. According to this model, 1-Si is expected to undergo nucleation in regions of the melt for which $T < T_n$ for a specified time t_n; these regions may be very small or more extended, depending on the temperature gradients. Two important conclusions resulted from the model calculations: (1) The release of latent heat by bulk nucleation before the melt front reaches its maximum penetra-tion (as determined by TEM), is essential to obtain agreement between calcu-lated and experimentally observed depths of melting; and, (2) this release of latent heat from bulk nucleation may result in the existence of buried, propagating molten layers of 1-Si in the interior of the sample, even after the surface has solidified.

Thompson et al. [6] have proposed a slightly different model for for-mation of the LG and FG p-Si regions: They suggest that the near-surface LG region is due to solidification of a primary liquid layer (produced by the pulsed laser), while the FG region is postulated to result from a secondary "explosive" crystallization process in which a thin, propagating liquid layer travels through the remaining a-Si. They do not discuss a nucleation mechanism, but it appears they have nucleation at the a-ℓ interface in mind. If bulk nucleation occurs in a thin layer near the melt front, the models are virtually indistinguishable. The models do differ in the assumed time sequence of events: Explosive crystallization assumes that the LG region begins to form first, while the bulk nucleation model to date has assumed that the FG region begins to form first. Both models recognize the impor-tance of the released latent heat in sustaining the melt front propagation and both postulate buried, propagating 1-Si layers.

In this and an accompanying paper [7] we present extensive new results of R measurements that were carried out using a-Si layers of three different thicknesses and at two different probe laser wavelengths (633 nm and 1.15 μm), in order to quantitatively study the formation of the undercooled liquid phase and its subsequent solidification process. The present paper focuses on changes in the characteristic shape of the R signals, with increasing pulsed laser E_ℓ, which provide in situ time-resolved information about the nucleation and crystallite growth process very near the under-cooled molten surface. Reflectivity measurements using the infrared probe laser have allowed us to look through the resolidified near-surface region: They reveal the presence of a buried liquid layer propagating beneath the recrystallized surface. New TEM cross-sectional micrographs, representative of melting and solidification in a-Si layers of different thicknesses, are also presented; these are correlated, at various E_ℓ, with the corresponding time-resolved R signals. In the accompanying paper [7], time-resolved R measurements of the surface melt duration and time of onset of melting of a-Si layers are compared with results obtained from a new formulation for melting model calculations, intended for use under conditions such that phase transitions do not occur at their normal (equilibrium) transition tem-peratures, such as the large undercoolings encountered in our experiments.

TRANSMISSION ELECTRON MICROSCOPY

A KrF (248 nm) excimer laser was used to irradiate a-Si layers that were produced by implantation of either 40-, 80-, or 180-keV Si ions into (100) c-Si substrates at 77 K; RBS showed the amorphous layer thicknesses to be 90, 190, and 440 nm, respectively. An excimer laser provides a much more homogeneous and reproducible laser beam than the Nd laser used in our earlier work [2]. In-situ E_ℓ monitoring of every laser pulse has reduced the uncertainty in individual E_ℓ measurements to ±2%.

We have used cross-sectional TEM micrographs at high resolution to image individual lattice planes and to study the morphology in the recrystallized FG and LG p-Si regions, and at lower resolution, using diffraction contrast to clearly distinguish between these regions. Figures 1 (diffraction contrast) and 2 (lattice image) are typical of the solidification morphologies observed following KrF laser (35 ns FWHM pulse duration) irradiation of the 190-nm and 440-nm thick a-Si layers, respectively. As shown in Fig. 1, following 0.14 J/cm² irradiation, the top 46 (±6) nm solidified as FG p-Si. At 0.2 J/cm², the FG region reached a depth of 106 (±15) nm, but now isolated grains of LG p-Si were also present, both at the surface and at greater depth. At 0.35 J/cm², two distinctly separated regions have formed: The thickness of the FG region has decreased to 81 (±11) nm, and it is now separated from the sample's surface by a 61 (±13) nm thick layer of LG p-Si. At 0.52 J/cm², the LG region continues to increase in thickness (to 122 ± 21 nm), while the FG material decreases again (to 36 ± 10 nm) and is displaced still deeper into the original a-Si layer. At 0.65 J/cm², the FG region is <6 nm thick (being indistinguishable from the heavily damaged back a-c boundary); well-defined columnar grains of LG p-Si, with an aspect ratio that varies from 2:1 up to nearly 10:1, almost fill the original a-Si layer. Finally, at 0.83 J/cm², complete epitaxial regrowth of a (100)

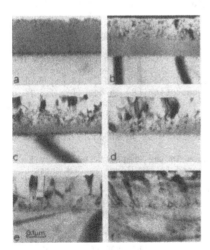

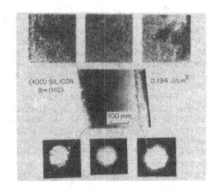

Fig. 1. TEM cross-sectional micrographs following irradiation of a 190-nm thick a-Si layer with 35-ns (FWHM) KrF laser pulses at (from top downwards) 0.14, 0.20, 0.35, 0.52, 0.65, and 0.83 J/cm².

Fig. 2. High resolution TEM micrographs and selected area diffraction patterns following irradiation of a 440-nm thick a-Si layer with a 35-ns (FWHM), 0.194 J/cm² KrF laser pulse: Left, FG p-Si; center, a-Si; right, a-c boundary.

single crystal occurs throughout the original a-Si layer, though some defects are still present.

Figure 2 illustrates the FG p-Si nature of the near-surface region following low E_ℓ melting and solidification of a 440-nm thick a-Si layer: (111) lattice planes are clearly visible in those individual fine grains that happen to be oriented with the (111) planes parallel to the beam direction. Microdiffraction patterns (obtained with the electron beam focused onto regions about 50 Å in diameter) show a clear progression from the "grainy" ring pattern expected for FG p-Si, to diffuse rings for the (unmelted) a-Si, and back to a typical defective crystalline pattern at the boundary of the (100) substrate. A microdiffraction pattern from an area that included the entire FG region showed that the FG p-Si microcrystals are approximately "randomly" oriented. Following irradiation of the 440-nm layer at 0.19 J/cm^2, only a 176 (±25) nm thick FG p-Si region was observed by TEM; however, at 0.37 J/cm^2, both a near-surface LG region (156 $\pm$ 15 nm thick) and a deeper FG region (191 $\pm$ 19 nm) were found. At 0.79 J/cm^2, the original a-Si layer was completely melted, with large, columnar grains (aspect ratio from 3:1 to 8:1) extending from the original a-c interface back to the surface. In this case, the regrown surface was found not to be flat but to be faceted by the slightly tilted faces of adjacent large grains that have slightly different orientations.

Detailed examination of TEM micrographs showed that, although the order of magnitude of the average grain size in the FG region is 100 Å (as described in earlier work [2,5]), the grains in this region actually vary widely in size and shape. Individual grains can be found in the FG region with dimensions ranging from fewer than 10 lattice planes up to several hundred Angstroms. The tendency to also find occasional large grains within the "FG" region becomes especially noticeable around the threshold E_ℓ for formation of the LG region. Nevertheless, there is still a clear boundary at the higher E_ℓ values between a deeper region that contains almost entirely FG p-Si and a near-surface region containing exclusively LG p-Si.

OPTICAL STUDIES OF NEAR-SURFACE NUCLEATION AND GRAIN GROWTH

In order to probe the melting and solidification behavior of both the near-surface region and the interior of an a-Si sample, we have carried out reflectivity measurements using red (633 nm) and infrared (1.15 μm) probe laser beams. Estimates of the optical absorption [8] in hot ($T > T_a$) FG or LG p-Si show that the 633-nm probe beam will be nearly completely attenuated by a few hundred Å thick surface layer of either material, while the 1.15-μm beam should be able to penetrate ~2 μm of FG or LG p-Si, to detect a buried liquid layer beneath. Both probe wavelengths will be strongly reflected by even a thin liquid layer.

Figure 3 shows characteristic changes in the shape of the 633-nm R signal, as a function of E_ℓ, that correspond to the lower E_ℓ TEM cross-sections shown in Fig. 1. At the lowest E_ℓ value, the R signal (Fig. 3) never achieves the flat-topped shape that is characteristic of pulsed laser melting of c-Si, but instead begins to decay to lower values almost immediately. One interpretation of this gradual decay is that it signals the occurrence of bulk nucleation, beginning within a few nanoseconds after the time when melting first occurs, in the highly undercooled melt that exists near the surface at the lowest E_ℓ. During this time, the near-surface region seems to be composed of a mixture of undercooled liquid and an increasing number of growing crystallites. Using an effective medium approximation to describe the optical properties of such a mixture, we have carried out calculations [8] that reproduce the continuously decaying R observed at low E_ℓ.

When E_ℓ is increased into the range (>0.2 J/cm^2) for which TEM first reveals solidification of LG p-Si in the near-surface region, the R signal

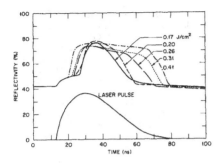

Fig. 3. Reflectivity signals vs time for selected KrF laser E_ℓ (633-nm probe laser).

(Fig. 3) also assumes a new shape: a shallowly sloping flat-top, followed by a slow fall back to the R value for hot c-Si. Effective medium calculations [8] strongly suggest that the very gradual decrease in R in the sloping flat-topped region is due to the growth of individual large grains back to the surface, the near-surface region during this time containing a mixture of liquid and isolated large grains. Planar regrowth of a number of large grains, regrowing at nearly the same rate and time, cannot reproduce the gradual decrease in R that is observed. The final, steeper drop in R corresponds to the remaining liquid volume eventually filling up more quickly with regrowing large grains. (Similar R behavior was observed over the same E_ℓ range during solidification of the 440-nm thick a-Si layer; see Ref. 8.) A possible explanation for the initial regrowth of only isolated large grains (assuming the bulk nucleation model) is that (1) it is difficult for these grains to initially nucleate and grow off of the FG p-Si that has already been formed in the lower temperature l-Si beneath [2,3,7]; and, (2) nucleation and growth of one isolated large grain will release latent heat, warming the adjacent l-Si and temporarily slowing the growth of neighboring large grains.

INFRARED REFLECTIVITY: BURIED LIQUID LAYERS

Figure 4 compares the R signals that were obtained using the red (633 nm) and infrared (1.15 μm) probe lasers, during irradiation of a 190-nm thick a-Si layer by 45-ns duration (FWHM) KrF laser pulses. (The melting threshold E_ℓ and the characteristic changes in the 633-nm R signal shape, shown in Fig. 3, were found to be displaced toward higher E_ℓ in Fig. 4 because of the longer KrF pulse duration.) Two striking differences between the red and infrared R signals in Fig. 4 are that the melt durations (duration of the HRP) measured using the infrared probe laser are much longer for most of the E_ℓ shown, and the infrared R signal is heavily modulated during the HRP by what appear to be interference maxima and minima. For example, the HRP durations at 0.18, 0.20, and 0.25 J/cm^2 are approximately 18 (35), 25 (37), and 34 (49) ns at 633 nm (1.15 μm). The longer melt duration at 1.15 μm apparently results from reflection of the 1.15-μm probe laser beam by a buried molten layer, after the surface has solidified. We have carried out computer simulations [8] of the reflected intensity from a wide variety of multi-layered structures composed of FG and LG p-Si, l-Si, and a- and c-Si, at both 633 nm and 1.15 μm. Several conclusions follow from these calculations: (1) The deep minimum observed in the 1.15-μm R signal is caused by the destructive interference of reflections from the sample's surface and from a buried l-Si layer; the simulations show that the zero reflectivity condition occurs when the top of a single liquid layer is 55–65 nm below the surface. No other zero of reflectivity occurs at 1.15 μm for a single liquid layer at greater depth, within the 190-nm thickness of the original a-Si layer. Thus, other incomplete R minima result as the single buried liquid layer propagates, or from reflections by a more complicated buried layer structure. (2) The shallow R minima observed at 633 nm for the two lowest E_ℓ in Fig. 4 are probably much-attenuated interference minima

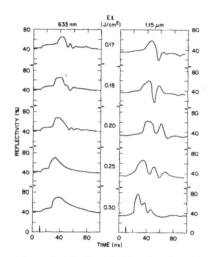

Fig. 4. Reflectivity signals at 633 nm and at 1.15 μm, during irradiation of a 190-nm thick a-Si layer by 45-ns (FWHM) KrF laser pulses.

that occur when the top of the buried liquid layer passes through a depth of about 70 nm. (3) The fact that no strong interference minimum is observed at 633 nm implies that the buried liquid layer is not formed by simply breaking away from a shallow solidifying melt at the surface: That would produce a deep minimum in R (633 nm) when the top of the liquid layer reached a depth of about 20 nm. Instead, the buried layer apparently is formed at a depth of about 45 nm or greater.

In summary, we find that time-resolved R measurements provide detailed information about crystallite nucleation and growth processes occurring in the near-surface region of an undercooled Si melt, and about changes in these processes with increasing pulsed laser E_ℓ. Changes in the shape of the R signal were shown to be correlated with corresponding changes (as a function of E_ℓ) in the solidification morphologies observed via TEM. R measurements using an infrared probe wavelength reveal the presence of a buried, propagating molten layer. We are currently using the new data obtained from these combined experimental techniques, together with model calculations [7], to try to distinguish between the explosive and bulk nucleation models for solidification in highly undercooled l-Si.

References

1. The melting temperatures of a-Si and a-Ge have been the subject of numerous recent publications. Please see the references to this literature that are given in our Refs. 2, 3, and 6 below.
2. D. H. Lowndes, R. F. Wood, and J. Narayan, Phys. Rev. Lett. 52, 561 (1984).
3. R. F. Wood, D. H. Lowndes, and J. Narayan, Appl. Phys. Lett. 44, 770 (1984).
4. D. H. Lowndes, R. F. Wood, C. W. White, and J. Narayan, Mat. Res. Soc. Symp. Proc. 23, 99 (1984).
5. J. Narayan and C. W. White, Appl. Phys. Lett. 44, 35 (1984).
6. M. O. Thompson, G. J. Galvin, P. S. Peercy, J. M. Poate, D. C. Jacobson, A. G. Cullis, and N. G. Chew, Phys. Rev. Lett. 52, 2360 (1984).
7. R. F. Wood, A. Geist, A. Solomon, D. H. Lowndes, and G. E. Jellison, Jr., these proceedings.
8. D. H. Lowndes, G. E. Jellison, Jr., and R. F. Wood (to be submitted to The Physical Review).

SPATIALLY AND TEMPORALLY RESOLVED REFLECTIVITY PROFILES OF
CRYSTALLINE SILICON IRRADIATED BY 48 PS PULSES
OF ONE-MICRON LASER RADIATION

STEVEN C. MOSS, IAN W. BOYD, THOMAS F. BOGGESS, AND ARTHUR L. SMIRL
Center for Applied Quantum Electronics, Department of Physics, North Texas
State University, Denton, Texas 76203

ABSTRACT

Time-resolved reflectivity measurements have been performed on single-
crystal silicon irradiated by intense 48 ps pulses of 1 micron radiation.
The temporal evolution of the spatial profile of the laser-induced reflectivity changes has been monitored using spatial correlation and surface
imaging techniques. These measurements resolve apparent discrepancies
between previous spatially and temporally averaged measurements at this
wavelength and similar measurements in the visible.

INTRODUCTION

Recent picosecond and femtosecond pump-and-probe measurements have
provided substantial support for purely thermal models of pulsed laser
melting of silicon within these time regimes. For example, the increased
self-reflectivity of a 20 ps, 0.53 μm pulse at a fluence above 0.2 J/cm^2 is
consistent with melting during the pulse and the correspondingly increased
reflectivity of liquid silicon [1]. In addition, the temporal evolution of
the reflectivity and transmission of silicon melted with these green light
pulses has been measured with a standard pump-and-probe geometry [2].
Here, again, the reflectivity rises in picoseconds to a value consistent
with the formation of a layer of liquid silicon. More recently, pump-and-
probe measurements of the reflectivity of silicon melted with 90 fs pulses
at 620 nm indicate that it can be melted or even evaporated within picoseconds of excitation [3]. These measurements, and the many others reported
to date, are all extremely consistent--when silicon is excited beyond the
melting threshold by picosecond or femtosecond pulses of visible light,
melting occurs within a few picoseconds of excitation.
    The experimental features are dramatically different when silicon is
excited beyond the melting threshold by picosecond pulses at 1.064 μm. In
contrast to the results with visible excitation, we observe no increase in
the self-reflectivity of a single pulse, even when the fluence is above the
melting threshold. Furthermore, the reflectivity of the excited region takes
hundreds of picoseconds to reach a value consistent with that of liquid
silicon [4]. These spatially-averaged measurements have been interpreted as
an indication of delayed melting and of the influence of a dense electron-
hole plasma [4].
    Here, we have performed temporally- and spatially-resolved reflectivity
measurements of single crystal silicon irradiated by intense picosecond
pulses of 1.064 μm light. Using spatial correlation and surface imaging
techniques, we show that it is essential to have an accurate knowledge of
the entire time-resolved spatial profile in order to interpret the spatially
averaged reflectivity data at 1 μm mentioned above. Finally, we demonstrate
that, although the spatially-averaged experimental features at 1 μm are very
different from those in the visible, they are also consistent with a rapid
melting of the Si.

EXPERIMENTAL APPARATUS

    An electro-optic shutter switched a single 48 ps (FWHM) pulse from a train of pulses produced by an actively/passively modelocked Nd:YAG laser. Part of the pulse was directed by a beamsplitter to a ratiometer that allowed us to determine the temporal pulsewidth shot-for-shot. The portion of the pulse transmitted by the beamsplitter was amplified and directed to a pump-and-probe apparatus. Here, the amplified pulse was split into two parts, a strong pump pulse and a weak probe pulse. The pump pulse was focused by a 1-m-focal-length lens at normal incidence onto the sample to a spot size of 320 μm (FWHM). The spot sizes were determined by horizontal and vertical beamscans using a scanning pinhole/photodetector apparatus. The spatial profiles at and on both sides of focus were determined to be Gaussian and to follow the usual Gaussian propagation formulae. The weak probe beam was directed to an optical delay line and then to the sample. During all but the spatial imaging measurements, the probe beam was focused by a 381-mm-focal-length lens to a spot size of 67 μm (FWHM) centered on the region of the sample illuminated by the pump beam. The peak probe fluence during these measurements was less than 3.6 mJ/cm^2. The light reflected by the sample and the light transmitted by the sample were collected by large-aperture, short-focal-length lenses and focused onto silicon PIN photodetectors. All photodetectors were isolated from luminescence and stray light by narrow bandpass filters placed in the beampath. The polarization of the probe beam was arranged perpendicular to the plane of incidence and also perpendicular to the polarization of the pump beam. We found it necessary to place prism polarizers in the reflected probe beam and transmitted probe beam to ensure that scattered light from the pump beam did not affect these measurements. For the spatial imaging measurements, the probe lens was removed, and the unfocused probe beam [spot size ~9.6mm (FWHM)] was centered on the region of the sample illuminated by the pump beam. The surface of the sample was then imaged onto a vidicon along the direction of the reflected probe beam. The magnification was x6, and the resolution of ~20 μm was limited by the resolution of the vidicon.

The sample was a 1-inch-diameter, 1-mm-thick flat cut from a boule of (111) Czochralski-grown, single-crystal, high-purity silicon. The sample was optically polished, and the second surface was AR coated to miminize Fabry-Perot effects.

RESULTS AND DISCUSSION

We used four separate techniques to measure the reflectivity of silicon when it was irradiated with picosecond pulses at one micron whose fluences were above the threshold (~1.6 J/cm^2) necessary to produce observable changes in the surface morphology. Varying the fluence from well below to factors of approximately four above threshold, we measured the self-reflectivity of the pump pulse. We then used standard picosecond pump-and-probe techniques, with the probe spot size much smaller than the pump, to measure the reflectivity of the silicon wafer. These measurements were performed both as a function of time delay with pump fluences fixed and as a function of fluence with time delays fixed. Next, we used a spatial correlation technique, scanning the pump pulse across the probe pulse, to map out the spatial profile of changes induced in the surface reflectivity. Finally, we used the surface imaging technique described by Downer et al. [5] to monitor shot-for-shot the changes in surface reflectivity. In all cases, the sample was translated following each shot, so that the sample was irradiated in a 1-on-1 manner.

The self-reflectivity of the pump pulse did not change from the crystalline value even for pump fluences approximately 4 times threshold. This result is in stark contrast to the same measurement performed with picosecond pulses of green light [1]. In those measurements, the increase in self-reflectivity at threshold has been interpreted as evidence for melting during the pulse. The fact that we observe no change in the self-reflectivity at 1 μm would <u>seem</u> to be consistent with a delayed melting.

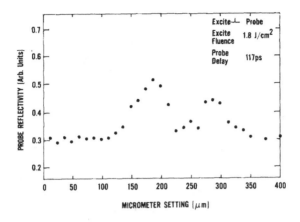

Fig. 1.   Spatial correlation measurement of probe reflectivity at 117 ps
delay. The horizontal axis shows the distance the lens in the
pump beam path was moved.

Furthermore, we performed time-resolved but spatially integrated
measurements of the reflectivity of crystalline silicon irradiated at fixed
fluences. Here the probe was focused onto the center of the excited region
to a spot size ~1/5 that of the pump beam. The spatially integrated
reflected probe signal was monitored. For a fixed fluence of 2.2 J/cm^2, the
apparent reflectivity rose to a maximum of approximately 0.6 only after a
delay of more than 400 ps. At higher fluences, the delay between excitation
and peak probe reflectivity was even longer. While these results are consistent with previous studies [4] and with the self-reflectivity measurements
described above, they do not appear to be consistent with rapid melting. In
contrast, when silicon is excited above threshold by 20 ps pulses of 0.53 μm
light, the optical properties of the surface become consistent with those of
molten silicon during the 20 ps melting pulse [2].
    We also measured the probe reflectivity as a function of pump fluence
for fixed delays between pump and probe pulses. At each fixed delay, the
probe reflectivity showed a dramatic variation with increasing fluence,
jumping to values near that expected of liquid silicon and then plunging to
values below that expected for crystalline silicon. Since each optical
pulse had a Gaussian spatial profile, that is, a spatial fluence profile,
these measurements suggested a strong spatial dependence to the reflectivity
change produced by each pulse. A complete understanding of the reflectivity
thus required that we spatially resolve the reflectivity profile of the
irradiated region.
    We first used spatial cross correlation techniques to spatially, as
well as temporally, resolve the reflectivity. The results of such a measurement are shown in Fig. 1. Here, the probe delay was held constant at
117 ps, and the pump fluence was held at 1.8 J/cm^2. The pump beam was
scanned across the probe beam by translating the lens in the pump beam path.
Care was taken to center the two beams vertically, and then the pump was
scanned in the horizontal dimension. Each data point shown is the average of
20 shots, with each shot irradiating a virgin piece of silicon. Notice that
the probe reflectivity dips dramatically near the center of the scan, rises
away from center on each side and then falls to the value of the crystalline
reflectivity in the wings. This spatial correlation technique graphically
displays the dramatic fluence dependence of the probe reflectivity, even at

110

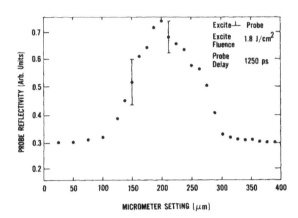

Fig. 2.   Spatial correlation measurement of probe reflectivity at 1250 ps
delay. The horizontal axis shows the distance the lens in the
pump beam path was moved.

fluences very near the melting threshold. The result of a scan at the same
pump fluence but a much longer time delay of 1250 ps is shown in Fig. 2.
Here, the reflectivity at the center of the spot has risen to values consistent with molten silicon. The rounded edges of the profile are probably an
artifact of the measurement technique, i.e. the correlation of the probe
pulse profile with the surface reflectivity features. Clearly, at fluences
very near threshold, the dip in reflectivity near the center of the irradiated region has fully recovered to the liquid reflectivity within one nanosecond. At higher fluences, however, the spatial correlation scans show
dips in the reflectivity profile even at 1250 ps delay. This spatial correlation technique is very time consuming and has limited usefulness
because of the cross correlation of the features of the reflectivity profile
with the spatial width of the probe pulse.
    A more elegant technique has been used by Downer et al. [5] to record
pictures of the surface of silicon melted with 90 fs pulses of 620 nm light.
The melted region was imaged onto a vidicon by 90 fs probe pulses at various
wavelengths and various delays. We use the same technique here to spatially
and temporally resolve the reflectivity of silicon melted by 48 ps pulses at
1.064 μm. The melted region is imaged onto the vidicon using the unfocused
probe pulses. A series of such profiles is shown in Fig. 3. as a function
of increasing fluence for three fixed time delays. Each of these represents
a narrow cross-section taken through the center of the irradiated region of
the sample. These results are completely consistent with the results of the
spatial correlation measurements described in the previous paragraph. At
each time delay, the dip in the center of the irradiated region becomes more
pronounced for higher pump fluences. At higher fluences, the dip lasts
longer. Even though the longest time delay shown here is 417 ps, the dip
lasts beyond 1250 ps delay for the highest fluence shown here. Keeping in
mind the finite width of our pulses, we believe that the reflectivity rise
during the pump pulse is indicative of rapid melting. Furthermore, the dip
in reflectivity that occurs thereafter may be caused by an overheating,
boiling, and evaporation of a thin surface layer of the liquid silicon. The
resulting evaporation then produces a cloud of material that scatters the
probe light and reduces the apparent reflectivity. These results are in

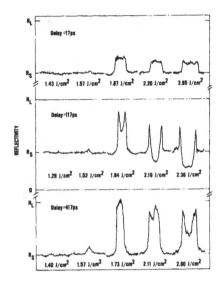

Fig. 3.   Sequence of images of
          the surface reflec-
          tivity for various
          fluences at three
          fixed delays.

qualitative agreement with those of Downer et al. [5]. An important distinction exists, however. We have found that it is extremely difficult to avoid
material ejection when melting the silicon surface with 1 μm picosecond
pulses. In fact, we observe the effects of material ejection even for
fluences as close to the threshold of ~1.6 J/cm^2 as we can experimentally
maintain. By contrast, no material ejection is observed for femtosecond
visible excitation until the fluence exceeds 2.5 times the threshold of
~0.1 J/cm^2 [5]. This disparity is a direct consequence of the more than
order-of-magnitude difference in the melting thresholds and the weak wavelength dependence of the liquid silicon absorption. We estimate that for
620 nm femtosecond pulses, once melting occurs, a 70% increase in fluence is
required to evaporate one skin depth of material. On the other hand, for
1 μm picosecond excitation, a fluence only 6% above threshold results in
evaporation [6]. Our observations of a constant self-reflectivity can be
readily understood in terms of this material ejection coupled with the
temporally-and spatially-integrated nature of the measurement. Moreover, in
view of the data in Fig. 3, the delayed rise in the probe reflectivity that
we observe in our pump-probe measurements can easily be interpreted in terms
of material evaporation without invoking a delayed melting model.

   In summary, we have used surface imaging techniques to temporally
resolve the spatially-dependent reflectivity of silicon melted by picosecond
pulses at 1 μm. These results are consistent with melting occurring during
the 48 ps pulse and explain the <u>apparent</u> delayed melting indicated by self-
reflectivity and pump-probe measurements [4] at this wavelength. In addition, the discrepancies between reflectivity measurements using 1 μm and
visible radiation have been explained in terms of the dramatically different
melting thresholds.

ACKNOWLEGEMENTS

   This work was supported by the U. S. Office of Naval Research, the
Robert A. Welch Foundation, and the North Texas State University Faculty
Research Fund.

REFERENCES

1. N. Bloembergen, H. Kurz, J. M. Liu, R. Yen, in "Laser and Electron Beam Interactions with Solids," edited by B. R. Appleton, G. K. Celler (Elsevier, NY, 1982), p. 3.

2. J. M. Liu, H. Kurz, N. Bloembergen, in "Laser-Solid Interactions and Transient Thermal Processing of Materials," edited by J. Narayan, W. L. Brown, R. A. Lemons, (North-Holland, NY, 1983), p. 3.

3. C. V. Shank, R. Yen, C. Hirliman, Phys. Rev. Lett. 50, 454 (1983).

4. K. Gamo, K. Murakami, M. Kawabe, S. Namba, Y. Aoyagi, in "Laser and Electron Beam-Solid Interactions and Materials Processing," edited by J. F. Gibbons, L.D. Hess, T. W. Sigmon (North Holland, NY, 1981), p. 97.

5. M. Downer, R. L. Fork, C. V. Shank, in "Ultrafast Phenomena IV," edited by D. H. Auston and K. B. Eisenthal (Springer-Verlag, NY, 1984), p. 106.

6. I. W. Boyd, S. C. Moss, T. F. Boggess, A. L. Smirl, Appl. Phys. Lett., to be published.

TIME-RESOLVED ELLIPSOMETRY AND REFLECTIVITY MEASUREMENTS OF THE OPTICAL PROPERTIES OF SILICON DURING PULSED EXCIMER LASER IRRADIATION*

G. E. JELLISON, JR. AND D. H. LOWNDES
Solid State Division, Oak Ridge National Laboratory,
Oak Ridge, Tennessee 37831

ABSTRACT

Several advances in time-resolved optical measurement techniques have been made, which allow a more detailed determination of the optical properties of silicon immediately before, during, and after pulsed laser irradiation. It is now possible to follow in detail the time-resolved reflectivity signal near the melting threshold; measurements indicate that melting occurs in a spatially inhomogeneous way. The use of time-resolved ellipsometry allowed us to accurately measure the optical properties of the high reflectivity (molten) phase, and of the hot, solid silicon before and after the laser pulse. We obtain $n = 3.8$, $k = 5.2$ (± 0.1) at $\lambda = 632.8$ nm for the high reflectivity phase, in minor disagreement with the published values of Shvarev et al. for liquid silicon. Before and after the high reflectivity phase, the time-resolved ellipsometry measurements are entirely consistent with the known optical properties of crystalline silicon at temperatures up to its melting point.

INTRODUCTION

Time-resolved reflectivity (TRR) measurements have been used for several years to examine the laser annealing process during and immediately after a laser pulse [1-3]. Early experiments used ruby [2] or Nd:YAG [1,3] lasers as the exciting lasers, and the 632.8-nm line or the 1152-nm line from a HeNe laser as the probing light. In most cases, only the reflectivity was observed, but the earliest work of Auston et al. [1] looked at the polarized reflectivity, and Lowndes et al. [2] observed the infrared transmission from a 1152-nm HeNe laser.

In this paper, we report recent time-resolved reflectivity measurements of crystalline silicon (c-Si) irradiated with light from a KrF (248 nm, 45-ns pulse width) excimer laser. Several improvements have been made in our measurement system. The use of an excimer laser simplifies the experiment, since it has very good pulse-to-pulse reproducibility (±3% in the energy density) and very good transverse homogeneity (also ±5% over a 3-mm spot). The second improvement involves the automation of the data acquisition system: The time-resolved reflectivity signal is fed into a digital oscilloscope (Tektronix 7912 A/D), which is connected to a desktop computer, allowing the data to be stored and manipulated later; signal averaging is then also possible. By using a beam splitter to monitor the incident excimer energy on every shot, the uncertainty in energy density can be further reduced to ~1%. An ultrastable HeNe laser (632.8 nm) was used for the probe beam. A final improvement is the use of wide-angle polarimetry measurements to supplement the near-normal incidence unpolarized reflectivity measurements.

---

*Research sponsored by the Division of Materials Sciences, U. S. Department of Energy under contract DE-AC05-84OR21400 with Martin Marietta Energy Systems, Inc.

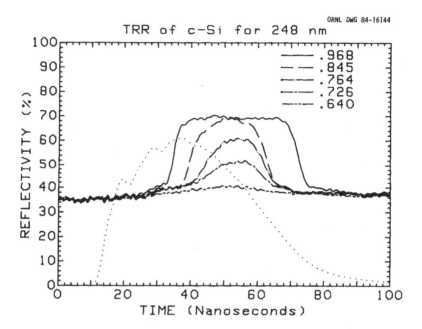

Fig. 1. The time-resolved reflectivity observed from silicon irradiated with various energy densities (shown in the upper right hand corner
in J/cm^2) from a KrF excimer laser. The probe laser was HeNe (632.8 nm). A
profile of the actual laser pulse is shown in arbitrary units by the dotted
line.

Normal Incidence Time-Resolved Reflectivity Measurements

Figure 1 shows TRR measurements for c-Si irradiated at several different energy densities (E_ℓ) near the melt threshold. The signal-to-noise
ratio is much better than in previously published results due to the
improvements mentioned above. With the improved signal-to-noise ratio,
several heretofore unseen features become apparent. For E_ℓ well above the
melt threshold, the reflectivity (normalized to the room temperature value
of 34.7%) rises dramatically (to 71% ± 1%) once enough energy has been
absorbed to melt the surface. The reflectivity continues at nearly this
value until it drops precipitously to ~41%, where it continues to decrease,
albeit much more slowly. For energy densities near the melt threshold (0.65
< E < 0.84 J/cm^2), the reflectivity increases, but never reaches 71%, indi-
cating that the surface is only partially melted. (This will be discussed
later.) For energy densities below the melt threshold (E < 0.64 J/cm^2), the
reflectivity rises but no sharp increase above ~41% is observed, indicating
that the surface is heated but no phase transition has occurred.

In all cases studied where a phase transition was observed, the sharp
increase of the reflectivity occurred at R = 41–42%. Detailed measurements
of the reflectivity of silicon up to 1000°C indicate that [4]

$$R = R_0 + aT \; , \tag{1}$$

where $R_0 = 0.347$ and $a = 4.6 \pm 0.1 \times 10^{-5}/°C$; therefore, $R = 41.5 \pm 0.5\%$ corresponds to $1480 \pm 110°C$ in agreement with the melting point of silicon ($1410°C$). As a result, the detailed time-resolved reflectivity measurements presented here provide yet another measurement of the lattice temperature and indicate that pulsed laser-annealed silicon is heated to the melting point of silicon before the onset of the high reflectivity phase, in agreement with the melting model for pulsed laser annealing.

A plot of the maximum reflectivity versus E_ℓ near the melting threshold (not shown) shows that the maximum reflectivity increases monotonically for $E_\ell > 0.64$ J/cm^2, until the maximum value of $71 \pm 0.5\%$ is obtained for energy densities greater than ~ 0.9 J/cm^2. Throughout this threshold region, the melt duration is never less than ~ 20 ns. It appears that the simple one-dimensional melting model [5] cannot be used to explain this result, since the melt duration would be expected to approach zero continuously, as the melt threshold energy density is approached from above. However, this behavior is consistent with the front surface region being only partially melted for energy densities between the threshold value and the value required for complete melting. This phenomenon has been observed by other techniques [6], and has recently been explained for irradiation with visible lasers by Combescot et al. [7]. It should be pointed out, however, that the calculations of Combescot et al. [7] may not be appropriate for the case of excimer laser irradiation since the optical properties of silicon do not change appreciably in going from the crystalline to the liquid phase at 248 nm.

Time-Resolved Ellipsometry

By incorporating a compensator and/or polarizers into a reflectivity experiment, it is possible to measure additional independent parameters of the reflected light. In our experiment, we have utilized the standard polarizer-compensator-sample-analyzer (PCSA) configuration with a large angle of incidence (68.2°). Depending upon the azimuthal orientation of the polarizer, compensator and the analyser, the light incident on the detector will be proportional to [8]

$$I \propto R_s = r_s r_s^* , \tag{2a}$$

$$I \propto R_p = r_p r_p^* , \tag{2b}$$

$$I \propto (1 \pm S) \quad \text{or} \tag{2c}$$

$$I \propto (1 \pm C) . \tag{2d}$$

The quantities R_s and R_p represent the reflectivities for light polarized perpendicular and parallel to the plane of incidence, respectively. The parameters N, S, and C are alternate ellipsometry parameters, given by

$$N = \cos 2\psi \tag{3a}$$

$$S = \sin 2\psi \sin\Delta \quad \text{and} \tag{3b}$$

$$C = \sin 2\psi \cos\Delta, \tag{3c}$$

and the common ellipsometric parameters ψ and Δ are expressed as

$$\frac{r_p}{r_s} = \tan\psi e^{i\Delta} , \tag{4}$$

where r_p and r_s are the Fresnel reflection coefficients (generally complex) for light polarized parallel and perpendicular to the plane of incidence, respectively. In general,

$$N^2 + C^2 + S^2 = \beta^2 , \tag{5}$$

where β represents the fraction of polarized light reflected from the surface; if $\beta < 1$, then the sample may be acting as a partial depolarizer.

Alternatively, the effective dielectric functions can be calculated from the values of N, S, and C. These are given by

$$\langle\varepsilon_1\rangle = \sin^2\theta_i \left[1 + \tan^2\theta_i \frac{N^2 - S^2}{(1 + C)^2}\right] \tag{6a}$$

$$\langle\varepsilon_2\rangle = \frac{\sin^2\theta_i\tan^2\theta_i \ 2NS}{(1 + C)^2} , \tag{6b}$$

where θ_i is the angle of incidence and it is assumed that the sample surface can be properly modeled by a single air-material interface (i.e., no oxide or other layer).

Figure 2 shows the time-resolved ellipsometric parameters for a sample of silicon irradiated with a series of 1.45 J/cm² pulses from a KrF excimer laser. The effective dielectric functions determined from the data in Fig. 2 are shown in Fig. 3. For both figures, the expected values for c-Si are

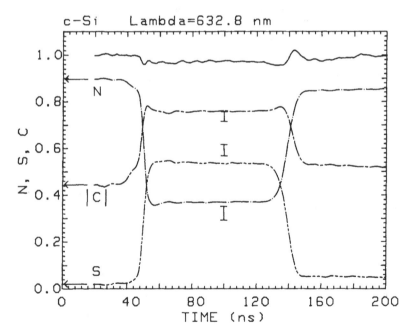

Fig. 2. The related ellipsometric parameters, N, S, and C plotted vs time for a silicon sample irradiated with a series of 1.45-J/cm² pulses from a KrF excimer laser (see text).

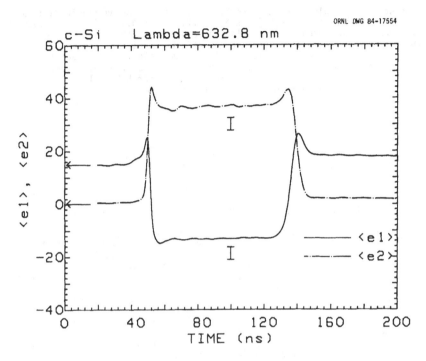

ORNL DWG 84-17554

Fig. 3. The effective dielectric functions determined from the N, S, and C parameters shown in Fig. 2 using Eqs. (6a) and (6b) (see text).

shown by arrows on the right, and the error bars indicate the values expected for liquid Si (l-Si) from an interpolation of the data of Ref. 9.

From Figs. 2 and 3, it is clear that the values of $\langle\varepsilon_1\rangle$ and $\langle\varepsilon_2\rangle$ obtained by time-resolved ellipsometry are in minor disagreement with the data of Ref. 9. Upon close examination, there are several possible errors in applying the data of Ref. 9 to TRR measurements of l-Si: (1) The probe wavelength (632.8 nm) was not measured in Ref. 9, and therefore an inter-polation must be used. (2) The authors of Ref. 9 used polaroid polarizers in their ellipsometer; because of the high leakage in these devices, severe errors can be introduced into the measurements. (3) The authors of Ref. 9 kept their l-Si sample at very high temperatures (>1410°C) for relatively long times; therefore, one must take into account the possibility that there was a film on the l-Si surface, even though the sample used in Ref. 9 was kept in a hydrogen atmosphere. A 4-nm film with n = 1.5 is sufficient to explain the difference between the two results.

In addition to measuring N, S, and C (Fig. 2), we also measured R_s and R_p, and R (θ_i = 0), all of which gave consistent results. Therefore, we obtained the following values of the optical functions for liquid silicon: ε_1 = -12.4 and ε_2 = 39.5 (±2.4) at 632.8 nm (n = 3.8, k = 5.2 ± 0.1), where the presence of a 9-Å layer of SiO_2 before, during, and after the laser pulse is taken into account.

There are several other features apparent from the effective dielectric functions presented in Fig. 3. (1) Both $\langle\varepsilon_1\rangle$ and $\langle\varepsilon_2\rangle$ increase monotoni-cally until melting occurs; the values of $\langle\varepsilon_1\rangle$ and $\langle\varepsilon_2\rangle$ at this point are consistent with values one obtains by extrapolating the available data to the melting point of Si [4,10] ($\langle\varepsilon_1\rangle$ = 20.7, $\langle\varepsilon_2\rangle$ = 4.4). (2) The values of $\langle\varepsilon_1\rangle$ and $\langle\varepsilon_2\rangle$ go through an oscillation before going to the values for l-Si; this is simply a result of interference effects created by the motion of the melt front going into the solid Si. (3) As the melt front recedes to the surface, the same oscillation is observed. (4) After the melt front has reached the surface, the values of $\langle\varepsilon_1\rangle$ and $\langle\varepsilon_2\rangle$ for hot, solid Si are again obtained, and continue to decrease slowly with time.

References

1. D. H. Auston, C. M. Surko, T. N. C. Venkatesan, R. E. Slusher, and J. A. Golovchenko, Appl. Phys. Lett. 33, 437 (1978).
2. D. H. Lowndes, G. E. Jellison, Jr., and R. F. Wood, Phys. Rev. B 26, 6747 (1982).
3. D. H. Lowndes, R. F. Wood, and R. D. Westbrook, Appl. Phys. Lett. 43, 258 (1983).
4. G. E. Jellison, Jr., L. K. Craven, and H. H. Burke, private communication.
5. R. F. Wood and G. E. Giles, Phys. Rev. B 23, 2923 (1982).
6. W. G. Hawkins and D. K. Biegelsen, Appl. Phys. Lett. 42, 358 (1983); J. E. Sipe, J. F. Young, J. S. Preston, and H. M. van Driel, Phys. Rev. B 27, 1141 (1983); J. F. Young, J. S. Preston, H. M. van Driel, and J. E. Sipe, Phys. Rev. B 27, 1155 (1983); J. F. Young, J. E. Sipe, and H. M. van Driel, Phys. Rev. B 30, 2001 (1984).
7. M. Combescot, J. Bok, and C. Benoit à la Guillame, Phys. Rev. B 29, (1984).
8. R. M. A. Azzam and N. M. Bashara "Ellipsometry and Polarized Light" (North Holland, New York, 1977).
9. K. M. Shvarev, B. A. Baum, and P. V. Gel'd, High Temp. 15, 548 (1977).
10. G. E. Jellison, Jr. and F. A. Modine, Appl. Phys. Lett 41, 1980 (1982).

TIME-RESOLVED X-RAY STUDIES DURING PULSED-LASER IRRADIATION OF Ge[†]

J.Z. TISCHLER*, B.C. LARSON* and D.M. MILLS**
* Solid State Division, Oak Ridge National Laboratory, Oak Ridge, TN 37831
** CHESS and Applied and Engineering Physics, Cornell University, Wilson Laboratory, Ithaca, NY 14853

ABSTRACT

Synchrotron x-ray pulses from the Cornell High Energy Synchrotron Source (CHESS) have been used to carry out nanosecond resolution measurements of the temperature distrubutions in Ge during UV pulsed-laser irradiation. KrF (249 nm) laser pulses of 25 ns FWHM with an energy density of 0.6 J/cm^2 were used. The temperatures were determined from x-ray Bragg profile measurements of thermal expansion induced strain on <111> oriented Ge. The data indicate the presence of a liquid-solid interface near the melting point, and large (1500-4500°C/μm) temperature gradients in the solid; these Ge results are analagous to previous ones for Si. The measured temperature distributions are compared with those obtained from heat flow calculations, and the overheating and undercooling of the interface relative to the equilibrium melting point are discussed.

INTRODUCTION

Time resolved x-ray studies of silion during pulsed laser irradiation have been previously reported in the literature [1,2]. This paper presents new results for similar experiments on germanium samples. We present here experimentally determined temperature profiles in a germanium crystal during pulsed laser irradiation. In contrast to the data on silicon, the measured temperature profiles in germanium do not agree with profiles from heat flow calculations. This difference is discussed with respect to variations of the thermal conductivity brought on by the very large (about 2%/μm) strain gradients, the uniaxial nature of the strain, and lattice impurities. By comparing liquid-solid-interface temperatures during melting and regrowth, we determined the difference between overheating and undercooling produced by the high interface velocities.

EXPERIMENT

The experimental procedure consists of measuring germanium rocking curves at a specific time after hitting the sample with a high power laser pulse. The shape of the rocking curve contains all of the information needed to obtain the strain gradient in the crystal. The angular range of intensity in the rocking curve indicates the maximum temperature, and the height of the rocking curve indicates the amount of material with a particular strain (the strain gradient for a uniformly strained crystal). A KrF (249 nm) eximer laser with a 25 ns pulse width was used to irradiate the germanium crystal. This short wavelength light is all absorbed in the first few 100 A (unlike ruby light on silion where there is considerable penetration of the laser beam). Also, the eximer laser's poor spatial coherence reduced diffration effects so that the beam uniformly heated the sample without the use of a quartz beam homogenizer.

[†]Research sponsored by the Division of Materials Sciences, U.S. Department of Energy under contract DE-AC05-840R21400 with Martin Marietta Energy Systems, Inc.

These measurments required a sharply pulsed source of x-rays; synchrotron radiation at CHESS provided an x-ray source that was both intense and sharply pulsed. The x-rays pulses from CHESS arrived every 854 ns with a pulse width of 150 ps. The laser was triggered so that its pulse hit the sample at a selected time before the x-ray pulse. Thus, the x-rays probed the sample at a particular time delay. To avoid bulk heating of the germanium sample, the laser was only fired once a second, thus most of the x-ray pulses were not used. The entire Bragg profile was measured on one spot on the sample, which developed a stable yellowish oxide 500 A thick after 10 laser shots.

The x-ray data consists of a simple rocking curve taken as a step scan at a fixed time delay between the laser and x-ray pulses. At each step in the scan, the reflectivity was measured using 9 laser shots. By taking rocking curves at different delay times, we accumulated a full set of data. Most of the data were taken at only two delay times, 15 ns and 45 ns after the laser pulse's peak. These times correspond to the maximum melting and regrowth velocities.

To measure the number of photons scattered during one x-ray pulse required digitizing the peak height from the x-ray detector because the diffracted beam detector recieved multiple (between 0 and 50) photons during each 150 ps x-ray pulse. Since the decay time in the NaI detector is longer than 150 ps, the output of the detector appeared as a single photon n times as large, where n is the number of photons collected.

To measure the reflectivity required accurate knowledge of the number of photons incident on the sample as well as the number scattered. To calibrate the incident x-ray beam we used one of the standard 6cm N_2 ion chambers available at CHESS. The calibration of charge per photon for this chamber was previously determined by others [3].

RESULTS

The results of the rocking curves are shown as the data points in figure 1. First, note that the scattering is weaker for the 15 ns data. This indicates that the temperature gradients are higher during melt-in, which is expected. Second, the scattering during melt-in extends to slightly higher angles than the scattering during regrowth. This implies a higher interface temperature during melt-in. This is also expected, since there should be some overheating during melt-in and undercooling during regrowth at these high interface velocities.

As mentioned above, most of the data were taken at delay times of 15 and 45 ns because these correspond to the times of maximum melt-in and regrowth, and also because the interface acceleration was a minimum. If the interface velocity is constant, the system is in a steady state, and so small errors in the timing only weakly affect the data.

To evaluate the data, we used a recursive procedure to find a temperature profile whose computed scattering matched the measured scattering. The comparison between fitted temperature profile and measured scattering was made with a program that calculated the scattering from a strained crystal using dynamical diffraction theory. The temperature profile was then adjusted until the calculated and measured x-ray scattering profiles agreed. The fitted temperature profiles are shown as the solid lines in figure 2. Note that the interface temperature for the 15 and 45 ns temperature profiles differ by about 100 K. This difference represents the sum of the overheating and undercooling. However, the interface temperature for the 45 ns data is higher than the equilibrium melting point indicating that some other effect has not been included. More will be said about this later.

For comparison with the experimentally determined temperature profiles, we calculated temperature profiles using a heat flow program written by M. O. Thompson and equilibrium thermodynamic data [4]. These calculations

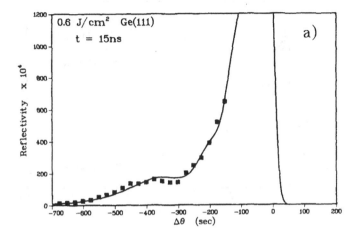

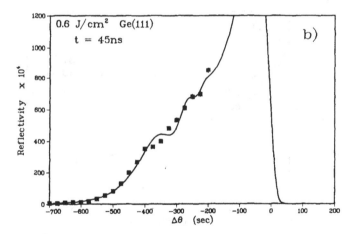

Figure 1. Measured and fitted Bragg profiles for the Ge (111) reflection. The solid line represents scattering calcualtions fitted to the measured data points. a) at 15 ns delay; b) at 45 ns delay.

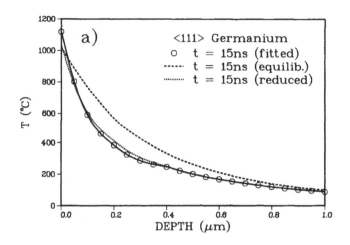

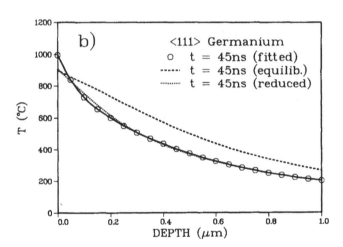

Figure 2. Measured temperature profiles (solid lines). Temperature profiles calculated using equilibrium thermal conductivity (dashed lines), and modified thermal conductivity (dotted lines). a) at 15 ns delay; b) at 45 ns delay.

are shown as the dashed lines in figure 2. Note that the theoretical curves have much smaller high temperature thermal gradients than the experimentally determined profiles.

Varying either the high temperature thermal conductivity or the laser power by a factor of two changes the maximum regrowth velocity by less than 15%. Furthermore, since the interface velocity is limited by the ability of the crystal to conduct away the latent heat of fusion, the interface velocity is,

$$v = -K\nabla T/\Delta H_m \tag{1}$$

where v is the interface velocity, K is the thermal conductivity, ∇T is the thermal gradient at the interface, and ΔH_m is the heat of fusion per unit volume. Using the numerically calculated regrowth velocity of 3.2 m/s, the measured thermal gradient of 0.24 K/Å, and the equilibrium latent heat of 3564 J/cm^3, equation 1 gives a thermal conductivity of 0.05 W/cm-K. This is 3.5 times smaller than the equilibrium value of 0.175. Since the error in the thermal gradient is 20%, and there is no basis for assuming that the velocity is 12 m/s or that the latent heat has changed, we conclude that the high temperature thermal conductivity is much smaller than the equilibrium value. To demonstrate the effect of a reduced high temperature thermal conductivity on the temperature profiles, we changed the thermal conductivity at temperatures above 900 K as shown in figure 3. Using this reduced thermal conductivity in the heat flow program produces the dotted curves in figure 2, which are much closer to the fitted temperature profile.

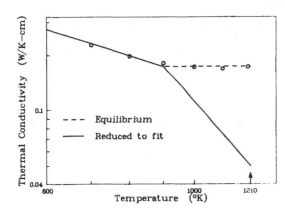

Figure 3. Log-log plot of thermal conductivity (only from 600 to 1210 K) used to compute the temperature profiles. The dotted line represents the equilibrium values, the solid line is the reduced value that better fits the data. The points are from reference [4].

DISCUSSION

These anomalously large temperature gradients near the interface were unexpected as they are not consistent with heat flow calculations using equilibrium values of the thermal conductivity; such an approach was successful for silicon. Three possibilites present themselves, first, the thermal conductivity might be reduced by an increase in phonon scattering brought on by the large strain gradient. Second, the electronic part of the thermal conductivity depends on the number of conduction electrons which is altered by the anisotropic nature of the strain. And third, the thermal conductivity may be reduced by the inclusion of impurities, particularly oxygen which might have diffused in from the thick oxide layer on the surface. We interpret this large strain gradient to arise from a change in the thermal conductivity and not from correspondingly large changes in the latent heat of fusion or specific heat. This conclusion comes directly from equation 1 and consideration of possible regrowth velocities.

Since most of the heat flow in germanium is conducted by phonons [5], the simplest way to achieve a large reduction in the thermal conductivity is to assume that phonon scattering is strain dependent. For large k phonons, the mean free path is about 10 Å and therefore still diffusive. Thus the large k phonons should not be affected by the strain gradient. However, at small k, the mean free path must be at least a few wavelengths long. And if the lattice constant changes over a phonon wavelength or there are many impurities, other phonon modes will mix in and so increase phonon scattering; this will add to the ordinary Umklapp process.

Not all expected effects will reduce the thermal conductivity. The electronic contribution to the thermal conductivity might be increased over its equilibrium value due to the one dimensional nature of the strain. Since the relatively cool crystal substrate keeps the hot layers from expanding laterally, the crystal's expansion is uniaxial, normal to the surface, and is larger than would be expected from the coefficient of thermal expansion. From other work on the band gap of germanium under uniaxial strain [6], we expect the band gap to drop from 0.3 eV to about 0.2 eV which should increase the number of conduction electrons by 50%. This increase in the number of conduction electrons will tend to increase the thermal conductivity, but apparently not as much as other effects have reduced it.

The increased conduction electron density would also have one other effect, as the electron density increases, the germanium lattice expands. This effect results in a larger expansion than expected from the thermal expansion coefficient, therefore, the peak temperatures at the interface may be less than those actually shown. Errors in the values used for the coefficient of thermal expansion (~5%) and temperature dependence of elastic constants must also be considered when interpreting the anomalously high interface temperatures.

CONCLUSION

This paper presents time resolved Bragg profiles for <111> germanium during pulsed laser irradiation. The most significant result is the observation of a thermal gradient during regrowth that is 3 to 4 times larger than expected. This implies that in the highly strained crystal, the thermal conductivity is 3 to 4 times smaller than the equilibrium value.

The thermal conduction effects observed here are under further investigation. Thermal conductivity under nonequilibrium conditions with high temperature and strain gradients has not been previously investigated, and may reveal new physical phenomena.

REFERENCES

1. B. C. Larson, C. W. White, T. S. Noggle, J. F. Barhorst and D. M. Mills, Appl. Phys. Lett. 42, 282 (1983).
2. B.C. Larson, C.W. White, T.S. Noggle, and D.M. Mills, Phys. Rev. Lett. 48, 337 (1981).
3. E. Matsubara, private communication.
4. Thermophysical Properties of Matter, Vol. 2, IFI/Plenum, 1970. Thermal conductivity of the liquid came from the Wiedeman-Franz Law.
5. C. J. Glassbrenner and G. A.Slack, Phys. Rev. 134, A1058 (1964).
6. Yet-Ful Tsay and B. Bendow, Solid State Commun. 20, 373 (1976.

ANISOTROPIC SECOND HARMONIC GENERATION INDUCED IN SILICON AND GERMANIUM BY PICOSECOND AND NANOSECOND LASER PULSES

JOHN A. LITWIN, HENRY M. VAN DRIEL AND JOHN E. SIPE
Department of Physics and Erindale College, University of Toronto,
Toronto, Ontario, Canada, M5S 1A7

ABSTRACT

We have performed experimental and theoretical studies of the origins of anisotropic second harmonic generation observed in reflection from (111) and (100) faces of crystal Ge and Si. Results, using 25 psec and 20 nsec 1.06 and 0.53 μm pump beams allow us to conclude that induced plasmas have no role in the effect and that the mechanism for the nonlinearity is associated with quiescent surface dipolar or bulk quadrupolar effects. Dispersion in the anisotropic behaviour is presented and the application of anisotropic harmonic generation to structural determinations of Si and Ge during laser annealing will be discussed.

INTRODUCTION

During the past year there has been a resurgence of interest in the phenomenon of second harmonic generation (SHG) in centro-symmetric crystals [1-3]. Observations of this effect in a variety of insulators, metals and semiconductors had been widely reported in the literature during the 1960's and 1970's[4,5] and "non-local" mechanisms such as magnetic dipolar and electric quadrupolar effects were used to account for the data since bulk electric dipolar contributions are forbidden by inversion symmetry. What was apparently not appreciated at the time is the fact that the conversion efficiency can be dependent on the crystal cut and orientation in many materials, including semiconductors. Guidotti et al. [1] last year reported anisotropic SHG in Si and Ge in a reflection geometry using 0.53 and 1.06 μm, 10 pisosecond pump pulses. Such effects had also been seen earlier albeit in transmission in the centrosymmetric insulator NaNO3[6]. Guidotti et al. were not able to account for the anisotropy using bulk quadrupolar effects alone and, since no signal was seen in Si at 1.06 μm where interband absorption is weak, they concluded that a high density electron-hole plasma was necessary. They proposed that a plasma-mediated inversion breaking mechanism occurred leading to "allowed" SHG. More recently Tom et al.[2] using nanosecond, 0.53 μm excitation, also reported anisotropic SHG in Si. Their data were apparently in conflict with that of Guidotti et al. but could be explained using a surface dipolar/bulk quadrupolar mechanism for quiescent Si. Most recently Shank and coworkers [3] have attempted to exploit the disappearance of the anisotropic effect in femtosecond pulse irradiated Si to signify the onset of melting. In this paper we will show that all the data taken to date, including ours presented here, can be accounted for by a quiescent model for Si with the plasma playing no role, even in the picosecond domain. We also show that a significant contribution to the anisotropy is a pure surface effect so that the use of SHG to obtain information about bulk structural changes in Si and Ge is not unambiguous.

THEORY

In a material possessing inversion symmetry where there is no bulk contribution to SHG of electric dipole symmetry, the dominant contributions are from terms of electric quadrupole or magnetic dipole symmetry. At the surface. where there is no inversion symmetry, electric dipole terms can exist. Phenomenologically one can then write the non-linear surface polarization density as

$$P_i^{2\omega} = \sum_{jk} \Delta_{ijk} E_j^\omega E_k^\omega \delta(z) \qquad (1)$$

where E_i is the amplitude of the field below the surface (located at $z = 0$), and Δ_{ijk} is the surface nonlinear susceptibility tensor. For cubic, homogeneous crystals with a plane wave incident, one can write the non-vanishing bulk part of the polarization density as,

$$P_i^{2\omega} = \gamma \nabla_i (\vec{E}^\omega \cdot \vec{E}^\omega) + \zeta E_i^\omega \nabla_i E_i^\omega \qquad (2)$$

with respect to the usual crystal axes of the diamond structure. The first term is isotropic in character while the second is anisotropic; they are phenomenologically characterized by the parameters γ and ζ respectively. Consider now light which is incident on a (100) surface. If the surface is assumed to have C_{4v} symmetry the field strength of the s-polarized SH light generated by a s- or p-polarized incident beam takes the form

$$E_s^{2\omega} = b_{s,p}^{(1)} \zeta \sin 4\phi \qquad (3)$$

where $b_{s,p}^{(n)}$ is a complex number dependent on the angle of incidence, the dielectric constant at the fundamental and second harmonic frequencies and the polarization of the incident light; it is proportional to $(E^\omega)^2$. Here ϕ is the angle that the plane of incidence makes with the <010> direction. For p-polarized SH light the corresponding expression is of the form

$$E_p^{2\omega} = b_{s,p}^{(2)}\zeta + b_{s,p}^{(3)}\gamma + b_{s,p}^{(4)} f_{s,p}(\Delta_{ijk}^{(100)}) + b_{s,p}^{(5)}\zeta \cos 4\phi \qquad (4)$$

where $f_{s,p}$ (and $g_{s,p}$ in the expression below) indicates a linear combination of elements of the surface tensor. For a (111) surface with assumed C_{3v} symmetry the corresponding expressions are

$$E_p^{2\omega} = b_{s,p}^{(6)}\zeta + b_{s,p}^{(7)}\gamma + b_{s,p}^{(8)} g_{s,p}(\Delta_{ijk}^{(111)}) + (b_{s,p}^{(9)}\zeta + b_{s,p}^{(10)}\Delta_{111}^{(111)})\cos 3\phi \qquad (5)$$

and

$$E_s^{2\omega} = (b_{s,p}^{(11)}\zeta + b_{s,p}^{(12)} \Delta_{111}^{(111)}) \sin 3\phi \qquad (6)$$

where now ϕ is the angle between the plane of incidence and the projection of the <100> vector on the (111) plane. These expressions will be used below where the intensity spectra for x-polarized input light and y-polarized output light will be denoted by I_{xy}.

EXPERIMENTAL RESULTS

Our experiments were carried out using single crystal wafers of Si and

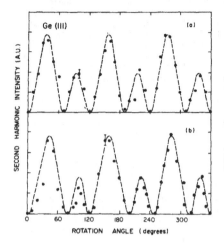

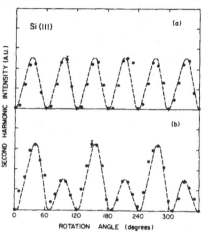

Fig.1. Angular spectrum of p-polarized SHG efficiency from (111) Ge induced by s-polarized 1.06 μm laser light with pulse widths (a) 25 psec and (b) 20 nsec. In this and the other figures, the origin of the rotation angle is chosen for convenience of display and the dotted lines are guides to the eye.

Fig.2. Angular spectrum of SHG efficiency from (111) Si for a 20 nsec 1.06 μm pump pulse with a s-polarized pump beam and (a) s-polarized and (b) p-polarized emission beams.

Ge, which were cut to reveal (100) and (111) faces. The samples were mechanically polished, with some being chemically etched; there was little difference in the results. Pump radiation at 1.06 and 0.53 μm was provided by Q-switched and mode-locked Nd:YAG lasers to provide pulses of 25 psec and 20 nsec respectively. The pulses were incident on the sample at an angle of incidence of 45 degrees on the sample face, which could be rotated about its normal. The weak SH signal (typically less than 10^3 photons) was detected using a monochromator and appropriate photomultipliers.

    Our first conclusion is that high density carrier effects did not influence any of our results and our observations are consistent with effects in quiescent semiconductors. This conclusion can be arrived at in several ways. For example, Figure 1 shows the angular rotation pattern of I_{sp} for 1.06 μm light on (111) Ge for both the nanosecond and picosecond excitation sources. In both cases the incident fluences was 50 mJ/cm^2, a factor of three below melting. To within experimental error (10%) the results are identical, not only with respect to the symmetry of the lobes as predicted by Equation (5) but also with respect to the relative heights of the lobes which are not determined by symmetry in this case. From previous reflectivity measurements [7] it had been determined that the induced carrier densities were 8×10^{19} and 5×10^{20} cm^{-3} for nanosecond and picosecond pulses respectively. The density independence of the results is further borne out by our observations of SHG in Si using nanosecond pulses at 1.06 μm for which the absorption depth is millimeters and the maximum induced carrier concentration at 50 mJ/cm^2 is much less than 10^{19} cm^{-3}. Typical results for different polarization pairs are shown in Figure 2. Both of these possess

128

patterns which are consistent with the equations above. For the same intensity of 1.06 μm light we find that Ge is only about a factor of 20 times more efficient than Si. Finally, and perhaps most convincingly, in both Si and Ge, with 1.06 μm incident light, the SHG intensity varied quadratically with the fundamental intensity below the threshold of melting. This tends to argue for a quiescent effect and against induced effects of any kind, including carrier density or lattice temperature effects. A similar conclusion was arrived at by Bloembergen and co-workers for nanosecond experiments. Shank and co-workers [3] apparently have noticed a plasma contribution using high intensity femtosecond pulses. To a first approximation, changes in the nonlinear and linear susceptibilities will become important at approximately the same density. On the basis of a Drude model for the free-carrier part of the linear susceptibility, this would occur at approximately 10^{21} cm^{-3} for the wavelengths of interest here.

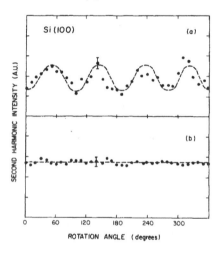

Fig.3. Angular spectrum of p-polarized SHG efficiency from (100) Si for a 20 nsec s-polarized pump pulse of wavelength (a) 0.53 μm and (b) 1.06 μm.

From the equations above it can be seen that only for second harmonic signals from the (100) face can one identify the anisotropy with bulk effects alone. From Figure 3 one sees that the bulk effects for 0.53 μm incident light on Si are not zero since there is significant modulation of the I_{sp} signal. For 1.06 μm light the modulation has virtually disappeared indicating a significant decrease, or dispersion, of the bulk anisotropic terms relative to the bulk isotropic or surface isotropic terms. In the same experiments conducted by Guidotti et al. [1] no modulation was observed for either sample or wavelength, but this may be due to poor signal-to-noise and low counting rates. Indeed in all cases where their observations differ from ours or those of Tom et al., experimental limitations seem to be the cause.

The surface terms obviously stand in the way of making anisotropic SHG a potentially powerful tool in the measure of bulk crystal short-range order in such areas as laser annealing. If one could relate their contributions to bulk effects progress could be made. In general, there are two contributions to these terms. One is associated with inversion breaking effects such as adsorbed molecules or interfacial layers which can give electric dipole contributions. The other is associated with quadrupolar-like contributions induced by the strong gradient (discontinuity) in the normal component of the electric field at the surface. A simple calculation shows that this latter effect's contribution to $\Delta_{j11}^{(111)}$ to be zero. If this were the only surface contribution, Equation (6) predicts that the I_{ss}/I_{ps} ratio of maximum intensities is determined only by the dielectric coefficient at the fundamental wavelength and the angle of incidence. For Si at 1.06 μm this ratio is predicted to be 0.615. Experimentally we find the ratio to be 0.37 indicating that other types of interfacial effects are present. It should be pointed out however that since our observations are virtually independent of surface preparation, such effects must be regarded as intrinsic properties of Si and Ge surfaces in air. Unfortunately, this means that any experiment which measures a change in characteristics of an

anisotropic signal, including that of Shank et al. who measure I_{pp} for (111)
Si cannot come to an unambiguous conclusion as to whether bulk, or surface,
or both contributions have changed. The exceptions are the pure bulk
sensitive cases (I_{ss} and I_{ps} on (100) surfaces); however they yield the
weakest signals because of geometrical effects and they (let alone their
changes) are difficult to observe. Experimentally, their signal strength
is more than two orders of magnitude below those reported here and are at
the limit of our apparatus' sensitivity.

ACKNOWLEDGEMENTS

We gratefully acknowledge financial support from the Natural Sciences
and Engineering Research Council of Canada.

REFERENCES

1) D. Guidotti, T. A. Driscoll and H. J. Gerritsen, Solid State Comm. 46,
337 (1983); T. A. Driscoll and D. Guidotti, Phys. Rev. B28, 1171 (1983).

2) H. W. K. Tom, T. F. Heinz, and Y. R. Shen, Phys. Rev. Lett. 51, 1983
(1983).

3) C. V. Shank, R. Yen and C. Hirlimann, Phys. Rev. Lett. 51, 900 (1983).

4) N. Bloembergen, R. K. Chang, S. S. Jha and C. H. Lee, Phys. Rev. 174,
813 (1968).

5) C. C. Wang and A. N. Duminski, Phys. Rev. Lett. 20, 668 (1968).

6) L. Ortmann and H. Vogt, Opt. Comm. 16, 234 (1976).

7) J. S. Preston and H. M. van Driel, Phys. Rev. B30, 1950 (1984); H. M.
van Driel, L.-A. Lompré, and N. Bloembergen, Appl. Phys. Lett. 44, 285 (1984).

EVOLUTION OF SILICON IRRADIATED WITH FEMTOSECOND PULSES

D. HULIN,[*] C. TANGUY,[*] M. COMBESCOT,[*] J. BOK,[*] A. MIGUS,[**] and A. ANTONETTI[**]
[*]Groupe de Physique des Solides de l'Ecole Normale Supérieure
[**]Laboratoire Optique Appliquée
ENSTA, Ecole Polytechnique, 91120 Palaiseau

ABSTRACT

The total energy reflected from a silicon single crystal irradiated by a 100 femtosecond laser pulse is measured. We observe a plasma resonance at wavelengths of 620 nm and 310 nm indicating electron-hole densities higher than 10^{22} cm . The result are interpreted using a highly non linear theory. Very short relaxation times are observed and attributed to electron-hole collisions. The study of the light scattered by the silicon surface shows a sharp decrease at high fluences that we interprete by a possible screening of irregularities by emitted electrons. A pump-test experiment is also reported showing the emission of Si particles. A possible mechanism for the extraction of these particles is proposed.

Laser pulses, of a duration of the order of 100 femtoseconds are a very unique tool to study the physical mechanisms of energy transfer from the electron-hole (e-h) plasma to the lattice in semiconductors. The incident photons are absorbed by the electrons, creating a hot and dense electron-hole plasma and breaking covalent bonds thus softening the lattice. After the pulse, the electron-hole pairs recombine, the plasma expands, and through electron-phonon interaction the energy is transferred to the lattice. Several experiments have recently been reported using femtosecond pulses to create a high density e-h plasma in silicon and study its time evolution [1,2,3]. The use of such intense and short pulses raises the possibility of breaking so many covalent bonds that the melting temperature of the crystal can be lowered [4,5] significantly. In a first period, a new phase is obtained, with atoms almost immobile (having a low kinetic energy) but imbedded in a dense hot plasma. In a time of the order of several electron-phonon relaxation times (τ_{e-p}) the energy is transferred to the atoms and the normal liquid phase is obtained. The understanding of the exact nature of the melting induced by very short pulses relies on a good knowledge of the energy transfer from the laser pulse to the sample. In this paper, we report measurements of the total amount of energy of a 100 fs, 620 nm and 310 nm of wavelengths, light pulse reflected by a silicon single crystal and its variation with pulse intensity (self reflectivity with no test beam). We also give measurement of the light scattered from the surface to see changes of the surface roughness. Finally, we give the result of a pump-test experiment showing the formation of a "blackhole" in the center of the incident spot, as already reported by other authors [6].

SELF REFLECTIVITY MEASUREMENTS

We have performed reflectivity experiments on a bulk crystalline silicon of orientation |100| with a laser pulse of 100 fs duration at 620 nm, energetic enough to melt (or even to evaporate) the crystal at each shot. The

132

penetration depth of this light in the unexcited silicon is 3 µm but can be reduced to less than 300 Å if a large number of electrons and holes are already present. The energy deposited per unit volume will thus be very different in these two extreme configurations and the knowledge of the e-h density profile is then of prime importance. For that purpose, we have measured the self-reflectivity of the laser pulse to obtain a time averaged information on the events inside this pulse, and consequently the mechanisms of the laser energy transfer towards the silicon. A detailed description of the experimental set-up was given in previous publications [7]. In the case of U.V. excitation, the second harmonic at 310 nm was generated in a 3 mm KDP crystal and focused on a 30 µm spot, yielding a maximum energy of about 10 mJ/cm². In the case of 620 nm light, the maximum incident fluence is 2 J/cm². The result of the experiments are shown on fig.1.

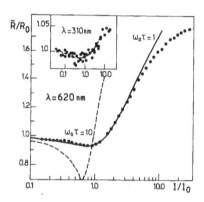

Fig.1 : Total reflectivity of Silicon points : experiments ; full line : theory using $\tau = 3 \ 10^{-16}$ s.

Let us first give a qualitative presentation of what is going on in the sample during the light pulse. The leading edge of the pulse finds a virgin silicon crystal, a part R_0 of the light is reflected, and the remaining part $(1-R_0)$ is absorbed through e-h pairs creation over a penetration depth of d = 3 µm, creating an e-h plasma with an exponentially decreasing profile. The plasma density increases in time over the same distance until the surface density reaches values near N_p, the plasma frequency density. Then the real part of the dielectric constant becomes negative and the beam transmitted into the sample becomes a vanishing wave and its penetration depth is reduced to a few hundred Ångströms, creating e-h pairs over that distance only. This explains the qualitative variation of the total fraction of energy of the beam reflected by the semiconductor with pulse intensity. A quantitative determination of the reflected energy R(t) during the pulse has been made by solving the non linear coupled equations for light propagation and e-h pairs generation. The light propagation is governed by Maxwell's equation :

$$c^2 \ \frac{\partial^2 E}{\partial z^2} = -\omega_0^2 \ \epsilon \ E \tag{1}$$

where E is the electric field of the plane wave propagating along the z direction, with frequency ω and ϵ is the dielectric constant of the medium. ϵ is given for a gas of free carriers by :

$$\epsilon(z,t) = \epsilon_0 \left| 1 - \frac{N(z,t)}{N_p(1+i/\omega_0\tau)} \right| + i \ \epsilon_0' \tag{2}$$

where N is the density of the e-h pairs, $\epsilon_0 + i \epsilon_0'$ is the static dielectric constant of the semiconductor ; the imaginary part ϵ_0' corresponds to interband absorption (i.e. electron-hole pair generation), τ is a relaxation time describing the dissipative processes in the plasma. The non linearity arises from the fact that N is a fonction of the absorbed power $\varepsilon_0'|E|^2$ and increases as :

$$\hbar \omega_0 \frac{\partial N}{\partial t} = \frac{1}{2} \varepsilon_0' |E|^2 \omega_0 \tag{3}$$

We thus have 3 equations allowing to compute $N(z,t)$ and $E(z,t)$. These equations are solved numerically using two boundary conditions : $E(\infty,t)=0$ and the continuity of the electric and magnetic fields at the incident surface. The instantaneous reflectivity is then easily obtained as $R(t) = |E(0,t) - E_0|^2/E_0^2$, E_0 being the amplitude of the electric field of the incident beam. Its averaged value R during the pulse duration is the quantity to compare with experiment. The experimental data (fig.1) can be very well fitted using a relaxation time $\tau = 3\,10^{-16}$ s. Values of N as high as $5\,10^{+2}$ cm^{-3} are obtained. We give the following interpretation for the very low value of τ. At low densities the relaxation time is due to electron-phonon (e-p) interaction. The e-p collision time τ_{e-p} is of the order of 10^{-13} s. For very dense plasma a new mechanism of absorption may appear due to electron-hole collisions. Collisions between electrons or between holes do not lead to dissipation because they conserve the total momentum of the electrons (or holes) and consequently the total electrical current. On the other hand, a collision between one electron and one hole, although it conserves the total momentum, gives a change in the relative velocity u and thus the electrical current eu of the e-h pair. We have estimated the e-h relaxation time using a screened coulomb interaction and the Rutherford scattering cross section |8| for the relative velocity. At Low plasma densities $1/\tau_{e-h}$ increases linearly with N and saturates at high densities to a value of $2\,10^{15}$ s^{-1}, in good agreement with the value used to fit the experimental data.

LIGHT SCATTERING MEASUREMENTS DURING THE PULSE

Figure 2 shows the light scattered from the surface and collected in the same experimental conditions as the reflected light, except that a small mask is introduced in the path of the reflected light. The ratio of scattered to incident intensity is almost constant up to an excitation 2 times larger than I_p (the laser energy corresponding to the minimum of reflectivity), and then drops rapidly to about 0.6 times its initial value. This decrease of scattering reflects a reduction of the surface roughness. Can this be due to atomic motion as for normal melting ? We don't think so because of the high velocity of atoms required. If we think of vanishing of irregularities of the order of 1000 Å in times of 10^{-13} s, this implies velocities of 10^6 m/s which are impossible to reach for atoms in a liquid. We may have an estimation of the lattice temperature by computing the energy transfer from the e-h

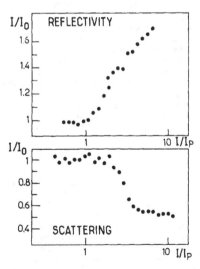

Fig.2 : Reflectivity and scattered intensity versus incident power.

134

plasma through phonon emission. This rate of energy transfer is given by : $dE/dt = E_{ph} \times N/\tau_{e-p}$ where N is the electron density and τ_{e-p} a characteristic electronphonon interaction time. For $E_{ph} = 50$ meV, N = 10^{22} cm^{-3} and $\tau_{e-p} = 10^{-13}$s, we find in 100 fs a maximum energy transfer of 80 J/cm^{-3} leading to a heating of the lattice of 40 K. This proves that almost all the energy absorbed by the sample is in the e-h gas. We may also estimate the electronic temperature T_e, which will be very high due to the low specific heat of the e-h plasma. The total energy density of the plasma is

$$U = N E_G + \int_0^\infty \rho_e(E) \, Ef_e(T_e)dE + \int_0^\infty \rho_h(E) \, Ef_h(T_h) \, dE \qquad (4)$$

where $\rho_{e,h}$ is the density of states of electrons (or holes) $f_{e,h}$ the distribution functions and $T_{e,h}$ the temperature of electrons (or holes). We take $T_e = T_h$ due to the fact that τ_{e-h} is of the order of 3.10^{-16} s and $f_{e,h} = A \, e^{E/kT}e$, A being a normalisation constant given by the density N. $\rho_{e,h}$ is taken from Cohen and Chelikowsky |9| to be αE. And we obtain

$$U = N E_G + 4 N K T_e$$

For an incident energy of 0.3 J cm^{-2}, an absorption depth of 300 Å, and N = 3 10^{22} cm^{-3}, we obtain $T_e = 40.000$ K. At these temperatures thermoelectronic emission is very strong and electrons are emitted from the silicon, leaving positive charges behind. A steady state situation is reached when the electric field created by this space charge stops the electronic motion. A kind of fluid dipôle layer is produced on the surface of the silicon crystal. The voltage V across this layer is of the order of KT_e/e. A tentative explanation for the decrease of scattering is that this electronic fluid screens out very rapidly all the electric field variations produced by the surface irregularities.

EVOLUTION OF THE REFLECTIVITY

A pump-test experiment was done using a test laser beam of low intensity with a variable delay. A typical result is given on figure 3 for an incident energy of 0.1 mJ focused on 100 μm diameter spot. The analysed reflected light comes from the central region of the spot. After an initial increase to a maximum which is less than the liquid value. The reflectivity drops to almost zero in 30 ps. This corresponds to a "black hole" in the center of the irradiated region. This black hole is due to the emission of small silicon particles. We argue that this is not thermal evaporation of silicon atoms. To fill a space of thickness of the order of 1000 Å in 10 ps, silicon atoms must have

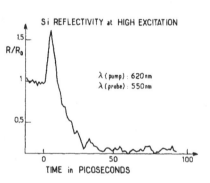

Fig.3 : Reflectivity versus time for the test beam.

a velocity $v_{Si} = 10^4$ m/s. If this is thermal evaporation, it corresponds to lattice temperature of the order of 10^5 K which is impossible. If it is charged particles accelerated by the electric field due to the dipole layer previously described, such velocities are reasonable. We thus think that the cloud of silicon particles are due to electrostatic extraction.

In conclusion, we have shown that

1) Irradiation by intense 100 fs laser pulse achieve e-h plasma densities up to $5 \ 10^{22} \ cm^{-3}$, relaxation times are due to electron-hole collsions, electronic temperatures up to $5 \ 10^{4}$ K are reached.

2) Electronic emission may perhaps explain the decrease of scattered light.

3) A tentative explanation for the formation of a cloud of silicon particles is electrostatic extraction.

References

1) C.V. Shank, R. Yen and C. Hirlimann, Phys. Rev. Lett. 50, 454 (1983).
2) C.V. Shank, R. Yen and C. Hirlimann, Phys. Rev. Lett. 51, 900 (1983).
3) D. Hulin, M. Combescot, J. Bok, A. Migus, J.Y. Vinet and A. Antonetti, Phys. Rev. Lett. 52, 1998 (1984).
4) J. Bok, Phys. Lett. 84A, 448 (1981).
5) M. Combescot, J. Bok, Phys. Rev. Lett. 48, 1413 (1982).
6) C.V. Shank, Proceedings of the Conference on "High excitation and short pulse phenomena", Trieste, July 84.
7) A. Migus, J.L. Martin, R. Astier, A. Antonetti and A. Orszag, in Picosecond phenomena III, Springer Series in Chemical Physics, 23, 6 (Springer, Berlin, 1982).
8) See for instance J.M. Ziman, Principles of theory of solids, Cambridge University Press (1965) p.187.

ULTRAFAST PHASE TRANSITION OF GaAs SURFACES IRRADIATED BY PICOSECOND LASER PULSES

J.M. LIU*, A.M. MALVEZZI**, AND N. BLOEMBERGEN**
*GTE Laboratories, Incorporated, 40 Sylvan Road, Waltham, MA 02254, USA
**Division of Applied Sciences, Harvard University, Cambridge, MA 02138, USA

ABSTRACT

Detailed analysis of the second harmonic signals in reflection from crystalline GaAs surfaces reveals that the structural transition associated with surface melting occurs in less than 2 ps. Preliminary results of picosecond time-resolved reflectivity measurements are also presented.

INTRODUCTION

Optical and photoelectric investigations of picosecond-irradiated silicon have shown that at ultrahigh excitation levels the surface is heated up to the melting point and a highly reflective metallic phase is induced [1]. These experiments probe the changes of electron density and energy. To study the structural disorder associated with melting, second harmonic (SH) generation has been recently exploited [2].

We present here a detailed analysis of the SH generation in reflection from crystalline GaAs under visible picosecond laser illumination. The SH energy dependence on the input laser fluence is monitored in single-shot experiments in a fluence range extending up to 30 times the critical value F_{th} where structural changes on the surface are first noticed. By comparing the results with model calculations, it is shown that the transition from the noncentrosymmetric ordered crystal surface to a centrosymmetric disordered structure occurs in less than 2 ps.

Preliminary results of picosecond time-resolved reflectivity measurements show heating of GaAs surfaces at pump fluences below the melting threshold and the existence of a highly reflective metallic molten phase at fluences sufficiently high above the melting threshold. Severe material evaporation induced by very high fluence pulses is also observed at time delays less than 100 ps.

SECOND-HARMONICS EXPERIMENT

SH generation from surfaces with microscopic symmetry, such as silicon, is principally a forbidden process in the dipole approximation. However, it has a very low efficiency, which is due to the symmetry breaking effect of quadrupole interactions. Noncentrosymmetric materials, such as GaAs, are per se more suitable for SH studies of laser-induced phase transition. The SH efficiency is in this case several orders of magnitude higher than in centrosymmetric systems, thus allowing greater sensitivity to a transition to a centrosymmetric disordered phase. A sharp decrease in SH efficiency has, in fact, been observed in experiments with nanosecond resolution if a threshold laser fluence is exceeded [3].

The frequency-doubled output from a passively mode locked Nd:YAG laser system is focused on a GaAs surface at an angle of incidence of 45°. The sample can be rotated around its normal. The pulse duration is $\tau = 12.5$ ps at the 1/e point of the Gaussian temporal profile. The time- and space-integrated UV signal is detected in the specular direction by a photomultiplier tube equipped with a suitable set of filters. The overall sensitivity of our apparatus is limited to ~ 200 photons generated on the target. The fluctuations of the laser pulse duration are taken into account by standard τ–A techniques [4], and the data from occasional double pulses are rejected. These precautions substantially reduce the scattering of the data.

At a threshold fluence $F_{th} = 30$ mJ/cm^2, structural changes of the GaAs crystal surface can be observed. This experimental value is in very good agreement with the calculated fluence necessary for melting the GaAs surface [6]. The SH signal below F_{th} depends on the polarization of the pump beam and the crystal orientation. Our data for (110) GaAs surfaces follow the predicted behavior [5] in this fluence range. A maximum efficiency $\eta_0 = 2.5$ x 10^{-18} cm^2/W is derived from the experiment when the laser electric field is parallel to a ⟨111⟩ crystal axis. Zero SH signal is observed with (100) GaAs samples for electric fields parallel to a ⟨001⟩ crystal axis.

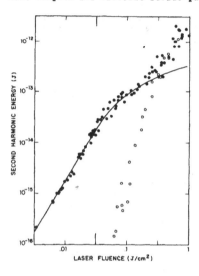

Fig. 1. Energy of the specular 266 nm emission vs incident laser fluence ($\lambda = 532$ nm) on GaAs. The laser spot diameter is 200 μm. (●): (110) surface, laser electric field parallel to the [111] crystal axis. (○): (100) crystal surface, laser electric field parallel to the [001] crystal axis. (–): calculated SH energy with $\tau_s = 0$, as discussed in the text.

The SH energy for the configuration of maximum efficiency is plotted in Fig. 1 in full circles versus laser fluence. The incident laser pulse is focused to a radius $\rho = 100$ μm on the crystal surface. A pure quadratic dependence is observed up to F_{th}, indicating that plasma or thermal processes do not affect the SH generation process. Above F_{th} the slope starts to decrease. This indicates that the noncentrosymmetric structure of the crystal is changed during the laser pulse. The solid curve refers to calculations in which the SH signal is assumed to drop to zero as soon as the threshold fluence for melting F_{th} is reached within a certain area. The curve fits the data up to a factor ∼4 above F_{th}. The UV signal received through the filter, however, increases again beyond this fluence. This additional emission has been studied by looking at the angular dependence and temporal behavior. At very high excitation levels the emission is accompanied by a blue spark on the surface of the target and appears to be isotropic. In addition, its duration is much larger than the temporal resolution of the detection system. By comparison, we studied the high fluence emission from (100) oriented GaAs surfaces with the laser electric field parallel to the [001] direction. In these conditions the SH efficiency is zero. The results, which are indicated with open circles in Fig. 1, show that this long-lasting emission process simply adds up to the SH signal in our experimental setup. Its origin is closely related to the evaporation of material and plasma formation in front of the target. We restrict ourselves in this contribution to the net SH signal in order to study the structural changes on the GaAs surface. For this purpose we have increased the spatial resolution of our measurements by simply introducing a diaphragm in the path

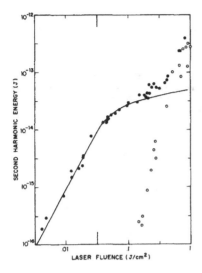

Fig. 2. Same as Fig. 1, but limiting the collection of the SH signal from a disk of 120 μm in diameter centered on the incident laser spot.

of the SH beam. Figure 2 shows the energy dependence of the SH emitted by the central portion of the excited area (120 μm in diameter) versus laser fluence.

As expected, the decrease of the SH efficiency above F_{th} is now more pronounced, indicating that at high fluences the central part of the irradiated area is less effective in the SH generation process. The data are compared with model calculations where the transition from an ordered noncentrosymmetric structure to a disordered liquid phase is simulated by a space- and time-dependent SH efficiency $\eta(r,t)$ decaying with a time constant τ_s.

$$
\eta(r,t) = \begin{cases} \eta_0 \text{ for } t < t_m(r) \\[2mm] \eta_0 \exp - \dfrac{t-t_m(r)}{\tau_s} \text{ for } t > t_m(r), \end{cases}
$$

with

$$
\int_{-\infty}^{t_m(r)} I(r,t)dt = F_{th}
$$

At each position r, η starts to decay as soon as the threshold pump laser fluence F_{th} is reached.

The solid curve in Fig. 2 shows the results of the calculations for the extreme case of $\tau_s = 0$. At fluences slightly above F_{th}, when $t_m(r)$ occurs on the trailing edge of the laser pulse, the data are in good agreement with the calculation. However, at higher fluences, when F_{th} is reached during the leading edge of the laser pulse, a noticeable deviation is observed. In Fig. 3 the same data are plotted in a linear scale and compared with calculations performed with different decay times τ_s of the SH efficiency η at the melting point. This fitting procedure demonstrates clearly that the transition from the crystalline surface structure to a centrosymmetric disordered phase occurs in less than 2 ps.

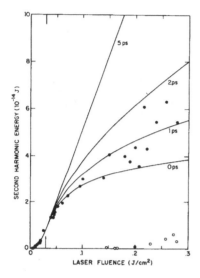

Fig. 3. Plot of the data of Fig. 2 on linear scales. The solid curves refer calculations with the values of τ_s shown.

TIME-RESOLVED REFLECTIVITY MEASUREMENTS

Picosecond time-resolved reflectivity measurements were carried out on undoped semi-insulating GaAs samples with (100) surfaces, using an experimental setup similar to that described in [7]. The GaAs surface is heated by a pump pulse of 20 ps FWHM (τ = 12.5 ps) at 532 nm and is probed by a delayed probe pulse derived from the pump pulse with a beamsplitter. The probe pulse is focused to one tenth the diameter of the pump spot in order to probe the reflectivity at the exact center of the pump spot. The pump and probe beams have orthogonal polarizations to suppress interference between these two beams. The probe beam is p-polarized and is incident at 26° to the normal of the sample surface. With this probe beam the calculated characteristic reflectivity is 33.8 % for GaAs at room temperature and is 61.2 % for molten GaAs at the melting point, 1511 K.

Preliminary results of the reflectivity measurements show many complicated features. In Fig. 4 the reflectivity of a GaAs surface probed at a time delay of 300 ps after the pump pulse is plotted versus the incident pump fluence. At fluences below the melting threshold the reflectivity increases gradually with increasing fluences due to heating of the GaAs surface by the pump pulse. As the pump fluence is increased above the melting threshold, 30 mJ/cm^2, the reflectivity continues to increase smoothly without an observable discontinuity. The reflectivity reaches a full maximum at 60 %, characteristic of the molten GaAs, only at pump fluences sufficiently high above the melting threshold. This behavior is very distinct from that observed in similar experiments on silicon [8] where a sharp rise of reflectivity occurs at the melting threshold. This may be an indication of a very thin molten layer on the heated GaAs surface. The exact cause of this phenomenon is still under investigation.

At still higher pump fluences the reflectivity starts to drop sharply. At fluences up to 0.15 J/cm^2 the reflectivity drops primarily because of overheating of the molten surface. It recovers to high reflectivity at time delays of 500 ps to 1 ns when the overheated molten surfce cools down. At the very high fluences (> 0.2 J/cm^2) the reflectivity drops to below 10 % and the data are scattered. This can be explained by a large amount of evaporated material which blocks and scatters the probe beam.

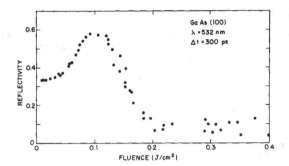

Fig. 4. Reflectivity of the GaAs surface versus pump fluence probed with a 532 nm pulse incident 26° to the normal of the surface, at 300 ps time delay.

The reflectivity versus probe time delay is shown in Fig. 5 for three different pump fluences. It is clear from these plots that the full liquid reflectivity is observed only at a fluence sufficiently high above the 30 mJ/cm^2 melting threshold after a sufficiently long time delay. At 50 mJ/cm^2 the reflectivity reaches a maximum of 47 % within 100 ps and decays within a few hundred picoseconds. At 25 mJ/cm^2, below the melting threshold, the reflectivity reaches a maximum of 39 %, indicative of a hot surface. Detailed accounts of these reflectivity signatures still need further investigation.

CONCLUSIONS

Second harmonic generation in reflection is very sensitive to the structural symmetry changes in a very thin surface layer. By comparison, time-resolved reflectivity measurements probe a thicker surface layer and give information about the lattice not as direct. However, time-resolved reflectivity does provide additional information about the electron-hole plasma, lattice heating, and material evaporation. We have used both techniques and have demonstratd ultrafast heating and melting of GaAs surfaces irradiated by picosecond laser pulses.

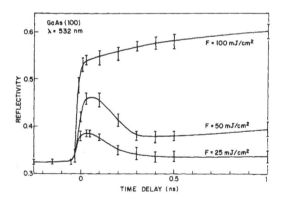

Fig. 5. Reflectivity of the GaAs surface versus probe time delay at three different pump fluences. The melting threshold is 30 mJ/cm^2.

ACKNOWLEDGEMENTS

This research was supported by the U.S. Office of Naval Research under contract N00014-83K-0030, and by the Joint Services Electronics Program of the U.S. Department of Defense under contract N00014-75-C-0648.

REFERENCES

1. L.A. Lompre, J.M. Liu, H. Kurz, and N. Bloembergen, Appl. Phys. Lett. 44, 3 (1984).
2. C.V. Shank, R. Yen, and C. Hirlimann, Phys. Rev. Lett. 51, 900 (1983).
3. S.A. Akhmanov, N.I. Koroteev, G.A. Paitian, I.L. Shumay, M.F. Galjaut-dinov, I.B. Khaibullin, and E.I. Shtyrkov, Optics Commun. 47, 202 (1983).
4. W.L. Smith and J.M. Bechtel, J. Appl. Phys. 47, 1065 (1976).
5. J. Ducuing and N. Bloembergen, Phys. Rev. Lett. 10, 474 (1963).
6. A. Lietoila and J.F. Gibbons, Mat. Res. Soc. Symp. Proc. 4, 163 (1982).
7. L.A. Lompre, J.M. Liu, H. Kurz, and N. Bloembergen, Appl. Phys. Lett. 43, 168 (1983).
8. J.M. Liu, H. Kurz, and N. Bloembergen, Mat. Res. Soc. Symp. Proc. 13, 3 (1983).

PART III

Fundamental Aspects of
Bulk Recrystallization

IMPURITY INCORPORATION DURING Si ULTRAFAST SOLIDIFICATION TO THE CRYSTALLINE AND AMORPHOUS PHASE.

SALVATORE UGO CAMPISANO
Dipartimento di Fisica dell'Università, Corso Italia 57,95129 Catania,Italy

ABSTRACT

Impurity segregation at the liquid-solid interface, interfacial instabilities due to constitutional supercooling and precipitation are compared
for the growth of a crystalline or of an amorphous phase from rapid solidifi
cation. Segregation at low concentration and interfacial instabilities at
high concentrations are shown to give the same information. The impurity
behaviour shows a distinct similarity between the crystalline and amorphous
phases.

INTRODUCTION

Pulsed laser irradiation of solid surfaces has provided a very powerful
tool to investigate melting and solidification phenomena occurring under far
from equilibrium conditions |1-3|. The step energy deposition profiles and
the very short time intervals involved in the process |4,5| produce very
large temperature gradients. If enough energy is supplied by the laser pulse the solid surface melts and solidifies with velocities that, for silicon,
can reach the value of 200 and 20 m/s respectively. Impurity segregation at
the liquid-solid interface |3,6| and liquid to amorphous transition |7-9|
are examples of far from equilibrium phenomena studied under such extreme
conditions.

Detailed |3,5,6| experimental investigations performed at v>1m/s have
shown that the redistribution of impurities at the l-s interface must be de
scribed by an interfacial distribution coefficient k' much larger than the
equilibrium value k_0. Moreover k' increases with velocity. Correspondingly
supersaturated solid solutions can be prepared by pulsed laser irradiation
of ion implanted semiconductors.

At very large velocities, however, it has been shown |8,10| that amor-
phous silicon is quenched from the melt. For (100) pure Si a solidification
velocity exceeding 15 m/s will result in the formation of the α-phase and
for (111) silicon such a threshold velocity is about 10 m/s. The usual
meaning of segregation coefficient is then lost when the solidification velocity exceeds the amorphization value.

The present paper reviews results concerning the redistribution of impu
rities in Si solidifying at velocities sufficient to grow an amorphous phase. The "segregation coefficient" at the liquid-amorphous interface k' has
been defined and measured for As,Bi and In in Si. The impurity profile moreover acts as a marker thus giving information on the α-l interface kinetics. Some properties of the α-Si- impurity system can be obtained.

EXPERIMENTAL

Samples used in the present experiments were single crystals of Si in
either (111) or (100) orientation. Several implanted species were investiga
ted: As,In,Te and Sb. Implantation energies and doses were in the 50-200
keV and 10^{14}-10^{16}ions/cm^2 ranges respectively. Ruby laser pulses of 2.5 ns
of either the fundamental or second harmonic wavelength were used for the
irradiation. The energy density varied between 0.1 and 0.6 joules/cm^2 and

a quartz homogeneizer was used to obtain a uniform energy distribution at
the surface. High resolution backscattering of 2.0 MeV He⁺ particles was
adopted to determine depth impurity profiles. Channeling effect technique
was used to determine the crystalline quality of the host Si and lattice location of impurities. Cross section electron microscopy (TEM) was used to
investigate the structure of the irradiated samples.

DOPANT INCORPORATION IN THE CRYSTALLINE PHASE

The directional solidification of a dilute solution has been studied since
many decades because of its importance in materials purification. A detailed
description of the process was given by Pfann |11| who pionereed such studies.
If the segregation coefficient is less than unity the impurities are rejected
into the liquid phase where they are allowed to diffuse. A solute rich layer
is thus formed in the liquid, just ahead of the interface and its concentration is determined by the competition between segregation and diffusion.
Such a layer, whose presence is detrimental for the purification process, is
responsible also for interface instabilities |12| through the constitutional
supercooling mechanism.
In a directionally solidified dilute solution the impurity depth profile
is determined by its initial distribution, by the interfacial segregation
coefficient and by the solidification velocity. For ion implanted laser irradiated Si the initial distribution can be measured, e.g. by RBS technique.
The solidification velocity has been either calculated or measured and
good agreement obtained|8,10|. The final impurity depth profile is thus uni-
quely determined by the value of the interfacial distribution coefficient|5,6|.
Using this approach the k' value has been measured for a large variety of
dopants in the velocity range of 1-7 m/s. As an example in Fig.1 is reported
the measured |13| profiles of Bi implanted in Si (111) after pulsing laser

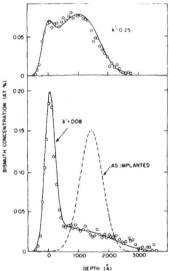

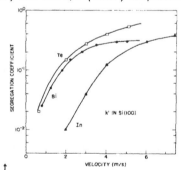

Fig.1. Experimental and calculated depth pro
files of Bi in Si (111) after laser irradiation. The solidification velocities are 2.3
(upper) and 0.9 m/s (lower) respectively.

Fig.2. Measured values of the interfacial se
gregation coefficient for Te,Bi and In in Si
(100) as a function of the solidification ve
locity.

irradiation so to solidify at velocities of 2.3 (upper) and 0.8 m/s (lower).
The depth profiles calculated using k'=0.25 and k'=0.08 at the larger and
the smaller velocities respectively are also reported. The velocity depen-
dence|14| of k' is reported in Fig.2 for Te,Bi and In in Si (100) in the ve-
locity range of 1-7 m/s. What is of importance to stress here is that the

measured and the fitted profiles must be those of impurities in the solid solution, i.e. the material must be free of precipitates and the dopant must
occupy a well defined lattice site.

Solute precipitates, usually in the form of a cell structure, can be present in a directionally solidified solid solution as a consequence of interface instabilities. These are explained in terms of constitutional super-
cooling, or more generally, in terms of morphological stability criteria|12|.
In simple words if the equilibrium solidification temperature of the solute
rich layer ahead of the interface is lower than the actual liquid temperature, the system becomes unstable developing a periodic array of precipitates.
Such effect is then observed |13| at high concentrations and it limits the
maximum amount of dopant that can be incorporated in substitutional lattice
sites. As an example in Fig.3 are reported the TEM and the RBS spectra in
presence of cells for In in Si. The aligned yield is about 80% the random
value and the non-substitutional In is precipitated in the cell structure.

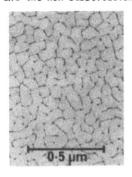

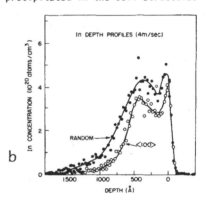

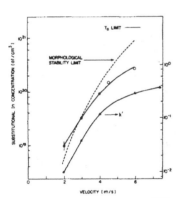

Fig.3. a) TEM image of a laser irradiated,
high dose In implanted Si; b) experimental
depth profile and channeling aligned yield
of a sample similar to that shown in part
(a).

←Fig.4. Measured maximum substitutional In
concentration as a function of the solidification velocity. In the same figure
are reported the measured values of k'(v)
and the calculated maximum concentration
giving rise to a planar solidifying interface.

In Fig.4 the measured|16| C_s^{max} versus velocity is reported for In in Si (100).
In the same figure are reported the k'(v) measured values. The dashed line
is the calculated maximum concentration obtained from the morphological stability limit and using the measured k' values. The observation of a cell
structure is then indicative that a solute rich layer is formed at the l-s
interface (i.e. k'<1).

We have then two criteria to say that k'<1: profile fitting for substitutionally diluted impurities and maximum substitutional concentration for a
two phase system consisting of a solid solution plus a periodic array of precipitates.

DOPANT INCORPORATION IN THE α-Si PHASE

The amorphous Si phase was first quenched from the liquid using picose-
cond|7| laser pulses. It was shown subsequently that the crystal-liquid-a-
morphous and the amorphous-liquid amorphous transitions can be obtained using
2-3 ns pulses with suitable wavelength in either pure|8| or doped|9| mate-
rial.

For the case of ion implanted amorphous Si we |17| have studied profile
modification of As, Bi and In. The as-implanted samples have been irradiated with ruby laser pulses of 2.5 ns duration. Energy densities between
0.18-0.60 j/cm² were used.
In the energy range 0.18-0.28 j/cm² the surface layer melted and solidified
in the amorphous phase. Above 0.28-0.30 j/cm², a transition to polycrystalline Si was observed. At these energies the laser pulse does not melt
through the amorphous layer. The melting was established by the gross changes that occurred in the implanted impurity depth distribution. The depth
profiles of In,As and Bi, measured before and after laser irradiation are
reported in Fig.5. A considerable change is observed in the In profile with
most of the In atoms segregated to a narrow band 400 Å beneath the surface.
A similar effect is observed in Bi, with lower accumulation at 250 Å beneath
the surface. No accumulation is observed for As.

Cross section TEM was performed on all samples. Fig.6a and 6b are the

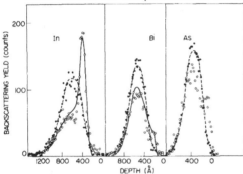

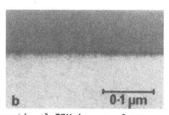

Fig.5. Experimental depth profiles of
In,Bi and As implanted in Si after ir
radiation with 0.24 J/cm² ruby laser
pulse of 2.5 ns duration. The full
line in the In profile is calculated
assuming v=7m/s and k'=0.25.

Fig.6.

Cross sectional TEM images of amorphous ion implanted and laser irradiated samples: a) In⁺ at 0.25 J/cm²
(note the buried In band); b) As⁺ at
0.25 J/cm².

cross section micrographies of laser irradiated In and As implanted samples
respectively. In the In case we observe a very narrow (<20 Å) band of segregated In in a featureless background of α-Si. No feature are seen in the As
case; in any case the as-implanted samples are featureless amorphous layers.

The formation of a buried metallic layer is explained by a solidification
scenario which is unusal for laser crystallization. The surface liquid layer
can reach such large undercoolings that nucleation and growth of α-Si phase
can start from the sample surface in addition to the inner liquid-amorphous
solidification. For both In and Bi cases the depth location of the buried
layer indicates that most of the implanted solute interacts with the inner
interface.

We can then assume that the mass transport at the l-α interface occurs as in the case of a dilute solution at the l-c interface. Profile fitting will determine the interfacial segregation coefficient. The l-α interface velocity has been determined by heat flow calculations |18| taking into account thermal properties of α-Si and it results 7±1 m/s. The resulting calculated In profile is reported in the same Fig.5 as a full line. The value $k_a=0.25$ fits well with the experimental depth profile.

This value is reported in Fig.7 together with the experimental values measured in the same velocity range at the l-c interface and for (111) and (100) oriented substrates. It is intriguing that we obtain comparable values of k' for the l-α and for the l-c interface. We must be sure that we are measuring a real segregation effect at the l-α interface. As in the case of crystallization a clear sign of k'<1 are interfacial instabilities. In Fig.

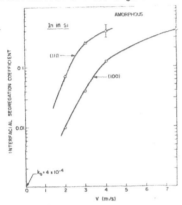

Fig.7. Values of the interfacial segregation coefficient of In in silicon as a function of the solification velocity for different crystal orientation and for amorphous Si.

Fig.8. TEM image of cellular In segregation pattern in laser irradiated amorphous Si.

is reported the cross section TEM micrography of a Si sample implanted with 170 keV In ions at a dose of 10^{16} at/cm^2 and irradited at 77 K with 2.5 ns ruby laser pulse of 0.4 J/cm^2. At the same depth of the In projected range ($\sim$ 700 Å) a band of segregated In is observed. Emerging from the In buried layer a periodic array of segregated In is observed. This observation is entirely consistent with the occurrence of constitutional supercooling[19].

The occurrence of interfacial instabilities is a clear indication that the interfacial segregation coefficient at the α-l interface is less than unity. More importantly is the fact that we can still use the concept of segregation coefficient for the solidification of an amorphous phase. The fact that we can use similar description for the l-c and for the l-α interfaces means that: i) the solidification of the α-Si occurs directionally and then is not bulk nucleated; ii) the α-Si phase is characterized by a limited solubility[20] for impurities such as In and Bi; and iii) the maximum solubility of such impurities in the amorphous and in the crystalline Si phase at the growth velocity of $\sim$7 m/s are comparable.

The effect of a limited solubility is that for an high solute concentration, exceeding the solubility value, we should observe the precipitation of a second phase if enough mobility is given to the atoms. For example high dose implantation in Si followed by high temperature annealing causes the formation of a supersaturated solid solution and then the excess of solute precipitates in small crystalline agglomerates|21|. We have investigated

150

the structure of ion implanted a-Si after low temperature annealing. Si samples have been implanted with 120 keV Sb^+ with doses of 1×10^{14} and 1×10^{16} at./cm^2 and annealed for 10 min. at 510°C. Results of high resolution TEM studies are summarized in Fig.9. Both the low dose and the high dose implanted samples show a partial regrowth of the a-Si.

Fig.9. High resolution TEM images of Sb^+ implanted Si after 10 min annealing at 500°C: a) 10^{14} at./cm^2 and b) 10^{16} at./cm^2.

The thickness of the regrown layer and its dependence upon the Sb concentration are well understood on the basis of the concentration dependence of velocity during solid phase epitaxy. The residual amorphous layer is however featureless in the low dose implant while in the high dose implanted samples we observe dark dots of about 50 Å diameter. These dark dots have been interpreted as amoprhous Sb agglomerates in the α-Si matrix and may be responsible of the rather complex behaviour of the SPE velocity at high dopant concentration. What is important to point out here is that we do observe a "precipitation" phenomenon in the amorphous Si matrix only at high dopant concentration as in the case of a crystalline phase.

The collection of all these experimental results clearly indicate that the a-Si phase is characterized by a limited solubility with respect to impurities such as In, Bi and Sb. Phenomena such as impurity segregation at the liquid solid interface, l-s interface stability during solidification and second phase separation during high temperature processing of a supersaturated alloy for an amorphous phase can be described with arguments similar to those used for a crystalline phase. Kinetic parameters such as interfacial segregation coefficient, and thus maximum solid solubility are of the same order of magnitude in the solidification velocity range around 7 m/s.

ACKNOWLEDGEMENTS

The results reported in this work have bee obtained during the course of a friendly and fruitfull collaboration among Catania, AT&T Bell Lab. and RSRE groups. The collaboration of the many scientists involved is gratefully acknowledged.

REFERENCES

1) J.M.Poate and J.W.Mayer, eds. "Laser Annealing of Semiconductors" (Academic Press, New York) 1982

2) J.M.Poate,G.Foti and D.C.Jacobson, eds. "Surface Modification and Alloying" (Plenum Press, New York) 1983

3) S.U.Campisano; Appl.Phys. A30,195(1983)

4) P.Baeri,S.U.Campisano,G.Foti and E.Rimini; J.Appl.Phys. $\underline{50}$,788(1979)

5) P.Baeri and S.U.Campisano:"Heat Flow Calculations"; in ref. 1, chap.IV

6) C.W.White,D.M.Zehener,S.U.Campisano,A.G.Cullis: "Segregation,Supersatura-ted Alloys and Semiconductor surfaces" in ref.2, chap.IV

7) P.L.Liu,R.Yen,N.Bloembergen and R.T.Hodgson; Appl.Phys.Lett. $\underline{34}$,864(1979)

8) A.G.Cullis,H.C.Webber,N.G.Chew,J.M.Poate and P.Baeri; Phys.Rev.Lett. $\underline{49}$,219(1982)

9) *S.U.Campisano, D.C.Jacobson, J.M.Poate,A.G.Cullis and N.G.Chew; Appl.Phys.Lett. (in press   Oct.1984)

10) M.O.Thompson,J.W.Mayer,A.G.Cullis,H.C.Webber,N.G.Chew,J.M.Poate and D.C.Jacobson; Phys.Rev.Lett. $\underline{50}$,896(1983)

11) W.G. Pfann: "Techniques of zone melting and crystal growing", in: "Solid State Physics" ed. by F.Seitz and D.Turnbull (Academic Press, New York,1957) p.424

12) R.F.Sekerka; J.of Cryst.Growth $\underline{3,4}$, 71(1968)

13) P.Baeri,G.Foti,J.M.Poate,S.U.Campisano,A.G.Cullis; Appl.Phys.Lett. $\underline{38}$,800(1981)

14) S.U.Campisano,G.Foti and J.M.Poate (unpublished data)

15) A.G.Cullis,D.T.J.Hurle,H.C.Webber,N.G.Chew,P.Baeri,G.Foti and J.M.Poate; Appl.Phys.Lett. $\underline{38}$,642 (1981)

16) S.U.Campisano and J.M.Poate; Appl.Phys.Lett. (in press)

17) S.U.Campisano,D.C.Jacobson,J.M.Poate,A.G.Cullis and N.G.Chew; Phys.Rev.Lett. (submitted)

18) H.C.Webber,A.G.Cullis and N.G.Chew; Appl.Phys.Lett. $\underline{43}$,669(1983)

19) A.G.Cullis,H.C.Webber and N.G.Chew; Appl.Phys.Lett. $\underline{40}$,998(1982)

20) S.Fischler; J.Appl.Phys. $\underline{33}$,1615(1962)

21) S.U.Campisano,A.E.Barbarino,R.Galloni and R.Rizzoli; Nucl.Instrum.& Meth. $\underline{209,210}$,645(1983).

A TEST OF TWO SOLUTE-TRAPPING MODELS*

M. J. AZIZ, J. Y. TSAO**, M. O. THOMPSON[+], P. S. PEERCY**, C. W. WHITE,
AND W. H. CHRISTIE[++], Oak Ridge National Laboratory, Oak Ridge, TN 37831.
**Sandia National Laboratories, Albuquerque, NM 87185.
[+]Department of Materials Science, Cornell University, Ithaca, NY 14853.

ABSTRACT

   Two solute trapping models are compared. They are shown to predict
identical behavior at any given end of a phase diagram but different behavior as the phase diagram is traversed. The segregation behavior of dilute
solutions of Ge in Si ($k_e < 1$) and of Si in Ge ($k_e > 1$) during regrowth
from pulsed-laser melting is being studied using transient conductance,
high resolution RBS, and SIMS. Our results to date suggest a significant
amount of solute trapping ($k \longrightarrow 1$) of Ge in Si and of Si in Ge. Such a
result would be inconsistent with the predictions of one of the models.

INTRODUCTION

   One of the most exciting aspects of pulsed-laser melting is that it
allows, for the first time in high-mobility systems, a quantitative study
of the kinetics of the fundamental atomic processes occurring at the solid-
liquid interface. Such a study is possible because these experiments have
reached the regime where, due to kinetic limitations, deviations from local
thermodynamic equilibrium are obvious and crystal growth is no longer heat-
flow limited. In this paper we describe an experimental test of models for
the departure from local compositional equilibrium at the interface — so-
called solute trapping models.
   There are a number of models [1-9] of the kinetics of the solidification reaction of a two-component system. A complete description of the
process requires the derivation of the two "interface response functions"
[10]; they predict (a) how fast the interface will move, and (b) the composition of the growing solid, in terms of the local conditions at the
interface, namely temperature and liquid composition.
   We will concentrate on two models for the following reasons. They are
"complete", in that they propose forms for both of the above response
functions and can therefore form a basis for our understanding of the entire
solidification process. They can also be readily applied by the experi-
mentalist to the particular problem at hand. In both of these models two
simultaneous, coupled reactions occur at the interface. In the model of
Jackson, Gilmer and Leamy [6,11], the two reactions are the solidification
of host atoms and that of impurity atoms. The barriers to solidification
of each species, Q_A and Q_B, respectively, are unknown constants that
influence the respective reaction rates. In the continuous growth model of
Aziz [7,12] the pair of reactions are a net alloy crystallization reaction
across a barrier Q_C, and a diffusive solute-solvent redistribution reaction
across a barrier Q_D. This pair of reactions is essentially a linear combination of the aforementioned pair. The differences between the models
arise from the choice of which barriers are the constant ones and the
manner in which the reactions are coupled. Solute trapping occurs in the

*Research sponsored by the Division of Materials Sciences, U.S. Department
of Energy under contract DE-AC05-840R21400 with Martin Marietta Energy
Systems, Inc.

former model (which we shall call "AB", for the two species) when $Q_A \ll Q_B$. Solute trapping occurs in the latter model (which we shall call "CD", for "crystallization and diffusion") when $Q_C \ll Q_D$.

Given a free choice of barrier heights, the models can predict virtually identical behavior for a dilute solution of any single dopant in any single host. This can be seen by the form of the resulting equations relating the segregation coefficient k to the growth velocity v. The CD model yields

$$k = \frac{(v/v_D) + k_e}{(v/v_D) + 1} ,$$ (1)

where k_e is the equilibrium segregation coefficient at the temperature T of the undercooled interface and v_D is the diffusive velocity, which depends upon the height of the barrier to redistribution Q_D. The prediction of the AB model can be put into the form

$$k = \frac{(v/v') + k_e}{(v/v') + 1}$$ (2)

where

$$v' = \exp\left[\frac{Q_B - Q_A}{RT} - \frac{\Delta S_f^B}{R}\right] / K_A^+ .$$ (3)

The entropy of fusion ΔS_f^B of species B is a positive number and K_A^+ is the velocity of the forward reaction alone across the barrier Q_A. (A lower limit on the magnitude of K_A^+ for Si (100) is 15 m/s, the largest measured crystal growth rate [13]; K_A^+ is probably significantly larger than this.) We see that in this form the AB model predicts a maximum value of k that should not be termed "saturation" but rather "truncation"; i.e. dk/dv does not approach zero at k < 1, but there is an upper limit to k which occurs at the maximum allowed velocity, $v = K_A^+$. Past confusion over apparent differences between the models on the dependence of k upon orientation and velocity has arisen due to the different assumptions for v(T) that were put into the models. With suitable choices of the pairs of fitting parameters (the Q's), the models give identical predictions if the same v(T) is input.

The two models differ substantially, however, in their composition dependence. If the barrier heights are constrained by forcing the models to predict the same behavior at one impurity concentration, then they need

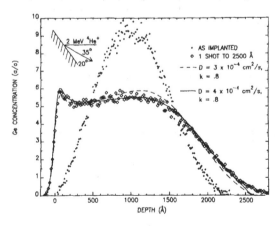

Fig. 1. Diffusion of Ge in Si. RBS spectra showing Ge profiles before and after pulsed laser melting. Solid line shows predicted spectrum for $D_L = 4 \times 10^{-4}$ cm^2/s; broken line shows predicted spectrum for $D_L = 3 \times 10^{-4}$ cm^2/s.

not predict the same behavior at other concentrations. In particular, the $\overline{CD}$ model predicts that if there is significant trapping of B in A then there should be significant trapping of A in B at a similar velocity. The AB model predicts that **if there is significant trapping of B in A then there cannot be significant trapping of A in B at any velocity.**

The presence of the diamond cubic crystal structure on both ends of the Si-Ge phase diagram allows a meaningful test to distinguish between these models. The solution behavior of both phases is well modelled, which reduces the likelihood of a chemical "surprise" and allows a surer test of the kinetics than has been previously attempted. The system exhibits complete miscibility of Si and Ge in both the liquid and solid phases. For a dilute solution of Ge in Si, $k_e{}^{Ge} \approx 0.3$; for a dilute solution of Si in Ge, $k_e{}^{Si} \approx 6$.[14] The phase diagram can be accounted for by ideal solutions in both the liquid and solid phases.[15] This extremely simple solution behavior is not necessary for distinguishing between the models; it does, however, make the interpretation of results much easier. If we could obtain data for $k_{Ge}(v)$ at the Si-rich end of the phase diagram, fit the parameters of both models to these data, use these parameters to obtain predictions for $k_{Si}(v)$ at the Ge-rich end of the phase diagram, and distinguish between them experimentally, we could rule out at least one of the models.

EXPERIMENTAL

Ge was implanted at 150 keV into silicon-on-sapphire (SOS), Si was implanted at 100 keV into Ge layers deposited on SOS, and crystallinity was restored in a vacuum furnace anneal. The samples were then irradiated with a pulsed-ruby laser with a pulse duration of 30 ns FWHM. The regrowth velocity was varied by changing the pulse energy density. The time evolution of the melt depth was measured with nano-second time resolution by the transient conductance technique.[13] The as-recrystallized and the final Ge impurity profiles were measured by glancing angle RBS and channeling. The Si impurity profiles were analyzed by glancing angle RBS and by SIMS depth profiling. To determine k with high accuracy the bulk liquid diffusion coefficient D_L must be known to high accuracy. For the case of Ge in Si, both were determined by numerical simulation. The measured as-implanted profile was broadened numerically using the measured melt depth as a function of time. Guesses were made for D_L and k; they were assumed not to depend upon v and T. The predicted final profile was convoluted with a 14 keV detector resolution and compared to the RBS data.

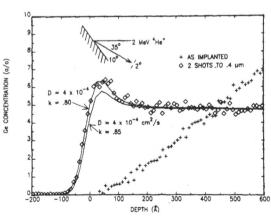

Fig. 2. Segregation of Ge in Si at 2.3 m/s. Glancing angle RBS spectrum showing surface peak of zone refined Ge. Peak height and area match simulation for k = 0.80. Predicted peak for k = 0.85 is too small.

RESULTS

Figure 1 shows an example of an RBS spectrum used to determine D_L. The value $D_L = = 4 \times 10^{-4}$ cm²/s provides excellent agreement with the data; in contrast, the simulation with $D_L = 3 \times 10^{-4}$ cm²/s simply does not provide enough broadening. It should be noted that the entire volume change upon melting will be taken up in the direction normal to the interface during pulsed laser melting. Numerical broadening programs like ours that do not account for the volume change will thus overestimate D_L. The resulting error is significant if a high precision determination is compared with the result from a conventional method of measuring D_L. This error does not affect our determination of k and we did not correct for it.

Once we know D_L we can determine k_{Ge} to high accuracy by a careful examination of the surface peak of zone-refined material. As shown in Fig. 2, we can easily discern between k = 0.80 and k = 0.85 by this method. This sample was shot twice in order to allow enough broadening of the Ge implant that the final profile is essentially flat beneath the surface. Uncertainties in identifying the peak area are thus reduced. We note that this surface peak is not detectable by normal geometry RBS.

The results for $k_{Ge}(v)$ are plotted in Fig. 3. The horizontal error bars represent fluctuations in the transient conductivity signal during the final 500 A of regrowth. The vertical error bars represent our estimate of the uncertainties in the results at the 90% confidence level, considering only uncertainties in detector resolution, D_L, and eyeballing the fit. Other systematic corrections may apply. The curve drawn through the data is obtained from the CD model with a diffusive velocity of $v_D = 1$ m/s. The same curve results from the AB model with $Q_{Ge}-Q_{Si} = 1$ eV and reasonable choices for the other parameters, $K_A^+ = 30$ m/s, T = 1645 K.

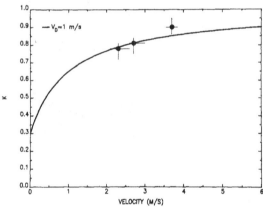

Fig. 3. Segregation of Ge in Si. Curve aises from CD model with VD = 1 m/s and from AB model with $Q_{Ge}-Q_{Si} = 1$ eV.

The AB model then predicts, with $Q_{Si}-Q_{Ge} = -1$ eV and $k_e^{Si} = 6$, $k_{Si}(v) = 6$ at all velocities. (Recall $v < K_A^+$.) This conclusion does not change even when all parameters are taken to their extreme limits of conceivability in order to maximize the predicted difference between $k_{Si}(v_{max})$ and k_e^{Si}. In contrast, the CD model predicts the familiar transition from equilibrium segregation to complete trapping centered at a diffusive speed which may be somewhat reduced due to a lower interface temperature on the Ge-rich side of the phase diagram. For $Q_D = 0.7$ eV, the CD model predicts k_{Si} (4 m/s) = 1.1.

We examined pulsed laser melted Ge layers with Si implants to see whether we could distinguish between $k_{Si} = 1.1$ and $k_{Si} = 6$. For most impurities in most semiconductors, $k < 1$. Thus during regrowth from pulsed laser melting a steady-state spike of zone refined impurity is set up in front of the interface. If C_s is the impurity concentration in the growing solid, the height of the spike above C_s is $C_s[(1/k)-1]$, its $1/e$ width is D_L/v, and its area is $(D_L/v)C_s[(1/k)-1]$. After solidification is completed, the spike appears at the free surface with a somewhat different shape; its area, however, is unchanged. The same analysis holds for Si in Ge where $k_{Si} > 1$, except that the Si spike becomes inverted, i.e. the liquid layer at the interface is purer in Ge than both the growing crystal and the bulk liquid. When this inverted spike reaches the free surface we are left with a surface deficit of Si, which can be examined in our samples by RBS. Figure 4 shows a high resolution glancing angle RBS spectrum of a Si-implanted polycrystalline Ge layer on SOS that solidified at approximately 4 m/s. The entire spectrum is shown in the inset at the right; the signal from Si at the free surface is shown in detail. The background counts presumably arise from dirt in the Ge layer; we see that they do not interfere with the identification of the Si edge. In order to identify the backscattered energy corresponding to Si at exactly zero depth, we translated into the beam a piece of virgin Si adjacent to the sample without changing the scattering geometry. We see the Si edge in the same location in the two cases. Curve (a) is the spectrum predicted from our simulation with the following assumptions: $D_L = 4 \times 10^{-4}$ cm^2/s, $k = 6$, detector resolution = 14 keV. If k is changed to 1.1 we obtain curve (b). The important feature of these curves is the location of the Si edge. We see that the surface deficit predicted by the AB model would manifest itself as an apparent 100 Å depression of the Si edge in the RBS spectrum. Within the resolution of our measurement, there is no evidence for such an effect. This result suggests that at 4 m/s k_{Si} differs significantly from k_e^{Si} for Si in Ge. Such a phenomenon, combined with a significant amount of trapping of Ge in Si, would rule out the AB model.

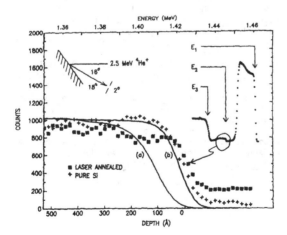

Fig. 4. Segregation of Si in Ge. Glancing angle RBS spectrum showing no surface deficit of Si. E_1: signal from surface Ge; E_2: surface Si; E_3: buried SOS. The virgin Si spectrum and the curves have been rescaled vertically. Depth scale for Si in Ge. Curve (a): AB model, $k_{Si} = 6$. Curve (b): CD model, $k_{Si} = 1.1$.

Our results are suggestive but not yet conclusive. Although the dependence of k upon orientation is expected to be small at these values near unity, the polycrystalline nature of the Ge layers introduces some uncertainty into our analysis. In addition, the presence of a surface oxide might obscure a surface Si deficit. Finally, the resolution of RBS is barely sufficient for these measurements. Before we draw a final conclusion, further work with better resolution on single crystal Ge in controlled atmospheres is desirable.

SUMMARY

High resolution RBS and SIMS measurements have been combined with the transient conductance measurement to determine the segregation coefficient for Ge as an impurity in Si and for Si as an impurity in Ge. We observe a significant degree of trapping of Ge in Si and of Si in Ge. Such a result is inconsistent with the model of Jackson and coworkers; it suggests that a superposition of an alloy crystallization reaction and a solute-solvent redistribution reaction might be a more appropriate basis for understanding binary alloy solidification.

REFERENCES

[1]   R. N. Hall, J. Phys. Chem. 57, 836 (1953).
[2]   A. A. Chernov, in Growth of Crystals, Vol. 3, eds. A. V. Shubnikov and N. N. Sheftal'. Consultants Bureau, New York, 1962.
[3]   J. C. Brice, The Growth of Crystals From the Melt, North-Holland, Amsterdam, 1965.
[4]   J. W. Cahn, S. R. Coriell, and W. J. Boettinger, in Laser and Electron Beam Processing of Materials, eds. C. W. White and P. S. Peercy, Academic Press, New York, 1980. pp. 89-103.
[5]   M. Hillert and B. Sundman, Acta Metall. 25, 11 (1977).
[6]   K. A. Jackson, G. H. Gilmer, and H. J. Leamy, in Laser and Electron Beam Processing of Materials, eds. C. W. White and P. S. Peercy, Academic Press, New York, 1980. pp. 104-110.
[7]   R. F. Wood, Appl. Phys. Lett. 37, 302 (1980).
[8]   M. J. Aziz, J. Appl. Phys. 53, 1158 (1982).
[9]   G. H. Gilmer, Mat. Res. Soc. Symp. Proc. 13, 249 (1983).
[10] J. C. Baker and J. W. Cahn, in Solidification, ASM, Metals Park, Ohio, 1970. pp. 23-58.
[11] K. A. Jackson, in Surface Modification and Alloying By Laser, Ion and Electron Beams, eds. J. M. Poate, G. Foti, and D. C. Jacobson. Plenum, Press, New York, 1983. p. 51.
[12] M. J. Aziz, Appl. Phys. Lett. 43, 552 (1983).
[13] M. O. Thompson, J. W. Mayer, A. G. Cullis, H. C. Webber, N. G. Chew, J. M. Poate, and D. C. Jacobson, Phys. Rev. Lett. 50, 896 (1983).
[14] M. Hansen, Constitution of Binary Alloys, 2d. ed., McGraw-Hill, New York, 1958. p. 774.
[15] C. D. Thurmond, J. Phys. Chem. 57, 827 (1953).

MODELING OF UNDERCOOLING, NUCLEATION, AND MULTIPLE PHASE FRONT FORMATION IN PULSED-LASER-MELTED AMORPHOUS SILICON*

R. F. WOOD,[1] G. A. GEIST,[2] A. D. SOLOMON,[2] D. H. LOWNDES,[1] AND
G. E. JELLISON, JR.[1]
[1] Solid State Division, Oak Ridge National Laboratory, P. O. Box X, Oak
Ridge, TN 37831
[2] Engineering Physics and Mathematics Division, Oak Ridge National
Laboratory, P. O. Box Y, Oak Ridge, TN 37831

ABSTRACT

Recently available experimental data indicate that the solidification
of undercooled molten silicon prepared by pulsed laser melting of amorphous
silicon is a complex process. Time-resolved reflectivity and electrical
conductivity measurements provide information about near-surface melting and
suggest the presence of buried molten layers. Transmission electron micrographs show the formation of both fine- and large-grained polycrystalline
regions if the melt front does not penetrate through the amorphous layer.
We have carried out extensive calculations using a newly developed computer
program based on an enthalpy formulation of the heat conduction problem.
The program provides the framework for a consistent treatment of the simultaneous formation of multiple states and phase-front propagation by allowing
material in each finite-difference cell to melt, undercool, nucleate, and
solidify under prescribed conditions. Calculations indicate possibilities
for a wide variety of solidification behavior. The new model and selected
results of calculations are discussed here and comparisons with recent
experimental data are made.

INTRODUCTION

It is now reasonably well established that T_a, the melting temperature
of amorphous (a) silicon, is approximately 200 ± 100 deg less than T_c, the
melting temperature of crystalline (c) silicon [1-3] and that the latent
heat L_a of a-Si is ~75% of L_c, the latent heat of c-Si [1]. Therefore, when
a-Si is melted by pulsed laser irradiation, highly undercooled liquid (ℓ)
silicon can be formed. Also, after the pulsed laser melting of c-Si under
some conditions, the kinetic rate processes at the liquid-solid (ℓ-s) interface may become so sluggish compared to the heat-flow rate that the liquid
undercools significantly before solidification, with the result that amorphous silicon is formed. Processes in which large undercoolings of the
liquid (or superheating of the solid) occur before the phase change takes
place are difficult to treat by the mathematical modeling techniques developed for heat conduction problems in which undercooling does not play a
significant role [4,6]. In order to treat such processes more adequately, a
computer program based on an enthalpy formulation of the heat flow problem
has been developed and tested. The program is being used to study the
problem of the pulsed laser-induced ultrarapid melting and solidification of
a-Si layers formed in c-Si substrates by self-ion implantation. In this
paper, we describe the new model and some of the results to date of an
ongoing study of melting of amorphous layers in semiconductors.
Recent time-resolved reflectivity measurements, transmission electron
microscopy (TEM) studies, and model calculations for silicon samples containing amorphous surface layers [5-9] have provided considerable insight

---
*Research sponsored by the Division of Materials Sciences, U.S. Department
of Energy under contract DE-AC05-84OR21400 with Martin Marietta Energy
Systems, Inc.

into the role played by such layers in the laser-annealing process. The work established that the thermal conductivity of a-Si is approximately an order of magnitude less than that of c-Si (see also Goldsmid et al. [9]). Furthermore, the calculations showed that the response of the amorphous layer to the annealing laser pulse is determined primarily by this greatly reduced thermal conductivity and that the reduction of T_a and L_a from T_c and L_c are comparatively unimportant. However, since the a-Si melts at T_a, laser melting of a-Si layers on c-Si substrates provides a unique method for forming highly undercooled molten silicon. Transient electrical conductivity measurements have also proved to be a powerful tool for studying the melting of a-Si layers [3].

The TEM carried out in connection with the time-resolved optical measurements shows that if the melt front does not penetrate through the a-Si layer, regions of fine-grained (FG) and large-grained (LG) polycrystalline (p) Si are formed, depending on the energy density E_ℓ of the laser pulse; this is illustrated schematically in Fig. 1. The random orientation and equiaxed nature of the fine grains suggests that bulk nucleation may have occurred at some $T < T_n$, the nucleation temperature, lying between T_a and T_c. When E_ℓ is low and the temperature of the undercooled liquid does not exceed T_n, the FG region on Fig. 1 is found at the surface and there is no LG material. As E_ℓ is increased the FG region moves into the sample and LG material appears at the surface. At higher E_ℓ the melt front penetrates through the a-c interface and liquid phase epitaxy ensues.

Three distinct cases may arise in the pulsed laser melting of a well-defined a-Si layer on a c-Si substrate.

Case a. E_ℓ is sufficiently low that the melt front does not penetrate through the a-Si layer. This means that a pool of highly undercooled molten Si is formed, separated from the c-Si substrate by an a-Si layer of low thermal conductivity.

Case b. E_ℓ is sufficient for the melt front to penetrate to the a-c interface. The melt front will pause at the interface until the heat flow adjusts to the differences in latent heats, $\Delta L = L_c - L_a$, and melting temperatures, $\Delta T = T_c - T_a$, in the a and c regions. Further complicating the calculations is the behavior of the thermal conductivity, which changes by about an order of magnitude at the a-c interface. In this case, as in case a, a pool of undercooled ℓ-Si will be present in the sample.

Case c. The melt front penetrates beyond the a-c interface. In this case, significant undercooling of the molten Si is not expected for any prolonged periods, and the physical conditions are closely similar to those that exist when c-Si melts and resolidifies. For a given E_ℓ, the difference in latent heat between a- and c-Si acts as an additional heat source to increase the melt-front penetration when an amorphous layer is present.

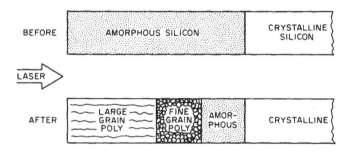

Fig. 1. Schematic of the results of pulsed laser melting and solidification of an a-Si layer on c-Si substrate.

It appears that conventional melting model calculations [4] can deal reasonably satisfactorily with case c, but it is not apparent how to deal with solidification from a highly undercooled melt in which either bulk nucleation or nucleation at the ℓ-a interface occurs. The major conceptual difficulty is to find a satisfactory way to include the effects of nucleation of the crystalline phase. The major computational difficulties are those of including solidification at a temperature other than the melting temperature and of introducing three-dimensional volume nucleation and growth effects into a calculation which, if it is to be tractable, must remain essentially one dimensional. In the following, we will describe how these effects have been approximately introduced into a one-dimensional treatment.

FORMULATION OF THE MODEL

The semiconductor sample is modeled as a slab composed of layers with various thermal and optical properties. All the material properties may be functions of temperature and state. The left boundary of the slab is assumed to be insulated because, for laser pulses of nanosecond duration, detectable convection and radiation losses do not occur at the surface [4]. The slab is semi-infinite, with the right boundary at x = infinity assumed to be at the ambient temperature. For practical purposes the slab must be assigned a finite width or the back boundary otherwise taken into account. The laser pulse is modeled by an internal heat generation function in the near-surface region of the slab. The amount of heat generated may be a function of position, time, temperature, and state.

a) Energy and Enthalpy Equations

We consider an elementary volume of the material and derive the enthalpy equation. For energy conservation the rate of change of the internal energy must equal the generation rate minus the rate at which energy is conducted away, i.e.,

$$\frac{d}{d\tau} \int_V \rho e \, dv = \int_V S \, dv - \int_A K\nabla T \cdot \vec{n} \, dA \ . \tag{1}$$

e is the internal energy, ρ is the density, S is the generation rate due to the laser pulse, K is the thermal conductivity, ∇T is the gradient of the temperature, dv and dA are volume and surface differentials and $\vec{n}$ is an outwardly drawn unit vector normal to dA. Recalling that the thermodynamic relationship for enthalpy h is

$$h = e + p\rho^{-1} \ , \tag{2}$$

we get

$$\frac{d}{d\tau} \int_V \rho h \, dv - \frac{d}{d\tau} \int_V p \, dv = \int_V S \, dv - \int_A K\vec{\nabla}T \cdot n \, dA \ , \tag{3}$$

in which p is the pressure. Since the pressure is constant during an isobaric process and the density change is reasonably small for melting of silicon, the second term on the left hand side can be neglected. By following straightforward manipulations found in many texts [11], we obtain

$$\rho \frac{\partial}{\partial \tau} h = \nabla \cdot K\nabla T + S \ , \tag{4}$$

162

which in the case of one-dimensional heat transfer reduces to

$$\rho \frac{\partial}{\partial \tau} h = \frac{\partial}{\partial x} \left(K \frac{\partial T}{\partial x} \right) + S . \tag{5}$$

This is the starting point for the discretization to be introduced in the finite-difference approach. Notice that if we apply the relation

$$dh = C_p \, dT , \tag{6}$$

in which C_p is the specific heat at constant pressure, the enthalpy may be eliminated, leaving us with the familiar parabolic differential equation for heat conduction without phase change,

$$\rho C_p \frac{\partial T}{\partial \tau} = \frac{\partial}{\partial x} \left(K \frac{\partial T}{\partial x} \right) + S . \tag{7}$$

Thus the formulations in terms of the time rate of change of h and T are identical.

We chose the enthalpy formulation because of the convenient and easily visualized properties it has when modeling change-of-phase problems. This is illustrated by Fig. 2 which shows for silicon what we shall refer to as a state graph. The state graph is nothing more than a plot of temperature versus enthalpy for the various states involved. For an isobaric process it shows clearly the thermodynamic conditions under which a material can undergo various changes of phase and state.

b) Source Term

In the calculations to be discussed here, the absorption of laser energy is assumed to have the usual exponential form appropriate to a constant absorption coefficient α. Thus, the amount of radiant energy

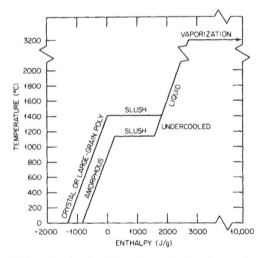

Fig. 2. State diagram for Si. The enthalpy is on the abscissa
because it is regarded as the fundamental quantity
and temperature as the derived quantity.

penetrating to a particular depth x is given by

$$S = P(t)\ \alpha e^{-\alpha x}\ , \tag{8}$$

where $P(t)$ is the variation in intensity of the laser pulse with time. In order to determine the internal heat generation in a finite region of the sample, Eq. (8) is integrated with respect to x to give

$$\int_{x_1}^{x_2} P(t)\ \alpha e^{-\alpha x}\ dx = -P(t)e^{-\alpha x}\ \bigg|_{x_1}^{x_2}\ . \tag{9}$$

The equations can easily be generalized to nonconstant absorption coefficients.

c) Discretization

Equation (5) was discretized using the classic backwards difference scheme. This gives an explicit method for updating the enthalpies at each new time step by way of the equation

$$\rho\ \frac{h_i^{n+1}-h_i^n}{\Delta \tau} = \frac{1}{\Delta x^2}\left[\frac{K_{i+1}+K_i}{2}\ (T_{i+1}^n-T_i^n) + \frac{K_i+K_{i-1}}{2}\ (T_{i-1}^n-T_i^n)\right] + S_i^n\ , \tag{10}$$

in which the superscripts and subscripts refer to the time and space increments, respectively. More complex discretization schemes were tried (for example Crank-Nicholson), but they did not seem to offer any significant advantages and actually caused some problems in implementing the state array described next.

d) The State Array

Multi-state capabilities are conveniently incorporated into the modeling by a state array. The rows and columns of the state array are labeled with the names of the various phases and states that may occur. The diagonal elements represent no change in state. The off-diagonal blocks can be used to prescribe the set of conditions under which any state can transform into another state. Since the model presently requires changes of state to occur essentially instantaneously (i.e., within one time step of 0.5×10^{-12} sec in the finite difference formulation), it was found desirable to explicitly introduce transition states through which the material of a given finite-difference cell passes during a phase change. For example, during a $\ell \rightarrow s$ phase transition the material may exist in a "mushy" or "slush" state for prolonged periods, with the cell containing solid and liquid in proportions dictated by the changing enthalpy content of the cell (see Fig. 2). It also proved useful to explictly recognize FG-, LG-, and c-Si as three distinct states. In fact, in its present form, the state array provides for nine different states, i.e., ċ, FG, LG, a, mushy amorphous (ma), mushy crystalline (mc), mushy FG (mFG), supercooled ($T<T_c$) liquid (scℓ), and normal ($T>T_c$) liquid. A complete description of the state array and the conditions required for each transition cannot be given here because of space limitations, but we will illustrate how the array functions by two examples.

The first example is the transition from a mushy crystal state to a single crystal state. The first condition to be met is that the enthalpy in the cell must be less than the minimum enthalpy required for a liquid fraction to exist in the cell. The second condition is that the neighboring cell to the right, i.e., toward the a-c interface (see Fig. 1) must already be in a single crystal state; otherwise, this cell does not have a crystalline interface on which to nucleate. If this second condition is not met the mc cell transforms to an LG cell.

In the second example we consider three of the transitions that a supercooled liquid cell can undergo. First is the change from supercooled to mushy amorphous state. This occurs when the enthalpy falls below the minimum enthalpy that a cell can have and still be entirely liquid. The second transition is from supercooled liquid to m FG material; this is the path taken when the liquid nucleates as FG p-Si. Two conditions must be met. First the enthalpy must be less than the enthalpy above which nucleation is improbable, and secondly a nucleation timer (described below) must be greater than the assigned nucleation time t_n. Physically t_n can be thought of roughly as the average time it takes for a nucleus of critical radius to form. In other words, the nucleation timer keeps track of the time a cell has been undercooled below T_n. The third transition is simply from undercooled to normal liquid, which occurs when the temperature of the liquid exceeds the melt temperature of the crystalline state.

e) Nucleation Timers

The model contains two timers which were introduced to simulate nucleation times. One timer has been used extensively in connection with bulk nucleation of FG material while the other has been used to control growth of LG material off the FG p-Si. Although there is a question as to whether nucleation occurs in the bulk (not necessarily homogeneously) or at the ℓ-a interface, we emphasize that as long as the melt-front does not contact the a-c interface, a nucleation event must occur for growth of the FG and/or LG material to proceed.

RESULTS OF THE CALCULATIONS

Two somewhat different models have been suggested to explain the solidification behavior of laser-melted a-Si. One model [5-7] emphasizes volume (bulk) nucleation over a spatially extended region at low values of E_ℓ. The other model [3] does not address the question of nucleation but emphasizes the important role of "explosive crystallization," a well-known phenomenon often discussed in the literature. Both models recognize the importance of the release of latent heat in determining the penetration of the crystallization front and both models predict that buried molten layers are a likely occurrence. Since some nucleation event is required to initiate conversion of a-Si to c-Si and since crystal growth can proceed very rapidly once nucleation occurs, it would seem that a synthesis of the two models is needed. The computer program described here is sufficiently flexible to provide such a synthesis.

The initial goal of the modeling has been a better understanding of the experimental and theoretical results described in Refs. 3-7. Figure 3 shows a schematic illustration of the complex behavior of melting and solidification revealed by a wide variety of test calculations. The details of this behavior depend critically on the energy density of the laser pulse and many of the parameters of the model; specific examples will be discussed shortly. The undercooled region on Fig. 3 occurs because of the melting of a-Si at a temperature $T_a < T_c$ by the laser pulse or by the release of latent heat. The region labeled "slush" is that part of the sample in which bulk nucleation has occurred in some cells but not others, so that the region is a mixture of liquid and small crystallites. We note that the release of latent heat due to nucleation in one cell may raise the temperature of neighboring

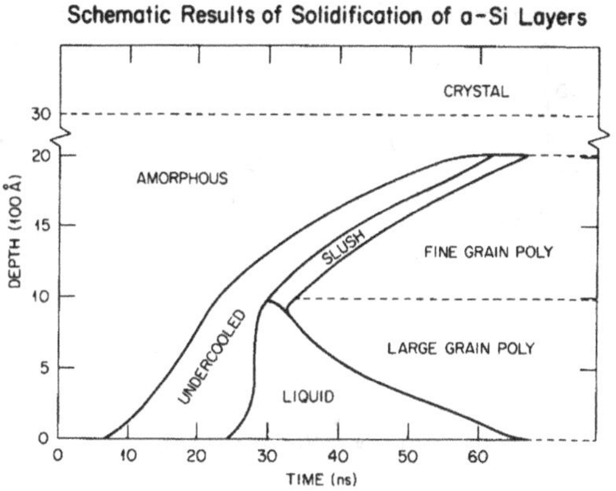

Schematic Results of Solidification of a-Si Layers

Fig. 3. Schematic illustration of the results of simulations of melting and solidification of an a-Si layer on a c-Si substrate. The diagram is not meant to represent strict relationships between the various states and phases.

cells above T_n and prevent nucleation from occurring in them. If then $T \simeq T_C$ in this region for any period, growth of crystalline material may be slowed or stopped entirely. The "liquid" region on Fig. 3 represents material in the molten state with temperature $> T_C$. For low values of E_ℓ this region may be absent. The region marked "Fine Grain Poly" is material which has solidified as a result of bulk nucleation or by explosive crystallization off of FG material. Large-grained p-Si can begin to grow off FG p-Si if the temperature in a cell of molten (undercooled or liquid) Si immediately adjacent to a FG p-Si cell is greater than T_n for a time $t_{\ell g}$. Roughly speaking, $t_{\ell g}$ may be viewed as a delay time required to establish growth of the randomly oriented fine grains in the preferred direction characteristic of the LG p-Si. We now turn to a consideration of some specific results.

Figure 4 shows the transient reflectivity signal of a 633 nm cw HeNe probe beam for a 0.2 J/cm^2 heating pulse incident on a 190-nm a-Si layer. It can be seen that the reflectivity signal, R, reaches a maximum value R_{max} where it persists for only a few nanoseconds before beginning to fall slowly to approximately its initial value R_0. The calculations for this case which seem to best agree with the reflectivity trace indicate that the near-surface region melts to a depth of 200-300 Å, nucleation events then occur in this region or at the ℓ-a interface, and the latent heat that is released subsequently forces the crystallization front into the solid while the initially melted material resolidifies. The molten zone that continues to propagate into the solid as a buried layer is only a few hundred angstroms wide and the process can probably be characterized as "explosive crystallization." The gradual fall of the reflectivity signal (Fig. 4) may be attributed to bulk nucleation, or to a slow receding of the buried layer from the surface. The existence of a buried molten layer is confirmed by experiments using an infrared probe beam reported by Lowndes et al. [12] in these Proceedings. We note here that these results are not inconsistent with the model proposed by Thompson et al. [3] if the requirement for nucleation events is explicitly taken into account. The accuracy of the experiments and modeling do not allow us to distinguish between volume and interface nucleation events.

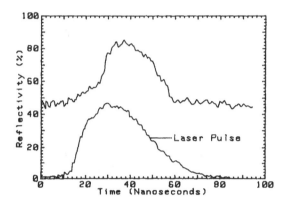

Fig. 4. Time-resolved reflectivity signal for a 0.2 J/cm² KrF
laser pulse incident on a 190-nm a-Si layer.

Figure 5 shows the reflectivity trace for a pulse with $E_\ell \simeq 0.7$ J/cm². The well-developed high reflectivity phase and sharply falling trailing edge of this trace are apparent. The model calculations for this value of E_ℓ show that (1) the liquid region of Fig. 3 is deep, (2) the undercooled region is very narrow because of the rapid heating by the laser pulse, and (3) the slush region is practically nonexistent. The melt front does not quite reach the a-c interface. The calculations give ~1400 Å of LG and 400 Å of FG p-Si after solidification, with only ~100 Å of a-Si remaining. These results are in satisfactory agreement with values obtained from TEM.

Figure 6 shows a comparison between the experimentally measured and calculated values of the time of onset of melting for both a- and c-Si that can be obtained with an appropriate choice of parameters. The most important parameters are K_a, the thermal conductivity, and R_a, the reflectivity of a-Si. We used $K_a = 0.015$ W/cm deg and $R_a = 0.58$, both values well within

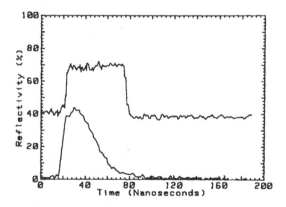

Fig. 5. Time-resolved reflectivity signal (upper curve)
for a pulse of ~0.7 J/cm² (lower curve).

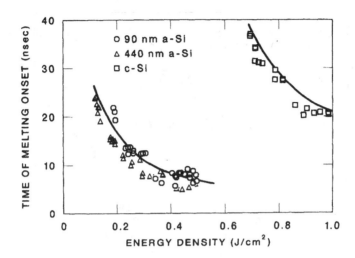

Fig. 6. Onset of melting as a function of energy density.
The solid curves are the calculated results obtained
for a 190-nm a-Si layer.

the ranges established for these quantities. A much more severe test for
the modeling is presented when the duration of surface melting must be
determined because the calculations then involve knowledge of the physical
properties of all of the various phases and states of the material (e.g.,
mush, FG p-Si, LG p-Si) that occur during melting and resolidification. As
can be seen from Fig. 7, the agreement between experiment and theory
requires further improvement, although the general features of the experimental data are reproduced by the calculations. In particular, the calculations indicate that the dip in the melt duration at $E_\ell \simeq 0.7$ J/cm^2 is caused
by the rapid increase in thermal conductance as the melt front approaches
the a-c interface and the amount of a-Si (with its low value of K_a) is
reduced. We anticipate that improvement in the fit between experiment and
modeling will be achieved as more data about the optical and thermal properties is obtained and the calculations are optimized.

SUMMARY AND CONCLUSIONS

We have briefly described a newly developed model for heat flow calculations and phase changes that is especially suited to conditions encountered during pulsed laser irradiation of a-Si layers. The model is extremely flexible and allows for the inclusion of such effects as undercooling,
bulk and interface nucleation of polycrystalline silicon, and explosive
crystallization. In spite of the flexibility of the model (or perhaps
because of it), we have not yet obtained a completely satisfactory fit to
all of the available experimental data. We have also not yet completed a
synthesis of the two models proposed to explain the solidification of laser-
melted a-Si. We believe that there is good evidence from the modeling that
explosive crystallization can occur under the conditions of the experiments.
Whatever the exact nature of the crystallization processes may be, nucleation events must initiate them. We recognize that metallurgists are reluc-
tant to invoke the concept of homogeneous bulk nucleation, except in most

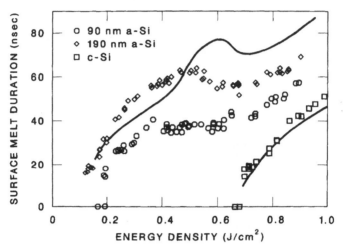

Fig. 7. Surface melt duration as a function of pulse energy density.

unusual and extreme situations. Although we feel that some of the experimental data and the modeling suggests that bulk nucleation occurs, it may be unnecessary to ascribe this to homogeneous nucleation.

Acknowledgments

We are indebted to G. E. Giles, J. B. Drake, and V. Alexiades for numerous discussions of various aspects of the modeling.

References

1. E. P. Donovan, F. Spaepen, D. Turnbull, J. M. Poate, and D. C. Jacobson, Appl. Phys. Lett. 42, 698 (1983).
2. G. L. Olson, S. A. Kokorowski, J. A. Roth, and L. D. Hess, Mat. Res. Soc. Symp. Proc. 13, 141 (1983).
3. M. O. Thompson, G. J. Galvin, P. S. Peercy, J. M. Poate, D. C. Jacobson, A. G. Cullis, and N. G. Chew, Phys. Rev. Lett. 52, 2360 (1984).
4. See, e.g., R. F. Wood and G. E. Giles, Phys. Rev. B 23, 2923 (1981).
5. D. H. Lowndes, R. F. Wood, and J. Narayan, Phys. Rev. Lett. 52, 561 (1984).
6. R. F. Wood, D. H. Lowndes, and J. Narayan, Appl. Phys. Lett. 44, 770 (1984).
7. D. H. Lowndes, R. F. Wood, C. W. White, and J. Narayan, Mat. Res. Soc. Symp. Proc. 23, 99 (1984).
8. J. Narayan and C. W. White, Appl. Phys. Lett. 44, 35 (1984).
9. H. C. Webber, A. G. Cullis, and N. G. Chew, Appl. Phys. Lett. 43, 669 (1983).
10. H. J. Goldsmid, M. M. Kaila, and G. I. Paul, Phys. Stat. Sol. (a) 76, K31 (1983).
11. See, for example, H. S. Carslaw and J. C. Jaeger, "Conduction of Heat in Solids," 2nd ed. (The Clarendon Press, Oxford, 1959).
12. D. H. Lowndes, G. E. Jellison, Jr., R. F. Wood, S. J. Pennycook, and R. W. Carpenter, these Proceedings.

EFFECTS OF ARSENIC DOPING ON THE SOLIDIFICATION DYNAMICS
OF PULSED-LASER-MELTED SILICON

MICHAEL O. THOMPSON*, P. S. PEERCY AND J. Y. TSAO
Sandia National Laboratories, Albuquerque, NM 87185
* On leave Dept. of Materials Science, Cornell University, Ithaca, NY 14853

ABSTRACT

The effects of arsenic doping on the solidification dynamics during
pulsed melting of silicon have been studied using the transient conductance
technique. At As concentrations below 1 at.%, the incorporation of As into
the Si lattice results in negligible differences in the solidification
dynamics. Between 2 and 7 at.% As, however, the interface velocity is
dramatically modified as the liquid-solid interface crosses the As
containing region. These velocity changes are consistent with a reduced
melting temperature for Si-As alloys. For concentrations of 11 at.% As, the
depression in the melting temperature is sufficient to allow the surface to
solidify while considerable melt remains buried within the sample. At
16 at.%, the melting temperature is drastically reduced and internal
nucleation of melt occurs prior to normal surface melting.

INTRODUCTION

The recent development of quantitative probes of the liquid-solid
interface motion during pulsed laser melting provides an opportunity to
study fundamental thermodynamic and kinetic effects during rapid
solidification. It has been well established that pulsed laser melting can
produce metastable alloys of silicon by the incorporation of impurity
concentrations substantially in excess of the equilibrium solid
solubility [1-2]. During this incorporation, the impurity segregation
coefficient varies dramatically from the equilibrium value. Several
theoretical treatments of impurity incorporation at high solidification
velocities exist which relate the impurity incorporation to a velocity
dependent segregation coefficient [3-4]. Experimental verification of these
theories has been hampered by the lack of direct information on the kinetics
of the solidification process. Previous kinetic analyses were limited to
measurements of impurity redistributions after melting and computer models
which assumed interface dynamics similar to pure Si. Preliminary studies of
impurity effects [5], however, indicated that the interface velocity during
impurity incorporation can be depressed from that for pure Si.

In this paper we present direct measurements, using the transient
conductance technique [6], of interface kinetics during pulsed laser melting
of silicon containing ~1-16 atomic percent arsenic. At concentrations below
1 at.%, the impurities have a negligible effect on solidification dynamics.
Between 2 and 7 at.%, however, the solidification dynamics show considerable
deviation from those for undoped Si. Several related phenomena are observed
at higher As concentrations, including surface solidification with a buried
melt and internal nucleation of melt. This observation of impurity induced
buried molten layers confirms previous indirect evidence for surface
nucleation inferred from anomalous impurity diffusion during laser melt [7].

EXPERIMENTAL

Samples for transient conductance measurements were patterned
photolithographically from 0.5 µm silicon on sapphire (SOS). The samples
were subsequently implanted with As ions at an energy of 250 or 275 keV to

170

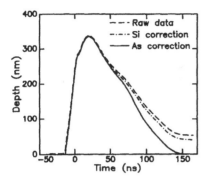

Fig. 1 The dashed curve shows the raw apparent depth for a 5×10^{16} As implanted sample. The dot-dashed curve includes a correction due to the conductivity of undoped Si and the solid curve shows the final melt depth after the conductivity of the solid doped Si has been subtracted.

fluences of 1.6, 5, 8 and $12 \times 10^{16}/cm^2$, corresponding to peak As concentrations of 1.7, 7, 11 and 16 at.%. Analysis of the As concentration by Rutherford Backscattering (RBS) indicated the peak As concentration at 160 nm with a full width at half maximum (FWHM) of approximately 100 nm. As implanted, the samples contained an amorphous layer ~250-300 nm thick.

The samples were irradiated through a quartz homogenizer [8] using a 28 ns FWHM 694 nm laser pulse. The transient conductance was monitored in the charge line configuration [9] while the surface reflectance was simultaneously monitored using a 488 nm probe laser [10]. Due to impurity diffusion in the molten phase, each sample was irradiated once and the final As profile was measured with RBS techniques.

In the charge line configuration, the conductance of a sample, designated $\eta(t)$, is given by

$$\eta(t) = \eta_s + \eta_l = \frac{1}{R_L} [V_b/V(t) - 2]^{-1} \qquad (1)$$

where η_s and η_l are the solid and liquid conductances, R_L is the charge line impedance, V_b is the bias voltage and $V(t)$ is the voltage observed at the oscilloscope. The molten layer thickness, $d(t)$, is linearly related to η_l by $d(t) = \rho(L/W)\eta_l$, where ρ is the liquid resistivity (80 $\mu\Omega$-cm) and L/W is the geometric aspect ratio for the samples. For pure SOS, η_s can be neglected; however, when As is quenched into electrically active sites, η_s can be significant. This solid contribution is treated as an effective molten depth, $d_{eff}(t)$, which must be subtracted from the apparent melt depth to yield the molten layer thickness as:

$$d(t) = \frac{\rho}{R_L} \frac{L}{W} [V_b/V(t) - 2]^{-1} - d_{eff}(t) \qquad (2)$$

In fig. 1, the steps to obtain $d(t)$ from the raw transient conductance data are shown for a 5×10^{16} As sample irradiated at 1.8 J/cm^2. The small ledge during melting at 280 nm results from the reduced melting temperature of amorphous Si relative to the crystalline phase [11]. The apparent molten thickness, neglecting solid conductance, is shown as the dashed curve and indicates a d_{eff} of 55 nm after complete resolidification at 150 ns. The contribution of undoped Si to the solid conduction can be estimated from the melt depth by assuming a conductivity ratio σ_l/σ_s of 30, and yields the dot-dashed curve. The remaining 41 nm of d_{eff} results from As doping and is subtracted by assuming a solid conductivity proportional to the As concentration. The approximate Gaussian profile of As gives a d_{eff} which is an error function of the solid thickness. The solid curve in fig. 1 shows the final determination of the melt depth using an As profile centered at 160 nm with a FWHM of 140 nm. Surface reflectance measurements indicate that samples above 7 at.% As do not solidify normally and thus deconvolution to a melt thickness is not possible for high concentrations.

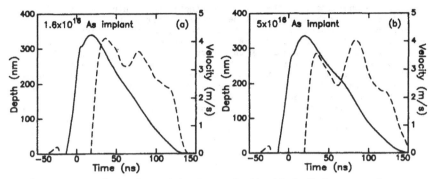

Fig. 2 Melt depth (solid) and interface velocity (dashed) for a sample
implanted with (a) 1.6×10^{16} As (1.7 at.%), (b) 5×10^{16} As (7 at.%).

RESULTS AND DISCUSSIONS

At low As concentrations (<2 at.%), the solidification dynamics are
only slightly modified from crystalline Si regrowth dynamics. Figure 2a
shows the melt dynamics of a 1.6×10^{16} As doped sample, with the velocity
shown as the dashed curve. While the effect of the As doping during
regrowth appears small in the melt depth curve, its effect can be readily
seen in the regrowth velocity. The solidification velocity of pure SOS
decreases monotonically with time corresponding to thermal diffusion of heat
into the substrate. The As doped samples, however, show a velocity decrease
as the melt interface approaches the impurity, and a velocity increase as
the interface passes the peak As concentration. Figure 2b shows the melt
depth and velocity for a similar incident energy density on a 5×10^{16} As
sample. At this concentration, As effects can be seen in the melt depth
trace, while the corresponding velocity changes are even more pronounced.

The observed velocity changes can result from several effects,
including: 1) a modification in the latent heat of melting, 2) a reduction
in the melting temperature of the Si-As alloy, and 3) modifications in the
kinetic undercooling relationship.

The heat of mixing for the Si-As alloy is positive, and thus the latent
heat required for melting, ΔH, is expected to decrease with increasing As
concentration. In the absence of other effects, a decrease in ΔH would
increase the interface velocity, v, since steady state requires

$$v\Delta H = K(\partial T/\partial z) \qquad (3)$$

where $(\partial T/\partial z)$ is the temperature gradient at the liquid-solid interface and
K is the thermal conductivity. A change in ΔH would yield a symmetric
increase (or decrease) in the velocity about the peak As concentration.
However, the data in fig. 2 indicate that the change in the interface
velocity is anti-symmetric about the peak concentration. Consequently, ΔH
changes are inadequate to explain the observed interface velocity.

Changes due to the melting temperature and interface kinetics can be
considered together, since both modify the interface temperature required to
support a given interface velocity. A reduced melting temperature requires
a reduced interface temperature to maintain a fixed undercooling and
velocity. Alternatively, if kinetic changes are invoked to explain the
velocity reductions, a greater undercooling and a lower interface
temperature would be required to maintain a given interface velocity. As
will be discussed, internal nucleation of melt at high As concentrations
suggests that the reduced melting temperature dominates the change in
velocity. For simplicity, we will consider the case of the melting

temperature changing with the interface kinetics remaining unmodified.

In fig. 2, both samples melt beyond the entire As profile. As solidification initiates, the required interfacial undercooling is established. In the region of the As impurities, the equilibrium melting temperature decreases, resulting in a reduced interfacial undercooling and consequently a reduced velocity. The difference between the heat released by the moving interface and the heat conducted into the substrate undercools the liquid layer further until the steady state of eqn. 3 is reestablished. When the interface passes the peak As concentration, the inverse occurs. The liquid passes into regions of increased melting temperature, increasing the interfacial undercooling and thus the interface velocity. The additional heat released at the interface is absorbed by the liquid layer to raise the interface temperature and reestablish steady state.

If quasi-steady state is maintained at all times, eqn. 3 can be modified to include the effect of these melting temperature changes by assuming an infinite thermal conductivity of liquid Si (uniform liquid temperature) and a thermal heat transport into the substrate which is independent of the interface temperature. The thermal conduction, $K(\partial T/\partial z)$, must then be balanced by latent heat released at the moving liquid-solid interface, $v\Delta H$, and by energy required to cool or heat the liquid layer to maintain a fixed undercooling. For a liquid of specific heat C_p, the heat flow due to a changing interface temperature is $C_p d(t)(\partial T_m/\partial t)$. The time derivative can be replaced by the convective derivative $-v(\partial T_m/\partial z)$ and thus eqn. 3 becomes:

$$K(\partial T/\partial z) = v_n\Delta H = v\Delta H - vC_p d(t)(\partial T_m/\partial z) \tag{4}$$

$$v/v_n = [1 - C_p d(t)(\partial T_m/\partial z)/\Delta H]^{-1} \tag{5}$$

$$\sim 1 + C_p d(t)(\partial T_m/\partial z)/\Delta H$$

$$(\partial T_m/\partial z) = (1 - v_n/v)\Delta H/C_p/d(t) \tag{6}$$

where v_n is the normal velocity with no impurities. As can be seen from eqn. 5, the velocity changes due to the impurities should be antisymmetric and follow the derivative of the impurity profile. Numerical integration of eqn. 6 for the traces of fig. 2 yields estimates of the peak melting temperature reduction as 40 K for the 1.6×10^{16} and 150 K for the 5×10^{16} As implants. For these estimates, v_n was assumed to decrease proportional to the square root of time. These estimates involve significant uncertainties in the form of v_n and thus are subject to a $\pm50\%$ possible error. Greater accuracy can be obtained with future studies at low concentrations.

Computer simulations were also performed to confirm these simple theoretical estimates. In fig. 3, the interface depth and velocity are shown for a computer simulation of a 340 nm melt. The As doping was approximated as a gaussian at 160 nm with a FWHM of 100 nm. The melting temperature reduction was assumed linear with As concentration with a peak depression of 100 K. Thermal and optical properties of crystalline SOS were assumed and a linear undercooling-velocity relationship with a slope of 17 K/(m/s) was used. The velocity exhibits the same derivative type relationship to the As profile as exhibited by the experimental data. Quantitative comparisons of the simulations with the experimental data confirm the estimates of the melting temperature reduction but also suggest that other effects may also be important. Further work is necessary to elucidate these effects.

Samples containing 8×10^{16} As failed to solidify normally, and instead exhibited nucleation of solid at the surface prior to full solidification of the internal molten layer. Results of simultaneous conductance and surface reflectance are shown in fig. 4a. Because of the high As concentrations, the traces in fig. 4 have not been corrected for solid conduction. The surface reflectance and transient conductance increase simultaneously

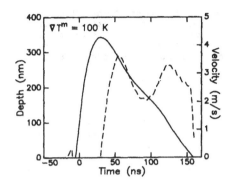

Fig. 3 Computer simulations of melt depth (solid) and interface velocity (dashed) versus time for simulated As profile. The melting temperature reduction follows a Gaussian centered at 160 nm with a FWHM of 100 nm and a peak temperature depression of 100 K.

indicating that melting occurs from the free surface. The surface reflectance, however, terminates at 50 ns while 100 nm of melt remains as indicated by the transient conductance. Since the reflectance is sensitive only to the surface while the conductance is sensitive to the bulk, this measurement indicates a buried molten layer 100 nm thick.

We interpret this surface solidification as further evidence for a reduction in the interface temperature during solidification. As the liquid-solid interface passes through the high As concentration region, the interface temperature is drastically reduced. The high thermal conductivity of the liquid phase requires the surface temperature to follow the interface temperature, and thus as the interface approaches the peak As concentration the surface is undercooled sufficiently to allow random nucleation of solid from the free surface.

As the As concentration is increased further, surface solidification occurs earlier in time. At sufficiently high concentrations, the melting temperature at the peak of the As profile is so severely depressed that internal nucleation of melt can occur, as shown in fig. 4b. Although the transient conductance indicates a peak melt thickness of 150 nm, surface reflectance fails to show any surface melt. A slight increase in the reflectance can be observed since the reflectivity probes significantly deeper than 20 nm in the solid phase. The internal nucleation of melt requires an actual reduction in the melting temperature for the sample

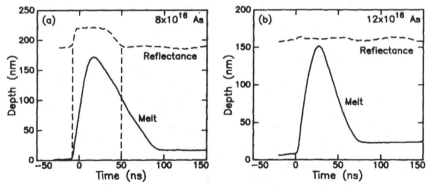

Fig. 4 Uncorrected melt thickness and surface reflectance for high concentration As implants. (a) Formation of a buried molten layer at 11 at.% As. Vertical lines indicate the duration of the surface melt. (b) Internal nucleation of melt at 16 at.% As.

interior compared to the surface. Changes in the undercooling-velocity relationship are unable to explain this internal melt nucleation without simultaneous reductions in the melting temperature. Thus, the existence of the internal melt strongly suggests the velocity effects observed at lower As concentrations are caused by melting temperature reductions rather than by interfacial kinetic effects.

CONCLUSIONS

Arsenic impurities in silicon can produce dramatic changes in the solidification dynamics during pulsed laser melting. Examination of the interface velocity indicate that these changes cannot occur solely from changes in the latent heat of melting. For peak As concentrations of 1.7 and 7 at.%, the velocity changes are consistent, respectively, with 40 K and 150 K reductions in the equilibrium melting temperature. At As concentrations exceeding 11 at.%, the melting temperature near the peak of the As is sufficiently reduced to allow nucleation of solid from the free surface and results in a buried molten layer. At 16 at.%, internal nucleation of melt is observed. Internal melt provides indirect evidence for reductions in the melting temperature rather than changes in the kinetic undercooling-velocity relationship.

ACKNOWLEDGEMENTS

We would like to acknowledge G. J. Galvin for valuable discussions and access to unpublished work. Work at Sandia was supported by U.S. Department of Energy under contract DE-AC04-76DP00789. Work at Cornell was supported by NSF through the Materials Science Center. Samples were prepared at the National Research and Resource Facility for Submicron Structures at Cornell supported by NSF.

REFERENCES

1. J. M. Poate in Laser and Electron Beam Interactions with Solids, edited by B. R. Appleton and G. K. Celler, (North Holland, New York, 1981), p. 121.
2. C. W. White, B. R. Appleton and S. R. Wilson, in Laser Annealing of Semiconductors, J. M. Poate, J. W. Mayer, Eds. (Academic Press, New York, 1982).
3. K. A. Jackson, in Surface Modification and Alloying by Laser, Ion and Electron Beams, edited by J. M. Poate, G. Foti and D. C. Jacobson (Plenum, New York, 1983), Chapter 3.
4. M. J. Aziz, J. Appl. Phys. 53, 1158 (1982).
5. G. J. Galvin, J. W. Mayer and P. S. Peercy, in Energy Beam-Solid Interactions and Transient Thermal Processing, edited by J. C. C. Fan and N. M. Johnson (North Holland, New York, 1984), p. 111.
6. Michael O. Thompson and G. J. Galvin, in Laser-Solid Interactions and Transient Thermal Processing of Materials, edited by J. Narayan, W. L. Brown and R. A. Lemons, (North Holland, New York, 1983), p. 57.
7. S. U. Campisano, D. C. Jacobson, J. M. Poate, A. G. Cullis and N. G. Chew, Appl. Phys. Lett. 45, 1216 (1984).
8. A. G. Cullis, H. C. Webber and P. Bailey, J. Phys. E. 12, 688 (1979).
9. Michael O. Thompson, Ph. D Thesis, Cornell University, Ithaca, NY, 1984.
10. D. H. Auston, C. M. Surko, T. N. C. Venkatesan, R. E. Slusher and J. A. Golovchenko, Appl. Phys. Lett. 33, 437 (1978).
11. Michael O. Thompson, G. J. Galvin, J. W. Mayer, P. S. Peercy, J. M. Poate, D. C. Jacobson, A. G. Cullis and N. G. Chew, Phys. Rev. Lett. 52, 2360 (1984).

LOW-ENERGY, PULSED-LASER IRRADIATION OF AMORPHOUS SILICON: MELTING AND RESOLIDIFICATION AT TWO FRONTS

W. SINKE AND F.W. SARIS
FOM-Institute for Atomic and Molecular Physics, Kruislaan 407, 1098 SJ
Amsterdam, The Netherlands

ABSTRACT

After low-energy pulsed-laser irradiation of Cu-implanted silicon, a double-peak structure is observed in the Cu concentration profile, which results from the occurrence of two melts. From Cu surface segregation we calculate the depth of the surface melt. Cu segregation near the position of the amorphous-crystalline interface gives evidence for a self-propagating melt, moving from the surface region towards the crystalline substrate. Measurements of As-redistribution and of sheet resistance as a function of laser energy density in As-implanted silicon are consistent with the crystallization model which is derived from the effects as observed in Cu-implanted silicon.

The results imply a large difference in melting temperature, heat conductivity and heat of melting between amorphous silicon and crystalline silicon.

INTRODUCTION

Recently, a controversy has arisen concerning the melting temperature of amorphous silicon (a-Si). While solid-phase regrowth can be obtained at temperatures very close to the melting temperature of crystalline silicon (c-Si) when heating is performed using a cw laser [1], experiments with pulsed electron beam annealing and pulsed-laser annealing indicate that a-Si melts at a temperature which is 200-500 K lower than the melting temperature of c-Si [2,3,4].

Moreover, after pulsed-laser irradiation at low energy densities, the originally amorphous film shows two kinds of polycrystalline structure [3,5, 6,7,8], coarse-grain at the surface and fine-grain underneath. The origin of the fine-grained region is not yet clear and two models have been proposed to explain this structure. It has been suggested [3] that a rapidly moving, self-sustaining melted layer produces the small grains. On the other hand, the structure may result from bulk-nucleation in the undercooled melt [6,7]. In the first model, there is a resolidification front moving inward while in the second case this is not the case.

In this paper we report on segregation effects in Cu-implanted Si after nanosecond pulsed-laser irradiation in the energy region below the threshold for epitaxial regrowth, ≈ 1 J cm^{-2}, which show that crystallization of the amorphous layer takes place via melting and resolidification at two fronts. A primary melt depth, produced by the absorbed laser light, is inferred from the surface segregation of copper. There is a plateau in the functional behaviour of melt depth versus laser energy density which is consistent with the suggestion that a-Si has a melting temperature and heat conductivity much lower than that of crystalline silicon (c-Si). In addition, copper segregation near the (a-Si) - (c-Si) interface gives evidence for a secondary melt propagating through the amorphous layer towards the crystalline substrate. As will be shown, the results are consistent with the mechanism proposed in Ref. 3.

In order to illustrate the crystallization processes as described above, we determined the maximum depth at which melting has occured and the sheet resistance as a function of energy density for As-implanted silicon.

EXPERIMENT

Si(100) was implanted at room temperature by either 170 keV Cu$^+$ ions to a dose of 2×10^{15} cm^{-2} or 100 keV As$^+$ ions to a dose of 3×10^{15} cm^{-2}.

Copper exhibits strong segregation effects and diffuses very fast in liquid silicon [9]. Therefore the as-implanted profile will be very sensitive to melting and crystallization within the amorphous layer. Changes in the Cu profile can thus be used to infer melt depths.

Arsenic has a very high solubility in silicon and therefore shows a different behaviour upon melting and resolidification in the amorphized layer. From changes in the concentration profiles it is only possible to estimate the maximum depth at which melting has occurred. On the other hand, the sheet resistance of the implanted layer as measured using a four-point probe, gives an indication of the conversion of amorphous material into (poly-)crystalline material and of changes in electron mobility [10] due to changes in average grain size.

Pulsed-laser annealing was performed using a Q-switched ruby laser, wavelength 694 nm, pulselength 20 ns. Laser energy density was varied between 0.1 J cm^{-2} and 2.0 J cm^{-2} using volume-absorbing neutral density filters. A guide diffuser [11] was used to obtain 5% uniformity over a 5 mm laser spot. The mean energy density was calibrated to within 5% using calorimetry. Laser irradiation was performed at room temperature. A new location on the implanted samples was used for each value of energy density.

Concentration profiles were determined by Rutherford backscattering spectrometry (RBS) [12] using 2 MeV He$^+$ in a random direction of the sample close to its surface normal. Reflected He particles were detected and energy analyzed by a surface barrier detector (energy resolution 12 keV) at a scattering angle of 110°, giving a depth resolution of about 10 nm. The depth of the (a-Si)-(c-Si) interface could be inferred from the high yield portion in the RBS spectra under channeling conditions and was found to be 225 nm for the Cu implantations and 138 nm for the As implantations. Channeling was also used to determine the threshold laser energy density for which epitaxial recrystallization of the originally amorphous layer was achieved. The threshold for surface melting could be inferred (± 0.01 J cm^{-2}) from a surface colour change after pulsed-laser irradiation, due to the conversion of a-Si to poly-crystalline Si.

RESULTS

Copper

Fig. 1 shows a comparison of Cu profiles after implantation at 170 keV and after irradiation at different laser energy densities. The threshold for surface melting was found to occur at 0.10 J cm^{-2}. Fig.1a shows that already at 0.12 J cm^{-2} the copper profile is changed drastically over a thickness of about 175 nm. Fig.1b shows the copper profile after irradiation at 0.14 J cm^{-2}; in this case changes can be observed to a depth of 210 nm. This result is remarkable because in the times involved in pulsed-laser heating the Cu implantation profile can only change due to liquid phase diffusivity, yet 0.14 J cm^{-2} is not sufficient to heat the entire layer of 210 nm thickness to its melting point and melt it [13,14].

A possible explanation for these changes far below the surface, as observed in Fig. 1a and 1b, is the following. At energies just above the threshold for surface melting, only a very thin layer of the amorphous film is melted but the temperature in the entire film is raised above its initial value. When crystallization starts, this will lead to the release of latent heat, ΔH_c. Because ΔH_c is larger than the heat needed to melt a-Si (ΔH_a) [2,13,14,15], the released energy may be used both to heat previously unmelted material to its melting point and to melt it, if the temperature gradient is not too large. During crystallization of this secondary melted silicon, again heat is re-

leased and the process proceeds, with a self-propagating layer of melted silicon moving from the surface region towards the (a-Si)-(c-Si) interface.
Depending on the temperature gradient in the amorphous material, this 'explosive' crystallization may either quench at a certain depth or proceed through
the entire layer. From the direction in which the secondary melt front moves
it is clear that this will lead to a segregation of copper toward the (a-Si)
-(c-Si) interface, which is indeed observed.

Both Fig.1a and 1b show very little surface accumulation of copper. From
this we conclude that initially only a thin surface layer has been melted by
the absorbed laser light, since melting in this region of low copper concentrations will not result in detectable segregation effects.

Cross-section TEM pictures of an a-Si film after laser annealing at low
energy densities show a dual grain structure which may be related to this two-
fold crystallization [3], while time-resolved conductivity measurements during
pulsed-laser irradiation indicate the presence of a buried melted layer [3].
To explain the difference in grain structure between the two poly-crystalline
regions, Lowndes et al. [6] and Wood et al. [7] suggest that bulk nucleation
occurs, however, according to their model there should be no copper accumulation at the (a-Si)-(c-Si) interface.

Fig.1c and 1d show the copper profile after irradiation at 0.21 J cm^{-2} and
0.31 J cm^{-2}, respectively. In both cases, a clear double-peak structure is observed, which for 0.31 J cm^{-2} coincides with the primary melt front reaching
to half of the amorphous layer thickness. An appreciable amount of copper
appears on the surface as a result of segregation in the primary melt. Copper is taken to the (a-Si)-(c-Si) interface because of segregation at the resolidification front which moves inward with the secondary melt. Fig.1e shows the result of a 0.53 J cm^{-2} pulse, the primary melt depth is 185 nm. As a result of the small width of the remaining amorphous layer (40 nm) and the low copper concentrations in this region the secondary melt causes only minor changes in the profile

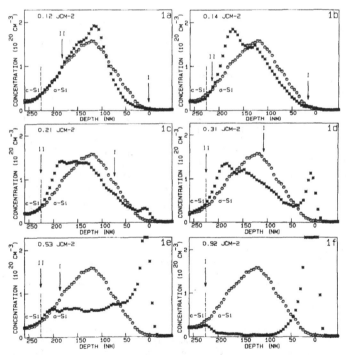

Fig.1. Cu concentration profiles in Cu-implanted Si, before
(∘) and after (∗) laser irradiation at various energy densities. Indicated are primary (I) and secondary (II) melt depth
(see text). Implantation: 2×10^{15} cm^{-2}, 170 keV.

near the interface.

The large amount of copper which is still present after crystallization of the primary melt (see Fig.1c - 1e) can be explained by solute trapping due to a very high speed of the resolidification front. When a-Si melts at a lower temperature than the melting temperature of c-Si, it forms a liquid which is undercooled with respect to c-Si and crystallization may occur very fast [16]. The same argument may hold for the copper which remains behind after crystallization of the secondary melt (see Fig.1a - 1d), but in this case there is a second possibility. When the liquid layer which propagates the secondary melt is thin, the copper concentration in the melt becomes very high and an appreciable amount of copper can be built in the crystal. However, at this moment, we cannot distinguish between these two effects.

For 0.70 J cm^{-2}, the copper concentration profile is found to be reduced over the full depth of the amorphous layer and segregated at the surface. From this we conclude that the primary melt has reached the depth of the (a-Si) - (c-Si) interface, which has been confirmed also by the observation of channeling in the recrystallized layer. Increasing the energy density further to 0.92 J cm^{-2} leads to improved channeling, but the spectra still reveal some dechanneling due to disorder at the position of the original interface. Between 0.70 J cm^{-2} and 0.92 J cm^{-2} the primary melt depth does not change, but the temperature in the liquid layer (the thickness of which is that of the originally amorphized layer) rises. This results in a lower rate of crystallization and a very effective segregation of copper to the surface at 0.92 J cm^{-2} (see Fig. 1f). Dechanneling at the original interface is absent when pulsed-laser irradiation is performed at 1.01 J cm^{-2}, at which energy density good epitaxial regrowth is obtained, indicating that melting beyond the depth of the (a-Si) - (c-Si) interface has occurred.

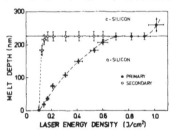

Fig. 2. Primary (●) and secondary (O) melt depth as function of laser energy density.

All measurements of the primary and secondary melt depth as a function of laser energy density are summarized in Fig.2. We have calculated the primary melt depth by subtracting the as-implanted spectrum from the spectrum which is taken after irradiation and integrating the difference-spectrum from the surface region down to some depth between the surface and the (a-Si) - (c-Si) interface. By varying the second boundary, the integral of the difference-spectrum is found to go to zero at a particular depth, which we have taken as the primary melt depth. All changes in the copper profile are assumed to be due to melting and segregation during resolidification. We also assume that after crystallization of the primary melt no copper is moved into or taken out of this region by the secondary melt. The secondary melt depth was calculated in a similar way.

As is shown in Fig.2, surface melting starts at 0.10 J cm^{-2}, while crystallization of the 225 nm amorphous layer is completed at an energy density as low as 0.17 J cm^{-2}! However, it takes 0.7 J cm^{-2} for the primary melt front to reach the (a-Si) - (c-Si) interface and over 0.9 J cm^{-2} to obtain epitaxial regrowth. Although from these measurements it is not possible to calculate directly the difference in melting temperature between a-Si and c-Si, the occurrence of a plateau in the relation between laser energy density and primary melt depth is in qualitative agreement with an appreciable melting point reduction in a-Si during pulsed laser annealing [13]. Both the very low onset for surface melting and the slope in the relation between energy density and melt depth close to the plateau indicate a large difference in heat conductivity between a-Si and c-Si [4,6].

Arsenic

Fig.3 shows a comparison of As profiles after implantation at 100 keV and after irradiation at various energy densities. The threshold for surface melting was found to be 0.16 J cm^{-2}. Fig.3a shows that after irradiation at 0.20 J cm^{-2} the profile is changed over a depth of more than 100 nm. Irradiation at 0.32 J cm^{-2} (Fig.3b) results in a change over the full depth of the amorphous layer. By increasing the energy density to 0.87 J cm^{-2}, the shape of the As profile after irradiation is changed (Fig.3c), but not the maximum depth at which changes in the as-implanted can be observed. Fig.3d shows that after irradiation at 1.02 J cm^{-2}, arsenic is clearly found beyond the (a-Si) - (c-Si) interface, indicating that melting has occurred beyond this interface.

Fig.4 shows the maximum depth at which changes in the As concentration profile can be observed after pulsed-laser irradiation, as a function of

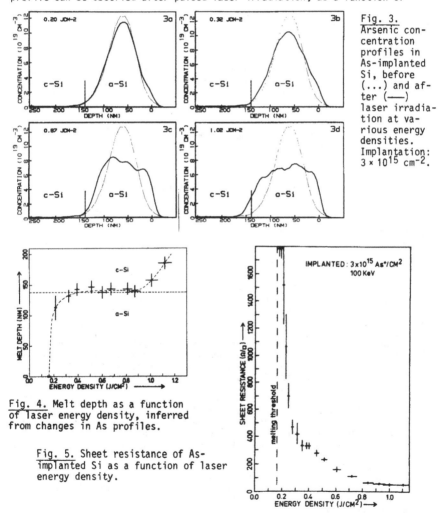

Fig. 3. Arsenic concentration profiles in As-implanted Si, before (...) and after (——) laser irradiation at various energy densities. Implantation: 3 × 10^{15} cm^{-2}.

Fig. 4. Melt depth as a function of laser energy density, inferred from changes in As profiles.

Fig. 5. Sheet resistance of As-implanted Si as a function of laser energy density.

laser energy density. We assume this depth to be the maximum melt depth.

Fig.5 gives the values of the sheet resistance after laser irradiation as measured with a four-point probe. The steep decrease in resistance around 0.2 J cm^{-2} corresponds with the (explosive) crystallization of amorphous material as indicated by the steep increase in melt depth around 0.2 J cm^{-2} in Fig.4. Between 0.3 J cm^{-2} and 0.9 J cm^{-2} there is a gradual decrease in resistance. We suggest that the first part of this decrease corresponds with the penetration of the primary melt into the amorphized layer, which results in an increase of the average grain size in the top (primary melted) layer. Once the primary front has reached the (a-Si)-(c-Si) interface, it does not penetrate further, but the temperature of the undercooled liquid rises until it reaches the melting temperature of c-Si. This temperature rise results in lowering of the crystallization velocity and hence, a further increase in average grain size. Between 0.9 and 1.0 J cm^{-2}, epitaxial recrystallization starts and the sheet resistance reaches a value which remains almost the same when irradiation is performed at higher energy densities.

CONCLUSIONS

We measured the redistribution of Cu and As as a function of laser energy density. From the concentration profiles after irradiation we conclude that besides a surface melt, there is a second, self-propagating melt which moves through the amorphized layer towards the (a-Si)-(c-Si) interface. Sheet resistance measurements on As-implanted Si are consistent with this crystallization model. From the changes in the impurity profiles we have derived quantitative information on the primary and secondary melt depths.

ACKNOWLEDGEMENTS

The authors wish to thank J. Politiek and R.J. Lustig of the Philips Research Laboratories for the Cu implantations, S. Doorn for the As implantations and M.O. Thompson and J.W. Mayer for fruitful discussions.

This work is part of the research program of FOM and is financially supported by ZWO.

REFERENCES

1. G.L.Olson, J.A.Roth, L.D.Hess and J.Narayan, Proc.of the US-Japan Seminar on S.P.E. and Interface Kinetics (1983) and references therein.
2. P.Baeri, G.Foti, J.M.Poate and A.G.Cullis, Phys.Rev.Lett. 45, 2036 (1980).
3. M.O.Thompson, G.J.Galvin, J.W.Mayer, P.S.Peercy, J.M.Poate, D.C.Jacobson, A.G.Cullis and N.G.Chew, Phys.Rev.Lett. 52, 2360 (1984).
4. H.C.Webber, A.G.Cullis and N.G.Chew, Appl.Phys.Lett. 43, 669 (1983).
5. J.Narayan, C.W.White, O.W.Holland and M.J.Aziz, J.Appl.Phys. 56, 1821 (1984).
6. D.H.Lowndes, R.F.Wood and J.Narayan, Phys.Rev.Lett. 52, 561 (1984).
7. R.F.Wood, D.H.Lowndes and J.Narayan, Appl.Phys.Lett. 44, 770 (1984).
8. J.Narayan and C.W.White, Appl.Phys.Lett. 44, 35 (1984).
9. P.Baeri, S.U.Campisano, G.Foti and E.Rimini, Phys.Rev.Lett. 41, 1246 (1978).
10. M.Miyao et al., J.Appl.Phys. 51, 4139 (1980).
11. P.Baeri and S.U.Campisano in Laser Annealing of Semiconductors - ed. by J.M.Poate and J.W.Mayer, Academic Press 1982, p.500.
12. W.K.Chu, J.W.Mayer and M.A.Nicolet, Backscattering Spectrometry, Academic Press, New York 1978.
13. See ref.11, p.75.
14. B.G.Bagley and H.S.Chen in Laser-Solid Interactions and Laser Processing - ed. by S.D.Ferris et al., A.I.P.Conf.Proceedings no.50 (1979) p.97.
15. F.Spaepen and D.Turnbull, ibid 50, 73 (1979).
16. See ref.11, p.32.

RELATION BETWEEN TEMPERATURE AND SOLIDIFICATION VELOCITY
IN RAPIDLY COOLED LIQUID SILICON

M. O. THOMPSON*, P. H. BUCKSBAUM**, AND J. BOKOR***
* Cornell University, Ithaca, NY 14853
** AT&T Bell Laboratories, Murray Hill, NJ 07974
*** AT&T Bell Laboratories, Holmdel, NJ 07733

ABSTRACT

A semi-empirical method to determine the undercooling-velocity relationship for laser induced melting is presented. The technique uses measurements of melt depth versus time to control numerical simulations, resulting in a map of the interface temperature as a function of time, and consequently as a function of the interface velocity. The results are independent of any model for the velocity-undercooling relationship. Results of the technique on simulated and experimental melt depth data are presented. Transient conductance data on 28 nanosecond 694 nm laser irradiation of silicon indicate an undercooling-velocity slope of 17$\pm$3 $K/(m/sec)$ near the melting point. Picosecond optical transmission data show a much smaller slope.

INTRODUCTION

Pulsed laser melting produces extreme thermal conditions that permit the study of phase transformations far from equilibrium. Of particular interest is the velocity of the solid-liquid interface when its temperature is elevated or depressed significantly from the equilibrium transition temperature T_m. The interface motion is governed simultaneously by molecular kinetics across the interface, and by the thermal gradients established during the irradiation. Under steady state conditions, the interface motion satisfies $v\Delta H = \kappa(\partial T/\partial z)_i$, where v is the interface velocity, ΔH the latent heat of melting, κ the thermal conductivity at the interface and $(\partial T/\partial z)_i$ is the temperature gradient at the interface. If the latent heat released by the interface velocity, $v\Delta H$, fails to equal the heat flow required by thermal conduction, the interface temperature will increase or decrease until the steady state condition is established. Assuming that the thermodynamic potentials and temperature can be validly defined, general kinetic arguments predict a simple relationship between the velocity of the liquid-solid interface and its temperature[1]:

$$v = f\lambda\omega \ e^{-\Delta Q/RT_i}[1 - e^{-\Delta G/RT_i}]. \tag{1}$$

Here f is a site fraction, ω is an attempt frequency, λ is an average jump distance, ΔQ is an activation energy, and ΔG is the change in Gibbs free energy during the phase change. At small undercoolings, the thermodynamic driving force can be approximated as $\Delta G = (1-T_i/T_m)\Delta H$. Likewise, the product $\lambda\omega$ can be approximated as the speed of sound in the liquid, v_s. Thus, for small undercoolings, the interface velocity is approximately linear with the undercooling $\Delta T = T_i - T_m$, and with slope:

$$\frac{v}{\Delta T} = fv_s \ e^{-\Delta Q/RT_i} \frac{\Delta H}{RT_i T_m}. \tag{2}$$

Using a sound velocity of 3000 m/sec and ΔH of 50530 $J/mole$, the interface velocity near T_m becomes:

$$v(m/sec) = 6.4 f \ e^{-\Delta Q/RT_i} \Delta T. \tag{3}$$

While the thermodynamic parameters are well known, values for the site fraction and the activation are essentially unknown. Limits on the values of f and ΔQ can be obtained from experimental measurements of the undercooling-velocity relationship.

Early numerical models for pulsed laser melting ignored the interfacial kinetics by assuming that melt and resolidification occurred at the fixed melting temperature T_m[2]. These simple models, however, adequately matched the early available experimental data of melt duration and impurity diffusion[8] since the expected superheating and undercooling are relatively small. As increasing accurate direct experiments of interface depth vs. velocity were reported, discrepancies between this simple model and data have become apparent. In experiments using picosecond laser pulses, the undercooling of the interface during solid phase regrowth is so large that this simple approximation is no longer adequate. More recent models have attempted to address the discrepancies by including either a linear relationship between the velocity and interface temperature, or by attempting to estimate the parameters in the full kinetic expression of eqn.1[4]. Although these have produced more satisfactory results, fits of numerical simulations to experimental data are relatively insensitive to the exact form of the undercooling equation. This difficulty arises because the interface temperature changes rapidly during a relatively brief part of the melt history, ie. during melting and near the peak of the melt depth. Comparisons of the experimental and numerical data are biased by the long regrowth regions, where the temperature is relatively constant and steady state conditions have been established. Deviations here are caused more by uncertainties in the thermal properties than in the undercooling-velocity relationship.

Several experimental groups have attempted to measure the interface temperature directly during pulsed melting. Of particular note are the synchrotron measurements by Larson et.al.[5] and direct thermocouple measurements by Campisano et.al.[6] Both experiments, however, are extremely difficult and uncertainties in the results prevent direct calculation of the undercooling-velocity relationship. Although reliable subnanosecond thermometry does not yet exist, the interface temperature may by estimated from careful measurements of interface position[7].

In this paper, we describe a numerical deconvolution algorithm for interface temperature, and present results obtained from experimental data. The procedure has been successful for both nanosecond and picosecond melting experiments, but gives different results in the two different time regimes. The discrepancy, if confirmed by future analysis and more experiments, may indicate interesting transient behavior at the solid liquid interface. The deconvolution is numerically stable for nanosecond melting of Si and yields a average slope for the kinetic equation (eqn.3) of $0.059 \ (m/sec)/K$ or $17 \pm 3 \ K/(m/sec)$. Picosecond data show a slope of $4 \pm 2 \ K/(m/sec)$.

NUMERICAL

Almost all numerical algorithms for the simulation of laser melting processes are based on the parabolic differential equation for heat transport:

$$C_p(T)\left[\frac{\partial T(z,t)}{\partial t}\right] = \frac{\partial}{\partial z}\left[\kappa(T)\left[\frac{\partial T(z,t)}{\partial z}\right]\right] + \alpha(z,t)I(z,t) + v\Delta H \tag{4}$$

$C_p(T)$ is the specific heat, $\kappa(T)$ is the thermal conductivity, $I(z,t)$ is the laser intensity as a function of depth in the sample and $\alpha(z,t)$ is the absorption coefficient for the laser energy. The differential equation is approximated as a difference equation in time and space and is numerically integrated to determine the sample temperature as a function of depth and time[8].

The numerical algorithm for a deconvolution of melt depth to interface temperature is essentially identical, with one exception, to the numerical codes for the direct simulation of melt depth. Both codes require the optical and thermal properties of the sample to be well known functions of temperature and time. However, in the direct simulation of the melt, the energy source term $v\Delta H$ is determined from a model equation of the interface-velocity relationship. Indeed, in the early calculations, the source term was determined by requiring that the interface temperature remain at the equilibrium melting point, and thus implying the interface velocity. In the deconvolution algorithm, the energy source $v\Delta H$ is obtained directly from the experimental measurement of melt depth versus time. In essence, the simulation is forced to follow the melt history determined from experiments, and the interface temperature is allowed to vary as necessary to satisfy the differential equation in eqn. 4. The output of the deconvolution consists, thus, of interface temperature as a function of time, or alternatively the interface temperature as a function of the experimentally determined interface velocity. Since the interface temperature can vary far more rapidly than melt depth, this deconvolution emphasizes any differences between the model calculations and the experimental data.

A standard spatial grid is imposed on the sample substrate. To reduce computation time, the grid spacing increases with depth, allowing fine spatial resolution near the surface and coarse resolution in the substrate. Simulations performed on silicon-on-sapphire (SOS) structures required continuity of the temperature and heat flow, $\kappa(\partial T/\partial z)$, across the silicon-sapphire interface. The liquid-solid interface is established at each time step in accordance with experimental data. Material properties are calculated based on the phase and temperature of each spatial grid. Since the temperature may exceed the thermodynamic range of each phase, the material parameters are analytically extrapolated from available data to all temperatures. The parameters used for all optical and thermal parameters are taken from the literature[9]. The liquid thermal conductivity was assumed to be 1.4 $W/(cm\ K)$ and a minimum absorption coefficient of 11,000 cm^{-1} for SOS was used to account for the defects in the Si film[10].

To test the deconvolution algorithm, a simulated melt depth profile was generated using the standard simulations with a linear undercooling-velocity relationship. The deconvolution algorithm accurately determined the interface temperature and reproduced the linear temperature-velocity relationship. The algorithm proved to be sensitive to the input experimental parameters, especially the incident laser energy and the exact timing of the laser pulse. These parameters can be iterated to the proper value by requiring that melting occur when the surface reaches the melting point, that the interface temperature pass through the melting point when the interface velocity passes through zero, and that the velocity-temperature curve be single valued. For 30 ns FWHM laser irradiation, the deconvolution is sensitive to incident energy errors of ±1% and pulse positions of ±1 ns. Despite the sensitivity, the shape of the resulting undercooling-velocity curve does not change significantly for large deviations in energy or pulse position. These changes only produce a shift of the temperature where the interface velocity passes through zero.

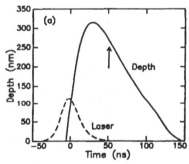

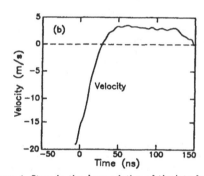

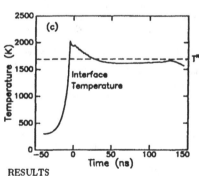

Figure 1. Steps in the deconvolution of the interface temperature from melt depth measurements. a) The melt depth and incident laser pulse as a function of time. The vertical arrow indicates the point at which steady state resolidification is reached. b) Interface velocity determined from a. c) Calculated interface temperature as a function of time from the data in b. The dashed line is drawn at the equilibrium melting temperature of Si.

RESULTS

Figure 1a shows an experimental melt depth versus time trace, determined from transient conductance[11], for 28 ns FWHM 694 nm irradiation of 520 nm SOS, with the corresponding laser pulse shown as the dashed curve. In fig. 1b, the interface velocity, determined by numerically differentiating the data in fig. 1a, is shown. The majority of the interface velocity-temperature information is obtained from the melting and the initial solidification, as indicated by the dotted lines in fig. 1a. The results of the interface temperature obtained from the deconvolution are shown in fig. 1c for an incident energy density of 1.614 J/cm^2 and a laser pulse offset of −0.6 ns. This figure shows the interface temperature passing through normal melting point at the peak of the melt depth. Since heat input from the laser tail is an important source of thermal energy, the deconvolution algorithm used the measured laser intensity. During the majority of the regrowth time, the interface temperature is at steady state, increasing slightly as the interface velocity decreases due to heat flow into the substrate.

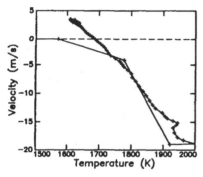

Figure 2. The interface velocity is plotted against the interface temperature from fig. 1c. Near zero velocity, the average slope is 17 K/(m/s). For negative velocities (melting), each point corresponds to a time step of 1 ns.

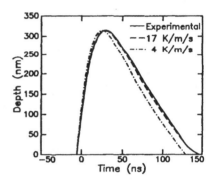

Figure 3. Comparison of the experimental data (solid curve) and calculations using a linear undercooling velocity relationship with slope of 17 K/(m/s) (dashed curve). In the region of the peak melt depth, the experimental data and the calculations agree extremely well. The dashed-dotted curve shows the discrepancy for the nanosecond data if an undercooling-velocity slope of 4 K/(m/s) is used instead.

This deconvolution technique is also extremely powerful in exhibiting differences between the material parameters used in the calculations and actual behavior of experimental data. For SOS calculations, the use of literature values of the thermal conductivity of polycrystalline sapphire for the SOS substrate resulted in the interface temperature continuing to rise during resolidification. This indicated that insufficient heat was being conducted into the sapphire substrate in the computer models. Increasing the high temperature thermal conductivity of the single crystal sapphire by 10% over the polycrystalline value resulted in a single-valued interface velocity versus undercooling curve. The need for this increase was suggested previously from indirect evidence based on the average resolidification velocity. In fig. 2, the interface temperature is replotted as a function of the interface velocity. The velocity-temperature relationship is approximately linear between -5 and 5 m/sec with a slope of 17 ± 3 $K/(m/sec)$. In the approximation of eqn.(3), this corresponds to a site fraction of 0.01 if the activation energy ΔQ is assumed to be zero, or an activation energy of 0.7 eV/atom assuming a site fraction of unity.

This undercooling-velocity slope of 17 K/(m/s) was used in a forward simulation to directly compare the results with the experimental data. Figure 3 shows the experimental data (solid curve) and the simulated data assuming a perfectly linear undercooling velocity relationship. In the region of interest for undercooling, ie. near the peak, the experimental and simulated data overlap extremely well. The differences in the regrowth result from the use of a linear undercooling while the experimental data shows a slight negative departure from linearity. For comparison, a simulation using a undercooling slope of 4 K/m/s is also shown in the figure as the dotted-dashed curve. The melt depth for this simulation rises more rapidly than the experimental data and turns over to resolidification more rapidly also.

The temperature of the solid-liquid interface produced in picosecond melting experiments has also been studied, using the same numerical procedure outlined above. Here, the experiments were performed on crystalline silicon substrates, using infrared optical transmission as a probe of depth of the liquid[4]. Each data set consists of between 1000 and 3000, 15 picosecond measurements of liquid depth before, during, and after irradiation with a 15 picosecond 248 nm laser pulse. In order to produce a stable numerical calculation, the data must be carefully smoothed to eliminate fluctuations due to noise in the experiment. Figure 4 shows the result of these smoothing operations on one data set. Note that the regime of greatest interest, near the maximum melt depth, is passed through very rapidly in the experiment. The interface acceleration in this area is much higher than for nanosecond data. The temperature range over the duration of the melt is also much greater. Although the velocity-undercooling relationship is not linear over the full range covered in picosecond melts, the value of the slope ($\Delta T/v$) evaluated at T_m, averaged over several different data sets, is $4.2K/(m/sec)$, approximately one-fourth as large as for the nanosecond SOS data (fig. 5). The analysis of the picosecond data is not yet complete, but the disagreement between the nanosecond and picosecond experiments has been seen at all levels of laser fluence.

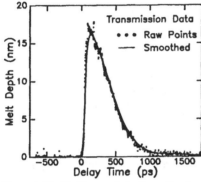

Figure 4. Liquid silicon melt depth vs. time for 15 psec, 248 nm melt pulses. This was obtained from approximately 1500 individual measurements of transmission through the liquid silicon layer by 15 psec pulses of 1.64 μm radiation. The smooth curve, which is the result of filtering this data with a 30 psec FWHM noise filter, was used in the numerical analysis.

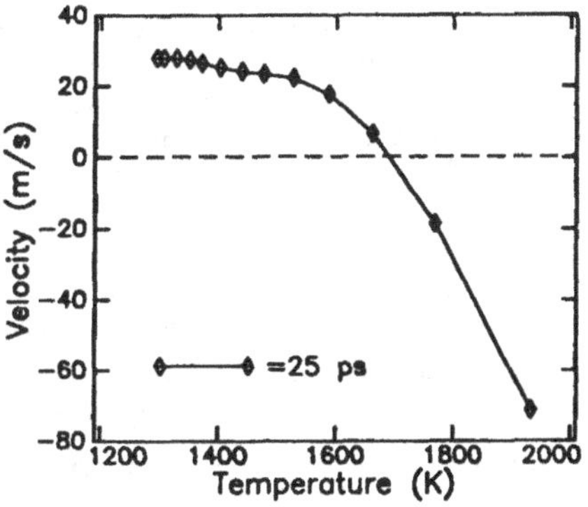

Figure 5. Result of numerical integration for data shown in fig. 4. Only the part of the data near the maximum melt depth is shown. The slope at $T = T_m$ is $4.2K/(m/sec)$. Points are separated by 25 psec.

The simple kinetic theory outlined above predicts a universal velocity-temperature relationship, independent of laser fluence, wavelength, temporal pulsewidth, or any other factors affecting the thermal gradients set up by the melting pulse. If the discrepancy is confirmed, it may point to a difference in the kinetics of the interface for picosecond and nanosecond laser melts. For example, interface roughness, which affects the regrowth site fraction and hence overall regrowth rate, may differ for the two regimes. The growth rate may also be affected by a less disordered liquid present in a melt which is only a few tens of picoseconds old. We cannot yet rule out measurement difficulties, however, such as an unexpectedly high free carrier absorption in the warm solid behind the melt at maximum melt depth.

CONCLUSION

We have developed a semi-empirical procedure for determining the temperature of the liquid-solid interface during pulsed laser melting. Time resolved melt depth measurements are used to control the position of the liquid-solid phase boundary in numerical heat diffusion simulations. This results in a calculation of the temperature of the interface as a function of time during the melt, and therefore produces a map of interface temperature vs. velocity. The stability of the procedure depends on high quality, low noise melting data. This analysis technique has been tested with silicon melting experiments, using both nanosecond data obtained by transient conductance and picosecond data measured by infrared transmission. Filtering of the noise resulting from differentiation of an experimental quantity was necessary to obtain stable solutions. Melts initiated by a 28 ns ruby laser pulse show 17_3K undercooling for each m/sec of interface motion. Picosecond data cover a much wider range of temperature, but near $T = T_m$, where nanosecond and picosecond data may by compared, numerical simulations of the latter show a much steeper temperature velocity relationship, $4\_2K/(m/sec)$. The origin of this discrepancy is not currently understood, and further investigations are underway.

We feel that these new techniques provide a superior means for interpreting the physics of the liquid-solid phase transition in pulsed laser melting experiments. Until direct, accurate time resolved interface temperature measurement techniques are developed, this analysis will be an important aid in determining the temperature of the moving liquid-solid interface.

This work was supported in part by an NSF Young Investigators Grant. (L. Toth)

REFERENCES

1. D. Turnbull and M.H. Cohen, in *Modern Aspects of the Vitreous State, Vol. 1*, J.D. MacKenzie, ed., (Butterworth, London, 1960) p. 38.

2. P. Baeri, S.U. Campisano, G. Foti and E. Rimini, J. Appl. Phys. *50*, 788 (1979); R.F. Wood and G.E. Giles, Phys. Rev. *B 23*, 2923 (1981).

3. J.M. Poate in *Laser and Electron Beam Interactions with Solids*, edited by B.R. Appleton and G.K. Celler, (North Holland, New York, 1982), p. 121; D.H. Lowndes, G.E. Jellison and R.F. Wood, Phys. Rev. *B 56* 6747 (1982).

4. P. H. Bucksbaum and J. Bokor, Phys. Rev. Lett. *53*, 182 (1984).

5. B. C. Larson, C. W. White, T. S. Noogle, J. F. Barhorst and D. M. Mills, in *Laser-Solid Interactions and Transient Thermal Processing of Materials*, edited by J. Narayan, W. L. Brown and R. A. Lemons, (North Holland, New York, 1983), p. 43.

6. S. U. Campisano, Appl. Phys. Lett. *45*, 398.

7. G. J. Galvin and P. S. Peercy, submitted Appl. Phys. Lett.

8. V. Vemuri and Walter J. Karplus, *Digital Computer Treatment of Partial Differential Equations* (Prentice-Hall, New Jersey, 1981).

9. *CRC Handbook of Chemistry and Physics* (Chemical Rubber Company, Boca Raton, 1980); *Thermal Conductivity of the Elements: A Comprehensive Review*, C. Y. Ho, R. W. Powell and P. E. Liley, J. Phys. Chem. Ref. Data *9* Supp. 1, 1-589 (1972). Optical constants are found in G.E. Jellison and F. A. Modine, Appl. Phys. Lett. *41*, 180 (1982); and D. Aspnes, A. A. Studna, and E. Kinstron, Phys. Rev. *B 23*, 768 (1984).

10. Michael O. Thompson, Ph.D. thesis, Cornell University, Ithaca, NY, 1984.

11. M. O. Thompson and G. J. Galvin, in *Laser-Solid Interactions and Transient Thermal Processing of Materials*, edited by J. Narayan, W. L. Brown and R. A. Lemons, (North Holland, New York, 1983), p. 57.

OVERHEATING AND UNDERCOOLING IN SILICON DURING PULSED-LASER IRRADIATION*

B. C. LARSON[+], J. Z. TISCHLER[+], AND D. M. MILLS[**]
[+]Solid State Division, Oak Ridge National Laboratory, Oak Ridge, TN 37831
[**]CHESS and Applied and Engineering Physics, Cornell University, Ithaca, NY 14853

ABSTRACT

We have used time-resolved x-ray diffraction measurements of thermal expansion induced strain to measure overheating and undercooling in <100> and <111> oriented silicon during pulsed laser melting and regrowth. 249 nm (KrF) excimer laser pulses of 1.2 J/cm² energy density and 25 ns FWHM were synchronized with x-ray pulses from the Cornell High Energy Synchrotron Source (CHESS) to carry out Bragg profile measurements with ±2 ns time resolution. Combined overheating and undercooling values of 120 ± 30 K and 45 ± 20 K were found for the <111> and <100> orientations, respectively, and these values have been used to obtain information on the limiting regrowth velocities for silicon.

INTRODUCTION

The use of short laser pulses has made it possible to carry out detailed studies of the physical processes associated with melting and solidification under large driving forces. The overheating and undercooling required for rapid melting and solidification has received considerable attention because of interest in the mechanism of amorphization in silicon.[1] Theoretical[2] and experimental[3-5] investigations have suggested 200—400 K undercooling for amorphization of silicon and limiting regrowth velocities of 13—30 m/s have been inferred for the <100> direction. In this study we have used time-resolved x-ray diffraction measurements of thermal expansion induced strain to directly measure interfacial overheating and undercooling in silicon during melting at ~6 m/s.

THEORY

The liquid-solid interface velocity during the regrowth phase of pulsed laser irradiation is governed by heat flow into the bulk of the crystal. Because of the very large melting enthalpy of silicon, the melt front velocity is governed by the crystal's ability to conduct the heat of crystallization at the liquid-solid interface into the bulk. Assuming a negligible temperature gradient in the liquid during regrowth, the interface velocity, V, is given by

$$V = -K\nabla T/(\rho \Delta H_m) \qquad (1)$$

where K is the thermal conductivity, ∇T is the temperature gradient in the solid at the interface, ρ is the density, and ΔH_m is the melting enthalpy. Since K is isotropic for cubic crystals, the regrowth velocity depends only on the gradient and is independent of crystal orientation.

*Research sponsored by the Division of Materials Sciences, U.S. Department of Energy under contract DE-AC05-840R21400 with Martin Marietta Energy Systems, Inc.

From the kinetic standpoint, the melt-front velocity can be written in microscopic terms as[6]

$$V = K_i d_{hkl} f_{hkl} (1 - e^{-\Delta G_c / k_B T}) \qquad (2)$$

where K_i is the reaction rate of liquid phase atoms, f_{hkl} is the fraction of surface sites available for crystal growth in the <hkl> direction, and d_{hkl} is the interface advance associated with one layer of growth. ΔG_c is the free energy associated with the thermodynamic driving force for crystal growth, k_B is Boltzman's constant, and T is the temperature. Assuming that ΔH_m is not temperature dependent, ΔG_c can be written in terms of interfacial undercooling, ΔT, as

$$\Delta G_c = \Delta H_m \; \Delta T / T_M \qquad (3)$$

where T_M is the equilibrium melting point of the crystal.

For large undercooling the exponential in Eq. (2) vanishes, yielding the limiting velocity

$$V_o{}^{hkl} = K_i d_{hkl} f_{hkl}, \qquad (4)$$

which carries with it microscopic information about the mechanism of crystal growth. However, little is known about V_o since K_i and f_{hkl} are difficult to specify and are expected to be temperature dependent as well.

Equations (1,2,3) can be combined to obtain

$$-K \nabla T / (\rho \Delta H_m) = V_o^{hkl}(T) \; (1 - e^{-\Delta H_m \Delta T / k_B T T_M}) \qquad (5)$$

which indicates that information about $V_o^{hkl}(T)$ can be determined from measurements of ∇T and ΔT. In addition, measurements of ΔT on different orientations provide information on the hkl dependence of $V_o{}^{hkl}$.

Time-resolved x-ray Bragg profile measurements of thermal expansion induced strain provide a means of determining ∇T and ΔT during pulsed laser irradiation. A lattice under thermal strain ε scatters x-rays at an angle $\Delta\theta$ relative to the normal Bragg angle, θ_B, according to the relationship

$$\varepsilon = -ctg(\theta_B) \Delta\theta \qquad (6)$$

and the strain ε, in turn, defines the temperature, T, through

$$\varepsilon = \int_{T_0}^{T} \alpha(T) \; \eta \; dT \qquad (7)$$

where T_0 is the ambient temperature, $\alpha(T)$ is the thermal expansion coefficient, and η is the enhancement due to the one-dimensionality of the strain. The interface temperature and the temperature distribution are obtained from the Bragg reflection profiles by fitting the x-ray intensity to scattering calculations for thermally strained crystals as described elsewhere [7].

EXPERIMENT

This experiment was performed using the pulsed time-structure of the Cornell High Energy Synchrotron Source (CHESS). Time-resolved x-ray Bragg profile measurements were made by synchronizing 1.2 J/cm^2, 249 nm excimer laser pulses with the arrival time of ~0.15 ns x-ray pulses as described

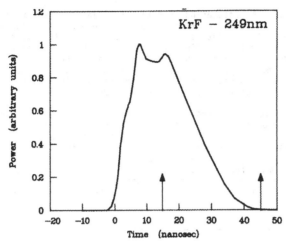

Fig. 1. Time structure of the excimer laser pulses used in this experiment.

previously.[7] The Bragg profile measurements provided time-resolved temperature distributions for the crystal (below the liquid-solid interface) through an analysis of the scattering in terms of thermal expansion induced strain. These Bragg profiles were measured with a time resolution of ±2 ns at times of 15 and 45 ns relative to the beginning of the laser pulses as shown in Fig. 1. The times were chosen to coincide with the maximum melt-in velocity and the maximum regrowth velocity as determined by numerical heat-flow calculations.

Measurements were performed on the (111) reflection of <111> oriented silicon and on the (400) reflection of <100> oriented silicon in a step scanning mode taking 18 laser shots at each angular setting. All measurements in a particular scan were made at the same position on the crystal; repeating measurements at selected angles after a scan showed no evidence of changes induced in the silicon as a result of the repeated laser shots.

RESULTS

The x-ray scattering profiles measured at the (111) reflection are shown in Fig. 2. The measured crystal reflectivity is reported as a function of the angular deviation, $\Delta\theta$, from the (111) Bragg angle, and the

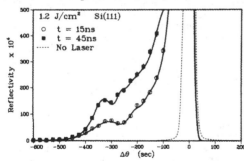

Fig. 2. Time-resolved Bragg profile measurements made at the (111) reflection of silicon during pulsed laser irradiation.

dotted line shows the scattering profile for no laser irradiation. X-ray
scattering at large negative $\Delta\theta$ indicates high temperatures (large thermal
strain); in general, the intensity as a function of $\Delta\theta$ is inversely proportional to the temperature gradients. Therefore, the higher intensity for
the 45 ns measurements signifies lower gradients during the regrowth phase
than during the melt-in phase. The solid lines represent least squares
fits of scattering calculations (for a thermally strained lattice) to the
data, where the temperatures as a function of depth were the fitting parameters.

The fitted temperature profiles are shown in Fig. 3 with the depth
scale measured from the liquid-solid interface; the thin (molten) liquid
layer produces no Bragg scattering and is essentially transparent to the x-
rays. As a result of overheating and undercooling, the interface temperatures during melting and regrowth are not the same; it can be seen that
the interface temperature measured during the melt-in phase (at 15 ns) was
120 ± 30 K higher than the interface temperature measured during the
regrowth phase (at 45 ns). This 120 K difference represents the sum of the
overheating and undercooling for <111> oriented silicon under these laser
conditions.

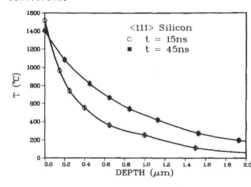

Fig. 3. Time-resolved lattice temperature profiles
for <111> oriented silicon
determined by fitting the
measured data in Fig. 1.

Figure 4 shows x-ray measurements made on <100> silicon using the (400)
reflection. The laser conditions and measuring times for these measurements were the same as those used in the (111) case. The solid lines again
represent fits to the Bragg profile measurements, and the temperature profiles producing these fits are shown in Fig. 5. These temperature profiles
are quite similar to those found for the <111> orientation (Fig. 3) with
the exception that the interface temperatures during melting and regrowth
differ by only 45 ± 20 K.

These results imply that a smaller driving force is required for
melting and regrowth on <100> oriented silicon than on <111> oriented silicon, and according to Eq. (5) they indicate that $V_0{}^{100} > V_0{}^{111}$. Using
Eq. (1) and the interface temperature gradients during regrowth (45 ns) in
Figs. 3 and 5, a regrowth velocity of ~7.5 m/s is obtained. The melting
velocity cannot be obtained from the 15 ns temperature profiles because
heat flow into the solid does not determine the rate of melting. Numerical
heat-flow calculations made for these conditions[8] predict a 10 m/s
melt-in velocity and a 6 m/s regrowth velocity. The experimental and
calculated regrowth velocities are, therefore, in reasonably good
agreement.

If the undercooling parts of the measured 120 K and 45 K temperature
differences could be determined accurately, $V_0{}^{hkl}$ could be determined for
the <111> and <100> orientations using Eq. (5). However, the uncertainties
in the temperature dependence of $\alpha(T)$ and n are such that absolute temperatures at the melting point may not be reliable to better than ±50 K.
The fact that the <100> regrowth temperature appears to be above the 1412°C

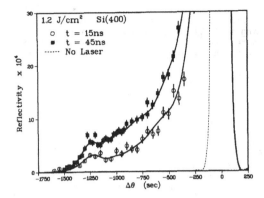

Fig. 4. Time-resolved Bragg profile measurements made at the (400) reflection during pulsed laser irradiation.

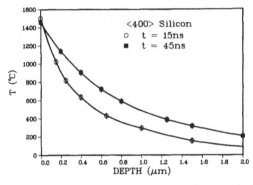

Fig. 5. Time-resolved lattice temperature profiles for <100> oriented silicon determined by fitting the measured data in Fig. 4.

equilibrium melting temperature bears out this uncertainty; regrowth cannot occur above the melting point. Fortunately the relative temperatures at the interface can be determined more accurately; the differences between the interface temperatures depend mainly on the fitting of the measured data.

Although it is not possible to make a direct determination of V_0^{hkl} for the regrowth process without specific knowledge of the undercooling, conditional values for V_0^{hkl} can be obtained by (1) assuming the overheating and undercooling to be prorated according to the melting and regrowth velocities, or (2) assuming the overheating to be negligible and ascribing all of the measured temperature differences to undercooling during regrowth. Using Eq. (5) and $\Delta T_{<111>} = 120$ K and $\Delta T_{<100>} = 45$ k we obtain V_0^{hkl} as indicated in Table I.

Table I. Limiting regrowth velocities inferred from the interface temperature differences on <111> and <100> silicon.

Assumption	V_0^{111} (m/s)	V_0^{100} (m/s)
Prorated as velocities	65 ± 16	195 ± 90
No overheating	36 ± 9	110 ± 50

Prorating the overheating and undercooling as the melting and regrowth velocities corresponds to assuming that V_0^{hkl} is the same for overheating and undercooling, and the alternate assumption of no overheating results in the minimum possible V_0^{hkl} for regrowth that would be consistent with these measurements.

DISCUSSION

The results of this study indicate that relatively small driving forces are required to melt silicon at 10 m/s and regrow it at 6 m/s. If the interface temperature differences are prorated for melting and regrowth, the undercooling at 6 m/s is 45 and 20 K on the <111> and <100> orientations, respectively. While the uncertainties in these values are quite large as a result of difficulties in measuring these small differences in temperature, large undercooling and overheating can apparently be ruled out for these conditions.

Comparing the results obtained here with theoretical calculations, we find that the ~2 times larger driving force calculated by Gilmer[2] for <111> regrowth compared to that for <100> regrowth is consistent with our results. However, the magnitudes of the driving force are quite different; Gilmer deduced that $V_0^{100} = 13.5$ m/s from segregation coefficient results, while we have found V_0^{100} to be >100 m/s (see Table I).

Bucksbaum and Bokor[3] inferred values of V_0^{hkl} of 25—27 m/s from their time-resolved measurements of laser melting and amorphization using picosecond laser pulses. Such low values seem to be inconsistent with the x-ray measurements made in the present study; however, Bucksbaum and Bokor's measurements were made in an entirely different regime, where amorphous silicon is formed and where heat-flow simulations implied an undercooling up to 700 K. Therefore, before a direct comparison of the results of the two experiments can be made, the temperture dependence of $V_0^{hkl}(T)$ must be considered.

Undercooling measurements are needed at higher velocities to directly relate our results to the amorphization of silicon at regrowth velocities >12 m/s; however, the results presented suggest that a non-linear undercooling versus velocity relationship is needed to support the 200—300 K undercooling proposed for reaching the liquid-amorphous transition temperature and the much larger undercoolings suggested[3] for even faster regrowth. Unfortunately, the short laser pulses needed to achieve faster regrowth velocity were not available in this experiment.

REFERENCES

[1] W. L. Brown, p. 9 in Energy Beam-Solid Interactions and Transient Thermal Processing, ed. by J. C. C. Fan and N. M. Johnson, North-Holland, New York (1984).
[2] G. H. Gilmer, p. 249 in Laser Solid Interactions and Transient Thermal Processing of Materials, ed. by J. Narayan, W. L. Brown, and R. A. Lemons, North-Holland, New York (1983).
[3] P. H. Bucksbaum and J. Bokor, Phys. Rev. Lett. 52, 182 (1984).
[4] M. O. Thompson, G. J. Galvin, J. W. Mayer, P. S. Peercy, J. M. Poate, D. C. Jacobson, A. G. Cullis, and N. G. Chew, Phys. Rev. Lett. 52, 2360 (1984).
[5] R. F. Wood, D. H. Lowndes, and J. Narayan, p. 141 in Ref. [1].
[6] F. Spaepen and D. Turnbull, p. 93 in Laser and Electron Beam Processing of Materials, ed. by C. W. White and P. S. Peercy, Academic Press, New York (1980).
[7] B. C. Larson, C. W. White, T. S. Noggle, J. F. Barhorst, and D. M. Mills, Appl. Phys. Lett. 42, 282 (1983).
[8] M. J. Aziz, private communication of results using the heat-flow program of M. O. Thompson.

LASER-BEAM-GLASS INTERACTION INDUCING
GLASS-CRYSTALLINE TRANSFORMATIONS

E. HARO[*] AND M. BALKANSKI[*]G. P. ESPINOSA[**] AND J. C. PHILLIPS[**]
*E. Haro and M. Balkanski Laboratoire de Physique des Solides, Associe au
C.N.R.S., Universite Pierre et Marie Curie, 4, Place Jussieu, 75230 Paris
Cedex 05, France
G. P. Espinosa and J. C. Phillips AT&T Bell Laboratories, Murray Hill, New
Jersey 07974, U.S.A.

INTRODUCTION

In recent years $GeSe_2$ glass has been intensively studied by different
methods. Light scattering experiments [1-10] have been used to investigate the molecular structure of the chalcogenide glasses. Diffraction
studies [16-17] have shown that the short range order of this glass is
similar to that of the crystalline $GeSe_2$. Raman [9-10] and Mössbauer [11-
15] spectroscopies have extended these results to include large number of
atoms (intermediate range order). Glass-crystalline transition induced by
laser irradiation has been investigated essentially by Raman scattering
[10-16].
    We present here an investigation of the glass-crystalline transition
in $GeSe_2$ glass induced by laser irradiation and studied by light
scattering. The results presented here show that the crystallization
kinetics depends on the irradiation power and the irradiation time.
    Two different methods of inducing crystallization are used to
investigate the dependence of the crystallization of the irradiation power
and or the irradiation time.
    With the first method structural changes are induced progressively by
slowly increasing the irradiation power. Using this procedure, one
observes three different steps in the glass-crystalline transformation.
The first one is characterized by a premisor effect indicating cluster
enlargement. The second is characterized by an intensity increase effect
indicating the transformation of clusters into microcrystallites. In the
third stage crystallites coalesce to form a crystalline material.
    With the second method the irradiation is delivered to the system by
high power pulses of short duration. In contrast with the "continuous"
irradiation method the precursor effect is not observed in this case.
However the final crystalline state obtained by both methods are
identical.

NONEQUILIBRIUM KINETICS

Initially the sample is irradiated using a very low power level i.e.,
the minimum permitting to obtain the Raman spectrum of the glassy state.
The irradiation power is then increased until a change is detected in the
spectrum. For low irradiation power crystallization does not take place,
even for long irradiation times (3 hours). It is necessary to reach a
threshold power level of irradiation in order to initiate crystallization. This is shown in Figure 1. At the threshold power level the
Raman spectrum shape changes as a function of time, eventually reaching an
equilibrium shape, as shown in Figure 2. Increasing the irradiation power
level beyond this point will cause drastic effects on the Raman spectrum
as shown in Figure 2.
    In Figure 1 the light scattering intensity for the peaks at 200 and
210 cm^{-1} is plotted as a function of the irradiation power. Up to a
certain power density the intensity of the peak changes very smoothly.
The system remains in a nonequilibrium state and is reversible. Above the

threshold power the intensity rises abruptly, the system undergoes an irreversible transition and reaches a much more nearly equilibrated state after a time of irradiation t_2. The intensity at the time t_1 is that of the spectrum taken immediately after the power density has been changed. The intensity at time t_2 is that of the spectrum taken after a time t_2 is that of the spectrum taken after a time t_2 beyond which further changes in the spectrum occur very slowly. The threshold energies for the narrowing and increase in intensity of the peak at 200 cm^{-1} P_{th}^{200} = 21 mW and that of the appearance and rapid increase in intensity of the peak at 210 cm^{-1} P_{th}^{210} = 25 mW are very close. This is understandable because the first is the precursor of the second and indicates the same trend in evolution. Neverthless, they correspond to two different stages of the evolution of the system. The first defines the nucleation and growth of clusters in a glass matrix determined by the increase of free volume and rotational degress of freedom in the frame of nonequilibrium kinetics. The second is characteristic of the coalescence of microcrystallites, randomly embedded in the glass matrix, into larger crystallites defining the initiation and growth of the crystalline phase which is now irreversible and tends toward an essentially stable state.

The spectrum in Figure 2a is recorded with an energy beam of low irradiation power, well below the threshold energy density, this spectrum reveals the glass structure.

In Figure 2b the irradiation power is very close to the threshold value P_{th}^{200} the spectrum is recorded at this density at a time t_1, immediately after the irradiation power has been changed.

The spectrum recorded at the same conditions after a time t_1' is presented in Figure 2c. A slight increase in the intensity and a narrowing of the 200 cm^{-1} band is clearly observed.

If the spectrum is recorded again under the same conditions but at a time t_2' beyond which changes in the spectrum occur very slowly, one obtains the spectrum presented in 2d. In this spectrum the broad band at 200 cm^{-1} has turned into a peak. This is the precursor effect, announcing the appearance of the crystalline mode at 210 cm^{-1}.

Further increase of the irradiation power leads to the spectrum presented in Figure 2e recorded at time t_1. The spectrum recorded at a time t_2' is shown in Figure 2f.

In this spectra the 210 cm^{-1} crystalline mode peak increases. At this point we have passed the power threshold value for the crystalline peak P_{th}^{210}, If the irradiation power is still increased the crystalline peak will grow at the expense of the cluster mode at 200 cm^{-1}. This process will continue until obtaining the spectrum characteristic of the crystalline state.

The Raman spectra presented in Figure 2 were recorded at the same beam power at which the sample was irradiated.

In Figure 3 are shown comparatively spectra recorded at a reference recording power, that does not induce structural changes in the sample. Starting from this reference irradiation power which is then increased progressively. The spectrum is recorded at the reference energy density as soon as shanges are observed. The operation is repeated until obtaining the crystalline peak.

PRECURSOR EFFECT

Figure 3 represents a succession of light scattering spectra taken in the following sequence: 3a is the reference spectrum for the glass state. The spectrum is recorded with a probe beam whose power is P_R = 9 mW. Repeated measurements at this power level show that the Raman spectrum in the glass phase is not perturbed by the probe beam. Two broad bands are observed in this spectrum centered respectively at 200 cm^{-1} and 215 cm^{-1}. The width of the band at 200 cm^{-1} is: Γ_{200}^g = 20 cm^{-1}. The following

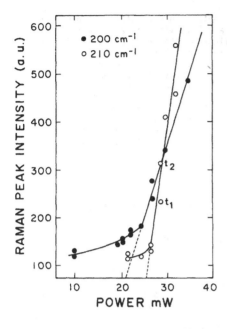

Fig. 1. Raman intensity vs. irradiation power showing two threshold powers: 22mW for cluster microcrystalline formation and 26 mW for crystallization

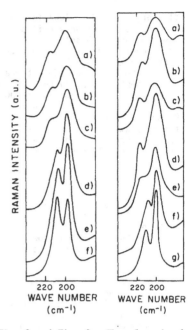

Fig. 2 and Fig. 3. The glass band at 200 cm^{-1} narrows and its peak height grows before the crystalline band at 200 cm^{-1} appears.

spectrum: 3b is the result of gradually increasing the power level until a change is detected in the spectrum, which first occurs at P_1 = 18 mW. The spectrum is recorded with a probe beam of P_R = 9 mW. There is practically no increase in the scattering strength of either of the two bands but one observes a slight narrowing of the 200 cm^{-1} band: $\Gamma_{200} \sim 18$ cm^{-1}. The next step is depicted in the spectrum 3c: the pump power now is P_1 = 23 mW and the probe beam remains at P_R = 9 mW. A dramatic increase and narrowing of the line at 200 cm^{-1} is observed. Here Γ^c_{200} = 4 cm^{-1} which is practically the linewidth of the normal mode of a crystal lattices. At the same time a new peak appears at 210 cm^{-1} (the crystalline value).

The important feature of this observation is the significant narrowing of the internal cluster mode at 200 cm^{-1} which signals two correlated effects: the enlargement of clusters and the precursor effect announcing the coalescence of clusters to form crystallites evidenced by the appearance of the crystalline normal mode peak at 210 cm^{-1}. The precursor effect is reasonable if we assume that at low energy densities the energy transferred to the system induces sufficient rotational freedom to nucleate clusters arising from the ordering of few molecular units. The broad band centered at 200 cm^{-1} characteristic of this stage corresponds to the envelope of the density of coherently scattering modes of the internal cluster vibrations.

As the input energy density increases clusters embedded in the glass become freer to rotate and undergo surface reconstruction. A large cluster, say 100 to 150Å in diameter, contains now a sufficient number of unit cells to define a Brillouin Zone and have vibrational branches obeying the $\vec{k} = 0$ selection rule for light scattering. Consequently the band moves towards a narrow spectral line which has the width of a Raman active normal mode $\Gamma_{200}^c = 4$ cm^{-1}. Hence the considerable narrowing from 20 cm^{-1} to 4 cm^{-1} of the internal mode A_1 demonstrates a significant enlargement of the cluster volume up to the formation of microcrystallites. The narrowing of the 200 cm^{-1} line, just before the crystalline normal mode at 210 cm^{-1} appears, constitutes a clear precursor effect. Precursor effects are characteristic of systems which are not in complete thermodynamic equilibrium, such as any crystal which contains domains. Here the precursor indicates that clusters are ready to fuse together forming the polycrystalline phase.

PULSED CRYSTALLIZATION

We have also investigated the effect, of a high irradiation power ($\simeq$ 100 mW) short time pulse (5 sec), on the structural transformation of GeSe$_2$ glass.
Once the laser beam power being fixed, the irradiation time is varied. All the Raman spectra are recorded at a low reference irradiation power.
An irradiation pulse of 100 mW and 5 sec length is sent onto the sample. After the pulse the Raman spectrum is recorded.
Results are presented in Figure 4. The most striking result for this type of inducing crystallization is the appearance and increase of the 210 cm^{-1} crystalline peak, without previous narrowing and increase of the 200 cm^{-1} band characterizing the precursor effect.
In Figure 4a the glass spectrum is obtained, after 25 sec of irradiation (5 x 5 sec) a slight change in the spectrum is observed.
After 16 pulses (90 sec) of irradiation the crystalline peak appears at 210 cm^{-1}. Note that in this spectrum the shape and intensity of the 200 cm^{-1} band has remain unchanged.
By further increase of the number of irradiation pulses one increases the intensity of the crystalline peak, until complete crystallization of the sample, as shown in spectrum. (4d and 4e)

DYNAMICAL REVERSAL EFFECT

The dynamical reversal effect shows two different types of relaxation dependent on the energy input level at which the system has been abandoned.
In Figure 5a we show a spectrum where the crystalline peak is clearly observable, if the irradiation is stopped, and the spectrum is recorded after 16 hours one obtains the spectrum presented in Figure 5B which is the spectrum of the original glassy state. The spectrum in Figure 5a has been obtained after 16 irradiational pulses of 5 l duration each.
One can obtain a similar result in the case of continuous crystallization if irradiation is stopped as soon as the precursor effect is detected.
Another type of relaxation is observed if one decreases the energy density gradually starting from the crystalline state. Results are shown in Figure 6.
When the irradiation power is progressively decreased, the intensity of the peak at 210 cm^{-1} decreases systematically and for low irradiation power $P_1 = 2.5$ mW, this intensity becomes comparable to the intensity of the peak at 200 cm^{-1}. The inverse transformation is illustrated in Figure 6.

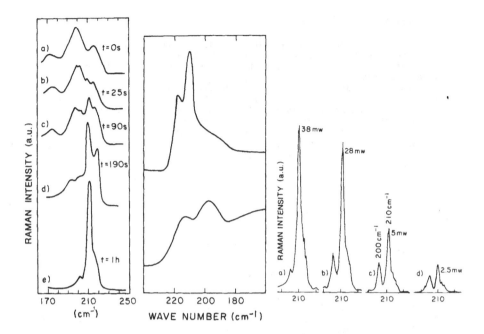

Fig. 4. Detailed dependence of crystallization on exposure time. Irradiation power 100 mW, recording power 38 mW.

Fig. 5. Complete revitrification, starting near threshold power for crystallization.

Fig. 6. Reversible partial revitrification with decreasing power level, starting from largely crystallized surface layer and power above threshold.

If the scattered intensity at 210 cm^{-1} corresponds to a certain density of crystallites whose size is d > 150 Å, a decrease of this intensity would have the meaning of a decrease in density of such crystallites. What is meaningful here is the ratio of the two peaks: 210 and 200 cm^{-1} attributed respectively to large crystallites and microcrystallites. Lowering the irradiation power reverses the trend of organization of the glass. Large crystallites disappear to the benefit of microcrystallites. Increasing the irradiation power leads to the glass to crystalline transformation. When this transformation is stopped at a certain level from which the irradiation laser power is now decreased, the system seems to reverse: the large crystallites are dissolved in the glass and the microcrystallites tend to dominate.

If one starts from a state at which crystallization has occurred and the peak at 210 cm^{-1} is the dominant feature of the Raman spectrum, the reversibility is towards a mixed state where crystallites and microcrystallites are both present.

CONCLUSION

The experiments discussed here demonstrate some of the possibilities for investigating structural transitions by Raman spectroscopy. A minimum irradiation power is defined for which the system is not perturbed by the laser irradiation.
Two methods are used to achieve crystallization by a laser beam. The first is continuous irradiation with stepwise changes in the laser power level. This spectra clearly reveal intermediate metastable states. The second method just described above is a time resolved procedure. The irradiation is applied in short intervals of time t and the recording of the Raman spectra is done after irradiation. A certain number of impulses are necessary to induce transformations analogous to that obtained by continuous irradiation. The duration of irradiation is a factor in the crystallization of glass. However, in this last method, the precursor effect is not observed.

REFERENCES

1. G. Lucovsky, F. L. Galeener, R. C. Keezer, R. H. Geils and H. A. Six, Phys. Rev. B10, 5134 (1974).
2. G. Lucovsky, F. L. Galeener, R. H. Geils and R. C. Keezer, in The Structure of Non-Crystalline Materials edited by H. Gaskell (Taylor and Francis, London 1977), p. 127.
3. R. J. Nemanich, S. A. Solin and G. Lucovsky, Solid State Commun. 21, 273 (1977) and references cited therein.
4. G. Lucovsky, R. J. Nemanich and F. L. Galeener, in Proceedings of the VIIth International Conference on Amorphous and Liquid Semiconductors, Edinburgh, 1977, edited by W. E. Spear (CICL, University of Edinburgh, Edinburgh, 1977) p. 130, and references cited therein.
5. G. Lucovsky, in Amorphous Semiconductors, edited by M. Brodsky (Springer, New York, 1979), vol. 36, p. 215.
6. A. Feltz, K. Zichmuller and G. Plaff, in ref. 10, p. 125.
7. R. J. Nemanich, G. A. N. Connell, T. M. Hayes and R. A. Street, Phys. Rev. B18, 6900 (1978); R. A. Street, R. J. Nemanich and G. A. N. Connell, Phys. Rev. B18, 6915 (1978).
8. K. Murase, T. Fukunaga, K. Yakushiji and I. Yunoki, J. Non-Cryst. Sol. 59/60 (1984).
9. P. M. Bridenbaugh, G. P. Espinosa, J. E. Griffiths, J. C. Phillips and J. P. Remeika, Phys. Rev. B20, 4140 (1979).
10. N. Kumagai, J. Shirafuji and Inuishi, J. Phys. Soc. Jpn. 42, 1261 (1977).
11. W. J. Bresser, P. Boolchand, P. Suranyi and J. P. de Neufville, Phys. Rev. Lett 46, 1689 (1981).
12. P. Boolchand, W. J. Bresser and M. Tenhover, Phys. REv. B25, 2971 (1978).
13. P. Boolchand, J. Grothaus, W. J. Bresser and P. Suranyi, Phys. Rev. B25, 2975 (1982).
14. P. Boolchand, H. Grothaus and J. C. Phillips, Solid State Commun. 45, 183 (1983).
15. M. Stevens, J. Grothaus, P. Boolchand and J. G. Hernandez, Solid State Commun. 47, 199 (1983).
16. J. E. Griffiths, G. P. Espinosa, J. P. Remeika and J. C. Phillips, Solid State Commun. 40, 1077 (1981); Phys. Rev. B25, 1272 (1982); J. E. Griffiths, G. P. Espinosa, J. C. Phikllips and J. P. Remeika, Phys. Rev. B28, 4444 (1983).
17. L. Cervinka and A. Hruby, in Amorphous and Liquid Semiconductors, eds. J. Stuke and W. Brenig (Taylor and Francis, London, 1974) p. 431.

LASER-INDUCED SURFACE RIPPLES:
WHAT IS UNDERSTOOD AND WHAT IS NOT

P.M. FAUCHET* and A.E. SIEGMAN**
*Department of Electrical Engineering and Computer Science, Princeton University, Princeton, NJ 08544
** Edward L. Ginzton Laboratory, Stanford University, Stanford, CA 94305

ABSTRACT

The most predominant ripple properties can be explained by existing perturbative theories. We present experimental results on novel types of ripples, which suggest that theory has to go beyond first-order perturbation and that the material's response is not well understood.

INTRODUCTION

It is now well known that relatively intense c.w. or pulsed laser beams can produce ordered surface structures on most materials. Although such laser-induced surface ripples had been observed for fifteen years, activity in the field started to peak up in the early 80's, when various groups published a detailed theoretical modelling of the electromagnetic interactions that lead to ripple's formation. We first describe and compare briefly the main features of some models and list their successes. Then, we show examples of nonstandard ripples which do not seem to be readily explained by the present theories. Finally, we suggest an explanation for some of these results and indicate how theory should be modified to account for these results.

THEORY

In all existing models, ripples result from interference between the incident beam and surface waves. Although it is believed that minute surface irregularities (of dimensions smaller than λ) are necessary for the production of the surface waves, there is a controversy on the exact nature of these waves. Among the candidates are surface polaritons [1] and radiation remnants [2]. In our theoretical work [3], we have described the interaction of the incident beam with surface diffracted waves on a corrugated surface (with a sinusoidal profile) or on a flat surface (where the dielectric function ε has a periodic sinusoidal variation).

The nature of the surface wave is revealed by applying the appropriate boundary conditions. After solving Maxwell's equations by first-order perturbation theory, we predict [4] ripples with a periodicity identical to that given by the surface polariton model, if Re $(\varepsilon) = \varepsilon_R$ is negative. For $\varepsilon_R > 0$, however, we find [5] that ripples can grow with a spacing Λ well described by $\lambda = \Lambda/n(1 \pm \sin\nu)$, where n is the refractive index of the incident medium (= 1 if air or vacuum) and ν is the angle of incidence. The ripple growth coefficient is then smaller than for metallic-like surfaces [4].

In all models, first-order perturbation theory describes the interaction (e.g., $h/\Lambda \ll 1$, where h is the height of the corrugation) and the properties of the substrate (e.g., crystalline orientation) play an insignificant role. Furthermore, theory has not combined the periodic variation of ε during illumination with the effect of surface corrugation. This variation is especially striking in the formation of aligned, coexisting liquid and solid lamellae during c.w. illumination [6]. Recently, permanent variations of ε have also been observed after U.V. illumination of metal films [7].

NONSTANDARD RIPPLES

Nonstandard ripples, which cannot readily be explained by an model, have recently been observed, especially during c.w. or repeated pulsed illumination. We concentrate on our results.

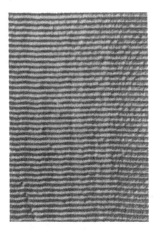 

Figure 1:   After forming deep ripples on Ge (left, $\Lambda = \lambda$), repeated illumination with normal-incident, vertically-polarized 1.06 μm pulses eventually produces nonstandard ripples, running at 90° and having the same spacing (right).

Once the aspect ratio (= h/Λ) of standard ripples becomes comparable to unity, additional illumination may lead to the formation of perpendicular ripples (running at 90° to the standard ripples). Figure 1 shows [8] typical results after repeated illumination of Ge above threshold for ripple formation. Here, the periodicity appears equal for the two types of ripples and the surface looks like a checkerboard. Often, the two spacings are different [9]. In some cases, instead of a regular pattern, we have observed [10] perpendicular stripes with a periodicity that is not very sharp and appears to vary with the intensity. Figure 2 shows such ripples on GaAs. In another experiment, we have observed [8] that repeated pulsed illumination of <111> Ge by normal-incident, circularly-polarized 1.06 μm light produced hexagonal patterns, reminescent of the crystal orientation, as shown in Figure 3. Others [11] have reported surface ripples with orientations at 45°.

Other observations are difficult to reconcile with theory. For example, repeated pulsed illumination with pulses well below melting threshold may eventually produce surface ripples [12] which, for an ion-implanted amorphized layer, appear as alternating stripes of recrystallized and amorphous material. In this case, however, the spacing and orientation are perfectly consistent with theory. All of these examples have some common features which clearly indicate the directions for future theoretical developments.

MODELS

It is quite clear that existing theories must fail for large aspect ratios. Our calculations [3] indicate that the modulation of absorbed power due to surface corrugation reaches 100% before $h/\Lambda = 1$. Then regions

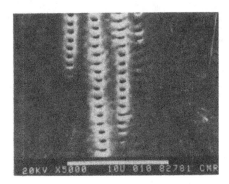

Figure 2: The superposition of standard ($\Lambda=\lambda$) and nonstandard ($\Lambda\neq\lambda$) ripples may lead to arrays of deep pits as shown in this SEM picture of GaAs after illumination with a normal-incident, vertically-polarized 1.06 μm beam.

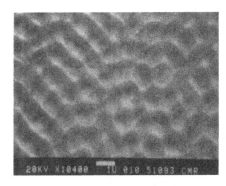

Figure 3: SEM pictures of hexagonal patterns produced on <111> Ge. Higher-resolution pictures indicate that the edges of the cells are quite sharp.

where the power flow is maximum alternate with regions where the power flow is zero. Consequently, interactions between different surface diffracted waves has to be considered. Furthermore, the dielectric function is also expected to vary strongly. Theory will break down where the magnitude of the surface wave becomes comparable to that of the incident beam as in the case of illumination of a surface with deep ripples. Since we have shown that the ripple's growth coefficient is larger for metallic-like surfaces ($\varepsilon_R < 0$), nonstandard ripples will also be produced more easily during U.V. laser-assisted growth of metal films. We have not developed our model beyond first-order perturbation theory. However, we have considered various mechanisms and attempted a phenomenological description of the surface interactions.

By analyzing the light diffraction pattern of the rippled surface illuminated by a probe beam at normal incidence, we may gain some insight to the surface wave mixing processes. The analysis was carried out in Ref. 8. If a screen is placed a distance d_o from the rippled surface (parallel to the x,y plane), we call $\vec{g}_T$ the total ripple wavevector normalized to the incident wavevector and r the ratio of the probe frequency to the excitation frequency. Then, the diffracted spot locus on the screen is given by

$$\frac{d_i}{d_o} = \frac{g_{Ti}}{\sqrt{r^2 - |g_T|^2}} \tag{1}$$

where $i = x$ or y. $\vec{g}_T$ describes a higher-order spatial frequency ripple, produced by interference among primary ripples $\vec{g}_k$ such that $\vec{g}_T = \Sigma \vec{g}_k$. Each primary ripple $\vec{g}$ obeys the ripple equation

$$(g_x \pm \sin\nu)^2 + g_y^2 = 1 \tag{2}$$

where (x,z) is the plane of incidence [13]. Such ripples may result from a nonlinear response of the surface when a weak ripple $\vec{g}_2$ begins to grow on a surface possessing a strong ripple $\vec{g}_1$. For example, Figure 4 shows the perfect fit between the observed and predicted patterns obtained from the hexagonal structures on <111> Ge. Here

$$\vec{g}_T = m \cos\left[\frac{l\pi}{3}\right] \hat{x} + m \sin\left[\frac{l\pi}{3}\right] \hat{y} + \vec{g}_o \tag{3}$$

where $m \in N$, $l = 0,1,\ldots,5$ and $\vec{g}_o$ is a primary ripple of random orientation. The figure shows diffraction patterns for $m = 0$, $m = 1$ and $m = 2$ (one branch only).

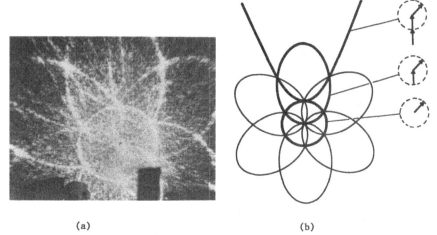

(a)   (b)

Figure 4: a) Hexagonal diffraction pattern from a <111> Ge surface similar to that in Fig. 3.
b) Simplified diffraction pattern predicted from Eq. (3). The reinforced diffraction traces are obtained by superposition of one, two or three ripples as shown in the inset.

This approach unfortunately does not shed much light on the nature of the surface interactions. A complete theory does not seem tractable. Clearly, the effect of the large aspect ratio of the ripples is important and is known to alter some results concerning the surface waves, obtained for small corrugation [14]. We are inclined to think that the surface should be considered as nonlinear, which also introduces new features in the surface waves [15]. Since power flow is very inhomogeneous, during illumination, the dielectric function ε of the solid could be written as

$$\varepsilon(x,y) = \varepsilon_o(x,y) + \varepsilon_2 |E(x,y)|^2 \tag{4}$$

where ε_0 is the dielectric function without illumination (which is already a periodic function in one direction for one type of primary ripple); ε_2 is the nonlinear coefficient and $|E(x,y)|^2$ is a periodic term proportional to the power flow. The surface waves which may exist at this interface are likely to differ from the usual solutions for a homogeneous interface [15].

The hexagonal structures reflect the symmetry of the substrate. It is conceivable that the six-fold anisotropy of the surface tension of <111> Ge is responsible for the observed hexagonal cells [16]. However, we present-ly have no firm evidence in favor of this hypothesis which raises many questions: for example, why is the hexagonal structure only observed for illumination by normal-incident, circularly-polarized light; why does the orientation of the pattern (as seen in diffraction) seem to be defined to $\pm$ 10 degrees only. The answer to these unresolved questions will come from a better description of the electromagnetic interactions and the material's response.

Finally, ripples produced by repeated subthreshold illumination can be explained by inhomogeneous energy deposition at the solid surface, leading to localized, periodic light-induced fatigue [17]. Here, the variation of ε produces the initial feedback mechanism. Absorption increases wherever the material yields, increasing the feedback until melting occurs at these sites. The result is the formation of ripples.

CONCLUSIONS

Although usually successful, existing models for laser-induced surface ripples cannot explain a significant fraction of the observed structures. Generalization of these models to account for the observations discussed here is difficult at best. We have indicated the directions that these future investigations should take.

We acknowledge support from AFOSR and IBM.

REFERENCES

1. S.R.J. Brueck and D.J. Ehrlich, Phys. Rev. Lett. 48, 1678 (1982).
2. J.E. Sipe, J.F. Young, J.S. Preston and H.M. van Driel, Phys. Rev. B 27, 1141 (1983).
3. Zhou Guosheng, P.M. Fauchet and A.E. Siegman, Phys. Rev. B 26, 5366 (1982).
4. P.M. Fauchet, Zhou Guosheng and A.E. Siegman, in Laser-Solid Inter-actions and Transient Thermal Processing of Materials, Narayan et al. editors, North-Holland (1983), pp. 205-210.
5. P.M. Fauchet, Zhou Guosheng and A.E. Siegman, in Picosecond Phenomena III, Eisenthal et al. editors, Springer-Verlag (1982), pp. 376-379.
6. R.J. Nemanich, D.K. Biegelsen and W.G. Hawkins, Phys. Rev. B 27, 7817 (1983).
7. R. Wilson, F.A. Houle and C.R. Jones, MRS meeting, Symposium B, paper B4.5 (Boston, 1984).
8. P.M. Fauchet and A.E. Siegman, Appl. Phys. A 32, 135 (1983).
9. R.M. Osgood, Jr. and D.J. Ehrlich, Optics Lett. 7, 385 (1982).
10. P.M. Fauchet and A.E. Siegman, Appl. Phys. Lett. 40, 824 (1982).
11. D. Haneman and R.J. Nemanich, Solid State Commun. 43, 203 (1982).
12. P.M. Fauchet, Phys. Lett. A 93, 155 (1983).
13. It can be checked that the very complex diffraction patterns obtained in most cases can be explained by this argument and not by just considering the nonsinusoidal profile of the ripples.
14. B. Laks, D.L. Mills and A.A. Maradudin, Phys. Rev. B 23, 4965 (1981).

15. W.J. Tomlinson, Optics Lett. $\underline{5}$, 323 (1980).
16. C. Herring, in Structure and Properties of Solid Surfaces, Gomer and Smith editors, Univ. of Chicago Press (1953), pp. 5-81.
17. P.M. Fauchet and A.E. Siegman, to be published in the NBS Proceedings of the 16th annual symposium on optical materials for high power lasers.

LASER IRRADIATION OF Ge(100): AN ASSESSMENT OF SURFACE ORDER WITH He DIFFRACTION

W. R. LAMBERT, P. L. TREVOR, M. T. SCHULBERG, M. J. CARDILLO, AND J. C. TULLY
AT&T Bell Laboratories, 600 Mountain Avenue, Murray Hill, New Jersey 07974

ABSTRACT

We investigate the effects of Nd:YAG and excimer laser irradiation on the Ge(100) surface under UHV conditions over a temperature range $140 < T(K) < 300$ using the surface sensitive probe of He atom diffraction. We study the effects of irradiation on surface damage and order using the apparent $(2 \times 1) \rightarrow c(2 \times 4)$ transition. We monitor surface contamination in situ. The temporal thermal response is modeled theoretically to aid in assessing the experimental results. The capability to maintain a Ge(100) surface at low temperatures free of contamination and well ordered is demonstrated.

I. INTRODUCTION

An important aspect of laser processing of materials is the potential to clean and anneal single crystal surfaces under ultra-high-vacuum (UHV) conditions. Experiments performed at either low temperatures or which require long signal averaging times are often severely restricted by surface contamination. Although in-situ laser cleaning and annealing of these surfaces has previously been implied,[1] to our knowledge no study of sufficient sensitivity has demonstrated the acceptability of this procedure with respect to surface order. Indeed recent high resolution electron spectroscopy suggests a general accompaniment of surface disorder.[2]

We present here a brief summary of our initial studies on laser cleaning and annealing of Ge(100) over a temperature range from 140K to 300K. We have used the technique of He diffraction complemented by in-situ Auger spectroscopy (AES) and Low Energy Electron Diffraction (LEED). The very high surface sensitivity of He diffraction makes it an ideal probe of surface order. We investigate the effect of both pulsed Nd:YAG (1.06 μm) and pulsed excimer laser (0.308 μm) irradiation on adsorbate contamination and surface order. To aid in the interpretation of our results, the time dependent surface temperature profiles have been modeled theoretically.

The Ge(100) surface is particularly well-suited for this investigation. Although the reciprocal net at room temperature has been described as a two-domain (2×1), based on LEED observations,[3] He diffraction consistently shows additional intensity in the form of streaks crossing the reciprocal net and connecting the half-order spots. These observations for Ge(100) are nearly identical to our previous results for Si(100).[4] For both crystals there have been reports of $c(2 \times 4)$ periodicities,[5,6] most recently for Ge(100) at low temperatures ($T \sim 100K$).[7] Total energy calculations suggest very small energy differences for possible arrangements of tilted dimer structural units corresponding to these different periodicities.[8,9] Our observation of disorder in the reciprocal net, manifested in the form of streaks in the diffraction intensity in addition to an underlying two-domain (2×1) surface periodicity for both surfaces, is consistent with this variety of experimental and theoretical results. We therefore utilize He diffraction measurements of the reciprocal net, which reflects the subtle inherent energetics of the surface reconstruction, as a sensitive indicator of the single crystal surface order of Ge(100) following pulsed laser irradiation.

II. EXPERIMENTAL

The vacuum apparatus, crystal manipulator, and He nozzle source have been previously described.[10] The minimum attainable sample temperature was 140K. For the majority of experiments the He atom beam and laser were incident at 50° and 40°, respectively, measured from the Ge(100) surface normal. Diffraction scans were obtained by rotating a differentially pumped mass spectrometer at a radius at 12 cm from the crystal. The angular resolution for He diffraction was primarily determined by the 2 mm lateral spread of the incident He beam at the crystal. From computer simulations of He trajectories, the effective angular resolution was determined to be 0.9°, which corresponds to a nearly symmetrical reciprocal space resolution (FWHM) of 10% of the spacing between integer diffraction peaks. The sensitivity of He diffraction to surface order will therefore be limited to domains with dimensions on the order of 100Å. In independent experiments with both the YAG and excimer lasers, the light beams were tightly focused, collimated, and then propagated into the far field to improve the spatial homogeneity, and then expanded to irradiate a 3 mm diameter region at the crystal surface. A collinear He-Ne laser beam provided a visual reference for the pulsed laser position at the crystal surface. Exact overlap of the laser and He beam at the crystal surface was confirmed by monitoring the change in He diffraction intensity for an adsorbate covered surface following laser irradiation. The power of the YAG laser was varied by adjusting the lamp voltage and pulse repetition rate with the knowledge and caveat that these parameters also influence the spatial qualities of the radiation. The power of the excimer laser was varied using wire mesh attenuators. The laser power was calibrated with a standard optical power meter. He diffraction, AES, and LEED measurements were made during independent experiments and the results were subsequently compared.

III. RESULTS AND DISCUSSION

In Fig. 1 we show He diffraction scans for a sputtered and annealed ($T_{max} \sim 900K$) Ge(100) surface taken at 300K along different azimuths and plotted against the parallel momentum transfer $\Delta K_{\parallel} = \frac{2\pi}{\lambda}(sin\theta_i - sin\theta_r)$. The angles of the diffraction scans in this plot are rotated to correspond to the appropriate azimuthal angles. This form of display provides a direct vertical correspondence between the normal view of the reciprocal net (e.g. as generally seen in LEED) and the pattern observed with He diffraction. The sharp two-domain (2×1) beams and the substantial extra intensity corresponding to streaking across the reciprocal net are clearly seen. The observation of this extra intensity is confined to a series of small areas in each reciprocal net as the mass spectrometer scans in one dimension. However, when folded back into the first Brillouin zone the extra intensity assumes the form of a streak. In the figure the streaks are shaded for emphasis. After several weeks of repeated processing of the crystal these streaks became sharper and eventually observable with LEED as well.

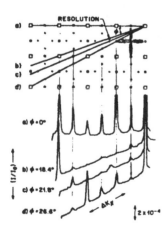

Ge (100) $\theta_i = 40°$ $T_s = 300K$ $\lambda_{He} = 1.0$ Å

Figure 1. He diffraction scans of Ge(100) at 300K for different azimuthal scattering angles. The abscissae ($\Delta K_{\parallel}$) are rotated to the azimuthal angle so as to allow a direct vertical comparison. The scans are also indicated on the c(4×2) reciprocal lattice. Streaking indicative of c(2×4) disorder is shaded for emphasis in the first Brillouin zone. The experimental resolution typical of the He diffraction experiment is indicated by the shaded circle. The diffraction intensities of the scans are directly comparable and the absolute scale is indicated.

Upon cooling the crystal surface the streaks can more readily be observed in LEED. He diffraction indicates that they become narrower and more intense but do not order into sharp spots. The cooling rate of the sample mount is sufficiently slow, however, that at 5×10^{-10} torr the background contamination rate renders the streak intensity unobservable before $T = 140K$ can be attained. As a result of this contamination the integer beams are also attenuated and within one hour are reduced to 20% of their clean, room temperature value. It is not clear from these observations if $T = 140K$ is not sufficiently cold for the surface to order into a $c(2\times4)$ or if the $c(2\times4)$ reconstruction is prevented by adsorbate contamination. In Fig. 2 we plot the diffraction intensity of the $(\overline{4}0)$ peak as a function of time after the onset of cooling. A curve representing the surface temperature is included. The initial rise is in accordance with the reduction of the Debye-Waller attenuation upon cooling and is consistent with a surface Debye temperature of $\theta_D^s = 190K$ which is $\sim \frac{1}{2}$ the bulk value ($\theta_D \sim 370K$), as represented by the open circles in the figure with subsequent leveling due to zero-point motion.[11] The subsequent fall-off with time in the $(\overline{4}0)$ diffraction intensity is ascribed to adsorbate contamination (primarily C, O, and perhaps H) as documented with AES.

At various times following the onset of cooling the YAG laser was fired at the surface with a repetition rate of 1 Hz and fluences between 0.3-0.8 J/cm^2. For these experiments the repetition rate is not physically significant for laser surface processing except that with the YAG laser it could not easily be decoupled from the spatial quality and laser power. In each case with repeated irradiation the $(\overline{4}0)$ diffraction intensity initially increased and then leveled off to a value which depended upon what point on this curve laser processing was initiated. These results are indicated schematically by the arrows drawn in Fig. 2. With continuing irradiation the diffraction intensity could be maintained at the recovered level for several hours. This variable level of diffraction intensity recovery can be most simply interpreted by two types of adsorbate sites with different binding energies: strong binding sites (steps and/or defects) which become irreversibly occupied by chemisorbed atoms with respect to laser induced desorption, and weak molecular adsorbate binding sites (terraces). Initial laser irradiation would then remove the weakly bound adsorbates and repeated irradiation prevents further accumulation of tightly bound species. This postulate requires the time for diffusion to, dissociation and capture at steps and defects to be slow compared to the laser repetition time.

When the laser irradiation was initiated at the peak of the $(\overline{4}0)$ intensity the extra intensity corresponding to a disordered $c(2\times4)$ periodicity remained evident for several hours. Although the streaks narrowed and sharpened with cooling they did not condense into sharp spots at $T_s = 140K$. Based on this persistence and narrowing of the streak intensity, the sensitivity of He diffraction to order and contamination, and the weak energeties of ordering, we conclude that residual disorder resulting from low power laser processing of Ge(100) is essentially negligible. It remains to be determined however whether with this processing technique, the surface can be completely ordered at $T < 140K$. It should be noted that the small energy differences postulated for the different periodic arrangements of tilted dimers may be less than the energies of the random step densities present even on well-cut crystals. Thus different crystals may have slightly different transition temperatures and perhaps different equilibrium periodicities.

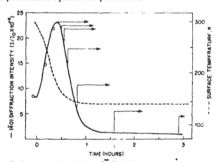

Figure 2. Diagram of the intensity of the $(\overline{4}0)$ diffraction peak following the onset of surface cooling and subsequent pulsed YAG laser irradiation. The degree of recovery in the scattering intensity following laser irradiation (fluence $= 0.3$ J/cm^2) is denoted by the arrows. Note the time dependence of the recovery maxima (arrows). The actual surface temperature is also presented as a function of time.

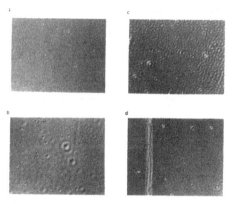

Figure 3. Phase contrast microscopy of Ge(100) following laser irradiation: a) following YAG irradiation at ∼0.3 J/cm², repetition rate = 1 Hz. Note the formation of terraces along the direction of the crystallographic axes; b) following YAG irradiation showing surface ripples; c) and d) following excimer irradiation. An example of a non-irradiated surface is visible at the left side of d).

In Fig. 3 we present photographs of the characteristic large scale features of the Ge(100) surface following laser irradiation at fluences between 0.3-0.8 J/cm². The YAG irradiated surface shows characteristic ripple patterns and the formation of terraces aligned along the principle crystallographic axes. The excimer irradiated surface shows a high density of "bubble" formation and can be directly compared in Fig. 3d to a non-irradiated portion of the same crystal. These rough surface features are a general indication of overall surface melting in combination with a widespread distribution of hotspots. Although our theoretical modeling confirmed that the Ge(100) surface was melted at these fluences, ordered recrystallization apparently occurred to the extent that the He diffraction intensity was not diminished. For both the YAG and excimer lasers, above a fluence of 0.8 J/cm², the Ge(100) surface was substantially damaged. The damage was inferred from the resulting irreversible deterioration of the He diffraction and LEED patterns, and extensive surface roughening was visually observed with phase contrast microscopy.

The effect of laser irradiation on the coverage of surface contaminants as determined using AES is summarized in Fig. 4. The C/Ge and O/Ge ratios following pulsed irradiation are presented as a function of laser fluence. Two points are included which indicate the degree of contamination shortly after sputtering and annealing and which correspond to a small fraction of a monolayer coverage. It is readily observed that both lasers are essentially equivalent in the ability to remove O at fluences in excess of 0.5 J/cm² and in their inability to remove C up to fluences which irreversibly damage the surface (> 0.8 J/cm²). Note that at fluences below ∼0.5 J/cm² the surface could still be maintained free of further surface contamination at low temperatures as evaluated using He diffraction (see Fig. 2). Thus the ability to maintain a clean and ordered surface is independent of the desorption parameters for apparently chemisorbed O and C. This interpretation is consistent with the He diffraction results (Fig. 2) and the postulation of weakly and strongly bound adsorbates with a slow interconversion time.

To aid in understanding the effects of laser irradiation on Ge(100) with both lasers, we theoretically modeled the temporal evolution of the surface temperature. We used a one-

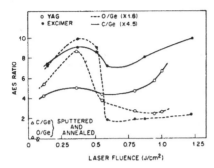

Figure 4.

The C/Ge and O/Ge ratios as determined by AES following laser irradiation. The clean surface values following sputtering and annealing are also shown.

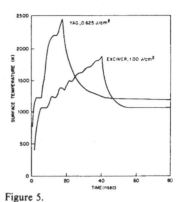

Figure 5.

Surface temperature profiles of Ge(100) following pulsed laser irradiation calculated as described in the text.

dimensional heat flow calculation for a series of 2000Å thick Ge slabs (12) with the parameters appropriate for our typical experimental conditions. The relevant optical and thermal transport properties for Ge were obtained from Refs. 12-16. Examples of calculated temperature profiles are shown in Fig. 5. With YAG irradiation (pulse width ~18 nsec, T = 150K) the calculated melting threshold (T = 1210K) occurs at a fluence of 0.21 J/cm² and the fluence necessary for the temperature to rise above 1210K is 0.33 J/cm². Corresponding values for excimer irradiation of the Ge surface (pulse width ~40 nsec, T = 150K) are 0.37 J/cm² and 0.57 J/cm², respectively. The fluence corresponding to substantial observable surface damage, as determined from He diffraction is ~0.8 J/cm², and for irradiation with the YAG laser corresponds to a maximum calculated surface temperature of ~3000K. At this temperature we estimate that ~1 monolayer of Ge is evaporated for each laser pulse. At a fluence of 0.5 J/cm² the estimated evaporation is approximately three orders of magnitude less. This implies that for laser fluences which were effective in maintaining the surface free from further surface contamination, extensive evaporation of the Ge surface layer is not a significant mechanism for adsorbate removal.

IV. CONCLUSION

We have demonstrated that a pulsed YAG laser can be utilized to maintain a Ge(100) surface at 140K free of contaminants for several hours without causing serious damage or disorder. At fluences from 0.3-0.8 J/cm² the surface is melted to various extents. Above these fluences irreversible damage occurs. It appears that at the low fluences (<0.8 J/cm²) for which the Ge(100) surface is melted but not damaged, and for which the surface is kept free from further contamination, no significant disorder is introduced. The intensity of the He diffraction streaks observed at room temperature become somewhat sharpened but do not order into spots at T = 140K. We think that for these Ge samples the equilibrium ordering of the c(2×4) structure occurs at a lower temperature, although it is possible that complete c(2×4) reconstruction is prevented by a slight laser induced increase in the surface defect density.

Results with an excimer laser are consistent with our observations using the YAG laser but are not as extensive or reproducible. In accord with our experimental observations, modeling of the surface temperature profile following pulsed irradiation suggests that the two lasers should not be significantly different in their effects on the surface at these fluences. However, the excimer laser proved to be significantly more difficult to control, perhaps due to inadequate spatial quality. Although both lasers show considerable promise for application to routine UHV single crystal experimentation, significant development requirements remain. In particular, refinements in laser power attentuation capability and improved beam spatial quality and control will be necessary. Further details of this study will be published elsewhere.

REFERENCES

[1] D. M. Zehner and C. W. White, "Laser Solid Interactions and Laser Processing" Ed. J. M. Poate and S. D. Ferris; (Academic Press, NY 1982) P. L. Cowan and J. A. Golovchenko, J. Vac. Sci. Technol., *17*, 1197 (1980).

[2] S. D. Kevan (private communication-to be published).

[3] F. Jona, H. D. Shih, D. W. Jepsen and P. M. Marcus, J. Phys. C., *12*, 1455 (1979).

[4] M. J. Cardillo and G. E. Becker, Phys. ReV. B., *21*, 1497 (1980).

[5] J. J. Lander and J. Morrison, J. Chem. Phys., *37*, 729 (1962).

[6] T. D. Poppendieck, T. C. Ngoc and M. B. Webb, Surface Sci., *75*, 287 (1978).

[7] Y. Kuk, private communication.

[8] D. J. Chadi, Phys. Rev. Let., *43*, 43 (1979).

[9] J. Ihm, D. H. Lee, J. D. Joannopoulos and J. J. Xiong, Phys. Rev. Lett., *51*, 1872 (1983).

[10] M. J. Cardillo, C. S. Y. Ching, E. F. Greene and G. E. Becker, J. Vac. Sci. Technol., *15*, 423 (1978).

[11] Simple Debye-Waller formulae generally do not fit He diffraction results well, in part as the result of scattering resonances from bound states. These, however, are weak for both the Si(100) and Ge(100) surfaces due to the presence of disorder.

[12] P. Baeri and S. U. Campisano, "Laser Annealing of Semiconductors" Ed. J. M. Poate and J. W. Mager (Academic Press, NY 1982).

[13] R. C. Smith, J. Appl. Phys. *37*, 4860 (1966) V. M. Glasov, S. N. Chizheuskaya and N. N. Clagoleva, 'Liquid Semiconductors', (Plenum, NY 1969).

[14] V. M. Glazov, A. A. Auivazov and V. G. Pavlor, Soviet Phys. Semiconductors, *5*, 1982 (1971).

[15] D. Aspnes (private communication).

[16] C. M. Surko, A. L. Simons, D. H. Auston, J. A. Golovchenko, R. E. Slusher and T. N. C. Venkatesan, Appl. Phys. Lett. *34*, 635 (1979).

EFFECTS OF IMPURITIES ON THE COMPETITION BETWEEN SOLID PHASE EPITAXY AND RANDOM CRYSTALLIZATION IN ION-IMPLANTED SILICON

G.L. OLSON*, J.A. ROTH*, Y. RYTZ-FROIDEVAUX*, AND J. NARAYAN**
* Hughes Research Laboratories, Malibu, CA 90265
** Microelectronics Center of North Carolina, Research Triangle
  Park, NC 27709

ABSTRACT

   The temperature dependent competition between solid phase epitaxy and random crystallization in ion-implanted (As^+, B^+, F^+, and BF_2^+) silicon films is investigated. Measurements of time-resolved reflectivity during cw laser heating show that in the As^+, F^+, and BF_2^+-implanted layers (conc.$\gtrsim 4 \times 10^{20} cm^{-3}$) epitaxial growth is disrupted at temperatures $\gtrsim 1000°C$. This effect is not observed in intrinsic films or in the B^+-implanted layers. Correlation with results of microstructural analyses and computer simulation of the reflectivity experiment indicates that disruption of epitaxy is caused by enhancement of the random crystallization rate by arsenic and fluorine. Kinetics parameters for the enhanced crystallization process are determined; results are interpreted in terms of impurity-catalyzed nucleation during the random crystallization process.

INTRODUCTION

   It is well known that amorphous silicon surface layers formed by either ion implantation or deposition onto atomically clean silicon substrates can be crystallized by solid phase epitaxy (SPE) and that the growth rate is thermally activated over a wide temperature range [1]. In addition to SPE, the nucleation and growth of randomly oriented crystallites can also occur during heating of amorphous silicon. The interplay between these competitive crystallization processes and the influence of this competition on the quality of the crystallized layer depend strongly on steady state temperature, amorphous film thickness, substrate orientation, and the type and concentration of impurity atoms in the film. In previous kinetics studies [1] we have shown that in intrinsic films or in films containing impurity atoms at concentrations considerably below the solid solubility limit, SPE is the dominant process over the entire range of temperatures (475°C to 1350°C) investigated. More specifically, measurements of time-resolved reflectivity obtained during cw laser heating of amorphous Si films on Si(100) substrates showed that epitaxy proceeds without interruption over that temperature range and that degradation of the interface which would be caused by the intervention of copious random crystallization does not occur. Those results are consistent with the temperature dependence of the rates of random crystallization measured in ultra-high vacuum deposited films [2] and the kinetics of SPE in ion-implanted and deposited layers [1].
   In contrast to those results we have found in this study that the presence of certain impurity atoms (i.e. As, F) at concentrations $\gtrsim 4 \times 10^{20} cm^{-3}$ can significantly enhance the rate of random crystallization such that at temperatures >1000°C nucleation and growth of polycrystalline material essentially dominate the crystallization kinetics of the film. These are to our knowledge the first quantitative measurements of the kinetic competition between SPE and random crystallization in thin amorphous layers in silicon. Such measurements provide a unique opportunity for examining the effects of impurities on the dynamics of structural transformations in silicon and are useful for determining conditions under which high quality crystalline layers can be formed by solid phase epitaxy.

To evaluate the effects of impurities on the dynamics of solid phase crystallization at high temperature we used samples implanted with a variety of different ions (As^+, B^+, F^+, BF_2^+) to concentrations $\geq 10^{20} cm^{-3}$. These impurities were selected because at reduced temperatures ($<950°C$) they modify the intrinsic SPE kinetics in different ways (i.e. enhance, retard or interrupt epitaxy) and thus would be expected to affect the competition between SPE and random crystallization differently as well. For example, boron significantly enhances the SPE rate over a wide temperature range [1,3], and can cause large deviations in interface planarity during SPE. Likewise, arsenic also enhances the SPE rate but in a manner which depends strongly on concentration and temperature [4]. In contrast, fluorine atoms retard the SPE rate over a wide temperature range with the amount of retardation decreasing with increasing temperature ($\sim 20x$ at $525°C$ and $\sim 2x$ at $950°C$) [5]; in BF_2^+-implanted samples the boron and fluorine each acts to compensate the rate altering effects of the other [5,6].

In this paper the time and temperature dependent changes in the crystallization kinetics in these samples are determined using measurements of time-resolved reflectivity (TRR) during cw laser heating; kinetics results are correlated with microstructural changes which occur during the crystallization process. The experimental data obtained in amorphous films which exhibit enhanced random crystallization during SPE are compared with computer simulations of the TRR experiments. The simulation includes the effects of both SPE and random crystallization on the net reflectivity of the film as a function of time and temperature and allows kinetics parameters for the random crystallization process to be determined.

EXPERIMENTAL

Samples were prepared either by implantation of the impurity atom into an initially crystalline Si(100) substrate or into a previously amorphized layer formed by $^{28}Si^+$ implantation (120 keV, $2x10^{15}cm^{-2}$; 80 keV, $1x10^{15}cm^{-2}$; 40 keV, $5x10^{14}cm^{-2}$ at $77°K$). A range of implantation energies and fluences were investigated for each impurity. Specific implantation conditions will be given when results are discussed in the next section.

Measurements of solid phase crystallization kinetics during cw laser irradiation were performed using TRR techniques as described in Ref. [1]. A cw Ar laser beam (4880/5145 Å) was used to heat the sample and a HeNe laser beam (6328 Å) was used to probe the changing reflectivity during crystallization. The spot sizes of the heating and probe beams were 150 μm and 14 μm (FWHM), respectively. During irradiation, the substrates were maintained at an elevated temperature ($300°C$ to $550°C$) on a resistively heated stage. Since the temperature dependence of the SPE rate in intrinsic and doped samples has been well characterized over a wide temperature range [1], the temperature during cw laser heating in these experiments could be readily obtained from the measured SPE rates.

A fast mechanical shutter was used to controllably gate the heating beam on and off the sample. This allowed samples to be heated for well-specified time intervals so that results of post-irradiation structural analysis could meaningfully be compared. Precise control of the irradiation interval also allowed us to partially crystallize the samples at a specified temperature and then vary the laser power or substrate temperature and continue the crystallization under different thermal conditions. Experiments of this type have been useful for determining the onset of random crystallization and for studying effects due to "pre-annealing" of the amorphous material. Structural characterization of the recrystallized material was performed using transmission electron microscopy (TEM), electron channeling analysis, and selective chemical (Wright) etching.

RESULTS AND DISCUSSION

Time-resolved reflectivity data obtained during crystallization of an arsenic-implanted (220 keV, $3x10^{15}cm^{-2}$; 150 keV, $5x10^{14}cm^{-2}$) amorphous film at five different temperatures are shown in Fig. 1. The original amorphous film thickness was 2400 Å; the substrate temperature was 450°C and the laser power was varied from 13W to 15.5W (similar results were obtained using a fixed laser power and different substrate temperatures). At a crystallization temperature of 960°C (SPE rate of $3.4x10^5$ Å/sec) a regular oscillation in the net reflectivity is observed. As previously discussed [1] this is caused by constructive and destructive interference between light reflected from the surface of the sample and from the advancing epitaxial growth interface. In contrast to the TRR result at 960°C which indicates unimpeded motion of the epitaxial interface, an increase in temperature is accompanied by a progressive diminution of the last interference peak. For example, during crystallization at 1040°C (SPE rate: $1.6x10^6$ Å/sec) the final interference peak is almost totally missing. As the temperature is increased further, this trend continues with each peak being sequentially eliminated.

This progressive loss of interference contrast suggests that the motion of the epitaxial interface is being disrupted by a competitive process having a temperature dependence which is different from that of epitaxial growth. We emphasize that this effect is not observed in intrinsic films or in films containing reduced arsenic concentrations. Likewise, extensive studies on the films produced by boron implantation (30 keV, fluence varied from $1x10^{15}cm^{-2}$ to $1x10^{16}cm^{-2}$) into previously amorphized Si substrates show that these materials can be crystallized epitaxially at very high temperatures without the detectable intervention of a competing crystallization process (the dependence of the crystallization rate on the crystal/amorphous interface position and the boron concentration at different temperatures will be discussed in a future publication).

As in the case of the arsenic-implanted samples the progressive loss of interference contrast was also observed in the TRR signals during crystallization of both the F^+ and BF_2^+-implanted films. In both cases the impurity was implanted into pre-amorphized silicon. The implantation conditions for the two impurities were as follows: F^+: 18 keV, $4x10^{15}cm^{-2}$; BF_2^+: 45 keV, $2x10^{15}cm^{-2}$. The onset of the change in interference contrast occurred in these samples in approximately the same temperature range as shown in Fig. 1 for the arsenic-implanted films.

In Fig. 2 we show results obtained from TEM analysis of the As^+-implanted sample crystallized at 1040°C. At this temperature the TRR result revealed significant disruption of the epitaxial interface at a distance of approximately 600 Å from the surface of the sample (see bottom trace in Fig. 1). The dark field micrograph in Fig. 2a shows the presence of a significant number of polycrystalline grains in the laser-irradiated zone. The existence of polycrystalline material is also confirmed by the ring pattern present in the electron diffraction result shown in Fig. 2b. These data are in good agreement with results of electron channeling analysis and selective chemical etching which clearly revealed the presence of polycrystalline material in regions crystallized at temperatures >1000°C.

We have developed a computer simulation of the time-resolved reflectivity experiment which incorporates reflectivity changes due to propagation of an epitaxial growth interface and temperature dependent changes in the volume fraction of polycrystalline silicon. Using this simulation together with the experimental TRR data we can readily extract kinetics parameters from the TRR signals. The program models the temperature dependence of the SPE rate using an Arrhenius expression ($v=v_0exp(-E_a/kT)$) with experimentally determined values of the pre-exponential term (v_0: $3.07x10^8$cm/sec) and activation energy (E_a: 2.68eV). The volume fraction of polycrystalline material (X) is calculated using a modified form of the

14595-6

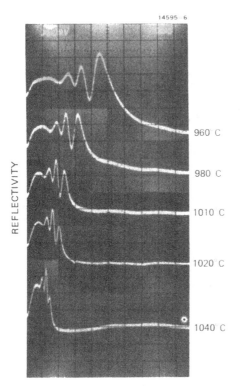

REFLECTIVITY

960 C

980 C

1010 C

1020 C

1040 C

TIME, 2 ms/div

Fig. 1. Time-resolved reflectivity data obtained
during crystallization
of As^+-implanted
(220 keV, $3 \times 10^{15} cm^{-2}$;
150 keV, $5 \times 10^{14} cm^{-2}$)
Si(100) at five dif-
ferent temperatures.

14595-7

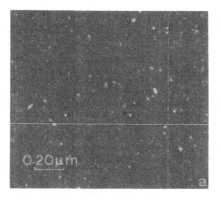

0.20μm

a

b

Fig. 2. Dark field transmission electron micrograph
and electron diffraction pattern showing
polycrystalline grains in As^+-implanted
(220 keV, $3 \times 10^{15} cm^{-2}$; 150 keV, $5 \times 10^{-14} cm^{-2}$)
Si(100) crystallized at 1040°C.

Avrami equation: $X=1-\exp[-(t/\tau)^4]$, where τ is a "characteristic" time for polycrystalline nucleation and growth which is related to the activation energy for random nucleation (E_a') as follows: $\tau=\tau_0\exp(E_a'/kT)$. The value of τ is determined by adjustment to give the best fit to the experimental TRR data at a given temperature. Additional features which are incorporated into the TRR simulation include effects due to: 1) SPE rate enhancement or retardation due to propagation of the interface through regions containing a graded concentration, 2) temperature dependence of the refractive index, 3) finite shutter response time and thermal risetime, and 4) changes in the relative sizes and positions of the probe and heating beams.

Computer simulations of the TRR experiment which incorporate effects due to enhanced random crystallization are shown in Fig. 3. There is good agreement between these results and the experimental data shown in Fig. 1 for crystallization at 960°C, 1010°C, and 1040°C. The effects of the finite shutter response time and thermal risetime are evident during the initial phases of epitaxial growth. As in the experimental TRR data, competition between SPE and random crystallization is apparent in the 1010°C and 1040°C simulations by the progressive reduction in the amplitude of the final interference peak. In order to determine kinetics parameters for the random crystallization process it is necessary to evaluate τ at the steady state growth temperature. This was accomplished in these simulations by "turning-on" the random crystallization process when a constant temperature was reached in the simulation (indicated by the arrow above each result in Fig. 3). At that particular time epitaxy had consumed approximately 800 Å of the original film thickness leaving a layer ~1600 Å thick to be crystallized by both SPE and random crystallization. Since a negligible fraction of the film converts to polycrystalline material before that time is reached, τ can be evaluated at a specific temperature thus permitting a meaningful comparison with other crystallization rate data to be made.

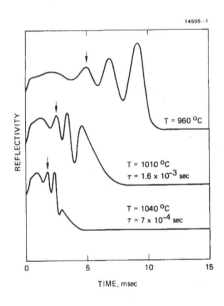

14595-1

REFLECTIVITY

T = 960 °C

T = 1010 °C
$\tau = 1.6 \times 10^{-3}$ sec

T = 1040 °C
$\tau = 7 \times 10^{-4}$ sec

0   5   10   15

TIME, msec

Fig. 3. Computer simulations of TRR data at three different temperatures which show the effects of enhanced random crystallization on the reflectivity signatures. These data are to be compared with the experimental results for the same temperatures in Fig. 1. Arrows indicate time at which random crystallization effects were initiated in program (time at which steady state temperature was reached).

216

A plot of the characteristic random crystallization time, τ, (time to form a polycrystalline volume fraction equal to $1-1/e$) vs $1/kT$ (T in °K), is given in Fig. 4 for the arsenic and fluorine-implanted layers. Also shown is the temperature dependence of the intrinsic random crystallization time determined previously in UHV-deposited amorphous films [2]. To give an indication of the competitive kinetics between random crystallization and SPE in these layers we also give the time to grow 1600 Å by SPE (1600 Å is the thickness for which crystallization occurs at constant temperature - see above). The data indicate that an order of magnitude enhancement in the random crystallization rate relative to the intrinsic rate occurs in the As^+ and F^+-implanted films. The differences between the activation energies (E_a') and pre-exponential factors (τ_0) for the random crystallization processes in the intrinsic and impurity-containing layers provide a quantitative measure of the random crystallization kinetics. The temperature dependent competition between SPE and random crystallization in these films is evident from the observation that at temperatures >980°C the characteristic time to form a polycrystalline volume fraction equal to $1-1/e$ is less than the time required to crystallize the film by SPE.

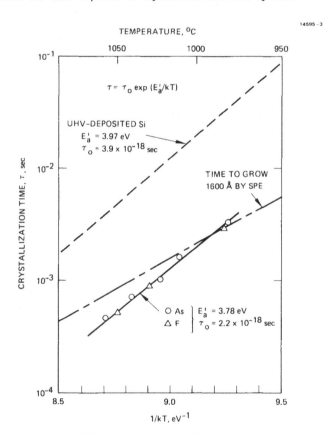

Fig. 4. Random crystallization time vs. $1/kT$ for intrinsic film and for films implanted with As^+ and F^+. Shown for comparison is the temperature dependence of time to grow 1600 Å by SPE (growth distance over which temperature was constant in these experiments).

In attempting to elucidate the mechanism responsible for the enhanced random crystallization in the films implanted with As^+, F^+, and BF_2^+ we addressed the issue of whether the instability of the SPE interface was the result of the nucleation and growth of polycrystalline material in the amorphous phase ahead of the interface or whether random crystallization occurred as a result of the interfacial instability. Experiments were performed in which we stopped the high temperature crystallization process at various times before the c/a interface reached the surface and then resumed crystallization at a (lower) temperature for which SPE was the dominant process. By comparing the TRR signals from the high and low temperature portions of the growth, the degradation of epitaxy due to formation of polycrystalline material ahead of the interface can be assessed. We found that even when SPE was interrupted at a time prior to that for which inhibition of epitaxy became evident, continued growth at reduced temperature (~680°C) was characterized by a significant reduction in interference contrast. The results obtained from these studies support the hypothesis that formation of polycrystalline material occurs at high temperatures in the amorphous film ahead of the advancing interface and the presence of the randomly oriented crystallites produce subsequent degradation of epitaxy.

Although these results provide information about the kinetics and spatial origin of the enhanced random crystallization process the detailed mechanism responsible for the impurity-enhanced process has not yet been established. However, the relative contributions of the nucleation and growth components to the enhanced random crystallization process can be determined from these data. We assume that solid phase epitaxy is the primary mechanism by which a crystallite expands after a nucleus of critical size has been formed. Since we have observed enhanced random crystallization in samples containing impurities that both reduce (F) as well as increase (As) the SPE rate relative to the intrinsic value, the rate enhancement is apparently not controlled by the growth component. Therefore, we suggest that the enhancement process is mediated by nucleation which is catalyzed by the As and F impurity atoms in the film. The effects that these impurities will have on the energetics of nucleation, e.g., reduction in the surface free energy contribution to the barrier for the formation of a stable nucleus, are currently under investigation.

SUMMARY

We have studied the temperature dependent kinetic competition between solid phase epitaxy and random crystallization in amorphous silicon films implanted with As^+, B^+, F^+, and BF_2^+. Time-resolved reflectivity measurements performed during cw laser heating of the films show that in the As^+, F^+, and BF_2^+-implanted layers at concentrations $\geq 4 \times 10^{20} cm^{-3}$ there is a progressive diminution in the amplitude of the TRR interference peaks as the temperature is increased above ~980°C. This effect is not observed in intrinsic (Si^+-implanted) layers or in the B^+-implanted films. By correlation with results of microstructural analysis and computer simulations of the TRR experiments this effect is ascribed to enhancement of the random crystallization rate by arsenic and fluorine during crystallization. Kinetics parameters were determined, and it was shown that over the temperature range from ~980°C to 1060°C the random crystallization rate is approximately ten times greater than the intrinsic rate. It was determined that polycrystalline material was formed ahead of the advancing interface and was responsible for interface degradation. The results suggest that an increased nucleation rate due to the presence of the impurities is responsible for enhancement of the random crystallization process.

218

ACKNOWLEDGEMENT

The authors gratefully acknowledge helpful discussions with L.D. Hess
and P.K. Vasudev and thank S.A. Kokorowski for his contributions to the TRR
simulation work. We also wish to acknowledge the technical assistance of
R.F. Scholl. One of the authors (Y. R.-F.) also wishes to thank the Swiss
National Foundation for Scientific Research for financial support.

REFERENCES

1. G.L. Olson, S.A. Kokorowski, J.A. Roth, and L.D. Hess, Mat. Res. Soc.
    Symp. Proc. 13, 141 (1983).

2. J.A. Roth, S.A. Kokorowski, G.L. Olson, and L.D. Hess, Mat. Res. Soc.
    Symp. Proc. 4, 169 (1982); J.A. Roth, G.L. Olson, and L.D. Hess, Mat.
    Res. Soc. Symp. Proc. 23, 431 (1984).

3. L. Csepregi, E.F. Kennedy, T.J. Gallagher, J.W. Mayer, and T.W. Sigmon,
    J. Appl. Phys. 48, 4234 (1977).

4. G.L. Olson, J.A. Roth, L.D. Hess, and J. Narayan, Mat. Res. Soc. Symp.
    Proc. 23, 375 (1984).

5. G.L. Olson, L.B. Roth, and P.K. Vasudev, submitted to Appl. Phys.
    Lett.

6. I. Suni, U. Shreter, M-A. Nicolet, and J.E. Baker, J. Appl. Phys. 56,
    273 (1984).

RAMAN MEASUREMENTS ON GRAPHITE MODIFIED BY HIGH POWER LASER IRRADIATION [1]

J. Steinbeck*, G. Braunstein*, M.S. Dresselhaus*, B.S. Elman**, T. Venkatesan***
*Massachusetts Institute of Technology, Cambridge, MA
**GTE Research Laboratory, Waltham, MA
***Bell Communications Research, Murray Hill, NJ

Abstract

The behavior of highly anisotropic materials under short pulses of high power laser irradiation has been studied by irradiating highly oriented pyrolytic graphite (HOPG) with 30 nsec Ruby-laser pulses with energy densities between 0.1 and 5.0J/cm². Raman spectroscopy has been used to investigate the laser-induced modifications to the crystalline structure as a function of laser energy density of the laser pulse. A Raman microprobe was used to investigate the spatial variations of these near-surface regions. The irradiation of HOPG with energy densities above ~ 0.6J/cm² leads to the appearance of the ~ 1360 cm⁻¹ disorder-induced line in the first order Raman spectrum. The intensity of the ~ 1360cm⁻¹ line increases with increasing laser energy density. As the energy density of the laser pulse reaches about 1.0J/cm², the ~ 1360cm⁻¹ line and the ~ 1580cm⁻¹ Raman-allowed mode broaden and coalesce into a broad asymmetric band, indicating the formation of a highly disordered region, consistent with RBS-channeling measurements. However, as the laser energy density of the laser pulses is further increased above 3.0J/cm², the two Raman lines narrow and can again be resolved suggesting laser-induced crystallization. The Raman results are consistent with high resolution electron microscopy observations showing the formation of randomly oriented crystallites. Raman Microprobe spectra revealed three separate regions of behavior: (i) an outer unirradiated region where the material appears HOPG-like with a thin layer of material coating the surface, (ii) an inner irradiated region where the structure is uniform, but disordered, and (iii) an intermediate region between the other regions where the structure is highly disordered. The changes in structure of the inner region are consistent with the behavior observed with RBS and conventional Raman spectra. The identification of an amorphous carbon-like layer on the outer region is consistent with a large thermomechanical stress at the graphite surface, introduced by the high power laser pulse, and known to occur in metals.

Introduction

Current work on the modification of the electronic and mechanical properties of graphite through the use of intercalation and ion-implantation has prompted research on the thermodynamic properties of graphite. While the low temperature regime has been extensively studied and is well understood, the properties in the high temperature regime still remain unclear. Through the use of a high power laser small portions of graphite can be brought into this regime to provide new information about the thermodynamic behavior of graphite, particularly the melting and sublimation on the graphite surface. These measurements also allow study of the effects of large thermal gradients upon graphite which can provide high temperature mechanical data.

Previous studies with pulsed laser irradiation at low pressure[1] and at high pressure[2] have shown that graphite can be melted at temperatures ~ 4300K. This melting process is accompanied by the formation of a disordered layer of randomly oriented crystallites on the graphite surface within the region of irradiation [1]. In addition to these processes, a sublimation process occurs at the surface leading to a loss of material from within the irradiated region [1]. Recent work has also shown that liquid carbon can also be formed by irradiating diamond with a high power laser [3].

Because graphite is a semi-metal, many of the results obtained from high power laser irradiation studies on metals can be applied to graphite. This is particularly true for in-plane processes where the electrical conductivity for graphite is on the order of $3.0 \times 10^4 \text{ohm}^{-1}\text{cm}^{-1}$. For this case, the study of thermomechanical stresses [4], heat transport [5], and shock waves [6] for metals irradiated with high energy laser pulses becomes valuable in determining the corresponding processes in graphite.

In the present study, Raman spectroscopy is used to determine the general behavior of the crystallinity of the irradiated and surrounding regions as a function of incident laser pulse fluence. A Raman Microprobe is used to obtain detailed information on the spatial variation of the structural changes induced in the crystal by the laser pulse. These two methods are complementary in nature and allow the exploration of both the macroscopic properties of laser-induced disorder (conventional Raman) and the microstructure of the graphite surface (Raman Microprobe). The Raman Microprobe technique is well suited to

[1]The MIT authors acknowledge NSF Grant #DMR 83-10482 for the support of their portion of the work.

the task of a spatial survey due to its high spatial resolution (area of the probing laser spot $\sim 4\mu m^2$) and the ability of the Microprobe to take a complete spectrum in ~ 5 minutes with a reasonably good signal to noise ratio. The short collection time is primarily due to the implementation of a linear photodiode array as the data collection device. This is in contrast to more conventional Raman techniques where the laser spot surveys ~ 0.1 mm^2, and collection times of ~ 30min are typical due to single wavelength sampling.

The Raman Microprobe is used to make an estimate of the changes in the degree of crystallinity across the boundary between the non-irradiated and the irradiated portions of the sample, by comparing the intensity of the disorder-induced ~ 1360cm^{-1} line (I_{1360}) with the intensity of the Raman-allowed ~ 1580cm^{-1} line (I_{1580}). The Raman technique has been shown to be effective in gaining valuable information on the general nature of the crystallinity in previous studies with ion-implanted graphite[8] and the technique is relatively simple to implement.

Experimental

The samples were prepared by cutting HOPG into thin 0.5cm^2 sections and irradiating them with a ruby laser operating at 6910Å. The irradiating pulses were generated by an oscillator (JK Lasers) and then directed through a quartz homogeniser before striking the sample surface, thus providing uniform irradiation over a spot ~ 0.25cm^2 in size. The duration of the irradiating pulses was 30nsec and laser pulse fluences in the range 0.1 to 4.9J/cm^2 were delivered to the sample surface. These fluences correspond to power densities on the order of 10^8W/cm^2.

Raman spectra were taken using two systems: a conventional system designed to maximize the signal to noise ratio and a second system consisting of a Raman Microprobe. The conventional system consisted of a Czerny-Turner double axis spectrometer with the illumination source being a continuous wave Argon laser operating at 4879.6Å. A phototube served as the data collection device.

The Raman Microprobe consisted of a similar Czerny-Turner double axis spectrometer with a micro-scope attachment. This enabled the area of the sample under investigation to be reduced to $\sim 4\mu m^2$. Backscattered light was collected by microscope optics and sent into the spectrometer where diffraction gratings served as the primary signal separator. The data were then collected by a linear photodiode array containing 512 photodiodes. The spatial resolution of the detector was 0.29Å/diode or approximately 1.01cm^{-1}/diode in the spectral region of interest. The illumination source was again a continuous wave Argon laser operated at 4879.6Å.

Surveys of the sample surfaces were conducted by taking a series of spectra across the entire sample with a spacing of $\sim 50\mu m$ between each spectrum. This allowed a fairly complete survey of the surface of the sample, and allowed many local structural fluctuations in the sample to be observed.

Results and Discussion

Raman spectroscopy and Raman microscopy both revealed several interesting characteristics of laser-irradiated graphite. All of the Raman spectra showed a definitive change in the crystallinity of the graphite samples as a function of laser pulse fluence. Raman Microprobe spectroscopy revealed that there are three regions which can be identified for study, each with a unique crystallographic structure. The three regions defined are the region well beyond the area of irradiation, the area of irradiation, and the area which lies in between.

The general trends in the degree of crystallinity which accompany increasing laser pulse fluence are based on the growth of the ~ 1360cm^{-1} disorder-induced line in the Raman spectra of laser-irradiated HOPG. As the laser pulse fluence is increased up to 0.6J/cm^2 the intensity of the Raman-active mode at ~ 1580cm^{-1} decreases, but no disorder-induced peak at ~ 1360cm^{-1} is observed. Above a threshold of 0.6J/cm^2 the disorder-induced peak becomes visible and grows with increasing laser pulse fluence. Accompanying this growth is a reduction in intensity and a broadening of the Raman-active mode at ~ 1580cm^{-1}. This behavior for several pulse fluences is illustrated in figure 1.

As the pulse fluence exceeds ~ 1.5J/cm^2 the Raman-active mode peak and the disorder-induced Raman peak begin to coalesce. The line broadening of both peaks continues until a laser pulse fluence of ~ 3.0J/cm^2 is reached. At this point the lineshapes begin to reverse their behavior. As the laser pulse fluence is further increased above ~ 3.0J/cm^2 to the upper limit of investigation at 4.9J/cm^2, the lineshapes narrow, become well-defined, and have intensity ratios (I_{1360}/I_{1580}) which remain constant, independent of further increases in fluence.

The Raman Microprobe results are consistent with these findings, but they also provide information on the spatial variation of the lineshapes. These spatial variations in turn can be related to changes in the microstructure on the graphite surface. The Microprobe surveys defined three regions on the sample surfaces with quite different characteristics. The spectra for the region well beyond the irradiated region

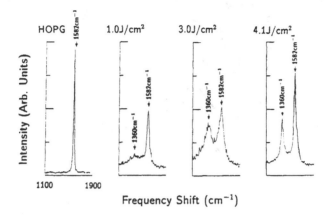

Figure 1: First order Raman spectra for laser irradiated graphite. Note the growth of the $\sim 1360\text{cm}^{-1}$ disorder-induced line for laser pulse fluences above 0.6J/cm^2.

on the sample were characteristic of those of HOPG. There was no evidence of any growth of a disorder-induced peak in the Raman signal for this region. Approaching the irradiated region a broad background signal is observed which grows as teh probe moves closer to the irradiated region. This signal grows at the expense of the Raman active mode and can be identified with an amorphous carbon Raman signal[7]. The peak for this signal occurs at $\sim 1580\text{cm}^{-1}$. This suggests that the disordered region extends far beyond the the irradiated region on the sample surface. This behavior is illustrated in figure 2. This background signal was typical for all laser pulse fluences above 1.0J/cm^2.

Within the irradiated zone there was a distinct change in the Microprobe spectra as the laser pulse fluence was varied. The behavior was similar to that observed in conventional Raman spectra, namely a growth of the disorder-induced ($\sim 1360\text{cm}^{-1}$) peak up to $\sim 3.0\text{J/cm}^2$, accompanied by a reduction and broadening of the Raman-active mode peak ($\sim 1580\text{cm}^{-1}$). The two peaks coalesced with increasing laser pulse fluence up to $\sim 3.0\text{J/cm}^2$. Above 3.0J/cm^2, the two peaks again separated and became more distinct. The relative intensities (I_{1360}/I_{1580}) of the peaks remained constant within the irradiated region. Exceptions to this behavior were observed within small islands within the irradiated zone which appeared very similar to HOPG or to amorphous carbon deposits on the basis of their Raman spectrum. These islands were small in size ($\sim 10\mu\text{m}^2$) and were randomly located within the irradiated zone and the region between the irradiated and nonirradiated zones.

An additional feature, unique to the irradiated zone, was the formation of smooth structures which resemble droplets. These structures appeared for all samples which had been irradiated with laser pulse fluences greater than $\sim 1.5\text{J/cm}^2$ and could be observed using the microscope of the microprobe. These structures appeared as darkened areas on the surface of the graphite. The Raman spectra of these areas contained peaks whose relative intensity (I_{1360}/I_{1580}) was the same as that of the irradiated region. However, the linewidth of both the disorder-induced and Raman-allowed peaks was smaller than for the regions outside these structures. The behavior of the Raman lineshapes within the structures with changing laser energy density was similar to that of the irradiated zone for those samples in which these structures could be observed.

The bounding region between the outer region and the irradiated region is mainly characterized by a series of spectra in which the ratio I_{1360}/I_{1580} sharply rises and then slowly descends to the level of I_{1360}/I_{1580} in the irradiated region. This behavior is illustrated in figure 3. The Raman observed peaks in this region are generally coalesced and somewhat reduced in intensity for all energy densities.

The processes which account for these phenomena can be related to those observed in other laser irradiation studies. From previous work with graphite implanted with arsenic [1] it is clear that melting occurs. The observation of droplet-like structures within the irradiated region confirms this results since these droplets have been attributed to melting in previous work with diamond [3].

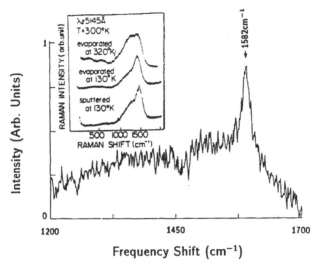

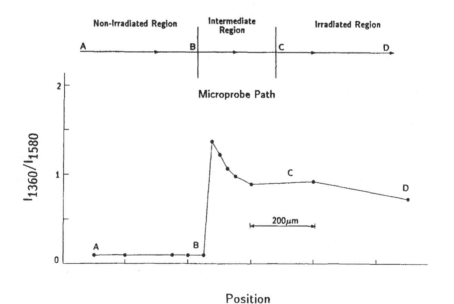

Figure 2: Growth of amorphous layer on the graphite surface in the region outside of the irradiated region. Also shown (inset) is a spectrum taken from Solin[7] for amorphous carbon.

Figure 3: Intensity ratio I_{1360}/I_{1580} versus Microprobe position

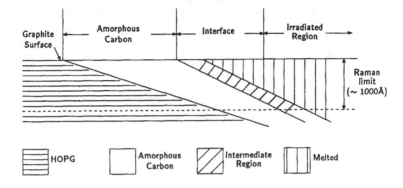

Figure 4: Speculative cross-section for laser irradiated graphite constructed from Raman Microprobe spectra of sample surface. Growth of Amorphous Carbon region and relatively large interface can be seen.

The structure outside the irradiated region can be understood by considering the thermomechanical stress within this region during irradiation. Musal [4] has derived an equation relating the mechanical constants of metals to the surface temperature change during irradiation which will cause permanent deformation of the surface of the metal in the limit of short pulse durations and large areas of irradiation. This equation is,

$$\Delta T = \frac{(1-\nu)}{Ea} Y \qquad (1)$$

where ΔT is the surface temperature change needed to cause permanent deformation, ν is Poisson's ratio, E is Young's modulus, a is the thermal expansion coefficient, and Y is the tensile yield stress of the material. The parameters for this experiment, pulse duration $30nsec$ and irradiation area $0.25cm^2$, fit the limits of validity for this equation proposed by Musal.

If we now treat graphite as a metal, use mechanical constants supplied by Reynolds[9], and apply Musal's equation for the in-plane thermomechanical stress, a surface temperature rise of $\sim 1200K$ is needed for permanent deformation of the surface. By comparison, the temperature change needed for a typical metal such as copper, which is used in laser optics, is only 20K. The approximation in treating graphite as a metal is appropriate since in this case we are concerned with in-plane deformations where the electrical conductivity is high ($3.0 \times 10^4 ohm^{-1}cm^{-1}$).

Since there is evidence to affirm that the graphite surface has melted under pulse fluences in excess of $0.6J/cm^2$, we can conclude that the surface temperature has risen above $4300K$. This temperature rise at the surface is more than sufficient to cause permanent plastic deformation of the surface of the graphite. Because of the rigid in-plane structure and small in-plane thermal expansion, this thermomechanical stress will be felt over relatively large ($1000\mu m$) distances. This accounts for the relatively large region ($\sim 400\mu m$ from the irradiated region) on the surface which has been transformed into amorphous carbon.

The small thermal conductivity coefficient and large coefficient for thermal expansion coefficient along the graphite c-axis combine to effectively isolate adjacent graphite planes in the lattice. We can then consider each plane in the lattice to be an independent surface. Since graphite planes farther from the surface will receive less energy than those near the surface, the in-plane stress generated by a smaller temperature rise of the planes farther from the surface. This means that the total area in the plane that is damaged by the stress will decrease for planes farther from the surface. This gives rise to a damage cross-section for irradiated samples as presented in figure 4.

The rapid rise and slow decent of the ratio I_{1360}/I_{1580} within the region between the irradiated region and the nonirradiated region (figure 3) can be accounted for by considering the cooling down phase after irradiation. For the pulse fluences where this behavior is observed, the irradiated region of the samples has been melted. As an interface between the cooling melted region and the outer amorphous region that is formed, high stress bonding will occur. These high stress bonds introduce a high degree of disorder for this interface region. This interface is seen in the I_{1360}/I_{1580} ratio within the region which lies between the irradiated and nonirradiated regions. The large size of this interface region ($\sim 100\mu m$ from the irradiated region) can be understood since this interface will occur closer to the irradiated region for successive graphite planes farther from the surface. Adding to this disordered region will be bonding occurring along the direction perpendicular to the surface.

As laser pulse fluences exceed the melting threshold ($\sim 0.6 J/cm^2$), the irradiated region becomes partially melted. The remaining portions of the region are exposed to a large thermomechanical stress which causes a large degree of disorder. For laser pulse fluences above ($\sim 3.0 J/cm^2$), an area larger than the conventional Raman probe laser spot is melted. This would account for the trends seen in the conventional Raman data, particularly the appearance of a leveling-off of the recrystallization behavior for energies above $3.0 J/cm^2$.

Conclusion

Both conventional Raman spectra and the Raman Microprobe spectra indicate that a melting process occurs within the irradiated region on the graphite surface. This is seen in the development of two well delineated peaks within the irradiated zone appearing at $\sim 1360 cm^{-1}$ and $\sim 1580 cm^{-1}$, corresponding to the disorder-induced peak and the Raman-allowed mode respectively. Evidence for melting also lies in the observation of droplet-like structures which have been attributed to melting in diamond melting studies. The ratio of the intensities (I_{1360}/I_{1580}) of these peaks is constant throughout the irradiated zone, consistent with uniform heating throughout this region due to the incident laser pulse.

Additional effects that are primarily seen in the Raman Microprobe spectra, due to the high spatial resolution of the Microprobe, include a region beyond the irradiated region which has become highly disordered. Furthermore, the Raman Microprobe spectrum can be identified with that of an amorphous carbon layer on top of HOPG. This layer grows as the irradiated region of the surface is approached. There is also an interface region between the irradiated and nonirradiated regions. This interface region is characterized by the sharp increase in the relative intensity of the disorder-induced peak at $\sim 1360 cm^{-1}$ to the Raman-allowed peak at $\sim 1580 cm^{-1}$, in passing from the nonirradiated zone to the irradiated zone, with this ratio gradually descending to the ratio of the intensities of the peaks within the irradiated zone. This behavior is consistent with the creation of a highly disordered zone which may be due to high stress bonding between the recrystallizing melted region and the highly damaged region.

There are more quantitative studies currently ongoing at M.I.T. to determine the exact nature of the observed behavior of the crystallinity for laser irradiated graphite. These studies will primarily focus on the growth of the region afected by the laser irradiation as a function of incident laser pulse fluence.

References

[1] T. Venkatesan, D.C. Jacobson, J.M. Gibson, B.S. Elman, G. Braunstein, M.S. Dresselhaus and G. Dresselhaus, *Phys. Rev. Lett.*.53, 360,(1984).

[2] R. Clarke and C. Uher, *High Pressure Properties of Graphite and its Intercalation Compounds*, Dept. of Physics, University of Michigan, Ann Arbor, Michigan, 1984.

[3] J.S. Gold, W.A. Bassett, M.S. Weathers, and J.M. Bird, *Science*, 225, 921, (1984).

[4] H.M. Musal, *Thermomechanical Stress Degradation of Metal Mirror Surfaces Under Pulsed Laser Irradiation*, NBS-SP-568, pp. 159.

[5] J.M. Poate, G. Foti, D.C. Jacobson, *Surface Modification and Alloying by Laser, Ion, and Electron Beams*, Plenum Press, New York, 1983.

[6] K. Mukherjee, J. Masumder, *Lasers in Metallurgy*, proceedings of a symposium by the Physical Metallurgy and Solidification Committees of The Metallurgical Society of AIME, 1981.

[7] S.A. Solin and R.J. Kobliska, *Proceeding of the 5th International Conference on Amorphous and Liquid Semiconductors*,(Taylor and Francis, London,1974); N. Wada, P.J. Gaczi, and S.A. Solin, J. Non-Cryst. Solids 35 and 36, 543 (1980).

[8] B.S. Elman, M.S. Dresselhaus, G. Dresselhaus, E.W. Maby, H. Mazurek, *Phys. Rev. B*24, 1027, (1981).

[9] W.N. Reynolds, *Physical Properties of Graphite*, Elsevier Pub. Co. LTD., New York, 1968.

ELECTRON BEAM ANNEALING OF PHOSPHORUS IMPLANTED CADMIUM TELLURIDE

C.B. YANG, M.L. PENG, J.T. LUE[+] and H.L. HWANG
Department of Electrical Engineering
+Department of Physics
National Tsing Hua University, Hsinchu, Taiwan 300, R.O.C.

ABSTRACT

The behavior of dopants in CdTe has been examined by Krogen and DeNobel and others. The n-type material is easy to dope with good electrical activity. The p-CdTe is more difficult to produce with concentration higher than $6x10^{16}$ cm^{-3}. For doping levels above this, the electrical activity of the dopant drops sharply and the hole mobility is reduced. The difficulty in doping p-CdTe stems from both strong compensation effects and low solubility of the usual dopant species.

The electron beam pulse method has been applied to annealing phosphorus implanted cadmium telluride. The threshold electron beam energy density necessary to give good electrical activation and mobility have been established in the range between 9.2-10.1 $J \cdot cm^{-2}$ for doses from 10^{14}-10^{16} ions cm^{-2}. A sheet resistance as low as $6.32x10^2$ $\Omega/\square$ and a carrier concentration as high as $3x10^{18}$ cm^{-3} have been obtained. The impurity profile of the annealed samples have been obtained by etching layer by layer with the etching rate calibrated by chemical techniques and the impurity concentration determined by van der Pauw/Hall technique.

INTRODUCTION

Because of its high bandgap (E_g=1.5 eV) and high average atomic number, CdTe can be used as excellent radiation detectors operated at room temperature [1-2] . Also, the low optical absorption of CdTe in 1-30 m and its large electro-optic coefficient makes it useful for the fabrication of electro-optic modulators [3]. Furthermore, CdTe is the only compound in II-VI semiconductors, in which p-n junction diode can be formed, CdTe is therefore an interesting material for optical integrated circuits [4].

Although CdTe has so many significant application potential, problems still exist in limiting its widespread applications as follows [5]:

(1) Lack of control over minority carrier behavior.

(2) Lack of doping control to low resistivity in P-type material.

The behavior of an impurity in II-VI compounds is not simple at all, the incorporated impurities are compensated electrically by native defects,

226

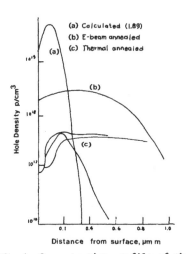

Fig.1. Concentration profile of the data in Table 1, and compared with the postimplantation thermal annealing depth profile.

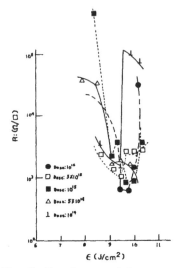

Fig. 2. The sheet resistance versus energy density for sample BS013

which are ionized with opposite sign. (self-compensation effect) [6]. Impurity often form complex centers with native defects [7]. There are many factors that might limit the formation of high hole concentration, such as the preimplantation annealing condition, implantation substrate temperature, post-implantation annealing temperature and Cd-over pressure. Almost a 100% doping efficiency could be achieved for a 300°C As+ implantation of 10^{15} ions/cm^2, if the CdTe had been previously annealed in Cd vapor at 400°C for 24 hr. [8].

The conventional thermal annealing techniques have many factors that can not be well controlled. It is our attempt to adopt the transient annealing method to eliminate whole or part of these factors.

CALCULATED IMPLANTS PROFILE

Taking a Thomas-Fermi atomic model, Lindhard, Schiott, and Scharff (LSS) derived a universal relationship for the nuclear stopping $(d\epsilon/d\rho)_n$ in terms of dimensionless length and energy parameters ρ and ϵ [9].

The theoretical ion distribution of p-implanted in CdTe is shown in Fig.1, which is approximated by the following Gaussian distribution [10]

$$N(x) = \frac{N_i}{(2\pi)^{2/3}[\Delta R_p]} \exp\{-(X-R_p)^2/2(\Delta R_p)^2\} \tag{1}$$

where R_p is the projected range, ΔR_p is standard deviation, x is the per-pendicular distance into the substance, and N_I is the ion dose. For a 100 KeV p+implanted in CdTe, R_p=875 Å, ΔR_p= 571 Å. [11]

EXPERIMENTAL

The CdTe single crystals were grown by travelling heater method [12] and sliced into wafers, Br_2-Methanol solution was used for chemical polish-ing to etch off the mechanical damage.

The energy of implanted p+ was 100 keV, and the dose was ranged between 10^{14} to 10^{16} ions/cm^2. The beam current was ranged between 5 to 150 μA. The cathode and anode tubes are connected to a main capacitor of extremely low inductance. The charging voltage of this capacitor determines the maxi-mum energy of the individual electrons of the beam. As the trigger pulse is on, a high current beam of electrons is extracted out of the discharge plasma, and the electrons pass through the barrier and hit the sample. In our system, the capacitance of the main capacitor is C=19.17 μF, the distance is R=6.5 cm, and the critical energy density at a voltage of 16.5 KV can be estimated to be

$$E = \frac{CV^2}{2} \cdot \frac{1}{2\pi R^2} = 9.83 \text{ J/cm}^2 \qquad (2)$$

Ultrasonic soldering of indium was used in this work to apply to ohmic con-tacts to p-type CdTe.

Since high temperature Ar plasma excited source (5500°C∿8000°C), the Inductively Coupled Plasma — Atomic Emission Spectrometry can excite almost all kinds of atoms, and its detection limit is in the p. p. b. range. It was used to determine the etching rate for CdTe, which is most important in the determination of impurity depth profile.

In this experiment, bromine-methanol was used as etchant, in which only selected area was etched, the rest areas were covered with wax. After a

constant duration of etching, the sample was taken out and the solution was heated in a water bath. The methanol was vaporized and only the Cd and Te with the bromine was remained, which was further dissolved in a 3.5% NHO_3 acid for the preparation of standard solution for I.C.P. measurement. From the measurement of the Cd and Te quantities, and in constant duration t, the etching rate can be determined as,

$$v= \frac{d}{t} = \frac{m_{Cd} + m_{Te}}{A \cdot \rho \cdot t} \qquad (3)$$

where A is the exposed surface area, and ρ is the CdTe density. The etching rate is a function of the concentration of bromine-methanol solution. To

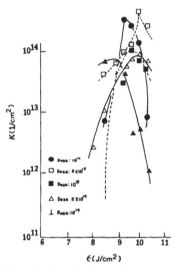

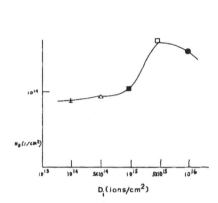

Fig. 3. The sheet carrier concentration versus energy density for p ion 10^{14} and 10^{16} cm$^{-\frac{1}{2}}$ annealed by pulse electron beam for sample BS013.

Fig. 4. The maximum sheet carrier concentration every ion dose from 10^{14} to 10^{16} ions/cm^2.

obtain uniform etching and more precise control of the etched depth, a solution of 0.05% bromine-methanol was used and the etching rate was estimated to be ~ 2.5 Å/sec.

The sheet resistance, sheet concentration and mobility were determined by the van der Pauw/Hall measurements.

RESULTS AND DISCUSSIONS

Critical range of energy density for E-beam anneal

In this work, when the CdTe was implanted by phosphorus ions having an energy 100 keV with a dose ranged between 10^{14} and 10^{16}ions/cm^2. The sheet resistance dependance on the energy density is shown in Fig. 2. Lower sheet resistance can be obtained in the energy density range between 9.2 to 10.1 J/cm^2. The minimum sheet resistance can reach 6.32×10^2 $\Omega/\square$. Fig. 3 shows the sheet concentration as a function of energy density. The maximum sheet concentration can reach 4.62×10^{14} 1/cm^2 for an implanted dose of 5×10^{15} ions/ cm^2 at an energy density of 10.1 J/cm^2.

Since using electron beam pulse of low energy density, the damage by ion implantation can't be completely removed. By raising the energy density, the electrical activation can be improved.

Once the energy density reaches the threshold, the sample surface will be melted and recrystallization will occur, in which most of the damage can then be removed. However, if electron beam energy density exceeds a certain value, damage could then be re-generated but by the electron beams. Therefore, a threshold for electron beam energy density should be found to obtain good electrical activation. The best annealing condition was with an energy density in the range of $9.2 \sim 10.1$ J/cm^2.

Doping efficiency

The doping efficiency, ϵ, is defined as follows:

$$\epsilon \stackrel{-}{=} \frac{N_s}{N_{ion}} \qquad (4)$$

where N_s is the maximum sheet carrier concentration, and N_{ion} is the implanted ion dose. The maximum sheet (hole) concentration and doping efficiency are plotted in Figs. 4 and 5, respectively, as a function of the implant doses.

The different volatilities of the component of CdTe may disturb its stoichiometry, and gives rise to intrinsic lattice defects, which can compensate the implanted impurity [13]. For example, Cd vancancy has the acceptor properties and can compensate the implanted donors [14]. When CdTe is doped with acceptors (e.g. phosphorus), the presence of V_{cd} makes some of the implanted impurities to occupy the Cd vanancy and to become a triple donor. This will reduce the doping efficiency [15].

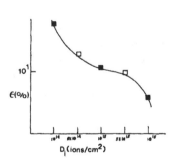

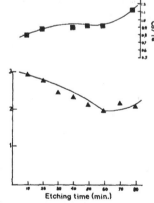

Fig. 5. The doping efficiency for every ion dose from 10^{14} to 10^{16} ions/cm^2.

Fig. 6. The etching rate and Cd/Te ratio for period of 10 minutes duration with 0.05% bromine-methanol.

Table 1. The concentration of bromine-methanol at 1%, 0.5%, and 0.25% with etching time 40 minutes.

concentra-tion	Cd(ppm)	Te(ppm)	d(Å)	Cd/Te	V(Å/sec)
0.251	47.3	56.3	20107	0.954	8.37
0.5 1	102	131	41026	0.484	17.09
1 1	294	304	92107	1.091	34.37

The doping efficiency was found to decrease with the implanted ion dose. This is because that the increase of the active carrier can't match with the increase in the implanted dose, since most the implanted ions are compensated by the crystal lattice defects.

The sheet carrier concentration can reach 10^{14} $1/cm^2$ only at a high dose larger than 10^{15} ions/cm^2. The saturation carrier concentration was found to occur at an ion dose of $5x10^{15}$ ions/cm^2, which means the implanted ions are compensated by the crystal defects.

Depth profile

Table 1 shows the results of a CdTe etched for a period of 40 minutes with a bromine-methanol solution of 0.25%, 0.5% and 1%. The high etching rate which stems from the high concentration of solution will cause problems in the etching control. It was, therefore, to choose 0.05% concentration in the investigation of the etching rate. The results for the same duration of a 10 minute ethching are shown in Fig.6. The etching rate was about 3 Å/sec in the first 10 minutes, and was changed to about 2 Å/sec in the last period.

The existance of defects in the surface layers where the Cd/Te ratio might not be stoichiometric and the etching rate being faster in the initial etching period is possibly due to the defect-assisted effect.

Using a pre-determined etching rate about 2.5Å/sec, the carrier concentration and mobility profiles could be obtained. The result is shown in Fig.1. The surface concentration can reach $2.97x10^{18}$ $1/cm^3$, which is the highest reported in literature. Also shown in the same figure is the impurity profile of the thermally annealed phosphorus implanted CdTe. [16]. Compared these results, the electron beam annealing clearly shows the potential to solve the p-type doping problems in CdTe.

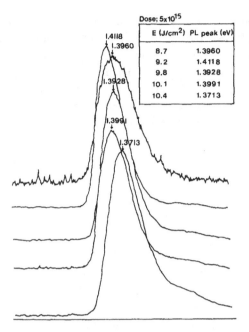

Fig. 7. The PL spectrum for different energy density electron-beam annealing.

Luminescence

Photoluminescence measurements provide information concerning the
localized levels associated with the impurities and defects. Using post-
implantation electron-beam pulse anneal, the 14°K emission spectra for
different energy density for the same ion dose 5×10^{15} cm^{-2} is shown in
Fig. 7. Only the 1.4 eV broad band was found, and it shifts from sample
to sample. This band was considered as the formation of V_{cd}-donor complexes
[17], however, its true identification has not yet been done. The shift in
this emission is correlated to the extent of implantation damages which are
removed by electron pulse. With increasing the energy density, this
emission band shifts to higher energy and then decreases. This is
correlated with the change of the sheet resistance and carrier concentration
as-depicted in Figs. 2 and 3.

CONCLUSIONS

Electron beam annealing has been demonstrated to be an effective method
to anneal the damage of p implanted CdTe. The range of the critical energy

density was determined between 9.2 and 10.1 J/cm^2 for the dose in the range between 10^{14} to 10^{16} ions/cm^2. I.C.P. was developed as an effective means to obtain the impurity profile of phosphorus in CdTe, and a hole concentration can reach about 2.9×10^{18} 1/cm^3, which is the highest reported in literature. It was also found that implanted ion dose must be larger than $10^{15}/cm^2$ to ensure the high impurity concentration. But higher doping efficiency could be obtained at lower ion doses.

REFERENCES

1. Mayer, J.W., Nucl. Instrum, Methods, 43, 55 (1966)
2. Akutagawa, W., Zanio, K., and Mayer, J.W., Nucl. Instrum. Methods, 55, 383 (1967).
3. Johnson, C.J., Proc. IEEE 56, 1719 (1968).
4. Popova, M. and Polivka, P., Czech. J. Phys. B23 (1973) 110.
5. F.V. Wald, Rev. Phys. Appl., No. 2, 277 (1977).
6. G. Mandel, Phys. Rev. 134, A1073, (1964).
7. B. Furgolle, et.al., Solid State Commun. 14, 1237 (1974).
8. J.C. Bean, J.F. Gibbons, et. al., in "Proc. Fourth International Conference on Implantation in Semiconductors",Osaka, Japan, 229 (1974).
9. J. Linhard, M. Scharff, and H.E. Schiott, Kgl. Danske, Videnskab Selskab, Mat. Fys. Medd. 33, No. 14 (1963).
10. W.K. Chu, et al., Backscattering Spectrometry (Academic Pr., N.Y.)(1978).
11. J.F. Gibbons, et al., "Projected Range Statistics in Semiconductors and Related Materials".
12. H.L. Hwang, M.L. Peng, M.H. Yang and C.Y. Sun, Tech. Report, May (1984).
13. E.N. Arkadeva, et al., Sov. Phys. Semicond. 9, 563 (1975).
14. N.V. Aqrinskaya, et al., Sov. Phys. Semicond. 5, 767 (1971).
15. R.B. Hall and H.H. Woodbury, J. Appl. Phys. 39, 5361 (1968).
16. M. Chu, et al., J. Electrochem. Soc., Vol. 127, No. 2, 483 (1980).
17. N.V. Agrinskaya, et al., Proc. Int. Symp. on CdTe, Strasbourg, 1971.

ELECTRIC CONDUCTIVITY OF VITREOUS SILICA
DURING PULSED-ELECTRON-BEAM IRRADIATION

STANLEY H. STERN AND RALPH B. FIORITO
Naval Surface Weapons Center, Silver Spring, MD 20910

ABSTRACT

We describe a measurement of the transient conductivity induced in amorphous SiO_2 during irradiation by an intense and energetic electron beam. Preliminary results of an experiment [1] we performed at the Pulsed High - Energy Radiographic Machine Emitting X Rays (PHERMEX) facility of Los Alamos National Laboratory indicate that a beam of electrons of energy 17.5 MeV, of current density 1 kA/cm^2, and of micropulse duration 3 nsec induce in a-SiO_2 an electric conductivity $\sim 6 \times 10^{-3}$ $(\Omega\text{-cm})^{-1}$. This value is 15 orders of magnitude larger than the ordinary (ionic) conductivity of the material [2].

INTRODUCTION

When a dielectric material is traversed by an intense beam of relativistic electrons, its electric conductivity increases by 10 to 15 orders of magnitude during the pulse [3]. In crystals, the increased, transient conductivity is related to promotion of electrons from the valence band to the conduction band and to the mobility and lifetime of those conduction-band charge carriers [4]. However, the precise role which beam energy-deposition plays in that promotion as well as the relationship of energy deposition to the electronic energy distribution of the promoted conduction electrons is just now beginning to be elucidated [5,6]. Moreover, beams of current density > 100 kA/cm^2 can alter media through a variety of mechanisms to enhance significantly stopping powers and beam energy-deposition [7]. A beam so intense is not a mere stream of individual particles which stop in matter according to the well-established principles of Bohr, Bethe, and Fermi. Rather, such a beam is a collective of particles that interacts synergistically with a medium it traverses [7]. It deposits energy in ways yet to be measured definitively and understood theoretically.

There are two motivations for our experiments. First, we want to establish a data base of values of electric conductivity induced in a variety of materials by electron beams of various energies, current densities, and pulse durations. By allowing tests of theories of induced conductivity and its dependence on material stopping-power and beam energy-deposition, such data would contribute to understanding basic physics of beam-material interactions. The second motivation is methodological. The thrust of our program is to study electron irradiation of materials at ultrahigh doses and dose rates. We need to develop a technique to measure energy deposited in such quantity and at such rates that it saturates the response of conventional dosimeters. It will soon be possible for an electron beam to deposit $\geq 10^9$ rad in ≤ 25 nsec to heat silica by $\sim 10^4$ K [8]. Such a dose would probably melt, vaporize, shock, or shatter most materials. Measurement of transient conductivity as a diagnostic of dose would allow in a nanosecond time-scale determination of energy deposited in materials that would be destroyed microseconds to milliseconds later. Such a technique could extend the useful range of dosimetry to values heretofore not directly measureable. Hence the focus of this paper is an analysis of the experimental configuration and data as they pertain to development of accurate measurement of electron-beam-induced transient conductivity of dielectrics.

EXPERIMENT

Configuration

Electric conductivity is determined experimentally from voltage signals $V_s(t)$ induced by the interaction of an electron beam with a dielectric sample fixed as a component of an electrical circuit. The configuration of the experiment is depicted schematically in Fig. 1.

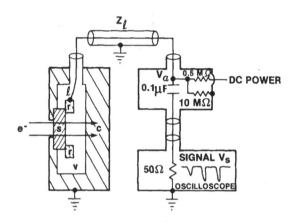

Fig. 1. Configuration of experiment: e^- electron beam, s sample, r high-voltage ring, ℓ lead to transmission line, c beam stop, v vacuum, V_a applied voltage, V_s signal voltage, Z_ℓ 50-Ω impedance of transmission line.

A sample whose thickness is smaller than the range of beam electrons traversing it is contained in a cylindrically symmetric stainless steel module. The symmetry axis is parallel to the direction of the electron beam. The module housing is grounded, as is the sample face serving as beam entrance, and so it forms a Faraday-shielded chamber admitting only the beam. Inside the chamber, the face of the sample from which the beam exits floats at high voltage through a retainer ring connected by a lead wire to a transmission line. Beam electrons which exit the sample are stopped in the grounded back wall of the module. The entire module is mounted on a rotatable wheel in vacuum of 4×10^{-5} Torr.

Before the pulse of the electron beam, a loaded power supply charges a 0.1-μF capacitor to bias the transmission line and silica sample to constant voltage V_a. Initially, the specimen resistance is $\sim 10^{17}$ Ω [2]. During the pulse of the electron beam through the material, the resistance $R(t)$ drops by ~ 15 orders of magnitude to a minimum $R_{min} \sim 10^2$ Ω, and the medium conducts a current driven by the voltage across it. Signals are recorded with a Tektronix 7104 oscilloscope in the nanosecond time-scale of the electron-beam pulse.

For an ohmic medium, one expects the following linear relationship between the (negative) peak signal $V_{s\ peak}$ and the bias voltage V_a applied to the sample [9]:

$$V_{s \text{ peak}} = \frac{Z_\ell R_{min}}{Z_\ell + R_{min}} I_{se} - \frac{Z_\ell}{Z_\ell + R_{min}} V_a. \tag{1}$$

Z_ℓ is the 50-Ω impedance of the transmission line, and I_{se} is the (maximum) current thought to arise predominantly from a net flux of knock-on secondary electrons deposited in the high-voltage electrode of the sample [10]. A fit of eq. (1) to the peak signals measured as a function of applied voltage yields values for I_{se} and for R_{min}, the minimum resistance induced in the silica during the beam pulse. The maximum electric conductivity σ_{max} can then be evaluated from

$$\sigma_{max} = w/A R_{min}, \tag{2}$$

where w is the thickness of the specimen, and A is the cross-sectional area of the beam in the specimen. Eq. (1) is only approximately true as it takes no account of capacitive and inductive elements of the sample-module system. These elements become significant when the rise and decay times of the electron-beam pulse are comparable to or shorter than the sample-module response time. Also, this approach would need further generalization should the response of the medium not be ohmic.

The experiment proceeded as follows: for a particular applied voltage V_a, a sample was shot with a beam pulse, and the signal $V_s(t)$ was recorded. After the sample cooled, the applied voltage was changed, and the sample was shot again, etc.

Beam parameters

Values of parameters which characterize the PHERMEX electron beam are listed in Table I. A single beam pulse is actually a "macropulse," i.e., a train of 10 to 11 distinct micropulses separated from each other in time by a 20-nsec period. The fast rise and decay times of the micropulses tend to distort the signals observed from a relatively slower-responding sample-module system.

TABLE I. PHERMEX ELECTRON-BEAM PARAMETERS

BEAM-ELECTRON ENERGY (MEAN)	17.5 MeV
BEAM CURRENT (MAXIMUM)	290 A
BEAM DIAMETER IN 1/16-IN a-SiO$_2$ SAMPLE (MEAN)	0.6 cm
BEAM CURRENT DENSITY IN SAMPLE (MAXIMUM FOR MEAN BEAM DIAMETER)	1 kA/cm^2
PULSE SHAPE: TRAIN OF 10-11 MICROPULSES, EACH MICROPULSE SEPARATED BY A PERIOD OF 20 NSEC	
MICROPULSE DURATION (FWHM)	3 nsec
MICROPULSE RISE-TIME AND DECAY-TIME	$\lesssim$ 1 nsec
DOSE TO SILICA PER PULSE TRAIN (MAXIMUM)	5 Mrad
TEMPERATURE RISE IN SILICA PER PULSE TRAIN, IN REGION PENETRATED BY BEAM (MAXIMUM)	70 K

Material properties

Amorphous silica specimens of optical-grade quality were purchased from
ESCO Products, Inc., Oak Ridge, NJ. They are disks of purity 99.7 % SiO_2,
fabricated from thermal fusion of quartz. Their diameters are one inch, and
their thicknesses range from 1/16 inch to 1/2 inch. Each sample face was
painted with several coats of a water-based silver conductor fixed with an
inorganic binder. After drying and baking in a furnace at 450 °F, the paint
on the sample faces formed a hard electrode of thickness ∼ 0.003 inch.
After repeated pulsing from the electron beam, there was no evidence of any
macroscopic mechanical or structural damage either to the silica or to its
silver electrodes. We report results for one specimen, the thinnest, whose
1/16-in thickness corresponds to 4 % of the range of a 17-MeV electron traversing this material.

RESULTS

The oscillograms recorded for two typical shots of run "C" are displayed
in Fig. 2. Beam currents were monitored with a time-integrating Ḃ-loop located upstream of the sample module. There was little variation in the beam
current from one shot to the next.

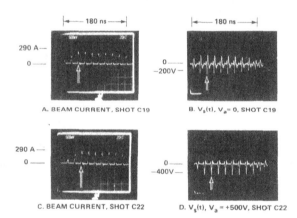

A. BEAM CURRENT, SHOT C19   B. $V_s(t)$, V_a= 0, SHOT C19

C. BEAM CURRENT, SHOT C22   D. $V_s(t)$, V_a = +500V, SHOT C22

Fig. 2. Oscillograms of beam current and signal voltage
for two shots of the PHERMEX electron beam through amorphous SiO_2, 1/16-in thick. Vertical arrows indicate third
micropulses.

Two distinct features are apparent in the $V_s(t)$ oscillograms: (1) There
is distortion of the signal that is evident in the bipolar nature of the
$V_s(t)$-micropulse waveforms, especially in the V_a = 0-V oscillogram, Fig. 2B.
One would expect an undistorted signal to follow -- perhaps not linerally --
the unipolar shape of the beam-current micropulses. (2) Despite distortion,
there is a discernible trend with applied voltage V_a: the more positive the
applied voltage, the more negative $V_s(t)$.

The trend of signal with applied voltage is illustrated in Fig. 3
by a plot of the (negative) peak values $V_{s\ peak}$ of third micropulses of the

trains vs. V_a.

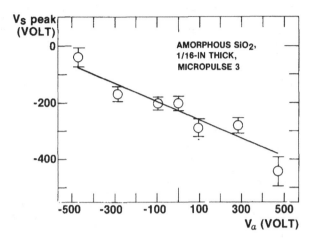

Fig. 3. Peak signal voltage of third micropulse vs. applied voltage. Straight line is a fit of eq. (1) to the data (see text).

A fit of eq. (1) to the points yields a value R_{min} = (110 ± 20) Ω for the minimum resistance of the silica sample during an electron-beam micropulse. From eq. (2), one deduces that the maximum induced conductivity is (0.006 ± 0.002) $(\Omega\text{-cm})^{-1}$. However, inasmuch as eq. (1) does not account for signal distortion, the values of uncertainty cited for the induced conductivity and resistance are inaccurate. Accurate assessment of the transient conductivity depends on an analysis of the sample-module circuit more general than that represented by eq. (1).

ANALYSIS

On the basis of time-domain-reflection measurements [11], we model sample-module reactance properties as schematized in Fig. 4.

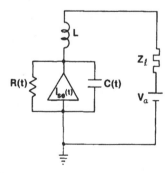

Fig. 4. Sample-module circuit.

238

The beam-material interaction is represented by a secondary-electron-generated current $I_{se}(t)$, by a beam-induced sample resistance $R(t)$, and additionally by a specimen capacitance $C(t)$. An inductive component L, whose value is a known constant, arises from the thin wire between the high-voltage retainer ring in the module and the transmission line. This model leads to the following relationship between applied voltage V_a and signal voltage $V_s(t)$:

$$V_s(t) + \frac{[Z_\ell R(t)C(t) + L + LR(t)\dot{C}(t)]\,\dot{V}_s(t) + LR(t)C(t)\ddot{V}_s(t)}{Z_\ell + R(t) + Z_\ell R(t)\dot{C}(t)} =$$

$$\frac{Z_\ell R(t)}{Z_\ell + R(t) + Z_\ell R(t)\dot{C}(t)}\,[I_{se}(t) - \frac{V_a}{R(t)}]. \tag{3}$$

Eq. (3) accounts for signal distortion and assumes ohmic behavior of the specimen. Its numerical solution may yield the dielectric response from $C(t)$ in addition to the medium conductivity from $R(t)$. If micropulse evolutions of beam current-density and beam energy are known as well, e.g. from measurements of transition radiation [12], time-correlation comparison would effectively broaden parameter space to encompass tests of dependences of beam-induced conductivity and dielectric behavior on time, beam current-density, and on beam energy.

This work was supported by the Directed Energy Program Office of the Naval Sea Systems Command.

REFERENCES

1. E.W. Pogue et al., Los Alamos National Laboratory Group Report M-4: GR-84-10, 1984 (unpublished).

2. Robert B. Sosman, The Properties of Silica (The Chemical Catalog Co., New York, 1927), p. 528.

3. D.I. Vaisburd and E.G. Tavanov, translation of Pribory i Tekhnika Eksperimenta (1), 215 (Tomsk Polytechnic Institute, 1976).

4. D.I. Vaisburd et al., Sov. Phys. - Dokl. 27, 625 (1982).

5. Oakley H. Crawford and Rufus H. Ritchie (private communication).

6. A. Lewis Licht (private communication).

7. D.W. Rule and Oakley H. Crawford, Phys. Rev. Lett. 52, 934 (1984).

8. T.P. Starke, IEEE Trans. Nucl. Sci. NS-30, 1402 (1983).

9. Simon Ramo and John R. Whinnery, Fields and Waves in Modern Radio, second edition, (John Wiley & Sons, New York, 1953), pp. 29-30.

10. Ralph B. Fiorito, Michael Raleigh, and Stephen M. Seltzer, Naval Research Laboratory Memorandum Research Report 5241, 1983 (unpublished).

11. Time-domain reflectometry was performed at the Naval Research Laboratory April 2 and July 13, 1984. We are grateful for the assistance of Michael Raleigh and for the support of J. Robert Greig.

12. D.W. Rule and R.B. Fiorito, Naval Surface Weapons Center Technical Report NSWC TR 84-134, 1984 (unpublished).

THRESHOLD ENERGIES FOR THE MELTING OF Si
AND Al DURING PILSED-LASER IRRADIATION

C. K. ONG, H.S. TAN AND E.H. SIN
Department of Physics, National University of Singapore, Kent Ridge,
Singapore 0511

ABSTRACT

The heat flow calculations have been performed to obtain the threshold
energies for the melting of Si and Al during pulsed-laser irradiations under
various laser conditions. The temperature dependent optical and thermal
properties of the solids are deduced from the available experimental data.
The melting threshold energies calculated for the solids are within the
accuracy of the experimental values.

INTRODUCTION

Many theoretical attempts have been made earlier to simulate the melting
and resolidifying processes in pulsed laser annealing of solids with heat
flow computation [1,2]. Some experimental results such as the laser-induced
dopants diffusion in ion-implanted silicon and the melt-history of laser
annealed silicon have been successfully explained. However, there are still
some uncertainties in the validity of the computational model and the
physical data employed. We begin to investigate these uncertainties by
calculating the threshold energies E_t required for the melting of solids
during pulsed laser annealing. It is because a reliable estimate of E_t of
a solid necessarily involved an accurate knowledge of the temperature
dependent optical and thermal properties of the solid which we can deduced
from the best experimental data available. Secondary, the calculations of
E_t were not complicated by the phase transition phenomenum and the much less
well known thermal and optical properties of the liquid phase. Silicon and
aluminium single crystals are the solids investigated in the present work as
the experimental thresold energies for these solids have been reported by
Peercy and Wampler [3], and Lowndes et al [4] and hence are available for
comparison.

CALCULATION MODELS AND RESULTS

The following one-dimensional heat diffusion equation is employed to
model the thermal evolution of the laser annealing processes:

$$\rho c \frac{dT}{dt} = \alpha(T)I(x,t) + \frac{d}{dx}(k(T)\frac{dT}{dx}) \tag{1}$$

where c is the specific heat capacity, ρ is the mass density of solid,
$T = T(x,t)$ is the temperature, $\alpha(T)$ is the optical absorption coefficient,
and $K(T)$ is the thermal conductivity. The absorbed radiation power density
at depth x is given by

$$I(x,t) = (1-R)I(o,t) \exp(-\alpha x) \tag{2}$$

Where R is the reflectivity at the interface between air and solid and $x = 0$
at the surface .
For optical and thermal properties of solids that varies with temperature, a numerical method has to be employed to solve eq. (1). In our

computation scheme, the irradiated region is divided into n layers of equal thickness Δx. The time t is chosen to be zero at the start of the laser pulse. By choosing a suitable time interval Δt, we can rewrite eq. (1) into a set of finite difference equations fulfilling the heat balance condition. For example, the change in temperature ΔT_i of the ith layer due to absorbed energy from the laser pulse and the thermal energy diffused between the nearest-neighbor layers, can be written as

$$I_i(1 - e^{-\alpha_i \Delta x})\Delta t + (k_{i-1} \frac{T_{i-1} - T_i}{\Delta x} + k_{i+1} \frac{T_{i+1} - T_i}{\Delta x})\Delta t = \rho c \Delta T_i \Delta x \qquad (3)$$

where k_{i-1} is the thermal conductivity between layer i and i-1 and

$$I_i = I_{i-1} e^{-\alpha_{i-1} \Delta x} \qquad (4)$$

and

$$I_1 = I(o,t)(1-R)\left|\frac{1-e^{-\alpha_1 \Delta x}}{\alpha_1 \Delta x}\right| \qquad (5)$$

In eq. (5), $I(o,t)(1-R)$ is the radiation power absorbed on the surface and I_1 is the radiation power at the center of the first layer. We have also assumed that no heat loss at the sample surface and the substrate temperature at x = 10μm was kept at room temperature at all time. In principle, we would like to choose the space steps Δx as small as desired to obtain accurate results. Unfortunately, it is restricted by the following stability criterion

$$\frac{k\Delta t}{c\rho(\Delta x)^2} \leqslant \frac{1}{2} \qquad (6)$$

In order to have some guidelines in choosing the size of space step Δx in the numerical calculation, we used an analytical solution of eq. (1) that assumed constant optical and thermal properties for the solid and employed a top-hat laser pulse. With the idealised conditions, the exact solution of eq. (1) is [5]

$$T(x,t) = \{(\frac{2I_0}{k}) \sqrt{Dt}\; ierfc\left|\frac{x}{2\sqrt{Dt}}\right| - (\frac{I_0}{\alpha k})e^{-\alpha x}$$

$$+ (\frac{I_0}{2\alpha k})\exp(\alpha^2 Dt - \alpha x)\; erfc\left|\alpha\sqrt{Dt} - \frac{x}{2\sqrt{Dt}}\right|$$

$$+ \frac{I_0}{2\alpha k}\exp(\alpha^2 Dt + \alpha x)\; erfc\left|\alpha\sqrt{Dt} + \frac{x}{2\sqrt{Dt}}\right|\}(1-R) \qquad (7)$$

where $I_0 = I(o,t)$ and

$$ierfc\frac{x}{2\sqrt{Dt}} = \int_{\frac{x}{2\sqrt{Dt}}}^{\infty} erfc\; \xi d\xi \qquad (8)$$

When comparing the temperatures of the Si surface at t = 40ns obtained with eq. (5) and with the present numerical method for various step sizes, it was found that even with Δx = 100nm, the numerical result only deviates 1.2% from the exact solution. We have checked the accuracy of our computation further by calculating E_t of ruby laser annealed silicon using temperature dependent optical and thermal parameters. The pulse-width employed was

15ns. The calculated E_t are 0.725, 0.716 J/cm² for Δx = 100, 20nm respectively. The difference in the calculated E_t is well within the uncertainty of the present experimental values for E_t.

In the present numerical computation, the value of E_t was identified as the minimum energy required to raise the surface (first layer) temperature of the solid to the melting point. The laser pulses were assumed gaussian in shape. The temperature dependence of the optical absorption coefficients of Si, especially in the high temperature domain were obtained from the calculated results by Sin et al [6] which are in excellent agreement with the experimental data obtained by Jellison and Lowndes [7] at optical wavelength of 1152nm, Jellison and Modine [8] at 694nm and Weaklien and Redfield [9] at 750nm. For the case of 532nm, we used the empirical expression

$$\alpha = 5.02 \times 10^3 \quad e^{T/430} \text{ cm}^{-1} \tag{9}$$

suggested by Jellison and Modine [8].

For the temperature dependence of the optical absorption coefficients of Al, Dreehsen et al [10] have determined the refractive index (n) and extinction coefficients (k) at elevated temperatures for CO_2 laser of 10.6μm. From the values of n and k, we deduced the following expressions that relate R and α to temperature:

$$R = 1.005 - 8.33 \times 10^{-5}T \tag{10}$$

$$\alpha = 6.816 \times 10^8 T^{-1.194} \text{cm}^{-1} \tag{11}$$

The reflectivity changes from 0.98 at 300k to 0.93 at 933k. It was noted that an accurate knowledge in the temperature dependence of R is essential in determining a reliable E_t for Al. Generally, the reflectivity of metal would also depend on the preparation and conditions of the surface. Since no experimental data are available for the temperature dependence of the optical constants for Al at other wavelengths. In those cases, we have treated α and R as constants for Al at all temperature. The optical absorption coefficient for Al at 0.693μm and 1.06μm are 1.27×10^6 cm^{-1} [11] and 10^6 cm^{-1} respectively. The later value is extrapolated from data given in ref. [11].

The temperature dependence of thermal conductivity of crystal Si is [12]:

$$k = 1521/T^{1.226} \text{ Wcm}^{-1}\text{K}^{-1} \quad 300K < T < 1200K$$

$$= 8.96/T^{0.502} \text{ Wcm}^{-1}\text{K}^{-1} \quad 1200K \leqslant T < 1683K \tag{12}$$

and the mass density ρ and average specific heat capacity c of silicon are 2.33 g/cm³ and 0.96 J/(gK) respectively.

The temperature dependence of thermal conductivity of crystal Al can be deduced from the available experimental data [13] and expressed as

$$k = 2.354 + 3.575 \times 10^{-4}T - 7.133 \times 10^{-7}T^2 \text{ Wcm}^{-1}\text{K}^{-1} \tag{13}$$

The mass density ρ of aluminium is 2.7 g/cm³ and the expression for specific heat capacity deduced from the available experimental data [14] is

$$c = 0.753 + 0.49 \times 10^{-3}T \text{ Jg}^{-1}\text{K}^{-1} \tag{14}$$

Our calculated E_t for Si and Al under different laser conditions, together with the reflectivity of solid and the available experimental E_t are given in table 1 and table 2 respectively. All the calculations were done for step size Δx = 100nm. Good agreement was found between the calculated and the available experimental values of E_t.

Table 1   Calculated threshold energies for melting
for silicon at various laser conditions

Type of laser	Pulsed width (ns)	Single crystal reflectivity	E_t (J/cm^2) Calculated	Experimental
Nd: YAG doubled-frequency 0.532μm	18	0.374 [15]	0.395	0.32 [4]
	30		0.474	-
Ruby 0.693μm	15	0.337 [15]	0.725	-
	30		0.805	0.80 [3]
Alexandria 0.750μm	15	0.328 [15]	1.09	-
	30		1.15	-

Table 2   Calculated threshold energies for melting
for aluminium at various laser conditions

Type of laser	Pulsed width (ns)	Single crystal reflectivity	E_t (J/cm^2) Calculated	Experimental
Ruby 0.693μm	15	0.894 [3]	2.44	-
	30		3.42	3.5 [3]
Nd:YAG 1.06μm	15	0.93 [16]	3.61	-
	30		5.09	-
Carbon dioxide 10.6μm	15	eq. (10)	4.72	-
	30		6.54	-

The calculated and measured threshold energies for melting of Si at 1.06μm will be reported shortly. At that wavelength, the absorption length of Si is large and hence the substrate will be heated substantially. Particularly the multiple reflection of radiation intensity at the interface must be included in the calculations of threshold energies for the melting of Si with an irradiated sample of finite thickness.

DISCUSSION

From table 1, it was noted that the value of E_t decreases with the wavelength of the laser employed. This is primarily due to the larger absorption coefficient and hence shorter optical absorption length at shorter wavelength. In general, the absorption coefficients for Al are much larger than those for Si at the wavelengths investigated. One might then expects smaller E_t for Al. However, firstly, the reflectivity of Al is about three times of that of Si and therefore only a small fraction of the laser energy incident on the Al surface is absorbed. Secondly, the thermal diffusion length of Al is larger than the absorption length of the radiation therefore energy was conducted to the substrate below more rapidly. Hence, larger E_t for Al were observed.

It is quite obvious that the magnitude of E_t of Al is primarily determined by its reflectivity in solid phase. Therefore, for an accurate calculated values of E_t, it is essential to have detailed experimental data on the temperature dependent reflectivity of Al. Nonetheless, we used a constant value of reflectivity of Al at 0.693µm irradition and found that the calculated E_t was in good agreement with the experimental value. It suggests that the temperature dependence of the reflectivity of Al at that wavelength is weak.

In summary, we have demonstrated that the temperature dependent optical and thermal properties we employed are quite realistic for the computation of E_t. The heat flow calculations are able to give the melt threshold energies for solid in agreement with the experimental values. This has given us the confidence to extend the present computation model to include the melting and solidification processes.

REFERENCE

1. P. Baeri and S.U. Campisano in Laser Annealing of Semiconductors. J.M. Poate and J.W. Mayer editors (Academic Press, N.Y. 1982) pp. 75-109.

2. E. Rimin in Surface Modification and Alloying by Laser, Ion, and Electron Beams. J.M. Poate, G. Foti and D.C. Jacobson editors (Plenum Press, N.Y. 1983) pp. 15-49.

3. P.S. Peercy and W.R. Wampler, Appl. Phys. Lett. 40, 768 (1982).

4. D.H. Lowndes, R.F. Wood, and J. Narayan, Phys. Rev. Lett. 52, 561 (1984).

5. H.S. Carslaw and J.C. Jaeger Conduction of Heat in Solids. (Oxford Univ. Press, London and N.Y., 1959).

6. E.H. Sin, C.K. Ong and H.S. Tan Phys. Stat Sol (a) 85, 199 (1984).

7. G.E. Jellison, Jr and D.H. Lowndes Appl. Phys. Lett. 41, 594 (1982).

8. G.E. Jellison, Jr and F.A. Modine Appl. Phys. Lett. 41, 180 (1982).

9. H.A. Weakliem and D. Redfield J. Appl. Phys. 50, 1491 (1979).

10. H.G. Dreehsen, C. Hartnich, J.H. Schaefer and J. Uhlembusch J. Appl. Phys. 56, 238 (1984).

11. AIP Handbook, 3rd Ed (McGraw Hill, 1972)   p6-125.

12. A.E. Bell, RCA Review 40, 295 (1979).

13. J. of Phys. and Chem. Ref. data 1, 305 (1972).

14. Ref. (11)   p4-108.

15. G.E. Jellison, Jr. & F.A. Modine ORNL/TM-8002, Oak Ridge National Laboratory (1982).

16. J.L. Brandt. Aluminium vol I. K.R. Van Horn ed. (American Society for metals, 1967).

CALCULATIONS ON MULTI-PULSE, LASER-INDUCED DIFFUSION AND
EVAPORATION LOSS OF ARSENIC AND BORON IN SILICON

H.S. TAN AND C.K. ONG
Department of Physics, National University of Singapore, Kent Ridge,
Singapore 0511

ABSTRACT

A dopant redistribution model that has the dopant diffusion equation
partially solved analytically before engaging numerical method in its final
solution was formulated. The model which included an evaporation loss
mechanism for the near-surface dopants was then employed in the theoretical
fitting of the measured As and B redistributed profiles in Si due to single
and ten-pulse laser irradiations. All the dopant profiles computed were in
good agreement with the measured SIMS data. The values of dopant diffusion
coefficients D_ℓ in liquid Si that were deduced presently from the dopant
profile fitting were $(3.2 \pm 0.3) \times 10^{-4} cm^2 s^{-1}$ for As and $(4.0 \pm 0.4) \times 10^{-4}$
$cm^2 s^{-1}$ for B respectively. Comparison to the reported experimental values
of D_ℓ was also made.

INTRODUCTION

The redistribution of ion-implanted dopants in Si due to liquid phase
epitaxial regrowth of the pulsed-laser irradiated surface is of great
significance to the performances of the devices so processed. However, any
attempt to compute a dopant redistributed profile must necessary involved
the melt-history of the laser irradiated region. Owing to the ultra-rapid
melt-in velocity, it is usually sufficient to characterise the melt-history
by the maximum melt-depth d_m and the average regrowth velocity v during the
solidification process. Nonetheless, it is only recently that reliable
measurements on maximum melt-depths [1,2] and full melt-histories [3,4] were
reported and that allow a model for dopant redistribution computation to
undergo more stringent test. The demand on the accuracy of a computation
model is even more stringent when it is required to fit the dopant redistributed profile due to multi-pulse laser irradiation.

The earlier dopant redistribution calculations [5,6] were based on the
first principle finite difference method that required the division of the
diffusion region into n layers of equal width Δx. Hence the redistribution
of dopant after a time-step Δt was obtained by the numerical solution of a
set of n finite difference equations of the form

$$C(x_i, t+\Delta t) = C_i(x_i, t) + \frac{D_\ell \Delta t}{(\Delta x)^2} \left[C(x_{i+1}, t) + C(x_{i-1}, t) - 2C(x_i, t) \right]$$

$$i = 1, 2, \ldots n \tag{1}$$

where $C(x_i, t)$ is the dopant concentration in the ith layer, D_ℓ is the dopant
diffusion coefficient in liquid Si, while $n\Delta x$ is the melt-depth at time t.
Since such an approach considered only the exchange of dopants between the
nearest neighbouring layers, the criteron $D_\ell \Delta t / \Delta x^2 < \frac{1}{2}$ must be satisfied for
reliable solutions. For good spatial resolution in the computed dopant
profile, Δt has to be small and consequently the necessity of long computation time.

The computation model proposed in the present work has the advantage of
having the mass diffusion equation partially solved analytically before
computing the final solution by numerical method. This allows the summing

of the dopants contributed to any layer from all the n layers during each time-step Δt and thus remove the restriction $D_\ell \Delta t/\Delta x^2 < \frac{1}{2}$ required in the earlier computation model. With the proposed model, the computation speed was shown to be faster with no loss in accuracy in the results.

THE COMPUTATION MODEL

Since the diffusion of dopants due to pulsed-laser irradiation occurs primarily in the liquid phase of Si, the mass diffusion equation to be solved is

$$D_\ell \frac{\partial^2 C}{\partial x^2} = \frac{\partial C}{\partial t} \tag{2}$$

D_ℓ is assumed a constant. The evaporation loss of dopants at the surface is allowed for with the boundary condition

$$D_\ell \frac{\partial C}{\partial x}\bigg|_{x=0} = HC \tag{3}$$

with the loss parameter $H = 0$ for zero loss and $H \to \infty$ for complete loss. The other boundary condition which is at the solid-liquid interface d is

$$D_\ell \frac{\partial C}{\partial x}\bigg|_{x=d} = 0. \tag{4}$$

It implies the dopant mass flux that incidents on the solid-liquid interface from the liquid side is totally reflected back into the liquid domain. This is a consequence of the fact that the diffusion coefficients of dopants in liquid Si are many orders of magnitude larger than their diffusion coefficients in solid. Secondly, dopants such as As and B exhibit no segregation effect in pulsed laser annealed Si. Therefore at the solid-liquid interface, one has $C_s = C_\ell$ with no segregation peak present on liquid side of the interface.

We first consider the diffusion of a dopant source initially at x' at time t_0. The solution of eq. (2) at time $t > t_0$ is

$$C(x,t) = C(x',t_0)(\pi\xi)^{-\frac{1}{2}}e^{-(x-x')^2/\xi} + B(x,t) \tag{5}$$

where $B(x,t)$ is the particular solution and $\xi = 4D_\ell(t - t_0)$. By performing Laplace transformation to C, one has

$$\bar{C} = \int_0^\infty e^{-pt'} C(x,t)dt' = \frac{1}{2D_\ell q} e^{-q|x-x'|} + Me^{-qx} + Ne^{qx} \tag{6}$$

where $t' = t - t_0$, $q = (p/D_\ell)^{\frac{1}{2}}$ and M, N were determined from the Laplace transformed boundary conditions (3) and (4). It can be shown by inverse Laplace transformation that

$$C(x, t) = C(x', t_0)\left[(\pi\xi)^{-\frac{1}{2}}Q - \frac{H}{D_\ell}R + \Theta(H^2) + \ldots\ldots\right] \tag{7a}$$

$$Q = e^{-Y^2/\xi} + \sum_{m=0}^\infty (e^{-[2md+Z]^2/\xi} + e^{-[2(m+1)d-Z]^2/\xi}$$

$$+ e^{-[2(m+1)d-Y]^2/\xi} + e^{-[2(m+1)d+Y]^2/\xi}) \tag{7b}$$

$$R = \text{erfc}(Z/\sqrt{\xi}) + \text{erfc}([2d-Y]/\sqrt{\xi}) + \text{erfc}([2d+Y]/\sqrt{\xi}) + 2\text{erfc}([2d+Z]/\sqrt{\xi})$$

$$+ \Theta(4d) + \ldots \tag{7c}$$

where $Y = x - x'$ and $Z = x + x'$. It should be noted that the analytical solution (7) is obtained assuming a stationary solid-liquid interface. In general, a numerical computation scheme must be adopted for moving solid-liquid interface. The liquid layer is divided into n segments of equal width Δx such that $d = n\Delta x$. The interface is assumed stationary for a duration $\Delta t_n = \Delta x/v$ where v is the speed of the interface motion. Thus for a given dopant profile at time $t - \Delta t_n$, the redistributed profile at time t can be computed using eq. (7). In particular for small loss, terms of $\Theta(H^2)$ may be neglected and the ith segment centered at $x_i = (i - \frac{1}{2})\Delta x$ has

$$C(x_i, t) \simeq \sum_{j=1}^{n} C(x_j, t-\Delta t_n)\Delta x \left[(\pi\xi)^{-\frac{1}{2}}Q - \frac{H}{D_\ell} R \right] \tag{8}$$

where $\xi = 4D_\ell\Delta t_n$. After obtaining the dopant profile $C(x_j, t)$, $j = 1, 2 \cdots n$, the solid-liquid interface is advanced instantaneously to the position $(n - 1)\Delta x$ or $(n + 1)\Delta x$ depending on whether the system is undergoing a melting or solidification process. The new melt-front is again assumed stationary for a duration of Δt_{n-1} or Δt_{n+1} and computation is repeat with the newly obtained $C(x_j, t)$ and with eq. (8) accordingly adjusted.

Generally, the values of d(t) and Δt_n to be employed in the above numerical computation are read from the melt-history curve of the pulsed-laser irradiated region. However, it is known that the melt-in velocity is many times faster than the regrowth velocity of a ns pulsed-laser irradiated region. Hence a "triangular approximation" to the full melt-history curve that assumes instantaneous melting of the region from the sample surface to the maximum melt-depth d_m was proposed and successfully applied in an earlier work [7]. With the approximation, the melt-history is sufficiently characterised with a value of d_m and an average regrowth velocity v. Then one would have $\Delta t_n = \Delta x/v$ independent of n. The computation scheme is thus simpler and the computation speed faster.

COMPUTATIONS AND RESULTS

Hill et al [1] had reported earlier some measured redistribution profiles of B and As implanted Si due to single or ten-pulse laser irradiations. The Q-switched ruby laser pulse-length was 25 ns and each pulse was of energy 2.5 J cm^{-2}. The dopant profiles before and after the 1 and 10 pulses of laser irradiations were measured with SIMS and the data were carefully calibrated.

Starting with the initial dopant profiles measured by Hill et al, the redistributed dopant profiles were computed with eq. (8) and using triangular approximated melt-histories. By comparing the unirradiated and single-pulse B profiles, the maximum melt-depth d_m due to the first pulse was estimated to be .58 μm. On the other hand, the steep portion of the final As profile indicated that d_m = .52 μm for each of the nine subsequent pulses. The extra melt-depth of .06 μm for the first pulse is consistent with the fact that less energy is required to melt the amorphous Si layer that only exists for the pulse. A value of the regrowth velocity v = 3 ms^{-1} was employed for the present computations. It is consistent with the v of 2.7 ms^{-1} measured earlier [3] by irradiating Si with a 30 ns ruby laser pulse. Computations with eq. (8) were performed by neglecting terms of $\Theta(4d)$ because of the rapid decay in the gaussian and erfc functions.

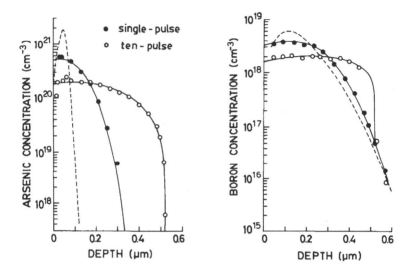

Figure 1. The dash curves are the initial unirradiated dopant profiles. The solid curves are the computed dopant profiles due to single-pulse and ten-pulse laser irradiations. The experimental data were obtained by Hill et al [1] with SIMS. See text for the details on computation parameters.

The initial and the computed As and B profiles are shown in figure 1. The computed and measured dopant profiles are in very good agreement when $D_\ell = 3.2 \times 10^{-4}$ cm^{2}s^{-1}, H = 3.2 cm s^{-1} are employed for As and $D_\ell = 4.0 \times 10^{-4}$ cm^{2}s^{-1}, H = 8.8 cms^{-1} are employed for B respectively. The computed results also show losses of about 4% and 17% (4% and 25%) in the initial Arsenic (Boron) dosage due to single and ten-pulse laser irradiations. It should be mentioned that the earlier As and B redistribution calculations [6, 8] were performed assuming no evaporation loss. It is likely because a loss of less than 5% may be easily overlooked as due to measurement error. However, with ten pulses of irradiation, the accumulated losses of 17% and 25% for As and B indicate that evaporation of these dopants were real.

The value of D_ℓ for As deduced presently is $(3.2 \pm .3) \times 10^{-4}$ cm^{2}s^{-1} which is in excellent agreement with the value $(3.3 \pm .9) \times 10^{-4}$ cm^{2}s^{-1} measured by Kodera [9]. The value of D_ℓ for B deduced presently is $(4.0 \pm .4) \times 10^{-4}$ cm^{2}s^{-1} is larger than the values of $(2.4 \pm .7) \times 10^{-4}$ and $(3.3 \pm .4) \times 10^{-4}$ cm^{2}s^{-1} measured by Kodera [9] and Shashkov and Gurevich [10] respectively. On one hand, the difference can be conviniently attributed to the interaction between the B and As ions that were present in the same sample prepared by Hill et al [1]. On the other hand, since B ion is much smaller and lighter than As ion, A larger D_ℓ for B than the D_ℓ for As as indicated by the present values obtained is consistent with a simple kinetic picture.

CONCLUSION

A dopant redistribution computational model which included a surface evaporation loss mechanism was formulated and employed in the calculations of As and B redistributed profiles in Si due to single-pulse and ten-pulse laser irradiations. Agreement is excellent between the measured and computed profiles if the values $D_\ell = 3.2 \times 10^{-4}$ cm^{2}s^{-1}, $H = 3.2$ cm s^{-1} for As and $D_\ell = 4.0 \times 10^{-4}$ cm^{2}s^{-1}, $H = 8.8$ cm s^{-1} for B are employed in the computations. The uncertainty in the values of D_ℓ and H is about 10%.

REFERENCES

1. C. Hill, A.L. Butler and J.A. Daly, Laser and Electron-Beam Interactions with Solids, edited by B.R. Appleton and G.K. Celler (Elsevier-North Holland, New York 1982) pp. 579-584.

2. Y. Hayafuji, Y. Aoki and S. Usui, Appl. Phys. Lett. 42, 702 (1983).

3. G.J. Galvin, M.O. Thomson, T.W. Mayer, R.B. Hammond, N. Paulter and P.S. Peercy, Phys. Rev. Lett. 48, 33 (1982).

4. M.O. Thomson, G.J. Galvin, Laser-Solid Interactions and Transient Thermal Processing of Materials, edited by J. Narayan, W.L. Brown and R.A. Lemons (Elsevier, New York 1983) pp. 57-67.

5. C.W. White, S.R. Wilson, B.R. Appleton and F.W. Young, Jr. J. Appl. Phys. 51, 738 (1980).

6. R.F. Wood, J.R. Kirkpatrick and G.E. Giles, Phys. Rev. B 23, 5555 (1981).

7. H.S. Tan and C.K. Ong, J. Phys. C: Solid State Phys. 16, 5063 (1983).

8. R.T. Young, R.F. Wood, W.H. Christie and G.E. Jellison, Jr. Appl. Phys. Lett. 39, 313 (1981).

9. H. Kodera, Jap. J. Appl. Phys. 2, 212 (1963).

10. Y.M. Shashkov and V.M. Gurevich, Russ. J. Phys. Chem. 42, 1082 (1968).

EXPLOSIVE CRYSTALLIZATION IN a-Ge FILMS
IRRADIATED WITH MICROSECOND LASER PULSES

R.K. SHARMA AND S.K. BANSAL
Department of Physics and Astro Physics, University of Delhi
Delhi - 110007, India.

ABSTRACT

     Explosive crystallisation has been observed in deposited
a-Ge films irradiated with pulsed Nd-glass laser beam of
400 micro second duration. Transmission electron microscopic
examination of the irradiated film shows a polycrystalline
region surrounded by radial dendrites of ∿ 8-10 μm size.
Analysis of the results is consistent with the duplex-melting
model of the explosive crystallisation mechanism. Heat released during the amorphous to crystalline transformation of a
localised region abruptly crystallises the surrounding area
resulting in dendritic growth. The crystallisation is self-
sustaining untill the temperature ahead of liquid/amorphous
interface drops below 775 K. These results are consistent
with earlier study of explosive crystallisation in unsupported
a-Ge films and further confirm that the heat loss to the underlying substrate governs the dynamics of the crystallisation
process.

Introduction

     One of the interesting phenomena frequently observed duri-
ng the laser processing of amorphous semiconductors is the run-
away or explosive crystallisation (EC) phenomenon [1-10]. It
has been observed during the pulsed as well as scanned CW laser
irradiation of a-Ge [1-5, 9, 10] and a-Si films [6-8]. It has
also been observed with a mechanical pin-prick [4] or a laser
shot [9] at same point on a several micron thick amorphous film.
The phenomenon manifests itself startlingly when following
initiation, crystallisation proceeds with a velocity of ⩾1
meter/sec across the film and is accompanied by heat liberati-
on. The latent heat liberated at the crystallisation front
serves to drive the reaction further. Crystallisation is self/
sustaining only at a substrate temperature T greater than some
critical value T_s^* [4,9]. With a scanning CW laser beam, a
periodic morphology consisting of oriented elongated crystal
grains has been observed[1-3]. The periodicity increases with
increasing background temperature. Zeiger et al.[2] theoriti-
cally analysed the EC phenomenon in terms of the solid state
process i.e. SPEC model whereas Gold et al.[3] proposed a
duplex-melting model for the same. Theoritical analysis of EC
phenomenon by Gilmer & Leamy[5] predicted that EC phenomenon is
intermediated via a thin liquid layer. A recent detailed study
of the EC phenomenon in a-Ge unsupported films [10] has
clearly shown that the explosive crystallisation propogates
via a melting step.

     In the present study, EC has been observed in a-Ge films
irradiated with μ-sec laser pulses. Electron microscopic
examination of the irradiated films show dendritic growth at
several places within the irradiated portion of the film

indicating explosive crystallisation. The present results
further confirm the duplex-melting model of the EC process and
its propagation is governed by the heat dissipation through
the underlying substrate.

Experiment And Results

        Amorphous films of 99.9999% pure Ge were deposited by
thermal evaporation onto freshly cleaved NaCl substrates in a
vacuum of $< 10^{-6}$ Torr. As deposited films were irradiated in
air at room temperature with Nd-glass laser (λ = 1.06 μm)
pulses (400 microsecond duration) of various power densities.
Microstructural features of the crystallised films were studied
with TEM observations. Figure 1a shows the electron micrograph
of an a-Ge film ($\sim$ 1800 $^{\circ}$A thick) crystallised with a single
laser pulse at 300 KW/cm^2 power density. It consists of a
central polycrystalline region ($\sim$3.0 μm dia, grain size $\gtrless$0,2 μm)
surrounded by radial dendritic growth ($\sim$15 μm-dia) all around
the polycrystalline region. Selected area diffraction (SAD)
pattern of the dendritic portion shows its crystalline nature
with a preferred orientation. These morphological features are
exactly identical with our earlier reported results on unsupported a-Ge films crystallised with a focussed electron beam
(reproduced in figure 1b for comparison from Sharma et al.[10],
J. Appl Phys. 55(2), 387, Jan. 1984). Figure 2 shows the electron micrograph of an a-Ge film ($\sim$1000 $^{\circ}$A thick) irradiated
with a single laser pulse of 100 kw/cm^2 power density. It
shows few crystal grains at the bottom left corner followed by
dendritic growth propagating towards the right upper corner upto
a certain distance ($\sim$6 μm) and stops abruptly showing a sharp
boundary.

        On irradiation with the laser pulse, the incident energy
is absorbed in a-Ge medium resulting in its uniform heating. A
temperature rise of 800K is calculated at 100 KW/cm^2 power
density. This is well above the crystallisation temperature of
a-Ge and amorphous to crystalline (a $\rightarrow$ c) transformation occurs
at random. A typical grain size of $\sim$0.2 μm in the polycrystalline region (figure 1a) is consistent with the extrapolated
crystal growth[11] of (100 $^{\circ}$A/min at 330°C) at the estimated
temperature rise of 800K. The transformation is exothermic[12,
13] with a consequent release of heat energy (H_c = 2.6 k Cal/mole).
This heat released would heat up the adjacent virgin amorphous
portion of the film. Using the thermal parameters of a-Ge and
NaCl substrate,calculations show that 80% of the heat released
is dissipated to the underlying NaCl substrate and the remaining
20% is consumed radially along the a-Ge film. Further calculations show that this 20% balance of energy is sufficient to
raise the temperature of an adjacent 1 μm wide annular portion
from 800K to 1000K. Direct solid phase processes can not account for the observed dendritic growth at the elevated temperature
of 1000K. The melting temperature (T_m^a) of a-Ge is reported to
be 970K [14,15] which is well below the melting temperature
(T_m^c = 1210K) of crystalline Ge. Melting of a-Ge,therefore,yields
a liquid which is substantially undercooled w.r.t. its crystalline phase so that crystallisation would be extremely rapid.
Dendritic growth has been satisfactorily explained in terms of
the duplex-melting model[10]. Duplex-melting crystallisation
front consists of two phase boundaries: (i) an endothermic

Figure 1. Transmission electron micrographic showing growth of dendrites around a polycrystalline region in the a-Ge film. (a) a-Ge film (1800 ōA thick) irradiated with Nd-glass laser ($\lambda = 1.06\,\mu m$) pulse of 400 μ.sec duration at 300 kw/cm2 power density. (b) a-Ge film 1000 ōA thick irradiated with a focussed electron beam [10]. Micrograph 1a also shows radial arrays of gaseous bubbles.

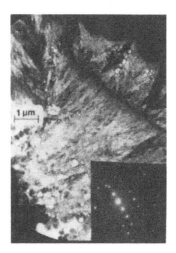

Figure 2. Electron micrograph of an a-Ge film (1000 Å thick) irradiated with the 400 μ. sec. duration laser pulse at 100 kw/cm² power density. Dendritic growth propogates from the bottom left corner toward the upper right corner. A dense array of gaseous bubbles is clearly seen.

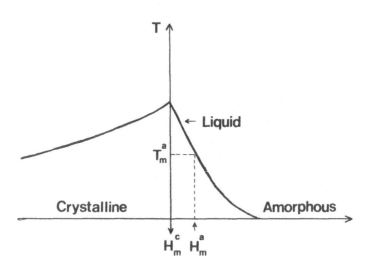

Schematic illustration of the duplex melting model showing temperature distribution pependicular to the crystallisation front during explosive transformation.

amorphous/liquid interface where melting is taking place at T^a_m and (ii) an exothermic liquid crystalline interface close behind. The heat dissipation radially outward along the a-Ge film yields a super cooled Ge melt which crystallises rapidly in the form of dendrites. Resolidification releases further heat. The difference (ΔH_m) in enthalpy of melting of crystalline and amorphous phases of Ge is estimated to ~ 2.0 K Cal/mole[14,15]. This energy released propagates radially outward and melts the further virgin amorphous portion which in turn crystallises and drives the reaction further. Thus a thin liquid layer propogates giving rise to radial dendritic growth all around the central polycrystalline region. Eventually this self-sustaining EC is quenched due to significant heat dissipation through the under-lying NaCl substrate such that the balance of the energy is insufficient to raise the temperature of a-Ge solid in front of the crystallisation front to T^a_m. Calculations based on the energy balance consideration of the heat liberated and heat lost to the underlying NaCl substrate show that the dendritic growth terminates when the temperature ahead of the crystallisation front ≤ 775 K.

A close examination of figure 1a and 2 shows the presence of dense arrays of gaseous bubbles radially arranged within the dendritic portion. The large bubbles typically 1000 ^{0}A in size are observed. This observation is consistent with the earlier reported results of Leamy et al.[9] during EC of supported a-Ge films . Gaseous impurities are present throughout the deposited amorphous film which penetrate the film through the inter connected network of voids. The formation of bubbles is due to the coaleseence of gas within the film upon melting and they have been frozen before the smoothing action of the surface tension upon the liquid layer has had time to be effective.

It can therefore be concluded that the EC phenomenon is intermediated by a thin liquid layer and the heat lost to the underlying NaCl substrate is responsible for the early termination of the dendritic growth and does not allow the formation of the annularly arranged periodic morphology[10] observed during the EC of unsupported a-Ge films.

Authors thankfully acknowledge the financial assistance from University Grant Commission, India during the course of this work.

References :

1.   Fan J.C.C., Zeiger H.J., Gale R.P., and Chapman R.L. (1980) Appl. Phys. Lett. 36, 158.

2.   Zeiger H.J., Fan J.C.C., Palm B.J., Galc R.P., and Chapman R.L. (1980) in "Laser and Electron Beam Processing of Materials" (C.W. White and P.S. Peercy eds.), p.234, Academic Press, New York.

3.   Gold R.B., Gibbons J.F., Magee T.J., Peng J., Ormand R., Deline V.R., and Evans C.A. Jr., (1980) in "Laser and Electron Beam Processing of Materials (C.W. White and P.S. Peercy eds.), p.221. Academic Press, New York.

4. Koba R. and Wickersham C.E. (1982) Appl. Phys. Lett. 40(8), 671.

5. Gilmer G.H. and Leamy H.J. (1980) in "Laser and Electron Beam Processing of Materials (C.W. White and P.S. Peercy eds.) p.227, Academic Press, New York.

6. Lemons R.A. and Bösch M.A. (1981), Appl. Phys. Lett 39, 343.

7. Narayan J. and White C.W. (1984), Appl. Phys. Lett 44(1), 35.

8. Cullis A.G., Webber H.C. and Chew N.G. (1980), Appl. Phys. Lett. 36, 547.

9. Leamy H.J., Brown W.L., Celler C.K., Foti G., Gilmer G.H. and Fan J.C.C. (1981), Appl. Phys. Lett. 38, 137.

10. Sharma R.K., Bansal S.K., Nath R., Mehra R.M., Bahadur K., Mall R.P., Choudhary K.L. and Garg C.L. (1984) J. Appl. Phys. 55(2), 387.

11. Csepregi L., Kullen R.P., Mayer J.W. and Sigman T.W. (1977), Solid state Commn. 21, 1019.

12. Chopra K.L., Randhawa H.S. and Malhotra L.K. (1977), Thin Solid Films 47, 203.

13. Chen H.S. and Turnbull D. (1969), J. Appl. Phys. 40, 4214.

14. Bagley B.G. and Chen H.S. (1979), AIP Conf. Proc. 50, 97.

15. Spaepen F. and Turnbull D. (1979) AIP Conf. Proc. 50, 73.

IMPLANTED IMPURITY INCORPORATION AND SEGREGATION PHENOMENA INDUCED BY PEBA IN SILICON

G. CHAUSSEMY, D. BARBIER and A. LAUGIER Laboratoire de Physique de la Matière (LA CNRS n° 358), Institut National des Sciences Appliquées de Lyon, 20 Avenue Albert Einstein 69621 Villeurbanne Cédex France

ABSTRACT

In this work, the PEBA induced thermal effects have been varied to study the diffusion of usual implanted impurities (P, As, Sb, In) and segregation phenomena in (100) and (111) silicon. The mean melt-front velocity has been adjusted between 1 and 4 m/s by modifying both the beam fluence and the sample starting temperature. A model for dopant redistribution has been developped, using a mean diffusion coefficient D and solving the one dimensional Fick's equation. Segregation and dopant evaporation are considered and introduced as limit conditions at the liquid-solid interface and at the wafer surface respectively. The impurity redistribution has been experimentally studied by SIMS profiling ; so that interfacial segregation coefificient Ki may be deduced from comparison between experimental and computed profiles.

I. INTRODUCTION

Dopant redistribution in silicon under pulsed laser irradiation has been widely studied over this five past years. Indeed, annealing in the liquid phase regime, with meter per second melt-front velocities offers new opportunities to access non equilibrium dopant incorporation in the regrowth layer (1,2). Also, this is especially the case of Pulsed Electron Beam Annealing (PEBA) for which a great variety of thermal effects can be achieved through the easy modifications of the beam parameters (3). Final impurity profiles generally depend on both the irradiation induced thermal effects (molten layer thickness, liquid phase duration, melt-front velocity) and the impurity properties (diffusivity in the liquid phase, segregation coefficient, solid solubility limit). In some cases, dopant evaporation out of the liquid phase must be taken into account to explain experimental results. Non-equilibrium segregation occurs at the regrowth interface when the rate of the dopant rejection from the solid to the liquid is exceeded by the host atom plane reconstruction rate (4). This phenomenon leads to solute trapping in the solidified material, it depends on both the melt-front velocity and the crystalline substrate orientation (2). Phenomenologically, solute trapping is well described by considering an interfacial segregation coefficient Ki which have been correlated to the equilibrium value Ko (5). In this work, incorporation of some usual implanted impurities in silicon (P, As, Sb, In), after Pulsed Electron Beam Annealing, has been studied. Secondary Ion-Mass Spectroscopy (SIMS) redistribution profiles have been fitted with computed profiles obtained from a numerical model given in § III.

Correlation between diffusivities, dopant losses, segregation coefficient and PEBA induced particular thermal effects will constitute the basis of our discussion.

II. EXPERIMENT

The SPIRE 300 Pulsed Electron Beam Processor used in this work has been previously described (6). It delivers polykinetic electron pulses of 50 ns in duration. Mean electron energies ($\bar{E}$) of 12 and 15 keV have been used with energy densities, or fluence, ranging from 0.7 to 1.4 J/cm^2. The substrate temperature was either 20°C or about 450-500°C to reduce the melt-front velocity. The PEBA induced melting process is characterized by an inhomogeneous latent heat deposition in a so-called "melting layer". Only the shallower part of this melting layer is expected to be fully molten and can even exceed the melting temperature T_m (at sufficiently high fluence) within the irradiation time, depending on the material thermal properties. During the freezing process the solid-melting zone interface velocity Vs is governed by the following expression (2) :

$$\eta (x_M) L \rho Vs = \lambda (dT/dx)_{x_M} \tag{1}$$

with : $0 < \eta(x_M) < 1$. $\eta(x_M)L$ denotes the latent heat fraction which must be evacuated at the interface through the action of the temperature gradient dT/dx. λ is the thermal conductivity. Immediatly after the pulse, the melt-front velocity, Vs exhibits a high value near the edge of the melting layer according to $\eta(x_M) << 1$ and equation /1/, at constant dT/dx. During recrystallization, the amorphous superheated fully molten zone acts as an heat reservoir and heat redistribution occurs into the underlying melting zone at $T = T_m$, increasing $\eta(x)$. This gives rise to a steady state velocity regime corresponding to $\eta(x)$ averaged during regrowth through this mechanism. For more details see ref.(3).

The beam parameters used in this work, combined with preheating of the sample at 450°C, have permitted to get Vs in the 1-4 m/s range.

Phosphorus diffusion has been studied in the case of a shallow implantation (10 keV -3.10^{15} cm^{-2}) in (111) silicon, producing a 500 Å thick damaged layer where complete amorphization is not achieved, as indicated by R.B.S. experiments. Arsenic has been implanted in (100) silicon at 140 keV at a dose of 10^{15}cm^{-2} with a subsequent 1500 Å thick amorphized layer. The same implantation conditions have been used with Sb in both (100) and (111) silicon and with In in (100) Si.

III. DOPANT REDISTRIBUTION : MODEL AND RESULTS

1. Model

According to § I, dopant redistribution in the molten layer is assumed to be governed by three mechanisms : (i) diffusion according to Fick's law, (ii) segregation at the melting solid interface, (iii) dopant evaporation accross the sample surface. All these phenomena have revealed an observable influence on the post-annealed SIMS profiles. Near the surface, the profile depends mainly on segregation effects and on dopant evaporation, whereas the tail is rather affected by diffusion properties of the melting layer. A numerical model has been developped neglecting diffusion motion in the solid phase, and convection in the liquid. The one dimensional Fick's equation has been solved by a finite difference method in a similar way than previously suggested by C.W. White et al. (7). The melting zone is divided into slices Δx of constant concentration Cj depending on x and time t. Diffusion is allowed to occur B times before the melt-front jumps from a slice into the next one (B is an integer taken equal to 5). Segregation is considered at the liquid-solid interface by programming :

$$C_k \leftarrow C_k + (1 - Ki) C_{k+1} \quad \text{and} \quad C_{k+1} \leftarrow Ki C_{k+1} \tag{2}$$

before the interface jumps from the $(k+1)^{th}$ to the kth slice. Modelization of the dopant evaporation is obtained through a simple control of the flux at the surface, via the concentration gradient ; an external but fictive layer with concentration $C_o = (1-P)C_1$ is introduced, with $0 \leq P \leq 1$ (C_1 is the concentration in the first slice). Two PEBA thermal kinetic parameters are introduced : maximum melting zone extent Xm and liquid phase duration θm. Then Vs is averaged by $Xm/\theta m$.

Table I - Coefficient for diffusion, segregation and dopant losses. The higher values for D are achieved with 12 keV mean energy ($\bar{E}$) electrons at a given fluence while the lower limit corresponds to $\bar{E} = 15$ keV. All diffusion coefficients D are in the 10^{-4} range, except for P which is a special case as discussed in the text.

Dopant and Si orientation	D (1E-4 cm2/s)		Ki	evaporation rate %
	this work	ref.(8)		
P Si(111)	.1 - .3	5.1±0.7	1	> 20
As Si(100)	2.5 - 3.0	3.3±0.9	1	6 - 7
Sb Si(100)	1.5 - 2.7	1.5±0.5	.2 - .7	0
Sb Si(111)	1.0 - 3.2		.4 - .9	0
In Si(100)	5.0 - 7.0	6.9±1.2	.011 .055	0

Astonishingly, diffusive properties of the material are well described by one mean diffusion coefficient D. The evaluation of D, Ki and P is obtained from the best fit of the experimental profiles. Although the model involves three adjustable parameters, this evaluation yields values lying within a relatively narrow range. The effective evaporation rate is obtained after normalization of the computed diffused profile.

2. Results

A significative set of results is given in table 1. It is clear that the whole melting zone presents liquid-like diffusive properties, but no evident correlation between fluence and D has appeared. However, raising the fluence increases diffusion length through the subsequent increase of the liquid phase duration (Fig.1). Generally, a misfit is observed at low concentration but this disagreement is not significative when occuring near the SIMS sensibility limit (i.e. 10^{18}at/cm^2). Increasing the fluence reduces the melt-front velocity Vs (3) then Ki is also reduced (Fig.1). At constant fluence, preheating of the sample strongly reduces Vs and leads to a smaller Ki (Fig.2). However, no segregation occurs with As and P dopants which exhibit a strong surface evaporation whereas losses can be ignored with In and Sb (see Fig.3).

IV. CONCLUDING REMARKS

The low diffusivity in the case of the shallow phosphorous implantation must be correlated with the particular thermophysical effects induced by PEBA in the melting layer. Indeed, heavy ions are likely to induce a rather thick and fully amorphized surface layer acting as a heat reservoir for the underlying part of the melting zone. Hence the average latent heat fraction absorbed in the melting zone is enhanced in the case of a fully amorphized surface layer, comparatively to crystalline silicon. The D values for As, Sb and In are close to previously published data (8), although slightly lower when using the 15 keV mean energy electron beam pulse. This is consistent with a lower latent heat fraction deposited in this latter case. So the "melt-like" diffusive properties of the PEBA induced "melting-layer"

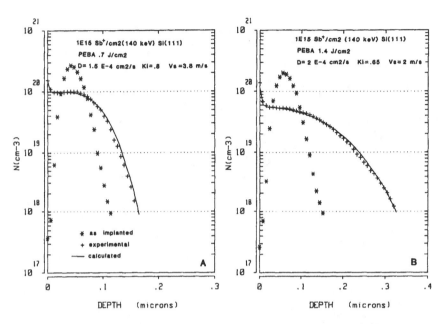

Fig. 1 – Sb concentration profiles before and after annealing. (A) Fluence = 0.7 J/cm^2 ; electron mean energy $\bar{E}$ = 12 KeV. (B) Fluence = 1.4 J/cm^2 ; $\bar{E}$ = 15 KeV

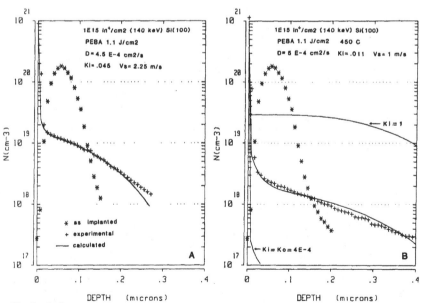

Fig.2 - Indium concentration profiles before and after annealing at constant fluence and $\bar{E}$ = 12 keV. (A) Wafer temperature 20°C ; (B) sample preheated at 450°C

are evidenced. Besides, they are correlated to the initial physical state of the processed layer.

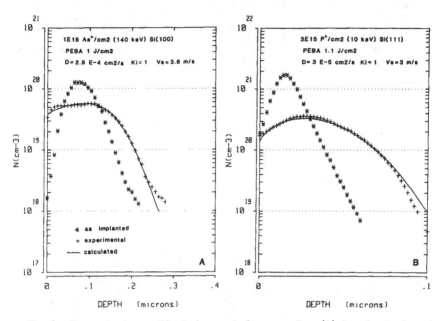

Fig. 3 - Concentration profiles before and after annealing. (A) Arsenic implanted silicon ; evaporation rate = 7 %. (B) Phosphorous implanted silicon ; evaporation rate = 35 %

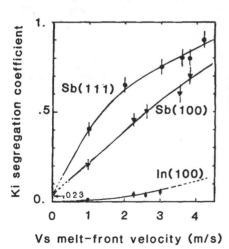

Fig. 4 - Measured segregation coefficient Ki versus melt-front velocity Vs plots, for In in Si (100), Sb in (100) and (111) silicon

262

Non equilibrium segregation effects are also noteworthy because these phenomena are likely to occur at a well defined liquid-solid interface. This is not exactly the case for the PEBA induced melting layer which includes a rather thick transition zone between solid and liquid. However, crystal regrowth and impurity incorporation can only be initiated at the moving edge of the melting layer where a high temperature gradient is able to drive the energy from the quasi-liquid to the solid. This point is supported by our results concerning the dependence of Ki on both Vs and the substrate orientation, as far as we have deduced Vs by locating the melt-front at the edge of the melting layer. Indeed, the Ki versusVs curves of Fig.4 are in good agreement with results obtained by some authors from laser annealing experiments (9).

In conclusion, the PEBA induced melting layer obviously exhibits "melt-like" properties considering diffusion and segregation of dopants. However, these properties are not well understood yet ; one can assume the existence of an "intermediate" and microscopically inhomogeneous state where diffusion would occur through a "percolation" mechanism.

ACKNOWLEDGEMENTS

This work was supported by A.F.M.E.

REFERENCES

(1)   R.F. WOOD, J.R. KIRKPATRICK, G.E. GILES. Phys.Rev.B 23, 10, p.5555 (1981)

(2)   P. BAERI, G. FOTI, J.M. POATE, S.U. CAMPISANO, E. REMINI, A.G. CULLIS. Laser and Electron Beam Solid Interactions and Transient Thermal Processing, J.F. Gibbons, L.D. Hess, T.W. Sigmon eds (Nort Holland, New York, 1981) p.67

(3)   G. CHEMISKY, D. BARBIER, A. LAUGIER. Journ. of Crystal Growth 66, p.215 (1984)

(4)   K.A. JACKSON, G.H. GILMER, H.J. LEAMY. Laser and Electron Beam Processing of Materials, C.W. White, P.S. Peircy eds (Academic Press, New York 1980) p.104

(5)   F. MOREHEAD. Laser and Electron Beam Processing of Materials, C.W. White, P.S. Peircy eds (Academic Press New York, 1980) 143

(6)   A. LAUGIER, D. BARBIER, G. CHEMISKY. Proceedings of the 4th E.C. Photovoltaic Solar Energy Conference, D. Reidel Publishing Company, p.1007 (1982)

(7)   C.W. WHITE, S.R. WILSON, B.R. APPLETON, F.W. YOUNG. J.A.P. 51, 1, p.736 (1980)

(8)   H. KODERA. Jps J.A.P. 2, p.212 (1965)

(9)   H. GILMER. Laser Solid Interactions and Transient Thermal Processing of Materials, J. Narayan, W.L. Brown, R.A. Lemons eds (North Holland, New York, 1983) p.249

-§-

PART IV

Rapid Thermal Processing

ELECTRONIC DEFECTS IN TRANSIENT,
THERMALLY PROCESSED SEMICONDUCTORS

N. M. JOHNSON
Xerox Palo Alto Research Center, Palo Alto, CA 94304

ABSTRACT

This paper summarizes the general observations that may be drawn from numerous studies of electronic defects in transient thermally annealed bulk single – crystal silicon and discusses the emerging subjects of electronic defect evaluation in beam – crystallized silicon thin films and in epitaxially – grown III – V semiconductors.

INTRODUCTION

Directed energy sources such as lasers and electron beams and incoherent light sources have been proposed as alternatives to conventional furnace annealing for (1) recrystallizing the amorphous layer created by or (2) activating dopants introduced by high – dose ion implantation in single – crystal semiconductors. These energy sources have also been used to crystallize silicon thin films on insulating amorphous substrates in order to obtain semiconducting material for electronic device fabrication. In both applications the purpose of transient thermal processing is to produce single – crystal material of high crystalline perfection. With both cw and pulsed energy sources in silicon, the recrystallized material is found to possess lower densities of extended defects than can be achieved by conventional furnace annealing. However, materials studies with techniques such as capacitance transient spectroscopy, luminescence, and electron – beam induced currents reveal high densities of residual defects in and near the recrystallized layers which give rise to deep levels in the silicon forbidden – energy band. Far less information is available on residual electronic defects in compound semiconductors. Structural defects in transient thermally annealed semiconductors are reviewed in Ref. 1. This paper summarizes the general observations that may be drawn from the numerous studies of electronic defects in transient thermally processed bulk single – crystal silicon and discusses the emerging subjects of electronic defect evaluation in beam – crystallized silicon thin films and in epitaxially – grown III – V semiconductors.

BULK SINGLE – CRYSTAL SILICON

Bulk single – crystal silicon has been the primary semiconductor of interest in research on energy – beam/solid interactions and transient – thermal processing since the first the Materials Research Society symposium was held on this subject in 1978. The subject of electronic defects in beam – processed silicon has been an integral part of the field since this first symposium and has received several reviews in recent years [2 – 5]. In this section the general observations that may be drawn from the numerous original studies of electronic defects in bulk silicon are summarized and illustrated with research which has indicated the microscopic origin of residual defects in transient thermally processed silicon.

General Observations

The following general observations have been drawn from numerous studies of electronic defects in transient thermally – processed bulk single – crystal silicon [5]:

(1) Residual electronic defects accompany all forms of transient thermal annealing.

(2) These defects are localized in the near – surface region.

(3) These include quenched – in defects.

(4) Beam – annealing conditions can be optimized for minimum defect densities. However, raster – scanned point sources always leave laterally – distributed residual defects.

(5) Laterally nonuniform recrystallization can result from wafer topography (e.g., patterned dielectric overlayers).

(6) Post – recrystallization processing (e.g., furnace anneals and hydrogen passivation) can be used to reduce the density of residual defects.

The first three of the above observations will be discussed with specific illustrations in the following subsections. The others are reviewed in Ref. 5.

N – Type Silicon

The existence of residual electronic defects and their near – surface localization after transient thermal annealing will be illustrated with results from an early study of cw – beam annealed silicon [6]. Silicon wafers of n – type conductivity were implanted at room temperature with SiH^+ to create

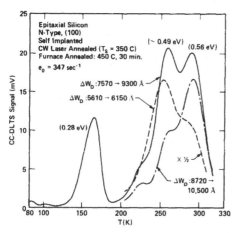
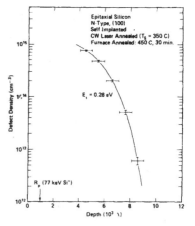

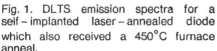

Fig. 1. DLTS emission spectra for a self – implanted laser – annealed diode which also received a 450°C furnace anneal.

Fig. 2. Spatial depth profile of the defect level at $E_C - 0.28$ eV in self – implanted laser – annealed silicon.

displacement damage of the form accompanying the implantation of dopant impurities without altering the original shallow dopant concentration; this permitted the fabrication of simple Schottky – barrier diodes for deep – level evaluation. The ions were implanted at 80 keV to a dose of 2×10^{15} cm^{-2}, which was sufficient to drive the silicon amorphous to a depth of ~0.12 μm. Then the material was recrystallized with a scanned cw Ar – ion laser beam, a rectifying metal contact was vacuum deposited, and the diode received a forming gas anneal (450°C, 30 min) in preparation for defect measurements by deep – level transient spectroscopy (DLTS). Electron emission spectra are shown in Fig. 1. The spectra are dominated by two midgap levels with activation energies for electron emission of ~0.49 and 0.56 eV. In addition, a prominent peak appears at 0.28 eV which is well – resolved from the other levels and therefore permits accurate depth profiling. This level appears only after the forming gas anneal and is stable to temperatures above 600°C. The spatial depth profile for the 0.28 eV level is shown in Fig. 2. The defect density decreases monotonically with depth below the laser – irradiated surface. Also shown is the projected range for the implanted silicon ions, which approximates the depth to which the material was amorphized by implantation. This reveals that the detected defects reside in material that was not driven amorphous by ion implantation. Such defects may arise from two sources. Ion implantation introduces displacement damage to depths beyond the amorphized surface layer. A beam annealing schedule can recrystallize the amorphous layer while only partially removing the more deeply – penetrating lattice damage. On the other hand, beam irradiation can generate both electronic and extended structural defects as a consequence of the high thermal gradients and rapid quenching which are consequences of the spatial and temporal features of beam recrystallization. Furthermore, in implanted material a clear distinction between these two sources can be obscured by defect interactions.

Unimplanted silicon most directly demonstrates that electronic defects can be quenched – in during transient thermal processing of silicon. This is illustrated in Fig. 3 with a DLTS spectrum for scanned electron – beam annealed (SEBA) silicon [7]. As – received n – type silicon was electron – beam annealed, under conditions that would be applicable for recrystallizing an As$^+$ – implanted layer, and used to fabricate Schottky – barrier diodes. The spectrum for the annealed diode is compared to that for unannealed material. The spectrum for the beam – annealed device is dominated by two emission peaks with activation

Fig. 3. DLTS spectrum for electron emission in unimplanted scanned electron – beam annealed silicon. Also shown is the spectrum for the unannealed material.

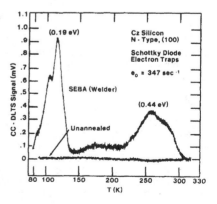

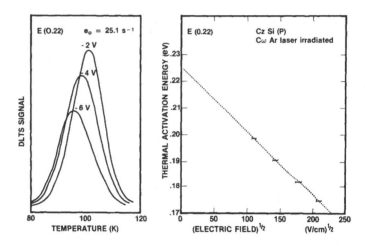

Fig. 4. Electric field effect on electron emission from E(0.22). (A. Chantre et al., Ref. 11.)

energies of 0.19 and 0.44 eV. In the unannealed control, no emission peak is detectable on the indicated scale of sensitivity. Similar quenched – in defects have been observed in laser irradiated silicon [8,9]. The 0.44 eV level has been identified as the phosphorus – vacancy complex based on the electron emission energy and effect of charge state on the defect annealing rate [7 – 10].

The low – temperature emission peak in Fig. 3 has been observed in both pulsed – and scanned – laser annealed silicon and found to have donor – like electron emission characteristics [2,11]. This is shown in Fig. 4 with DLTS spectra for different bias voltages in (a) and the resulting electric – field dependence for thermal emission in (b) [11]. It is evident that reducing the reverse bias moves the DLTS peak to higher temperatures, which corresponds to a reduction in the electron emission rate at constant temperature. The apparent thermal activation energy for electron emission, E_a, is found to vary with the junction field F as follows:

$$E_a = E_0 - q[qF/(\pi \varepsilon)]^{\frac{1}{2}} \text{ with } E_0 \approx 0.225 \text{ eV}, \tag{1}$$

where E_0 is the zero – field activation energy, q is electronic charge, and ε is the silicon permittivity. This identifies a Poole – Frenkel emission process and confirms the donor character of the defect. In addition, it has been proposed that the defect is a chromium atom at an interstitial site (Cr_i) [11].

P – Type Silicon

Transient thermal annealing studies in p – type boron – doped silicon have revealed additional, identifiable quenched – in defects with energy levels in the lower half of the silicon bandgap. A representative DLTS spectrum for boron – doped silicon is shown in Fig. 5 [12]. The material was implanted with As^+, annealed with a scanned cw Ar – ion laser, and used to fabrication $n^+ - p$ junction diodes for capacitance transient measurements. The spectrum is dominated by two hole traps with energy levels at $E_v + 0.10$ eV and $E_v + 0.45$ eV;

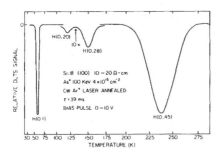

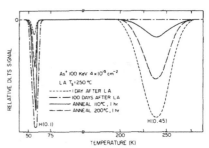

Fig. 5. Typical capacitance transient spectrum of p – type Si implanted with As⁺ and annealed by a cw Ar – ion laser. This spectrum was obtained ~4 days after sample preparation, so that the H(0.10) and H(0.45) peaks are of comparable strength. (N. H. Sheng and J. L. Merz, Ref. 12.)

Fig. 6. The transmutation between the deep level H(0.45) and the shallow level H(0.10) during room – temperature anneal. The restoration of H(0.45), and consequent annihilation of H(0.10), can be stimulated by annealing at 100 – 200°C. (N. H. Sheng and J. L. Merz, Ref. 12.)

the level at $E_v + 0.28$ eV is discussed in the next subsection. The concentrations of the two dominant defects are strong functions of time after beam annealing and subsequent thermal treatments. This is shown in Fig. 6 [12]. The H(0.45) level decays with time and transmutes into the shallower level H(0.10). The H(0.45) level can be recovered by low – temperature annealing and by minority – carrier injection. Rapid quenching of furnace – annealed samples produced the same defects as in the laser – annealed material. By correlating the results with published information, the laser – induced quenched – in defects were identified as interstitial iron for H(0.10) and the Fe – B pair for H(0.45) [12].

Rapid Thermal Annealing

Of the several transient thermal processing techniques that have been investigated, rapid thermal (or isothermal transient) annealing is the most promising as a replacement for conventional furnace annealing in silicon integrated – circuit technology. It also most closely resembles a furnace anneal in its thermal treatment of a silicon wafer. In this case, a wafer is irradiated with an incoherent light source (e.g., tungsten halogen lamp, arc lamp, or graphite heater), and the ion – implantation damage is annealed on a time scale of seconds. The combination of large – area illumination and long dwell times (as compared to pulsed – or scanned – beam annealing) ensure that the wafer is heated essentially uniformly.

Residual electronic defects have also been detected in rapid thermally annealing silicon [13,14]. A typical trap – level spectrum for an unimplanted boron – doped epitaxial silicon wafer after rapid thermal annealing is shown in Fig. 7 [14]. After an anneal at 900°C for 20 sec, the DLTS spectrum is dominated by a single peak which has an activation energy of 0.29 eV. This appears to be the same as the $E_v + 0.28$ eV level that was discussed above. The level is not detectable in the starting (reference) material, but is present in low concentration

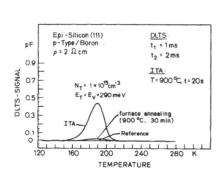

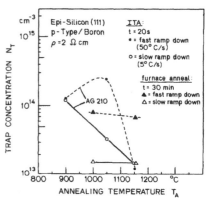

Fig. 7. DLTS signal versus temperature for a boron – doped epi – silicon sample after thermal pulse annealing (TPA).

Fig. 8. Comparison of residual trap concentrations of the 290 – meV – level after slow and fast ramp – down rates. The circles and triangles represent measurement points of TPA and furnace – annealed samples, respectively.

after a conventional furnace anneal. After rapid thermal annealing, the level was not detectable in unimplanted Czochralski – grown or float – zone silicon but was present in As^+ – implanted Cz – grown silicon. Also, in n – type silicon no electrically active deep levels with concentrations greater than 10^{11} cm^{-3} were found in the upper half of the bandgap after rapid thermal annealing.

The quench rate after rapid thermal annealing was found to be critical in determining the residual trap density [14]. Results are presented in Fig. 8. The open circles represent measurement points for a slow ramp down of 5°C/sec, and the solid circles are points taken at a ten times faster rate; furnace anneals were performed for comparison. The results reveal that the production of traps is strongly reduced by a slow ramp down of the temperature, independent of the anneal method.

Two defect models have been proposed for the level at $E_V + 0.29$ eV. In the study of rapid thermally annealed silicon it was suggested that the level involves a complex consisting of boron atoms and lattice vacancies [14]. It was speculated that such a complex could provide the defect kinetics responsible for the dependence of trap concentration on annealing temperature that is shown in Fig. 8. The model involves two chemical reactions: (1) boron atoms and vacancies react to form the complex and (2) vacancies and silicon interstitials react to form an ideal lattice. The competition for vacancies in these reactions should give rise to a peak in the defect density versus anneal temperature as is observed in Fig. 8. However, essentially the same energy level and capture cross section have been reported in scanned laser – annealed silicon and ascribed to pairing of an interstitial chromium atom and a substitutional boron atom [11]. Given that H(0.29) is the dominant deep level in rapid thermally annealed silicon, and is also detectable after a conventional furnace anneal, its identity should be firmly established.

Thermal Donor

The above subsections were concerned with defects that are quenched – in by transient thermal processing. This subsection demonstrates the apparent removal of an existing defect by rapid thermal annealing. The "thermal donor" is an oxygen – related defect which is produced in single – crystal silicon at temperatures from 300 to 525°C and has the electrical characteristics of a shallow donor impurity [e.g., Ref. 15]. In Czochralski – grown silicon, with oxygen concentrations of typically 10^{18} cm^{-3}, thermal – donor formation can substantially decrease the resistivity of lightly – doped n – type material and produce conductivity – type conversion in (initially) p – type silicon. Even though the thermal donor is one of the most extensively studied semiconductor defects, its microscopic structure has yet to be identified.

It has been demonstrated that the changes in electrical conductivity that are associated with the formation of thermal donors can be readily removed by transient thermal annealing. This was first shown with a scanning electron beam [16] and more recently by rapid thermal annealing [17]. In Fig. 9 the resistivity of a wafer of Cz – grown n – type silicon is plotted after specified stages of furnace and rapid thermal annealing. The as – grown material (point # 1) was first annealed at 1000°C for 10 sec. The resistivity increased to point # 2 and was essentially unchanged by additional similar exposures (points # 3 and 4), which indicates saturation. The wafer was then annealed at 450°C for 64 hr which substantially reduced the resistivity (point # 5), as expected from thermal – donor generation. Again, a rapid thermal anneal at 1000°C for 10 sec returned the resistivity to the former high value (point # 6), and the process is reproducible.

The changes in the resistivity shown in Fig. 9 can be explained by the low – temperature furnace generation of thermal donors and their rapid annihilation at high temperatures. Indeed, this model is fully plausible and has been proposed to account for the phenomenon [16,17]. However, electrical resistivity is a non – specific sensor of deep levels, so that the model remains to be confirmed by a direct spectroscopic technique. More importantly, the temporal features of transient thermal annealing should enable useful studies of the annealing kinetics of the thermal donor, for example, under conditions in which impurity diffusion is suppressed. Such information should be helpful in evaluating proposed models for the thermal donor, especially if annealing stages can be identified as has been suggested [16].

Fig. 9. Typical resistivity variation for an n – type (100) – oriented seed – end silicon wafer after various thermal annealing processes. (S. Hahn et al., Ref. 17.)

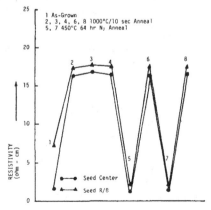

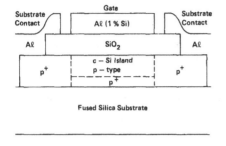

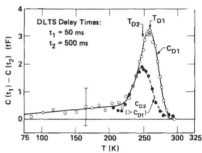

Fig. 10. Schematic diagram of an MOS capacitor fabricated in a laser – crystallized silicon thin film on bulk fused silica.

Fig. 11. DLTS spectra for a laser – crystallized thin – film silicon MOS capacitor.

BEAM – CRYSTALLIZED SILICON THIN FILMS

Beam – crystallized semiconducting thin films are of technological interest for large – area fabrication of high – performance thin – film transistors, photodetectors, and logic circuits. The field of semiconductor – on – insulator (SOI) technologies and device applications is reviewed in Ref. 18. To date only a few studies have been reported which specifically characterize electronic defects in crystallized thin films of silicon, and there are no reports on the study of defects in any other crystallized thin – film semiconductor. All of the existing studies have relied on the metal – oxide – silicon (MOS) structure and concluded that the $Si - SiO_2$ interface is the dominant source of residual electronic defects in this material [19 – 22]. This is illustrated in the present section with results from a study of residual electronic defects in fully processed cw laser – crystallized silicon thin films on fused silica [22].

Transient capacitance spectroscopy was performed on MOS thin – film capacitors of p – type conductivity. A schematic cross section of the test device is shown in Fig. 10 [22]. The silicon thin film was crystallized on fused silica with a scanning CO_2 laser and used to fabricate MOS capacitors by conventional microelectronic processing techniques. The buried p^+ layer, which is situated directly beneath the gate electrode, was found to be essential to minimize the series resistance of the silicon substrate in these laterally – contacted devices. The gate area was 40 μm $\times$ 50 μm. Hole – emission spectra are shown in Fig. 11 and were obtained by recording the capacitance transient at fixed temperatures [22]. At each temperature, the capacitance – voltage characteristic of the MOS capacitor was recorded and used to select the gate bias required to maintain the same equilibrium device capacitance in depletion, C_D, for all temperatures. This simplified the interface – state analysis and, more importantly, permitted a determination of the origin of the high – temperature emission peak in Fig. 11. This emission peak could be due to defects in the bulk of the silicon film which have a discrete energy level located near midgap, or it could arise from minority – carrier generation through a continuous distribution of deep levels as exists at the $Si - SiO_2$ interface [23]. The second spectrum in Fig. 11 was recorded with an equilibrium depletion capacitance C_{D2} chosen such that $C_{D2} >$

C_{D1}. This resulted in (1) a lower peak height and (2) the peak is shifted to a lower temperature (i.e., $T_{D2} < T_{D1}$). Increasing C_D reduces the magnitude of the electric – field intensity within the silicon depletion layer [24]. Therefore, if the shift of the peak temperature with C_D were due to Poole – Frenkel emission, the peak temperature would increase, rather than decrease, with an increase in C_D; the Poole – Frenkel mechanism was discussed in connection with Fig. 4. The dependence of the high – temperature emission peak on C_D is consistent with the effect of surface generation on the DLTS measurement of interface states near the silicon midgap [23]. Therefore, the entire DLTS spectrum in Fig. 11 may be ascribed to a continuous distribution of deep levels at th Si – SiO$_2$ interface, and the interface – state density can be estimated with a standard analysis of the DLTS signal which can be applied at temperatures below the emission peak where hole emission dominates [22]. For example, the DLTS signal just below the peak (i.e., near 210 K) yields a value of ~ 6×10^{10} eV^{-1} cm^{-2} for the density of interface states near the silicon midgap.

For the detection of bulk deep levels, interface states contribute a large background signal. Against this high background, the density of a spatially uniform discrete level would need to have been > 1×10^{14} cm^{-3} to the detectable in the capacitors examined in this study. In halogen – lamp recrystallized silicon films on SiO$_2$, an upper limit of ~ 5×10^{11} cm^{-3} for bulk deep levels has been reported [21]. It is suggested here that transient conductance spectroscopy can be combined with multiple – terminal MOS devices (e.g., thin – film transistors) to achieve high sensitivity for bulk defect detection, in part by excluding any contribution from the interface; such techniques have been used on bulk silicon devices [25]. Indirect evidence for the high crystalline perfection and low defect density of laser – crystallized silicon films on fused silica has recently been presented with the demonstration of highly photosensitive transistors which display optical gains > 10^3 and photocurrent decay times of the order of 10 μs [26]. The high technological potential of SOI devices and circuits should motivate the continued study of electronic defects, and their dependence on processing, in these materials.

COMPOUND SEMICONDUCTORS

No systematic study has yet been reported in the open literature on electronic deep levels in a compound semiconductor which has been treated by transient thermal processing. The effect of beam processing on band – edge luminescence in III – V semiconductors has been reported and provides indirect information on deep levels [e.g., see Ref. 27]. The absence of detailed studies is all the more remarkable in light of the substantial progress that has been achieved during the last few years in the application of rapid thermal annealing for transistor fabrication in GaAs and its alloys; this topic is reviewed in Ref. 28.

An initial survey of electronic defects in (unimplanted) metalorganic chemical – vapor deposited GaAs has revealed that rapid thermal annealing changes the densities of existing defects and introduces new levels [29]. Similarly, photoluminescence reveals changes in the densities of donor – and acceptor – related defects after rapid annealing [30]. In both cases, the rapid thermal annealing conditions were the same as those which have been generally identified as optimal for dopant activation in GaAs. It is anticipated that the variety of deposition techniques (e.g., molecular – beam epitaxy, metalorganic CVD, and liquid – phase epitaxy), the complexity of the defect chemistry (as compared to elemental semiconductors), and the large variability in the quality of the substrate material, both in terms of structural perfection and impurity content, will contribute to a wide disparity of results under otherwise identical rapid thermal annealing conditions. It will therefore be even more importantn in compound semiconductors than in silicon to combine several complementary techniques for meaningful and detailed materials characterization and electronic defect identification. From the viewpoint of fundamental studies of electronic defects, rapid thermal annealing offers a controlled means for systematically perturbing both shallow and deep levels in compound semiconductors to correlate results from different measurement techniques.

FUTURE DIRECTIONS

In bulk single – crystal silicon from the viewpoint of electronic defects, rapid thermal annealing is the most promising as a replacement for conventional furnace annealing in integrated – circuit processing. Directed energy sources, such as lasers and electron beams, may offer unique advantages for materials modifications other than recrystallization. In addition, they provide a controlled means of introducing defects in silicon for fundamental studies.

For electronic – defect studies in SOI materials, multiple – terminal thin – film devices combined with transient conductance techniques should be considered. This approach eliminates the practical difficulties, and reduced sensitivity, of capacitance measurements on small – geometry devices and permits at least partial separation of the effects of bulk versus interface defects. Although not discussed in this paper, the effects of grain and sub – grain boundaries on SOI device operation may continue to be an issue, in which case techniques that can isolate and characterize electronic defects at grain boundaries in a thin film should be identified and applied.

In epitaxially – grown III – V semiconductors, rapid thermal annealing shows great promise for transistor processing. However, detailed systematic studies of electronic defects in these processed materials remain to be conducted.

REFERENCES

1. J. Narayan, these proceedings.
2. L. C. Kimerling and J. L Benton, in *Laser and Electron Beam Processing of Materials* (Academic Press, New York, 1980), eds. C. W. White and P. S. Peercy, pp. 385 – 396.
3. J. L. Merz, M. Mizuta, and N. H. Sheng, Inst. Phys. Conf. Ser. No. 67, 115 (1983).
4. A. Chantre, J. de Phys. C5, 269 (1983).
5. N. M. Johnson, in *CW Beam Processing of Silicon* (Academic Press, New York, 1985), ed. J. F. Gibbons, chap. 4.
6. N. M. Johnson, R. B. Gold, and J. F. Gibbons, Appl. Phys. Lett. 34, 704 (1979).
7. N. M. Johnson, J. L. Regolini, D. J. Bartelink, J. F. Gibbons, and K. N. Ratnakumar, Appl. Phys. Lett. 36, 425 (1980).
8. A. Chantre, M. Kechouane, and D. Bois, Mat. Res. Soc. Symp. Proc. 4, 325 (1982).
9. A. Chantre, M. Kechouane, G. Auvert, and D. Bois, Appl. Phys. Lett. 43, 98 (1983).
10. A. Chantre, M. Kechouane, and D. Bois, Physica 116B, 547 (1983).
11. A. Chantre, M. Kechouane, and D. Bois, Mat. Res. Soc. Symp. Proc. 14, 547 (1983).
12. N. H. Sheng and J. L. Merz, J. Appl. Phys. 55, 3083 (1984).
13. J. L. Benton, G. K. Celler, D. C. Jacobson, L. C. Kimerling, D. J. Lischner, G. L. Miller, and Mc. D. Robinson, Mat. Res. Soc. Symp. Proc. 4, 765 (1982).

14. G. Pensl, M. Schulz, P. Stolz, N. M. Johnson, J. F. Gibbons, and J. Hoyt, Mat. Res. Soc. Symp. Proc. 23, 347 (1984).
15. G. S. Oehrlein and J. W. Corbett, Mat. Res. Soc. Symp. Proc. 14, 107 (1983).
16. C. J. Pollard, J. D. Speight, and K. G. Barraclough, Mat. Res. Soc. Symp. Proc. 13, 413 (1983).
17. S. Hahn, D. Hung, S. C. Shatas, and S. Rek, in *VLSI Science and Technology/1984*, (The Electrochemical Soc., New York, 1984), eds. K. Bean and G. Rozgonyi, pp. 85–92.
18. B.-Y. Tsaur, these proceedings.
19. N. M. Johnson, M. D. Moyer, and L. E. Fennell, Appl. Phys. Lett. 41, 560 (1982).
20. N. M. Johnson, M. D. Moyer, L. E. Fennell, E. W. Maby, and H. Atwater, Mat. Res. Soc. Symp. Proc. 13, 491 (1983).
21. D. P. Vu, A. Chantre, H. Mingam, and G. Vincent, J. Appl. Phys. 56, 1682 (1984).
22. N. M. Johnson and M. D. Moyer, Mat. Res. Soc. Symp. Proc. 33, 101 (1985).
23. N. M. Johnson, J. Vac. Sci. Technol. 21, 303 (1982).
24. S. M. Sze, *Physics of Semiconductor Devices, 2nd Ed.* (Wiley, New York, 1981), chap. 7.
25. M. G. Collet, Solid–State Electron. 18, 1077 (1975).
26. N. M. Johnson and A. Chiang, Appl. Phys. Lett. 45, 1102 (1984).
27. D. Kirillov and J. L. Merz, these proceedings.
28. J. S. Williams, these proceedings.
29. N. M. Johnson, (unpublished).
30. R. A. Street, (unpublished).

DYNAMIC ANNEALING PHENOMENA AND THE ORIGIN
OF RTA-INDUCED "HAIRPIN" DISLOCATIONS

W. MASZARA,[1] D.K. SADANA,[1,2] G.A. ROZGONYI,[2,1] T. SANDS,[3] J. WASHBURN,[3] AND J.J. WORTMAN[1]
[1]North Carolina State University, Raleigh, North Carolina 27695-7916
[2]Microelectronic Center of North Carolina, Research Triangle Park, North Carolina 27709
[3]Lawrence Berkeley Laboratory, Berkeley, CA 94720

ABSTRACT

The geometry, origin, and diffusion along hairpin defects in Si were investigated using TEM and SIMS techniques. The defect that grows from the amorphous-crystalline (a/c) interface following solid phase epitaxy growth front was found to be a perfect dislocation with a/2(101) Burgers vector. Misoriented microcrystallites within the a/c transition region are proposed to be nucleation sites for the hairpin dislocations. The density of the crystallites increases with an overall coarsening of the interface which occurs during dynamic annealing processes stimulated by implantation or post-implantation low temperature annealing. Hairpin dislocations were found to pipe-diffuse boron at much higher rates than bulk processes significantly shifting dopant profiles. The diffusion coefficient of boron pipe diffusion at 1150°C was found to be about 10^4 times higher than the bulk one.

INTRODUCTION

The ion implantation process has become a routine means of doping semi-conductors. An electrical activation of the dopants requires an annealing at elevated temperatures. Although temperatures as low as 600-650°C can provide proper activation [1-3] in an amorphized substrate, temperatures of the order of 1000°C are necessary to fully activate the impurity when the structure of host crystal has not been rendered amorphous before or during implantation. It is also well known that implantation into a preamorphized substrate eliminates channeling of ions along major crystallographic directions assuring tight control of the dopant concentration profile and junction location. Silicon self-implantation has been chosen by several workers [3-5] as a preamorphization process to avoid any chemical interaction with the dopant. While the regrowth of the deep amorphous layer accomplished through the solid phase epitaxy (SPE) process during annealing removes any residual damage caused by the shallower dopant implantation, it may introduce certain characteristic defects of its own. A heavy dose of the dopant may induce precipitation and microtwin formation [6,7] or polycrystallinity [8] during regrowth. Regardless of the dose the recrystallized layer contains interstitial type dislocation loops near the original location of the amorphous-crystalline (a/c) interface. Also, it has been observed [5,6,9-11] that elongated, V-shaped "hairpin" dislocations span the regrowing layer (Fig. 1). The latter dislocation follows the regrowing a/c interface and in the case of an amorphization extending to the surface of the wafer terminates at that surface. The hairpins have been observed in samples amorphized with arsenic [9] and silicon [6], as well as in those preamorphized with Si [9-11] or Ge [12] and subsequently implanted with shallow BF_2. In our experiments we have also observed the defects in the Si preamorphized samples implanted with boron.

Krimmel et al. [9] who observed "dislocation pairs" or hairpins in As implanted and electron beam annealed silicon postulated that the defects originate from small dislocation loops acting as nuclei. Two segments of dis-location loops [6], or half-loop [5] intersected by the a/c interface have been proposed in other works as an origin of the hairpin dislocation. The form and specific location of the dislocation-along the gradient of implanted dopant(s) and hence across the interface(s) of p-n junction(s) suggests its potentially significant and detrimental role in device performance. To our knowledge, no detailed information about the role and relation of the defect's existence and population to the implantation process has been published. Early studies of ionic transport in pipe diffusion process along dislocations in Si [13,14] indicate a four to six orders of magnitude increase in diffusion coefficients of such ions as Sb, Ag, In, and P in the 900-1200°C temperature range. In this investigation the origin, the relation to the morphology of a/c interface, and the role in the dopant pipe diffusion of hairpin dislocations will be examined.

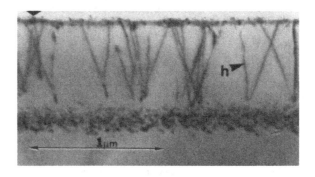

Fig. 1. Cross-section TEM showing hairpin dislocations (h) spanning the region where amorphous layer existed before annealing.

DYNAMIC ANNEALING AND HAIRPIN DENSITY

Our previous investigation [5] showed that there is a strong correlation between the concentration of hairpin defects in the preamorphized samples implanted with BF_2 and annealed via rapid thermal annealing (RTA), and the temperature of the wafer during dopant implantation. Samples implanted at nominal room temperature (RT), without proper heat sinking, at a power density of about $0.1W/cm^2$ exhibited a high density of hairpins. An etch pit count for the sample subjected to a Secco etch for 30 seconds revealed about 7×10^4 dislocations per cm^2. Almost no hairpin defects were observed in the sample implanted at liquid nitrogen temperature (LN) with heat conducting silver paste applied between the back of the wafer and a cooled sample holder. Since the temperature of the secondary implant was the only parameter different for both samples, beam heating was suspected as the cause of the increase in hairpin density, N_d. In order to simulate heating effects during BF_2 implantation, Si^+-preamorphized samples were given furnace heat treatments (250-350°C for 1h) in inert atmosphere prior to RTA. There was no appreciable increase of hairpin density over that of the preamorphized sample with LN BF_2 implant ($N_d < 10^4 cm^{-2}$). However, samples which received a furnace heat treatment at 500°C for 1 hour prior to RTA were found to contain hairpins whose density was comparable to that of preamorphized sample with BF_2 implanted at room temperature. These results suggest that the existing a/c interface can be modified in similar fashion by beam and furnace heating and that this process can lead to an increase in the population of hairpin dislocations. The modification is presumably a ripening process of very small crystallites [11] in the part of amorphous layer in close proximity to the a/c interface which otherwise may be annihilated during high temperature RTA process. Cross section TEM of both BF_2 implanted samples observed prior to RTA process has shown that the RT BF_2 implant produced significant coarsening of the interface. Figure 2 shows the TEM micrographs of the preamorphized sample before and after BF_2 implantation at RT. The sample with BF_2 implantation performed at LN exhibits similar features to those of a preamorphized only sample.

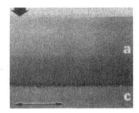 

a   b

Fig. 2. Morphology of a/c interface in the sample (a) preamorphized with Si^+ at LN and (b) subsequently implanted with BF_2 at RT (0.1 W/cm^2). a – amorphous, c – crystalline layer. Distance marker in this and following micrographs indicates 0.5 μm.

Although beam heating of the a/c interface during the second (doping) implantation can have, as was just shown, remarkable impact on the dislocation density, the quality of the interface due to amorphization process will play a dominant role in determining the density of the hairpins. If the temperature of the sample during amorphizing implantation is low enough to prevent radiation induced defects from significant diffusion (recall that the negative vacancy in Si has been observed to become mobile above 100K [15]), the appearance of the a/c interface can be sharp and smooth, with possible undulations due to ion straggling effect only. The TEM micrographs of Fig. 3 show a LN cooled Si sample implanted with Si^+ at 300 keV, 1E16 cm^{-2} dose. These conditions produce a buried amorphous layer with two a/c interfaces. Note that the a/c interface closer to the surface appears rougher than the deeper one (Fig. 3a) and was observed to generate about an order of magnitude more hairpins after RTA than the latter (Fig. 3b). The difference is probably associated with a different gradient of damage energy density at the depths corresponding to each interface (Fig. 3c). The steeper the curve, the less spatial straggle of the amorphous-to-crystalline threshold zone would be expected. A preamorphization under similar implantation conditions with heavier ions like germanium yields steeper damage energy density curves, sharper a/c interfaces and fewer hairpins after RTA.

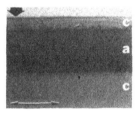

a

b

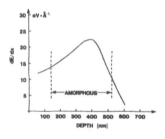

c

Fig. 3. (a) Buried amorphous layer after single 300 keV, 1E16cm^{-2} Si^+ implant into LN cooled Si substrate, (b) after RTA, 950°C/10s, annealing hairpin dislocations (h) grow from each a/c interface toward the middle of amorphous layer. The density in the upper layer is about one order of magnitude higher than in the lower one, (c) corresponding damage energy density curve [16].

If the temperature of the substrate during implantation is sufficiently high to cause the diffusion of radiation induced defects, part of the damaged area regrows. This so-called dynamic regrowth does not necessarily reproduce the crystalline structure in its original form. Instead, a highly damaged, although not amorphous, transition layer develops between amorphous and crystalline regions. Amorphization may not occur at all if the heating is sufficient. This in effect overcomes the residual ion straggling nonuniformity of a/c interface and becomes a dominating factor determining its roughness. Figure 4 shows cross section TEM micrograph of a Si sample implanted with Si_2^+ under the same conditions as the sample of Fig. 2a (i.e., 300 keV 1E16cm^{-2}, 150 keV 5E15cm^{-2}, 70 keV 5E15cm^{-2}), but at RT. The surface amorphous layer was reduced from about 0.6 μm to 0.15 μm, and the a/c interface became coarse and fragmented (Fig. 4a). The RTA at 950°C/10s produced about 10^7cm^{-2} hairpins, evaluated from TEM observations in Fig. 4b.

 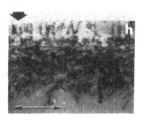

Fig. 4. TEM micrograph of silicon amorphized with Si[+] at nominal room temperature with the same energies and doses as samples in Fig. 2a, (a) before and (b) after rapid thermal annealing at 950°C/10s. d - heavy damage layer.

HAIRPIN GEOMETRY AND ORIGIN

More detailed information about hairpin geometry and its origin was obtained by tilting experiments on plan-view and cross section TEM samples. The dislocations were found to be perfect with Burgers vector b=a/2<101> inclined by 45° to the <100> sample surface. Observation of a partially SPE regrown amorphous layer indicates that the hairpin tips are located in the region corresponding to the coarse a/c transition region (Figs. 3 and 4). The transition region can be quite broad. High resolution imaging of the sample in Fig. 2b (Fig. 5) reveals the presence of misoriented crystallites surrounded by amorphous material. The misorientation is thought to be a result of stresses in the transition region which result from the difference in density [17] between amorphous and crystalline silicon.

Fig. 5. Cross-sectional, high magnification image of the a/c interface of the sample shown in Fig. 2b. Microcrystallites imbedded in amorphous surroundings show a 4° misorientation relative to the substrate.

Based upon these observations the following model for nucleation of hairpin dislocation is proposed (Fig. 6). During RTA, the advancing and recrystalliz- ing a/c growth front intersects a small misoriented crystallite in the a/c transition region. A perfect dislocation segment is formed at the intersec- tion to accommodate the misorientation. As the combined growth front con- tinues to move, the dislocation segment wraps around the misoriented material to form a half loop. As the annealing continues, the hairpin arms diverge since the growth front protrusion of misoriented material is spreading later- ally as well as vertically. During further annealing the dislocation shortens its length by moving toward the surface along the glide cylinder defined by the inclined Burgers vector and line direction. The model clearly implies that the density of hairpins is related to availability of the misoriented crystallites within a/c transition region, i.e., the narrower and sharper the region, the lesser are the chances for hairpin nucleation.

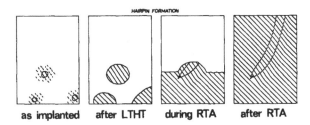

HAIRPIN FORMATION

as implanted   after LTHT   during RTA   after RTA

Fig. 6. Schematic diagram illustrating model for hairpin nucleation during RTA (see text for description). The hairpin is shown in almost edge-on orientation.

PIPE DIFFUSION ALONG HAIRPINS

Two Si^+ preamorphized samples with identical BF_2 implants, which were found to have significantly different densities of hairpins, have been used for boron concentration profile examination in secondary ion mass spectroscopy (SIMS). The densities N_d were $<10^3$ and about $2.5 \times 10^7 cm^{-2}$. A boron SIMS profile of the sample with a higher density of hairpins, annealed via RTA at 1150°C for 10 seconds, shows a significant shift toward the bulk with respect to the profile of the other sample (Fig. 7). We attribute this shift to the enhanced diffusion along the defects. An approximate evaluation of diffusion coefficient, D_d, of boron along hairpins based upon preliminary experimental results is given below.

Following derivations by Sedgwick et al. [18] and Seidel and MacRae [19] for assumed Gaussian dopant density the ratio of diffusion coefficients of two samples annealed in identical conditions can be expressed in terms of profile shifts Δx_1, Δx_2 between the as implanted and annealed profiles for each sample at some fraction, Q, of the peak concentration:

$$\frac{D_v}{D_{eff}} = \frac{\Delta x_1^2 + 2\Delta x_1 \Delta R_p (2\ln 1/Q)^{1/2}}{\Delta x_2^2 + 2\Delta x_2 \Delta R_p (2\ln 1/Q)^{1/2}}$$

Where: D_v = bulk diffusion coefficient, D_{eff} = effective diffusion coefficient for the sample with high concentration of dislocations, ΔR_p = straggling of implanted ion distribution. We can safely neglect dislocation density of the first sample ($N_d < 10^3$) and consider its $D = D_v$. For Q=0.01 and ΔR_p=0.019 μm (BF$_2$ implant energy was 42 keV, 9.5 keV of which is shared by boron) D_v/D_{eff}=0.52. Measurement at Q=0.001 produced a result within 2% of the previous one.

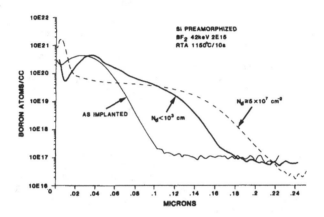

Fig. 7. SIMS profiles of B in Si$^+$ preamorphized, BF$_2$ implanted, RTA 1150°C/10s annealed samples showing different densities of hairpin dislocations.

Assuming that the diffusion coefficient for pipe diffusion, D_d, is much greater than bulk diffusion coefficient, D_v, we can neglect any flux of ions taking place beside the dislocation pipes. Therefore, it can be concluded from Fick's first law for directional flow of the point defects that the effective diffusion coefficient, D_{eff}, as calculated from a concentration profile (SIMS measurement represents average concentration over 250 x 250 μm raster area) is related to the actual pipe diffusion coefficient as follows:

$$D_d = D_{eff}A/A_d$$

Where: A = unit area perpendicular to the flux (here assumed equal to 1 cm^2), A_{eff} = cross section area of all pipes crossing the unit area. The value of the radius of dislocation pipe was assumed to be r_o = .01 μm after ref. [20]. It is also assumed that the dislocations are perpendicular to the sample surface (actual angle between the hairpin arm and the surface was evaluated in TEM studies to be about 70°). Since the hairpin consists of two arms, the total concentration of dislocation pipes is N_d=5 x 10^{7}cm^{-2} and A_d=$N_d\pi r_o^2$, hence D_d=6.4 x 10^3 D_{eff}=1.2 x 10$^4 D_v$. This result is, of course, only approximate and if, contrary to our assumption, the mass transported through bulk diffusion process is not totally negligible with respect to the pipe diffused quantity, the value of D_d will become somewhat smaller than estimated. The discrepancy becomes more significant when samples with lower dislocation density are considered in such an estimate.

CONCLUSIONS

Whether it is routine dopant implantation (in the case of heavy ions) or doping preceeded by a preamorphization process, the resulting amorphous layer can potentially contain nuclei at its interface with crystalline zones which would initiate the growth of hairpin dislocation during subsequent annealing process. Misoriented small crystallites present in a/c transition region have been identified as the nuclei. Roughness of this region due to ion straggling determines the density of hairpins with more nonuniformity leading to a higher density. Dynamic annealing occurring due to ion beam heating causes coarsening and broadening of the transition region thus contributing to a further increase of the hairpin density. Post-implantation, low temperature annealing or beam heating of the a/c interface by a secondary, shallower implantation process can also induce increase of the density. A pipe diffusion of doping ions along the dislocations occurs at a highly accelerated rate and can cause a significant shift of the dopant concentration profile toward the bulk of the sample. The pipe diffusion coefficient, D_d, for boron diffusion along hairpin dislocation was evaluated to be about 10^4 times higher than its bulk value, D_v, at 1150°C.

ACKNOWLEDGMENTS

The authors would like to express their thanks to Dr. Dieter Griffis for performing SIMS measurements and to Jiann Liu for assistance in implantation of our samples. We would also like to thank Dr. O.W. Holland and Dr. David Poker for helpful discussions and providing an access to TRIM 84 computer software. This research was partially supported by a grant from the Semiconductor Research Corporation.

REFERENCES

1. B.L. Crowder and F.F. Morehead, Appl. Phys. Lett. 14, 133 (1969).

2. D.G. Beanland, Solid State Elec. 21, 537-547 (1978).

3. T.E. Seidel, IEEE Electron. Dev. Lett., EDL-4 10, 354 (1983).

4. M.Y. Tsai and B.G. Streetman, J. Appl. Phys. 50(1), 183-187 (1979).

5. W. Maszara, C. Carter, D.K. Sadana, J. Liu, V. Ozguz, J. Wortman, and G.A. Rozgonyi in: Energy Beam-Solid Interactions and Transient Thermal Processing, J.C.C. Fan and N.M. Johnson eds. (North-Holland, New York 1984), Mat. Res. Soc. Symp. Proc. 23, 285-291.

6. J. Narayan, O.W. Holland, and B.R. Appleton, J. Vac. Sci. Technol. B1(4), 871-887 (1983).

7. T. Sands, J. Washburn, R. Gronski, W. Maszara, D.K. Sadana, and G.A. Rozgonyi, Appl. Phys. Lett. 45(9), 982-984 (1984).

8. J.S. Williams, C.E. Christodoulides, and W.A. Grant, Radiat. Eff. 48, 157-160 (1980).

9. E.F. Krimmel, H. Oppolzer, H. Runge, and W. Wondrak, Phys. Stat. Sol. (a) 66, 565-571 (1981).

10. T.E. Seidel, R. Knoell, F.A. Stevie, G. Poli, and B. Schwartz in: VLSI Science and Technology/1984, K.E. Bean and G.A. Rozgonyi eds. (Proc. of Electrochem. Soc. Meeting, 1984, vol. 84-7) pp. 201-210.

11. T. Sands, J. Washburn, R. Gronsky, W. Maszara, D.K. Sadana, and G.A. Rozgonyi, J. Electronic. Mat. Dec. 1984, in press.

12. D.K. Sadana, W. Maszara, J.J. Wortman, G.A. Rozgonyi, and W.K. Chu, J. Electrochem. Soc. 131(4), 943-945 (1984).

13. V.A. Sterkov, V.A. Panteleev, P.V. Pavlov, Sov. Phys.-Solid State 9, 533 (1967).

14. G.V. Dudko, M.A. Kolegaev, and V.A. Panteleev, Sov. Phys.-Solid State 11, 1097 (1969).

15. G.D. Watkins in: Radiation Effects in Semiconductors, F.L. Vook ed. (Plenum Press, New York 1968) p. 67.

16. E.g. see computer programs TRIM 84, MARLOWE for nuclear and electronic stopping profiles of implanted ions.

17. J. Wilson, Metal Rev. 10, 381 (1965).

18. T.O. Sedgwick, S.A. Cohen, G.S. Oehrlein, V.R. Deline, R. Kalish, and S. Shatas, Ref. 10, pp. 192-200.

19. T.E. Seidel and A.U. MacRae, Trans. of the Metallurgical Soc. of AIME 245, 491-498 (1969).

20. P.V. Pavlov, L.V. Lainer, V.A. Sterkhov, and V.A. Panteleev, Sov. Phys.-Solid State 8(3), 580-584 (1966).

A STUDY OF ELECTRICALLY ACTIVE DEFECTS INDUCED BY PULSED ELECTRON BEAM ANNEALING IN (100) N-TYPE VIRGIN SILICON

M.S. DOGHMANE, D. BARBIER and A. LAUGIER
Laboratoire de Physique de la Matière (LA CNRS N° 358), Institut National des Sciences Appliquées de Lyon, 20 Avenue Albert Einstein 69621 Villeurbanne Cédex France

ABSTRACT

Au/Si Schottky contacts have been used as test structures to investigate defects induced in virgin C.Z (100) N-type silicon after irradiation with a 12 to 20 KeV mean energy electron beam pulse. A thin and highly damaged surface layer was observed from a fluence threshold of 1 J/cm^2. In addition electron traps were detected in the PEBA induced melting layer with concentrations in the 10^{12}-10^{13} cm^{-3} range. Their depth profiles have been related to the PEBA induced melting layer thickness. Quenching of multidefect complexes is the most probable mechanism for electron trap generation in the processed layer.

I. INTRODUCTION

The electrical characteristics of Implanted and Energy Beam Annealed P-N junctions are generally limited by excessive reverse currents. In fact, annealing in the liquid phase epitaxy regime using pulsed lasers has soon been recognized as leaving defect states in the space charge region of devices (1). The origin of these states has been discussed by some authors in term of implantation tail defects (2). However, recent experiments have shown that deep centers are induced in virgin silicon as well by various annealing procedures operating either in the liquid phase or in the solid phase regime (3-4). A previous paper was devoted to Pulsed Electron Beam Annealing (PEBA) induced defects in virgin P-type (100) silicon (5). In this work defect generation by PEBA in N-type virgin silicon has been investigated using gold Schottky contacts directly deposited on the annealed surface. Electrical characterization of the diodes and DLTS experiments have been carried out for a number of annealing conditions in order to investigate the defects state origin and their possible generation mechanisms.

II. EXPERIMENT

Irradiations of C.Z. boron doped (100) silicon, about 3 Ω.cm in resistivity have been performed by means of a SPIRE 300 Pulsed Electron Beam Processor.

A description of this annealing device has been given elsewhere (6).After HF etching 1 cm^2 silicon wafers have been processed in a vaccum of 5.10^{-6} Torr using 50 ns duration single pulses. The energy densities (fluences) were ranging from 0.7 to 1.4 J/cm^2 and the electron mean energy has been varied between 12 KeV and 20 KeV in order to produce differences in the thermal effects. In these conditions, the physical state of the surface layer is characterized by a rather penetrating electron energy deposition profile which produces a so-called melting layer (7).

At the end of the pulse only the top of this melting layer of thickness x_M may be turned into liquid state up to a depth x_{FM} depending on the electron mean energy and on the fluence value. For example at the end of a 1.2 J/cm^2 PEBA shot the fully melted zone in crystalline silicon is only 0.3 x_M for a 20 KeV mean energy beam and raises up to 0.7 x_M when the mean electron energy is lowered down to 12 KeV. From the depth x_{FM} to the edge of the melting layer at x_M the deposited fraction of latent heat decreases from 100 % to 0. So, in this transition zone, crystalline order is likely to subsist although silicon atoms are turned highly mobile as indicated by PEBA induced impurity redistribution (8).

The above described details will be considered for the discussion in section IV.

After irradiation gold Schottky contacts were evaporated at several places on the processed area. The DLTS experimental set-up uses a double pulse generator and has a sensitivity $\Delta C/C \simeq 10^{-6}$. A full description of the aparatus is given in ref.(9). The defect state depth-profiles have been established by differential capacitance measurement in the DDLTS mode (9).

III. RESULTS

Electrical characterization

The I-V characteristics of the Schottky diodes were systematically recorded as a function of the beam fluence. From 0.7 to 1.0 J/cm^2 a slight increase of both the ideality factor n and the saturation current density J_{SC} is observed. From 1.0 J/cm^2 a strong degradation occurs whatever the mean electron energy of the pulse, with significant inhomogeneities over the pulsed annealed area. The degradation effect is also characterized by a high increase of the diode capacitance which indicates that donor centers are generated with concentrations as high as $10^{16} cm^{-3}$ for wafers processed at 1.2 J/cm^2 (see ref. (10)). In the same time departure from linearity is observed from 1.0 J/cm^2 in the variation of $1/C^2$ as a function of the reverse bias voltage. Furthermore a temperature scan from 300 K to 173 K produces only a slight decrease of the saturation current density J_{SC} compared to the one measured on non-PEBA processed diodes. This result suggests that J_{SC} is dominated by tunneling effects which might be due to the presence of a thin surface damaged layer. The regular decrease of the barrier height with the beam fluence also supports the conclusion that surface defects are generated even if the beam fluence is not sufficiently high to induce a fully melted zone inside the material. Moreover surface damage should also contribute to the high increase of J_{SC} by leakage around the gold contacts as far as we did not use mesa structures. However after repetitive removal of silicon layers by oxyde stripping, it appeared that the damaged layer do not extend deeper than 0.1 µm from the surface even after PEBA at 1.2 J/cm^2. At this depth the capacitance of a reference diode was almost completely restored and the $1/C^2$ versus V curve recovered a straight line shape as shown on fig.1.

Defect state introduction

Fig.2 shows typical DLTS spectra obtained with silicon wafers processed with a 20 KeV mean energy electron beam pulse. They are dominated by two peaks E_1 and E_2 attributed to electron traps at $E_C-0.29$ eV and $E_C-0.48$ eV respectively. A third structure E_3, attributed to an electron trap near $E_C-0.42$ eV, appeared from 1.0 J/cm^2 and the peak heights were growing when increasing the beam fluence. The samples processed with a 12 KeV mean energy pulse exhibited again the E_1 (0.48) and E_2 (0.29) levels but not below 0.8 J/cm^2. However, the E_3 (0.42) level is replaced by another one E'_3 (0.38) which appeared from 0.9 J/cm^2. We must note that under zero bias condition, no deep level located at less than 0.1 µm from the surface could be detected. The depth-profiles of E_1 and E_2 are shown on Fig.3 as a

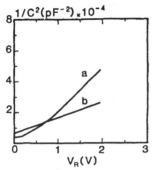

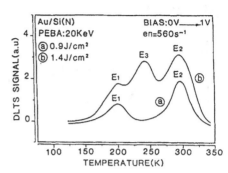

Fig.1- $1/C^2$ versus V curve of a Au/Si N Schottky diode :
a)Gold contact directly deposited on the silicon surface after irradiation with a 15 KeV mean energy electron beam pulse at 1.2 J/cm^2.
b) Gold contact deposited after removal of a 90 nm surface layer from the same PEBA processed sample as case a).

Fig.2- DLTS spectra recorded after silicon wafer processing with a 20 KeV mean energy electron beam pulse

function of the beam fluence in the case of a 20 KeV mean energy pulse. Even at 1.4 J/cm^2 the E_1 level remained confined within about 0.3 μm from the surface. This depth almost coincide with the computed thickness of the fully melted layer in these conditions. In the contrary E_2 is always distributed beyond the fully melted zone ($x_{FM} < x < x_M$) where the silicon lattice was brought at the melting temperature but not completely destroyed. The same remarks could be made for the depth-distributions of E_1 and E_2 when induced by a 12 KeV mean energy pulse. As for E_3 and E'_3, they are distributed around the edge of the fully melted layer. Their profiles were regularly shifted along with the fully melted zone penetration inside the material when the energy density was raised. The maximum concentrations of all the observed defect states were located between 10^{12} and 10^{13}cm^{-3} with rather higher values in the case of a 12 KeV mean energy pulse.

Defect state stability

Thermal treatments of 30 mn under H_2 atmosphere have been carried out from 300 to 800°C on irradiated samples in order to investigate the PEBA induced defect state stability. The gold contacts were chemically etched before thermal treatment and afterwards deposited again at the same place on the wafer surface. As shown on Fig.4, only E_1 exhibited a reverse annealing behaviour up to 700°C and was completely annealed by thermal treatment at 800°C. The other defect state concentrations were under the detection limit at only 500°C.

IV. DISCUSSION

Surface damage

The strong degradation of the Schottky diode electrical characteristics was not homogeneously observed on a given sample and appeared at only 1 J/cm2. Besides,

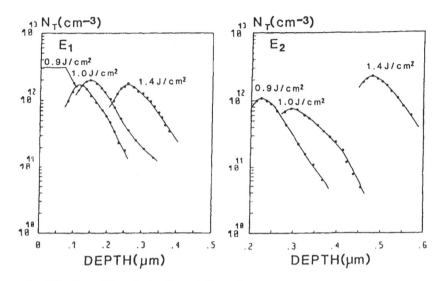

Fig.3 Depth-profiles of the E_1 and E_2 defect states as a function of the beam fluence for silicon wafers processed with a 20 KeV mean energy electron beam pulse

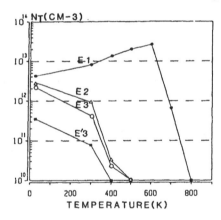

Fig.4 Defect state concentrations as a function of the furnace annealing tempe - rature (30 mn under H_2 atmosphere). PEBA : 1.2 J/cm^2. $\bar{E}$ = 12 KeV.

the capacitance anomalous increase was no longer observed after removal of a thin surface layer and thus could not be directly correlated with the electron traps observed much deeper in the material in the same annealing conditions. We therefore conclude that a shallow highly damaged layer can be created after PEBA at the wafer surface. Indeed the wafer surface may be acting as a source of defects during the melting process producing defect states in high concentration. Heat flow calculation predicted that the liquid phase duration was about 220 ns after a

1.2 J/cm^2 PEBA shot using a 15 KeV mean energy electron beam pulse. This allows impurity diffusion up to 0.1 μm from the surface assuming diffusivities in the 10^{-14} cm^2/s range. This depth is in good agreement with the measured highly damaged layer thickness in these conditions.

Origin of the PEBA induced electron traps

The following identification is mainly based on comparisons between already observed electron traps and must therefore presently be considered as tentative. The E_1 (0.48) electron trap appeared whatever the annealing kinetics with a shallow profile compared to the melting zone extent x_M. The high annealing temperature of this defect state leads to suspect extended defects to be responsible of its formation. Dislocations have already been detected in Pulsed Electron Beam Annealed Silicon (11). They are created during crystal regrowth, from the edge of the melting layer up to the surface, through the action of high thermal stresses. A level at $E_C - 0.48$ eV has been reported by Kimmerling et al. following deformation of N-type C.Z. (100) silicon at 770°C (12). In section III, we have shown the slight shift of the E_1 level toward the bulk of the material along with the increase of the liquid phase duration. This behaviour seems to indicate that a diffusion process is involved in the generation of E_1. However, unlike the creation of the highly damage surface layer, the depth-profile of E_1 cannot be explained by in-diffusion from the surface taking into account liquid phase durations of a few hundred of nanoseconds. We rather suggest interaction of defects with dislocation cores during resolidification of the fully melted surface layer. The defect responsible for the E_2 (0.29) electron trap being distributed beyond the fully melted zone could be compared to defects produced in the solid phase epitaxy regime. The high quenching rate of the partially melted zone (10^9 K/s) may account for the creation of E_2. Indeed, at the melting point the high vacancy concentration provides a source for multidefect complex formation. Considering this mechanism the thermal signature of E_2 has been found close to the one of a level at $E_C - 0.3$ eV reported by Kimmerling et al. after 1 MeV electron irradiation of N-type silicon (1). This electron trap has been attributed to a multivacancy center. The low annealing temperature of E_2 also supports our proposed identification. Furthermore regarding the very similar thermal signature of the E_3 (0.38) and E'_3 (0.42) levels we are tempted to make no difference between the defect from which they are originating, more especially as their annealing behaviours are also very similar (see Fig.4). The distribution tail of E_3 and E'_3 approximately follows the one of E_2 and a first assumption is that these two defect states are also related to multidefect complexes, generated in the silicon lattice at the melt temperature, and subsequently quenched during the freezing process. However, a late investigation has shown that E_3 is absent from the DLTS spectrum when PEBA is performed on pre-heated samples at 450°C whereas E_1 and E_2 are still clearly detected. This result seems to indicate that either the cooling rate or the melt front velocity play a significant role in the generation of E_3.

CONCLUSIONS

We have shown that at a relatively low energy density (> 1 J/cm^2) a thin damaged surface layer is induced by PEBA in bare (100) N-type silicon. Electron traps have been detected much deeper than the damaged layer in PEBA processed samples with concentrations ranging from 10^{12} to 10^{13} cm^{-3}. One of them at $E_C - 0.48$ eV has been tentatively attributed to defect interaction with dislocations. The other electron traps are likely to be originating from quenched-in multidefect complexes. This mechanism is mostly probable in the PEBA induced partially melted zone where the silicon lattice is not destroyed although at the melting temperature.

ACKNOWLEDGMENTS

This work was supported by A.F.M.E.

REFERENCES

1   J.L. BENTON, L.C. KIMMERLING, G.L. MILLER, D.A.H. ROBINSON. Laser-Solid Interaction and Laser Processing, S.D. Ferris, H.J. Leamy and J.M. Poate eds (A.I.P. Conf. Proceedings, New York 1979), p. 543

2   A. MESLI, J.C. MULLER, P. SIFFERT. Laser-Solid Interaction and Transient Thermal Processing of Materials, Journ. de Phys., Coll.5, p. C5-281 (1983)

3   L.C. KIMMERLING, J.L. BENTON. Laser and Electron-Beam Processing of Materials, C.W. White and P.S. Peircy eds (Academic Press, New York, 1980) p. 385

4   A. CHANTRE. Laser-Solid Interaction and Transient Thermal Processing of Materials, Journ. de Phys., Coll.5, p. C5-269 (1983)

5   D. BARBIER, M. KECHOUANE, A. CHANTRE, A.LAUGIER. Laser-Solid Interaction and Transient Thermal Processing of Materials, J. Narayan, W.L. Brown, R.A. Lemons eds (North Holland, New York, 1983), p. 449

6   A. LAUGIER, D. BARBIER, G. CHEMISKY. Proceedings of the 4 th E.C. Photovoltaic Solar Energy Conference, D. Reidel Publishing Company, p. 1007 (1982)

7   G. CHEMISKY, D. BARBIER, A. LAUGIER. Journal of Crystal Growth, 66, p.215 (1984)

8   G. CHAUSSEMY, D. BARBIER, A. LAUGIER. This symposium

9   M.S. DOGHMANE, Thesis (June 1984)

10   M.S. DOGHMANE, D. BARBIER and A. LAUGIER. Laser Solid Interaction and Transient Thermal Processing of Materials, Journ. de Phys., Coll.5, p. C5-297 (1983)

11   M. PITAVAL, M. THOLOMIER, M. AMBRI, G. CHEMISKY, A. LAUGIER. Inst.Phys.Conf.Ser. No. 67: section 3, p.173 (1983)

12   L.C. KIMMERLING, J.R. PATEL. Appl.Phys.Lett. 34(1), p.73 (1979)

A COMPARISON OF DEEP LEVELS GENERATED
IN InP BY PROTON AND DEUTERON BOMBARDMENT

A. T. MACRANDER AND B. SCHWARTZ
AT&T Bell Laboratories, Murray Hill, New Jersey, USA

ABSTRACT

A predominant deep level which has an activation energy of 0.4 eV was found to have been formed in the region between the surface and the end-of-range of 200 keV ions. In the fluence range $10^{12}-10^{13}$cm^{-2} essentially no difference between the deep level spectra for proton and deuteron bombardments was found.

INTRODUCTION

For many years, III-V compound semiconductors in general, and GaAs in particular, have been known as the materials of the future. Numerous GaAs- and GaP-based discrete devices (e.g. lasers, FET's, photodetectors, LED's, etc.) have moved into production, and a complex material and fabrication technology has begun to develop. Following very close behind are the InP-based devices, and a required technology is developing here as well.

Among the interesting properties of these III-V materials is the ability to make the material semi-insulating by various means. Deliberate doping of bulk crystals with impurities such as chromium, iron, cobalt, or oxygen, or by careful growth of undoped (i.e. high purity, meaning $\sim 10^{14}$ cm^{-3} or less of background dopant) has been one of the main means of preparing large volumes (i.e. ingots) of material that is semi-insulating. One other means that has proven to be very useful for making high resistivity GaAs and GaP is by ion bombardment which damages the crystal lattice. The most widely used ion species for this purpose, but by no means the only one, has been the proton. Because of its low atomic number and low ionic mass, it is the most deeply penetrating of the possible ion choices, and thus finds use in the fabrication of devices requiring localized isolation and/or current confinement. Specific application for this technology is in the fabrication of short wavelength semiconductor lasers and GaAs IC's.

It was natural, then, to apply a similar approach to the fabrication of InP-based devices, and, indeed, it was found that InP could be made highly resistive by proton bombardment,[1] and some proton-bombarded longer wavelength lasers have been fabricated with this technology. However, the ability to control the generation of high resistivity, without type conversion, has proven to be a difficult problem. Recently, it was observed that deuterons, when used as the impinging particle on p-type InP, reproducibly and over a very wide dose range (wider than for proton bombardment), resulted in high resistivity material.[2]

One of the questions that thus comes into focus is, why the difference? It was this question that led us to study the energy levels generated in InP by bombardment with ^{1}H$^+$ or ^{2}H$^+$. Specifically, we address the question of whether or not the basic defects generated at a low fluence (relative to the fluence needed for the maximum resistivity) are different between proton and deuteron bombardment of n-type InP. We find that the deep level spectra are quite similar and conclude that the differences between the two types of ions vis-a-vis the type conversion of p-InP at large fluences are due to phenomena not occurring for low fluences or are related to the dopant species. We note that this similarity between the proton and deuteron bombarded cases is analogous to that found for n-GaAs.[3]

EXPERIMENTAL RESULTS

Gold Schottky barrier diodes 250 μm or 500 μm in diameter were formed by evaporation of Au through a contact mask onto wafers consisting of a single nominally undoped 6-7 μm thick homoepitaxial n-type layer grown by vapor phase epitaxy on an n^+ ($\sim 10^{19}$ cm^{-3}) S doped substrate which was oriented along (100). The substrate side was covered with a full surface evaporation of Au for the ohmic contact. Wafers were cleaved, mounted on TO-18 transistor packages, and wire-bonded using a single component silver epoxy. The epoxy was cured at 150°C for 1 hr. This cure constituted a radiation damage anneal since it was performed after ion bombardment.

For unirradiated material the carrier concentration obtained from capacitance-voltage (C-V) measurements varied over the range 10^{15} to 10^{16} cm^{-3} over the wafer and over the layer thickness. Thermally stimulated capacitance (TSCAP) curves and deep level transient spectroscopy (DLTS) curves revealed the presence of either 2 or 3 traps at concentrations up to 10^{14} cm^{-3}.[4,5]

Both proton irradiated (10^{13} cm^{-2}) and deuteron irradiated (10^{12} cm^{-2}) material were studied via TSCAP and DLTS. Irradiations were performed at 200 keV and at 7° from the (100) direction to avoid channeling.

TSCAP curves for the proton case are shown in Fig. 1 and for the deuteron case in Fig. 2. In both cases the capacitance was found to decrease dramatically with temperature and an easily recognized TSCAP step which is characteristic of a deep level was observed at a temperature of $\sim$100 K. The large capacitance change occurring at $\sim$250 K is also the result of a trap. It was determined via admittance spectroscopy that the electron trap responsible is likely the same as the one leading to the TSCAP step. This trap shows up clearly in the DLTS spectra shown in Fig. 3 and Fig. 4 and has been labeled a D-center because it was first observed in the deuteron irradiated material.[5] The D-center was found to have an electron emission activation energy of 0.4 eV.

Capacitance-voltage data for the irradiated cases are shown in Figs. 5 and 6. A strong temperature dependence was observed. The C-V data and the TSCAP and DLTS data can all be understood[5] by modeling the D-center as an acceptor which is formed in the region between the surface and the end-of-range of the ions (2 μm). In addition one must invoke a shallow (in energy) donor which is formed in this same region. The C-V data fall into three categories which correspond to the three curves shown in Figs. 5 and 6: i) At very low temperatures the D-center is frozen out but the shallow donors are not, so that the region in which D-centers occur is compensated and has a low charge density ($\sim 10^{14}$ cm^{-3}). ii) At higher temperatures the D-center becomes thermally activated. However, the temperature may still be too low for the D-center to be able to respond to the frequency used to measure the capacitance (1 MHz). In this case, an increased bias will ionize more D-centers but the capacitance will be determined by the crossing point of the shallow donor level and the Fermi level. This point is pinned at the end-of-range, and the result is a constant capacitance until the D-center ionization punches through to the end-of-range at $\sim$20 volts bias. iii) At high temperatures the D-center can respond to the 1 MHz signal and an ordinary Schottky-diode-like C-V curve was obtained with a reasonable built-in potential. D-center concentrations of $\geq 4.6 \times 10^{16}$ cm^{-3} and $\geq 4 \times 10^{15}$ cm^{-3} were obtained for the proton and deuteron cases, respectively. We note that these concentrations have roughly the same ratio as do the fluences.

In addition to the D-center another deep level was observed after proton and deuteron irradiations. This is labeled IE2 and can be seen at 250-300 K in Figs. 3 and 4. The high temperature portions of these spectra are irregular due to the large change in impedance between high and low temperatures of the diodes. The spectra in Figs. 3 and 4 were obtained by balancing the capacitance bridge[6] at low temperatures. To resolve high temperature DLTS features one must do this at high temperatures. We have done this, and we find a smoothly varying DLTS peak. The concentration observed after proton bombardment was 4×10^{15} cm^{-3} and after deuteron bombardment was 3×10^{14} cm^{-3}. The electron emission activation energy was in the range 0.6-0.7 eV. We note that these concentrations also have roughly the same ratio as do the fluences.

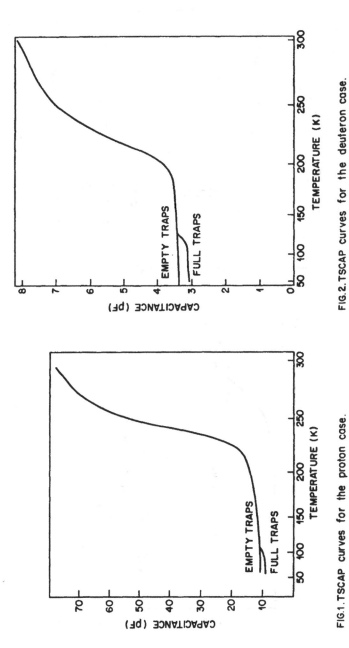

FIG.2.TSCAP curves for the deuteron case.

FIG.1.TSCAP curves for the proton case.

296

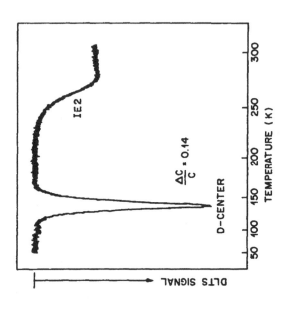

FIG.4. DLTS curve for the deuteron case.

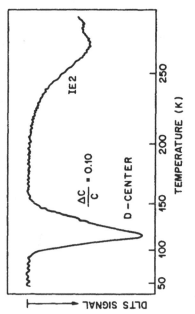

FIG.3. DLTS curve for the proton case.

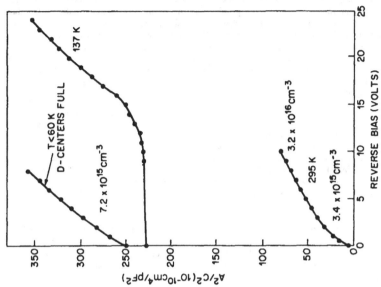

FIG.6. C-V data for the deuteron case.

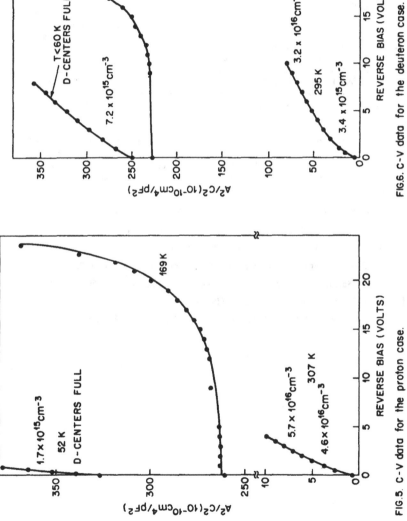

FIG.5. C-V data for the proton case.

DISCUSSION

The difference between the resistivities of p-InP at higher fluences for the proton and deuteron cases may be due to a sluggishness of the Fermi level to move through midgap as a function of increasing deuteron fluence. For the proton case the Fermi level shifts rapidly as the fluence is raised and the resistivity drops off markedly. The sluggishness in the case of deuterons is likely due to deep levels other than the D-center. We note that we did find a deep level (IE2) which is situated near midgap. Although no significant difference between the introduction rates of this level for the proton and deuteron cases was observed at the low fluences used presently it may well be that at higher fluences ($\geq 10^{14}$ cm^{-2}) more are introduced upon deuteron bombardment than upon proton bombardment, and this might explain the lack of rapid type conversion for heavily deuteron bombarded p-InP.

Finally, we note that recently Loualiche et al.[7] have reported DLTS measurements of n-InP bombarded with 100 keV protons to a fluence of 5×10^{11}cm^{-2}. After room temperature annealing they find two main electron traps (their A3 and A4) having activation energies of 0.36 and ~ 0.7 eV, respectively. We associate these with the two traps we have detected. Furthermore, these workers found a uniform distribution of A3 throughout the ion range which is in agreement with our findings for the D-center.[5] These results strongly suggest that the electronic energy loss (10 eV/Å/ion) which predominates over the nuclear energy loss (≤ 0.2 eV/Å/ion) and is constant up to the end of range is responsible.

CONCLUSIONS

In the fluence range studied, namely, $10^{12}-10^{13}$cm^{-2}, we have found essentially no difference between the deep level spectra after proton and deuteron bombardment. Our results can be understood if the primary deep level formed is taken to be an acceptor. This deep level center is formed in the region between the surface and the end-of-range of the ions.

REFERENCES

[1] J. P. Donnelly and C. E. Hurwitz, Solid-State Electron. *20*, 727 (1977).

[2] M. W. Focht and B. Schwartz, Appl. Phys. Lett. *42*, 970 (1983).

[3] P. Blood, Inst. Phys. Conf. Ser. 56, p. 251 (1981).

[4] A. T. Macrander and W. D. Johnston, Jr. J. Appl. Phys. *54*, 806 (1983).

[5] A. T. Macrander, B. Schwartz, and M. Focht, J. Appl. Phys. *55*, 3595 (1984).

[6] D. V. Lang, J. Appl. Phys. *45*, 3014 (1974).

[7] S. Loualiche, P. Rojo, G. Guillot, and N. Nouailhat, Revue Phys. Appl. *19*, 241 (1984).

CHARACTERIZATION OF DEFECTS IN PROTON-IMPLANTED GaAs

H. A. JENKINSON*, G. N. MARACAS**, M. O'TOONI*, and R. G. SARKIS*
*U. S. Army Armaments Research and Development Center, Dover, NJ 07801-5001
**Arizona State University, Center for Solid State Electronics Research,
  Tempe, AZ 85287

ABSTRACT

    Studies have been made of n^+-GaAs implanted with 300 keV protons to
fluences of 10^{14}-10^{16}/cm^2. Electrical studies included I-V analysis, DLTS,
and thermally stimulated current measurements made on implanted FET-like
structures fabricated from MBE GaAs epi-layers. Optical infrared reflectance
spectra and high resolution transmission electron micrographs were obtained
for as-implanted material and for specimens annealed at 300 °C and 500 °C.
Results show as-implanted material can be characterized by a band of traps
lying about 0.2 eV below the conduction band with a relatively uniform
distribution throughout the implanted region. Thermal processing causes
a significant alteration in the density and distribution of these defects.

INTRODUCTION

    The use of proton implantation for electrical isolation of devices in
doped GaAs optoelectronic circuits is attractive because it allows a higher device packing density than can be achieved with mesa isolation technologies [1].
This procedure also alters the refractive index of the material making it suitable, under certain conditions, for use as an infrared optical waveguide [2].
The ability to easily fabricate on-chip optical waveguides in this manner
greatly facilitates development of monolithic integrated optical circuits.

    Both of these applications rely on the reduction of the free carrier con-
centration in the implanted region. Reducing the number of carriers increases
the electrical resistivity of the material. Concurrently, the free carrier
plasma depression of the dielectric function is decreased, increasing the refractive index with respect to more heavily doped regions. Proton bombardment
creates a thin layer on the wafer surface containing large concentrations of
simple defects in the crystalline material. Energetically, this can be described by the creation of a band of levels below the conduction band. The
population of these newly formed levels reduces the free carrier concentration
in the conduction band. It is important therefore to characterize the energy
distribution and density of these levels throughout the damaged layer. Moreover, the effects of post-implantation thermal processing on structural,
electrical and optical properties must be correlated.

    This paper reports recent results in continuing investigations of these
effects. Experiments consisted of infrared reflectance measurments to optically characterize the implanted layer, a series of electrical measurments to
attempt to identify the radiation-induced traps, and high resolution electron
microscopic examination to determine structural changes. Due to the differing
requirements of the anticipated experiments, it was not feasible to use identical samples for each type of characterization. These differences are discussed
below. The implantations, however, were all done with 300 keV protons, and
were performed at room temperature with the samples oriented 8° with respect to
the beam to minimize the effects of ion channeling. To preclude effects due to
wafer heating during implantation, the samples were bonded with a thermal paste
to the target assembly and the beam current was held to 10^{-6} amps/cm^2. It is
predicted from the LSS projected range calculations [3] and it has been verified by SIMS measurements [4] that for these conditions the proton distribution

in GaAs is nearly Gaussian with a projected range of 2.4-2.7 μm and a standard deviation of about 0.25 μm.

ELECTRICAL CHARACTERIZATION

Four n-type, Si-doped epitaxial layers were grown on semi-insulating, Cr-doped GaAs substrates by Molecular Beam Epitaxy (MBE). Each layer was 3.5 μm thick, uniformly doped to a carrier concentration of $2 \times 10^{17}/cm^3$ as determined by capacitance-voltage measurements, and isolated from the substrate by an undoped layer 1 μm in thickness. This carrier concentration was considered to be as large as practical for these measurements. One sample was kept as a control and the others processed to fabricate test structures for material evaluation. The basic test structure employed was a mesa-isolated MESFET (Figure 1). This was chosen because of its accurately defined geometry and adaptibility to conductance Deep Level Transient Spectroscopy (DLTS) measurements.

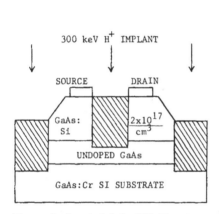

Figure 1. Ungated GaAs FET Structure

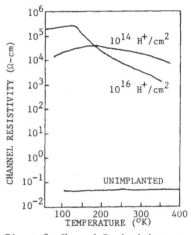

Figure 2. Channel Resistivity vs. Temperature of Implanted FETs

Ohmic contacts were deposited by photolithographically defining contact windows and evaporating metal in an e-beam system. A total of 2200 Å of Au/Ge/Ni was deposited in layers at a background pressure of ~5x10⁻⁷ Torr. The contacts were alloyed at ~490 °C for 30 seconds in a dry nitrogen atmosphere. This high temperature process was performed before implantation so there would be no effects due to unintentional annealing of the bombarded samples. Samples were then implanted to fluences of $10^{14}, 10^{15}$, and 10^{16} protons/cm². It was expected the bombardment would compensate the channel region between the source and the drain, as well as serve to completely isolate the mesa structures. Although the implantation would penetrate the electrodes for the source and the drain, compensating the underlying material somewhat, these regions were expected to remain conducting with respect to the channel. Conductance would therefore be across the channel and the conductance versus depth profile should reflect the defect distribution in depth.

The four samples were mounted in individual 40 pin ceramic flat packages and the FET terminals bonded to the package pins with 0.7 mil Au wire in a thermocompression bonder. All processing temperatures after implantation were kept below 90 °C to prevent annealing from altering the implantation-induced damage.

Electrical measurements were performed on the as-inmplanted samples using a microcomputer-based data acquisition/analysis system similar to the one described in reference [5]. Conductance DLTS measurements were performed on the samples by optically filling traps with an AlGaAs laser and a GaAs LED. A small drain bias was applied and the transient channel conductance was monitored after the filling pulses. The polarity of the transients indicated the presence of majority carrier (electron) traps. It was observed that instead of a few discrete defects, a band of defects was created by the proton implantation. Conventional DLTS analyses could not extract an accurate activation energy for the band but it was found to have an energy of about 0.2 ±0.1 eV from the conduction band. Such a complicated defect structure is not easily characterized by DLTS so the temperature dependence of the resistivity was measured. Figure 2 shows the results for the unimplanted sample and for samples implanted to fluences of 10^{14} and 10^{16} protons/cm^2. Reliable contacts could not be made to the sample implanted to 10^{15} protons/cm^2. The resistivity of the unimplanted sample was essentially constant in the 100-400 °K temperature range. All implanted samples showed a rise in resistivity with temperature and then a decrease. The resistivity peaks occured in the range of 110-150 °K for the samples tested. The temperature of the peak decreased with increasing fluence. Also, the total resistivity change was larger and the slope of the high temperature rolloff was sharper as the fluence increased.

This behavior can be explained by considering the effect of the trap band as a function of temperature. At low temperatures (<150 °K), the Fermi level is between the deep traps and the shallow Si donor level. Most of the free carriers are in the trap states and the resistivity is high. As the temperature increases, the Fermi level passes through the trap level which causes it to ionize, releasing carriers and decreasing the resistivity. The transition is not abrupt, as would be expected for a single state, indicating that a distribution of states determines conduction in the material. However, the $\rho(T)$ curve is more abrupt for the sample implanted to 10^{16} protons/cm^2 than for the one implanted to 10^{14} protons/cm^2. This suggests the high fluence sample has a high concentration of a narrow band of states while the lower fluence sample has a relatively lower concentration of a wide band of states. A tentative explanation for this behavior is that the initial effect of the bombardment is to cause lattice damage. This is evidenced by the broad band of energy levels composed of overlapping discrete levels associated with a specific point defect. As the fluence increases, states associated with the presence of hydrogen (as an impurity) begin to dominate the energy band structure and, as manifested by the sharpening of the resistivity transition. The temperature dependence of the Schottky barrier height using I-V measurements was performed and it was found that for the high fluence samples, the barrier height decreased from 0.6 eV at room temperature to ~0.3 eV at ~150 °K and remained constant at lower temperatures. This number agrees well with the trap activation energy estimated by DLTS and also corresponds to the breakpoint temperature in the $\rho(T)$ curve. One can conclude that the Fermi level is being pinned under the Schottky barrier because of the presence of high concentrations of hydrogen.

OPTICAL CHARACTERIZATION

The samples used for optical characterization were obtained commercially and were heavily doped with Si ($n \sim 3\times10^{18}$/cm^3) in order to obtain large optical effects. The wafers were cut to expose (100) surfaces which were polished by the manufacturer. No further processing was performed prior to implantation. Following implantation to a fluence of 10^{15} protons/cm^2, measurements were made of the differential infrared reflectance spectra for wavenumbers between 800 cm^{-1} and 4000 cm^{-1} (2.5 μm and 12.5 μm). The measurements were made on a Perkin-Elmer model 180 spectrophotometer fitted with dual specular reflectance attachments and were relative in that the ratio of the reflected intensity of the implanted sample to that of an unimplanted piece of the same wafer was recorded. These measurements were made on as-implanted material and on two samples annealed for 30 minutes at 300 °C and 500 °C respectively. Results are shown in Figure 3.

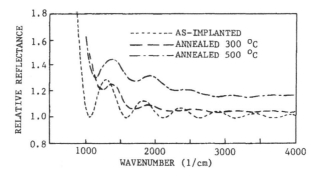

Figure 3. Measured Reflectances of Implanted GaAs Wafers

Several things are immediately apparent from these spectra. For the as-implanted sample, the location of the fringe minima are very periodic and indicate the implanted layer can be optically characterized by a single layer on the substrate. The decay in amplitude with wavenumber is consistent with the free carrier model for the dielectric function. Using an analysis reported previously [6,7] it is possible to show from these spectra that the implanted layer has a higher refractive index than the substrate (since the reflectances are greater than 1.0), that the thickness is about 3.25 μm, and the carrier concentration is about a factor of 100 less, or $3\times10^{16}/cm^3$.

The fringes of the sample annealed at 300 °C are reduced considerably in amplitude indicating some of the compensating defects have been destroyed. The major fringe minima are not as periodic and are lower in amplitude. In addition the fringe minima at lower wavenumbers are moving upward from 1.0. These factors indicate the layer can no longer be modelled as a single slab on a substrate. Computer simulations using a multilayer reflectance model indicate this fringe structure may be explained by the formation of a graded-index transition region between the layer and the substrate.

For the sample annealed to 500 °C, the fringes are larger in amplitude again. However the fringe minima are here considerably above 1.0. These factors indicate the structure has a basic thickness of about 3.0 microns and is still well compensated, but has an index gradient to the substrate.

ELECTRON MICROSCOPY

Samples similar to those used in the optical characterizations and annealed to 300 °C were prepared for examination in a high resolution electron microscope. In contrast to previously reported efforts with cross sectional specimens [7], these samples were prepared for transverse examination. A series of sections were bonded together using a low melting point adhesive, and then progressively thinned by mechanical abrasion, electrolytic polishing, and ion milling. Examination was then performed in both image and diffraction modes. Figures 4 and 5 are an electron micrograph and the associated electron

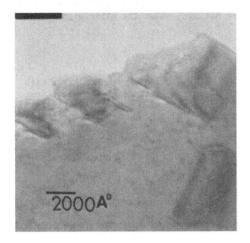

Figure 4. Transverse Electron
Micrograph of Implanted GaAs

Figure 5. (100) Electron
Diffraction Pattern of
Implanted GaAs

diffraction pattern for one of these specimens. The photomicrograph shows an array of dislocations at a depth of approximately 1250 Å from the irradiated surface. The diffraction pattern is oriented in the (100) direction and shows the material is highly crystalline.

DISCUSSION

Proton bombardment of GaAs has been known to produce a variety of compensating defects in GaAs. The mechanism of this effect is not known however, nor has the stability of the defect distribution been determined. The research reported here was performed in an attempt to begin to answer these questions.

Measurments on as-implanted material show a defect distribution which is much more uniform than would be expected based on energy loss models for hydrogren implanted in GaAs. Although the optical measurements are not sensitive to the energy band structure of the compensated material, the electrical characterizations reported here do suggest that this structure is composed of many overlapping levels and is dose dependent in an unusual way. At low fluences, the energy levels observed agree with those reported for electron irradiated material [8]. All of these observations are consistent with a model that includes ionization effects in defect creation. This would explain the high degree of compensation observed near the surface of the samples.

As the fluence is increased, a defect strongly associated with hydrogen is apparent in the electrical measurments. Although time did not permit the repitition of these experiments with annealed samples, it would be informative to determine the trap distribution following high temperature treatment. Presumably simple point defects would be annealed out, causing their associated energy levels to disappear. If there are levels assocated with hydrogen as an impurity, they should remain until the hydrogen diffuses away from its implanted location. The reflectance curves suggest a process like this may be occuring as hydrogen diffuses into the substrate. Further studies will be necessary to confirm this hypothesis.

Electron microscopy has shown the existence of defect agglomerations in as-implanted and annealed material and has also shown the material remains crystalline following bombardment.

ACKNOWLEDGEMENTS

The authors would like to thank Dr. Wyn Laidig of North Carolina State University for growing the MBE layers, Dr. Harry Dietrich of the Naval Research Laboratory for the implantations, and M. Tischler of North Carolina State University and T. Schaeffer of Motorola for aid in preparing the electrical samples.

REFERENCES

1. J.D. Speight, P. Leigh, N. McIntyre, I. G. Groves, and S. O'Hara, Electron. Lett. 10, 98-99, (1974).
2. E. Garmire, H. Stoll, A. Yarive, and R. G. Hunsperger, Appl. Phys. Lett., 21, 87-88, (1972).
3. J. F. Gibbons, W. S. Johnson, and S. W. Mylorie, Projected Range Statistics, 2nd Edition (Dowden, Hutchinson, and Ross, Stroudsburg, PA 1975).
4. J. M. Zavada, D. K. Sadana, R. Wilson, R. G. Sarkis, and H. A. Jenkinson, Proceedings of SPIE Conference 530 - Advanced Applications of Ion Implantation, Paper 530-26, Los Angeles (1985).
5. G. N. Maracas, W. D. Laidig, and H. R. Wittman, J. Vac. Sci. Tech., JVST-B2 July/Sept. #3, (1984).
6. H. A. Jenkinson and D. C. Larson, NASA Conference Publication 2207, 231-240, Hampton, VA (1981).
7. H. A. Jenkinson, M. O'Tooni, J. M. Zavada, T. J. Haar, and D. C. Larson, Proceedings of the Fall Meeting of the Materials Research Society, Paper E7.22, Boston, MA (1983).
8. D. V. Lang, R. A. Logan, and L. C. Kimerling, Phys. Rev. B, 15, 4874-4882, (1977).

STRAIN/DAMAGE IN CRYSTALLINE MATERIALS BOMBARDED BY MeV IONS:
RECRYSTALLIZATION OF GaAs BY HIGH-DOSE IRRADIATION

C. R. WIE, T. VREELAND, JR., AND T. A. TOMBRELLO
Divisions of Physics, Astronomy, and Mathematics and Engineering and Applied
Science, California Institute of Technology, Pasadena, California 91125

ABSTRACT

MeV ion irradiation effects on semiconductor crystals, GaAs(100) and Si
(111) and on an insulating crystal CaF_2 (111) have been studied by the x-ray
rocking curve technique using a double crystal x-ray diffractometer. The results on GaAs are particularly interesting. The strain developed by ion
irradiation in the surface layers of GaAs (100) saturates to a certain level
after a high dose irradiation (typically $10^{15}/cm^2$), resulting in a uniform
lattice spacing about 0.4% larger than the original spacing of the lattice
planes parallel to the surface. The layer of uniform strain corresponds in
depth to the region where electronic energy loss is dominant over nuclear
collision energy loss. The saturated strain level is the same for both p-type
and n-type GaAs. In the early stages of irradiation, the strain induced in
the surface is shown to be proportional to the nuclear stopping power at the
surface and is independent of electronic stopping power. The strain saturation phenomenon in GaAs is discussed in terms of point defect saturation in
the surface layer.
    An isochronal (15 min.) annealing was done on the Cr-doped GaAs at tempe-
ratures between 200° C and 700° C. The intensity in the diffraction peak from
the surface strained layer jumps at 200° C $< T \leq 300^{\circ}$ C. The strain decreases
gradually with temperature, approaching zero at $T \leq 500^{\circ}$ C.
    The strain saturation phenomenon does not occur in the irradiated Si.
The strain induced in Si is generally very low (less than 0.06%) and is interpreted to be mostly in the layers adjacent to the maximum nuclear stopping
region, with zero strain in the surface layer. The data on CaF_2 have been
analysed with a kinematical x-ray diffraction theory to get quantitative
strain and damage depth profiles for several different doses.

INTRODUCTION

    Sadana and coworkers observed that GaAs amorphized by MeV ion irradiation
at liquid nitrogen temperature could be reconverted to a single crystal by a
subsequent MeV ion irradiation at room temperature [1]. The following experiments show that higher substrate temperature and dense electronic ionization
around the path of a bombarding MeV ion can be effective in the suppression of
amorphization and in amorphous-to-crystalline reconversion. In an x-ray rocking curve study, a GaAs crystal implanted at room temperature with 300 keV Si
ions (where the energy loss process is mainly due to nuclear cascade collisions with comparable energy to this nuclear stopping deposited into the target electronic system only near the surface) gives a small diffraction signal
for a $1.2 \times 10^{15}/cm^2$ beam dose [2]. A 220 keV ion implantation at room temperature produces an amorphous layer in GaAs at a beam dose of $10^{13}/cm^2$, whereas
the implantation performed at 150° C does not produce an amorphous region at a
$10^{15}/cm^2$ dose [3].
    Hodgson and coworkers used a pulsed 200 keV proton beam to anneal Si
amorphized by 100 keV As ion implantation [4]. Even though the 200 keV proton

beam loses most of its energy into the target electronic system, there is a phenomenological difference between the typical pulsed beam annealing and the keV ion implantation or the MeV ion bombardment. In the pulsed beam annealing, the typical beam current density is 10^{20}-10^{21} ions/cm^2 sec for a proton beam and 10^{21}-10^{22} electrons/cm^2 sec for a pulsed electron beam, so that the cylindrical regions of dense ionization around the particle path heavily over-lap instantaneously, and the irradiation region becomes laterally uniform, so the irradiation effect results in the melting of the target from thermal heating. In the MeV ion irradiation, however, the typical particle current density is 10^{11} to 10^{12} ions/cm^2 sec so that the cylindrical ionization regions around each ion path do not overlap appreciably within the relaxation time of the ionized electrons.

MeV ions leave damage tracks [5] in and cause enhanced sputtering from the surface of insulators and some compound semiconductors [6]. The cylin-drical damage track structure is ascribed to processes following the dense ionization around the ion path in the target [7]. Elemental semiconductors do not show enhanced erosion by electronic ionization/excitation from MeV ion bombardment [8]. The electronic ionization/excitation effect on the solid surfaces induced by MeV ion bombardment is used for surface modification [9], interfacial changes for the enhancement of thin film adhesion, and for the modification of electrical contact properties [10]. We used a CaF$_2$ single crystal, a track forming material, to study the strain and damage in the sur-face layers induced by MeV ions [11].

Among the III-V compound semiconductors, simple defects are essentially immobile near room temperature in GaAs and well above room temperature in GaP; however, the primary defects in GaSb, InSb, and InAs and those in elemental semiconductors, Si and Ge, are all mobile below 200 K [12]. Lang and co-workers have shown that the isochronal recovery stage versus Debye tempera-tures within the III-V group is proportional to Θ_D^2 based on the previously reported data [12 and references therein].

The x-ray rocking curve technique [13] has been used to measure the damage and strains in ion implantation [5], in MeV ion irradiated garnet single crystals [14], and to analyze supperlattices [15]. In the next section, we show rocking curve data on various crystals bombarded with MeV ions.

EXPERIMENTAL RESULTS AND DISCUSSION

GaAs

Cr-compensation doped p-type semi-insulating GaAs$_3$(100) and Si$_3$ and Te-doped n-type GaAs (100), of resistivities of 1.4×10^{-3} and 8×10^{-3} Ohm-cm, respectively, were irradiated with Cl and O ions (3 to 15 MeV) to various doses (10^{13} to 5×10^{15} ions/cm^2) using the tandem accelerator at Caltech. The irradiation current density was typically 10^{11} to 10^{12} ions/cm^2 sec. The irradiation was done approximately normal to the sample surface which is about 2-3 degrees off the (100) plane. No channeling effect was observed. The irradiated samples were analyzed by the x-ray rocking curve technique described elsewhere [11] using a double-crystal x-ray diffractometer. The Bragg reflections of FeK 1 radiation from (400), (422), (511), and (333) planes have been recorded at each angle, which is typically changed in .001 degree steps.

The (400) rocking curves taken for a Cr-doped GaAs single crystal irra-diated with a 15 MeV Cl ion beam (Range = $\sim$ 5.8um in GaAs) are given in Fig. 1 in order of increasing beam dose. The small single peak at zero angle is the diffraction signal from the undamaged substrate beyond the ion range. The diffraction pattern deviates toward negative angle from the substrate peak, varying with the ion beam dose, and corresponds to the strain development in

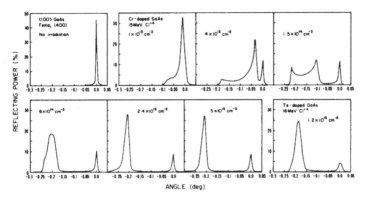

ANGLE (deg)

Fig. 1. (400) Bragg reflections for a Cr-doped
GaAs (100) crystal are shown for increasing doses
of 15 MeV Cl^{+4} ions. The peak at zero angle
corresponds to the diffraction by the substrate
crystal unaffected by the ion beam. The diffraction pattern deviates toward negative angle,
and corresponds to the strain development in the
surface layers by the ion irradiation. The single
peak around .22 degrees after a high dose
($\geq 1.2 \times 10^{15}$ cm^{-2}) irradiation indicates a
uniform lattice spacing like a single crystal in
the suface layers.

the surface layers affected by the ion beam. After a high dose irradiation
(e.g., D $\geq 1.2 \times 10^{15}$ ions/cm^2 for 15 MeV Cl) a single diffraction peak
develops at a well defined angular separation from the substrate peak, with a
small satellite of vanishing intensity for higher doses on the lower angle
side, which means that the strain level in the surface layer saturates to a
uniform lattice spacing like a single crystal, with a gradually amorphizing
layer corresponding to the maximum nuclear energy loss region sandwiched between the surface crystal and substrate crystal. The strain perpendicular to
the sample surface, in the surface layer can be calculated from the Bragg con-
dition $n\lambda = 2d\sin\theta_B$ to be $\varepsilon^\perp = \Delta d/d = \Delta\theta_B \cot\theta_B$ for a symmetric reflection.
Thus, the measured perpendicular strain, which is uniform after high dose
irradiation, is $\sim 0.4\%$. The strain parallel to the sample surface (defined
as "parallel strain" hereafter) can be measured by an asymmetric reflection.
This strain saturation phenomenon from high dose irradiation is the same for
other heavily doped GaAs crystals with approximately the same strain level (as
shown in the last diagram in Fig. 1 for Te-doped GaAs). Cr-doped GaAs crystals were bombarded with Cl and O ions at different energies to clarify the
strain production mechanism in the surface layers. The final strain level is
the same for all the ions at different energies. The strain at the sample
surface can be measured directly from the rocking curve data because the
strain level at the surface is the lowest in the strain vs. depth distribution. The surface strain perpendicular to the surface is plotted versus increasing ion dose in Fig. 2a. The saturation dose is lower for ions with
higher nuclear stopping power, for example, the saturation dose is
$\sim 1.5 \times 10^{14}$/cm^2 for 3 MeV Cl and $\sim 1.2 \times 10^{15}$/cm^2 for 15 MeV Cl. Fig. 2b
shows that at doses below the saturation dose the surface strain is propor-
tional to the nuclear stopping power, S_n for a given ion, and is independent
of the electronic stopping power, in contrast with the fact that MeV ions give
rise to an enhanced sputtering yield from the surface of compound semiconductors [6].

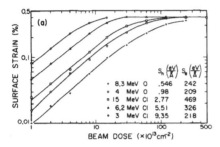

Fig. 2a. The perpendicular strain at the surface of GaAs (100) is given as a function of beam dose of various ions. The final "saturated" strain level is about the same for all the ion beams used. The higher the nuclear stopping power, the earlier the strain saturates.

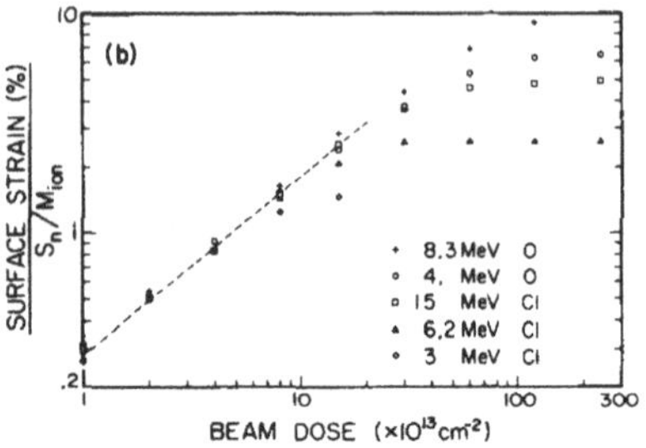

Fig.2 b. This plot was obtained by dividing the surface strain in (a) by S_n/M_{ion}. The data indicate that strain is proportional to $(S_n/M_{ion})D^{0.85}$.

To obtain the parallel strain in the surface layer, rocking curves were measured from asymmetric 422, 511, and 333 reflections for a Cr-doped GaAs (100) crystal bombarded with a 15 MeV Cl ion beam to a dose of $1.2 \times 10^{15}/cm^2$ and a 8.3 MeV O ion beam to $2.4 \times 10^{15}/cm^2$. The incident angle of the x-ray beam was $\theta_B - \phi$ with respect to the sample surface for the reflections measured, where ϕ is the reflecting plane angle with respect to the sample surface. For an asymmetric reflection, the angular separation of the strain peak from the substrate peak is related to the parallel and perpendicular strain by the following formula [15].

$$|\Delta\theta_B| = \varepsilon^{\perp}(\cos^2\phi\tan\theta_B \pm \sin\phi\cos\phi) + \varepsilon''(\sin^2\phi\tan\theta_B \mp \sin\phi\cos\phi) \qquad (1)$$

where the upper and lower signs refer to the incident angle of $\theta_B - \phi$ and $\theta_B + \phi$ with respect to the sample surface, respectively. Using Eq. (1) with the perpendicular strain found in the symmetric 400 reflection and the angular separation measured in the asymmetric rocking curves we find a parallel strain of essentially zero. This indicates a strong coupling of the strained surface layer to a much thicker substrate. The coupling does not permit the surface layer to expand laterally.

Fig. 3 shows the change of perpendicular strain level with temperature after 15 min. of isochronal annealing in vacuum for 15 MeV Cl doses of $1.2 \times 10^{15}/cm^2$. For the annealing, the GaAs sample was capped with another GaAs wafer to prevent As out-diffusion. There is an intensity jump of the reflecting power at a temperature 200° C $< T \leq 300^{\circ}$ C. This intensity jump is consistent with the 500 K recovery stage of electron irradiation-induced defects in GaAs [16]. Lang, Logan, and Kimmerling assigned the defects which recover at 500 K to Ga atom displacements [12] based on their work on the orientation dependence of defect production in GaAs. However, in more recent work, Pons and Borgoin reversed this assignment and proposed that it was due to As atom displacements (from their work on orientation dependence of the introduction rates of electron traps E1, E2, and E3 as a function of electron irradiation energy) [17]. Pons and Bourgoin demonstrated that for electron energies below 0.6 MeV, there are more As-displacements than Ga-displacements, but above 0.6 MeV there are more Ga-displacements, consistent in trend (but not in magnitude) with the work of Lang, Logan, and Kimmerling. Since in the heavy MeV ion irradiation case the elastic process in the ion-solid collisions may correspond to higher energy electron irradiation, the 500 K recovery stage in MeV ion bombarded GaAs is more likely due to Ga-vacancies. But in contrast to the relatively well defined recovery stage of radiation induced vacancies, the strain level in the surface layers decreases rather smoothly in the temperature range 200° C to 500° C, approaching zero at 500° C. The absence of a sharp recovery stage of the strain indicates that the origin of uniform surface strain in GaAs is not due to vacancies alone. If the heavily damaged layer in the nuclear stopping region recrystallizes epitaxially upon annealing, then it will grow in direction both from the surface crystal layer and from the substrate crystal [1]. It can be seen that the surface crystal layer does not recover epitaxially to the original crystal from the fact that the reflecting power of the strain peak does not decrease by increasing the substrate reflecting power, but only the relative angular separation changes upon annealing.

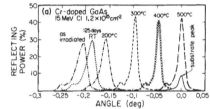

Fig. 3a. (400) Bragg reflection was taken for 15 MeV Cl ion bombarded GaAs (100) after a successive 15 min. isochronal annealing at the temperatures indicated. The broken curves represent the diffraction pattern from the surface strained layer. The solid curve at zero angle represents the diffraction pattern from the substrate, and is common for all the broken curves except the one taken after 500° C annealing. The broken curve 'as irradiated' is the rocking curve taken after ion irradiation, and the curve '125 days RT' after 125 days room temperature aging.

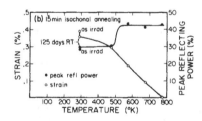

Fig. 3b. The perpendicular strain and the peak reflecting power of GaAs surface layer, obtained from (a), are given as a function of annealing temperature. The jump around 500 K in the peak reflecting power is consistent with the 500 K recovery stage of the radiation induced defects in GaAs. The strain decreases gradually with temperature.

The angular separation of the two peaks is not due to misorientation of the surface layer with respect to the substrate crystal as it did not change position after a 180 degree rotation of the sample about the surface normal.

The strain saturation in the surface layer indicates a saturation in the point defects responsible for the strain. There seem to be criteria for the strain saturation in room temperature MeV ion irradiation, i.e., (1) the point defects must be immobile near room temperature, (2) the electronic process in the energy loss of ion does not contribute to the damage production in the material, and (3) the energy loss into nuclear collisions is small enough so that no significant number of isolated disordered regions are created. If the point defects are mobile near or below room temperature as in Si and Ge, then the surface strain does not appear. If the electronic process contributes to the damage production as in the track forming materials, then the high dose irradiation eventually converts the surface layer completely to the amorphous state. The above criteria are all satisfied in the MeV ion irradiation of GaAs. For a Cl ion in GaAs, for example, the electronic process is dominant in energy loss for ion energies greater than about 0.3 MeV [20]. The nuclear stopping power of a Cl ion in GaAs is about 36 eV/Å for 0.3 MeV, and less for higher energies (see Fig. 2a for higher energies), calculated from the Kr-C formula [21]. Assuming a threshold displacement energy of about 14 eV in the semiconductors [22], it can be seen that only isolated point defects will be created in the GaAs surface layer by the MeV ion irradiation. The saturation of disordered regions in keV ion implantation leads to a complete amorphous layer because the individual disordered regions overlap sufficiently [23]. The saturation of point defects in MeV ion irradiated GaAs, however, leads to a uniformly strained layer, i.e., a single crystal layer with a high uniform concentration of point defects. This point defect saturation phenomenon in GaAs can be considered in terms of the competition of the creation and the recovery processes of point defects, which are simultaneously going on during the irradiation. If the dense electronic ionization around the ion path plays any direct role in the recovery process, then the local final saturation level will be governed by the local ratio of the electronic stopping power to the nuclear stopping power. This does not seem to be the case because the ratio is the largest at the surface and decreases along the ion path in the present energy range. Fig. 2a also shows that the initial strain level is independent of the electronic stopping power. It is more likely that the recovery process is governed by the temperature in the material during the irradiation, which may be uniform along the ion path.

A detailed strain vs. depth distribution, which may be very useful in comparing the strain level with the nuclear stopping power, can be obtained by the analysis of the x-ray rocking curve data by a suitable x-ray diffraction theory. When the reflecting power of the surface strained layer does not exceed about 6% (as in 15 MeV Cl irradiation CaF$_2$), a kinematical diffraction theory can be used. When the reflecting power of the strained layer is larger than about 6%, a dynamic theory must be used. We are currently developing a capability to analyse the rocking curves using the dynamic theory.

Si and CaF$_2$

A 15 MeV Cl ion beam was used to irradiate n-type Si (111) of resistivity .005-.02 cm and CaF$_2$ (111) single crystals to various doses. In Fig. 4 the rocking curve data for Si show that the strain level in Si is generally very low and that the maximum strain reached in the maximum nuclear stopping layer is around 0.06% before that layer becomes amorphous at a dose of 1.25 × 10^{14}/cm^2. The data also show that no appreciable strain is induced in the surface layer except for a very low strain of about 0.02%, which may be from the layers adjacent to the maximum nuclear stopping layer. The zero strain in the surface layer can be ascribed to the fact that the point defects in Si, vacancies and interstitials, are mobile at room temperature [12] and that the elemental semiconductors do not have antisite defects, unlike the compound semiconductors. The antisite defects generally have a higher annealing temperature than vacancies. The strain in the higher nuclear stopping region around the region of maximum nuclear stopping power can be from the defect complexes in Si. Isolated amorphous regions can also contribute to the strain measured by x-rays by the strain field around the isolated disordered region. The influence of this effect on the measured total strain has not been analysed.

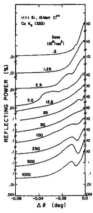

Fig. 4. (333) Bragg reflection from Si <111> bombarded by a 15 MeV Cl ion beam is shown. The substrate peak is at zero angle. The angular deviation of the diffraction pattern is small compared with GaAs or CaF$_2$. The strain giving rise to the diffraction pattern is in and around the region of maximum nuclear stopping power.

The rocking curves taken for a CaF$_2$ (111) single crystal are shown as a function of beam dose in Fig. 5a. When the maximum reflecting power of the strained layer is less than ∿ 6%, the strain and damage depth profile can be obtained by a kinematical x-ray diffraction theory analysis of the rocking curve data [11,13]. The strain/damage depth distribution (Fig. 5b) obtained by this analysis shows that the nuclear collision process, which is the dominant energy loss process near the end of ion range, is effective in producing strain and damage. The nearly constant strain level toward the surface is consistent with the fact that the nuclear stopping power in the high ion-velocity region $v \leq e^2/\hbar$ is nearly constant [18] and that the radial dimension of the nuclear damage track in an insulator is proportional to $(dE/dx)^{1/3}$ [19]. The radiation induced defects which are responsible for the strain production have not been determined.

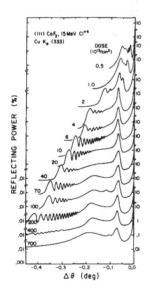

Fig. 5a. (333) Bragg reflection from a 15 MeV Cl bombarded CaF$_2$ <111> single crystal. The substrate peak is at zero angle.

312

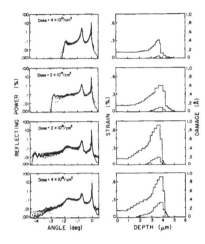

Fig. 5b. The dots are the experimental reflecting power and the solid curves are the calculated reflecting power using a kinematical x-ray diffraction theory, assuming the strain/damage depth distribution given on the right-hand side. The damage is defined by the mean atomic displacement (Å) from the lattice site, which appears in a Debye-Weller factor in the structure factor.

CONCLUSION

A high dose irradiaton of semi-insulating and highly doped GaAs (100) single crystals with MeV ions has been shown to produce a uniformly strained single layer in the surface region. The strain saturation in GaAs seems to be due to the point defect saturation in the surface layer. In the early stage of irradiation (i.e., at low dose), the strain produced at the surface is proportional to the nuclear stopping power for a given ion beam. The surface strain saturates at the same level (i.e., $\sim 0.4\%$ for the perpendicular strain) regardless of the ion beam used and the beam energy. The annealing behavior shows that there is a recovery stage of radiation induced vacancies at 200° C $< T \leq 300^\circ$ C in the surface single crystal layer and the strain is not due to vacancies alone and that, unlike the rather well defined recovery stages of the radiation-induced specific defects, the strain decreases gradually with temperature. The strain goes to zero at $T \leq 500^\circ$ C recovering the original lattice spacing. This annealing behavior of the strain indicates that the point defects in GaAs, i.e., vacancies, interstitials and antisite defects (the recovery stage of the antisite defect, As_{Ga}, is around 500° C [24]), are responsible for the strain in the surface layer. An EPR measurement of MeV ion beam induced antisite defects in GaAs, and some independent measurement of specific point defects as a function of beam dose in relation to the rocking curve measurement of lattice strain may be important in determining the importance of each point defect type in producing the strain.

MeV ion induced strain in Si is generally very low with no appreciable strain in the surface layer, consistent with the fact that no electronic ionization-induced enhanced erosion exists in the MeV ion irradiated Si. The strain induced in an insulating single crystal CaF_2 has been measured with the x-ray rocking curve technique, and quantitative strain and damage depth distributions were obtained by a kinematical diffraction theory analysis.

ACKNOWLEDGEMENTS

This work was supported in part by the National Science Foundation [DMR83-18274] and the Caltech President's Fund.

REFERENCES

1. D. K. Sadana, H. Choski, J. Washburn, and N. W. Cheung, Appl. Phys. Lett. 44, 301 (1984).
2. V. S. Speriosu, B. M. Paine, M-A. Nicolet, and H. L. Glass, Appl. Phys. Lett. 40, 604 (1982).
3. J. S. Harris, F. H. Eisen, B. Welch, J. D. Haskell, R. D. Pashly, and J. W. Mayer, Appl. Phys. Lett. 21, 601 (1972).
4. R. T. Hodgson, J. E. E. Baglin, R. Pal, J. M. Neri, and D. A. Hammer, Appl. Phys. Lett. 37, 187 (1980).
5. R. L. Fleischer, P. B. Price, and R. M. Walker, Nuclear Tracks in Solids, (Univ. of California Press, Berkeley, 1975).
6. Y. Qiu, J. E. Griffith, and T. A. Tombrello, Nucl. Instr. Meth. B1, 118 (1984); T. A. Tombrello, Int. J. Mass Spec. Ion Phys. 53, 307 (1983).
7. Y. Qiu, J. E. Griffith, W. J. Meng, and T. A. Tombrello, Rad. Eff. 70, 231 (1983).
8. T. A. Tombrello, Nucl. Instr. Meth, B2, 555 (1984); T. A. Tombrello, C. R. Wie, N. Itoh, and T. Nakyama, Phys. Lett. 100A, 42 (1984); T. A. Tombrello, Nucl. Instr. Meth. B1, 23 (1984).
9. T. A. Tombrello, Nucl. Instr. Meth 218, 679 (1983); T. A. Tombrello, J. Phys. Soc. Jap. in press (1984).
10. M. H. Mendenhall, Ph.D. Thesis, California Institute of Technology (1983); T. A. Tombrello, J. Mat. Sci. Eng., in press (1984); C. R. Wie, C. R. Shi, M. H. Mendenhall, R. P. Livi, T. Vreeland, Jr., and T. A. Tombrello, Nucl. Instr. Meth., in press (1984); S. Paine, C. R. Wie, M. H. Mendenhall, R. P. Livi, T. Vreeland, Jr., and T. A. Tombrello, Mat. Res. Soc. Symp. Proc., submitted (1984).
11. C. R. Wie, T. Vreeland, Jr., and T. A. Tombrello, Nucl. Instr. Meth., submitted (1984).
12. D. V. Lang, R. A. Logan, and L. C. Kimmerling, Phys. Rev. B 15, 4874 (1977).
13. V. S. Speriosu, J. Appl. Phys. 52, 6094 (1981); G. L. Miller, R. A. Boie, P. L. Cowan, J. A. Golvchenko, R. W. Kerr, and D. A. H. Robinson, Rev. Sci. Instr. 50, 1062 (1979).
14. B. Strocka, G. Bartels, and R. Spohr, Appl. Phys. 21, 141 (1980).
15. V. S. Speriosu and T. Vreeland, Jr., J. Appl. Phys. 56, 1591 (1984).
16. K. Thommen, Rad. Eff. 2, 201 (1970).
17. D. Pons and J. Borgoin, Phys. Rev. Lett. 47, 1293 (1981).
18. J. Lindhard and M. Scharff, Phys. Rev. 124, 128 (1961).
19. T. A. Tombrello, Nucl. Instr. Meth. B1, 23 (1984).
20. J. W. Mayer, L. Erikson, and J. A. Davis, Ion Implantation in Semiconductors (Academic Press, New York and London, 1970, p. 23).
21. W. D. Wilson, L. G. Haggmark, and J. P. Biersack, Phys. Rev. B 15, 2458 (1977).
22. A. Sosin and W. Bauer, Studies in Radiation Effects, ed. G. J. Dienes, Vol. 3 (Gordon and Breach, New York, 1969).
23. J. W. Mayer, L. Erikson, and J. A. Davis, Ion Implantation in Semiconductors (Academic Press, New York and London, 1970, p. 99).
24. R. Wörner, U. Kaufman, and J. Schneider, Appl. Phys. Lett. 40, 141 (1982).

ACTIVATION OF Si-N MODES IN SILICON BY PULSED LASER ANNEALING*

H. J. STEIN, P. S. PEERCY and C. R. HILLS
Sandia National Laboratories
Albuquerque, NM 87185

ABSTRACT

Retention and bonding of nitrogen implanted into crystalline Si were examined by infrared absorption (ir) and transmission electron microscopy
(TEM) after furnace and pulsed laser annealing. Localized Si-N vibrational
modes for N-N pairs are observed, and the associated ir band intensities
increase upon pulsed annealing. Furnace annealing above 600°C decreases
the ir intensity for N-N pairs and fine structure defects appear in TEM.
Subsequent laser annealing removes most of the fine structure and reacti-
vates the pair spectrum which we interpret as dissolution of N precipitates
and pair formation upon quenching from the melt. Any realistic model for
N in Si must include the formation and consequences of N-N pairs.

INTRODUCTION

Physical and chemical characteristics of nitrogen in silicon have
been of scientific interest for a number of years. Recently, mechanical
strengthening of Si wafers by N doping in the melt [1] and dielectric
layer formation by implantation of N into Si [2] have been widely explored
for technological applications. From studies using various techniques
[1-7], it is apparent that more than one kind of center is produced by N
doping and implantation and that the centers are strongly dependent upon
thermal history.

Nitrogen bonds to Si and local vibrational modes are observed by infrared absorption in both melt-doped [1] and N-implanted Si [8]. Combined
implantations of ^{14}N and ^{15}N have shown that the predominant local modes
are associated with N pairs [8] which are bonded to Si. Pairing occurs
during implantation and ir band intensities increase due to additional
pairing and/or annealing of implantation-induced displacement damage [9]
upon heating to 600°C. The present laser annealing study was undertaken
to obtain additional information about the formation and characteristics
of Si-N centers in Si. Studies of N-N pair formation after the implanted
layers are melted by pulsed laser irradiation gives evidence for pair
formation in the melt and for solute trapping of up to $\sim 10^{20}$ pairs/cm^3.
In contrast, a previous study of oxygen-implanted layers quenched from the
melt showed oxygen in dispersed interstitial sites [10].

EXPERIMENTAL DETAILS

Nitrogen was implanted 7° off axis at $\sim$50°C into both sides of polished
<111> float zone Si samples. Ion fluences between 2 x 10^{14} and 5 x 10^{15}/cm^2
at energies from 60 to 190 keV were utilized in the study. Laser annealing
was performed with $\sim$ 25 ns pulses at 2.0 to 2.2 J/cm^2 from a ruby laser.
This energy density was sufficient to produce a melt duration of $\sim$ 200 nsec
at the surface. Furnace annealing was performed in flowing N$_2$ for 1 hour
periods. Infrared absorption spectra were measured at 2 cm^{-1} resolution

*This work performed at Sandia National Laboratories supported by the U.S.
Department of Energy under contract number DE-AC04-76DP00789.

with a Nicolet SX60 FT-IR spectrometer. Transmission electron microscopy was performed using a JEOL model JEM 200CX microscope.

RESULTS AND DISCUSSION

Absorption spectra for the spectral range between 650 and 1100 cm^{-1} are shown in Fig. 1 for samples with sequential 60, 100 and 180 keV implantations of 1 x 10^{15} ^{14}N/cm^2. Spectra after subsequent 600°C furnace or pulsed ruby laser annealing are also shown. The maximum calculated implanted N concentration is 6 x 10^{19} cm^{-3} which is approximately four orders of magnitude greater than solid solubility and one order of magnitude greater than the reported liquid solubility.

Absorption bands at 767 and 963 cm^{-1} associated with Si-N bonds in Si are observed in both implanted [8] and melt doped Si [1]. These bands are associated specifically with N-N pairs [8]. Possible one and two vacancy sites for N-N pairs are sketched in Fig. 1. Annealing at 600°C significantly increases the intensities for these pair bands, and even larger increases are produced by laser annealing.

Intensities for the 767 and 963 cm^{-1} bands are correlated; therefore we use the intensity of only one band to illustrate fluence dependences. Peak intensity for the 963 cm^{-1} band is plotted in Fig. 2 as a function of 100 keV ^{14}N fluence after implantation and after subsequent furnace and laser annealing. The slope of the intensity versus fluence after implantation is greater than unity for fluences up to 2 x 10^{15}/cm^2; at higher fluences the intensity decreases. Annealing increases these band intensities for all fluences; however, the slope of the intensity versus fluence decreases.

Several factors could combine to determine the fluence and annealing dependence for the 963 cm^{-1} Si-N band: statistics and energetics of N-N pair formation, displacement damage, N mobility in Si and sinks which compete with pair formation for available N. Bimolecular reaction-limited formation of N-N pairs would give a slope of 2 for intensity versus fluence. Displacement damage which accompanies implantation, however, will cause a reduction in band intensity [9] and can explain the decrease in intensity at high fluences. Displacement damage is removed by furnace or laser annealing, and additional N-N pairs can form during annealing if N is mobile. Since some N-related defects are removed by annealing below 600°C, at least limited motion of N must occur at this temperature. Diffusion is expected to be ≤ 300 Å during the <200 ns melt produced by laser annealing.

The separation between damage and ion profiles is illustrated for the single energy implant in the inset in Fig. 2. A significant fraction of the implanted N comes to rest in a region of low displacement damage which is more readily annealed at 600°C than is the higher damage level present

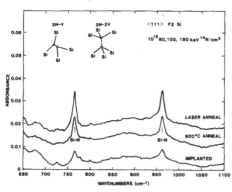

Figure 1. Si-N absorption bands in Si: 1) after ^{14}N implantation, 2) after implantation and 600°C furnace annealing, and 3) after implantation and laser annealing, along with sketch of possible one-vacancy/2N and two-vacancy/2N defects.

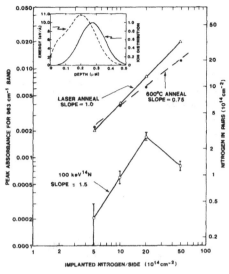

Figure 2. Intensity vs. fluence for 963 cm^{-1} Si-N absorption after implantation and subsequent furnace annealing. Inset illustrates separation of N and damage profiles.

in multiple energy implantation. This difference can explain the smaller differences between band intensities observed after furnace and laser annealing for this single energy implant compared to the multiple energy implant discussed for Fig. 1.

Complete pairing of N is a possible explanation for the unity slope for intensity versus fluence after laser annealing (Fig. 2). While small concentrations of other N centers are formed under laser annealing [3], approximately 35 percent of the implanted N can be accounted for based upon previously published data relating peak absorption coefficient, α, for the 963 cm^{-1} band and the N concentration, C; $\alpha = 1.3 \times 10^{-17} \cdot C$. The right ordinate on Fig. 2 gives the N atoms/cm^2 in pairs deduced from this relationship. The pair concentration after 5×10^{15} N/cm^2 and laser annealing is $\approx 10^{20}$ cm^{-3} near the peak of the distribution.

The N pairing upon quenching from the melt is in contrast to dispersed interstitial O centers observed when quenching O-implanted layers from the melt [10]. It is interesting to note that the relative dissociation energies for diatomic molecules are N-N > Si-N > Si-Si, whereas Si-O > O-O > Si-Si. If the bonding sequence upon implantation or quenching from the melt goes in the order of these energies then N-N bonds would form first in N-doped Si, whereas Si-O bond would be favored in O-doped Si.

Perhaps the best evidence that pairs are formed in the melt by breaking previously existing Si-N and N-N bonds is obtained from laser annealing of samples that were furnace annealed above 600°C. Absorption spectra for samples implanted with 10^{15} N/cm^2 at 60, 100 and 180 keV after furnace annealing at 750 and 900°C are shown by the lower traces in Figs. 3 and 4, respectively. The upper traces show the spectra for the same samples after subsequent laser annealing. Absorption peaks in the spectra after annealing at 750°C are close in frequency to strong absorption bands in crystalline silicon nitride so that N is probably forming precipitates. No Si-N bands are resolved after annealing at 900°C; however, subsequent laser annealing regenerates pairs in samples annealed at both 750 and 900°C. We infer from these observations that N is redissolved in the melt and solute trapped in pairs. Previous SIMS measurements [11] suggested that N movement toward free or internal surfaces occurs near 850°C. Such behavior can explain the weaker intensity for N-N pairs bands after 900 than after 750°C furnace annealing and subsequent laser annealing.

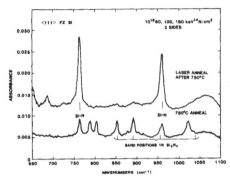

Figure 3. Si-N absorption after 750°C furnace annealing and after subsequent laser annealing.

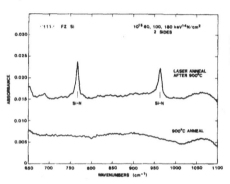

Figure 4. Si-N absorption after 900°C furnace annealing and after subsequent laser annealing.

TEM micrographs of N-implanted Si were taken in dark field using {220} weak-beam condition with foils oriented near the [111] zone. No defect structures are observed in TEM after annealing at 600°C. Micrographs in Fig. 5 show fine rod-like defects elongated in the <110> and <112> directions, and small white dots after 750°C, whereas 900°C anneals created large rod defects and dislocation loops with fewer fine structure defects. TEM thus supports the suggestion from ir data that small (fine) precipitates form at 750°C but are removed at 900°C. Small precipitate defects must dissolve rather than form Si_3N_4 precipitates at 900°C. Formation of N_2 within the layer would explain the loss of ir bands upon annealing at 900°C and the availability of N for reactivation of Si-N modes upon subsequent laser annealing. Other workers [13] have shown that laser annealing of layers implanted with fluences $> 10^{17}$ cm^{-2} causes blistering of the layers, presumably due to molecular N_2 formation.

TEM for samples subjected to furnace and furnace plus laser annealing are compared in Fig. 6. Laser annealing removes most of the fine defect structures as well as the loops and large rods. A few stacking faults were observed after laser annealing. The removal of the fine defect structure parallels the regeneration of N-N pairs in the ir spectra. Thus there appears to be an inverse relationship between fine defect structure, loops and rods observed in TEM and N-N pairs observed in ir for N-implanted Si.

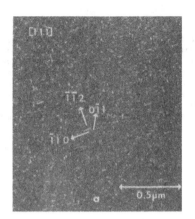
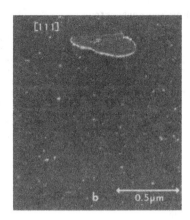

Figure 5. Defect structure in nitrogen implanted silicon after annealing for 1 hour at (a) 750°C and (b) 900°C.

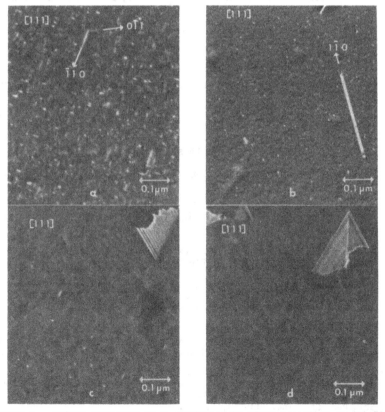

Figure 6. Defect structure in N- implanted Si after annealing at (a) 750°C and (b) 900°C, followed by laser annealing for (c) 750°C, and (d) 900°C.

SUMMARY AND CONCLUSIONS

Efficient pairing of N in Si is found when quenching N-implanted layers from the melt. Intensities for N-N pair bands decrease upon furnace annealing and extended defects are formed. Reactivation of N-N pairs and removal of extended defects upon subsequent laser annealing is consistent with dissolution, or at least partial dissolution, of N precipitates and reformation of N-N pairs in the melt. Pairing of N is in contrast to dispersed interstitial O centers which were observed after laser annealing of O-implanted Si layers. While the energetics which favor N-N pairing are not understood, pairing and low solubility are probably related consequences of effective bond energies for N in Si. Models for the properties of N in Si must explain pairing since it occurs over a wide range of doping conditions.

REFERENCES

1. T. Abe. K. Kikuchi, S. Shirai and S. Murako, "Semiconductor Silicon 1981," edited by H. R. Huff and R. J. Kriegler, The Electrochemic. Soc. Inc., Pennington, NJ (1981), p. 54.

2. G. Zimmer and H. Vogt, IEEE Trans. on Electr. Devices, ED-30, 1515 (1983).

3. K. L. Brower, Phys. Rev. B26, 6040 (1982).

4. M. Tajima, T. Masui, T. Abe and T. Nozaki, Jap. J. Appl. Phys. 20, 6423 (1983).

5. Y. Tokumara, H. Okushi, T. Masui and T. Abe, Jap. J. Appl. Phys. 21, L443 (1982).

6. H. Ch. Alt and L. Tapfer, Appl. Phys. Letts. 45, 426 (1984).

7. M. Sprenger, E. G. Sieverts, S. H. Muller and C.A.J. Ammerlaan, Solid State Comm. 51, 951 (1984).

8. H. J. Stein, Appl. Phys. Letts. 43, 296 (1983); H. J. Stein, Proc. of 13th Intl. Conf. on Defects in Semiconductors, Coronado, CA, 8/12-17/84 (to be published).

9. H. J. Stein, Electrochem. Soc. Extended Abstract #117, Vol. 84-1, accepted for publication in J. Electrochem. Soc.

10. H. J. Stein and P. S. Peercy, Materials Research Society Symposium Proceedings, Vol. 13, edited by J. Narayan, W. L. Brown and R. A. Lemons, Elsevier Sci. Pub. Co., Inc. (1983), p. 229.

11. C. I. Drowley and T. I. Kamins, Materials Research Society Symposium Proceedings, Vol. 13, edited by J. Narayan, W. L. Brown, R. A. Lemons, Elsevier Sci. Pub. Co., Inc., (1983), p. 511.

12. T. P. Smith, III, P. J. Stiles, W. M. Augustynaik, W. L. Brown, D. C. Jacobson and R. A. Kant, Materials Research Society Symposium Proceedings, Vol. 23, edited by J.C.C. Fan and N. M. Johnson, Elsevier Sci. Pub. Co. Inc., (1984), p. 453.

CHEMICAL REACTION PRODUCTS FROM INTERACTION OF HIGH-ENERGY BEAMS WITH COMPOSITES

J. S. PERKINS, S. S. LIN, P. W. WONG, D. J. JAKLITSCH, and A. F. CONNOLLY
Army Materials and Mechanics Research Center, Watertown, MA 02172

INTRODUCTION

In a study to examine the effect of irradiating a number of silica- and carbon-reinforced composites with high-energy beams, surface analytical techniques have proved especially effective in revealing reactions between undoped or doped matrices and silica or carbon reinforcements, respectively.

Because of selective photoexcitation of and multiphoton absorption by molecular bonds, high-intensity lasers can produce extremely high electronic excitation of solids on or near the surface with consequent possibilities for endothermic reactions between reinforcements and certain matrix constituents. Both silica and carbon, being very stable materials, are ideal reinforcements for shielding vehicles exposed to hostile environments. When highly energized, these reinforcements and their matrices may interact to yield products of high internal energy and thus dissipate very significant amounts of the absorbed energy.

Endothermic reactions are revealed by sharply defined borders between the thin layers in chars produced by high-energy beams; the nature of the reactions is determined by identifying the reactants and products in the adjacent layers. When crystaline products are produced in high-melting liquid phases during irradiation, identification is simplified; when vapor species are formed, special techniques reveal such gaseous products, e.g., bubbles trapped in resolidified melts or solids condensed in cooler areas.

A CO_2-laser (10.6-μm wavelength) has a photon energy of less than 3 kcal/mole. This low infrared energy can only produce rotation and vibration of molecular bonds unless the photons are absorbed so rapidly that the acquired energy cannot be dissipated by vibrations alone. This situation leads to electronic excitation which can even extend to bond breaking and realignment (chemical reaction) or electron and ion ejection (plasma formation). With the high-intensity beams provided by focusing lasers or increasing the current of electron beams, both chemical reactions and plasma formation have been observed.

The objective of this paper is to prove the existence of highly endothermic chemical reactions as significant pathways to dissipate the energies deposited by either a high-intensity laser or a high-energy electron beam on two composites.

RESULTS AND DISCUSSION

Lasers on silica-polyimide composites

The chars in the silica-reinforced polyimide samples shown in cross-section (Figure 1) were produced by focusing a 7.6 kW continuous-wave (CW) CO_2-laser beam on 2.90 and 0.33 square-centimeter surface areas, respectively. In the first case the cavity produced was found by X-ray diffraction to be lined with SiC. In the second case, no SiC was found, only traces of a high energy form of silica (cristobalite), but there was evidence of gaseous SiO production along with CO. The two reactions:

$$SiO_2(l) + 3\ C(s) \longrightarrow SiC(s) + 2\ CO(g) \qquad \Delta H_{1800°K} = +140 \text{ kcal} \qquad Eq(1)$$

$$SiO_2(l) + C(s) \longrightarrow SiO(g) + CO(g) \qquad \Delta H_{2000°K} = +156 \text{ kcal} \qquad Eq(2)$$

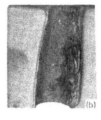

Figure 1. Cross-sectional views of cavities produced in a silica-polyimide composite by a 7.6 kW CO_2-laser focused in (a) to 2.90 cm^2 and in (b) to 0.33 cm^2.

indicate the greater energy sink provided in the second case[1]. The carbon for these reactions came from the charring of the polyimide matrices.

Electron beams and lasers on tungsten-doped carbon/carbon composites (WC/C)

The material used in our second example was a laminate of carbon cloth impregnated with a tungsten-containing resin. The consolidated material was processed at temperatures exceeding 2500°C. Under these conditions, X-ray diffraction indicated that the "fired" resin contained W_2C and WC, the laminate, W_2C, β-WC_{1-x}, and WC (Figure 2). The phase diagram of the tungsten-carbon system[2] shows that WC melts peritectically to release carbon and becomes enriched in tungsten (Figure 3). Hence the fired resin is rich in W_2C. In the laminate, β-WC_{1-x} predominates and WC increases because of a carbon cloth and liquid W_2C reaction.

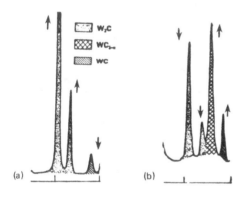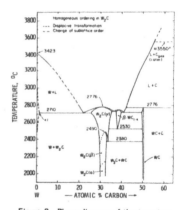

Figure 2. X-ray diffraction patterns of (a) a tungsten containing resin fired to 2800°C; (b) a composite of carbon cloth and the resin similarly treated (WC/C).

Figure 3. Phase diagram of the tungsten-carbon system after E. Rudy.

Thermodynamics[3] for the system indicate that at 3050°K:

$$2\ WC(1) \longrightarrow W_2C(1) + C(g) \qquad\qquad \Delta H = -5\ kcal \qquad\qquad Eq(3)$$

$$W_2C(1) + C(s) \longrightarrow 2\ WC(1) \qquad\qquad \Delta H = +175\ kcal \qquad\qquad Eq(4)$$

$$C(s) \longrightarrow C(g) \qquad\qquad \Delta H = +170\ kcal \qquad\qquad Eq(5)$$

The first reaction is slightly exothermic, hence favored at 3050°K; the second is highly endothermic. The net reaction requires 170 kcal which is equivalent to 60 kJ/g of carbon vapor produced if monoatomic carbon (C_1) is the principal product.

We present evidence for C_1-formation and C_1-vaporization from the tungsten-doped carbon/carbon composite as follows:

 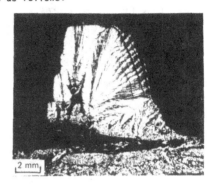

Figure 4. Pyrolytic carbon deposit produced downstream from WC/C when irradiated with a 30 kW cm^{-2} CW CO$_2$-laser in a 0.1 Mach wind stream.

    a. **Pyrolytic graphite formation.** Evidence that a vapor form of carbon is produced is shown by the laminar growth of pyrolytic graphite downwind from a laser-produced cavity (Figure 4).

    b. **Enthalpy-intensity plot.** That C_1 is a major product is indicated by the work of Brown, Gellert, and Hilton[4] who found that the ablation rate of the composite leveled off at about 70 kJ/g of carbon at high intensities of a CO$_2$-laser (Figure 5).

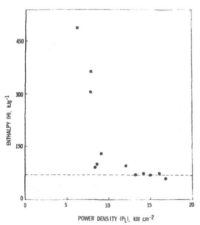

Figure 5. Ablation enthalpy versus laser power-density for interaction of a CW CO$_2$-laser with WC/C.

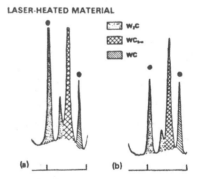

Figure 6. X-ray diffraction patterns from burn surfaces produced by different intensities of the CO$_2$-laser interacting with the material in Figure 2.(a) 15 kW cm^{-2}; (b) 30 kW cm^{-2}. The higher energy flux intensifies the reaction of W$_2$C with carbon cloth.

c. WC to W$_2$C ratio. Increased laser intensity increased the ratio of WC to W$_2$C in the quenched melt (Figure 6) as would be anticipated if atomic carbon is released from the cloth layers, traverses the molten tungsten-carbide phase, and emerges from the melt surface. The higher the beam intensity, the faster the release of atomic carbon, the larger the C to W ratio on quench.

d. Back-face temperatures. That a significant endothermic reaction occurs is evident in Figure 7 which shows the back-face temperature of a sample during irradiation. A powerful energy-absorbing reaction alters the slope of the temperature profile.

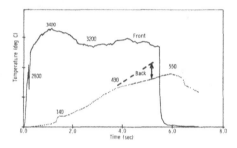

Figure 7. Front and back face temperatures of 1/8"-thick WC/C during a typical CW CO$_2$-laser run. The sharp change in rate of temperature rise at 430° (back face) indicates a substantial heat-sink reaction on the front face.

Value of surface analytical techniques

Scanning electron micrographs (SEM) (Figures 8a, 8b, 9a, and 10a) clearly reveal (1) the thin surface areas where reaction occurred and (2) the liquid phase involved. Use of energy dispersive analysis of X-rays (EDAX) in the same areas (Figures 8c, 9b) shows that the element W is concentrated in the burn areas. Scanning Auger micrographs (SAM) can map many light elements and extend our knowledge of the distribution of C, N, O, etc. Figure 10b shows that C-distribution is widespread and coextensive with the W-coverage found in the platelets in the burn areas indicating the presence of tungsten carbides (cf. Figure 8c).

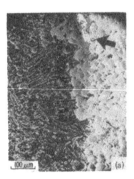  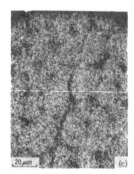

Figure 8. (a) and (b) SEM's of the edge and center of an electron-beam burn on WC/C and (c) W-map of the same area as (b) by EDAX. An impurity is marked in (a).

Figure 9. (a) Enlargement of the impurity in Figure 8a with (b) W-map and (c) Ca-map proving it is a Ca salt.

The burns produced by either an electron beam or the laser resemble fish scales which on close examination result from contraction of the shallow surface layers of tungsten carbides upon cooling. The fragmented films are centered over the exposed ends of the carbon yarns in the cloth (Figures 8b and 10a).

The two surface analysis techniques, EDAX and SAM, also show sequestering of impurities near the edges of the electron beam burn and a laser burn. Calcium and potassium were initially detected in a general EDAX scan and in an Auger spot analysis. Mapping by activating these elements separately showed the location of these impurities, a Ca salt (Figure 9) and a K salt (Figure 10).

One final SEM is particularly revealing of the field of reaction. Figure 11a shows that the solidified fluid phase sealed a portion of the upper of two thin 3-ply composites to the second 3-ply layer. The enlargement in Figure 11b of the center portion of Figure 11a clearly shows erosion by the liquid phase of an open side of one carbon fiber and a cauliflower-like growth by vapor deposition of carbon on the end of a cooler fiber. The EDAX view (Figure 11c) shows at (1) heavy W-concentration from escape of carbon vapor, at (2) a dilution of W as the melt reacts with the fiber, and at (3) no W where carbon vapor has deposited on the end of a cooler fiber. These corroborate the postulated reactions in Equations (3) to (5).

 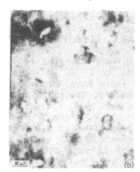 

Figure 10. Back-scattered electron micrograph near the edge of a CO_2-laser burn on WC/C; (b) a C-map and (c) a K-map of the same area. Note the solid carbon filament growing from the K-rich pool and diffuse carbon over all fragments of the tungsten carbide film free of K.

326

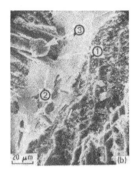

Figure 11. SEM's and W-map of an area showing attack of molten tungsten-carbide alloy on one fiber, deposit of solid carbon on another, and the region where the intermediary C-vapor has been released.

SUMMARY

Endothermic reactions (Equations 1, 2, and 5) are produced by high-intensity and hence high-energy fluences and provide chemical energy sinks which minimize damage from high-energy sources. However, the reactions are only initiated when energy fluences are sufficiently great to exceed the energy barriers separating low-energy reactants from high-energy products.

The principal stages in high-energy-beam - solid interactions are:

- selective photon activation,
- multiphoton absorption,
- initiation of endothermic chemical reactions, and
- plasma formation (not discussed in this paper).

Shallow, near-surface char layers are observed which are formed by successive endothermal chemical reactions. Fluorescent and phosphorescent radiation and slower thermal distribution mechanisms provide less exoergic pathways than these chemical reactions to dissipate the remainder of the initially-absorbed high-energy radiation.

ACKNOWLEDGMENTS

The authors wish to acknowledge the superior work of the following colleagues: T. P. Sheridan for X-ray diffractograms, R. M. Middleton for optical photomicrographs, and L. Elandjian for infinite care in reducing data and keeping it accessible.

The CW CO_2-lasers used were operated by K. Choy, Ford-Aeronutronics Division, Newport Beach, CA; by C. J. Oblinger and J. O. Bagford, Acurex Corp., Wright-Patterson Air Force Base, OH; and by J. H. McDermott and R. E. Beigel, Naval Research Laboratory, Washington, DC. Initial arrangements for use of these facilities were made by R. Fitzpatrick of our laboratory.

We are grateful to Dr. J. R. Brown, Materials Research Laboratory, Melbourne, Australia, for permission to use Figure 5; to Dr. M. B. Frish, Physical Sciences, Inc., Andover, MA, for use of the electron-beam-irradiated WC/C sample he produced; and to many other unnamed investigators in this field for comparisons of data and constructive discussions.

REFERENCES

[1] Thermodynamic data taken from JANAF Thermochemical Tables, 2nd Ed., D. R. Stull and H. Prophet (Project Directors), NSRDS-NBS 37, 1971.
[2] E. Rudy, Ternary Phase Equilibria in Transition Metal-Boron-Carbon-Silicon Systems, Part V, AFML-TR-65-2, Air Force Materials Laboratory, 1969, p. 192.
[3] H. L. Schick, _Thermodynamics of Certain Refractory Compounds_, (Academic Press, New York, 1966), Vol. 2, pp. 83, 85.
[4] J. R. Brown (private communication).

RAPID THERMAL ANNEALING IN SI

T.E. SEIDEL[+], C.S. PAI[*], D.J. LISCHNER, D.M. MAHER,
R.V. KNOELL, J.S. WILLIAMS, B.R. PENUMALLI, and
D.C. JACOBSON
AT&T Bell Laboratories, 600 Mountain Avenue, Murray Hill,NJ 07974

ABSTRACT

Certain aspects of Rapid Thermal Annealing (RTA) are reviewed.
Temperature considerations are discussed. The implant disorder removal
rate is measured (5eV removal energy for As induced damage). Shallower
defect-free junctions are obtained using RTA. Results of a "Round Robin"-
RTA annealing are presented, transient enhanced diffusion is not prominent
for As. New results for the concentration enhanced diffusion of As are
presented. Diffusion from the channeling-tail region of shallow boron
diffusions is noted as a limiting factor for producing shallow
p^+-junctions. Other issues are briefly discussed.

INTRODUCTION

Traditional annealing processes for semiconductor fabrication use
heating times of about one hour. Recently, various radiative heating
sources (e.g. tungsten-halogen, graphite heaters and arc-lamps) have been
developed to give controlled annealing times of 1-100 seconds. This
"Rapid Thermal Annealing" (RTA) allows effects to be studied using high-
temperature, short-time combinations.[1,2] Diffusion effects are limited
because of the shorter times, while thermally activated processes such as
dislocation removal or reaction rate limited processes such as silicide
reactions or oxynitride formation take place efficiently at higher
temperatures.

This paper discusses: (I) temperature determination and control;
optical coupling by free carrier absorption has an effect on the heating
rate.[3-5] (II) implant disorder removal; here we report the removal of
the "end of range" dislocation-array which exists just beyond the As-
implanted, non-annealed amorphous/crystalline interface ("α/c" disorder).
[6] Arsenic concentration profiles are compared under the constraint that
the "α/c" disorder is removed in minimal time at various temperatures.[7]
These defect-free profiles are shallower when higher temperature-shorter
time anneals are used to remove the "α/c" dislocation disorder array.
This result indicates an RTA advantage for producing defect-free shallow
arsenic junctions. (IIIa) the issue of enhanced diffusion for As; results
of a "Round Robin" experiment, where anneals at 1100°C for 1-50 sec, done
by five laboratories, are reported. Transient-enhanced diffusion is not
observed for As.[8] (IIIb) further studies of arsenic diffusion are
discussed: the importance of oxide capping to establish a no-arsenic-loss
boundary condition, and a comparision of Ghez [9] and two-Gaussian Seidel-
Mac Rae [10] solutions is made. (IIIc) arsenic diffusion data for 850 -
1200°C, and 10^{20} -10^{21}/cm^3 concentrations C are reported. The diffusion
D values show very strong concentration enhancement for average
$C > 3 \times 10^{20}$/cm^3 and $T > 1100$°C. (IIId) the strong concentration enhancement is accompanied by metastable doping. Reversible resistivity changes

---

[+]At J. C. Schumacher Co. Oceanside, CA.   [*]At UCSD, LaJolla, CA.

are shown by repeated 1100°C - 800°C anneal cycles. (IV) RTA of high dose
boron implants; these indicate a more effective dislocation defect removal
when compared with traditional annealing, under the normalized constraint
of equal junction depths.(11) These dislocations are distributed over the
region of maximum displacement disorder. When very shallow boron profiles
($x_j \sim 0.1$ μm) are considered, diffusion is dominated by enhanced
diffusion of the channel-tail region.[12] Preamorphization gives limited
tail diffusion and allows electrical activation under low temperature
solid phase epitaxial regrowth.[13-17] However, another class of defects:
"α/c", hairpins or spanning dislocations and other defect arrays have
potentially adverse effects on junction leakage.[18,19] (V) defect
studies; especially fabrication and control of localized defect arrays,
impurity interactions with these defect arrays by gettering, impurity
diffusion during silicide formation [20] and from silicide contacts are
briefly discussed. The RTA effects concerning slip, [21] glass flow [22]
and oxynitridization [23] appear interesting. Compatibility with an
entire technology sequence is left as an open question.

(I) TEMPERATURE CONSIDERATIONS

A schematic of a modern RTA system is shown in Fig. 1. The wafer is
supported by low-mass quartz pins and enclosed by a quartz chamber which
allows ambient control via the input and output pipes. A tungsten-halogen
heating source supplies a blackbody-like radiation spectrum to the wafer,
the lamps are driven with a power P(t). An optical pyrometer can be used
to measure the temperature response of the wafer. Quartz cuts off beyond
$\lambda \sim 4.0$ μm, and the wafer will absorb radiation for $\lambda < 1.1$ μm. If the
pyrometer is sensitive between 1.1 and 4.0 μm, then the filament emission
is detected immediately, while the wafer becomes hot only after a several
second delay dependent on the heat capacity of the wafer and its
absorption properties. An optical window, e.g. CaF_2, can be used if the
pyrometer is sensitive beyond 4 μm.[24]

Two schemes for power drive can be used; "open loop", where the drive
and temperature response are independent, or "closed loop" where feedback
is used to over-drive the power at short times and the result is a more
constant temperature, T(t) response.[25] Equivalent physical effects may
be obtained from either approach. (However some effects due to the nature
of the activated process may be anticipated. Athermal processes, if
present, may be excited by a closed-loop configuration. Furthermore, a
low activation energy process (~1.0ev) will be undergoing significant
changes on the heat-up, while a higher activation process (~5.0ev) will be
undergoing changes only near the peak of the temperature cycle. These
possible caveats remain to be demonstrated.)

If an open-loop P(t) arrangement is used and the wafer is heavily
doped (n^+, p^+, nn^+ or pp^+), then extrinsic free carrier absorption results
in a faster heating rate than a lightly doped sample.[4] See Fig. 2a. In
Fig. 2b, the overlap of absorption and blackbody spectrum are shown. The
rate of temperatures rise, is given by energy considerations:
$c(dT/dt) = P_a - P_e$, where c is the heat capacitance and $P_{a,e}$ is
the power absorbed, and emitted respectively. The power absorbed is
approximated by $\int d\lambda\, I_\lambda (1 - e^{-\alpha_\lambda d})$ (for a reflectance which is constant
with λ.) I_λ is the blackbody intensity and α_λ is the absorption, d is the
thickness of the wafer. For heavily doped Si ($n,p \gtrsim 10^{19}$) (cm^3), one can
see that the initial P_a and heating rate can be about twice as large as
the lightly doped Si.[4] The problem is non-linear in temperature since
heating to intermediate temperatures (300-660°C) results in intrinsic free
carriers which also produce significant free carrier absorption. Detailed
numerical solutions and experiments remain to be done.

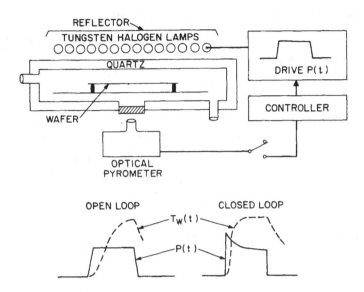

Fig. 1: Schematic of Tungsten-Halogen RTA system. Temperature T(t) are measured with an optical pyrometer. The lamps can be driven with power pulse P(t) in an open or close-loop configuration. (See Text)

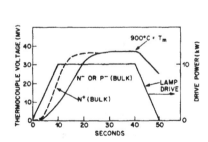

Fig. 2a: Thermocouple output voltage plotted against time for an open-loop lamp drive in AG-210M tungsten-halogen system for N⁻ and N⁺ silicon.

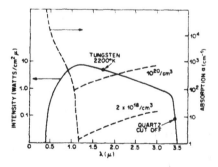

Fig. 2b: Normalized black body intensity spectrum for Tungsten lamps at 2200°K. Absorption coefficient for silicon showing fundamental edge and free carrier absorption.

(II) DEFECT REMOVAL WITH LIMITED DIFFUSION (ARSENIC)

RTA can be used to remove ion damage while obtaining minimal dopant diffusion. Arsenic implanted samples were annealed between 850°C - 1200°C, for a variety of times at each temperature (e.g., at 1200°C, 0.7, 1.3, 1.8, 2.2 sec ...). The shortest times at which no "α/c" dislocation arrays were observed are plotted as open circles in Fig. 3. The 5.0eV energy is the same as for Si self-diffusion, showing the movement of Si atoms are involved in this defect removal process.[26] The profiles of As corresponding to the shortest time at which no "α/c" dislocations exist are shown in Fig. 4, showing that more shallow defect-free junctions are obtained using RTA. The profiles in Fig. 4 show surface peaks (As_xO_y) and are discussed below in connection with retention of arsenic and the useful boundary condition for diffusion analysis. The defect-free profiles are more shallow using higher temperatures because defects are removed with ~5.0eV, while arsenic diffusion proceeds with <4.0eV.[27,28]. If the defect and diffusion activation energies were the same, there would be no high temperature, short time advantage.

The defect-free condition may be taken as a necessary condition for the fabrication of low leakage junctions. These results are consistent with an earlier scenario discussed by Sedgwick [1] and recent leakage measurements by Kamgar et al.[29]

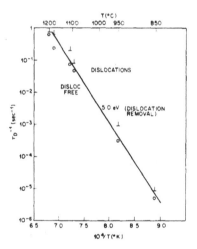

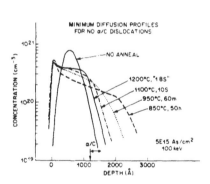

Fig. 3: Dislocation removal rate (τ_0^{-1}) plotted against reciprocal temperature (open circles). The bar symbols are for the longest times where dislocations are still observed.

Fig. 4: Arsenic profiles for the condition that the "α/c" disorder has just been removed. Higher temperature-shorter times give shallower defect-free profiles.

(III) ARSENIC DIFFUSION

a. Enhanced Diffusion (Search for Transient Effects)

Arsenic diffusion under RTA has been controversial.[7] Round Robin experiments were carried out to try to settle the issue of transient effects, and to sample the RTA community's collective ability to measure temperature. A group of 4-inch silicon wafers were batch-implanted at Bell Labs (B) and four wafers each were distributed to Tamarack (T), IBM (I), AG Assoc. (G), and Eaton (E) for annealing at 1100°C. Annealed samples were measured at Bell Labs for profiles, the results are shown in Fig. 5a, with the Tamarack profiles shown in Fig. 5b. The agreement of the data between different annealing locations is good, especially for I, B and T. Note that no individual set of data (e.g. all E points) shows strong evidence of transient enhanced diffusion. A strong transient diffusion would show about the same diffusion in 1-2 sec. as 10 sec.[8] Fig. 5b shows a surface peak of As, just as Fig. 4.

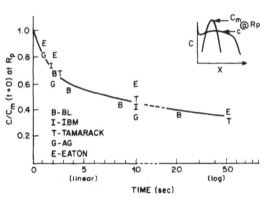

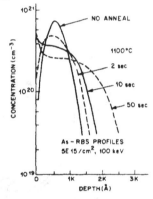

Fig. 5a: Plot of normalized concentration for the Round Robin studies.

Fig. 5b: Concentration profiles for the Tamarack (T) data of Fig. 5a.

b. Boundary Condition and Analysis of Diffusion

The pile-up of a small amount of As at the surface is due to an oxide or oxidized surface. The surface As is non-substitutional (channeling RBS measurements), and can be removed by an HF etch of the oxide implying As bonding to the oxide: As_xO_y. The oxide acts as a barrier to out diffusion, as shown in Fig. 6. When the oxide is present essentially all the dose remains in the silicon, with about 10-15% precipitated at the surface.

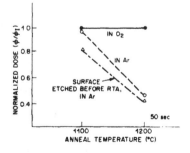

Fig. 6: Plot of normalized retained dose after 50 sec. RTA's at high temperatures. (100keV, 5E15/cm^2)

334

(c) Ghez and Two Gaussian Approximation

Ghez, et al [9] recently developed a general solution for a diffusion profile, characterized by a constant diffusion coefficient. They compared the general solution with "no-out diffusion" as a boundary condition with the approximate one-Gaussian solution by Seidel and Mac Rae (SM) [10] for a surface condition which allows out diffusion. Here we compare Ghez and the published SM two-Gaussian approximation for the same boundary condition, no-out diffusion. See Fig. 7. The results are experimentally indistinguishable, the Ghez theory confirms that the SM two-Gaussian approximate solution is nearly exact. We use this two-Gaussian approximation to extract $2Dt/\Delta Rp$ values (where $\bar{D}$ is an average diffusion constant, t the time of diffusion, and ΔR_p is the initial straggle), from the Round Robin data. The results are plotted in Fig. 8. The slope gives an average D value for concentrations between $4\text{-}8 \times 10^{20}/cm^3$. Extrapolation to zero time implies no measureable transient enhanced diffusion.

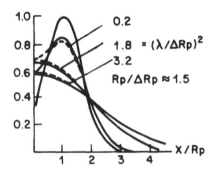

Fig. 7: Normalized concentration profiles for exact Ghez, et al (solid) solution and Seidel-Mac Rae for the same boundary condition.

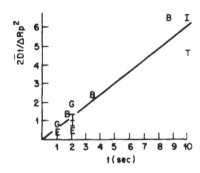

Fig. 8: Plot of average diffusion parameter $\bar{D}t/\Delta R_p^2$ against time for the Round Robin samples. No evidence for transient behavior exists.

Similiar analysis give $\bar{D}$ values for other doses $10^{15}\text{-}10^{16}/cm^2$ and energies 50-200keV. They are plotted in Fig. 9. The solid curves are from the literature, the average result from the Round Robin is plotted, and the open points are new RTA data. The interesting feature is that $\bar{D}$ rises more rapidly than linearly with C above 1100°C, while a saturation and reduction in $\bar{D}$ with $\bar{C}$ is observed at and below 1000°C. The saturation is associated with "clustering" and precipitation; [30-33] and the higher temperature results are associated with "no clustering," and possibly a doubly charged vacancy mechanism. We demonstrate the "clustering" by sequential, repeated high (1100°C)-low (800°C) anneals where declustering and clustering are induced.[5]

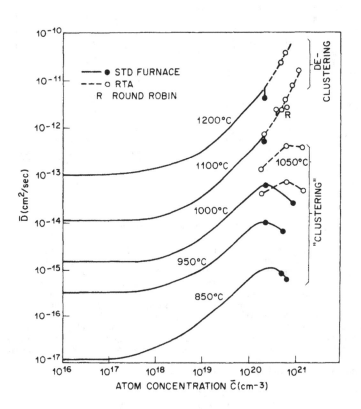

Fig 9: Diffusion constant $\bar{D}$ vs concentration $\bar{C}$ for As implants. For $\lesssim 1000°C$, the concentration enhancement is limited by saturation ("clustering") for $\gtrsim 1100°C$, the concentration enhancement is very strong.

Fig. 10: Sheet resistance of As →
implanted wafers, following RTA and
standard anneals.

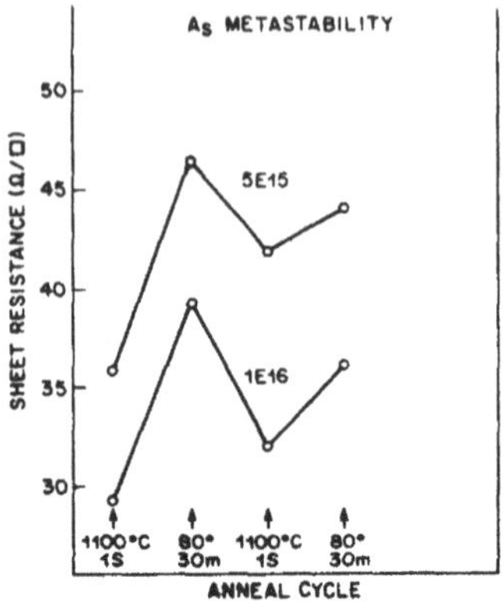

(d) Metastability

The electrically active arsenic
after RTA ($1100°C$, ~1 sec.) can be
"cluster-precipitated" by a standard
($800°C$, 30 min.) anneal cycle. The
effect is partially reversible, see
Fig. 10. Thus, high diffusivity is
accompanied by metastable doping.

(e) Discussion of Arsenic Diffusion

The test of the validity of the data in Fig. 9 requires a self-
consistent numerical analysis which uses the $D(C)$ data, including a doubly
charged vacancy $D^=$ term which successfully calculates $C(x,t)$ under RTA
conditions. Preliminary work appears to do this without transient
effects. We do not have, at present, a detailed understanding of the
electrical properties, i.e. how free carrier density depends on C [31] and
RTA thermal history which may include metastable conditions.

(IV) DEFECT REMOVAL AND DIFFUSION FROM BORON IMPLANTS

Defect removal of ion damage from high dose B implants is more
effective using RTA than standard furnace annealing under the normalizing
condition of the same diffused junction depth.[11] See Fig. 11. In this
case the residual dislocations are distributed near the region of the
maximum displacement damage. For very shallow boron profiles, the
behavior of boron diffusion under RTA and standard anneals is dominated by
diffusion in the channel-tail region of the profile.[12-17] Preamorphization leads to reduced diffusion under RTA for the tail diffusion and
allows electrical activation using solid phase epitaxy. See Fig. 12.
However, defects near or beyond the junction region can exist to various
degrees depending on the thermal history.[34]

One of the more intriguing effects of
residual disorder is the interaction
of impurities with dislocation defect
arrays. Fig. 13 shows F (Fluorine)
decoration of the α/c disorder in a
preamorphized Si layer which was
annealed at various temperatures.
[35,36] The double peak is commonly
observed, a minimum is proposed to be
due to a minimum in strain due to
dislocation-dislocation interactions.
Hairpin-spanners can also getter F.

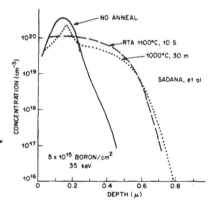

Fig. 11: B profiles for 35keV, →
5×10^{15} B/cm^2, normalized to the
same X_j (0.8 μm). Damage-
precipitated B is reduced using RTA.

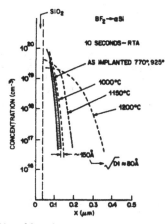 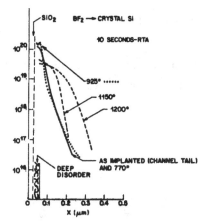

Fig. 12: Boron profiles for BF$_2$ implants into preamorphized (left) and crystalline (right) silicon. Samples (α and c) were annealed together for the same temperature labels.

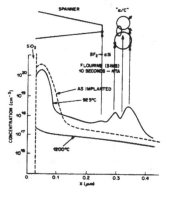

Fig. 13: Fluorine concentration from SIMS analysis, for BF$_2$ implants into preamorphized Si. The gettered F correlates to the disorder.

(V) OTHER ISSUES

Shallow junctions, defect array formation and control, and defect-impurity interaction [37] continue to be studied. Shallow junctions may be obtained by implanting into silicides, [20] where ion damage in silicon and ion channeling effects in silicon are avoided, followed by a limited thermal cycle. Glass flow, [38] slip, [21] and reaction rate limited effects such as oxynitridization [23] appear interesting. Glass flow may be implemented with B-P-SiO$_2$ at lower flow temperatures, so RTA may not make a commercial impact unless diffusion must be further limited, or dopant activation is better by quenching from high temperatures leading to reduced contact resistance. Slip needs to be defined, slip can be nucleated by poor edge rounding, and can be limited by "proper" oxygen control. Processes such as oxynitridization, which are reaction rate limited, should limit the diffusion of nitrogen to the Si/SiO$_2$ interface. Finally, compatibility with an entire technology sequence is an important consideration for any RTA use. For example, the high temperature RTA results in higher arsenic doping density but the doping is metastable. Further work will show if arsenic doping obtained by 1100°C-RTA survives lower temperature process sequences.

338

ACKNOWLEDGEMENTS

We thank our colleagues: T. O. Sedgwick, A. E. Michel, R. E. Sheets, S. Shatas and J. Gelpey for cooperating on the Round Robin experiment. We also thank J. M. Poate, W. L. Brown, S. S. Lau for their helpful comments and J. W. Mayer and G. A. Rozgonyi for their encouragement pursuant to this work.

References

1. T. O. Sedgwick, J. Electrochem. Soc. 130(2), p. 484 (1983).

2. C. Hill, in Laser and Electron-Beam Solid Interaction and Material Processing, p. 361, J. F. Gibbons, L. D. Hess, and T. W. Sigmon, Eds., (North Holland, NY, 1981).

3. R. T. Fulks, in Rapid Thermal Processing for Integrated-Circuit Applications, Notes-Continuing Education in Engineering, College of Engineering, UC, Berkeley - Oct 15-16, 1984.

4. T. E. Seidel, D. J. Lischner, C. S. Pai and S. S. Lau, Proc. of the 2nd Internal Sympos. on VLSI Sci. and Technol., Vol. 84-7, p. 184 (1984)., and J. Appl. Phys. Jan. 15, 1985.

5. S. R. Wilson, W. M. Paulson, R. B. Gregory, A. H. Hamdi, and F. D. Mc Daniel, J. Appl. Phys. 55, (12), p. 4162 (1984).

6. R. T. Hodgson, J. E. E. Baglin, A. E. Michel, S. Mader, and J. S. Gelpey, Mat. Res. Soc. Symp. Proc. 13, p. 355 (1983).

7. T. E. Seidel, D. J. Lischner, C. S. Pai, R. V. Knoell and D. M. Maher, to be published, Conf. Proc. Ion Beam Modif. of Mat. '84, Ed B. Manfried Ullrich and J. Nucl. Instr. and Meth. in Phys. Res. - Sec. B Beam Interactions with Materials and Atoms, Eds. H. H. Anderson and S. T. Picreaux, Mar. 1985.

8. R. Kalish, T. O. Sedgwick, S. Mader, and S. Shatas, Appl. Phys. Lett. 44(1), p. 107 (1984).

9. R. Ghez, G. S. Oehrlein, T. O. Sedgwick, F. F. Morehead and Y. H. Lee, Appl. Phys. Lett. Nov. (1984).

10. T. E. Seidel and A. U. Mac Rae, Trans. of Met. Soc. of AIME, 245, p. 491 (1969).

11. D. K. Sadana, S. C. Shatas and A. Gat, Inst. Phys. Conf. Sec. No. 67; Sec. N3, p. 143 (1983).

12. W. K. Hofker, Phillips Res. Repts. Suppl., No 8 (1975).

13. T. E. Seidel, in VLSI Technology, Ed. S. M. Sze, p. 234, 260, (1983).

14. B. L. Crowder, J. F. Ziegler, and G. W. Cole, Ion Implantation in Semiconductors and Other Materials, Yorktown Heights, NY, Plenum, p. 257, (1973).

15. M. Y. Tsai and B. G. Streetman, J. Appl. Phys. 50, p. 183 (1979).

16. T. M. Liu and W. G. Oldham, IEEE Electr. Dev. Lett. EDL-4, 3, p. 56 (1983).

17. T. E. Seidel, IEEE Electr. Dev. Lett. EDL-4, p. 353 (1983).

18. W. Marzara, C. Carter, D. K. Sadana, J. Liu, V. Ozguz, J. J. Wortman and G. A. Rozgonyi, Energy Beam-Solid Interaction and Transient Thermal Processing, Eds., J. C. C. Fan and N. M. Johnson, North Holland, NY (1984).

19. T. E. Seidel, D. M. Maher and R. V. Knoell, 13th Int. Conf. Def. in Semicond., J. Electr. Matl, Dec. 1984.

20. F-C Shone, K. C. Saraswat and J. D. Plummer, Stanford Electronics Lab. Rept ICL17-79, July 1984.

21. C. Russo, 5th Int. Conf. Ion Impl. and Tech. Smugglers Notch, July 1984, to be published in Nuc. Inst. and Meth.

22. T. Hara, H. Suzuki, and M. Iurukawa, Japan J. Appl. Phys. 23 (7), L452 (1984).

23. J. Nulman, J. P. Krusius and L. Rathbun, IEDM 1984, IEEE Electron Device Meeting.

24. R. E. Sheets, (private communication), Tamarack Scientific Co., Inc. Anaheim, CA.

25. S. Shatas (private communication), AG Associates, Palo Alto, CA.

26. J. M. Fairfield and B. J. Masters, J. Appl. Physics, 38, p. 3148 (1967).

27. R. N. Ghostagore, Phys. Rev. B3, p. 397 (1971).

28. R. B. Fair and J. C. C. Tsai, J. Electrochem. Soc. 122 p. 1689 (1975).

29. Avid Kamgar, W. Fichtner, T. T. Sheng, and D. C. Jacobson, Appl. Phys. Lett. 45 (7), 1, p. 754 (1984).

30. S. M. Hu, Atomic Diffusion in Semiconductors, Ed. D. Shaw (Plenum, London) p. 217 (1973).

31. R. B. Fair and G. R. Weber, J. Appl. Phys. 44, p. 583 (1973).

32. M. Y. Tsai, F. F. Morehead, J. E. E. Baglin, and A. E. Michel, J. Appl. Physics 51 (6), p. 3230 (1980).

33. N. R. Wu, D. K. Sadana and J. Washburn, Appl. Phys. Lett. 44, 8 p. 782 (1984).

34. T. Sands, J. Washburn, R. Gronsky, W. Maszara, D. K. Sadana and G. A. Rozgonyi, Appl. Phys. Lett. 45 (9), 1 p. 982 (1984).

35. W. Maszara, C. Carter, D. K. Sadana, J. Liu, V. Ozguz, J. Wortman, and G. A. Rozgonyi, Proc. Matl. Res. Soc.; Energy Bean-Solid Interactions and Transient Thermal Processing, Boston, MA, Nov 13-17, 1983 (in press).

36. T. E. Seidel, R. V. Knoell, F. A. Stevie, G. Poli, B. Schwartz, VLSI Science and Technology/1984 Eds. Beam and Rozgonyi, p. 201, The Electrochem. Soc., Vol. 84-7 (1984).

37. T. E. Seidel, D. M. Maher, and R. Knoell, J. Electronic Materials, Dec. 1984.

38. J. E. Tong, K. Schertenlieb and R. A. Carpio, Sol. State Technology, p. 161 (1984).

A SIMPLE MODEL FOR THE TRANSIENT, ENHANCED DIFFUSION
OF ION-IMPLANTED PHOSPHORUS IN SILICON

F. F. MOREHEAD and R.T. HODGSON
IBM Thomas J. Watson Research Center, Yorktown Heights, N.Y. 10598

ABSTRACT

Unlike As, B as well as P implanted into Si exhibits transient, enhanced diffusion. For example, when P implants are annealed for times of ~1 s at temperatures > 900°C, we observe a large movement of dopant toward the furnace of the Si wafer which is nearly independent of temperature 1050-1200°C. Once the temperature rises above 1200-1250°C the diffusion is similar to that normally observed. We model the experimental results as a transient, enhanced diffusion of a mobile component, about half the total phosphorus implant, distributed deeper in the bulk than the total P distribution. This mobile component may be linked to a large super-saturation of self-interstitials produced by the 50 keV implantation, which are expected to be left deeper in the bulk than the total dopant profile.

Accurate knowledge of any transient diffusion phenomena resulting from defects produced by ion implantation is essential if shallow junctions and high doping levels required of very large scale integrated devices (VLSI) are to be controlled. Laser annealing usually melts the wafer surface and yields complementary error function profiles for the dopants because of the very high diffusion constants for molten Si. Rapid thermal annealing with high power, incoherent radiation and effective heating times of a few seconds or less may, for some maximum temperatures, allow the transient effect to be the dominant contribution to the net motion of the dopant. Fortunately for our VLSI goals, arsenic does not appear to suffer from this effect [1]. Earlier reports [2,3] of an "enhanced" diffusion of arsenic appear to be due to an underestimation of the temperature to which the wafers were heated during RTA. Boron, however, has long been known to show transient, enhanced diffusion even with 30 minute furnace anneals of implanted wafers [4]. A recent report on the RTA of phosphorus implants in silicon clearly demonstrated the effect for this dopant [5]. We describe here additional corroborating experiments with phosphorus with a different high power lamp for heating [6] and a simple model which accounts for the striking transient changes in the phosphorus profile.

P-type, 10 ohm-cm, (100) silicon wafers were implanted at room temperature with 50 keV phosphorus ions at beam currents of 10 microamps cm^{-2} or less for doses of 10^{14} and 10^{15} ions cm^{-2}. A 100 kW argon arc lamp [6] was used for heating. Temperature-time measurements for the argon arc lamp are made with an optical pyrometer focussed on the back of the sample [6]. The latter system normally indicates maximum temperatures within 10-20°C of those actually attained. In contrast, the halogen lamp system employed in the earlier work [5] consistently gave a "nominal" maximum temperature over 50°C lower than that actually achieved, as indicated by the extent of "normal" dopant diffusion following any transient, enhanced motion [7]. This error in temperature measurement is almost certainly the origin of early [2,3] and current [8] reports of "enhanced" diffusion in arsenic.

The temperature-time profile of these experiments is similar to that published earlier [1,6] and corresponds roughly to 1 s at the indicated maximum temperature. These isochronal "fast" anneals were carried out for both doses for "peak" temperatures of 850-1250°C at 50°C intervals. Despite the- fact that the higher dose amorphized the silicon [9] while the lower one did not, the pattern for both sets of SIMS profiles (oxygen beam) is essentially the same. Extrapolation of results obtained at lower temperatures [10] indicates that epitaxial regrowth of the amorphized layer occurs in less than 30 ms, even at 850°C. Sheet resistivity measurements indicate 85% and 95% of the expected electrical activity at 850°C for the 10^{14} and 10^{15} cm^{-2} doses respectively, which is consistent with earlier annealing studies [9].

For both sets of profiles, there is almost no change from the initial, as-implanted profile at 850°C. Enhanced diffusion occurs between 900 and 1050°C, as shown in Figures 1 and 2 for both doses. No significant further motion occurs after 1050°C until 1250°C, when the expected "normal" diffusion becomes dominant. Although anneals for longer times at 900, 950, and 1000°C would be helpful here, this pattern definitely establishes the transient character of the observed motion. Reasonable agreement of the dopant motion at 1250°C with that expected from furnace annealing studies indicates that the temperatures are good estimates.

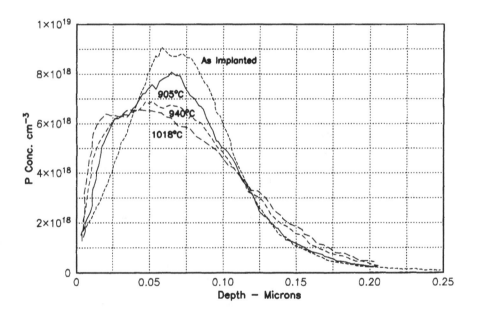

Figure 1. Linear plot of the concentration profiles of 10^{14} cm^{-2} keV implanted P before and after rapid thermal annealing (RTA) of ~1 s at the indicated temperatures.

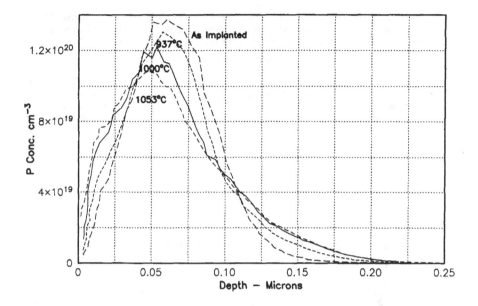

Figure 2. Linear plot of the concentration profiles of 10^{15} cm^{-2} 50 keV implanted P before and after 1 s RTA at the indicated temperatures.

The most striking feature of both Fig. 1 and Fig. 2 is the rather massive movement of material to the surface. The initial implanted profile is not simply uniformly broadened, as might be expected for a uniform for a uniform, transient supersaturation of vacancies and interstitials created by the implantation process. Interstitials are known from silicon oxidation experiments to enhance P diffusion [11]. A model, suggested for transient boron diffusion [8], in which normal diffusion is maintained in a "damaged" region near the impurity peak, and enhanced, defect assisted diffusion occurs outside this region, both toward the surface and the interior of the silicon crystal, does not explain the observations. Nor does any reasonable distribution of impurity-trapping defects at the crystal surface. A remarkably simple model which "explains" the results surprisingly well follows.

The basic assumption is that only a fraction of the P is made mobile by the defects produced by the implantation. For simplicity this fraction is assumed to be distributed as a reflected Gaussian whose initially narrow standard deviation (or straggle) is spread by the isochronal anneals at 900-1050°C. The difference between these two reflected Gaussians when added to the initial profile produces the measured annealed profile. One simply chooses (1) the fraction of the total that is assumed mobile, (2) the location of the peak of its distribution, and (3) the initial and final straggles. How well this can be done is illustrated in Figures 3 and 4 for the two doses at two different annealing temperatures. For both doses the location of the peak of the mobile fraction is 800 Å, (compared to 650 Å for the peak of the impurity distribution), and the initial straggle of the mobile component is 250 Å (most of the dopant is contained within a Gaussian with a 350 Å straggle). Roughly 30% of the 10^{14} cm^{-2} dose is mobile while 40% of the 10^{15} cm^{-2} dose moves. From the increase in

344

the straggle at the different 1 s isochronal annealing temperatures, an effective diffusion constant can be computed [12,13]. These values are given in Table I. The temperature dependence of the calculated diffusion constants for the mobile P is given by

$$D_{eff} = 0.007 \exp(-2.2 \text{ eV}/kT).$$

The 2.2 eV activation energy may well represent the migration enthalpy of a phosphorus-self-interstitial pair. The location of the distribution of the mobile component of the phosphorus to the right of the impurity distribution is the expected location of self-interstitials generated by the ion bombardment [14]. Vacancies, on the other hand, are expected to peak nearer the surface.

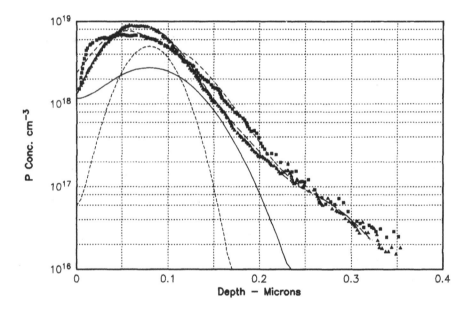

Figure 3. A mobile component model for the 1 s annealing of 10^{14} cm^{-2} 50 keV P at 940°C. The triangles (as implanted) and squares (after RTA) are from SIMS data. The narrow inner reflected Gaussian contains 3×10^{13} cm^{-2} P. Its straggle is spread by the RTA from 250 to 455 Å to give the solid curve. When the difference is added to the curve through the triangles, the outer dashed curve results, agreeing well with the squares.

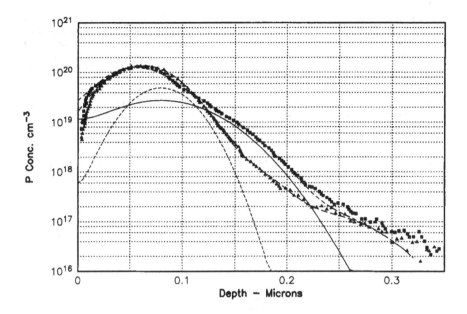

Figure 4. The mobile component model for 1 s annealing of 10^{15} cm^{-2} 50 keV P at 937°C. See Figure 3. Here the mobile component is 4×10^{14} cm^{-2} P and its straggle is spread by the RTA from 250 to 417 Å.

TABLE I

Effective Diffusion Constants for the Mobile Phosphorus Component

Temp.	Dose	Range	Mobile P Peak	Initial ΔR_i	Final ΔR_f	Total Mobile	D_{eff}
°C	cm^{-2}	Å	Å	Å	Å	cm^{-2}	cm^{2}s^{-1}
905	10^{14}	650	800	250	312	3×10^{13}	2.0×10^{-12}
940					455		7.0×10^{-12}
1018					625		1.6×10^{-11}
905	10^{15}				338	4×10^{14}	3.0×10^{-12}
937					417		6.0×10^{-12}
1000					550		1.2×10^{-11}

The present results are quite consistent with the earlier report on the RTA of phosphorus [5]. The published data support the observation of a transient effect for the 2×10^{15}cm^{-2} dose quite like that illustrated in Fig. 2 if a 50°C discrepancy between "actual" and reported temperatures is assumed. The reported transient for a P dose of 10^{14} cm^{-2} is also consistent with the present case. At any rate it is clear that the presence of an amorphous layer does not eliminate the transient motion, as [5] concludes, and does not even alter its nature significantly.

346

Our assumption of a "mobile component" is not a significant departure in representing transient enhanced diffusion. In a very real sense all diffusion in semiconductors like silicon is the result of motion by a mobile component, consisting of impurity-defect pairs. The motion is usually represented by calculating an appropriate diffusion coefficient and applying it to the total impurity population. This approach works well for normal diffusion, but not when the defects have local distributions which differ from that of the impurities. Consider, for example, Si uniformly doped with B or P, which is bombarded with Ar or Si ions. The implantation produces a Gaussian-like distribution of defects related to the range and straggle of the implanted Si or Ar ions. Application of the defect-enhanced time-dependent diffusion constant suggested in [8] to a high temperature annealing process produces *no* net impurity redistribution whatever since there is no concentration gradient in the impurity. The mobile component, formed by pairing defects and impurity ions, does have a concentration gradient and a high diffusivity, producing the observed redistribution [4, p.100 &ff.]. This redistribution is well characterized by our mobile component model, which gives a large dip in the formerly flat profile at the peak of the defect concentration with pile-up on either side. We have tried more elaborate kinetic models with pairing and unpairing constants, diffusivities for the pairs, etc., and can, of course, produce the same profile as the simple mobile component model. The simple model gives the same diffusion constant and activation energy (Equation 1) as the kinetic model, so the increased computational effort is not justified.

REFERENCES

1. R.T. Hodgson, V. Deline, S.M. Mader, F.F. Morehead, and J. Gelpey, Mat. Res. Soc. Symp. Proc., *23*, 254 (Elsevier, 1984).
2. R. Kalish, T.O. Sedgwick, S. Mader, S. Shatas, Appl. Phys. Lett. *44*, 107 (1984).
3. J. Narayan, O.W. Holland, R.E. Eby, J.J. Wortman, V. Ozguz, and G.A. Rozgonyi, Appl. Phys. Lett. *43*, 957 (1983).
4. W.K. Hofker, Philips Res. Repts. Suppl. 1975, No. 8.
5. G.S. Oehrlein, S.A. Cohen, and T.O. Sedgwick, Appl. Phys. Lett. *45*, 417 (1984).
6. R.T. Hodgson, V.R. Deline, S. Mader, and J.C. Gelpey, Appl. Phys. Lett. *44*, 589 (1984).
7. F. Morehead, unpublished.
8. R.B. Fair, J.J. Wortman, and J. Liu, J. Electrochem. Soc. *131*, 2387 (1984).
9. B.L. Crowder and F.F. Morehead, Appl. Phys. Lett. *14*, 313 (1969).
10. L. Csepregi, J.W. Mayer, and T.W. Sigmon, Appl. Phys. Lett. *29*, 92 (1976).
11. T.Y. Tan, U. Goesele, and F.F. Morehead, Appl. Phys. A, *31*, 97 (1983).
12. T.E. Seidel and A.U. MacRae, Tras. AIME *245*, 491 (1969).
13. R. Ghez, G.S. Oehrlein, T.O. Sedgwick, F.F. Morehead, and Y.H. Lee, Appl. Phys. Lett. *45*, 881 (1984).
14. M.D. Giles, private communication.

ACKNOWLEDGMENT

The authors would like to thank V. Deline of this laboratory for the SIMS measurements and J. Gelpey, Eaton Ion Implantation Systems, for the implantations and RTA's.

A COMPARISON OF MILLISECOND ANNEALING OF B IMPLANTS
AND ISOTHERMAL ANNEALING FOR TIMES OF A FEW SECONDS

R. A. McMAHON*, D. G. HASKO*, H. AHMED*, W. M. STOBBS** AND D. J. GODFREY+
*Microcircuit Engineering Laboratory, Cambridge University, Cambridge
Science Park, Cambridge, CB4 4BH, U.K.
** Department of Metallurgy and Materials Science, Pembroke Street,
Cambridge CB2 3QZ, U.K.
+ GEC, Hirst Research Centre, East Lane, Wembley, Middlesex HA9 7PP, U.K.

ABSTRACT

The annealing behaviour of B implants in the millisecond time regime
using a combination of swept line beam and background heating is compared
with isothermal annealing with heating cycles of a few seconds. Carrier
concentration profiles derived from spreading resistance measurements show
that under annealing conditions which restrict diffusion, millisecond
processing gives higher activation of B implants than isothermal heating.
Transmission electron microscopy shows that millisecond annealing also
results in a lower defect density than that following an equivalent
isothermal anneal.

INTRODUCTION

The need for further restriction of diffusion during implant
activation arises from a continuing general reduction in device dimensions.
The annealing of ion implants in the millisecond time regime has been
studied and compared with rapid isothermal annealing. Millisecond annealing
is attractive because it activates implants with markedly less diffusion
than isothermal processes, and by reaching a higher temperature removes many
residual defects. The millisecond anneals were performed using a dual
electron beam system, with one beam providing isothermal background heating
and the other beam formed into a line, which was scanned across the specimen
to give localised heating. [1]

Isothermal heating is well established as a means of annealing ion-
implants; in this work anneals were performed using the multiple-scan,
electron-beam isothermal annealing method. [2] With thermally isolated
wafers, the shortest time for a heating cycle is set by the available power
that can be coupled to the wafer, which determines the time taken to reach
the peak temperature, and an irreducible time taken for the wafer to cool by
radiation. For implants such as As, which form amorphous layers, it is
possible to regrow the damage layer, activate the dopant fully and restrict
diffusion to less than 50nm using this kind of heating cycle. However, for
B implants, which do not form amorphous layers, the temperatures required
for full activation cause diffusion significantly greater than 50nm and this
kind of heating cycle is not satisfactory. The combinations of time and
temperature that restrict diffusion to 50nm are given by Sedgwick, [3], but
substantially less diffusion than 50nm is necessary for shallow junction
formation for fine dimension CMOS or hot electron structures. [4] Studies
of the microstructure show that higher temperatures above those required for
activation are required to remove point defects resulting from implantation,
and even higher temperatures for the removal of extended defects. [3]

Dual beam processing overcomes the time and temperature constraints of
isothermal annealing, and avoids the problems of conventional thermal flux
techniques. Sedgwick [3] has reviewed the reported results from these

thermal flux systems, operating with characteristic time-temperature cycles lasting for a few milliseconds or less. Scanning spot systems, such as cw lasers, with background specimen heating up to 400°C, can activate implants but the process window is narrow, the throughput is low and there are inter- ference effects with lasers. More seriously, the high temperature gradients around the spot can cause additional damage. Line beam systems have higher throughput, but still suffer from high temperature gradients under the line. [5,6] This study concentrates on the activation of B implants, these being the most difficult to anneal, using a line beam and relatively high background temperatures, 500°C and 800°C, to avoid temperature gradient induced slip. The line beam of 6 to 10mm sweeping at 0.5m/s is compatible with good through- put. The electrical activation and diffusion have been measured by spreading resistance, and the microstructure has been examined with cross- section TEM. Isothermal anneals were performed for comparison, and to con- firm the relationship between activation, diffusion and the heating time and temperature.

DUAL BEAM HEATING AND TEMPERATURE MODELLING

The dual e-beam method combines isothermal background heating with a second e-beam which gives a short temperature transient above the background temperature. The lowest usable background temperature is set by the need to avoid crystal damage from stresses resulting from the temperature gradients under the line. A background temperature of 500°C was found to be adequete for peak temperatures under the line up to the melting point, as no associated damage was seen with this background temperature in the form of macroscopic slip lines, nor was any lattice strain seen by TEM. The highest background temperature used was 800°C, which was expected to cause less than 1nm diffusion for a typical background heating cycle lasting up to 20s. Higher background temperatures will cause unacceptable diffusion.

The line beam is formed by rapidly scanning a round electron beam spot at a frequency of 100kHz. The temperature rise under the beam, for a given spot size, hence linewidth, depends on the sweep speed and the power per unit length, which are independently adjustable over a wide range. The peak temperature under the beam has been estimated from a one-dimensional heat flow model, using a temperature dependent specific heat. The accuracy of the model was checked by measuring the power required just to start melting and comparing it with the predicted values, which were in close agreement. The isothermal processing was performed with another e-beam system, using the multiple scan method. The specimens were heated with a constant power density, 35 W/cm^2, and the temperature was monitored by an IRCON two colour pyrometer. Heating was stopped once the desired peak temperature was reached; cooling was by radiation. Heating curves are shown in Fig. 1.

CARRIER CONCENTRATION PROFILES

Double implants of differing energy and dose were used for this study, with B implanted at doses at 3.10^{14}/cm^2 at 25keV and 5.10^{14}/cm^2 at 120keV and also a single implant of 10^{16}/cm^2 at 25keV. The samples used for isothermal annealing had a capping layer of 0.5μm of undoped deposited SiO$_2$ to prevent loss of dopant during annealing. The samples for line beam annealing had no capping layer, as the shorter heating cycle was not expected to cause signi- ficant out-diffusion of dopant. The double implant conditions were chosen to provide a minimum in the atomic distribution below the surface which acts as a diffusion marker for high doping concentrations. The double implants were chosen not to exceed solid solubility over the range of anneal tempera- tures used. The isothermally annealed samples were heated to peak tempera- tures of 800°C to 1200°C in steps of 50°C and the time taken to reach the maximum temperature ranged from 3 to 5s.

Carrier profiles after annealing were measured by the high resolution spreading resistance method on a bevelled sample, which gives the peak carrier concentration, the junction depth, and the sheet resistance. [7] The activation behaviour of B is shown in Figs. 2 to 4 , which show carrier profiles after 850°C, 1000°C and 1150°C isothermal anneals.   An 850°C anneal produces partial activation, and the dip in the as-implanted atomic profile is not clearly seen.   A 1000°C anneal gives about twice the peak carrier concentration, and there is a marked dip in the profile corresponding to the one in the as-implanted concentration.   There is no movement of

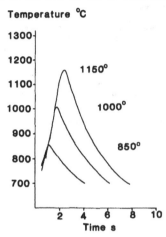

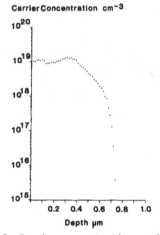

Fig.1 Experimental Temperature-time curves for isothermal heating of Si at 35W/cm^2 to peak temperatures of 850, 1000 and 1150°C.

Fig.2 Carrier concentration against depth for a 3.10^{14} at 25keV + 5.10^{14} at 120keV B/cm^2 implant after isothermal annealing at 850°C.

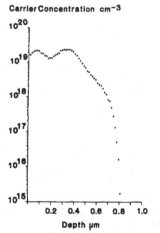

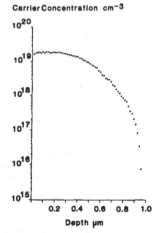

Fig. 3 Profile for implant of Fig.2 after annealing at 1000°C.

Fig. 4 Profile for implant of Fig.2 after annealing at 1150°C.

of the junction within experimental measurement errors. At 1150°C, the peak carrier concentration has fallen as significant diffusion has occurred, the dip in the profile has disappeared and the junction depth has deepened considerably.

Implants of B at the doses used for the isothermal anneals were processed in the dual electron beam system, using two background temperatures, 500°C and 800°C. The higher background temperature causes some dopant activation, as seen from the isothermal annealing results, but negligible diffusion, and reduces the power required in the line beam for a given maximum wafer temperature. All line beam anneals were performed with a 160µm wide line, sweeping at 50cm/s, with a power per unit length of three quarters of the power required just to melt the surface visibly. The line power was 50W/mm, giving an estimated peak temperature of 1170°C with 500°C background heating and 21W/mm giving 1260°C with 800°C background. Results from the B implants are shown in Figs. 5 and 6 and show carrier profiles as determined from spreading resistance measurements. The concentration profile for 1000°C isothermal heating shows the highest peak to valley ratio of 1.8 for the dip in the implanted profile. For the dual beam annealed sample the ratio is more than twice as much, indicating that much less diffusion has taken place even in the high concentration region.

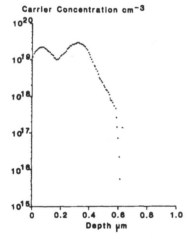

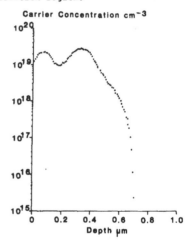

Fig.5 Profile for implant of Fig.2 after line annealing with 500°C background temperature.

Fig.6 Profile for implant of Fig.2 after line annealing with 800°C background temperature.

TRANSMISSION ELECTRON MICROSCOPY

Cross-sectional transmission electron microscopy, XTEM, at 120keV was used, on ion beam thinned samples, to examine the defect structure after annealing. Rapid isothermal annealing results in a reduced defect density compared to conventional furnace treatments. [8] The microstructure resulting from a rapid anneal which causes negligible diffusion is shown in Fig.7. The implanted region contains a large number of small dislocation loops, 20 to 30nm in diameter. Annealing at a higher temperature results in some diffusion, but greatly reduces the defect density, leaving only a few loops of 100nm diameter.(Fig.8) The microstructure resulting from millisecond annealing of a 3.10^{14} at 25keV plus 5.10^{14} at 120 keV B implant is substantially different. With 500°C background temperature, the micro-structure is as shown in Fig.9. Two bands of small loop dislocations, ∿10nm in diameter, are observed, corresponding to the two peaks in the boron distribution. With 800°C background heating two defect bands are again observed, but in this instance containing a mixture of loops 20-30nm diameter, and line defects. (Fig.10) The generally paired line defects, ∿150nm long, were largely parallel to the surface, but some other distinct orientations are also present. The overall density of defects is much less than in the previous case. No lattice strain or other defects arising from the tempera-ture gradients experienced during line beam processing could be observed further down from the implanted layer.

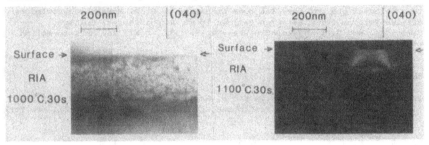

Fig.7 XTEM of 10^{16} at 25keV B/cm^2 implant after isothermal annealing at 1000°C for 30s (bright field).

Fig.8 XTEM of 10^{16} at 25keV B/cm^2 implant after isothermal annealing at 1100°C for 30s (dark field).

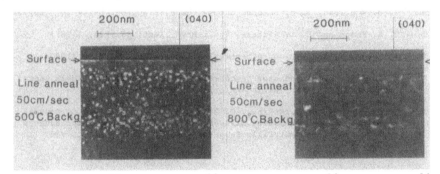

Fig.9 XTEM of 3.10^{14} at 25keV + 5.10^{14} at 120keV B/cm^2 implant after line beam annealing at 50cm/s with 500°C background temperature (dark field).

Fig.10 XTEM of 3.10^{14} at 25keV + 5.10^{14} at 120keV B/cm^2 implant after line beam annealing at 50cm/s with 800°C background temperature (dark field).

DISCUSSION

This work confirms that rapid isothermal annealing gives increasing activation with temperature for a boron implant, with the peak carrier concentration occurring at about 1050°C after which diffusion of the boron causes the peak value to fall for higher temperatures. Diffusion of about 10nm is present following annealing at 1000°C. With millisecond line beam annealing superimposed on background heating, the highest background temperature gave the highest activation, similar to that from the best isothermal anneals, but with much less than 10nm diffusion. The lower background temperature gave less activation but the peak temperature under the beam was, of course, lower. Annealing at the higher background temperature gives a lower overall defect density. The results of this study provide evidence that for boron implants short, high temperature anneals can combine to give good activation, low diffusion and low defect structure. The use of the dual beam method avoids damage, gives readily controllable heating and good throughput compared with conventional thermal flux techniques.

ACKNOWLEDGEMENTS

G. F. Hopper, GEC Hirst Research Centre, is thanked for help with temperature calculations for the line beam annealing. R. A. McMahon acknowledges British Telecom for a Research Fellowship at Corpus Christi College, Cambridge, and D. G. Hasko acknowledges the SERC, U.K., for an Information Technology Fellowship.

REFERENCES

1. R. A. McMahon, J. R. Davis and H. Ahmed in Laser and Electron Beam Solid Interactions, eds. J. F. Gibbons, L. D. Hess and T. W. Sigmon, New York: North Holland (1981).

2. R. A. McMahon, H. Ahmed, D. J. Godfrey and K. J. Yallup, IEEE, Trans. ED $\underline{30}$, 1550 (1983).

3. T. O. Sedgwick, Proc. Electrochem. Soc. $\underline{82}$ (7),130 (1982).

4. G. B. McMillan, J. M. Shannon and H. Ahmed, Electron. Lett. $\underline{20}$ (1984).

5. J. A. Knapp and S. T. Picraux, Appl. Phys. Lett. $\underline{38}$, 873, (1981).

6. T. Yu, K. J. Soda, B. C. Streetman, J. Appl. Phys. $\underline{51}$ 4399 (1980).

7. M. Pawlik in Semiconductor Processing ASTM STP850 ed. C. Dinesh Gupta, American Society for Testing Materials (1984).

8. D. J. Godfrey, R. A. McMahon, D. G. Hasko, H. Ahmed and M. G. Dowsett to be published in Impurity Diffusion and Gettering in Semiconductors, Proceedings of the Materials Research Society, Fall Meeting 1984, Boston, Mass.

RAPID THERMAL ANNEALING OF PRE-AMORPHIZED
B AND BF$_2$-IMPLANTED SILICON

I.D. CALDER*, H.M. NAGUIB*, D. HOUGHTON**, AND F.R. SHEPHERD**
* Northern Telecom Electronics, Box 3511, Stn. C, Ottawa, Ontario, K1Y 4H7
** Bell-Northern Research, Box 3511, Stn. C, Ottawa, Ontario, K1Y 4H7

ABSTRACT

Shallow p$^+$n junctions have been formed through a combination of pre-amorphization of the silicon surface by implantation of ^{28}Si, ^{33}Ar, or ^{73}Ge, low energy implantation of boron or BF$_2^+$, and rapid thermal annealing (RTA) in a tungsten halogen lamp system. Both pre-amorphization and RTA are required to form a shallow (< 0.25 μm) junction, for either boron or BF$_2^+$. Argon pre-amorphization results in poor electrical activation of the boron, while germanium gives the lowest sheet resistivity, but is responsible for a deep boron tail during implantation. The residual damage is characterized by a plane of dislocation loops centred either close to the boron concentration peak, for B$^+$ implantation into a crystalline substrate, or at the original amorphous-crystalline interface, for pre-amorphized specimens.

INTRODUCTION

As the planar dimensions of semiconductor devices are scaled down for VLSI, an accompanying reduction in junction depths is necessary [1]. To form shallow junctions both the doping and annealing steps must be optimized. Low energy ion implantation is necessary [2], but steps must be taken to eliminate ion channelling [2], either by the use of a heavy molecular beam such as BF$_2$ (for p-type doping) [3] or by a prior amorphization of the surface layer [4-6]. Diffusion can be contained by using either a lower temperature or shorter anneal time. This transient (short time) method has proven very successful at reducing diffusion while producing a high degree of dopant activation [4,5,7-9].

The objective of this study was to develop a technology for the fabrication of shallow (< 0.25 μm) p$^+$n junctions. A combination of pre-amorphization, low energy boron or BF$_2^+$ implantation, and rapid thermal annealing (RTA) techniques was investigated. The depth of amorphization, the crystal quality, and the residual damage were characterized by Rutherford backscattering/ion channelling (RBS) and transmission electron microscopy (TEM). The dopant distribution was measured by secondary ion mass spectroscopy (SIMS) and a spreading resistance probe (SRP) technique. SRP, in conjunction with four point probe measurements of sheet resistivity, also provided information about the degree of electrical activation.

JUNCTION FORMATION

The starting substrates were 4-6 Ω-cm n-type (100) silicon wafers containing 10^{15} phosphorus ions/cm^2. Most wafers were then oxidized in dry O$_2$ to form 400Å of thermal oxide. Pre-amorphization was performed by implanting 2 x 10^{15} ^{28}Si$^+$/cm^2 @ 50 keV, or 2 x 10^{15} ^{33}Ar$^+$/cm^2 @ 60 keV, or 1 x 10^{15} ^{73}Ge$^+$/cm^2 @ 100 keV. The dopant was introduced by the implantation of 3 x 10^{15} B$^+$/cm^2 @ 10 keV, or 3 x 10^{15} BF$_2^+$/cm^2 @ 45 keV. BF$_2^+$ was

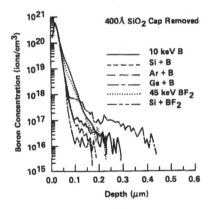

Figure 1. SIMS concentration profiles for boron in unannealed specimens implanted through 400Å SiO_2, showing the effects of various pre-amorphization schemes. The doses were 3×10^{15} ions/cm^2 for B^+ and BF_2^+, 2×10^{15} Si^+/cm^2 @ 50 keV, 2×10^{15} Ar^+/cm^2 @ 60 keV, and 1×10^{15} Ge^+/cm^2 @ 100 keV.

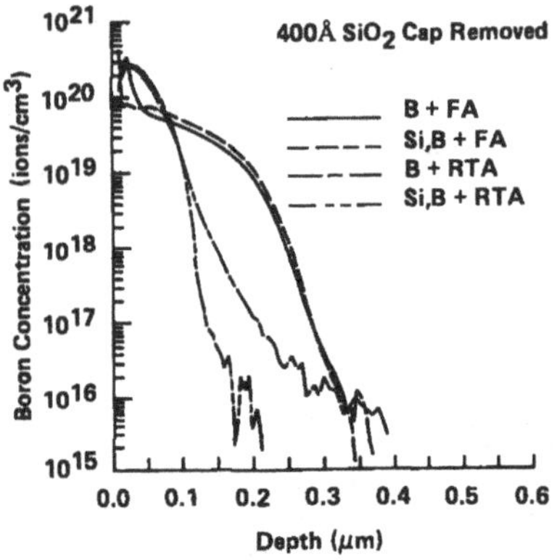

Figure 2. Boron concentration profiles after furnace annealing (FA) at 925°C for 30 minutes, or rapid thermal annealing (RTA) at 1050°C for 10 seconds, with and without Si pre-amorphization.

not implanted into the Ar- or Ge-amorphized specimens. Annealing was carried out either in a furnace at 925°C for 30 minutes, or by Heatpulse rapid thermal annealing (RTA) at a temperature of 1050°C, over times ranging from 4 seconds to 30 seconds. The ambient was dry N_2 in all cases. A few pre-amorphized samples also received a cw laser anneal (LA) with an argon laser at 6 W in a 42 μm spot scanned at 400 cm/s while the substrate was maintained at 500°C.

RESULTS AND DISCUSSION

The Profile As/Implanted

Figure 1 shows SIMS profiles of the boron concentration after various preamorphization treatments and implantation through 400Å of SiO_2. The metallurgical junction depth was essentially the same (~0.25 μm) for either boron or BF_2 implanted into crystalline substrates. The profile is much sharper for the preamorphized samples, although a small tail is still visible for boron implanted into Si or Ar amorphized substrates. Optimization of these amorphization implants is still required. The germanium implanted samples were found to contain a long boron tail with a concentration and depth much greater than can be attributed to channelling, as seen in the unamorphized case. Some form of low temperature damage enhanced diffusion during implantation may be responsible. There is a correlation between the mass of the amorphizing species as it increases from ^{28}Si to ^{33}Ar to ^{73}Ge and the extent of the tail (Figure 1).

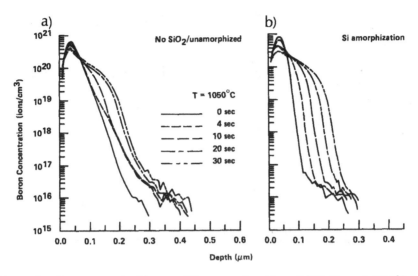

Figure 3. Boron concentration profiles as a function of RTA time at 1050°C, for (a) unamorphized and (b) Si pre-amorphized specimens. The implants were into bare silicon.

Rapid Thermal Annealing, Diffusion, and Activation

A comparison between furnace annealing and RTA is seen in Figure 2, for crystalline and pre-amorphized specimens. The junction depth x_J is ~0.36 μm in all cases except for the pre-amorphized, rapidly annealed case where $x_J \simeq 0.2$ μm. For implantation into bare Si, a 1050°C/20 sec RTA resulted in $x_J = 0.22$ μm and $\rho_\square = 52$ $\Omega/\square$. Therefore both of these technologies are required to form a shallow junction.

The diffusion of boron as a function of time in crystalline and Si-amorphized substrates is illustrated in Figure 3. For short times in the crystalline specimens (Fig. 3a) there is a very fast initial diffusion in the tail, perhaps due to interstitial processes, and negligible diffusion at the peak, where some precipitation or clustering is expected to take place above the solubility limit of 4 x 10^{20} cm^{-3}. As the anneal continues for longer times there is some diffusion in the peak and shoulder because of concentration enhancement effects, eventually forcing movement of the tail as the concentration gradient increases. On the other hand, diffusion in the amorphized specimens (Fig. 3b) takes place in a more conventional manner, with no obvious time dependent effects. There is also much more diffusion at the peak. Specimens which received a laser pre-anneal experienced slightly reduced diffusion, junction depth, and sheet resistivity.

A comparison between SIMS and SRP results revealed that an unactivated peak remains for Ar pre-amorphization, while anomalously high diffusion occurred for the Ge case. The activation is reflected by measurements of sheet resistivity (Figure 4), which give much higher values for Ar pre-amorphization than for any other case; Ge gives the best results.

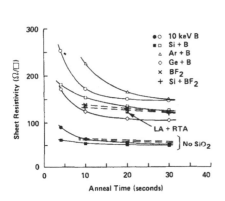

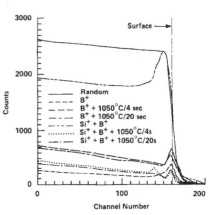

Figure 4. Four point probe measurements of sheet resistivity for various amorphization and annealing conditions. The RTA temperature was 1050°C.

Figure 5. RBS/ion channelling data showing residual damage before and after RTA. The depth scale is 48Å/ channel, with a resolution of 9 channels. x_{min} was reduced to 4.7% for the pre-amorphized RTA case.

RBS/ion channelling (Figure 5, Table 1) and TEM measurements (Figure 6, Table 1) show that residual damage remains after annealing. For pre-amorphized specimens, ion channelling measurements show an initial damage (amorphous) peak of width 1100Å (x_{min} = 75%) which is eliminated after RTA (x_{min} = 4.7%). However a peak formed of a dislocation network of width 850Å remains at the original amorphous-crystalline interface, with depth = 1100Å measured by cross-sectional TEM. This dislocation band has also been observed previously for deeper pre-amorphized structures [10]. For the annealed samples that were not pre-amorphized, a damage peak distinct from the surface peak is not resolved (Figure 5), but TEM reveals a layer of dislocations centred at a depth close to the B^+ projected range of 440Å [11].

CONCLUSIONS

Both pre-amorphization and rapid thermal annealing are required to form a shallow p^+n junction in silicon, for either boron or BF_2; a final junction depth of 0.22 μm and sheet resistivity of 52 Ω/□ were obtained with Si preamorphization. Argon pre-amorphization is not beneficial, while Ge pre-amorphization gives the lowest resistivity, but can cause anomalous low temperature diffusion, resulting in a much deeper junction. For crystalline substrates residual damage occurs at a dislocation network close to the boron implantation peak, and diffusion is time dependent. Residual damage in pre-amorphized specimens occurs as a dislocation network near the original amorphous-crystalline interface, and diffusion takes place in the recrystallized substrate in an essentially conventional manner.

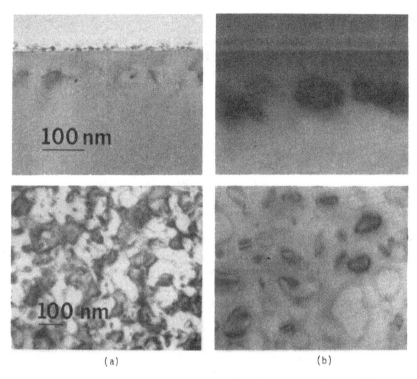

(a) (b)

Figure 6. TEM cross-sectional and plan view micrographs for the 20 second RTA samples described in Figure 5. There is a layer of dislocation loops near the boron peak, for the un-amorphized specimens (a), or near the original amorphous-crystalline interface for the pre-amorphized samples (b).

Table 1. Summary of RBS and TEM results for various pre-amorphization and annealing conditions.

Amorph.	RTA (°C/s)	x_{min} (%)	Damage Peak Depth (Å) RBS	TEM	Width (Å) TEM
–	–	5.7	surface	–	–
–	1050/4	17	surface	550	500
–	1050/20	15	surface	450	450
Si^+	–	75	500	–	–
Si^+	1050/4	4.7	1100	1100	850
Si^+	1050/20	5.0	950	900	900
Ge^+	1050/20	3.3	1150	–	–

ACKNOWLEDGEMENTS

We are grateful to W. Vanderworst, C. Slaby, M. Normandin, and A. Naem for many useful discussions. Particular thanks are due to Surface Science Western (University of Western Ontario) for the SIMS analysis, to Queen's University for the RBS, to Solid State Measurements Inc. for the SRP results, and to CANMET and R. Mallard (Bell-Northern Research) for the TEM work.

REFERENCES

[1] Y. El-mansy, IEEE Trans. Electron Dev. ED-29, 567 (1982).

[2] H. Maes, W. Vanderworst, and R. van Overstraeten, in "Impurity Processes in Silicon", ed. by F.F.Y. Wang, Chap. 8 (North-Holland, Amsterdam, 1981).

[3] T.M. Liu and W.G. Oldham, IEEE Electron Dev. Lett. EDL-5, 299 (1984).

[4] T.E. Seidel, IEEE Electron Dev. Lett. EDL-7, 353 (1983).

[5] D.K. Sadana, E. Myers, J. Liu, T. Finstead, and G.A. Rozgonyi, in "Energy Beam-Solid Interactions and Transient Thermal Processing", p. 303 (NorthHolland, Amsterdam, 1984).

[6] F.R. Shepherd, W.H. Robinson, J.D. Brown, and B.F. Phillips, J. Vac. Sci. Technol. A1, 991 (1983).

[7] T.O. Sedgwick, R. Kalish, S.R. Mader, and S.C. Shatas, in "Energy Beam-Solid Interactions and Transient Thermal Processing", p. 293 (North-Holland, Amsterdam, 1984).

[8] R.T. Hodgson, V. Deline, S. Mader, and J. Gelpey, Appl. Phys. Lett. 44, 589 (1984).

[9] J.B. Lasky, J. Appl. Phys. 54, 6009 (1983).

[10] C. Carter, W. Maszara, D.K. Sadana, G.A. Rozgonyi, J. Liu, and J. Wortman, Appl. Phys. Lett. 44, 459 (1984).

[11] M. Simard-Normandin and C. Slaby, to be publ. in Can. J. Phys.

LIGHT-INDUCED, SHORT-TIME ANNEALING OF SILICON-IMPLANTED LAYERS

D. WOUTERS AND H.E. MAES
ESAT Laboratory- K.U.Leuven
Kardinaal Mercierlaan 94
B-3030 Heverlee, Belgium

ABSTRACT

Short time anneal experiments were performed in a specially developed annealing system based on a bank of Tungsten-Halogen lamps. The annealing process during the temperature transient was studied as a function of the illumination time with the system operating in the constant maximum power regime. By using this procedure it was shown that electrical activation of all elements (B, P, As) considered in this study is completed within the transient period without noticeable diffusion. Redistribution of the implanted profiles was only observed for longer times after the steady state temperature was reached. Pre-amorphization with Argon is found to significicantly alter and retard the annealing behaviour.

INTRODUCTION

Short time anneal (STA) is rapidly gaining interest for the annealing of ion-implanted profiles, as present and future integrated circuit technologies require better control of the implant distribution [1]. For this reason, much research is done on new annealing techniques using laser and electron beam or incoherent light heaters.

A STA system was developed in our laboratory consisting of a bank of 6 Tungsten-Halogen lamps of 6kW, 480 V each, resulting in a power density of 40 W/cm². The wafers are positioned on quartz pins inside a quartz tube in an Argon flow.

Conventionally [2-4], STA and more specifically the degree of activation and possible diffusion are reported in terms of the actual time the wafer remains at a certain steady state temperature (isothermal procedure) disregarding possible annealing effects during the transient period itself. With our system however it was possible to anneal wafers using different procedures. In the present study a constant maximum power regime was used and the illumination time could be varied to any possible value, allowing the examination of the annealing behaviour during temperature transient. Lamp ignition occured after 4 seconds. Typical measured transients of the reactor temperature and of the temperature on the wafer as obtained with a thermocouple attached to the wafer, are shown on Fig.1.

EXPERIMENTAL PROCEDURE

For this study a large set of implanted samples were prepared. Boron, Phosphorus and Arsenic were implanted either in bare silicon of the opposite type or in silicon through a SiO₂ layer. Different combinations of energies and doses were used. For some samples the silicon surface was pre-amorphized with Argon. For this purpose two different energies were used resulting in a shallow (Ar1) or a deep (Ar2) amorphous surface layer. The sample reference, implantation parameters and conditions are given in Table 1.

The evaluation of the STA effects was done using four point probe (FPP)

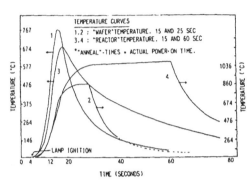

Fig. 1 Time dependence of the wafer and reactor temperature for different anneal time settings

measurements, the spreading resistance (SRP) technique and secondary ion mass spectrometry (SIMS). Isochronal STA experiments at 35 and 60 sec with different temperature set points were performed on samples B1 and P1. For both dopants no significant difference in the sheet resistances at each set temperature after 35 or 60 seconds was found as shown in Fig.2. For the limited available power these two times correspond to a steady state regime in case of a low set point but to a steady state (60 sec) and only onset of steady state (35 sec) for a high set point. The results of Fig.2. therefore suggest that the major annealing process occurs during the transient period, well before the steady state temperature is reached. Therefore the conventional isothermal STA procedure at elevated temperatures is not appropriate to reveal the actual annealing progress. In this paper we report on the results of STA annealing in this transient temperature regime when using the constant maximum power.

EXPERIMENTAL RESULTS

The sheet resistances of annealed samples obtained with FPP are shown in Figs.3-5 as a function of the illumination time. In the case of P and As implants (Fig.3. and 4.) the initial as-implanted sheet resistance was found to be unexpectedly low corresponding to the substrate value and type. After a few seconds of annealing a sharp increase of the sheet resistance was observed followed by an exponential decrease for longer anneal times (as expected for solid state regrowth [3], [5]) until a low steady state value is reached. So the sheet resistance versus anneal time shows a characteristic peak. Only for the B implants (Fig.5.) a normal behaviour for the sheet resistance is found except when a deep Argon pre-amorphization precedes the B implantation (see B3 on Fig.5.). Since an Argon pre-amorphization alone (without further implants) showed no change in sheet resistance neither before or after annealing and since a low dose As implant (below the self-amorphizing dose) also gives rise to the peak effect, this phenomenon is not related to the presence of an amorphous layer.

TABLE I : IMPLANTATION PARAMETERS

REFERENCE	DOSE (cm^{-2})	ENERGY (kV)	R_p (µm)	ΔR_p (µm)	CONDITION
ARGON :					
Ar1	1x10^{15}	70	0.071	0.029	
Ar2	1x10^{15}	160	0.212	0.056	
BORON :					
B1	6x10^{15}	50	0.157	0.055	Ar1
B2	2x10^{15}	50	0.157	0.055	Ar1
B3	2x10^{15}	50	0.157	0.055	Ar2
B4	2x10^{15}	50	0.157	0.055	Bare Si
B5	2x10^{15}	70	0.215	0.065	50 nm SiO$_2$
PHOSPHORUS :					
P1	2x10^{16}	60	0.072	0.033	Ar1
P2	5x10^{15}	60	0.072	0.033	Ar1
P3	5x10^{15}	60	0.072	0.033	Bare Si
P4	5x10^{15}	120	0.150	0.057	50 nm SiO$_2$
ARSENIC :					
As1	1x10^{15}	50	0.034	0.012	Ar1
As2	1x10^{15}	50	0.034	0.012	Bare Si
As3	5x10^{13}	50	0.034	0.012	Bare Si
As4	1x10^{15}	180	0.110	0.036	50 nm SiO$_2$

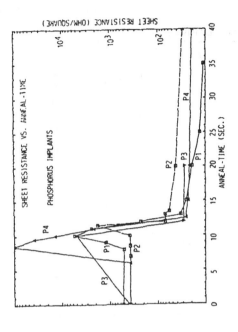

Fig. 3

Sheet resistance of Phosphorus-implanted substrates vs. time.

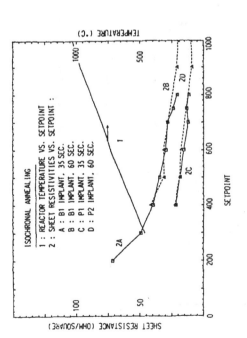

Fig 2.

Reactor temperature and sheet resistance for B1 and P1 vs. setpoint for 35 and 60 sec. isochronal annealing.

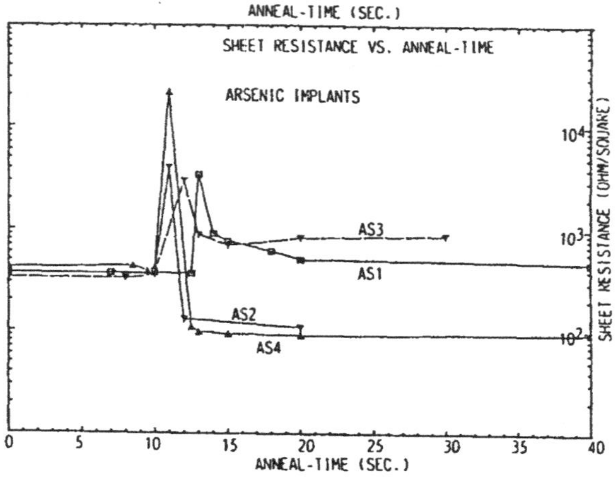

362

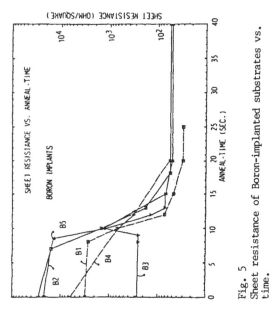

Fig. 4
Sheet resistance of Arsenic-implanted substrates vs. time.

Fig. 5
Sheet resistance of Boron-implanted substrates vs. time.

Excessive probe penetration could also be ruled out. The reason for this
behaviour was expected to be revealed from complementary SRP measurements.
However in most cases these measurements were in contradiction with the FPP
results. Indeed a low initial sheet resistance could never be predicted
from the electrical active profiles obtained with SRP. However before any
SRP measurement a low temperature (400 C, 15 min) oxide (Silox, 80 nm) is
deposited on the wafer in order to allow a sharp detection of the actual
surface and to eliminate bevel rounding effects. This "low" temperature
however was then found to sufficiently electrically activate the implanted
profile such that the deviation between the results of both techniques could
be explained. Indeed if the FPP measurement was repeated on a wafer that
also received a Silox treatment, the low initial sheet resistance and even
the peak were no longer observed and a close agreement between measured FPP
and the sheet resistance calculated from the SRP results, was obtained
(Fig.6.). From comparative SRP measurements on wafers with and without
Silox treatment it could be concluded that the low initial sheet resistance
corresponding to the bulk value was obtained in FPP because of the complete
depletion of the insufficiently active implanted layer. The peak value

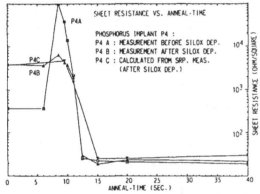

Fig. 6   Sheet resistance
of Phosphorus implanted
substrates before and after
Silox treatment and com-
parison with sheet resis-
tance as computed from SRP
measurements

therefore corresponds to the onset of junction activation and can be used as
an indicator of annealing. The time at which the peak occurs is strongly
dependent on the used dopant and on the implantation condition. For Boron
a peak is only observed when a deep Argon amorphized layer is present
(Fig.5.) and Silox deposition always eliminates the peak effect. For Phosphorus a peak is observed in all cases before a Silox deposition whereas the
latter always removes the peak effect (Fig.6.). Finally for Arsenic a peak
is observed under all circumstances and this peak is not always removed
after a Silox deposition. From these observations and from Figs.3-6 it
follows that activation is slower and requires higher temperatures for As
than for B and P.

Pre-amorphization with Argon is found to result in a lower initial
activity of the implanted dopants and a retarded activation as compared to
implantation in bare silicon or through SiO2 (Figs.3. and 5.). The final
electrical activity is always lower for the amorphized samples. These effects are most outspoken for As. These observations are in qualitative
agreement with previous studies [5], [6]. The steady state sheet resistance
and electrical activation are given for all the samples in Table II.

Figs.7-9 show the electrical profiles for P, As and B respectively at
different times during the anneal progress as obtained from SRP and the SIMS
profile corresponding to a 15 sec anneal. The zero time SRP profile was
obtained after Silox deposition. For all dopants a 15 second anneal, well

364

within the transient period, is seen to be sufficient for "complete" electrical activation. The anneal progress is very fast between 11 and 15 seconds. From SIMS measurements and from the shown SRP results no redistribution was observed for anneal times below 20 seconds. Diffusion however becomes important for As and P after 40 second anneal.

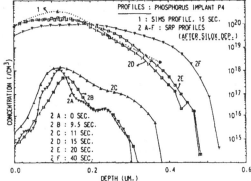

Fig. 7 Phosphorus :
Electrical profiles for different anneal times (from SRP) and SIMS profile after 15 sec anneal

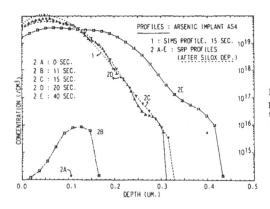

Fig. 8 Arsenic : Electrical profiles for different anneal times and SIMS profile after 15 sec anneal

Fig. 9 Boron : Electrical profiles for different anneal times and SIMS profile after 15 sec anneal

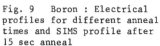

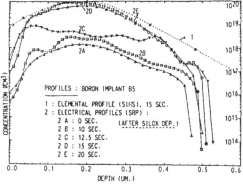

TABLE II : STEADY-STATE SHEET RESISTANCE
AND ELECTRICAL ACTIVATION

REFERENCE	TIME	R_{SH}	R_{SH}	ACTIVATION
	(sec)	(measured) $\Omega/\square$	(computed,SRP) $\Omega/\square$	%
B1	20	30	46	43
B2	20	55	77	75
B3	15	70	92	60
B4	20	56	85	65
B5	20	51	71	80
P1	20	20	30	15
P2	20	42	52	32
P3	20	27	44	38
P4	20	23	26	66
As1	20	500	1.16k	4
As2	20	108	152	52
As3	20	850	1.95k	30
As4	20	90	92	87
B1*	60	31	-	63
P1*	60	12	-	36

* INCREASE OF ACTIVATION LEVEL BUT WITH
SIGNIFICANT REDISTRIBUTION !

CONCLUSIONS

Short time anneal experiments in the transient temperature regime have
been shown to provide more information on the actual annealing progress
than the conventional isothermal STA procedure. "Complete" electrical activation of the implanted dopants occurs within this transient temperature
period without diffusion.

References

[1] C. Hill, "The contribution of beam processing to present and future
integrated circuit technologies", presented at the Navo Advanced
Research Workshop, Jan 83, Mons, Belgium

[2] R.B. Fair, J.J. Wortman, J. Liu, Proc. of the IEDM 83, p. 658

[3] T.E. Seidel, IEEE EDL-4, 10,353 (1983)

[4] R. Kalish, T.O. Sedgwick, S. Mader, S. Shatas, Appl. Phys. Lett. 44,107
(1984)

[5] J. Kato, S. Iwamatsu, J. Electroch. Soc., 131,1145 (1984)

[6] P. Spinelli, M. Bruel, Nuclear Instruments and Methods 209/210, North-
Holland Publ. Co, 751 (1983)

HIGH-QUALITY BORON AND BF_2^+-IMPLANTED P^+ JUNCTIONS IN Si USING SOLID PHASE EPITAXY AND TRANSIENT ANNEALING

P.K.VASUDEV, A.E.SCHMITZ AND G.L.OLSON
Hughes Research Laboratories, Malibu, CA 90265

ABSTRACT

We report on a systematic study of the doping profiles resulting from rapid thermal annealing of boron and BF_2^+-implanted silicon samples that were preamorphized by Si^+ implantation. A two-step process consisting of an initial solid phase epitaxial regrowth followed by a brief (~5 sec) high temperature (1050°C) anneal produces extremely shallow (<1500Å) junctions with low defect concentrations. The quality of the epitaxial regrowth is very sensitive to implant conditions and impurity effects as deduced from time-resolved reflectivity measurements. Using the best conditions for implantation and solid phase crystallization, we have obtained boron-doped regions with sheet resistivities of ≤ 40 $\Omega/\square$ and BF_2-doped regions of resistivity ≤ 60 $\Omega/\square$.

INTRODUCTION

The scaling down of MOS device channel lengths to submicrometer dimensions will require a concomitant reduction in source/drain junction depths (<0.2 μm) to minimize short channel behavior such as threshold and punchthrough shifts [1]. Problems with shallow source/drain junctions are particularly acute for p-channel MOS devices fabricated using boron implantation and conventional furnace annealing. This is due to the large penetration depth of boron when it is implanted at conventional energies (≥ 15 keV) and to the substantial diffusion of boron during annealing. The use of reduced energies is not very effective in achieving shallow junctions, since the combination of low mass and low energy results in the production of a significant channeling "tail" in the boron profile. Furthermore, the high diffusivity of boron also leads to increased junction depths (>0.3 μm) during conventional annealing at the high temperatures (~1000°C) and prolonged times required to achieve full activation. Although the use of rapid thermal annealing has recently been effective in reducing the boron junction depths [2-4], a significant tail remains in the annealed profiles. Likewise, in boron-doped layers produced by BF_2^+ implantation there is generally a high residual defect concentration present after rapid thermal annealing. This can result in the degradation of junction quality.

An approach that we have used to minimize these problems is to introduce boron or BF_2^+ into preamorphized silicon substrates. The amorphous layer is then recrystallized using a combination of solid phase epitaxy (SPE) and high temperature transient annealing. This approach prevents ion channeling and restricts the junction depths to below 0.2 μm. The sharpness of the dopant boron distribution profile is strongly influenced by the degree of amorphicity of the matrix, as well as by the conditions used for solid phase epitaxy. The optimization of these parameters is thus necessary for production of a hyperabrupt dopant distribution.

EXPERIMENTAL

Boron was implanted either as ^{11}B or $^{49}BF_2$ into n-type (100) silicon. The ion dose was $2x10^{15}cm^{-2}$ in both cases; the energy of the BF_2 implant (45 keV) was selected to give a boron projected range equivalent to that of the 10 keV ^{11}B implant. Preamorphization of the silicon was accomplished by implantation of Si^+ at an energy of 100keV and a dose of $2x10^{15}cm^{-2}$. Implantation was performed at substrate temperatures of 300°K and 77°K. Rapid thermal annealing (RTA) of the implanted layer was performed using an AG210T system in a dry N_2 ambient. Atomic depth distribution profiles were determined using secondary ion mass spectrometry (SIMS). An O_2^+ primary beam was used for sputtering in the SIMS studies. Electrical resistivities were measured with a conventional four-point probe measurement system. The rate of solid phase epitaxial crystallization was monitored in situ using time-resolved reflectivity (TRR) techniques [5]. The TRR results were correlated with the spatial distribution profiles of the dopant.

RESULTS AND DISCUSSION

The production of sharp implanted dopant profiles is strongly dependent on the nature of the amorphized region. Two variables which are particularly important in the formation of an amorphous layer are the ion species and the substrate temperature. The effect of these variables on the amorphization of Si is shown in Figure 1. The various curves compare the spatial variation of RBS yield for a number of samples aligned in a channeled orientation. The regions exhibiting 100% backscattered yield correspond to the amorphous zones (which regrow by solid phase epitaxy upon annealing), while the remaining regions correspond to damaged (subcrystalline) zones which recrystallize randomly without epitaxy. For the Si^+ implants, the data show that when the substrate is maintained at 77°K a uniform amorphous region of thickness ~2100Å is produced. In contrast, when the substrate is held at a temperature of 300°K, an identical implant produces a subcrystalline region in the sample. This is due to self-annealing of the implantation damage. Although the high mass of a BF_2^+ molecular ion is sufficient to amorphize silicon at either temperature, the temperature plays a major role in determining the "sharpness" of the crystal/amorphous (c/a) interface. It is seen from the

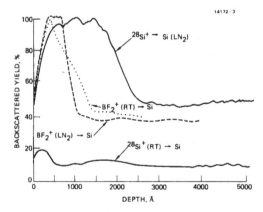

Figure 1. Effect of ion species and substrate temperature on the spatial variation of the RBS yield.

figure that by reducing the substrate temperature to 77°K a uniform
amorphous zone of thickness ~800A can be produced for BF_2 implants, while
at 300°K the same implant leads to a very thin and graded amorphous region
that ·appears to have undergone considerable self annealing.

The effect of the amorphicity of the Si substrate on the sharpness of
the implanted boron profiles is shown in Figure 2. The typical as-
implanted boron profile exhibits a rather broad distribution revealing
considerable channeling into the crystal. Reducing the substrate
temperature to 77°K has only a minor effect on reducing the extent of the
channeling tail, since there is no appreciable blockage of the channels in
the crystalline substrate upon cooling. However, by implanting boron into
a substrate that was preamorphized by Si^+ implantation, a considerable
sharpening of the profile is observed. Even for a 300°K Si^+ implant, the
profile undergoes some sharpening, but still shows evidence of channeling
near the tail region due to incomplete blocking of the (110) channels.
The sharpest dopant distribution corresponding to an idealized LSS profile
is produced by implanting boron or BF_2 into an amorphous region that is
much thicker than the width of the boron or BF_2-implanted zone.

Since the redistribution of the dopant profile upon annealing is
strongly dependent on the shape of the initial as-implanted profile, we
compared the various implanted profiles following identical anneals in an
RTA system. The results are shown in Figure 3. A two-step annealing
sequence which consisted of an initial low temperature (600°C) solid phase
epitaxial regrowth, followed by a "short" (~5 sec), high temperature
anneal at 1050°C was used. During the 600°C anneal, the amorphous region

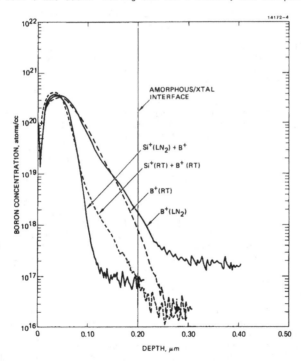

Figure 2.   Effect of substrate temperature and preamorphization
             on the as-implantled boron concentrating profiles
             determined from SIMS measurements.

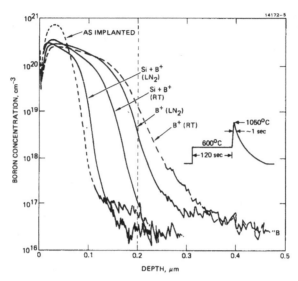

Figure 3. Boron concentration profiles after RTA
for samples produced under different
implantation conditions.

regrows by SPE, causing very little dopant redistribution, while the high
temperature step causes significant diffusion into the material. This
depends on the distribution and concentration of residual defects retained
in the material following SPE. It is seen that the sharpest distribution,
corresponding to the smallest amount of diffusion, is produced by implantation of the boron into a substrate which was preamorphized at 77°K.
This result is consistent with the observation [6] that a low temperature
implant produces the narrowest damaged zone below the amorphous region,
thereby leading to a very abrupt c/a interface. This region undergoes SPE
regrowth as a planar moving interface, with a minimum of residual defects
being accumulated. All other implantation conditions produce
distributions having significant tails.

The effect of various RTA conditions on the redistribution of boron
implanted into a substrate preamorphized at 77°K is shown in Figure 4.
With a low temperature (600°C) anneal only, very little impurity
redistribution during SPE is observed. However, a reduction in the peak
concentration to the solid solubility limit ($\sim$3x10^{20} cm^{-3}) is evident.
These results suggest that during SPE only lattice ordering occurs and
that there is insufficient energy for dopant diffusion. However, the
quality of the SPE regrowth is critical in determining the nature and
distribution of residual defects retained in the regrown material. The
defect density is strongly influenced by the presence of impurities such
as fluorine and boron which have a marked effect on the SPE growth rate
[5,7,8]. The sharp peak near the surface in all of the profiles is due to
the accumulation of inactive boron in regions of high residual defect
density [9] (this region also shows a corresponding pile-up in the
fluorine distribution). When only a short ($\sim$5 sec) high temperature
anneal is used a large redistribution in the boron profile is observed.
Under these conditions, the presence of the high defect density peak in
the profile indicates that the sample has retained the highest concentration of residual defects following annealing. The use of the two-step
anneal results in a profile that is intermediate between the above two

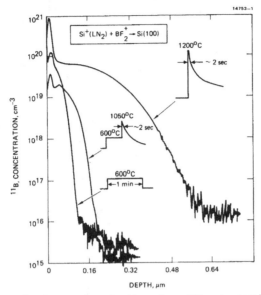

Figure 4. Boron concentration of profiles in samples
produced by BF_2^+ implantation into preamorphized substrates (77°K) after annealing
with three different RTA conditions.

extremes and also shows the best compromise between electrical activation
and dopant redistribution. In addition of results in the smallest
residual defect density peak near the surface and the least amount of
diffusion during the high temperature anneal. This clearly shows the
beneficial effects of a two-step sequence which combines both the SPE and
the high temperature annealing steps.

The effect of the various anneal conditions on the electrical
resistivity of both boron and BF_2^+-implanted samples is shown in Figure 5.
We note that in all cases the two-step anneal leads to the lowest sheet
resistivity, corresponding to nearly complete activation of the implanted
boron and a carrier mobility that is nearly equal to the theoretical
limit. Combining the results of Figures 4 and 5, we conclude that both
abrupt and low resistivity p^+ junctions can be produced by a combination
of low temperature implants and two-step rapid thermal anneals.

The effect of impurities on the SPE growth rate obtained from TRR
measurements during crystallization and the correlation of the SPE rate
with the distribution of implanted impurities are shown in Figures 6 and
7, respectively. As discussed in Ref. 5, the nonuniform temporal spacing
of the interference features in the TRR data is directly related to the
variations of the SPE rate during crystallization. In Figure 6, TRR data
are presented for crystallization of films at 620°C that contain the
following implants: (1) Si^+ only (uniform amorphous layer); (2) $B^+ + Si^+$;
and (3) $F^+ + Si^+$. In cases (2) and (3), the boron and fluorine implantation
energies (10 keV and 18 keV, respectively) were selected so that the
impurities would be located in a region near the surface, well within the
preamorphized layer. In the sample implanted with Si^+ only, the temporal
spacing of the interference peaks is constant, indicating a uniform
epitaxial growth rate throughout the entire layer. In the boron-implanted
sample there is a significant increase in the SPE rate [5,7] when the c/a

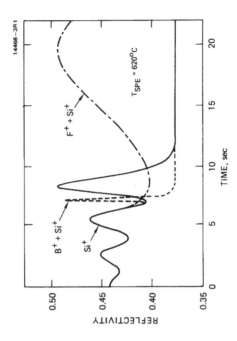

Figure 6. Time-resolved reflectivity data for crystallization at 620°C of samples implanted with: Si$^+$ only, Si$^+$ + B$^+$ (10 keV, 3x10^{15}cm^{-2}), and Si$^+$ + F$^+$ (18 keV, 6x10^{15}cm^{-2}).

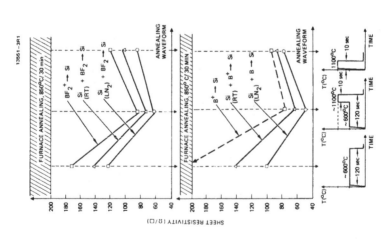

Figure 5. Effect of RTA conditions on sheet resistivity for doped films produced by B$^+$ and BF$_2^+$ implantation into amorphized and non-amorphized substrates.

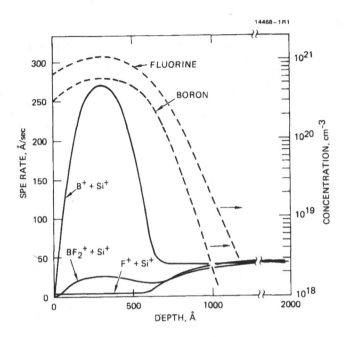

Figure 7. Dependence of the SPE rate on interface position
for crystallization at 640°C of amorphous layers
implanted with B⁺, F⁺, and BF₂⁺. The concentration
profiles of the boron and fluorine are given for
comparison.

interface propagates through the region containing the boron atoms. This
is evidenced by the dramatic increase in the oscillation frequency of the
TRR signal as the growth interface proceeds through the boron-containing
zone. In marked contrast, in the fluorine-implanted sample there is a
substantial reduction in the SPE rate as the interface moves through the
near-surface region containing the fluorine atoms [5,8]. The enhancement
or retardation of the SPE growth rate by ion-implanted impurities can have
a significant effect on the concentration of residual defects retained in
the regrown material. We suggest that this is due to the competition
between the interface velocity and the growth rate of extended defects
such as dislocation loops and stacking faults.

The dependence of the SPE rate on interface position for samples
implanted with B, F, and BF_2 is compared with the spatial distribution
profiles of the implanted impurities in Figure 7. At these temperatures
(~640°C) the rate retardation due to the fluorine is much more pronounced
than the rate enhancement due to the boron [5]. Consequently there is an
overall reduction in the SPE rate when both impurities are present
simultaneously in the film. To obtain regrown material with a minimum of
residual defects, the SPE growth rate must be faster than the growth rate
of extended defects. In addition, the growing interface should be planar,
thereby minimizing the number of sites for the nucleation of defect
clusters. The low temperature SPE growth conditions for minimizing the
residual defect content have not yet been fully optimized in this study.
However, this work must be carried out before the formation of
hyperabrupt, high quality p⁺-n junctions is realized.

CONCLUSIONS

A detailed study of the shape of the impurity atom profile and electrical effects in ion implanted and transient annealed silicon samples has been performed. It has been shown that implantation of impurity atoms into a preamorphized layer and the use of a two-step transient annealing cycle are effective in reducing the diffusion of boron and producing high quality, spatially abrupt p^+-doped regions. The effect of impurities on the SPE growth conditions can play a significant role in determining the quality of the recrystallized material and thereby influence junction quality.

ACKNOWLEDGMENTS

The authors gratefully acknowledge R.G. Wilson and D. Jamba for assistance with ion implantation. We also wish to thank R.F. Scholl for his technical assistance.

REFERENCES

1. J.R. Brews, W. Fitchner, E.H. Nicollian and S.M. Sze, IEEE Electron Dev. Letters, EDL-1,2 (1980).

2. K. Yamada, M. Kashiwagi and K. Taniguchi, Pro. 14th Conf. on Solid State Devices, Jap. J. Appl. Phys., 22, Suppl 22-1, 157 (1983).

3. T.E. Seidel, IEEE Elect. Dev. Lett, EDL-4, 353 (1983).

4. T.O. Sedgwick, R. Kalish, S.R. Mader and, S.C. Shatas, Mat. Res. Soc. Symp. Proceedings, 23, 293 (1984).

5. For recent reviews, see G.L. Olson, these proceedings, G.L. Olson, S.A. Kokorowski, J.A. Roth and L.D. Hess, Mat. Res. Soc. Symp. Proc. 13, 141 (1983), and references therein.

6. L.D. Glowinski, K.N. Tu and P.S. Ho, Appl. Phys. Lett. 28, 312 (1976).

7. L. Csepregi, E.F. Kennedy, T.J. Gallagher, J.W. Mayer and T.W. Sigmon J. Appl. Phys. 48, 4234 (1977).

8. I. Suni, U. Shreter, M-A. Nicolet and J.E. Baker, J. Appl. Phys. 56, 273 (1984).

9. T. Sands, J. Washburn, R. Gronsky, W. Maszara, D.R. Sadana and G.A. Rosgonyi, Appl. Phys. Lett. 45, 982 (1984).

FORMATION OF SHALLOW P+ JUNCTIONS USING TWO-STEP ANNEALS

C.I. DROWLEY*, J. ADKISSON**, D. PETERS*, AND S.-Y. CHIANG*
*Hewlett-Packard Laboratories, 3500 Deer Creek Road, Palo Alto, CA 94304
**Department of Electrical Engineering and Computer Science, Massachusetts
Institute of Technology, Cambridge, MA 02139

ABSTRACT

Shallow (0.15-0.2 μm deep) p+ junctions have been formed using boron
implanted into silicon which was pre-amorphized using a silicon implant. The
implants were annealed using a two-step process; initially the wafers were
furnace annealed at 600 °C for 100 min., followed by a rapid isothermal
anneal (RIA) at 950-1100 °C for 10 sec. For comparison, some wafers were only
given a single-step rapid isothermal anneal at 950-1100 °C for 10 sec.

The shallowest junctions were formed when the amorphous silicon layer
was deeper than the boron implant, because of the suppression of channelling.
When the amorphous/crystalline interface was shallower than the tail of the
boron implant, some channeling occurred. This channeling tail exhibited an
enhanced diffusion during the single-step RIA which was reduced significantly
by the two-step anneal. When the amorphous layer was deeper than the boron
implant, the single-step and two-step anneals gave identical results.

INTRODUCTION

As MOS device dimensions are scaled down into the sub-micron range,
source/drain junction depths will have to be reduced to the 0.1-0.2 μm range
in order to minimize short-channel and punchthrough effects. This reduction
in junction depth is comparatively easy for n+ junctions, since arsenic, with
its small implant range and low diffusivity, can be used. The problem is much
more difficult with p+ junctions, since the only major candidate is boron.
Boron tends to channel during implantation, resulting in a relatively deep
"tail" to the implant; moreover, it diffuses much more rapidly than arsenic,
resulting in significant redistribution during implant annealing.

An approach which minimizes these two problems is to first amorphize the
silicon surface with a silicon implant, implant the boron, and then perform a
short (∿10 sec.) isothermal anneal at elevated temperatures. The presence of
the amorphous silicon layer reduces or eliminates the boron channeling, while
the rapid isothermal anneal attempts to minimize the redistribution of the
boron.

This approach has been studied previously [1,2]. However, those studies
concentrated on a relatively deep amorphous layer (deeper than the junction
depth) which may cause excessive leakage in the annealed junctions [3]. Also,
no detailed examination of the diffusion of the boron was performed.

Other studies of shallow boron junction formation using RIA [4-6] have
suggested that boron diffusion is enhanced when RIA is used without a
pre-amorphized surface layer. This enhancement has been attributed to an
interaction of point defects with the diffusing species [4,5,7].

The study presented in this paper examines the formation of shallow boron junctions with two different pre-amorphization conditions. Both shallow (less than the junction depth) and deep (greater than the junction depth) amorphous layers were used. The implants were subject to either a single-step anneal using RIA or a two-step anneal using a 600 °C, 100 min. low-temperature step to regrow the amorphous layer followed by a high-temperature RIA step. The high-temperature steps included temperatures up to 1100 °C in order to investigate the diffusion of the boron.

EXPERIMENTAL

P(100), 16-24 ohm-cm silicon substrates were used for the annealing study. The wafers were oxidized in dry oxygen to produce a 20 nm thick oxide, and implanted with Si_{28}^+ at 77 °K to produce an amorphous layer at the surface. Two sets of silicon implants were employed: the first, which we will refer to as the "shallow" case, used an implant of 3×10^{15} cm^{-2} at 25 keV, producing an amorphous layer 50 nm thick at the surface of the silicon. The second set, which we will refer to as the "deep" case, used a series of implants: 3×10^{15} cm^{-2} at 40 keV, 5×10^{15} cm^{-2} at 70 keV, and 5×10^{15} cm^{-2} at 150 keV. This group of implants produced an amorphous layer 330 nm thick. After amorphization of the surface, all wafers were implanted with 1.5×10^{15} cm^{-2} of B_{11}^+ at 15 keV.

The wafers from each set of silicon implants were split into two groups. The first group was given a 600 °C anneal in dry nitrogen for 100 min. to regrow the amorphous layer, while the second set was not annealed. One wafer from each silicon implant/anneal condition group was selected (four wafers total) and split into quarters. The quarter wafers were given rapid isothermal anneals in an AG Associates 210T Heatpulse annealer. The temperatures were monitored by a thermocouple mounted on a chip next to the wafers, and ranged from 950-1100 °C. These temperatures are estimated to be accurate to ±30 °C [8]. The annealing times were uniformly 10 sec. The annealing ambient was dry nitrogen. Quarter wafers from the same silicon implant group but different low-temperature annealing conditions were annealed simultaneously to minimize run-to-run differences in temperature.

Following the rapid isothermal anneals, the boron profiles were examined using a Cameca IMS 3f secondary ion mass spectrometer with an oxygen beam [9]. The junction depths were obtained using spreading resistance. Sheet resistivity was measured using a four-point probe.

DISCUSSION

Sheet resistivity. The sheet resistance measurements indicated that the boron was essentially completely activated by the 600 °C anneal of the two-step anneal for either amorphous layer thickness. The subsequent high-temperature RIA left the sheet resistivity essentially unchanged. The boron was completely activated by the lowest RIA (950 °C) in the single-step anneal for either amorphous layer thickness.

Junction depth. The junction depths were measured for the two-step anneals only. The junction depths ranged from 0.15 to 0.2 µm for the deep amorphous layer case. For the shallow amorphous layer, the junction depths were between 0.25 and 0.30 µm, with evidence of a channeling tail below a concentration of 10^{17} cm^{-3} extending from 0.15 µm depth to the junction.

SIMS profiles. The SIMS profiles for the different splits are shown in Figs. 1-4. Significant diffusion of the boron can be seen for temperatures at and above 1050 °C. The boron profiles after the 1100 °C RIA are compared for the single-step and two-step anneals and for the shallow and deep amorphous layer cases in Figs. 5 and 6. The two-step anneal reduces the diffusion of the boron when the shallow amorphous layer is used; however, essentially no difference between the two anneals is seen with the deep amorphous layer.

The Dt product was obtained from the boron profiles for concentrations equal to 1/10 of the peak (Q=1/10) using the method outlined in Ref. [3] (Figs. 7 and 8). For comparison, the expected diffusion for a 10 sec. anneal using standard diffusion theory is also shown [4]. The activation energies are all relatively close to values in the literature [4,10]; the estimated errors in the activation energies are large because of the uncertainties in the measured profiles, particularly at temperatures at and below 1000 °C.

The magnitude of the diffusion in all of the cases shown is considerably larger than expected from standard diffusion theory. It is unlikely that this difference is related to an error in temperature measurement; a shift of the temperatures upward by 50 °C still results in a Dt product 2-4 times greater than "normal". If the diffusion occurs in much shorter times than 10 sec., as has been observed in other cases [4,5], the enhancement is even greater than that shown in Figs. 7 and 8. No measurements of the time dependence were attempted in this experiment.

The nature of the diffusion in the case of the shallow amorphous layer appears to be different from the deep amorphous layer case. Examination of the boron profiles for the shallow amorphous layer case reveals several points. First of all, some retrograde diffusion is evident near the initial amorphous/crystalline (a/c) interface for RIA temperatures at and below 1050 °C (particularly pronounced for the 600 °C, 100 min. + 950 °C, 10 sec. case). This behavior has been observed by others [2,11] and has been attributed to diffusion into dislocations immediately below the a/c interface. Secondly, the profiles of the single-step anneals show a change in slope slightly below the a/c interface, suggesting that the diffusivity of the boron is enhanced below this point. This change in slope is not evident with the two-step anneal, indicating that the enhancement is reduced by the 600 °C anneal.

The two-step anneal has no significant effect with the deep amorphous layer. This observation indicates that any enhancement in diffusion is probably not related to the crystallization of the amorphous layer.

The boron concentrations in this experiment are large enough to cause concentration enhancement of the diffusivity [9]. Concentration effects are expected to be similar in all of the cases considered, however, so that differences in behavior between the various amorphization/annealing conditions are significant. Concentration enhancement may account in part for the overall enhancement of the diffusion (figs. 7 and 8), depehding on the time in which the diffusion occurs.

CONCLUSIONS

The two-step anneal is effective in reducing the diffusion of boron below the initial amorphous/crystalline interface in the case where the amorphous layer is shallower than the junction. Where the amorphous layer extends deeper than the boron implant, no effect of the two-step anneal is seen.

378

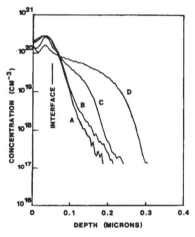

Fig. 1. Boron profiles for the shallow amorphous layer case. All wafers had a single RIA. (A) as implanted. (B) 950 °C, 10 sec. anneal. (C) 1050 °C, 10 sec. anneal. (D) 1100 °C, 10 sec. anneal.

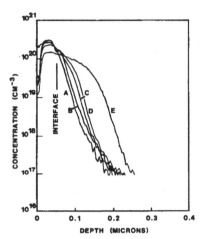

Fig. 2. Boron profiles, shallow amorphous layer case. (A) as implanted. The others are for two-step anneals of 600 °C, 100 min. + (B) 950 °C, 10 sec., (C) 1000 °C, 10 sec., (D) 1050 °C, 10 sec., or (E) 1100 °C, 10 sec.

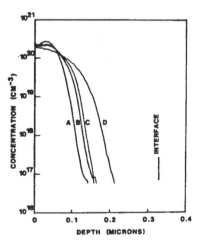

Fig. 3. Boron profiles for the deep amorphous layer case. All wafers had a single RIA. (A) as implanted. (B) 950 °C, 10 sec. anneal. (C) 1050 °C, 10 sec. anneal. (D) 1100 °C, 10 sec. anneal.

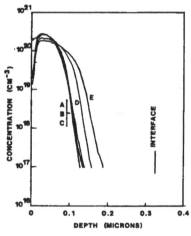

Fig. 4. Boron profiles, deep amorphous layer case. (A) as implanted. The others are for two-step anneals of 600 °C, 100 min. + (B) 950 °C, 10 sec., (C) 1000 °C, 10 sec., (D) 1050 °C, 10 sec., or (E) 1100 °C, 10 sec.

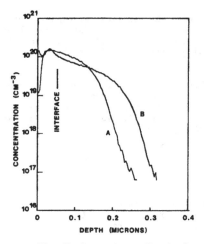

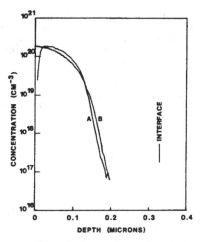

Fig. 5. Comparison of single- and two-step anneals for the shallow amorphous layer case. Both samples received a 1100 °C, 10 sec. RIA. (A) 600 °C, 100 min. + RIA. (B) RIA only.

Fig. 6. Comparison of single- and two-step anneals for the deep amorphous layer case. Both samples received a 1100 °C, 10 sec. RIA. (A) 600 °C, 100 min. + RIA. (B) RIA only.

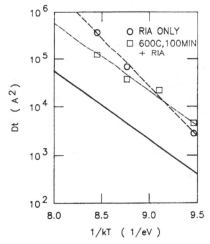

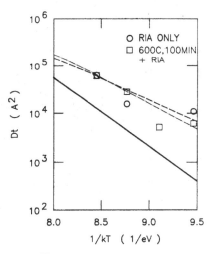

Fig. 7. Diffusion, shallow amorphous layer case. Solid line: standard diffusion theory [3]; $E_a = 3.35$ eV. Dashed line: fit to single-step anneal data; $E_a = 4.5\pm1.0$ eV. Dot-dash line: fit to two-step anneal data; $E_a = 3.1\pm1.0$ eV.

Fig. 8. Diffusion, deep amorphous layer case. The solid line is the same as in Fig. 7. Dashed line: fit to single-step anneal data; $E_a = 2.1\pm1.5$ eV. Dot-dash line: fit to two-step anneal data; $E_a = 2.4\pm1.1$ eV.

The diffusion of the boron is larger with the shallow amorphous layer than the deep amorphous layer, even for the two-step anneal. The boron diffusion is particularly enhanced in the region below the initial amorphous/crystalline interface for the single-step anneal of the shallow amorphous layer.

In all cases, the diffusion of the boron is significantly enhanced over standard diffusion theory. This enhancement may be due in part to concentration effects; however, the effect of the two-step anneal in the shallow amorphous layer case and the shapes of the diffusion profiles in that case indicate that defect interactions are significant.

ACKNOWLEDGMENTS

The authors would like to acknowledge the assistance of the staff at Charles Evans Associates, particularly Craig Hopkins, Paul Chu and Clive Jones. Jan Turner's help with the experiments is also greatly appreciated.

REFERENCES

1. T.E. Seidel, IEEE Electron Dev. Lett. EDL-4 , 353 (1983)
2. T.E. Seidel, R. Knoell, F.A. Stevie, G. Poli, and B. Schwartz, in VLSI Science and Technology/1984 , ed. by K.E. Bean and G.A. Rozgonyi, (Electrochemical Society, Pennington, NJ, 1984) 201
3. K. Yamada, M. Kashiwagi, and K. Taniguchi, Proc. 14th Conf. (1982 International) on Solid State Devices, Jap. J. Appl. Phys. 22 , Suppl. 22-1, 157 (1983)
4. T.O. Sedgwick, R. Kalish, S.R. Mader, and S.C. Shatas, Mat. Res. Soc. Symp. Proc. 23 , 293 (1984)
5. T.O. Sedgwick, S.A. Cohen, G.S. Oehrlein, V.R. Deline, R. Kalish, and S. Shatas, in VLSI Science and Technology/1984 , ed. by K.E. Bean and G.A. Rozgonyi, (Electrochemical Society, Pennington, NJ, 1984) 192
6. R.T. Hodgson, V. Deline, S.M. Mader, F.F. Morehead, and J. Gelpey, Mat. Res. Soc. Symp. Proc. 23 , 253 (1984)
7. R.B. Fair, J.J Wortman, and J. Liu, International Electron Dev. Meeting Tech. Digest, (IEEE, New York, 1983) 658
8. A. Gat, AG Associates, private communication. See also ref. [5].
9. The SIMS measurements were performed at Charles Evans Associates.
10. R.B. Fair, J. Electrochem. Soc. 122 , 800 (1975)
11. G.B. McMillan, J.M. Shannon, and H. Ahmed, IEEE Electron Dev. Lett. EDL-5, 280 (1984)

IMPURITY DIFFUSION DURING RTA

R. B. FAIR
Microelectronics Center of North Carolina, Research Triangle Park,
North Carolina 27709

ABSTRACT

Enhanced dopant diffusion during RTA depends upon whether the following physical phenomena occur individually or in combination: (1) amorphization of the Si, (2) damage-induced dislocation formation, (3) damage annealing, (4) self-interstitial trapping, (5) solubility enhancement. RTA of B in crystalline or preamorphized Si presents significantly different environments for enhanced diffusion. In preamorphized Si, enhanced B diffusion is modeled as increased B solubility following SPE. In addition, a different intrinsic diffusivity is observed which corresponds to B diffusion in preamorphized Si. Anomalous diffusion of P and As from high dose implants can be modeled with the same mechanism -- self-interstitial trapping following SPE.

INTRODUCTION

Pulse and short-time annealing of ion-implanted Si have been receiving attention as potentially useful techniques for achieving shallow junctions free of extended defects. However, several observations have been made of significant dopant diffusion occurring as a result of rapid thermal annealing (RTA) using incoherent light sources [1-4]. And, it has been observed that this enhanced diffusion occurs during the initial stage of the RTA cycle (4,5). Hence, enhanced diffusion during RTA is observed to be a transient process, and therefore requires computer modeling in order to understand the important mechanisms that are operating.

Observations of enhanced diffusion during RTA will be described in this paper, and models proposed to explain the observations. In addition to the point defect transient models previously discussed [5], two new mechanisms will be described: point defect trapping by Group V impurities and solubility enhancement of B implanted into preamorphized Si substrates. It will be shown that enhanced diffusion during RTA depends upon whether the following physical phenomena occur individually or in combination:

*amorphization of the implanted layer

*damage-induced dislocation formation

*damage annealing

*self-interstitial trapping

*solubility enhancement

OBSERVATIONS AND MODELS

Boron Implanted in Crystalline Substrates

Ion implantation of B^+ or BF_2 into crystalline Si results in B ions being scattered into <110> channeling directions. These channeled ions form exponential tails on the deep side of the implanted distributions. It has been observed that during RTA enhanced diffusion occurs in these tails [6]. It has been proposed that this diffusion is a result of interstitial boron atoms in the tail which move rapidly until they become substitutional.

Recently, Fair et al [5,7] have proposed an alternate model whereby point defects generated by dislocation movement during RTA are responsible for enhanced tail diffusion. The observations and model apply to high dose B implants where the peak implant concentration exceeds the solid solubility of B. These can be summarized as follows:

1. The B^+ implant is not sufficient to drive the Si surface amorphous. Thus, no solid phase epitaxial regrowth occurs.

2. For doses greater than $10^{15}B/cm^2$ networks of dislocations occur at the peak of the profile following annealing due to the high B concentration.

3. Point defects are generated during dislocation network stabilization which occurs within the first 30 seconds of RTA.

4. The lifetime of generated point defects is very low within the network of dislocations. This dislocation region is an efficient generator and recombination area for point defects, and as such the crystal can easily maintain point defect equilibrium.

5. The lifetime of point defects is relatively high outside of the dislocated network. Thus, generated point defects can diffuse significant distances away from the network before they recombine.

6. Enhanced B diffusion occurs outside of the dislocation network where generated point defects can reach supersaturated levels.

A schematic diagram of the situation during RTA of implanted B is shown in Fig. 1. It is assumed that the peak B concentration exceeds the solid solubility of B at the RTA temperature. The dislocation network is indicated by the cross-hatched area in the figure. It is assumed that within this area diffusion can proceed in conjunction with B precipitation, but with no point defect supersaturation [8]. However, during the first 30 seconds of the creation of this network, a significant population of point defects is generated which diffuses out into the non-dislocated crystal. These point defects contribute to enhanced diffusion outside of the cross-hatched area. The time dependence of the density, C_p, of this point defect supersaturation is represented by an equation of the form

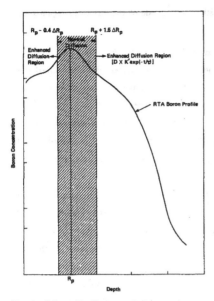

Fig. 1. Schematic diagram of the transient point defect model for B during RTA. The cross-hatched region corresponds to the implant-induced dislocation network.

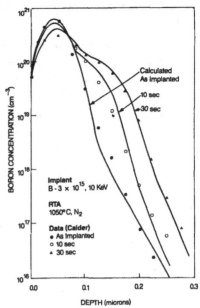

Fig. 2. Calculated and measured (Calder, et al) RTA profiles for B$^+$ implanted into crystalline Si.

$$C_{pd} = K(T)exp(-^t/\tau). \tag{1}$$

Inside the dislocated band it is assumed that

$$C_{pd} = C_{equil}, \tag{2}$$

where C_{equil} is the equilibrium concentration of point defects. The establishment of the dislocation network involves nonconservative processes such as climb which can be efficient vacancy generators. The intrinsic diffusion coefficient is multiplied by an enhancement factor of the form

$$D_B = D_i(1+K'(T)exp(-^t/\tau)). \tag{3}$$

Over the RTA furnace setpoint range of 1000-1100°C it was found that

$$K'(T) = 1x10^{-6}exp\left[\frac{2.0eV}{kT}\right], \tag{4}$$

$$\tau = 4.4sec. \tag{5}$$

giving B diffusion coefficient enhancements of 20-80 at t=0 outside of the dislocated region shown in Fig. 1.

Comparisons between measured and calculated profiles of $3\times10^{15}B/cm^2$, 10 keV into crystalline Si are shown in Figure 2. The experimental results by Calder [9] were obtained after 10 and 30 second RTA processing 1050°C. To simulate the profiles the bulk-side depth of the dislocated region is assumed to be 400Å, which agrees with measurements. In the peak region little diffusion occurs. This is modeled by assuming a 12-atom B cluster model as proposed by Ryssel, et al [10]. Since the concentration at the peak of the implant is $>>C_{sol}$, diffusion during RTA is limited by B precipitation. For x > 400Å, diffusion is assumed to be enhanced by K'(T) given in Equations (3) and (4). Agreement with the SIMS data is reasonable.

Implantation of BF_2 into crystalline Si is modeled differently from B^+ implants. No damage dislocation networks are observed at the implant peak, although some disorder is produced [11, 12]. Calculations show that no enhanced diffusion of BF_2 implants takes place beyond what can be predicted by thermally-assisted diffusion processes. This is believed due to the fact that the BF_2 implants studied were of sufficient dosage to drive the surface region in the Si amorphous. Regrowth of amorphous layers in the presence of fluorine have been shown to produce unique defect structures [13] which will change the conditions for point defect generation during RTA.

Boron Implanted in Amorphous Substrates

The defects generated when B^+ is implanted into preamorphized Si have been previously discussed [12]. The diffusion behavior of B in preamorphized Si was first studied by Crowder, et al. [14]. Boron was implanted into two wafers and one wafer was then implanted with Si to amorphize the surface. The two wafers were then annealed at 900°C for 30 minutes in a neutral furnace ambient. The resulting measured profiles are shown in Fig. 3. Diffusion of B was enhanced in the amorphized sample, even though the layer was epitaxially regrown to single crystal Si in milliseconds at 900°C. The calculated curves were determined from thermally-assisted diffusion models [15]. The intrinsic diffusion coefficients used were:

$$D_i = 1.2\times10^{-15}cm^2/sec \text{ (crystalline Si)}$$
$$D_i = 8.4\times10^{-15}cm^2/sec \text{ (preamorphized Si)}$$

The factor-of-seven difference is unexplained but is consistent with RTA studies as discussed below.

Examples of RTA of low energy B^+ implants into preamorphized Si are shown in Figures 4a and 4b. Shown are $3\times10^{15}B/cm^2$, 10keV B^+ implants annealed at 950, 1000, 1050 and 1100°C for 10 seconds [9]. The preamorphization was accomplished with a 2×10^{15} Si/cm^2, 50keV implant. The depth of this layer is estimated to be in excess of 1000Å.

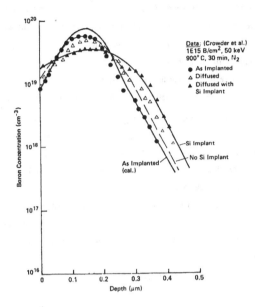

Fig. 3. Implanted B^+ diffusion in crystalline Si with and without a post Si implant

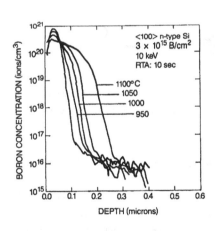

Fig. 4a. RTA of B^+ implanted into preamorphized Si data from Calder, et al

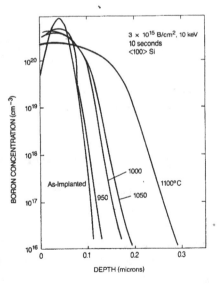

Fig. 4b. Calculated RTA profiles for the conditions in Fig. 4a

Comparison with the profiles in Figures 2a and 2b show significant differences. In the peak of the B profile, the diffusion of B is significantly greater in the preamorphized samples. At the deep side of the B profile, a more uniform broadening is observed following annealing. The calculated curves in Figure 4b were obtained by assuming:

    -- enhanced solid solubility of B

    -- no damage dislocation networks are formed at the B peak

    -- enhanced intrinsic diffusion coefficient consistent with B diffusion in preamorphized Si.

The enhanced solid solubility of B following RTA of preamorphized Si is shown in Tables I and II where room temperature sheet resistance measurements performed on samples after RTA are shown [9]. It can be seen that R_s of the preamorphized samples is significantly less for all samples except the 1050°C, 30 second sample as well as the 1100°C sample. At 1050°C, the difference in the R_s values in Table II show an exponential time delay with $\exp(-t/\tau_1)$, where $\tau_1 = 6.5$ seconds. This dependence is assumed to be due to the build up of B precipitates. Following epitaxial regrowth of the preamorphized layer it is believed that all of the B is electrically active. Thus, enhanced concentration-dependent diffusion is possible because the maximum diffusion coefficient is

$$D_{max} = D_i C_{sol}/n_i. \qquad (6)$$

The calculated curves in Figure 4b are based on this assumption and the use of the intrinsic diffusivity

$$D_i = .002\exp\left|\frac{-2.72eV}{kT}\right| \text{ (preamorphous Si)} \qquad (7)$$

Good agreement with the measured profiles is obtained. This section for D_i is very similar to the equation of Ghostagore for buried layer B diffusion [16]. The expression for D_i in crystalline Si is [17]:

$$D_i = 3.17\exp(-3.59eV/kT) \qquad (8)$$

Pennycook, et al [18] studied Si-B alloys in float-zone Si which was first amorphized by Si^+ implantation (175keV,$1.5\times10^{16}cm^{-2}$) at liquid nitrogen temperature, followed by B^+ implantation ($35keV,1\times10^{16}cm^{-2}$ or $1.5\times10^{16}cm^{-2}$). These samples were then SPE grown by furnace annealing at 575°C for 90 min. Subsequent furnace anneals were then performed at higher temperatures. The results of Pennycook's work show that B diffusion during furnace annealing of preamorphized Si does not exhibit any anomalous enhanced diffusion effects other than an enhanced D_i. Solubility enhancement only occurs during the first 10-20 seconds of the anneal, which makes this an important mechanism for RTA, but not longer time anneals. Also, no trapping of self-interstitials by B was observed by Pennycook [18], a mechanism which may have significant implications for Group V element diffusion in Si [19].

RTA of Implanted Arsenic

The defect morphology introduced by high-dose As implants is significantly different from the high dose B case. It is known that at room temperature, a 2×10^{14}As/cm^2 implant is sufficient to create an amorphous surface layer in Si [20]. In addition, such an amorphous layer can epitaxially regrow to a single crystalline layer in milliseconds at 1000°C [21]. Solid phase epitaxy (SPE) of this amorphous layer produces defect-free material, although secondary defects below the amorphous/crystalline interface may still be present [3].

Observations of enhanced diffusion of implanted As during RTA are mixed and controversial. Kalish, et al [22] observed anomalous enhanced diffusion during RTA for temperatures above 1000°C. The diffusion was characterized by a low activation energy of 1.8eV and was active during 1 second. However, at a later time this group decided that the enhanced diffusion could all be accounted for by concentration enhancement effects [23]. We have simulated their results with PREDICT with no transient enhancements [15] and conclude that some enhanced diffusion did occur in their data during RTA (see Figure 5), provided the temperature measurement was accurate.

Powell et al [24] annealed a 100keV, 1×10^{15}cm^{-2} dose As implant for 10 seconds at 1000°C and saw no measurable dopant redistribution. However, Narayan et al [25] and Fair, et al [5, 7] reported anomalous enhanced diffusion of As during RTA for doses ranging from 2×10^{14}cm^{-2} to 1×10^{16}cm^{-2}. Kwor, et al [26] also noticed enhanced diffusion during 2 second, 1170°C RTA of 1×10^{15}cm-2 As$^+$ implants.

Modeling of enhanced diffusion of As during RTA has been performed [5, 7] by overlaying a transient point defect model onto the steady state diffusion models in PREDICT [15]. These models assume As diffusion occurs primarily by V$^-$ and V^x vacancies, thus diffusion coefficients were made to be directly proportional to the vacancy transient. At high concentrations, diffusion of electrically active As is reduced by clustering, and clustering was assumed to occur instantaneously and independently of the vacancy transient. Diffusion coefficients of the form of Eq. (3) were used with

$$K(T) = 1.3\times10^{-6}(E/200)\exp\left[\frac{2.0\text{eV}}{kT}\right], \tag{9}$$

$$\tau = 4.4\text{sec}$$

388

for As doses $> 2\times10^{14}\text{cm}^{-2}$ and temperatures in the 1000-1150°C range. The parameter E is the As implant energy and is included to account for the increased width of the amorphous region with implant energy for $50 \leqslant E \leqslant 200\text{keV}$. It is assumed, unlike the case of B, that enhanced diffusion occurs over the entire profile since no dislocation network is created. Examples of calculated profiles using the transient model are shown in Figure 6. It should be pointed out that the "activation" energy associated with K'(T) is somewhat fictitious, since we are dealing with temperature transients.

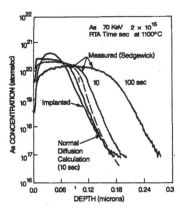

Fig. 5. Comparison of 10 sec. calculated profile (no transient RTA effects) with measured data of Sedgwick, et al.

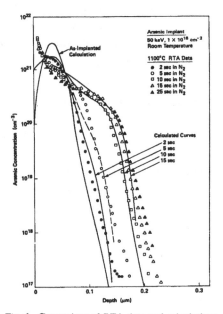

Fig. 6. Comparison of RTA data and calculations showing the time-dependent evolution of enhanced As diffusion during RTA.

The origin of enhanced diffusion of As and other Group V dopants in Si has been investigated by Pennycook, et al [18, 19]. In studying supersaturated Si alloys produced by As and Sb implantation with subsequent low temperature SPE, they observed enhanced diffusion during furnace annealing. In particular, for the Sb implants, they observed the formation of interstitial loops concurrent with enhanced Sb diffusion. These results are shown in Figure 7. In similar alloys produced by liquid-phase epitaxial regrowth with a pulsed laser, much lower diffusion coefficients were obtained and no loops were observed.

The authors concluded that the SPE samples produced excess Si self-interstitials which assisted Sb diffusion via an interstitialcy mechanism and which also condensed into the observed loops. Interstitials were trapped at the Sb atoms during SPE regrowth. It was proposed that Sb atoms are electronically saturated in the amorphous Si, having five neighbors. As the crystallization interface passes, trapped self-interstitials are created which results in supersaturation. The Sb diffusion activation energy from Figure 7 is 1.8eV which corresponds to the migration enthalpy of Sb via the interstitialcy mechanism.

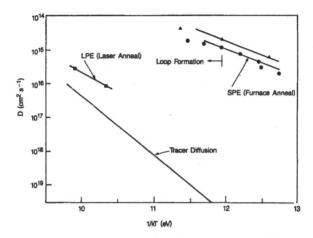

Fig. 7. Antimony diffusion in Si deduced from 20 minute anneals - SPE grown and LPE grown implanted samples (Pennycook, et al).

Transient effects with As implanted samples have also been shown to have a 1.8eV activation energy [22]. If Eq. (9) is used in Eq. (3) with $D_i = 22.9\exp(-4.1eV/kT)$, and using for As

$$D \alpha D_i K'(T) n/n_i, \qquad (10)$$

then the activation energy for D is 1.43eV for $n > n_i$ and 2.1eV for $n < n_i$, in reasonable agreement.

To study the effect of SPE on RTA, a 550°C anneal for 30 min. was performed on as-implanted samples. Such an anneal is sufficient to regrow approximately 90% of the amorphous layer associated with a high dose As implant (27). The effect of a low temperature, recrystallization anneal on enhanced diffusion of As is demonstrated in Fig. 8. A $1 \times 10^{16} As/cm^2$, 50keV implant was performed in two Si wafers. One wafer was preannealed at 550°C for 30 min. in N_2. RTA was then performed on both wafers at 1100°C for 10 sec. It can be seen that the effect of the preanneal was to reduce enhanced diffusion to a level that can be accounted for by thermally-assisted diffusion. Kwor, et al [26] reported no reduction in enhanced As diffusion by the preanneal step.

The effective temperature required to account for the observed level of diffusion (ignoring transient effects) in Fig. 8 is 1130°C. In addition, the preannealed sample should heat up to a higher temperature than the other sample due to higher free carrier absorption of the SPE grown layer [11]. Nevertheless, the sample with no preanneal showed the enhanced diffusion.

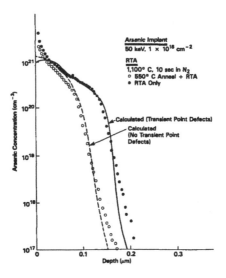

Fig. 8. Comparison of RTA data and calculations for 550°C preannealed and no preannealed As-implanted Si

RTA of Implanted Phosphorus

The diffusion of ion-implanted P in Si during RTA has been reported by Oehrlein [28]. The existence of enhanced diffusion depends upon the P dose. At high doses $(2\times10^{15}\mathrm{cm}^{-2})$ enhanced diffusion is observed as shown in Fig. 9 for 1050°C anneals at 1 sec. - 100 sec. The calculated curves are based upon the transient diffusion model for As. This model with $K(T)$ given in Eq. (9) was overlaid on the thermal diffusion models for P with the good results shown. For comparison, the calculations for a 30 min. furnace anneal are compared with data in Fig. 9.

For the low dose case Oehrlein, et al studied $1\times10^{14}\mathrm{P/cm}^2$ implants. They found a profile redistribution at 900°C for RTA times of 10 sec., but which was temperature independent in the 800-1150°C range. Since this dose is below that required to amorphize the surface, no enhancement mechanism discussed so far applies. However, this effect is consistent with transient enhanced diffusion caused by damage annealing. Such an effect would be enhanced as the anneal temperature is reduced.

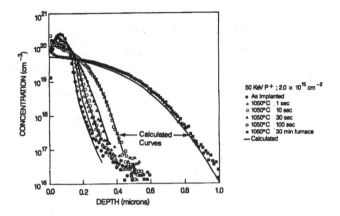

Fig. 9. Comparison of calculated and measured time dependence of RTA of phosphorus profiles (data of Oehrlein, et al).

CONCLUSIONS

Modeling the diffusion of As, P and B during RTA presents a unique challenge, since point-defect generation mechanisms are believed to be operating during the short annealing intervals. In the case of high-dose B implants into crystalline silicon, damage clusters created by the high concentration of B in the peak of the implant distribution grow into well-defined cross-grid dislocation networks during the first 10-30 seconds of the anneal. Quantifying the concentration of point defects generated during this process requires knowledge of the types of dislocation reactions, climb processes, and cross-slip processes that may occur simultaneously. For the case of B implantation into amorphous Si, no damage dislocation networks are formed. Enhanced B diffusion has been modeled as enhanced solid solubility of B following SPE. In addition, the intrinsic diffusion coefficient of B is larger in Si that was preamorphized. No satisfactory explanation exists at this time.

Enhanced diffusion of P and As occurs by a similar mechanism when the implant dose is sufficient to amorphize the surface. Diffusion calculations were performed by overlaying a transient point defect model on thermally-assisted diffusion models associated with steady-state furnace processes. A weighted exponential function was assumed to describe the transient process, and diffusion coefficients were directly multiplied by this function. Numerical integrations were performed over 200 millisecond constant temperature intervals to approximate the RTA environment.

The mechanism whereby high dose Group V dopant diffusion is enhanced is believed to be self-interstitial trapping following SPE. An analogous vacancy trapping may occur in the case of B implanted into preamorphized Si. However, this conjecture needs to be investigated.

392

REFERENCES

1. J.L. Benton, G.K. Celler, D.C. Jacobson, L.C. Kimberling, D.J. Lischner, G.L. Miller and Mc. D. Robinson, *Mat. Res. Soc. Sym. Proc.*, "Laser and Electron Beam Interactions with Solids", eds. B.R. Appleton and G.K. Celler, vol. 4, p. 765 (Elsevier Publishing Co., 1982).

2. R.B. Fair, J.J. Wortman, J. Liu, M. Tischler, N.A. Masnari, and K.Y. Duh, Device Research Conference, Burlington, VT, 1983.

3. D.K. Sadana, S.C. Shatas and A. Gat, *Proc. of Microscopy of Semiconducting Materials*, Inst. of Phys. London, 1983 (in press).

4. R. Kalish, T.O. Sedgwick, S. Mader, S. Shatas, *Applied Phys. Lett.*, 44, 107 (1984).

5. R.B. Fair, J.J. Wortman and J. Liu, *1983 International Electron Device Meeting, Tech. Dig.*, p. 658 (1983).

6. R.T. Hodgson, V. Deline, S.M. Mader, F.F. Morehead, and J. Gelpey in "Energy Beam-Solid Interactions and Transient Thermal Processing" (ICC Fan and N.M. Johnson, eds. New York) 1984.

7. R.B. Fair, J.J. Wortman and J. Liu, *J. Electrochem. Soc.*, 131, 2387 (1984).

8. V.A. Panteleev, R.S. Baryshev, L.V. Lainier, A.G. Zinina and E.F. Pakkutina. Fiz. Tuerd. Tela, 16, 502 (1974)

9. I.D. Calder, H.M. Nagub, W. Vandervorst, D. Houghton, and F.R. Shepherd, Materials Research Society, Fall 1984 Meeting, paper A5.4

10. H. Ryssel, K. Hoffman, F. Haberger, G. Prinke and K. Müller, Meeting of the Electrochemical Society, Spring 1980, St. Louis (unpublished).

11. T.E. Seidel, R. Knoell, F.A. Stevie, G. Poli and B. Schwartz in "VLSI Science and Technology/1984," (eds. K.E. Bean and G.A. Rozgonyi, The Electrochem. Soc., Pennington, N.J. 1984) p. 201.

12. D.K. Sadana, private communication

13. W. Maszara, C. Carter, D.K. Sadana, J. Liu, V. Ozguz, J. Wortman and G.A. Rozgonyi, Proc. of the Materials Research Society; Energy Beam-Solid Interactions and Transient Processing, Boston, Mass., Nov. 13-17, 1983, p. 303.

14. B.L. Crowder, J.F. Ziegler and G.W. Cole, Ion Implantation in Semiconductors and Other Materials (Yorktown Heights), Plenum, New York, 1973, p. 257.

15. PREDICT computer program - PRocess Estimater for the Design of IC Technologies.

16. R.N. Ghostagore, *Phys. Rev. B.*, 389 (1971).

17. R.B. Fair, *J. Electrochem Soc.*, 122, 800 (1975).

18. S.J. Pennycook, J. Narayan and O.W. Holland, to be published.

19. S.J. Pennycook, J. Narayan, and O.W. Holland, *J. Appl. Phys.*, 55, 837 (1984).

20. F.F. Morehead and B.L. Crowder in *First Inter. Conf. on Ion Implantation, Thousand Oaks*, eds. F. Eisen and L. Chadderton (Gordon and Breach, New York, 1971) p. 25.

21. L. Csepregi, J.W. Mayer and T.W. Sigmon, *Appl. Phys. Lett*, 29, 92 (1976).

22. R. Kalish, T.O. Sedgwick, S. Mader and S. Shatas, *Appl. Phys. Lett.*, 44, 107 (1984)

23. T.O. Sedgwick, S.A. Cohen, G.S. Oehrlein, V.R. Deline, R. Kalish and S. Shatas, in "VLSI Science and Technology/1984, (eds. K.E. Bean and G.A. Rozgonyi, The Electrochem. Soc., Pennington, N.J. 1984) p. 192.

24. R.A. Powell, T.O. Yep and R.T. Fulks, *Appl. Phys. Lett.*, 39, 150 (1981).

25. J. Narayan, O.W. Holland, R.E. Eby, J.J. Wortman, V. Ozguz and G.A. Rozgonyi, *Appl. Phys. Lett.*, 43, 957 (1983).

26. R. Kwor, D.L. Kwang, Y.K. Yeo, *Appl. Phys. Lett*, 45, 77 (1984).

27. J.S. Williams and R.G. Elliman, *Appl. Phys. Lett.*, 37, 829 (1980).

28. G.S. Oehrlein, S.A. Cohen and T.O. Sedgwick, *Appl. Phys. Lett.*, 45, 417 (1984).

SOLID-PHASE EPITAXIAL REGROWTH OF ION-IMPLANTED SILICON ON SAPPHIRE USING RAPID THERMAL ANNEALING

A M HODGE, A G CULLIS AND N G CHEW
Royal Signals and Radar Establishment, St. Andrews Road, Malvern, Worcs.
WR14 3PS, UK

ABSTRACT

Solid phase epitaxial regrowth of silicon on sapphire is used to improve the quality of as-received silicon films prior to conventional device processing. It has been shown that this is necessary, especially for layers of 0.3μm and thinner, if the full potential of this particular silicon on insulator technology is to be realised. Si^+ ions are implanted at an energy and dose such that all but the surface of the silicon film is rendered amorphous. In this study, the layer is regrown using a rapid thermal annealer operated in the multi-second regime. A second shallower implant followed by rapid thermal annealing produces a further improvement. Characterisation of the material has been principally by cross-sectional transmission electron microscopy. The structures observed after different implant and regrowth treatments are discussed.

INTRODUCTION

Silicon on sapphire (SOS) is recognised as the most advanced technology currently available for the production of VLSI silicon on insulator circuits. If smaller, faster devices are to be produced, thinner epitaxial silicon will be required. However, the quality of thinner layers (less than about 0.3μm), is inferior principally because of defects close to the interface arising from lattice and thermal expansion mismatch between the silicon film and sapphire substrate. Various techniques have been devised to improve the silicon quality after deposition (for example [1,2]). Solid-phase epitaxial regrowth (SPEG) involves implanting Si^+ ions in the silicon with a peak projected range such that all but the surface of the single crystal film is rendered amorphous, followed by regrowth from this good surface silicon by a suitable heat treatment. In this study we have used rapid thermal annealing in place of a conventional longer furnace treatment both to make the process compatible with the use of transient annealing in subsequent device fabrication and to minimise the risk of aluminium diffusion from the sapphire into the silicon film. This problem of auto-doping has been reported by other workers [3,4] and techniques have been devised to minimise it [5,6]. Double solid-phase epitaxial regrowth (DSPEG) [7] has also been investigated whereby amorphous silicon produced by a second shallower Si^+ implant is regrown. The quality of the various silicon films has been investigated at each stage by cross-sectional transmission electron microscopy.

EXPERIMENTAL

Commercially produced 3" SOS wafers supplied by Union Carbide were used throughout this study. The epitaxial silicon film of (100) orientation was nominally 0.3μm thick. $Si29^+$ ions were implanted into the as-received wafers with energies approaching 200keV in a Varian 350D ion implanter. Ion doses were set at values between 5E14 and 5E15cm-2. The wafers were held at room temperature during implantation. The $Si29^+$ isotope was used in order to minimise the possible implantation of contaminant species with the same mass. The conditions were chosen so that the peak range of the implanted ions lay close to the silicon-sapphire interface and that the silicon film thickness was rendered amorphous except for a surface layer. After implantation the wafers were annealed in argon at a temperature between 600 and 1000°C in a Heatpulse incoherent light source rapid thermal annealer (A G Associates). The isothermal anneal period was 10s in most experiments. DSPEG requires a second implant after the first anneal. For this 1E15 Si29+ ions cm-2 were implanted at either 100 or 120keV. The layers were regrown using the same conditions as for the first anneal.

The structure of both implanted and regrown samples was examined in a transmission electron microscope (TEM) operated at 120keV: cross-sectional specimens were thinned to electron transparency by mechanical polishing followed by low voltage argon ion milling. For comparative purposes, certain layers were regrown in argon using a conventional furnace at 900°C for 30min. In addition, the effect of conventional furnace treatment for device fabrication after DSPEG was also investigated. The aluminium content of the recrystallised films was assessed by SIMS analysis of selected samples.

RESULTS

The structure in an as-received 0.3μm wafer is shown in a cross-sectional micrograph in figure 1; 2 sets of inclined microtwins on {111} planes emanating from the sapphire interface and decreasing in density away from it are clearly apparent. Defects close to the sapphire interface lie in the active device areas, whereas in thicker films devices are fabricated in less defective material away from the interface.

The Si+ ion doses and energies were chosen to amorphise the film in a considerable part of its depth. Figures 2, 3 and 4 show cross-sectional micrographs of various implanted wafers. In figure 2a most of the silicon has been amorphised with only an extremely thin surface layer remaining completely crystalline. On annealing such a structure, regrowth from the surface is very irregular and for the 10s anneals in these experiments, recrystallisation is incomplete, as illustrated in figure 2b. The opposite extreme is exemplified in figure 3. The amorphous silicon formed during implantation did not reach the sapphire interface or the external silicon surface so that during regrowth, epitaxy occurs from crystalline silicon remaining on both sides. Complete regrowth of the layer occurs within 10s at 800°C but the layer is still traversed by numerous twins emanating from both surfaces.

Figure 1: Cross-sectional transmission electron micrograph of an as received SOS wafer. Two sets of microtwins on {111} planes are in contrast showing that they emanate from the sapphire interface but decrease in density away from it.

Figure 2: Micrographs of an implanted silicon film: (a) as implanted, (b) regrown at 800°C for 10s. Recrystallization is incomplete because of the implant conditions.

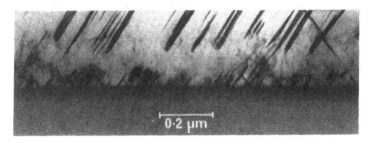

Figure 3: Micrograph of a silicon film, regrown at 800°C for 10s, where the implant used was such that residual crystalline silicon remained at the sapphire interface.

The conditions for implantation and annealing were optimised for the sample in figure 4. The as-implanted material is shown in figure 4a - the silicon has been amorphised to the sapphire interface with a layer of the original single crystalline material (with a few twin remnants) left at the surface. Epitaxial regrowth proceeds from this crystalline seed layer so that (as shown in figure 4b) the twin density close to the sapphire interface has been reduced considerably in comparison with as-received wafers. However, the few twins remaining in the crystalline layer have propagated through the film during regrowth. A second implant and anneal is therefore used to reduce the number of these surface defects.

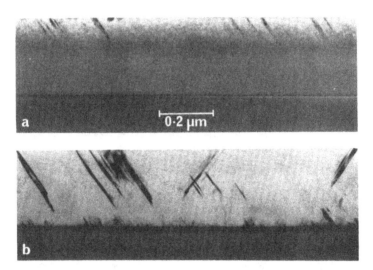

Figure 4: Micrographs of a silicon film with optimised first implant: (a) as implanted, (b) regrown at 800°C for 10s.

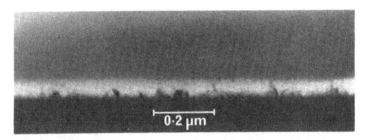

Figure 5: Micrograph of a silicon film improved by a deep implant and regrowth (as shown in figure 4b) and subsequently subjected to a second shallower implant for DSPEG.

The second implant is such that the surface silicon is amorphised but the improved silicon film closest to the sapphire interface (about a quarter of the total film thickness) is unaffected, as shown in figure 5. Micrographs of the regrown material are shown in figures 6 and 7. It is evident from figure 6 that the layers have regrown relatively free of twins. However, as shown in figure 7 (the same specimens but with different TEM imaging conditions), the layers contain dislocation arrays emanating from the substrate interface. In addition, for the sample implanted at 120keV (figures 6a and 7a) there are some residual defects at the silicon surface indicating that the implant range was slightly too great. Reducing the energy to 100keV (figure 6b and 7b) produced the best quality material observed and in this case there are no surface defects. For comparison, figure 8 shows a wafer subjected to the same optimised implant conditions but regrown after each in a conventional furnace. No significant differences have been observed.

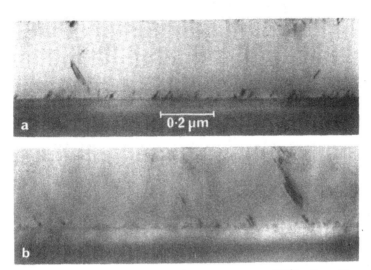

Figure 6: Micrographs of silicon films regrown at 800°C for 10s after a shallow implant for DSPEG at energies of: (a) 120keV, (b) 100keV. The few residual twins are in contrast.

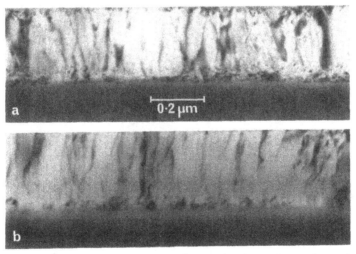

Figure 7: Micrographs of the same samples as in figure 6 but with the TEM conditions optimised to reveal the presence of dislocations: (a) 120keV implant, (b) 100keV implant.

The micrographs show that rapid thermal annealing results in relatively low defect density single crystalline silicon on sapphire. The amount of aluminium in the silicon after the ion implantation and regrowth must also be considered. SIMS analyses of selected samples are therefore shown in figure 9. Curve (i) is for an as-received wafer while (ii) and (iii) are

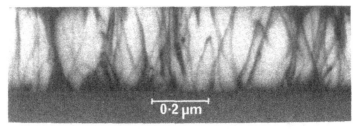

Figure 8: Micrograph of a silicon film with optimised first and second implants but regrown after each in a conventional furnace at 900°C for 30min.

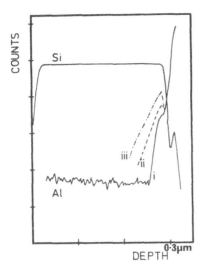

Figure 9: SIMS analyses of selected implanted and annealed samples. (i)as-received wafer, (ii) DSPEG wafer with both anneals in the rapid annealer, (iii) DSPEG wafer annealed twice in a conventional furnace.

after DSPEG. Some aluminium has migrated into the film but the amount is considerably less following rapid thermal (ii) than for conventional furnace (iii) treatment. SIMS analyses of the wafers subjected to furnace treatments during device fabrication after DSPEG show little further migration of the aluminium.

DISCUSSION

The results have shown that solid phase epitaxial regrowth using rapid thermal annealing produces good quality single crystalline SOS. Commercially produced 0.3μm films contain high densities of twins emanating from the silicon-sapphire interface (figure 1). These lie in the active device regions and degrade their performance. In comparison figure 6b shows the material produced by optimised implants and regrowth. The great reduction in the number of twins is clearly apparent although dislocation arrays remain (figure 7b). It is, however important to note that only slight changes in implant range or dose will affect the results significantly. Too great a surface dose during the first implant can render most or all of the film amorphous so that recrystallisation is poor (figure 2) or, conversely, too low a dose does not amorphise the film to a sufficient depth so regrowth occurs from both surfaces with little reduction in twin density (figure 3). Relatively twin-free silicon can be obtained by the use of optimised implant conditions. Further work is in progress to reduce the number of dislocations remaining in the material.

The control of implant energy and dose cannot, however be considered in isolation - they are only optimised for a particular epi-layer thickness. The wafers supplied by the major manufacturers of SOS quote a thickness tolerance of +/-10% in their standard specification. In this study we have seen that layer·thicknesses do vary within the limits of the specification both within one layer and also from wafer to wafer. This means that implant dose and range may be optimised over only part of the wafer. It is also important to control the temperature of the wafer during implantation - lower doses are required for amorphisation at reduced temperatures and may be beneficial [6].

Cross-sectional transmission electron microscopy is an essential technique for the assessment of the crystalline quality of the silicon film but is unsuitable for routine analyses. An alternative non-destructive, rapid but sensitive method is required for characterising SOS films routinely prior to device fabrication. Techniques such as UV reflectometry [8] or X-ray pole figure analyses [9] may be suitable.

CONCLUSIONS

We have shown that DSPEG of SOS using rapid thermal annealing after each amorphisation implant is an effective way of producing good quality, relatively twin free material prior to the production of devices. The aluminium diffusion from the sapphire substrate into the silicon is considerably reduced when using the rapid heating in place of longer furnace treatments during DSPEG. In order to achieve this improvement, the implant conditions and epi thickness uniformity must be controlled.

ACKNOWLEDGEMENTS

The authors are grateful to Mr F G H Smith for preparing the ion implanted specimens and Mr D Sykes of Loughborough Consultants who performed the SIMS analyses.

REFERENCES

1. S.S.Lau, S.Matteson, J.W.Mayer, P.Revesz, J.Gyulai, J.Roth, T.W.Sigmon and T.Cass, Appl. Phys. Lett., 34, 76 (1976)
2. I.Golecki and M-A.Nicolet, Solid State Electr., 23, 803 (1980)
3. I.Golecki, G.Kinoshita and B.M.Paine, Nucl. Instr. and Methods, 182/183, 675 (1981)
4. R.E.Reedy and T.W.Sigmon, J. Cryst. Gr., 58, 53 (1982)
5. R.E.Reedy, T.W.Sigmon and L.A.Crystel, Appl. Phys. Lett., 42, 707 (1983)
6. I.Golecki, R.L.Maddox and K.M.Stika, J. Electr. Mat., 13, 373 (1984)
7. T.Yoshi, S.Taguchi, T.Inoue and H.Tango, Jap. J. Appl. Phys., 21, Suppl. 21-1, 175 (1982)
8. M.T.Duffy, J.F.Corboy, G.W.Cullen, R.T.Smith, R.A.Soltis, G.Harbeke, J.R.Sandercock and M.Blumenfield, J. Cryst. Growth, 58, 10 (1982)
9. R.T.Smith and C.E.Weitzel, J. Cryst. Growth, 58, 61 (1982)

IMPROVING THE QUALITY OF A HETEROEPITAXIAL CaF$_2$ OVERLAYER BY RAPID POST ANNEALING

Loren Pfeiffer, Julia M. Phillips, T. P. Smith, III*, W. M. Augustyniak, K. W. West
AT&T Bell Laboratories, Murray Hill, NJ 07974

ABSTRACT

We show that post anneals of short duration at high temperature can markedly improve the quality of CaF$_2$ films grown by molecular beam epitaxy (MBE) on Si(100). Anneals at 1100°C for 20 sec in an Ar ambient improved χ_{min}, the ratio of backscattered 1.8 MeV He^4 ions in the aligned to random direction, from as-grown values of .07 to .26, to post post-anneal values of .03 to .045. This is the best χ_{min} yet reported for the CaF$_2$:Si system. The post-annealed films also show improved resistance to chemical etching and mechanical stress, and increased dielectric breakdown voltages.

The heteroepitaxial quality of molecular beam deposited CaF$_2$ films on Si(100) has been shown to be a sharply peaked function of the substrate temperature during the deposition.[1] This is illustrated by the filled circles of Fig. 1 which show the aligned minimum Rutherford Backscattering (RBS) ratio of these as-grown films as a function of the deposition temperature. Deviations of as little as 25°C from the optimum seriously degrade the heteroepitaxial quality. The sharpness of the optimum is believed to arise from a tradeoff of the following temperature dependent effects. At low growth temperatures the molecular mobility of CaF$_2$ on Si(100) is insufficient to ensure good CaF$_2$:Si heteroepitaxy; on the other hand if the growth temperature is too high the CaF$_2$ tends to form islands on the Si substrate, or worse can react chemically with it during the growth. These effects also degrade the heteroepitaxy.

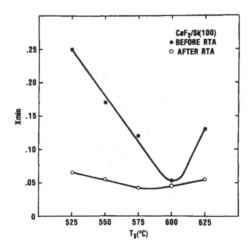

Fig. 1. Minimum backscatter yield before and after post annealing as a function of substrate temperature during MBE growth.

These considerations suggest that a very brief high temperature anneal after growth might be useful if it could heat the sample to a temperature sufficiently high for a long enough time to increase the atomic mobility and improve the epitaxy, but not long enough[2] to activate chemical reactions. In this paper we show there does exist at least for CaF_2:Si(100) a set of time and temperature values for which post growth annealing is useful.

Figure 2 shows random and (100) aligned RBS and channeling spectra from a 5000Å MBE CaF_2 film over Si(100) before and after a post anneal at 1100°C for 20 sec in Ar gas. The channeling minimum yield, χ_{min}, improved from .26 to .03 as a result of the post anneal. As measured by channeling the heteroepitaxy of the post annealed film is nearly ideal, except for a slight indication of dechanneling at the back interface of the CaF_2 film. This dechanneling may be evidence of disorder of the CaF_2:Si interface possibly due to strain induced by the differential thermal contraction of the two materials as the sample cooled after the post anneal.

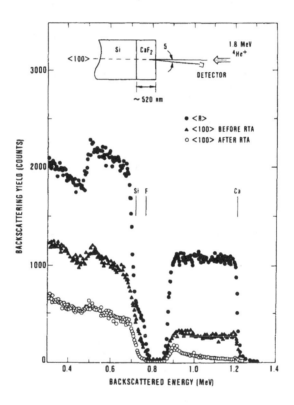

Fig. 2. RBS Spectra of CaF_2 MBE films on Si(100) before and after a rapid post anneal treatment.

Figure 1 shows the χ_{min} of several samples before and after the 20 sec 1100°C post anneal, plotted as a function of the original CaF_2 growth temperature. In all cases the χ_{min} of the post annealed films is improved, but the improvement is especially striking for films grown under non-ideal conditions. This ability to repair films of initially poor epitaxy is useful because it allows some relaxation of the sensitive conditions during MBE growth.

As might be expected the success of the post anneal is a sensitive function of the peak annealing temperature, the time at that temperature and the ambient gas. Our initial experiments were done at 10^{-6} torr in a graphite strip heater chamber using a base temperature of 770°C, a graphite upper heater wire temperature of 2240°C, and a wire scan speed of 1.5 min/sec. Strip heater post annealing also improved the CaF_2:Si(100) epitaxy; it also produced films with values of χ_{min} as low as .03. However, because the directionality and speed of the scanned line source proved to have no effect on the heteroepitaxy, subsequent experiments were done in a commercial AG 210 flash lamp machine[3] using an Ar gas ambient.

The purity of Ar ambient proved to be important. Contamination of the Ar with as little as 1% air during the high temperature processing induced an oxygen reaction at the CaF_2:Si interface and ruined the stoichiometry of the adjacent material as well as the epitaxy. Short anneals in pure Ar gas for which the 1100°C peak temperature was maintained less than 10 sec showed less improved epitaxy presumably due to insufficient mobility of the CaF_2 atoms. Anneals in pure Ar gas held at 1100°C for 30 sec or longer showed evidence of chemical reactions between the Si and the CaF_2 at the interface.

The morphology of the films before and after annealing is shown in Fig. 3. The as-

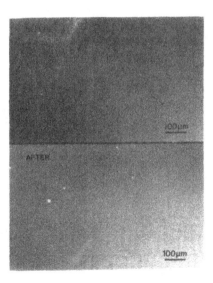

Fig. 3. Surface Morphology of CaF_2 MBE films on Si(100) before and after post anneal.

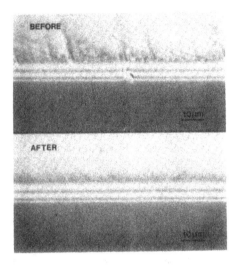

Fig. 4. Angle Lap (10:1) of epitaxial CaF_2:Si(100) before and after post anneal.

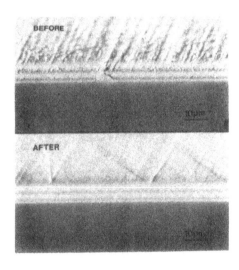

Fig. 5. Defect etched angle lap (10:1) of MBE CaF_2:Si(100) before and after post anneal.

grown films are uniform, but show a definite micro-roughness in Nomarski contrast. After annealing this micro-roughness disappears but a system of lines in an ashlar pattern running along <110> and <101> appears on the sample surface in Nomarski contrast. These lines may be stress cracks due to the differential thermal contraction after the anneal. However, if these are cracks, we note they seem to have no deleterious effect on the observed dielectric breakdown properties of the films. (See discussion of electrical properties below.)

Figure 4 is a pair of Nomarski photomicrographs of 10 to 1 angle lap sections of the film before and after the post anneal. During the preparation of this sample we observed that the post-annealed film is clearly more mechanically robust. The as-grown sample sometimes chipped or eroded at the exposed angle lapped edge during mechanical polishing; the post annealed film did not. Both films stood up to the final Syton polish well. The sections through the films show a residual roughness in the as-grown film when compared to the annealed film, indicating that the as-grown films is in some way inhomogeneous. This is brought out more clearly in fig. 5 which shows these same 10:1 angle sections after a chemical defect etch of 90 sec using $HC\ell:H_2O$ at 4:1. We believe this effect is due to misoriented CaF_2 microcrystalites embedded in the as-grown film which chemically etch at a different rate than the surrounding (100) film.

Figure 6 shows the 1 MHz capacitance and conductance of an $A\ell-CaF_2-Si$ capacitor made from these post annealed films. As expected, the capacitance drops and the conductance peaks as the surface of the semiconductor is driven from accumulation to depletion. From these measurements the surface state density is calculated to be $\sim 5 \times 10^{11}/cm^2$ eV. This is typical of most good films prior to RTA, indicating that post annealing does not significantly change the electronic properties of the CaF_2-Si interface. There is some hysterisis in the capacitance-voltage characteristic if the bias is changed slowly. This is probably due to ionic transport in the CaF_2.[4] We note, however, that the RTA films are as good or better than unannealed films in this regard.

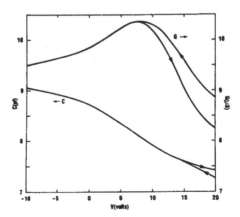

Fig. 6. Capacitance-Voltage and Admittance-Voltage characteristics of post annealed $CaF_2:Si(100)$.

The breakdown voltage of the RTA films is in general better than the as-grown films. Most unannealed films of this thickness can support only $2\text{-}5 \times 10^5$ V/cm, however, RTA films are typically able to sustain fields of $1\text{-}2 \times 10^6$ V/cm. This observation remains true even for annealed films that are known to have ashlar lines as in fig. 3 running across the breakdown test regions.

We conclude that a rapid post anneal treatment of MBE grown CaF_2 films on Si(100) is indeed useful. With the exception of the possible microcracking shown in fig. 3, all measured properties of the post-annealed films are at least as good as, and often clearly superior to the as-grown starting films. The CaF_2:Si system is not as far as we know especially or uniquely suited to this post anneal treatment. We believe the improvement in epitaxial quality that we report here is likely to be a quite general effect. The generalized prescription would be to deposit an overlayer of mediocre epitaxial quality at a substrate temperature low enough to avoid all possible chemical reactions. The epitaxial quality of these films would then be improved in a post anneal of suitable duration and peak temperature.

REFERENCES

* Permanent address Department of Physics, Brown University, Providence, R.I. 02912.

1. J. M. Phillips, Paper D3.1 of this conference. Symposium on Layered Structures, Epitaxy and Interfaces.

2. This idea is somewhat similar in spirit to some recent RTA-damage removal experiments. See T. E. Seidel, D. J. Lischner, C. S. Pai, R. V. Knoell, and D. M. Maher in Nucl. Instr. and Meths. March (1985). In that work there are also two competing processes with different activation energies. The intention is to use RTA to remove crystal damage by promoting Si to Si epitaxy, without bringing up the sample temperature long enough to allow substantial redistribution of an As dopant.

3. The Model AG Heatpulse 210 is manufactured by AG Associates, Inc., Pal Alto, CA 94303.

4. R. People, T. P. Smith, III, J. M. Phillips and W. M. Augustyniak, Paper D3.5 of this conference. Symposium on Layered Structures, Epitaxy and Interfaces.

TRANSIENT ANNEALING OF NEUTRON-TRANSMUTATION-DOPED SILICON

C J POLLARD*, K G BARRACLOUGH** and M S SKOLNICK**
*VLSI Technology Division, British Telecom Research Laboratories, Martlesham Heath, Suffolk, England
**Royal Signals and Radar Establishment, Malvern, Worcestershire, England

ABSTRACT

Scanning electron beam annealing has been used to study the annealing of NTD silicon in the temperature range 600-1180°C. The annealing process has been found to be very rapid above 800°C, a 30 second anneal producing highly uniform n-type material. Below this temperature, an initial drop followed by a progressive rise in sample resistivity with increasing anneal time is consistent with phosphorus dopant activation and the formation of a defect-related acceptor complex.

INTRODUCTION

Neutron Transmutation Doping (NTD) is now a well established technology for the production of uniform n-type phosphorus-doped silicon material. The radiation damage inherent in the process is normally annealed at temperatures above 750°C using conventional furnaces to produce silicon ingots suitable for the manufacture of power devices such as thyristors. However, the mechanisms involved in the annealing of the radiation-induced defects are not completely understood, and are dependent on the the interaction of many factors such as the neutron dose, the thermal-to-fast neutron ratio (cadmium ratio), the annealing time and temperature.

The doping technique has associated with it a number of damage mechanisms (1). These arise from both the initial irradiation recoil events from both slow and fast neutrons (the ratio of which depends on reactor type), and radiation damage produced by energetic decay products emanating from the subsequent nuclear reaction. Crystal damage will therefore range from single atom displacements to large and relatively stable complex damage structures, and the thermal excursion required to anneal these fully will vary accordingly.

From this, it is evident that, unlike ion implantation damage, several classes of fundamentally different crystal defect will co-exist in the crystal following neutron irradiation, each of which is likely to have different annealing characteristics.

The purpose of the series of experiments described here was to investigate the early stages of annealing of NTD silicon. The temperatures and times were chosen to allow the characteristics of partially annealed samples to be examined with the aim of establishing the minimum anneal conditions necessary for the production of device-worthy material.

SAMPLES

The material examined in this series of experiments were Syton polished wafers cut from a dislocation-free 30-34mm diameter <111> single crystal silicon ingot grown by the vertical float-zone method. Pre-irradiation resistivity was in the range 1 - 1.5 kohm-cm. Infra-red absorption measurements indicated an interstitial oxygen content of $<10^{16}$ cm^{-3} (standard: DIN 50438/1) and substitutional carbon content of 5 x 10^{15} cm^{-3} (ASTM F123-74). The ingot was irradiated, following a sulphuric/peroxide clean and HF dip, in a reactor with a cadmium ratio of 800:1. The neutron dose for these experiments was 9.65 x 10^{17} cm^{-2} to give a final phosphorus concentration of 1.79 x 10^{14} cm^{-3}. This corresponds to a final bulk resistivity of 28 ohm-cm, which was confirmed by a control furnace anneal at 1000°C for one hour in argon. Uniformity of the slice was also measured, and found to be better than 5% over the whole wafer area.

EXPERIMENTAL

In order to reduce the errors inherent in short period furnace annealing due to the uncertainties in the heating and cooling stages of the thermal cycle, the technique of scanning electron beam annealing (SEBA) was used for these experiments. This method, described in detail elsewhere (2), employs a high power electron beam which is scanned in a high frequency raster pattern over the surface of the sample, which is held in thermal isolation in vacuum, thus raising the temperature of the whole sample to the required temperature very rapidly. This technique allows very short anneal times to be studied repeatably under clean conditions. Following heat treatment, the samples were examined using four point probe and spreading resistance measurements to assess the electrical activation. Photoluminescence (PL) spectra were taken from selected samples to assess changes in the relative intensities of damage related and phosphorus-bound exciton PL features as a function of anneal time and temperature. Photoluminescence was measured at 4.2K excited by 100mW of 488nm light from an argon ion laser.

The experiments were split into two groups. The aim of the initial group was to assess the minimum anneal conditions required for full conversion of the irradiated material to its target resistivity. In order to achieve this, the samples were subjected to annealing treatments ranging from 600°C to 1180°C for times of 10 to 30 seconds. Samples were considered to be fully annealed when the resistivity and uniformity of the samples matched that of the furnace controls. The time taken to reach 95% of the target temperature did not exceed 5s in any of the anneals, and radiative cooling of the wafer to less than 400°C was achieved at rates in excess of 150Ks^{-1}.

The second group of experiments was designed to investigate the progression of the annealing process as a function of time for temperatures ranging from 650 - 815°C. Wafers were heated to a peak temperature in this range for 15 seconds, allowed to cool in vacuum, removed from the annealing chamber and their resistivity measured. The cycle was then repeated, giving total anneal times of up to 200s.

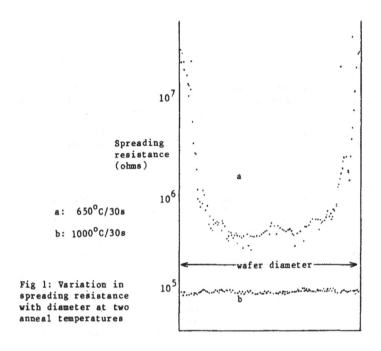

Spreading
resistance
(ohms)

a: 650°C/30s

b: 1000°C/30s

Fig 1: Variation in
spreading resistance
with diameter at two
anneal temperatures

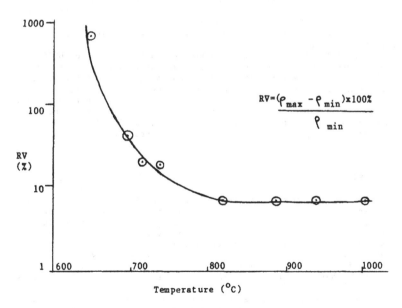

$$RV = \frac{(\rho_{max} - \rho_{min}) \times 100\%}{\rho_{min}}$$

Fig 2: Radial resistivity variation vs. anneal temperature (30s anneals)

RESULTS

Following anneals at 800 - 1180°C for 10 - 30s, the samples were found to be fully annealed, their resistivity having fallen from the post-irradiation value of greater than 10^5 ohm-cm to the target of 28 ohm-cm, with high resolution spreading resistance measurements indicating a uniformity of approximately 5%. The samples which were annealed at temperatures less than 750°C for 30 seconds exhibited much higher resistivity (fig 1) and a significant radial variation in resistivity, characteristic of partially annealed material (fig 1, fig 2). A comparison of the photoluminescence spectra from a partially annealed sample (650°C for 30s) and a fully annealed sample (815°C for 30s, this being the minimum conditions found to be necessary to return the sample to its target resistivity) is shown in fig 3. The partially annealed sample shows features associated with substitutional phosphorus dopant atoms (marked y), and also features relating to residual crystal damage (marked x); the strongest line at 1.138eV is the K-spectrum reported by Tajima (3) in similar, partially annealed NTD material. By comparison, the 815°C/30s sample spectrum shows stronger phosphorus bound exciton features (y), indicating a high level of substitutional dopant atoms, and a corresponding decrease in the intensity of the damage related features. It is interesting to note, however, that although the higher temperature treated sample showed full recovery to the target resistivity, a weak broad band at approximately 800mV, possibly related to residual damage, is still evident in the spectra.

Results from the progressively annealed samples showed unexpected behaviour. The resistivity of all the samples fell rapidly after the first 15 seconds from >10^5 ohm-cm to 400-700 ohm-cm. Further annealing, however, caused the resistivity to rise again in a repeatable and smooth fashion. This increase in resistivity is illustrated in fig 4 for five temperatures in the range 610-800°C, and clearly shows that the rate of increase of resistivity rises at higher annealing temperatures; further annealing at the temperatures shown continues this upward trend. To eliminate the possibility of contamination doping, some samples were then given a higher temperature anneal. This subsequent heat treatment (900°C for 15s) returned the samples to the target resistivity and uniformity, indicating that the previously observed behaviour was not due to progressive contamination effects introduced by the annealing method.

DISCUSSION

Our results on the transient annealing of NTD silicon show that extremely rapid dopant activation can be achieved with a very low thermal budget - typically 815°C for 30 sec - which is compatible with the requirements of low temperature device processing. For example, the combination of low temperature epitaxy, neutron transmutation doping and transient annealing would allow the preparation of n-type epitaxial layers with superior uniformity and interface sharpness compared with conventional layer growth and doping techniques.

The transient annealing technique used here has also allowed us to study the initial stages of dopant activation. The extremely rapid

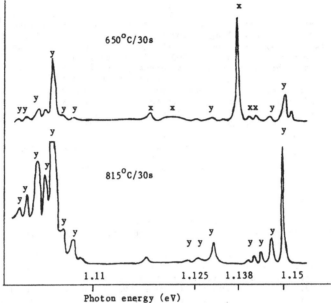

Fig 3. Comparative photoluminescence spectra

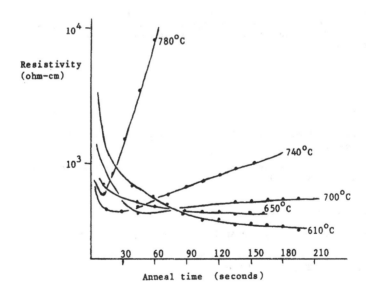

Fig 4: Wafer resistivity vs anneal time

decrease in resistivity in the first 15 seconds is consistent with
rapid vacancy annealing at low temperatures which has been reported
elsewhere (4). The subsequent change in resistivity with annealing time
for temperatures less than 800°C (fig 4) could be associated with
mobility changes arising from defect generation. Alternatively, there
may be two competing dopant activation mechanisms; one due to the
activation of the phosphorus dopant, and the other a temperature-time
dependent generation of an acceptor defect complex. Evidence for the
generation of such an acceptor defect in this temperature range has
been reported previously (5) at concentrations greater than the
background level of boron in our samples ($<<10^{13}$cm^{-3}). The
significant difference in annealing behaviour between samples treated
at 780°C and 815°C would therefore indicate that the acceptor
complex becomes unstable at temperature in excess of 800°C.

SUMMARY

1. Isothermal scanning electron beam annealing has demonstrated
extremely rapid phosphorus dopant activation of NTD-silicon for
temperatures greater than 800°C.
2. The combination of such rapid annealing, neutron transmutation
doping and low temperature epitaxy offers the potential for the
preparation of abrupt device interfaces.
3. Partial annealing at temperature below 800°C resulted in
resistivity changes which can be related to the simultaneous generation
of both donors and acceptors.

ACKNOWLEDGEMENTS

The authors are grateful to the Director of Research, British
Telecom Research Laboratories, for permission to publish this work, and
to A D Pitt, R W Series and D Kemp for assistance with sample
characterisation.

REFERENCES

1. H Herzer in Semiconductor Silicon '77 (ed. R Huff and E Sirtl),
Electrochem Soc, 1977, p106

2. C J Pollard, J D Speight and K G Barraclough   Mat Res Soc Symp
Proc, Vol 13 (1983), p413

3. M Tajima   Proc of 3rd International Conference on Neutron
Transmutation Doping, Aug 27-29 1980, ed J Guldberg, Plenum 1981, p377

4. K Sahu and T Rs Reddy   J Appl Phys, Vol 54 part 2 (1983), p706

5. J M Meese, D L Cowan and M Chandrasekhar   IEEE Trans on Nuclear
Science, Vol NS-26 part 6 (1979), p4858

PULSED-CO$_2$-LASER ANNEALING
OF ION-IMPLANTED SILICON

R.B. JAMES*,+ J. NARAYAN**,++ W.H. CHRISTIE***, R.E. EBY***,

O.W. HOLLAND*,**, and R.F. WOOD*
*Solid State Division, Oak Ridge National Laboratory, Oak Ridge, Tn 37831
** Microelectronics Center of North Carolina, Research Triangle Park,
   NC 27709
*** Analytical Chemistry Division, Oak Ridge National Laboratory, Oak Ridge,
    TN 37831

Abstract

From studies of time-resolved reflectivity and microstructural changes, we have obtained direct evidence of CO$_2$ laser-induced melting above a threshold energy density. Results of the optical measurements, transmission electron microscopy, and secondary ion mass spectrometry are reported. The measurements show that melt depths as deep as 1 μm can be achieved with pulsed CO$_2$ laser radiation. By using differential absorption between layers with different free-carrier densities, we find that a CO$_2$ laser can be used to melt regions which are embedded in the material. It is likely that this observed phenomenon is impossible to obtain with a visible or ultraviolet laser.

Introduction

Although there exists an extensive list of experimental and theoretical reports on pulsed laser annealing of ion-implanted silicon, only a few investigations have been conducted with a laser which has a photon energy smaller than the bandgap[1-4]. For CO$_2$ laser radiation ($\lambda \sim 10$ μm), the absorption of light by single-photon interband transitions is not energetically allowed, and another absorption mechanism is required to optically heat the material. In doped Si the dominant absorption mechanism is by free-carrier transitions. Since free-carrier absorption is an extrinsic absorption property in most heavily doped Si wafers at room temperature, one can control the penetration depth of the laser radiation by varying the carrier concentration. Control of the absorption coefficient is especially important for applications where one would like to achieve a deeper penetration of the pulse energy than is achievable with a visible or ultraviolet laser. Furthermore, one can use the extrinsic absorption property to preferentially deposit energy in heavily doped layers that are encapsulated by lightly doped layers.

In this paper we report time-resolved reflectivity measurements of ion-implanted silicon surfaces during and immediately after CO$_2$ laser annealing. We find that by varying the energy density up to several joules/cm^2, the near-surface region of the material melts via free-carrier absorption. Transmission electron microscopy (TEM) and secondary ion mass spectrometry (SIMS) measurements were performed on the irradiated samples to obtain information on the microdefects and the redistribution of the implanted species.

Experimental

A Q-switched, TEA CO$_2$ laser was used to generate the annealing pulses at a wavelength of 10.6 μm. The laser was operated with a low nitrogen mix so

---

+ Present Address: Sandia National Laboratories, Livermore, CA 94550
++ Also at North Carolina State University, Raleigh, NC 27650

that the amplitude of the long tail on the pulses could be greatly suppressed. About 80% of the energy in each pulse was contained in the form of a Gaussian-like peak with a duration of 70 ns (FWHM), and the remainder of the energy was in a long tail following the spike which lasts for several hundred nanoseconds. The output pulse was diverged by a spherical convex mirror with a 1-m radius of curvature. The diverging beam was later collimated by a spherical concave mirror with a 2-m radius of curvature. The collimated beam then impinged on a CO_2 laser beam integrator which spatially homogenized the pulse to within ±10%, according to the specifications by the manufacturer. The laser pulses had a size of 12x12 mm in the target plane of the integrator. A photon-drag detector and a volume absorbing calorimeter were used to measure the intensity and energy of the laser pulses, respectively.

A cw 633-nm He-Ne laser was used to measure the time-resolved optical reflectivity of the ion-implanted surface during and immediately after ir-radiation with a CO_2 laser pulse. The angle of incidence of the unfocused, He-Ne probe beam was $30°$ from the surface normal, and the CO_2 laser pulses impinged on the sample at normal incidence. A silicon avalanche photo-diode and fast oscilloscope were used to monitor the probe reflectivity. Narrow bandpass He-Ne filters were used to attenuate scattered radiation from the Si surface. The probe reflectivity was calibrated by using a chopper in the path of the He-Ne laser and equating the signal without the CO_2 laser with the known reflectance of amorphous Si.

Results and Discussion

Time-resolved reflectivity measurements were performed on 340-μm thick, boron-doped Si (111) wafers to determine the onset of the high-reflectivity phase and the duration as a function of the energy density of the laser pulses. The samples had a resistivity of 0.0073 Ω-cm at room temperature, which corresponds to a hole concentration of about $2-3 \times 10^{19}$ cm^{-3}. The wafers were amorphized on one side by implantation of ^{75}As at an energy of 180 keV to a dose of 1×10^{16} cm^{-2}.

The observed transient reflectivity signals of the probe laser are similar in shape to those reported by Auston et al.[5]. For energy densi-ties exceeding about 2.9 J/cm^2, the signals consist of a flat top with a dura-tion t_h and a decaying tail. The reflectivity during the high-reflectivity phase is approximately two times its initial value (R_0), which corresponds to a reflectivity of about 70%. The measured values of the melt durations are shown in Fig. 1 as a function of the energy density of the CO_2 laser pulses. The melt durations were taken by adding the time (t_h) of the high-reflectivity phase and the time required for the reflectivity R to fall to a value where $(R-R_0)=0.5R_0$. Each data point in Fig. 1 is an average of the melt durations from ten independent probe-reflectivity signals, where each reflectivity signal was recorded for a fixed value (to within ±2%) of the averaged energy density of the CO_2 laser pulse. [The value for an averaged energy density of a CO_2 laser pulse is obtained from a measurement of the total pulse energy through a 2-mm aperture at the position of the sample hold-er.] The melt durations are found to vary by ±15% from shot to shot, of which less than 5% can be attributed to variations in the averaged energy density between independent shots. The remaining variation in the melt durations is primarily caused by spatial inhomogeneities in the laser pulses that are not completely averaged out by the beam integrator.

The melt durations are found to be as long as 1 μs, which is somewhat longer than the observed melt durations using a laser which has a photon en-ergy greater than the intrinsic absorption edge[7]. The longer recrystalli-zation times strongly suggest that the melt depths are comparatively large, which may be advantageous for some applications of pulsed laser processing.

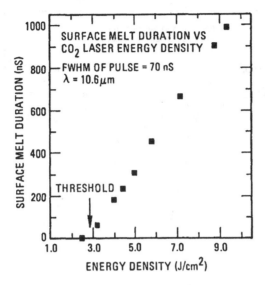

Figure 1. Measured duration of high-reflectivity phase versus the energy density of the laser pulses.

SIMS measurements were also conducted on the laser-annealed wafers to study the redistribution of the implanted arsenic and to investigate the possibility of controlling the dopant profiles by varying the energy density of the laser pulses. Since the diffusion of the implanted dopants primarily occurs during the time the near-surface region is molten, the SIMS measurements were also useful in obtaining approximate values for the melt depths. The results for the redistribution of the implanted arsenic are shown in Figure 2 for several different averaged energy densities of the CO_2 laser pulses. The implanted arsenic is found to diffuse to depths of over 7000 Å after single-shot irradiation, and there was no segregation behavior observed in the samples.

Cross-section TEM was used to further investigate the melt depths in the material and the defects in the recrystallized layer. Lightly doped silicon (111) wafers with a resistivity of 2-6 Ω-cm at room temperature were implanted with [11]B at several different ion energies in order to produce a uniform concentration (~1.0×10^{20} cm^{-3}) of boron to a depth of 0.7 μm. The implantation produced dislocations which were distributed to a depth of about 0.9 μm from the surface. Initial attempts to laser anneal the ion-implanted samples were found to be unsuccessful due to the high transparency of the CO_2 laser radiation in the lightly doped silicon. In order to increase the coupling of the laser light to the material, the samples were heated to 610 °C for five minutes prior to irradiation and were removed from the substrate heater immediately after irradiation. [The dislocation loops induced by the boron implantation are stable for thermal treatments at 610 °C for five minutes[6].] The preheating of the boron-implanted Si significantly increases the absorption coefficient by electrically activating a fraction of the implanted dopants and by increasing the temperature-dependent intrinsic carrier concentration. Figure 3(a) shows a cross-section TEM micrograph of a sample which has been irradiated by a single shot with an averaged energy density of 7.3 J/cm^2. The complete removal of extended defects in the

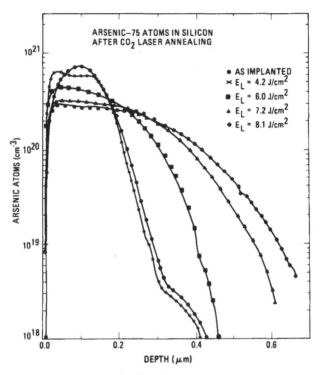

Figure 2. Depth profiles of $^{75}As^+$ emission in silicon before and after laser annealing at several different energy densities. Each of the samples were heavily doped with boron and had a resistivity of 0.0073 Ω-cm at room temperature.

annealed region and the abrupt change in the density of dislocation loops are consistent with the melting of the near-surface layer. Defect-free regions ranging up to a depth of about 8000 Å were observed in the lightly doped samples which were preheated to 610 °C and annealed with a CO_2 laser pulse.

Since the absorption of the laser radiation is dominated by free-carrier transitions, one expects the energy deposition to be considerably different for lightly and heavily doped Si. A heavily phosphorus-doped Si wafer with a resistivity at room temperature of 0.0026 Ω-cm was implanted with ^{11}B at an energy of 185 keV to a dose of 1.5×10^{16} cm^{-2}. Although the implanted region remains crystalline, electrical measurements on the as-implanted samples show that most of the free electrons near the surface are trapped by implantation-induced defects. Thus, the absorption coefficient of the CO_2 laser radiation is smaller in the implanted region than in the un-damaged substrate, and consequently, much of the energy in the laser pulse will be deposited in the lower-resistivity material below the implantation-damaged layer. The results of the cross-section TEM measurements on the heavily doped, ion-implanted Si samples show that one can melt regions near the interface of the undamaged and implantation-damaged layers, without melting

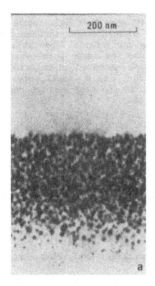

Figure 3. Fig. 3(a) is a cross-section TEM micrograph which shows the defect-free recrystallization of a lightly doped, boron-implanted Si wafer after CO_2 laser irradiation. The sample was heated at 610 °C for five minutes and then irradiated with a pulse having an energy density of 7.3 J/cm^2. Fig. 3(b) shows a cross-section TEM micrograph of a heavily doped Si wafer which has been implanted with ^{11}B at an energy of 185 keV to a dose of 1.5×10^{16} cm^{-2}. The sample was irradiated by a laser pulse with an energy density of 6.1 J/cm^2.

through the implantation-damaged layer near the surface [Figure 3(b)]. The observed melting of embedded regions further demonstrates how one can use free-carrier absorption to control the energy deposition of the laser radiation in the material. Furthermore, it is unlikely that this type of melting phenomenon can be obtained using a laser where the absorption is dominated by an intrinsic absorption mechanism.

From Figure 3(b) we see that microcracks are formed at a depth of about 9000 Å as a result of the pulsed CO_2 laser annealing. Since the density of both heated and liquid Si is different than crystalline Si at room temperature, rapid heating and subsequent melting of the embedded layer causes large mechanical stress in the material. The fracturing of the sample is due to the laser-induced heating of the embedded layer and the lack of a free surface to relieve the mechanical and thermal stress.

Summary and Conclusions

For CO_2 laser-energy densities greater than a few joules/cm^2, the energy deposited by free-carrier absorption can melt the near-surface region of ion-implanted Si. The melt durations are found to be as long as 1 µs, which is comparable to or larger than the melt durations which have been reported using a laser with a photon energy in the visible or ultraviolet spectrum. The results of SIMS measurements show that the implanted dopants can diffuse to depths of greater than 7000 A after pulsed CO_2 laser irradiation. TEM studies of boron-implanted crystalline Si show that melt depths in excess of 7500 A from the surface are obtainable without laser-induced defects in the annealed region. In addition, by using differential absorption between layers with different free-carrier densities, the laser pulse energy can be preferentially deposited in regions which are embedded in material that is relatively transparent to the laser radiation. As a result, one can use a CO_2 laser to melt layers which are embedded in the near-surface region without melting through the layer that encapsulates it.

Acknowledgments

This research was sponsored by the U. S. Department of Energy. We would like to thank R. T. Young, K. van der Sluis, D. H. Lowdnes, J. R. Hogan, R. J. Bastasz, and A. J. Antolak for many useful discussions. One of us (RBJ) would like to thank M. I. Baskes and W. D. Wilson for their encouragement and helpful comments while this work was completed at Sandia National Laboratories.

References

1. G. K. Celler, J. M. Poate, and L. C. Kimerling, Appl. Phys. Lett. 32, 464 (1978).
2. K. Naukkarinen, T. Tuomi, M. Blomberg, M. Luomajarvi, and E. Rauhala, J. Appl. Phys. 53, 5634 (1982).
3. M. Blomberg, K. Naukkarinen, T. Tuomi, V. M. Airaksinen, M. Luomajarvi, and E. Rauhala, J. Appl. Phys. 54, 2327 (1983).
4. M. Hasselbeck and H. S. Kwok, J. Appl. Phys. 54, 3627 (1983).
5. D. H. Auston, C. M. Surko, T. N. C. Venkatesan, R. E. Slusher, and J. A. Bolovchenko, Appl. Phys. Lett. 33, 437 (1978).
6. J. Gyulai and P. Revesz, Conf. Ser.-Inst. Phys. 46, 128 (1979).
7. See, for example, D. H. Lowdnes, R. F. Wood, and J. Narayan, Phys. Rev. Lett. 52, 561 (1984).

DOPANT INCORPORATION IN SILICON DURING NONEQUILIBRIUM SOLIDIFICATION: COMPARISON OF TWO PROCESSES*

E. P. Fogarassy,[+] D. H. Lowndes,[**] J. Narayan,[++] and C. W. White[**]
[+]Guest scientist from the Centre National de la Recherche Scientifique, Strasbourg, 67037, FRANCE
[**]Solid State Division, Oak Ridge National Laboratory, Oak Ridge, TN 37831.
[++]On sabbatical at the Microelectronics Center of North Carolina, Research Triangle Park, NC 27709 and the Materials Engineering Department, North Carolina State University, Raleigh, NC 27650.

ABSTRACT

The incorporation properties of implanted or deposited Sb into the silicon lattice during laser irradiation with a UV laser has been studied. For both implanted or deposited Sb, we find a maximum substitutional concentration of $2.1 \times 10^{21}/cm^3$ following laser melting and solidification at $V \simeq 6$ m/sec. In both cases, substitutional solubility is limited by interfacial instabilities which develop during regrowth. For the deposited case we observe in addition a much larger cellular microstructure which may result from convection induced instabilities.

INTRODUCTION

The rapid deposition of energy from a pulsed laser into the near surface of silicon leads to melting of crystal, followed by liquid phase epitaxial regrowth from the underlying substrate. During regrowth, velocities of several meters per second can be achieved [1]. Implanted or deposited elements from group III and group V can be incorporated into substitutional sites in the silicon lattice with concentrations far in excess of their equilibrium solubility limit during the rapid solidification process [2,5]. In this study, we have compared the incorporation of implanted or deposited Sb into the Si lattice as a result of irradiation with a pulsed KrF ultraviolet laser. Measurements have been made of the surface melt duration and the total and substitutional distribution profiles of the dopant. In both cases, mechanisms limiting dopant incorporation at high concentration have been determined. Finally we discuss the formation of a new type of cellular structure in surface of deposited samples.

EXPERIMENTAL CONDITIONS

The samples used in this study were single crystal silicon of <100> orientation. A part of these crystals were implanted at room temperature with antimony ions of 150 keV energy. The implantation doses ranged from 2×10^{16} to 5×10^{16} cm^{-2}. Thin films of antimony (50-100 Å) were deposited on similar substrate surfaces by electron gun evaporation in vacuum ($P \sim 10^{-6}$ Torr). These samples were irradiated with KrF excimer laser pulses ($\lambda = 248$ nm) of 35 ns FWHM. The laser energy density could be varied between 0.5 and 1.3 J/cm^2. The total and substitutional dopant distribution profiles were determined by Rutherford backscattering and ion-channeling measurements performed along the <110> direction. A 2.5 MeV ^{4}He$^+$ ion beam (diameter $\simeq$1 mm) was used, and the energy of the backscattered particles was measured by a cooled detector. This permitted a depth resolution of about 150 Å to be achieved.

*Research sponsored by the Division of Materials Sciences, USDOE under contract DE-AC05-840R21400 with Martin Marietta Energy Systems, Inc.

Time-resolved optical reflectivity measurements to determine the onset and duration of surface melting were performed using a cw HeNe probe laser beam (λ = 633 nm) incident at 10° to the surface normal and focussed to a 35 μm (e^{-1}) spot diameter. Reflected light was detected using a silicon avalanche photodiode and a storage oscilloscope. Transmission electron microscopy (TEM) results were obtained using a Phillips (EM-400) analytical microscope.

RESULTS

RBS Experiments

Figure 1 shows the measured total and substitutional concentration profiles of implanted Sb (150 keV, 5 × 10^{16} cm^{-2}) after laser annealing at energy density E = 1 J/cm^2. The as-implanted profile is approximately Gaussian with a peak at ~750 A. Laser annealing causes a broadening of the Sb distribution, both toward the surface and deeper in depth, which is characteristic of a diffusion in liquid Si. The melt depth, ~2000 A, has been deduced from channeling spectra. The substitutional concentration of antimony reaches to a value of 2.1 × 10^{21} cm^{-3}, and then becomes constant or decreases slightly in the near surface region. This concentration, equal to that found by using pulsed ruby laser, exceeds the retrograde maximum solutiblity limit at thermal equilibrium by a factor of about 30 (6 × 10^{19} cm^{-3} at T = 1200°C) [6]. The value of 2.1 × 10^{21}/cm^3 is the solubility limit of Sb in Si for interface velocities (deduced from heat flow calculations) close to 6 m/s. Figure 2 shows results obtained after irradiating an Sb film (~80 A) deposited on Si. Following laser irradiation (E = 1 J/cm^2), the dopant distribution is composed of two distinct regions, an in-depth profile and a rich antimony surface peak. The in-depth profile results from diffusion of Sb into liquid silicon during liquid phase epitaxial regrowth. The Sb substitutional distribution profile increases continuously to the surface, reaching a maximum concentration of 2.1 × 10^{21} cm^{-3}, a value which is somewhat higher than values found after pulsed ruby laser irradiation of deposited Sb layers (~1.1 × 10^{21} cm^{-3}). There is no loss of Sb by evaporation during laser irradiation. We find ~34% of the initial Sb in the surface peak and ~66% in the in-depth profile following the laser treatment. The proportion of incorporated Sb can be increased by repeating the laser irradiations, as shown in Figs. 3 and 4 for 10 and 20 shots respectively (E = 1 J/cm^2). During repeated shots, the Sb surface peak acts as a diffusion source. For 20 shots, we measure an incorporation of ~83%, with only 17% of the initial deposited antimony remaining in the

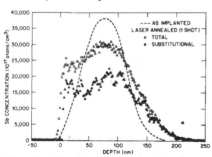

Fig. 1. Total and Substitutional Concentrations for ^{122}Sb (150 keV, 4 × 10^{16} cm^{-2}) in <100> Si after Excimer Laser Annealing (1 Shot)

residual surface peak. In Fig. 4, the depth of the plateau in the Sb distribution profile provides an appropriate measure of the depth of melting (~2500 A) during the laser induced diffusion process. This value is considerably larger than the depth predicted by calculations for crystalline Si (~1000 A). The in-depth Sb concentration in Fig. 4 is fully substitutional since the maximum concentration is less than the solubility limit of 2.1 × 10^{21}/cm^3 due to repeated melting and diffusion in the liquid.

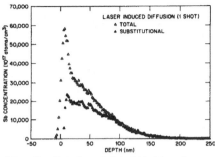

Fig. 2. Total and Substitutional Concentrations for a Thin Film of Sb (80A) Deposited on <100> Si After Excimer Laser Induced Diffusion

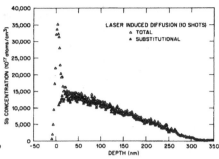

Fig. 3. Total and Substitutional Concentrations for a Thin Film of Sb (80A) Deposited on <100> Si After Excimer Laser Induced Diffusion

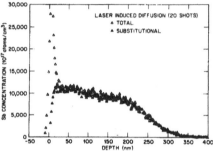

Fig. 4. Total and Substitutional Concentrations for a Thin Film of Sb (80A) Deposited on <100> Si After Excimer Laser Induced Diffusion

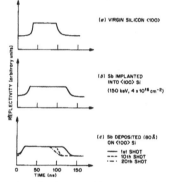

Fig. 5. Surface Reflectivity During Melting by a 35 ns Pulsed KrF UV Laser (E = 1 J/cm^2)

Reflectivity Measurements

Time-resolved optical reflectivity measurements were used to obtain better insight into the surface melt dynamics resulting from laser irradiation [7]. Time resolved reflectivity measurements were performed on (a) virgin silicon, (b) Sb implanted samples and (c) Sb deposited layers during the UV laser treatment (E = 1 J/cm^2). The results are reported in Fig. 5. In all cases, the sharp transition to the high reflectivity phase (R $\simeq$ 72% at λ = 633 nm) is characteristic of molten silicon. Surface melt durations measured on Sb implanted and deposited layers (t $\sim$ 95 ns) are significantly larger than for virgin silicon (t $\sim$ 55 ns). However, for the deposited case, we measure a decrease of surface melt duration when we repeat the laser treatment several times at the same energy density. Surface melt duration decreases from $\sim$95 ns on the first shot to $\sim$64 ns on the 20th shot, a value which is only marginally greater than that measured on virgin Si. This behavior may be related to the decrease of the antimony contained in the surface peak with increasing number of pulses, as observed by RBS experiments. This means that the deposition of Sb strongly modifies the melt dynamics of laser treated crystalline silicon. At the KrF wavelength (λ = 248 nm), the absorption coefficient of crystalline silicon is very high ($\alpha \sim 10^6$ cm^{-1}), and it is reasonable to assume that this parameter is not significantly modified in presence of a thin antimony layer. In addition, we measure a decrease of the solid state reflectivity (from $\sim$65% to 38.6%) on the as-deposited sample. This result is not well understood, but it may be related to interference effects in the thin layer. A decrease of reflectivity increases the amount of energy available to melt the sample. Another important parameter is the melting temperature of the surface layer. Since the melting point of antimony (T_M = 630°C) is lower than for silicon (T_M = 1410°C), the melting of the Sb deposited layer does not require as much energy as for crystalline silicon. This is suggested by the earlier time of the onset of melting measured on Sb deposited layers (Fig. 5c). This means that for identical irradiation conditions more energy is available to melt the underlying silicon when it is covered with a thin film of antimony through the formation of a Sb-Si eutectic alloy. The main consequence of this is to increase thickness of the melted silicon layer and the surface melt duration, in good agreement with experimental results.

Transmission Electron Microscopy

Figure 6(a) is a plan-view micrograph showing the formation of cells in the ion implanted and laser annealed (1.0 J/cm^2, 1 shot) specimens. Cell formation starts at a depth ~1300 Å, in agreement with ion channeling results of Fig. 1 which indicate differences between substitutional and total concentrations at this depth. The average cell size in Fig. 6a is ~ 500 Å and cell walls are perpendicular to the surface. Figure 6(b) shows a selected-area-diffraction pattern in which the extra spots correspond to the pure antimony phase. The TEM results from deposited and laser annealed (1.0 J/cm^2) samples are displayed in Figs. 7 and 8. Figure 7 shows the existence of two types of cells following the initial laser pulse: small cells with an average size of 350 Å are associated with the walls of large cells of average size 3000 Å. Figure 8 shows the formation of cells as a function of number of laser pulses in deposited samples. After one pulse, we observed small and large cells including denuded zones in the middle of large cells. The electron micrograph in Fig. 8a was taken under kinematical diffraction conditions, where small cells near the wall are clearly delineated. The formation of cells starts at a depth of 850 Å, in agreement with ion channeling results of Fig. 2. After irradiation with 10

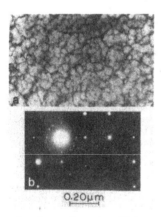

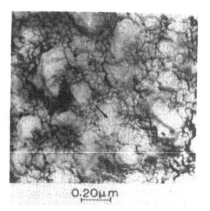

Fig. 6. Formation of cells in Sb implanted (150 KeV, 5 x 10^{16} cm^{-2}) and laser annealed silicon: (a) plan view micrograph, (b) diffraction pattern showing Sb spots.

Fig. 7. Formation of two types of cells in Sb deposited (80 Å) and laser treated silicon.

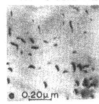

Fig. 8. Cells in Sb deposited layers as a function of number of laser shots (E = 1.J/cm²) (a) 1 shot, (b) 10 shots, (c) 20 shots.

pulses (Fig. 8b) the presence of cells was observed in the top 200 Å only. The width of this region decreased to 150 Å after irradiation with 20 pulses (Fig. 8c). The residual structure in Figs. 8(b) and (c) seems to be associated with large cells.

The formation of small cells both in surface of implanted and deposited layers is presumably the result of interfacial instability caused by constitutional supercooling during solidification [8,9]. During solidification of a dilute binary alloy, a planar liquid-solid interface can become morphologically unstable above a certain solute concentration at a given velocity of solidification. The sizes of small cells are in good agreement with calculations based upon the Mullins-Sekerka [10] theory modified to include the dependence of distribution coefficient upon the velocity of solidification. The concentration of Sb corresponding to the development of interface instability is close to 2×10^{21} cm^{-3}, as deduced from distribution profiles after 1 pulse irradiation (Figs. 1 and 2). However, for the deposited samples the small size cellular structure disappears after 10 laser pulses (Fig. 8b) since the maximum substitutional Sb concentration does not exceed 1.5×10^{21} cm^{-3} (Fig. 3), due to the redistribution in depth of the dopant during repeated melting and liquid phase diffusion. This value is less than that required for the development of interface instability. The formation of large cells seems to be related to the convection induced instability leading to segregation of solute near the large cell walls. The formation of small cells in certain middle regions of the large cells is suppressed (as shown by the arrow on Fig. 7) because the transport of the solute from the middle of the large cells reduces the concentration to a value which is less than that required for the formation of cells by constitutional supercooling.

DISCUSSION

To better understand these results, we must consider the thermodynamic process which occur during the irradiation of silicon with a uv laser. The absorption coefficient for either crystalline or amorphous Si at $\lambda = 248$ nm is $\sim 10^6$ cm^{-1}, which is much higher than that at the fundamental frequency of the ruby laser ($\lambda = 694$ nm) where $\alpha \simeq 5 \times 10^3$ cm^{-1} in crystalline Si and $\alpha \simeq 5 \times 10^4$ cm^{-1} in amorphous Si. Thus, at the uv wavelength, the radiation is absorbed in the first few hundred angstroms generating very high temperature gradients both in crystalline and amorphous Si and consequently higher regrowth velocities, as deduced from calculations ($V_{c-Si} \simeq V_{a-Si} \simeq 6$ m/sec)[11]. This is higher than calculated regrowth velocities resulting from ruby laser irradiation of similar pulse duration ($V_{c-Si} \simeq 3$ m/sec., $V_{a-Si} \simeq 4.5$ m/sec.). Therefore, the substitutional solubility is similar for deposited or implanted Sb because the regrowth velocity is approximately the same. At the ruby wavelength, the solubilities are different presumably because the velocity of regrowth is different. The value of $2.1 \times 10^{21}/\text{cm}^3$ is limited by interface instabilities which develop during regrowth and lead to the formation of a well defined small size cellular structure, which is observed in surface of both deposited and implanted samples. However, the experimental results have shown the presence of a second type of cell in the surface of the deposited layer. These large cells, with a size of 3000 A, which is comparable to the liquid layer thickness might be attributed to convection induced instability forming "Bénard cells", as observed by van Gurp et al. [12], and by Possin et al. [13] on laser treated thin metallic films of cobalt and nickel deposited on silicon. The thin liquid films are known to be dominated by surface tension forces. There is a well known tendency for the liquid film to ball up and form droplets on the surface of the substrate. In case of a thin liquid film of Sb on liquid silicon, there is a large difference in surface tensions for these two materials ($\delta_{Si} \sim 750$ dynes/cm, $\delta_{Sb} \sim 350$ dynes/cm). This difference can lead to a large surface tension gradient over a small lateral distance. This results in the fluid film breaking up into small convective cells of size comparable to the depth of the melted layer.

CONCLUSION

We have demonstrated, in this study, that the substitutional incorporation of antimony in laser treated silicon does not depend on deposition or implantation, but is only a function of the velocity of liquid-solid interface during resolidification of the molten layer. In the two cases (implanted and deposited) the solubility limit of Sb is 2.1×10^{21} cm^{-3}, and is limited by interface instabilities which develop during regrowth at $V \simeq 6$ m/sec and lead to the formation of a small size cell structure in the near-surface region. However, in the Sb deposited and laser annealed samples, we observe a second type of cellular structure, an order of magnitude larger in size, (Bénard cells). This larger structure is believed to be related to convective movements which take place in the surface of deposited samples during laser treatment.

REFERENCES

1. See for example, Laser Annealing of Semiconductors, ed. by J. M. Poate and James W. Mayer, Academic Press, New York (1982).
2. C. W. White, S. R. Wilson, B. R. Appleton, and F. W. Young, Jr., J. Appl. Phys. 51, 738 (1980).
3. C. W. White, J. de Physique T44, 145 (1983).
4. E. Fogarassy, R. Stuck, J. J. A. Grob, and P. Siffert, Laser and Electron Beam Processing of Material, ed. by C. W. White, and P. S. Peercy, Academic Press, New York (1980) p. 117.
5. E. Fogarassy, R. Stuck, M. Toulemonde, D. Salles, and P. Siffert, J. Appl. Phys. 54, 5059 (1983).
6. F. Trumbore, Bell Syst. Tech. J. 39, 205 (1960).
7. D. H. Auston, C. M. Surko, T. N. C. Venkatesan, R. E. Slusher, and J. A. Golovchenko, Appl. Phys. Lett. 33(5) 437 (1978).
8. J. Narayan, H. Naramoto, and C. W. White, J. Appl. Phys. 53, 912 (1983).
9. J. Narayan, J. Appl. Phys. 52, 1283 (1981).
10. W. W. Mullins and R. F. Sekerka, J. Appl. Phys. 35, 444 (1964).
11. M. Aziz, private communication.
12. G. J. van Gurp, G. E. J. Eggermont, Y. Tamminga, W. T. Stacy, and J. R. M. Gijsbers, Appl. Phys. Lett. 35, 273 (1979).
13. G. E. Possin, H. G. Parks, and S. W. Chiang, P. 73 in Ref. 7.

PART V

Compound Semiconductors

RAPID ANNEALING OF GaAs AND RELATED COMPOUNDS

J.S. WILLIAMS, Microelectronics Technology Centre, Royal Melbourne Institute of Technology, Melbourne 3000 Australia.

S.J. PEARTON, A.T. & T. Bell Laboratories, Murray Hill N.J. 07974, U.S.A.

ABSTRACT

In recent years, rapid annealing of GaAs has been employed to activate ion implanted dopants and to produce metal-semiconductor contacts. More recently, rapid heating techniques have been used to anneal heterostructures without degrading requisite high mobilities of SDHT devices. This paper reviews the various annealing regimes which are useful for heat treating GaAs and related compounds. Examples are chosen which illustrate unique and potentially useful features of rapid annealing.

INTRODUCTION

The thermal processing of GaAs and other compound semiconductors, necessary for the fabrication of devices and circuits, is accompanied by considerable complication [1-3] in comparison with silicon processing. Difficulties experienced with conventional furnace processing of GaAs include: surface dissociation and associated As loss at temperatures above $500^\circ C$, inability to electrically activate ion implanted n-type dopants above $3 \times 10^{18} cm^{-3}$, sometimes excessive dopant diffusion during annealing to temperatures ($>850^\circ C$) which are required for best activation of ion implanted dopants, and lack of reproducibility for alloyed metal-semiconductor contacts. Such processing difficulties have led to a strong interest in more rapid forms of heating, as illustrated in the previous GaAs review papers presented at this forum [4-8] and elsewhere [9 - 12].

Early studies of rapid annealing of GaAs concentrated on two areas: i) the removal of ion implantation damage and corresponding activation of dopants; and ii) the alloying of metals with GaAs for ohmic contacting applications. These studies invariably used either pulsed lasers (electron beams) to melt the resolidify the near-surface regions of the material in nanosecond time scales [12-18] or CW lasers (electron beams) operating in the 10-100 ms time regime to locally heat small volumes of the material to temperatures below the GaAs melting point [19-23]. Although some success was obtained in these studies, the electrical properties were not as promising as might have been expected and these methods have not been pursued as useful processing techniques. However, rapid thermal annealing (RTA) in the time regime of 1-100 seconds has received increasing attention in recent years as a more promising method for device processing [24-28]. In this time regime isothermal heating of the entire wafer is obtained and the shorter anneal time compared with furnace annealing is found to have particular advantages. This review briefly outlines the behaviour during rapid annealing of GaAs, covering time regimes from nanoseconds to several minutes. In particular, both the annealing of ion implanted layers and ohmic contacting applications are discussed. Examples are given which illustrate unique and potentially useful features of rapid annealing for device applications.

OVERVIEW

Annealing Regimes

Fig. 1 summarises the various annealing regimes for GaAs with respect to a heating scale spanning 10^{-9} to 10^3 s. As indicated in the upper

428

portion of the figure, temperatures in excess of the GaAs melting point must be employed to induce significant surface modification (interdiffusion or removal of implantation damage) in times <10^{-6} s. For longer times, the magnitude of solid phase diffusion coefficients and crystallisation kinetics are consistent with inter mixing of components and damage removal. In order to rapidly heat and cool the near-surface region of the sample in times less than about 1s, some form of local (rather than bulk) heating must be employed. In contrast, heating for times greater than about one second usually ensures near-isothermal heating of the entire wafer. The nsregime requires use of a pulsed energy source, the most rapid solid phase regime (10^{-6} to 1s) requires local heating with either pulsed or CW sources, and isothermal heating in the RTA regime (>1s) can be accessed using, for example, a strip heater [24, 25] or an incoherent light source [26-29].

In the lower portion of Fig. 1 we illustrate the useful annealing regimes for the two major heating applications of GaAs, namely removal of ion implantation damage and formation of metal-semiconductor contacts. For the former process, pulsed melting and rapid solidification can remove ion implanted damage by liquid phase epitaxial growth (LPEG). The useful upper limit to this annealing time scale is about 10^{-7}s, above which surface melting is accompanied by excessive dissociation and evaporative loss of As and Ga from the surface, as indicated more fully in the following section. Several studies [7-9,21-23] have now indicated that local solid phase heating for times less than 1s is not useful for damage removal since, at temperatures of >1000°C (which is necessary to totally remove defects) the severe thermal gradients lead to slip and surface cracking. Therefore, RTA is the only useful rapid solid phase annealing regime for damage removal in ion implanted GaAs. In terms of metal-semiconductor contacts, alloying or eutectic melting normally requires temperatures in the range 400-700°C, depending on the system under study.

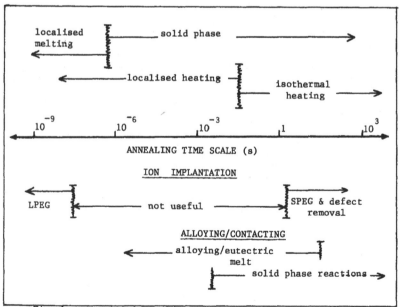

Fig. 1.   Localised heating regimes for GaAs

All annealing regimes have been used for contact formation although furnace annealing is usually restricted to times not exceeding a few minutes to minimise undesireable morphological features (balling up and spiking) which can result from longlived alloy melts. Thus, rapid annealing would appear to be particularly well suited to alloying of contacts : local heating and thermal gradients are not a problem if maximum temperatures are kept below about 800°C for annealing times of 10^{-6} to 1s. Solid phase reactions [30] of metals (e.g. Ni, Pt, Ti) with GaAs provide further possibilities for contact applications, although these have not been the subject of major interest. The relevant time scales available are indicated on Fig. 1.

Pulsed Annealing (<50 ns time scale)

Fig. 2 gives a schematic illustration of the relevant features of pulsed annealing for both (a) ion implantation damage removal and (b) metal-semiconductor alloying.

In terms of ion implanted GaAs most pulsed beam annealing studies have concentrated on the selection of optimum annealing conditions which give good crystal recovery and high dopant activation while minimising surface degradation. The following features have been the subject of most detailed study.

i) The removal of ion implantation disorder by near-surface melting and liquid phase epitaxial growth (LPEG) [12] is illustrated in Fig. 2. The quality of the crystal is good but the optimum energy window (typically 0.4 - 0.6 Jcm^{-2} for a 15ns ruby laser) is narrow [9,18,31], compared with pulsed annealing of silicon. At low powers (>0.6 Jcm^{-2}), deep defects are not removed and at high powers (>0.6 Jcm^{-2}), surface damage occurs (as a result of surface dissociation). However, extended-defect-free growth can be obtained under optimum conditions [31].
ii) Surface dissociation and evaporation, attendant surface damage and accumulation of free surface Ga have been examined in some detail by serveral workers [31-34]. These observations are consistent with melting models for GaAs during pulsed annealing [12,35]. Time-resolved reflectivity measurements [35,36] have been employed as more definitive evidence for surface melting.
iii) Electrical measurements on ion implanted and pulse annealed GaAs have not been as promising for device applications as was initially expected [7]. For example, low dose (<10^{13}cm^{-2}) n-type implants could not be electrically activated [37,38], presumably as a result of quenched-in point defect complexes [39]. In addition, the carrier mobilities of both n-type and p-type layers were lower than for corresponding thermally annealed GaAs [10]. Finally, although high dose implants (especially n-type) could be electrically activated by pulsed annealing to higher levels than with furnace annealing [4,10], such layers were metastable and moderate heating (to ~ 300°C) resulted in a marked reduction in electrical activity [40].

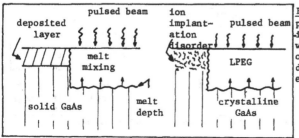

Fig.2 Schematic showing pulsed annealing for mixing of deposited layers with GaAs and removal of ion implantation damage via liquid phase epitaxial growth.

Other, often conflicting, observations regarding dopant redistribution during pulsed melting and high solid solubility following solidification have been made in pulse annealed GaAs. For example, some authors have reported redistribution of impurities in the melt [41,42] or segregation to the surface on solidification [43,44], whereas other studies have given no evidence for redistribution [13,18]. Similarly, high impurity solubility (supersaturation) to levels above those reported for furnace annealing have been obtained with pulse annealing [41] but no comprehensive investigations of this topic have been forthcoming. In addition, a range of fundamental phenomena which have been investigated in some details for pulse annealed Si [45] (e.g. suppressed melting point of amorphous silicon under ultra-rapid heating, the kinetics of LPEG, thermodynamic properties of rapidly melted and solidified silicon, and explosive crystallisation processes) have not yet been observed, let alone studied, in GaAs.

Finally, several studies of alloying of deposited metal-GaAs systems by pulse annealing (c.f. Fig. 2) have demonstrated the capabilities of the technique for fabricating successful contacts which are more reproducible and exhibit better morphology [23,46]. It has also been demonstrated that less complex metal overlays may be employed (e.g. Ge [47] or Au-Ge [46]) when using pulse alloying compared with those often required for furnace alloying (e.g. Au-Ge-Ni layers for ohmic contacts to n-layers). However, the elemental composition of pulse-alloyed layers has not often been analysed nor has the identification of alloy phases received any serious consideration.

Rapid Termal Annealing (>1 sec time scale)

In terms of device applications, RTA offers more potential than pulsed annealing for both activation of ion implanted dopants in GaAs and the formation of metal-GaAs contacts. The technique appears to provide electrical properties at least as good as those achievable by conventional furnace heating and has other more specific advantages over furnace heating, which are derived from the shorter heating time of RTA. Some of these advantages are discussed with reference to the solid phase annealing summary (principally for furnace annealing) given in Fig. 3.

The top portion of Fig. 3 illustrates the various annealing regimes for removal of ion implantation damage in GaAs [5,48]. Ion implantation damage can consist of either amorphous layers or extended crystalline defects (loops and stacking faults) depending on the implantation

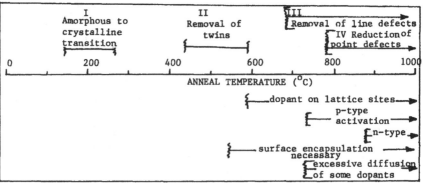

Fig. 3 Summary of ion implantation damage removal in GaAs by furnace annealing.

conditions. Amorphous layers recrystallise epitaxially during annealing at 150-250°C (stage I) but the recrystallised layer is invariably highly defective, consisting of twins, stacking faults and other defects. These defects anneal out (stage II) to leave essentially dislocation loops in the range 400-550°C. The loops grow and annihilate above about 700°C (stage III) and the remaining point defect clusters begin to anneal out (stage IV) above about 750°C.

Other features of the annealing of GaAs are indicated in the lower part of Fig. 3. Although dopant atoms may be observed to occupy the mandatory Ga or As lattice sites (i.e. substitutionally soluble in GaAs at temperatures above 600°C [1-3]) optimum electrical activity does not ensue until significantly higher temperatures [10]. Interestingly, n-type dopants are more difficult ot activate than p-type dopants. During furnace annealing the surface of the GaAs must be protected against dissociation and As loss for annealing temperatures > 500°C [49] (either by capping the surface with a suitable material such as Si_3N_4 or using an As overpressure). In addition, some dopants (e.g. S, Sn and Zn) are reported to be subject to excessive diffusion during high temperature annealing [10]. RTA offers the possibility of minimizing these latter, undesirable features of furnace annealing, namely dissociation and dopant diffusion.

Early applications of RTA [20,24-28] for the removal of ion implantation damage in GaAs and the consequential promotion of dopant activation showed considerable promise and stimulated much research interest. A variety of rapid heating methods have now been employed, including strip heaters [24,25], incoherent light sources [26-28] and scanning electron beams [20]. To date the following attributes of RTA have been established: i) the electrical properties (activation and mobility) following optimum RTA are at least as good as those obtained by furnace annealing [11]; ii) high solubility of dopants can be achieved by RTA [7]; iii) less diffusion has been claimed under optimum RTA conditions [7], and iv) it is possible to activate both n-type and p-type dopants by RTA without recourse to capping of the surface, although optimum n-type activation requires higher temperatures and/or times which may often necessitate preventative measures to minimise surface dissociation [50]. More detailed accounts of these attributes are given in published reviews [7,9-11], some specific illustrations are given later in this review.

Not all aspects of RTA of ion implanted GaAs have been the subject of detailed investigation. For example, the damage/defect removal processes and the relationships between dopant solubility, diffusivity and electrical activity have not been established. Such studies, along the lines of those currently being undertaken for RTA of ion implanted Si [51], are needed to provide more insight in the annealing behaviour of GaAs. In addition, device characteristics, device yield and uniformity of annealing across large wafers have not previously been examined for RTA of GaAs.

With regard to metal-semiconductor ohmic contacts, recent studies have indicated the ability of RTA to provide reproducible, morphologically improved alloyed contacts with low specific contact resistance to both p-type and n-type layers [52,53]. Examples of uses in GaAs device fabrication are beginnning to appear in the literature [53]. However, there are few studies of solid state reactions induced by RTA. Very recent work of Palmstrom et al [55] for various metals on GaAs and Soloman and Smith [56] for Ni on GaAs have indicated the potential of rapid heating for inducing solid state reactions.

432

EXAMPLES FROM RECENT RTA STUDIES

Si$^+$ Implants for Active Layers

Two important considerations in the assessment of RTA for the activation of Si$^+$ implants are: i) The surface dissociation problem; and ii) control over profile broadening. We indicated earlier that it is possible to activate dopants during RTA without taking steps to protect the surface from dissociation. However, although measurements of the extent of dissociation during RTA have indicated that annealing for 10s at 850°C causes only ~ 30 Å of surface dissociation, 10s at 1000°C can result in As loss from a depth of > 800 Å [32]. The extent of dissociation of a face up sample during RTA is indicated in Fig. 4. Since best activation in n-type dopant layers requires annealing at higher temperatures [10], it is necessary to protect the surface in some way during annealing. Often simplified measures can suffice, as indicated in Fig. 4., where the placing of the GaAs wafer face down on a Si or a GaAs wafer during annealing [50] can allow optimum activation to be achieved before excessive dissociation occurs. Normally, more conventional capping procedures such as SiO$_2$ or Si$_3$N$_4$ layers are utilized [10]. Even when such caps are employed, RTA offers major advantages since the quality of the cap, so important for longer time furnace annealing, is not crucial during RTA [26]. This feature can greatly improve the reproducibility and simplicity of the annealing steps in GaAs technology.

Recent measurements at Bell Laboratories [57] have investigated RTA of Si$^+$ implanted GaAs in some detail. Both the degree of electrical activation and the active doping profile obtained using RTA, in comparison with furnace annealing, are indicated in Fig. 5. This example shows a typical FET implant of 3 x 10^{12} cm^{-2} of 60 keV Si$^+$. Two important features are illustrated. Under conditions (RTA and furnace) which give the same profile shape (i.e. minimal profile broadening), the RTA case exhibits a considerably higher peak activity. Average mobilities in RTA samples have also been measured to be about 10% higher than for furnace annealed wafers. The apparent broadening shown by both profiles in comparison with the LSS profile is most probably not indicative of the diffusion during annealing, but of the shape of the original implanted profile with a deep 'channeling' tail.

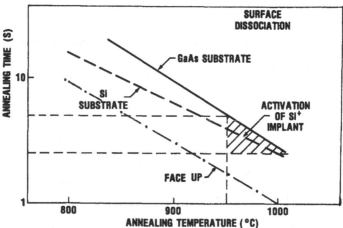

Fig. 4. Surface dissociation of GaAs under RTA. Data partially taken from Ref. 50.

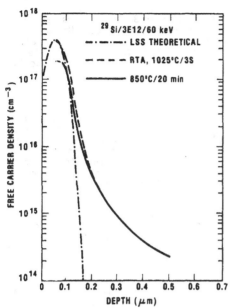

Fig. 5. Electrical profiles of
Si$^+$ Implanted (100)
GaAs; annealing under
various conditions.

It is significant to note that for Si$^+$ implants in GaAs, the Si is not
observed to diffuse excessively during furnace annealing. However, for
cases where shallow junctions are needed any diffusion results in a
departure from ideal behaviour. In this regard, benefits of RTA are
indicated in Fig. 6 which illustrates typical dual implants of Si used to
form n/n$^+$ in source/drain contact region of FET'S. These doping profiles
were measured following three annealing schedules. Furnace annealing after
each implant clearly results in the greatest diffusion and the lowest peak
activation in comparison with the RTA cases. It is also significant that
one RTA annealing step after both implants does not give as high an
activation as separate RTA steps following each implant. This presumably
results from different defect structures and associated dopant-defect
interactions during annealing.

Fig.6 . Electrical profiles
of dual implants
of Si$^+$ into (100)
GaAs for n/n$^+$
layers. Various
RTA and furnace
annealing treat-
ments are shown.

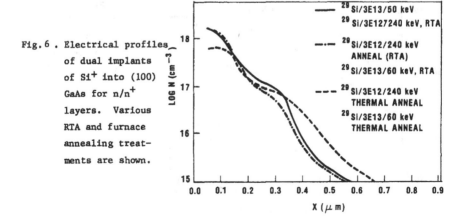

434

Finally, single and dual implants of Si$^+$ over 3" GaAs wafers have been checked for lateral uniformity of the electrical properties. Uniformities are comparable with those from furnace annealed wafers (i.e. average mobilities within 5% and sheet activities within 7% across the wafers). Thus, RTA appears to be an emanently suitable method for activating ion implanted n-type dopants in GaAs FET technology.

High Dose Implant Studies

Furnace annealing studies indicate that high dose n-type implants into GaAs cannot be fully electrically activated and that the degree of activity depends on the implant temperature [10]. Early RTA studies [26] suggested that solubility of dopants (e.g. Te) may be higher following RTA than furnace annealing. It is interesting to pursue this feature to determine if RTA can indeed provide higher electrical activity than furnace annealing. We illustrate some recent results [58] for the electrical activation of Sn implants into GaAs in Figs. 7 and 8.

Fig. 7 shows the sheet carrier concentration as a function of Sn dose following both furnace annealing and RTA. For the lowest dose (10^{13}cm^{-2}), RTA gives slightly higher electrical activity than furnace annealing. In this case, the implanted layer is almost certainly not amorphous and the electrical results following annealing are consistent with those for Si$^+$ implants discussed in the previous section. However, for the higher Sn doses, the electrical measurements in Fig. 7 indicate incomplete activity for both annealing methods but higher activity for furnace annealing than RTA. Significantly, these higher implant doses generate amorphous layers and, presumably, higher temperatures and longer annealing times are needed to completely remove disorder and activate dopants. On Fig. 7, we have also shown the furnace and RTA results for a 10^{15}Sn cm^{-2} implant at 400°C. The electrical activities are significantly higher for these cases in comparison with a room temperature implant. In addition, RTA gives slightly higher activity than that for furnace annealing. This again indicates that implants which do not generate amorphous layers (as is certainly the case for a 400°C implant) exhibit higher activities following RTA.

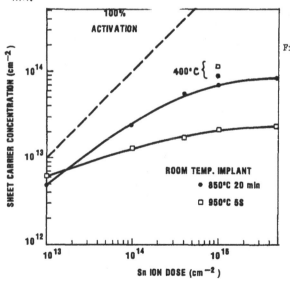

Fig. 7. Sheet carrier concentration vs Sn dose for RTA and furnace annealing treatments.

In Fig. 8, we illustrate the substitutional Sn fraction (as measured by ion channeling) as a function of anneal temperature for room temperature and 400°C implants of 1 x 10^{15}Sn cm^{-2}. The implant temperature is observed to have a strong influence on the substitutional (soluble) Sn fraction. For room temperature implants, the substitutional fraction is initially high after a 600°C anneal and then drops dramatically after annealing at 950°C. In constrast, 400°C implants are about 60% substitutional after implantation and this fraction increases on annealing at higher temperatures. As suggested previously [59], this may indicate strong interactions between solute atoms and residual defects, the soluble fraction being dictated by the particular defect configuration which accompanies implantation and annealing. The RTA results (solid symbols) on Fig. 8 show that these shorter time anneals at 950°C behave somewhat like a lower temperature furnace anneal. Consequently, RTA of the room temperature implant exhibits higher substitutionality than the corresponding furnace anneal whereas the opposite behaviour is observed for the 400°C implant.

It is interesting to compare the higher temperature (~950°C) anneal data from Fig. 8 with the 1 x 10^{15}Sn cm^{-2} electrical data on Fig. 7. There is clearly only a weak correlation between the often high substitutional fraction and the usually much lower electrically active fraction; only for the furnace anneal of the room temperature implant is the electrically active fraction (~7%) at all comparable with the substitutional Sn fraction (~10%). Further measurements show that the electrically active dopant profiles (as measured by chemical stripping and CV techniques) do not correlate well with the total Sn profile (obtained by Rutherford backscattering). Although the electrical profiles following RTA show much less diffusion broadening than those following corresponding furnace anneals, both profiles indicate considerable diffusion of the active Sn fraction, whereas the Rutherford backscattering data show that most (>90%) of the Sn is immobile during both RTA and furnace annealing. This result suggests that only the active dopant fraction is mobile and the inactive fraction is presumably immobilised at defects.

In summary, RTA of high dose Sn implants has not provided higher activity in comparison with furnace annealing although shallower dopant profiles are observed following RTA. The implant temperature has a major influence on both the substitutional fraction and the electrically active fraction. Furthermore, solute-defect interactions appear to control the level of solubility and activity for both RTA and furnace annealing. There is clearly a need for such correlations to be studied in more detail before the attributes of RTA for activating high dose implants are fully realised.

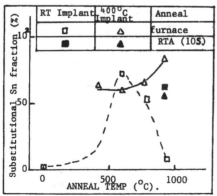

	RT Implant	400°C Implant	Anneal
	□	△	furnace
	■	▲	RTA (10S)

Fig.8 Substitutional Sn fraction as a function of anneal temperature for 1 x 10^{15} cm^{-2} implants into GaAs

436

Annealing of Heterostructures

Selectively doped AlGaAs/GaAs heterostructure transistors (SDHT's) offer much promise for supercomputer applications in view of their fast switching speeds (<10ps). However, in integrated circuits these devices need to be able to withstand high temperature implant-activation steps, to provide, for example, low source-drain resistance. In Fig 9a we illustrate the important electrical properties (mobility and sheet electron concentration) for RTA and furnace annealing of the heterostructure shown in Fig. 9b. These results are taken from the work of Tatsuta et al [60]. Normal furnace annealing causes a dramatic fall off in electron mobility (measured at 77K) and this results mainly from diffusion of Si ions across the buffer spacer layer (undoped AlGaAs) into the undoped GaAs region of the 2 dimensional electron gas (2 DEG). The additional impurity scattering caused by these ions at 77K gives rise to the fall off in electron mobility. However, rapid annealing at temperatures and for times needed to activate ion implanted layers causes significantly less diffusion of the Si and, consequently, much less degradation in electron mobility. The electron concentration does not change significantly during RTA whereas it increases with temperature for furnace annealing (again as a result of Si diffusion).

The annealing of heterostructures by RTA is therefore the only feasible method of performing a crucial step in the processing of a very promising technology.

CONCLUSIONS

i)   For the annealing of GaAs device structures, RTA is a most promising regime. For activation of Si implants (n and n^+ layers), RTA gives high activation, less diffusion and is less restrictive of the cap material. The alloying of metal-semiconductor contacts is more reproducible with RTA compared with furnace annealing and the annealing of heterostructures is only feasible using RTA methods.

ii)   In both the pulse anneal and RTA regimes, there is a need for a better understanding of the systematics of defect removal, impurity diffusion, solubility and the interrelationships between such parameters and electrical activity.

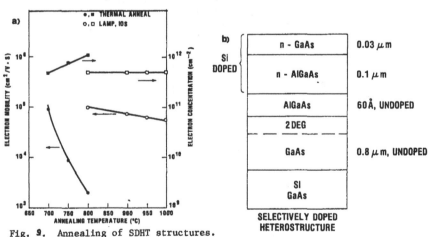

Fig. 9.   Annealing of SDHT structures.

iii) There is much scope for research into novel alloying and solid state reactions produced in pulse anneal and RTA regimes.

ACKNOWLEDGEMENTS

One of us (JSW) wishes to acknowledge financial support from the Commonwealth Special Research Centres Scheme and the ARGS.

REFERENCES

1   J.P. Donnelly, Inst. Phys. Conf. Ser. 33, 167 (1977)
2   F.H. Eisen, in "Ion Beam Modification of Materials", J. Guylai, T. Lohner and E. Paztor, eds. (Hung. Acad. Sci. Budapest 1978) p. 147.
3   K.G. Stephens, Nucl. Instr. Meth. 209/210, (1983).
4   F.H. Eisen, in "Laser and Electron Beam Processing of Materials," C.W. White and P.S. Peercy, eds (Acadmeic Press, New York, 1980) p. 309
5   J.S. Williams and H.B. Harrison, Proc. Mat. Res. Soc. 1, 209 (1981)
6   C.L. Anderson, Proc. Mat. Res. Soc. 4, 653 (1982)
7   J.S. Williams, Proc. Mat. REs. Soc.13, 621 (1983)
8   J.C.C. Fan, Proc. Mat. Res. Soc. 1, 261 (1981)
9   J.S. Williams, ch. 11 in "Laser Processing of Semiconductors", J.M. Poate and J.W. Mayer, eds. (Academic Press, New York 1982) p. 383.
10 F.H. Eisen, Ch. 10. in "Ion Implantation and Beam Processing", J.S. Williams and J.M. Poate, eds. (Academic Press, Sydney, 1984).
11 B.J. Sealy, Microelect. Journal 13, 21 (1982).
12 B.H. Lownes in "Pulsed Laser Processing of Semiconductors", R.F. Wood, C.W. White and R.T. Young, eds (Academic Press, New York, 1985).
13 J.A. Golovchencko, and T.N.C. Venkatesan, Appl. Phys. Lett. 32, 464 (1978).
14 P.A. Barnes, H.J. Leamy, J.M. Poate, S.D. Ferris, J.S. Williams and G.K. Celler Appl. Phys. Lett. 33, 965 (1978)
15 M.H. Badawi, J.A. Akintunde, B.J. Sealy, and K.G. Stephens, Electron. Lett. 15, 447 (1979).
16 J.L. Tandon and F.H. Eisen, in "Laser - Solid Interactions and Laser Processing", S.D. Ferris et al, eds. (AIP, New York, 1979) p. 616.
17 C.L. Anderson, H.L. Dunlop, L.D. Hess, G.L. Olson, and K.V. Vaidyanathan, in "Laser and Electron Beam Processing of Materials", C.W. White and P.S. Peercy, eds. (Academic Press, New York, 1980) p. 334.

18 S.U. Campisano, G. Foti, E. Rimini, F.H. Eisen, W.F. Tseng, M.A. Nicolet, and J.L. Tandon, J. Appl. Phys. 51, 295 (1980).
19 J.C.C. Fan, R.L. Chapman, J.P. Donnelly, G.W. Turner, and C. Bolzer, App. Phys. Lett. 34, 780 (1979)
20 N.J. Shah, H. Ahmed, P.A. Leigh, Appl. Phys. Lett. 39, 322 (1981).
21 G.L. Olson, C.L. Anderson, L.D. Ness, H.L. Dunlap, R.A. McFarlene and K.V. Vaidyanathan, in "Laser and Electron Beam Processing of Electronic Materials", C.L. Anderson, G.K. Celler and G.A. Rozgonyi, eds. (ECS Princeton, 1980) p. 467.
22 Y.I. Nissim and J.F. Gibbons, Proc. Mat. Res. Soc. 1, 275 (1981)
23 G. Eckhardt in "Laser and Electron Beam Processing of Materials", C.W. White and P.S. Peercy, eds. (Academic Press, New York, 1980) p. 467.
24 B.J. Sealy and R.K. Surridge, IBMM-78, J. Guylai et at, Eds. (Budapest 1979) p. 487.
25 R.L. Chapman, J.C.C. Fan, J.P. Donnelly, B.Y. Tsaur, Appl. Phys. Lett. 40, 805 (1982).
26 H.B. Harrison, F.M. Adams, B. Cornish, S.T. Johnson, K.T. Short and

438

J.S. Williams, Proc. Mat. Res. Soc. 13, 641 (1983).

27 M. Arai, K. Nishiyama and N. Watanabe, Jap. J. Appl. Phys. 20, L124 (1981).

28 D.E. Davies, P.J. McNally, T.G. Ryan, K.J. Soda, and J.J. Comer, Inst. Phys. (Lond.) Conf. Ser. No. 65, 619 (1983).

29 A. Gat, IEEE Elect. Dev. Lett. 2, 85 (1981).

30 A.K. Sinha and J.M. Poate, Ch 11 in "Thin Films - Inter - diffusion and Reactions", J.M. Poate et al, eds. (John Wiley, New York 1978) p. 407.

31 J. Fletcher, J. Narayan and D.H. Lownes, Proc. Mat. Res. Soc. 2, 421 (1981).

32 A. Rose, J.T.A. Pollock, M.D. Scott, F.M. Adams, J.S. Williams and E.M. Lawson, Proc. Mat. Res. Soc. 13, 633 (1983).

33 A. Pospieszczyk and M. Abdel Harith, J. Appl. Phys. 54, (6) (1983)

34 D.O. Boerma, H. Hasper, and K.G. Prasard Phys. Lett. 92A, 253 (1983).

35 D.H. Lowndes and R.F. Wood Appl. Phys. Lett. 38, 971 (1980)

36 D.H. Auston, J.A. Golovchenko, A.L. Simons, R.E. Slusher, P.R. Smith, C.M. Surko and T.N.C. Venkatesan, Am. Inst. Phys. Conf. Proc. 50, 11 (1979)

37 S.G. Liu, C.P. Wu and C.W. Magee, Am. Inst. Phys. Conf. Proc. 50, 603 (1979).

38 K. Gamo, Y. Yuba, A.H. Onaby, K. Murakami, S. Namba, and Y. Kawasaki, in "Laser and Electron Beam Processing of Materials", C. White and P.S. Peercy, eds. (Academic Press, N.Y. 1980) p. 322.

39 D.E. Davies, T.G. Ryan and J.P. Lorenzo, Appl. kPhys. Lett. 37, 612 (1980).

40 P.A. Pianetta, C.A. Stolte, and J.L. Hanson, ibid, p. 328.

41 P.A. Barnes, H.J. Leamy, J.M. Poate, S. D. Ferris, J.S. Williams, and G.K. Celler, Am. Inst. Phys. Conf. Proc. 50, 647 (1979).

42 B.J. Sealy, S.S. Kular, M.H. Badawi and K.G. Stephens, ibid p. 610.

43 H.B. Harrison and J.S. Williams, in "Laser and Electron Beam Processing of Materials", C.W. White and P.S. Peercy, eds. (Academic Press, N.Y., 1980) p. 481.

44 R.F. Wood, D.H. Lowndes, and W.H. Christie, Proc. Mat. Res. Soc. 1, 231, (1981).

45 A. G. Cullis, these proceedings.

46 R.B. Gold, R.A. Powell and J.F. Gibbons, Am. Inst. Phys. Conf. Ser. 50, 635 (1979).

47 G. Badertscher, R.R. Salathe and W. Luthy, Elect. Lett 16, 113 (1980).

48 S.S. Kular, B.J. Sealy, K.G. Stephens, D.K. Sadana, G.R. Booker, Sol. State. Elect. 23, 831 (1980).

49 S.T. Picraux, Rad. Eff. 17, 261 (1973).

50 H. Kehzu, M. Kuzuhara and Y. Takyama, J. Appl. Phys. 54, 4998 (1983).

51 T.E. Seidel, D.J. Lischner, R.V. Knoell, D.M. Maher, C.S. Pai and J.S. Williams, this symposium.

52 S.K. Tiku, J.B. Delaney, N.S. Gabriel and H.T. Yuan, this symposium.

53 R. Zuleeg, private communication.

54 J.C.C. Fan, this symposium.

55 C. Palmstrom, K.L. Kavanagh and J.W. Mayer, Symposium D, this conference

56 J. Soloman and S. Smith, private communication

57 S.J. Pearton, unpublished.

58 S.J. Pearton, D.O. Boerma, J.M. Poate, J.S. Williams, S.T. Johnson and K.G. Orrman-Rossiter, unpublished.

59 K.G. Orrmon-Rossiter, S.T. Johnson and J.S. Williams, Nucl. Instr. Methods (in press).

60 S. Tatsuta, T. Inata, S. Okamura and S. Hiyamizu, Jn. J. Appl. Phys. 23, L147 (1984).

EPITAXY AND DEFECTS IN LASER-IRRADIATED, SINGLE-CRYSTAL BISMUTH

AUBREY L. HELMS, JR.,[*] CLIFTON W. DRAPER,[**] DALE C. JACOBSON,[***]
JOHN M. POATE,[***] and STEVEN L. BERNASEK[*]
[*]Department of Chemistry, Frick Chemistry Laboratory, Princeton University,
Princeton, NJ 08554; [**]AT&T Technologies Engineering Research Center,
P.O. Box 900, Princeton, NJ 08540; [***]AT&T Bell Laboratories, 600 Mountain Avenue, Murray Hill, NJ 07974.

ABSTRACT

The (0001), ($\bar{1}$010), and (2$\bar{1}$$\bar{1}$0) faces of Bi have been pulse laser melted at 0.5 J/cm^2 with a Q-switched Ruby laser. Nomarski Interference Contrast Microscopy, Channeling, and selective chemical etching have been used to investigate the response to the laser irradiation. The response of the material and the level of damage is shown to be strongly correlated to the critical resolved shear stress characteristics in the particular crystallographic direction studied.

INTRODUCTION

Over recent years numerous studies, many detailed in the proceedings of this conference, have shown that elemental semiconductors can be epitaxially recrystallized following pulsed laser melting. Futhermore, the regrown surface layers are free of extended defects. To a lesser extent transition metals have been similarly investigated and although epitaxial regrowth is observed, all metals studied to date recrystallize with incorporated defect densities greater than the unirradiated starting crystals [1]. In most of the cases studied, a crystallographic orientation dependence was observed with the most densely packed surfaces generally showing epitaxial regrowth of poorer quality. The extent of the introduced disorder was found to correlate well with the laser fluence [1].

An interesting class of elements yet to be investigated is Group VB (As, Sb, and Bi). The Group VB semimetals have rhombohedral crystal structures in which the atoms are arranged in double layers such that each atom has three nearest neighbors and three at a slightly greater distance [2]. The bonding within the layers consists of strong, directional covalent bonds, while the bonding between layers is often described as changing from Van der Waals to metallic in nature as one increases in atomic number from As to Bi. This atomic arrangement results in very anisotropic bonding character in different crystallographic directions which is further evidenced by a large anisotropy in the thermal, physical, electrical, and mechanical properties. For example, the thermal conductivity for Bi is twice as high perpendicular to the trigonal axis as compared to parallel to the axis [3]. Another manifestation of the large bonding anisotropy is the fact that Bi has only one major slip system, parallel to the (0001) planes. Bismuth is also an interesting material because of its low melting point (271°C) and, similarly to Si, it contracts upon melting.

The results of laser melting and epitaxial regrowth of Bi single crystals are reported here. Crystal quality has been examined by Rutherford Backscattering and channeling, Nomarski Interference Contrast Microscopy, and selective chemical etching. The results indicate that the quality of epitaxial regrowth may be ordered (0001) > ($\bar{1}$010) > (2$\bar{1}$$\bar{1}$0). The (0001) face can be laser surface melted and recrystallized epitaxially with no increase in disorder with the Q-switched Ruby laser much the same as Si.

EXPERIMENTAL

Single crystal slices of Bi, oriented in the $\langle 0001 \rangle$, $\langle \bar{1}010 \rangle$, and $\langle 2\bar{1}\bar{1}0 \rangle$ directions, were electro-mechanically polished to a mirror finish. To insure a strain-free surface it was necessary to mount the samples on a specially designed teflon holder which held the crystal for polishing without damage. The electrolyte used for the polishing was a mixture of 10 parts methanol, 2.5 parts HCl, 2.5 parts H_2SO_4, and 1 part glycerin [4]. The current density was held a 1 amp/cm^2 and the samples were washed in distilled water followed by an ethanol rinse and dried with air.

The samples were irradiated with a Q-switched Ruby laser with a pulse width of 30 nsec. A bent waveguide was used to produce a homogeneous circular output 7 mm in diameter [5]. The fluences ranged from 0.5 J/cm^2, which produced a well defined melt spot, to 0.9 J/cm^2 which resulted in significant evaporation of material.

The crystalline quality of the samples was evaluated by both channeling [6] and selective chemical etching. A solution of 2% I_2 in methanol was used to decorate the intersection of screw dislocations on the Bi(0001) face as triangular pits [7]. The pits could then be counted and served as an indication of the extent of deformation or annealing.

RESULTS

The rhombohedral structure may be referenced to a hexagonal basis set. In this notation, the Bi(0001) surface is perpendicular to the C-axis, that is, parallel to the planes formed by the covalent double layers. The other two surfaces studied are perpendicular to this face. They were chosen because they form a set of surfaces displaying the anisotropic properties of Bi [FIG. 1].

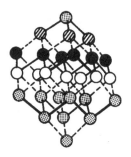

Figure 1. The rhombohedral structure of Bi showing the bonding within and between the bouble layers.

The regrowth in the $\langle 0001 \rangle$ direction is epitaxial and without an increase in the defect density as evidenced by both channeling and selective etching techniques [FIG. 2]. The slight decrease in the X_{min} for this surface from 9.5% for the virgin surface to 8.7% for the laser irradiated surface and the scarcity of etch pits observed inside the laser irradiated area are attributed to the release of stress, incorporated by polishing, during the liquid phase.

a)

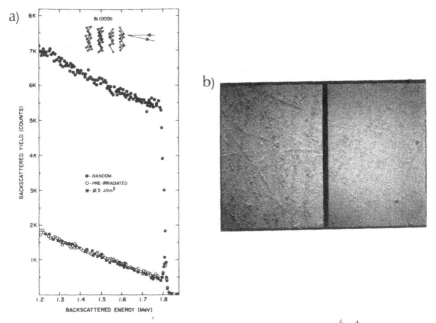

b)

Figure 2. a) RBS/Channeling data, using a 2.0 MeV ^{4}He$^+$ ion
beam on Bi(0001) showing pre- and postirradiated
spectra. b)Nomarski micrograph for Bi(0001) showing pre- (left) and postirradiated (right) surfaces, showing triangular etch pits at the intersection of screw dislocations with the surface.

The experimental χ_{min} value of 8.7% as compared to a calculated
minimum yield of 4.4% [8] indicates that the crystal was of good
overall quality. The <0001> half-angle [FIG. 3] was measured to be
0.54° in comparison with the calculated value of 0.46° obtained using
Moliere's screening function as outlined in Mayer and Rimini [8].

Figure 3. Half-angle spectra for pre-
and postirradiated Bi(0001).

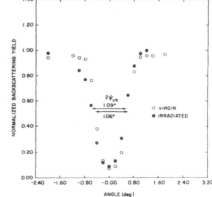

In the $\langle\bar{1}010\rangle$ and $\langle2\bar{1}\bar{1}0\rangle$ directions the channeling results show epitaxial regrowth but with an increased dechanneling rate which is indicative of increased disorder in the form of dislocations [FIG.4]. The dechanneling rate does not return to a value equal to that of the virgin crystal suggesting the crystal is deformed over depths of at least 9000 Å. For laser fluences of 0.5 J/cm^2 the $(\bar{1}010)$ surface showed a χ_{min} of 16.3% after laser irradiation as compared to 16.1% for the virgin surface. The χ_{min} for the $(2\bar{1}\bar{1}0)$ surface increased from 7.0% to 11.1% upon irradiation. The observed increase in the χ_{min} for these surfaces is indicative of damage and follows trends as a function of laser fluence observed in fcc, bcc, and hcp metals [1]. Figure 4b is a Nomarski Interference Contrast Micrograph showing the $(\bar{1}010)$ face before and after laser irradiation above the melt threshold. The (0001) basal slip planes are readily apparent. The number of slip lines did not show any correlation with laser fluence. Trends in the channeling results on crystals irradiated at fluences greater than 0.5 J/cm^2 are consistent with the above results.

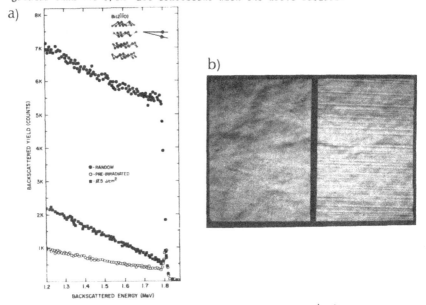

Figure 4. a) RBS/Channeling data, using a 2.0 MeV ^{4}He$^+$ ion beam, on Bi(2110) showing pre- and postirradiated spectra. b)Nomarski micrograph for Bi(1010) showing pre- (left) and postirradiated (right) surfaces at a laser fluence of 0.5 J/cm^2.

DISCUSSION

The response of a material to rapid melting and resolidification is a strong function of the thermal and mechanical properties in the material. This correlation has been used to explain the differences in the behavior observed in semiconductors and metals in response to pulsed laser melting. It is well documented that in the semiconductors, the regrowth is epitaxial with no increase in disorder [13], while the metals regrow epitaxially with an increase in disorder [1]. The rhombohedral crystal structure and large anisotopy in Bi provide a unique opportunity to study the influence of these properties and crystallographic orientation on the response to pulsed laser melting in the same element.

The increased dechanneling rate observed in the channeling spectra of laser melted metals has been shown to be due to an extended dislocation network [9]. The extent of this network often manifests itself in the form of slip lines depending on the particular crystallographic orientation [10,11]. It is well known from dislocation theory that slip lines are due to the activation of dislocation sources within the slip planes [12]. The multiplication of dislocations and their movement within the slip plane is a function of the applied stress relative to the critical resolved shear stress (CRSS) for the particular slip system in question [12].

The geometry of a pulsed laser experiment where the spot diameter is very large compared to the melt depth allows the problem to be approximated by a semi-infinite slab with one-dimensional heat flow [13]. This implies that the thermal and mechanical properties perpendicular to the surface are the important properties that determine the evolution of the system. The damage introduced is due to compressive and tensile stresses incurred during the heating and cooling phases of the pulsed laser melting [15].

The slip lines and increased dechanneling observed for the ($\bar{1}010$) and ($2\bar{1}10$) surfaces are consistent with the aforementioned model. For these two surfaces, the basal slip system of Bi is perpendicular to the surface. The slip lines appear where the (0001) basal planes intersect these surfaces. The slip within the melt puddle must be due to compressive stress introduced during the cooling of the hot solid because the liquid melt puddle is unable to support either shear or tensile stresses. The appearance of slip lines and the increased dechanneling rate indicates that these stresses were well above the CRSS for dislocation sources lying within the (0001) basal planes. The deep bulk damage, below the melt depth, is due to stress introduced during the heating, melted, and cooling phases of the experiment. The introduced stress is perpendicular to the surface because of the one dimensional heat flow. The CRSS is a very strong function of temperature for Bi [16]. This may explain why the slip lines on these two surfaces terminate very quickly outside the laser melted region. The heat flow is poor parallel to the surface so the temperature profile decreases rapidly as a function of distance beyond the edge of the melt puddle. As the temperature decreases, the CRSS increases dramatically and the introduced stress may not be sufficient to activate dislocation sources responsible for the slip lines. The behavior perpendicular to these surfaces is similar to that observed for metals and the increased disorder introduced by pulsed laser melting is consistent with this analogy.

The abscence of slip lines on the (0001) surface after pulsed laser melting is due to the fact that the main slip system is parallel to this surface. The stress introduced by the thermal gradients is not sufficient to activate the movement of dislocations perpendicular to the surface. The CRSS for the $\{0\bar{1}11\}$ family of planes, which slip along the $\langle 0\bar{1}1\bar{2}\rangle$ direction, has been reported to be 50 times higher than the CRSS for slip along the (0001) basal planes [17]. This also explains the lack of increased disorder observed in the channeling spectra. The behavior perpendicular to this surface is similar to that observed in the elemental semiconductors, and the high quality epitaxial regrowth of this surface is also consistent with this analogy.

444

CONCLUSIONS

Epitaxial regrowth was observed for three crystallographic directions for the semimetal, bismuth. A correlation was found between the thermal mechanical character and the quality of the epitaxial regrowth. In the <0001> direction, the epitaxial regrowth is good and without any detectable increase in disorder. The ($\bar{1}010$) and ($2\bar{1}\bar{1}0$) surfaces, regrow epitaxially, but with some increased disorder visible in the form of slip lines on the irradiated surfaces. The χ_{min} and half-angle were measured for the <0001> direction and compare well with calculated values.

ACKNOWLEDGEMENTS

The expertise and guidance of Mr. John Garno of AT & T Bell Labs in the preparation of high quality single crystal Bi surfaces is greatfully acknowledged. We also gratefully acknowledge the helpful discussions and suggestions of Dr. Everett Canning, Jr. of AT & T Technologies.

REFERENCES

1. L. Buene, E.N. Kaufmann, C.M. Preece, and C.W. Draper in LASER AND ELECTRON-BEAM SOLID INTERACTIONS AND MATERIALS PROCESSING, Gibbons, Hess, and Sigmon eds., (Elsevier North Holland, New York, 1981), pp. 591-597.
2. J. Donohue, THE SRUCTURES OF THE ELEMENTS, (Wiley, New York, 1974), pp. 311-316.
3. C. Ho, R. Powell, and P. Liley, THERMAL CONDUCTIVITY OF THE ELEMENTS: A COMPREHENSIVE REVIEW, (J. Phys. Chem. Ref. Data, Vol 3, Suppl. 1, 1974), pp. 79-93.
4. Mr. John Garno, AT & T Bell Labs, private communication.
5. A.G. Cullis, H.C. Webber, and P. Bailey J. of Physics E, 12, 688 (1979b).
6. E. Rimini, in MATERIALS CHARACTERIZATION USING ION BEAMS, Thomas and Cachard eds., (Plenum, New York, 1978), pp. 455.
7. L.C. Lowell, J.H. Wernick, J. Appl. Phys. 30, 234 (1959).
8. B.R. Appleton and G. Foti in ION BEAM HANDBOOK FOR MATERIALS ANALYSIS, Mayer and Rimini eds., (Academic, New York, 1977), pp. 67-107.
9. L. Buene, J.M. Poate, D.C. Jacobson, C.W. Draper, and J.K. Hirvonen, Appl. Phys. Lett., 37, 385 (1980).
10. F. Haessner and W Seitz, J. of Mat. Sci., 6, 16 (1971).
11. J.O. Porteus, M.J. Soileau, and C.W. Fountain, Appl Phys Lett., 29, 156 (1976).
12. C.E. Birchenall, PHYSICAL METALLURGY, (McGraw-Hill, New York, 1959), chapter 6.
13. P. Baeri, S.U. Campisano, G. Foti, and E. Rimini, J. Appl. Physics. 50, 788 (1979).
14. G. Foti and E. Rimini in LASER ANNEALING OF SEMICONDUCTORS, Poate and Mayer eds., (Academic, New York, 1982), pp. 203-245.
15. H.M. Musal, Jr., Symp. on Optical Materials for High Power Lasers, Boulder, Colorado (1979).
16. R.E. Reed-Hill, PHYSICAL METALLURGY PRINCIPLES, (D. Van Nostrand, New York, 1964), pp. 121.
17. S. Otake, H. Namazue, and N. Matsuno, Jap. J. of Appl. Phys., 19, 433 (1980).

BORON REDISTRIBUTION DURING TRANSIENT
THERMAL METAL SILICIDE GROWTH ON Si

C.J. SOFIELD[*], R.E. HARPER[+], and P.J. ROSSER[++]
*UKAEA Harwell, Didcot, Oxon., U.K.
+ Dept. of Electrical Engineering, University of Surrey, Guildford, Surrey, U.K.
++ Standard Telecommunication Laboratory Ltd., Harlow, U.K.

Abstract

We have used the heavy ion elastic recoil technique to study B distribution changes during Co and Ti di silicide formation on B implanted single crystal Si wafers. B diffuses to the interface of $TiSi_2$ and Si and to the surface of $CoSi_2$.

Introduction

The concentration of dopants at the junction between a metal silicide and silicon influences the Schottky barrier height and contact resistance of such a junction. As the most common method of producing such a junction is the thermally driven solid state reaction between a deposited metal film and a silicon substrate considerable interest has arisen in changes to Si dopant concentrations consequent upon this reaction. The element Boron is the most frequently used P type dopant in such situations. This light element is also more difficult to analyse, i.e. measure depth versus concentration than for example As which is ammenable to depth profiling by Rutherford backscattering [1].

The elastic recoil depth analysis (ERDA) technique had been used previously [2] to study Boron dopant levels in a-Si:H semiconductor material while simultaneously profiling the H level. This experience suggested this technique would be suitable for the study of boron redistribution following metal silicide formation. Although several metal Si systems were investigated (Ti, Co, Ta, W) the data presented here will concentrate on Boron redistribution following Titanium and Cobalt disilicide formation. These two disilicides have about the lowest resistivities observed for any of the metal silicides. Perhaps more interesting from the point of view of dopant redistribution studies they are believed to be formed by Si diffusion to form $TiSi_2$ and Co diffusion to form Co_2Si [3]. Thus one might see the consequences, if any, of these different influences on dopant redistribution.

Rapid thermal annealing (RTA) was used to produce the metal silicides, this method being chosen because of the increasing technological importance of RTA arising from the trend towards small dimensions in VLSI structures.

Experimental Methods

(a) Sample preparation

Commercially available 3" diameter <100> Si wafers of 30 Ωcm resistivity were implanted at 7° to the normal with 10keV $^{11}B^+$ ions to a dose of 1×10^{16} ions/cm^2. The implantation was carried out with a current of 5μA in a cryopumped chamber at a pressure of about 2×10^{-6} torr. This high dose was chosen to suit the sensitivity range of the ERDA technique. These wafers were then chemically cleaned in 1:10 HF solution for 10s, washed in super Q water and blow dried. Ti or Co metal was deposited in a Leybold Hereus Z550 D.C. magnetron sputter unit at pressures around 10^{-7} torr. The thickness of metal deposited, as determined by Rutherford backscattering was 235±10Å for Ti and 180±10Å for the Co films. The Boron concentration determined by ERDA analysis agreed with the implant dose and also indicated a B range R in Si of 380±40Å and a range straggling $< R^2 >^{\frac{1}{2}} = 240 \pm 20$Å. It is interesting to note these values are in agreement with those observed by Wach and Wittmaack [4] when SIMS profiling 10keV ^{11}B implanted into amorphous Si. Though thicker metal films would have ensured total consumption of the B implanted layer upon disilicide formation, some compromise was necessary to optimise the information obtainable from the ERDA technique: the thickness chosen permitted simultaneous profiling of B and any contaminants (H, C, O, N) present in the film or at the interfaces.

The silicides were prepared primarily by raster scanned 30keV electron beam annealing (EBA) using a 'SEZA' system (Lintech Instruments Co., Cambridge, U.K.). The steady state vacuum in this apparatus is typically $< 5 \times 10^{-6}$ torr. Details of temperature calibration and specimen mounting are given elsewhere [5]. The temperature rise time is dependent on beam power varying from about 2.5s to reach 1000°C at 40W to 10s to reach ~600°C at 5W. The final annealing temperatures we quote are reasonable to no worse than ±50°C.

Two other methods of annealing were used in the temperature range covered by the EBA method. A commercial flash lamp driven RTA device (Model 210.T-10 HEATPULSE, A.G. Associates, Sunnyvale, California, U.S.A.) was used to anneal at 900°C for 30s in an N_2 atmosphere. The temperature rise time to 900°C was 2.5s. Samples were also annealed at 700°C in a vacuum furnace (~2 10^{-6} torr) for 30 minutes.

The thickness and stoichiometry of the metal disilicides produced by RTA were determined using Rutherford backscattering of 1.5 MeV $^4He^+$ ions. The apparatus used [6] enabled film thicknesses to be determined to better than 35Å.

(b) Boron and light element profiling

The ERDA analysis, which is basically the inverse of Rutherford backscattering, was carried out by recoiling elements out of the specimens using incident 30 MeV Cl ions. The technique is fully described elsewhere [2] and is only summarised here. The collimated heavy ion beam was allowed to impinge on the specimen at an angle of 20° to its surface. A Melinex foil of sufficient thickness (> 8.9μm) to stop elastically scattered Cl ions was placed in front of a surface barrier detector situated at 30° to the beam direction. Any nuclei scattered out of the specimen with sufficient energy to pass through the Melinex could be detected: i.e. elements from the specimen surface up to a mass of 19AMU. Measurements of implanted H, B, Li in Si [2] samples prepared as standards give implant doses at the level of 1×10^{16} cm^{-2} to an accuracy of 5% and depth information to a similar accuracy. As the technique is sensitive to light elements the interaction of such elements present as residuals in the vacuum system, especially H, C, O, with specimen surface (perhaps chemically activated by the incident beams) can produce troublesome background signals in the form of surface peaks (H, C, O).

As an example of the range of information provided by such ERDA analysis the spectra of recoil elements from a Ti film deposited on B implanted Si before and after annealing is shown in Figure 1.

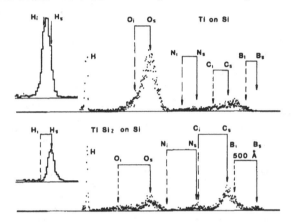

Figure 1: ERDA spectra of Ti/Si and the TiSi$_2$/Si film formed by EBA are shown. The position of the various elements at the film surface(s) and the Si interface (i) are indicated.

The various elements are indicated and their differing depth scales are shown for the Ti or TiSi$_2$ layer. These data show the dramatic reduction in H and O content of the film brought about by annealing to form TiSi$_2$. One can also see the difficulty experienced in positively identifying some elements, e.g. surface N and C at the TiSi$_2$ interface. The depth resolution available with this method is primarily limited by energy-loss straggling in the Melinex absorber. Typical values, [FWHM], for monolayers of H, B, C, O at the surface are 300Å, 140Å, 140Å, 150Å.

Results

1.   Boron redistribution upon TiSi$_2$ formation

The B distributions resulting from the various different annealing regimes are compared in Figure 2. If the B had not moved the volume change occuring during TiSi$_2$ formation would have led to an apparent re-scaling to give a B peak at about 250Å. Clearly therefore B has redistributed during the annealing. In general terms it is seen that the B distribution piles up below the TiSi$_2$/Si interface becoming broader with longer annealing and diffusing further into the TiSi$_2$. The pile up of B in the Si substrate was checked by etching the TiSi$_2$ off and re-measuring the remaining B level.

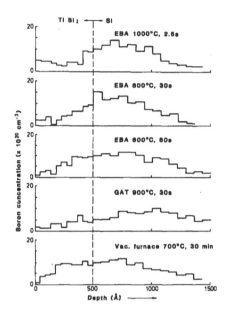

Figure 2: B distribution in TiSi₂ formed by various annealing procedures. The TiSi₂/Si interface is indicated. The surface B peak has a width set by the depth resolution.

The light element data showed that all the TiSi$_2$ films had O, N, C incorporated to varying degrees near the TiSi$_2$ surface. The furnace anneal produced O and C levels of 4.4×10^{16} cm^{-2} and 2.5×10^{16} cm^{-2} respectively. The RTA anneal in a N$_2$ atmosphere produced O (4.2×10^{16} cm^{-2}), N (2.4×10^{16} cm^{-2}) and C (1.5×10^{16} cm^{-2}). In general EBA produced C (1×10^{16} cm^{-2}) and O (1×10^{16} cm^{-2}). Background levels of C and O resulting from the ERDA analysis are in general less than or equal to 5×10^{15} atoms/cm^{-2}. The original Ti film contained about 6.5 atom % H through its bulk, O at the level of 6×10^{16} cm^{-2} and negligible C and N. These values are not surprising in view of the high reactivity of Ti. A comparison with the impurities observed in the EBA cases indicates a dramatic decrease in the O content of the films upon RTA in vacuum. In all cases hydrogen was driven out of the films upon annealing; decreasing by at least an order of magnitude in the bulk. The light element impurity levels at the TiSi$_2$ surface appear to show an inverse correlation with the amount of B found in the first 100Å of the TiSi$_2$.

2. Boron redistribution upon CoSi$_2$ formation

A comparison of the B distributions in CoSi$_2$ prepared by the different annealing regimes are given in Figure 3. B is always seen at the CoSi$_2$ surface but the magnitude of the broad B peak near the CoSi$_2$/Si interface and the amount of B extending into the Si substrate vary strongly with annealing conditions. In the case of the two highest temperature shortest time EBA treatments the deeper B peak virtually disappears. These distributions are in marked contrast to those found upon TiSi$_2$ formation.

Light element impurities in the CoSi$_2$ films were at much lower levels than in the TiSi$_2$ films. The starting material, 180Å Co on Si, showed evidence of O in the near surface region (~1.8×10^{16} cm^{-2}) and C at just above background levels (~7×10^{15} cm^{-2}). On annealing by any of the various methods the O near the surface decreased to ~9×10^{15}

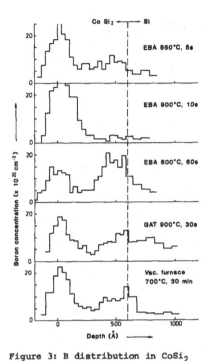

Figure 3: B distribution in CoSi$_2$ formed by various annealing procedures. The CoSi$_2$/Si interface is indicated. The surface B peak has a width set by the depth resolution.

cm^{-2} and the C increased to around 1.5×10^{16} cm^{-2}. No evidence of N incorporation into the surface or bulk was found for the sample annealed in N$_2$, suggesting that in the TiSi$_2$ case the nitrogen was taken up by reaction with to Ti rather than Si.

Discussion

In a study of B re-distribution during TiSi$_2$ formation Chow, Katz and Goehner [7] determined B profiles using SIMS and sputter profiling. They implanted B upto 6×10^{15} cm^{-2} i.e. slightly lower than the dose we used. After B implantation damage was annealed at 900°C for 30 min in N$_2$ to activate the B. Ti was then deposited and further annealing at either 650°C or 700°C for 30 mins in H$_2$ was carried out to form TiSi$_2$. The data they present for 700°C 30 min anneal of 6×10^{15} cm^{-2} B implanted Si indicates B redistribution to the Si side of the TiSi$_2$/Si interface. This result is very much as we observe. They however see a surface peak of B which contains up to 10% of the implanted B. The data for an implant dose of 5×10^{14} cm^{-2} shows a relatively stronger surface peak (around 60% of the B) and the rest of the B again located below the TiSi$_2$/Si interface.

It is thus difficult to reconcile their SIMS data with their conclusion that the B redistributes into the TiSi$_2$ film and accumulates at the TiSi$_2$ surface.

In the course of a solid-solid reaction different steps [8] such as nucleation, transfer of atoms across phase boundaries and most importantly diffusion in the reaction product occur. The transport of matter involved in these steps is essentially due to the mobility of point defects. It is known that B diffuses in Si [9] via a vacancy point defect coupling rather than interstitially. Thus one might expect point defects produced as a consequence of the metal/silicon reaction to affect the mobility of B near the reaction front. The presence of point defects in the Si produced by the low energy B implant may also contribute to both the diffusion of B and the diffusing species in silicide formation. Consideration of solid solubilities of the B in the different phases should also be important.

The B distribution we see in TiSi$_2$ and CoSi$_2$ are consistent with B redistribution in the direction of vacancy motion during silicide formation. As Si is the diffusing species [3] in the Ti/Si reaction vacancies will move away from the reaction front into the Si. Coupling of B diffusion to these could lead to a pile up of B below the $TiSi_2$/Si interface such as we observe. Subsequent thermal diffusion would tend to wash this distribution out; again as our results for longer time annealing suggest.

In the case of $CoSi_2$ formation it is not so certain whether Co and/or Si diffuser [3]. Thus Co diffusion might give rise to a surface peak of B while Si diffusion gives rise to B peak at the Si side of the $CoSi_2$/Si interface. Precipitation of B at interface grain boundaries may also contribute to B pile up near the $CoSi_2$/Si interface. However, the detailed knowledge of diffusion constant and solid solubilities required to model these processes is not yet available so that these mechanistic suggestions must be regarded as highly speculative.

Several possibilities exist for the major difference in B distribution results in $TiSi_2$ observed by Chow et al and those reported here, namely the presence of a strong B surface peak in their samples. SIMS sputter profiling may have difficulties with Oxygen present as an impurity near the film surface (possibly causing enhanced ion yield) or with differential Ti/Si sputtering rates. Another possibility may lie in the implant damage left in the Si by our B implants; Chow et al annealed to remove this before Ti deposition.

Obviously the final distribution of a dopant such as B following annealing to form metal silicide can affect the characteristics of a device produced by such techniques. The empirical data provided by this study is useful in this context. However it is also clear that there is a need for a deeper understanding of the mechanisms by which dopants diffuse during solid state reaction of metal and silicon.

Acknowledgements

We wish to thank, A.M. Stoneham, G. Dearnaley and J.M. Shannon for helpful discussion. One of us (REH) would like to thank Phillips Research Labs. for funding of a research fellowship.

1.   E.A. Maydell - Ondrusz, R.E. Napper, I.W. Wilson and K.G. Stephens to be published in 'Vacuum'.
2.   P.M. Read, C.J. Sofield, M.C. Franks, G.B. Scott, M.J. Thwaites. Thin Solid Films, 110 (1983), 251.
3.   G. Ottaviani. Materials Research Society, Symposium Proceeding, Vol. 26 ed. J.E.E. Baglin, E.R. Cambell, W.K. Chu. North Holland, New York 1984, p21.
4.   W. Wach and K. Wittmaack, Nucl. Instr. Met. Phys. Res. 194 (1982), 113.
5.   E.A. Maydell-Ondrusz, R.E. Harper, A. Abid, P.L.F. Hemment and K.G. Stephens. Materials Research Society, Symposium Proceeding. Vol. 25 ed.
6.   P.L.F. Hemment, J.E. Mynard, E.A. Maydell - Ondrusz and K.G. Stephens, Proc. IV Int. Conf. on Ion Implantation Equipment and Techniques (Berchtesgadin 1982).
     T.P. Chow, W. Katz, and R. Goehner. Electrochem Society, Extended Abs. Vol. 84-1.
8.   Treatise on Solid State Chemistry Vol. 4 Reactivity of Solids Ed. N.B. Hannay, Plenum Press, New York 1986.
9.   R.B. Fair in 'Impurity Doping Processes in Silicon'. Ed. F.F.Y. Wang, North Holland, New York 1981, p320.

OPTICAL FURNACE ANNEALING

NEIL J. BARRETT, D.C. BARTLE, A.G. TODD AND J.D. GRANGE
GEC Research Laboratories, Hirst Research Centre, Wembley, Middlesex, UK

ABSTRACT

The rapid annealing of Be implanted GaAs has produced electrical activations of 70% for doses of 5×10^{14} cm^{-2} and hole profiles similar to the as implanted distribution. Si implanted GaAs has also been investigated using doses of 1 and 2×10^{14} cm^{-2} with sheet resistivities of 40 Ω/square after rapid thermal annealing. GaAs has been annealed with W-Si on the surface for the application of self aligned gate FET technology. Temperature cycles upto 850°C are required to activate the implanted dopant. Such cycles do not cause inter diffusion between the W-Si and GaAs or deterioration of the metallisation surface morphology.

INTRODUCTION

In many circuit applications it is imperative that high dopant activation efficiencies are achieved for both p- and n-type implants. Ion implantation annealing of Be implanted GaAs has generally been carried out using a conventional furnace anneal between 700 and 900°C. However, to regain crystallinity an anneal at 900°C is necessary [1,2] which results in the loss of Be by out-diffusion at the surface and a broadening of the Be atomic distribution [3]. A solution to this problem is rapid thermal annealing [4-7].

Carrier concentrations for a post ion implantation anneal of Si implanted GaAs generally do not exceed 2×10^{18} cm^{-3} [8-10]. For high atomic concentrations, this has been attributed to the amphoteric nature of Si, i.e. its location on both Ga lattice sites as a donor Si_{Ga} and As lattice sites as an acceptor Si_{As}. Experimental work [11] is indicating that the Si_{Ga}-Si_{As} neutral pair is the dominant defect once carrier concentrations saturate at 2×10^{18} cm^{-3}. This Si defect model is frequently used to interpret experimental results; Greiner and Gibbons [12] for ion implantation and Hurle [13] for crystal growth are recent examples. The formation of this Si complex is also temperature dependent with transitions from Si donors to Si acceptors beginning to take place above 900°C [14]. The situation is complicated further when ion implantation induced crystal damage is taken into consideration [10] in the presence of other impurities [15] and vacancies [16]. There have been a number of publications on the annealing of Si implanted GaAs by incoherent light sources [17-22]. Davies et al [18] achieved a 6.5×10^{18} cm^{-3} carrier concentration after an illumination time of 2½ s for a dose of 4×10^{14} Si cm^{-2} at 200 keV. The short anneal cycle may be of considerable importance for obtaining high carrier concentrations for Si implanted GaAs.

If a self-aligned GaAs MESFET process is used during the post ion implantation anneal the gate metal may be still present. In order to anneal the implanted dopant, without affecting the gate metal properties [23] it is necessary to keep the anneal time to a minimum. A suitable refractory alloy system is W-Si which has been used in this study.

EXPERIMENTAL

The optical furnace consists of twelve 1½ kW linear tungsten halogen lamps, positioned above and under the wafer chamber (Figure 1). (Further

452

details are given in Reference 7). GaAs slices were loaded onto a carbon disc susceptor for annealing. A thermocouple was secured to the carbon disc to measure the sample temperature in a nitrogen ambient. $9Be^+$ ions generated from a BeQ ceramic source [24], and $28Si^+$ ions, were implanted nominally at room temperature at 7° to the normal into semi-insulating undoped LEC(100) GaAs. Unimplanted substrate samples maintained a sheet resistivity of greater than 10^7 Ω/square after an 850°C/20 minute anneal in an arsine ambient [25]. 40 and 75 keV Be was implanted at a dose of 1×10^{15} cm^{-2} and 5×10^{14} cm^{-2} respectively. Post implantation annealing in a conventional furnace was undertaken using 20 minute anneal cycles with a plasma deposited Si$_3$N$_4$ encapsulant [26] or capless in an arsine ambient [25]. Rapid anneal cycles were carried out in an optical furnace with a Si$_3$N$_4$ encapsulant or with a silicon proximity encapsulant positioned on the GaAs. 360 keV and 400 keV Si was implanted at doses of 1 and 2×10^{14} cm^{-2} respectively. Optical furnace annealing was used with a Si$_3$N$_4$ encapsulant. Differential Hall and step measurements [27] were used to determine electrical hole and carrier concentrations and Hall mobility profiles. Dopant atomic distributions were obtained by SIMS.

W-Si alloy layers were deposited by co-sputtering from elemental targets. The deposits, which were approximately 1000 A thick were either homogeneous or were multilayer with a layer thickness of a few nm. The resistance between two indium contacts along two edges of a square was used to measure the films sheet resistivities. Auger electron spectrometry and X-ray florescence analysis were used for film characterisation.

Fig 1: Side view of the 18 kW optical furnace

The anneal times quoted are the dwell times at temperature, for example 850°C/30 s. Temperature rise rates of about 150°C/s were used for Be implanted GaAs, from a background temperature of 600°C which was maintained for a few seconds. All other anneals were within a maximum temperature rise rate of 40 and 100°C/s to approximately 800°C and to 1000°C respectively. Thermal pulse anneals, straight to the set temperature and down again, had a dwell time of zero. In these cases the anneal is quoted as, for example 950°C/0s.

RESULTS

Be: A zero dwell time was necessary at 950°C to limit diffusion and obtain a maximum electrical activation of 26% for a 40 keV Be implant at a dose of 1×10^{15} cm^{-2} (Figure 2). A Si$_3$N$_4$ or a Si proximity encapsulant gave very similar results with hole concentrations of 1×10^{19} cm^{-3}. A higher hole concentration was measured for a 60 s dwell time at 850°C rather than a standard anneal for 20 minutes which saturated with a hole concentration between 1 and 2×10^{18} cm^{-3}.

The maximum activation obtained for a 5×10^{14} cm^{-2} dose at 75 keV was 70%. Figure 3 emphasises the disadvantage of a long anneal time, such as a standard furnace anneal for 20 minutes compared with the as implanted atomic

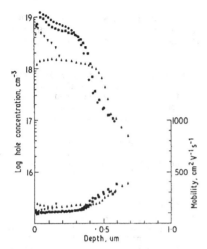

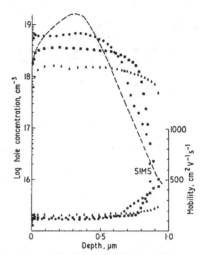

Fig 2: OFA for 40 keV Be with doses of 1 x 10^15 cm^-2 compared with a standard furnace anneal using a Si3N4 encapsulant. Temperature/times were ● 950°C/0 s, ■ 900°C/5 s, ▼ 850°C/60 s and ▲ 850°C/20 minutes.

Fig 3: OFA for 75 keV Be with doses of 5 x 10^14 cm^-2 compared with a standard anneal using a Si3N4 encapsulant. Temperature/times were ● 1000°C/0 s, ■ 900°C/25 s, and ▲ 850°C/20 minutes. SIMS is the as implanted atomic distribution.

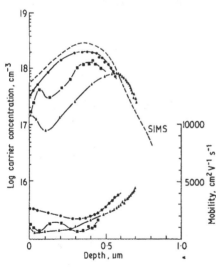

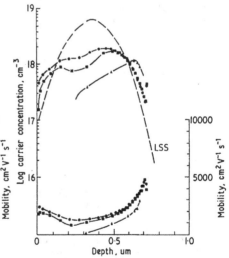

Fig 4: OFA for 360 keV Si with doses of 1 x 10^14 cm^-2 using a Si3N4 encapsulant. Temperature/times were ● 950°C/0 s + 800°C/30 s, ■ 950°C/0 s, and ▲ 800°C/30 s. SIMS is the as implanted atomic distribution.

Fig 5: OFA for 400 keV Si with doses of 2 x 10^14 cm^-2 using a Si3N4 encapsulant. Temperature/times were ● 950°C/0 s + 800°C/30 s, ■ 950°C/0 s and ▲ 800°C/30 s. LSS is the calculated as implanted atomic distribution.

distribution. Even a dwell time of 25 s as compared with 5 s at 900°C (Figure 2) is detrimental to the peak hole concentration. For the 850°C/20 minute anneals SIMS profiles were the same as the electrical profiles both for capped or capless in an arsine ambient.

Si: To obtain high electrical activations with doses of 1 and 2 x 10^{14} cm^{-2} a thermal pulse anneal to 1000°C is necessary. If a pre anneal is carried out before the thermal pulse anneal, it is possible to reduce this temperature to 950°C for a dose of 1 x 10^{14} cm^{-2} and obtain similar results. Pre anneals at 800°C/30 s (Figures 4 and 5) were an advantage combined with a thermal pulse anneal to 950°C. Pre anneals at 850°C/30 s compared with 800°C/30 s were not an advantage, generally producing lower carrier concentrations. Temperature rise rates of 40°C/s and 100°C/s to 800°C and 950°C respectively were used for all three anneal cycles in Figures 4 and 5. The maximum electrical activation was 70 and 40% for doses of 1 and 2 x 10^{14} cm^{-2} respectively.

WSi: The surface morphology of the W-Si deposits after optical furnace annealing has proved to be far superior to that produced by conventional furnace annealing. Auger electron spectrometry with a resolution of about 100 A showed no diffusion at the W-Si:GaAs interface (Figure 6) for 850°C/30 s anneals. For multilayered W-Si (Figure 6) a few percent of oxygen through the film, increasing at the W-Si:GaAs interface, was observed. Homogeneous layers with a deposition rate of approximately 100 A/min (2-3 times that of multilayered films) had no detectable oxygen levels through the film (<1 atomic %) with reduced levels at the interface.

Figure 7 shows the results of resistivity measurements after annealing for 30 s at temperature upto 950°C for $W_{60}Si_{40}$ and 850°C for $W_{90}Si_{10}$. Films with Si concentrations of 20 atomic % or less tended to have stress lines on the surface of the layer even before annealing. Resistivities decreased with anneal temperature to approximately 1 x 10^{-4} Ωcm and 6 x 10^{-3} Ωcm for $W_{60}Si_{40}$ and $W_{90}Si_{10}$ respectively.

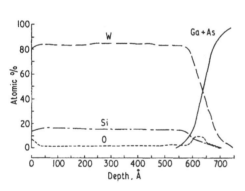

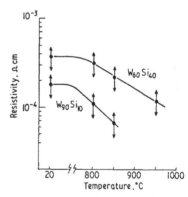

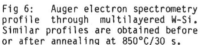

Fig 6: Auger electron spectrometry profile through multilayered W-Si. Similar profiles are obtained before or after annealing at 850°C/30 s.

Fig 7: Resistivity of WSi before OFA and after a 30 s anneal at temperature

DISCUSSION

There was very little difference between the electrical profiles of Be implanted GaAs annealed with a Si_3N_4 encapsulant or in an arsine ambient.

SIMS profiles after annealing have indicated that the majority of the Be has been lost from the surface during a standard furnace anneal. However, if the dwell time is reduced to 60 s or less (Figure 2), high anneal temperatures are an advantage. The encapsulants Si_3N_4 or Si proximity, made very litte difference to the electrical profiles using an optical furnace anneal if the dwell time was no more than a few seconds. Be is electrically activated in less than 5 s at 900°C and at 950°C it was only necessary to go to temperature with a zero dwell time. If the dwell time is longer than necessary, the results indicate that Be is lost from the surface in agreement with McLevige et al [3].

Pre anneal cycles at 800°C increased the electrical activation of high doses of Si. However, 850°C pre anneal cycles were not so advantageous. Results suggest that it is preferable to have some residual damage left after the pre anneal. Thus, at annealing temperatures above 900°C, when the Si prefers the As lattice site, there may be higher concentrations of Ga vacancies with an 800°C pre anneal as opposed to an 850°C pre anneal. With additional Ga vacancies the activation of Si will be higher. The alternative explanations are that there is out diffusion of compensating impurities or encapsulation failure. At present impurity diffusion can not be ruled out but there is no evidence at present to suggest that a pre anneal at 850 instead of 800°C for 30 s would cause a large enough change in the impurity distribution to account for the decrease in electrical activation. The Si_3N_4 after annealing and the GaAs surface were of high quality. This was reflected in the capped and capless Be results where any surface loss would have been observed as a shift in the electrical profiles.

W-Si films with a high tungsten content produced the lowest resistivities (Figure 7). However, films with 80 atomic % or more of tungsten produced strain lines before or after annealing. The resistivity increased with the Si content of the film. 30 to 40 atomic % of Si with resistivities of 1 and 2 x 10^{-4} Ω cm after annealing is a suitable compromise.

CONCLUSIONS

The three methods used to prevent the GaAs surface from degrading have produced very similar electrical profiles, indicating that the activation or diffusion of Be is not affected by surface strain generated at the Si_3N_4/GaAs interface for the atomic concentrations used. The results from SIMS suggest that there is Be out diffusion from the GaAs surface as well as through the Si_3N_4. Hole concentrations of upto 1 x 10^{19} cm^{-3} were obtained. Pre anneal cycles at 800°C before the thermal pulse anneals are advantageous to Si implanted GaAs. WSi withstands an OFA at temperatures which will electrically activate Si implanted GaAs. This is due to the short anneal times which are not possible with conventional furnace anneals.

ACKNOWLEDGEMENT

This work has been carried out with the support of the Procurement Executive, Ministry of Defence, sponsored by DCVD.

The authors wish to thank Dr Colin Wood for his continued support and technical interest in this optical furnace programme. Thanks are also due to P G Harris for the Auger electron spectrometry, R Nicholls for Hall effect measurements, A J Vale for X-ray florescence measurements and A J Taylor for W-Si deposition. M G Dowsett of the City of London Polytechnic provided the SIMS profiles.

456

REFERENCES

1   Chatterjee, P.K., Vaidyanathan, K.V., McLevige, M.V. and Streetman, B.G., Appl. Phys. Lett. 27, (1975) p 567
2   McLevige, W.V., Helix, M.J., Vaidyanathan, K.V. and Streetman, B.G., J. Appl. Phys. 48, (1977) p 3342
3   McLevige, M.V., Vaidyanathan, K.V., Streetman, B.G., Comas, J. and Plew, L., Sol. State. Comm. 25, (1978) p 1003
4   Banerjee, S.K., Dejale, R.Y., Soda, K.J. and Streetman, B.G., IEEE Elect. Dev. ED-30. (1983) p 1755
5   Tabatabaie-Alavi, K., Choudbury, A.N.M., Kanbe, H., Fonstad, C.G. and Gelpey, J.C., Appl. Phys. Lett. 43 (1983) p 647
6   Asbeck, P.M., Miller, D.L., Babcock, E.J. and Kirkpatrick, C.G., IEEE Elect. Dev. Lett. EDL-4, (1983) p 81
7   Barrett, N.J., Bartle, D.C., Nicholls, R. and Grange, J.D., Proc. 11th International Symposium on GaAs and related compounds, Biarritz (1984)
8   Kamalov, M.N., Kolesnik, L.I., Milvidskii, M.G., Rakov, V.V. and Shershakova, I.N., Sov. Phys. Semicond. 12, (1978) p 340
9   Tandon, J.L., Nicolet, M.A. and Eisen, F.H., Appl. Phys. Lett. 34, (1979) 165
10 Masuyama, A., Nicolet, M.A., Golecki, I., Tandon, J.L., Sadana, D.K. and Washburn, J., J. Appl. Phys. Lett. 36, (1980) p 749
11 Bhattacharya, R.S., Rai, A.K., Yeo, Y.K., Pronko, P.P., Ling, S.C., Wilson, S.R. and Park, Y.S., J. Appl. Phys. 54, (1983) p 2329
12 Greiner, M.E. and Gibbons, J.F., Appl. Phys. Lett. 44, (1984) p 750
13 Hurle, D.T.J. J. Cry. Growth 50, (1980) p 638
14 Hwang, C.J., J. Appl. Phys. 39, (1968) p 5347
15 Kanber, H., Feng, M. and Whelan, J.M., Appl. Phys. Lett. 40, (1982) p 960
16 Onuma, T., Hirao, T. and Sugawa, T., J Electrochem. Soc.: Solid State Sc. and Tech. 129, (1982) p 837
17 Arai, M., Nishiyama, K. and Watanabe, N., Jpn. J. Appl. Phys. 20, (1981) p L124
18 Davies, D.E., McNally, P.J., Lorenzo, J.P. and Julian, M., Elect. Dev. Lett. EDL-3, (1982) p 102
19 Kuzuhana, M., Kohzu, H. and Takayama, Y., Appl. Phys. Lett. 41, (1982) p 755
20 Ito, K. Yoshida, M., Otsubo, M. and Murotani, T., Jpn. J. Appl. Phy. 22, (1983) p L299
21 Badawi, M.H. and Mun, J., Elec. Lett 20, (1984) p 125
22 Kohzu, H., Kuzuharam, M and Takayama, Y., J. Appl. Phys. 54, (1983) p 4998
23 Ohnishi, T., Yamaguchi, Y., Inada, T., Yokayama, N. and Nishi, H., IEEE Elect. Dev. Lett. EDL-5, (1984) p 403
24 Sugata, S., Tsukada, N. Nakajima, M., Kuramoto, K. and Mita, Y., Jpn. J. Appl. Phys. 22, (1983) p L470
25 Grange, J.D. and Wickenden, D.K., Sol. State. Elect. 25, (1983) p 313
26 Bartle, D., Andrews, D.C., Grange, J.D., Harris, P., Trigg, A.D. and Wickenden, D., Vacuum 34 (1984) p 315
27 Stewart, C., Medland, J.D. and Wickenden, D.K., Semi-insulating III-V Materials Conference, Oregon (1984)

RAPID THERMAL ANNEALING OF TiSi$_2$ FOR INTERCONNECTS

P.J. ROSSER* AND G.J. TOMKINS**
*Standard Telecommunication Laboratories Limited, London Road, Harlow, Essex, UK
**Standard Telephones and Cables, Maidstone Road, Foots Cray, Kent, UK.

ABSTRACT

A self-aligned polycide gate interconnect process has been developed at STL in order to reduce the RC delays in existing and planned MOS circuits. This process has been chosen due to its ease of implementation into an existing process line. Incoherent lamp annealing is used to form the silicide after the deposition of titanium over patterned polysilicon.

This paper will discuss the various materials and processes considered, outlining the significant attributes of each. The factors which must be controlled to achieve a practical process are discussed, together with the degree of redistribution of dopants and the role of other impurities.

INTRODUCTION

As device sizes are reduced so the resistivity of the interconnects becomes the dominant limiting factor in the speed of the device.

The first section of this paper briefly reviews the various alternative materials which have been considered for gate level interconnects. The choice is discussed in the context of the proposed process implementation.

The second section describes the various process options considered, concentrating on the STALICIDE process developed at STL.

Finally, the interaction between titanium and silicon during rapid thermal processing in a nitrogen ambient is considered in more detail.

Material Options

Heavily doped polysilicon gate level interconnects have been accepted and used in the semiconductor industry for many years. As device geometries are reduced to below 3 μm so the propagation delay of the polysilicon interconnects becomes a significant, if not dominant, factor in the overall device delay (Figure 1).

Many alternative materials have been considered, both for use in the 'short term', where the requirement is for minimum changes to existing processes and in the 'long term' where major process changes are envisaged. For the latter case refractory metals such as molybdenum or tungsten have been seen as prime candidates. Some process modifications would be required as clearly an

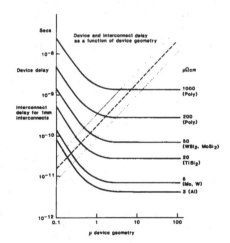

Fig. 1

insulating oxide could not be grown on these materials. Indeed, oxidising ambients must generally be avoided due to the high vapour pressure of the metal oxides. Aluminium is a further candidate for the 'long term' applications, but only if a process can be designed around its poor tolerance to high temperatures.

For 'short term' applications a 'drop-in' replacement for polysilicon is needed and it was in order to fulfil this role that work started on the refractory metal silicides. These compounds have very low resistivities, are readily deposited and etched and, what is most important can be oxidised to form a high quality silicon dioxide layer of the same high quality as that used in the polysilicon based processes. It is this property that sets the silicides apart from all other materials as a drop-in replacement for polysilicon.

The need for stability during subsequent high temperature anneals limits the choice to the refractory metal silicides, notably WSi_2, $MoSi_2$, $TaSi_2$, and $TiSi_2$. Of these, tungsten and molybdenum disilicide can suffer from the 'pest' reaction during oxidation due to the volatility of the metal oxides.[1] Titanium disilicide etches rapidly in hydrofluoric acid based etches and tantalum is more difficult to etch,[2] making a tantalum based self-aligned process impossible if only etches compatible with silicon processing are considered.

For this work the aim was a silicide process to be used in a 1 μm MOS process using dry etching. The silicide need not be exposed to hydrofluoric acid. This combined with the significantly lower resistivity of titanium disilicide compared with the other refractory metal silicides resulted in titanium disilicide being selected for this application.

Process Options

The three most commonly quoted silicide based processes are the silicide, polycide and SALICIDE[3] processes. The silicide process avoids the complication of etching a two-layer structure, as is required in the polycide process. The polycide process however retains the accepted polysilicon over silicon dioxide gate structure. The SALICIDE process avoids the complication of etching a new material, but does need sidewall spacers and all the silicon exposed after spacer etching (or polysilicon etching) is silicided.

For our application it is not permissible to silicide all the exposed silicon after the polysilicon etch and no sidewall spacers are used. The self-aligned approach is very promising however and so the STALICIDE process was developed (Figure 2). The only new etch step required is the selective wash for titanium. The result is a self-aligned polycide process with only the gate interconnect being silicided.

SILICIDE FORMATION

During silicide formation it is inevitable that the regions of silicon dioxide on which titanium has been deposited will begin to be reduced since titanium dioxide is thermodynamically favoured over silicon dioxide. In the STALICIDE process, it is essential to ensure that the 'gate oxide' over source and drain regions is not reduced.

Figure 3 shows three curves obtained from 100 nm thick titanium films evaporated onto polysilicon films or onto 30 nm thick oxide films, both on silicon substrates. Two of these curves show data from the titanium over polysilicon samples. The first indicates the point at which some samples became completely silicided while the second indicates the point at which all samples became silicided. The region between these two is considered to be the area in which slight differences in film interfaces and purity dominate. A third curve shows the results

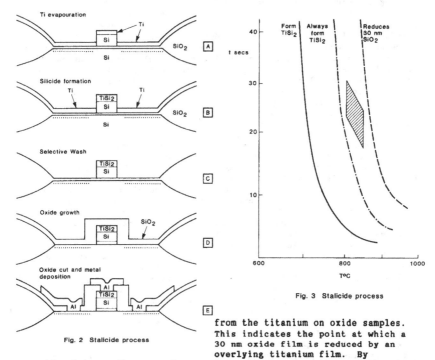

Fig. 2 Stalicide process

Fig. 3 Stalicide process

from the titanium on oxide samples. This indicates the point at which a 30 nm oxide film is reduced by an overlying titanium film. By operating between the curve for complete silicide formation and that for oxide reduction it is possible to silicide routinely polysilicon interconnects without reducing the 60 nm gate oxide too severely. Further improvements are possible using a very short, high temperature anneal to overcome interface effects followed by a lower temperature anneal to enable complete siliciding without further reduction of the oxide.

Impurities, notably oxygen, in the evaporated titanium film and nitrogen which diffuses into the titanium during the anneal are snowploughed ahead of the disilicide as the silicon diffuses into the titanium. Figure 4 shows how the initial oxygen concentration is first homogenised throughout the titanium film and then caused to peak again as the silicide forms. Meanwhile the nitrogen ambient causes a nitride to form at the surface, resulting in a characteristically gold coloured nitride layer over an oxide layer over the disilicide. Subsequent annealing does not alter this. The last diagram on Figure 4 shows the same film after the selective wash used to remove unreacted titanium. The oxide and nitride layers have been removed by the wash leaving only a thin silicon oxide on the stoichiometric disilicide.

Dopant Diffusion

Polysilicon, when used for interconnects in existing MOS processes, is very highly doped in order to reduce its resistivity. During the silicide formation anneal it is inevitable that either some dopant is incorporated in the silicide or that the doping level in the silicon increases. Phosphorus redistribution has been studied using 4PP, AES, RBS and SIMS analysis. Phosphorus from the polysilicon diffuses into the growing silicide region and on into the unreacted titanium at the surface. Its distribution peaks just below the surface, in a similar position to the oxygen peak. The vast majority of the phosphorus is incorporated in this impurity layer during the anneal (Figure 5) and subsequently removed in the selective wash. Of the remaining phosphorus, most is to be found in the silicide layer with only between one and five per cent being left in the polysilicon. Figure 6 shows the phosphorus remaining in the silicide even after subsequent oxidation.

This dramatic change in doping level does not, however, appear to affect the threshold voltage significantly, the resulting phosphorus concentration being relatively high. The sheet resistance of the polysilicon film rises dramatically to several hundred ohms per square, but this change is not significant as the polysilicon resistivity is shunted by the low resistivity silicide. The phosphorus in the silicide does not noticeably affect its resistivity.

Arsenic redistribution has been studied using 4PP, AES and RBS. In this case, single crystal silicon was chosen as the source of silicon for the silicide in order to emulate the silicidation of heavily doped source and drain regions, as would occur in a fully self-aligned process. During the silicide formation anneal the arsenic diffuses into the silicide. Very little arsenic diffuses into the unreacted titanium (Figures 7 and 8 show the arsenic distribution before and after the anneal). After annealing titanium deposited over a silicon substrate which had been implanted with 10^{16} arsenic ions per sq. cm. at 40 keV, the arsenic level in the silicide was at least four times that in the silicon. This loss of dopant could prove significant. The problem could possibly be overcome by implanting higher doses into the silicon, but other workers[4] have reported that titanium does not form a silicide on very heavily arsenic implanted silicon.

Boron redistribution has been reported elsewhere[5].

In the STALICIDE process (Figure 2), it is important to note that lines of polysilicon over an oxide of varying thickness are being silicided, and it is in considering the formation of silicides on these non-planar regions that the anneal ambient becomes of great relevance. For example, by annealing in argon, a sheet resistance of 0.8 $\Omega/\square$ can be achieved, comparing very favourably with the value of 1.2 $\Omega/\square$ achieved with a nitrogen anneal. The high sheet resistance of the film annealed in nitrogen is due to the formation and subsequent removal of a titanium nitride layer at the surface. By annealing in an inert ambient such as argon the 'unreacted titanium' removed from above the silicide is comprised solely of titanium reacted with impurities incorporated during the deposition. This leaves more titanium available for siliciding, hence the lower resistance.

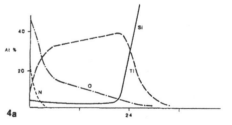

4a

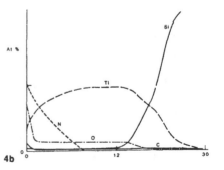

4b

Despite its relatively high resistance, all anneals are carried out in nitrogen in order to avoid severe problems with lateral diffusion, sidewall siliciding and gate oxide reduction. If the selective wash were capable of removing all titanium oxides whilst leaving titanium silicides untouched then an oxygen anneal ambient would in fact be preferable to an inert ambient.

Lateral diffusion of titanium disilicide is caused by silicon diffusing horizontally and reacting with the titanium over the silicon dioxide. If the silicide formation anneal is carried out in nitrogen then this titanium is nitrided completely before the silicon can diffuse any significant distance. In the case of anneals in argon this is not the case and silicides can be formed spreading several microns from the original polysilicon line. Similarly with sidewall silicide formation, by depositing titanium with some degree of

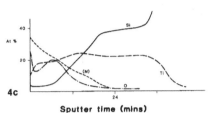

4c

Sputter time (mins)

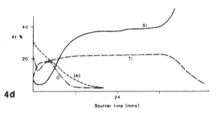

4d

Sputter time (mins)

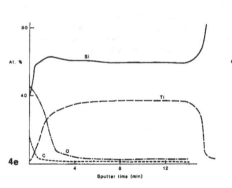

4e

Sputter time (min)

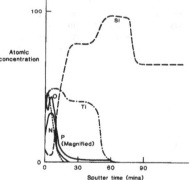

Sputter time (mins)

Fig. 5 Phosphorus distribution after silicide formation

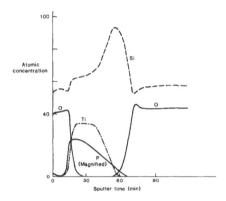

Fig. 6 Phosphorus distribution after selective
wash and oxidation

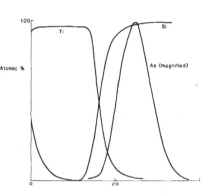

Fig. 7 Arsenic re-distribution as deposited film

anisotropy it is possible to avoid the formation of any sidewall
silicides by nitriding all of the available titanium. Figure 9 shows the
thickness of titanium nitrided to be largely independent of the deposited
titanium thickness, while Figure 10 shows some of the undesirable effects
of a sidewall silicide on line profile after subsequent oxidation.
Clearly a sidewall silicide is unavoidable using an inert anneal ambient.

Reduction of the gate oxide is a severe problem in the STALICIDE
process. By forming titanium nitride rapidly over any silicon dioxide
region the free energy change of any subsequent reduction of silicon
dioxide is significantly reduced.

Finally, the films annealed in nitrogen are significantly less rough
than those annealed in inert ambients, possibly again due to the
competition between the two reactions, siliciding and nitriding.

It is worth noting that in this application the selective wash is
not used to remove unreacted titanium, but unsilicided titanium.

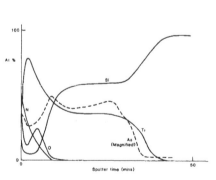

Fig. 8 Arsenic re-distribution after silicide formation

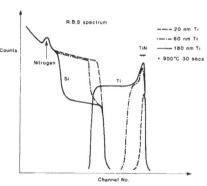

Fig. 9 Silicide formation: Ti thickness

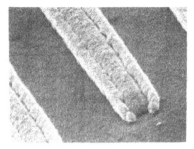

11,000X

Fig. 10

The beneficial role of nitrogen in annealing titanium to form titanium disilicide has been outlined, but even the oxygen in the original titanium plays a useful part in enabling reproducible silicide formation. Whilst a layer of titanium oxides remains between the surface nitride and the titanium disilicide (Figure 4) the titanium disilicide can continue to be annealed and is not significantly changed by the anneal. If this oxide layer is not present and titanium nitride is in intimate contact with the titanium disilicide, then any further anneal in a nitrogen ambient will enable the nitride film to grow at the expense of the silicide film. This nitriding of the silicide is undesirable during the silicide formation, but could prove of great use in forming a titanium nitride diffusion barrier layer over the silicide to prevent the diffusion of underlying silicon into subsequently deposited aluminium layers.[6],[7]

CONCLUSIONS

A novel self-aligned polycide process, the STALICIDE process, has been outlined. This process has been designed in order to enable implementation into existing and planned MOS processes using polysilicon interconnects with a minimum of process perturbation.

Dopant redistribution during silicide formation has been recognised as an area for concern, but in this particular application doping levels over the gate region remain high enough to retain the same threshold voltage as degenerately doped polysilicon controls. Impurities such as oxygen and nitrogen incorporated in the titanium film are shown to perform useful roles in controlling silicide formation and enabling the subsequent formation of self-aligned diffusion barrier layers.

ACKNOWLEDGEMENT

The authors thank STC plc for permission to publish this paper.

REFERENCES

1)   A.K. Sinha, J. Vac. Sci. Technol. $\underline{19}$, 778 (1981).

2)   J.L. Vossen & W. Kern, <u>Thin Film Processes</u>, 473, (Academic Press, Orlando, Florida)

3)   T.P. Chow, A.J. Steckel, <u>IEEE ED-30</u>, 1480 (1983)

4)   H.K. Park, J. SWachitano, M.McPherson, T Yamaguchi, G. Lehman. J. Vac. Sci Technol. $\underline{A2(2)}$, 264, (1984)

5)   Ibid.

6)   Ibid.

7)   H. Norstrom, T. Donchew, M. Ostling, C.S. Peterson. Physica Scripta $\underline{28}$, 633, (1983).

TRANSIENT PROCESSING OF TITANIUM SILICIDES
IN A NON-ISOTHERMAL REACTOR

C.S. WEI, J. VAN DER SPIEGEL, J. SANTIAGO* AND L.E. SEIBERLING**
Center for Chemical Electronics, Department of Electrical
Engineering, University of Pennsylvania, Philadelphia, PA 19104
* Physics Department, University of Puerto Rico, Rio Piedras,
  PR 00931
** Department of Physics, University of Pennsylvania,
   Philadelphia, PA 19104

ABSTRACT

Transient processing of titanium silicides on single-
crystal Si in a non-isothermal reactor provides high quality
films. Heat from quartz-halogen tungsten lamps and a small
temperature gradient act as driving forces for the reaction. The
temperature gradient, small compared to the concentration
gradient, shows negligible influence on the formation process.
The influence of sample reflectivity on the other hand is
appreciable. From Xe+ marker experiments, Si atoms are found to
be the moving species either up or down the temperature
gradient. Small amount of TiSi as an intermediate phase is
found to be coexistent with TiSi2. The silicide formation of
the implanted wafers is somewhat slower than that of the
unimplanted wafers.

INTRODUCTION

Rapid thermal annealing, using incoherent light sources,
has been employed successfully for the formation of refractory
and near noble metal silicides [1-4]. The reaction chamber
usually consists of high intensity lamps, which heat the samples
isothermally in a relatively short time of 1 to 10 s. In the
system used for the experiments described in this paper, a non-
isothermal situation is created by positioning the wafers
between the lamps and a water-cooled heat sink. As a result a
small temperature difference over the wafer is present. The
silicide layer is formed by the inter-diffusion of silicon and
metal atoms. The driving forces for the solid-state reaction
comes from the intense radiant energy of the lamps and a
temperature gradient across the wafer ($\sim 200^\circ$C/cm). In this
paper, we studied the influence of the temperature gradient and
sample surface reflectivity on titanium silicide formation.

APPARATUS

The apparatus consists of a bank of six parallel 1.2 kW
quartz-halogen tungsten lamps and a vacuum reactor with water-
cooled wall and base. The reactor can be pumped down to a
pressure of 2E-7 Torr. The samples are put on three thin
conical quartz pins between the lamps and the base. Details of
the apparatus have been reported elsewhere [5,6].

SAMPLE PREPARATION

Two series of experiments are conducted in this paper. In the first, 100 nm thick Ti films are E-gun deposited on both sides of (100), 10Ω·cm, n-type, double-side polished single crystal Si wafers. This is done to study the influence of a temperature gradient on the silicide formation.

In a second series of experiments, 100 keV Xe+ ions with a dose of 2E15 /cm2 are implanted at one side of the double-side polished wafers prior to Ti deposition as a marker to determine the moving species. The projected range is 45 $\pm$ 15 nm below the Si surface. After ion implantion, a 60 nm wet SiO2 film is grown at 860°C to anneal the surface damage and also to reduce the amount of contaminants. The oxide is then removed by diluted HF. Prior to Ti deposition, all the samples are dipped in 2% HF for 2 min to remove the native oxide. The deposition of Ti on the implanted side of the wafer is done in vacuum under a pressure of 3E-7 Torr. These samples are prepared to investigate the influence of surface reflectivity.

The samples are processed in the vacuum reactor under a constant radiant power dinsity of about 20 W/cm2 for times ranging from 3 to 20 s. This power density is chosen to complete the silicide formation within about 10 s.

TEMPERATURE GRADIENT

Si atoms have been reported as the moving species during the formation of Ti silicides [3,7,8]. Therefore, we will take only a flux of Si atoms into consideration in the following model.

In the geometry shown in Fig.1, a vertical temperature gradient is present across the sample. Let Jn1 and Jn2 be the atom flux induced by the concentration gradient at the two interfaces, and Jt1 and Jt2 be the corresponding flux induced by the temperature gradient. Depending on whether the atoms are driven up or down the temperature gradient, the total flux across the interface can be described as [9].

$$|J1| = |Jn1| - |Jt1| = D(T)*(dN/dx)|1 + S(T)*(dT/dx)|1 \qquad (1)$$

$$|J2| = |Jn2| + |Jt2| = -D(T)*(dN/dx)|2 - S(T)*(dT/dx)|2 \qquad (2)$$

where the diffusion coefficient D and the Soret coefficient S are both functions of temperature. Since the temperature difference between the two interfaces is only a few degrees, we can assume that

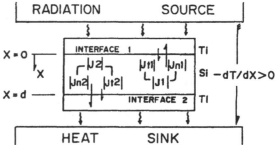

Fig.1 the geometry for understanding the temperature gradient at both interfaces

$$|Jn1| \cong |Jn2| \quad , \text{ and } \quad S(T)|1 \cong S(T)|2 \tag{3}$$

and therefore,

$$\Delta J = |J2| - |J1| \cong |Jt2| + |Jt1| = -S(T)*(dT/dx|2 + dT/dx|1) \tag{4}$$

The effect induced by the flux difference ΔJ, on the silicide formation will be appreciable only when $|Jt|$ is not too small compared to $|Jn|$. If the effect is appreciable, depending on the sign of the Soret coefficient, the temperature gradient will assist or retard the silicide formation.

EXPERIMENTS AND RESULTS

Effect of Temperature Gradient

This series of expeiments are conducted to understand the influence of the temperature gradient on silicide formation. Ti thin films are deposited at both sides of the unimplanted Si wafers and subsequently processed in the reaction chamber under a vacuum of 2E-7 Torr. The side facing the lamps is called "front surface", and that facing the base is called "back surface".

A comparison of the sheet resistance as a function of processing time is shown in Fig.2. A slight increase in sheet resistance in the first three seconds has been explained by oxygen diffusing into the Ti film prior to silicide formation [3,10]. After 5 s the sheet resistance at both sides decreases rapidly as a result of the silicide formation. The sheet resistance shows the same trend at both sides, illustrating that the influence of the temperature gradient is negligible on the silicide formation. The final sheet resistance is 0.94 $\Omega/\square$. The corresponding thickness as measured by a stylus-based thickness analyzer is 200 nm, which corresponds to a film resistivity of 18.8 $\mu\Omega$ cm.

The x-ray diffraction peak intensities as a function of processing time for both cases are shown in Fig.3. No appreciable difference is observed in the spectra except for the TiSi2 signal which appears somewhat faster at the back surface. TiSi appears after 5 s, while TiSi2 starts to form after 6 s and grows at the expense of TiSi phase. TiSi was not found in our

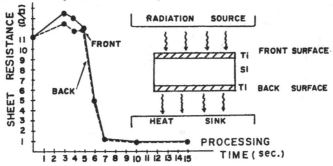

Fig. 2 the sheet resistance as a function of processing time for both sides

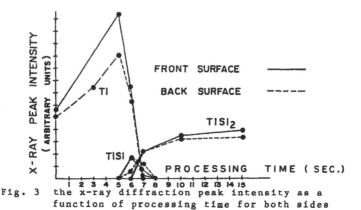

Fig. 3 the x-ray diffraction peak intensity as a
function of processing time for both sides

previous results [3] where a higher radiant power density of 30
W/cm2 was used. It is expected that a small amount of TiSi was
present in the beginning but that the transformation into TiSi2
was very fast at higher power levels. As a result, no TiSi was
detected. The presence of TiSi as an intermediate phase has
also been reported for convetional furnace annealing [10-12].

Effect of Reflectivity of Sample Surface

Xe+ ions are implanted at one surface of the double-side
polished Si wafers prior to Ti deposition as a marker. Low
temperature wet oxidation was done after implantation. Without
this additional step, nonuniform silicide growth occurs with
considerable amount of Ti-rich silicides present even after long
processing time. In some samples the Ti film flakes off during
or after the processing. 100 nm thick Ti is deposited onto the
implanted side of the wafers. The wafers are then processed
with either the deposited Ti or the polished Si surface facing
the lamps (see Fig.4). The reflectivity of a shiny Ti surface is
higher than that of a polished Si surface [13], so that more
energy is absorbed in the latter case. As a result, the
substrate temperature will be higher for the same incident
energy.

A comparison of sheet resistance for the two cases is

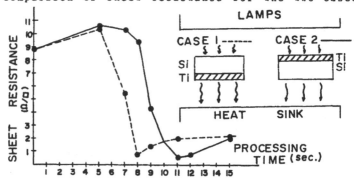

Fig. 4 the sheet resistance as a function of processing
time with either Ti or Si facing the lamps

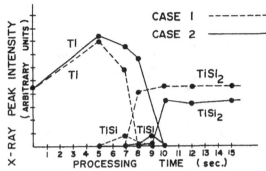

Fig. 5 the x-ray signal as a function of processing
time with either Ti or Si facing the lamps

shown in Fig.4. A lag of about 2 s is clearly observed for case
2 where Ti surface is facing the lamps. The x-ray results are
shown in Fig.5. TiSi being an intermediate phase is again
observed in both cases. The formation rate of silicide is
higher in case 1 as predicted.

Rutherford backscattering spectrometry (RBS) results shown
in Fig.6 are obtained using 16 MeV 0^{16} beam at a backscattering
angle of 110°. The spectra for the two cases are taken at
slightly different detection angle. Si atoms being the dominant
moving speicies in both cases can be seen from the moving of the
Xe+ ions. This is in agreement with previously reported results
for conventional furnace anneal. The evolution of the Ti signal
in the spectra suggests that TiSi and TiSi2 are coexistent.

By comparing case 2 in Fig.4 (Ti layer facing the lamps)
with the results obtained in the first experiments, we can see
that the presence of Xe+ implants at the interface retards the
formation process by about 3 s.

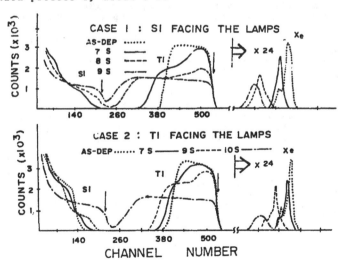

Fig. 6 Rutherford backscattering spectra with either
Ti or Si surface facing the lamps

CONCLUSIONS

Transient processing of Ti silicides on double-side polished single crystal Si wafers using quartz-halogen lamps is successful in a non-isothermal reactor. The resistivity of the TiSi2 films is 18.8µΩ cm. The small temperature gradient across the wafers has negligible influence on the silicide formation. The concentration gradient provides the main driving force in the reaction.

The influence of surface reflectivity on the formation process is appreciable. The case with Si surface facing the lamps has a higher formation rate because of its lower reflectivity and a resulting higher temperature. Si atoms are found to be the moving species either up or down the temperature gradient. Small amount of TiSi as an intermediate phase is observed in all the experiments and to be coexistent with TiSi2. There is an indication that the presence of the implanted Xe+ ions retard the reaction.

ACKNOWLEDGEMENTS

The authors would like to thank Dr. Zemel for valuable discussions. This work was supported in part by the National Science Foundation MRL program under grant no. DMR-8216718 and the Ben Franklin Partnership of the Advanced Technology Center of Southeastern Pennsylvania.

REFERENCES

1.R.T. Fulks, R.A.Powell and W.T. Stacy, IEEE Electron Device Lett. 7, 179 (1983)
2. C.G. Hopkins, S.M. Baumann and R.J. Blattner, " Thin Films and Interfaces II ", eds. J.E.E. Baglin, D.R. Campbell and W.K. Chu, Elsevier Science, N.Y., 87 (1984)
3.J. Santiago, C.S. Wei and J. Van der Spiegel, Materials Lett. 2, 477 (1984)
4.T.O. Sedgwick, F.M. d'Heurle and S.A. Cohen, J. Electrochem. Soc. 131, 2446 (1984)
5.C.S. Wei, J. Van der Spiegel and J. Santiago, Thin Solid Films 118, 155 (1984)
6. C.S. Wei, J. Van der Spiegel, J. Santiago and L.E. Seiberling, Apply. Phys. Lett. 45, 527 (1984)
7.W.K. Chu, S.S. Lau, J.W. Mayer, H. Müller and K.N. Tu, Thin Solid Films 25,393 (1975)
8. A.P. Botha and R.Pretorius, Thin Solid Films 93, 127 (1982)
9. P.G. Shewmon, " Diffusion in Solids ", McGraw-Hill, N.Y. (1963)
10.S.P.Murarka and D.B. Fraser, J. Appl. Phys. 51, 342 (1980)
11.H. Kato and Y. Nakamura, Thin Solid Films 34, 135 (1976)
12.L.S.Hung, J. Gyulai, J.W. Mayer, S.S. Lau and M-A. Nicholet J. Appl. Phys. 54, 5076 (1983)
13.Eds., Y.S. Touloukian, C.Y. Ho, " Thermophysical Properties of Matter ", Vol.7,8, Plenum Press, N.Y. (1972)

TITANIUM SILICIDATION BY HALOGEN LAMP ANNEALING

T. OKAMOTO, M. SHIMIZU, K. TSUKAMOTO, and T. MATSUKAWA
LSI R&D Lab., Mitsubishi Electric Corp.
Mizuhara, Itami, Hyogo, 664, JAPAN

ABSTRACT

Silicidation of titanium on silicon is carried out with the halogen lamp annealing. It is found that the lamp annealing is quite effective in forming an oxide-free and homogeneous titanium disilicide layer with resistivity of 15-17 μohm·cm. Rutherford backscatterring and X-ray diffraction studies show that the halogen lamp annealing over 650 °C for only 60 sec results in disilicide. By a silicidation reaction, arsenic and boron atoms at silicon beneath a titanium layer are incorporated into a formed silicide layer. Arsenic atoms initially in a titanium layer are swept toward the surface as silicidation reaction proceeds. Arsenic atoms in titanium have an effect to retard silicidation reaction.

INTRODUCTION

Recently, in MOS VLSI circuits, refractory metal silicides have been used to reduce sheet resistances of gate electrodes and interconnect lines as a replacement for conventional polycrystalline silicon [1-3], and also to reduce sheet resistances of source/drain regions. Especially, titanium silicide is the most promising material among various refractory metal silicides because of its lowest resistivity [4-6]. But titanium is known to be a reactive metal and to oxidize easily during heat treatment in atmosphere containing redisual oxygen gas. Therefore, when titanium on silicon is silicided with a conventional furnace, care must be taken in order to prevent oxidation of titanium and to form a homogeneous silicide layer.

We have investigated the short time annealing process with the halogen lamp for titanium silicidation. This method is found to be quite effective in forming oxide-free and homogeneous titanium silicide on silicon.

In this paper, we describe characteristics of silicidation of sputter-deposited titanium on silicon by the halogen lamp annealing; comparison of the halogen lamp annealing with the furnace annealing, silicidation process at various annealing temperature, impurity (arsenic , boron) redistribution with silicidation and the influence of arsenic atoms in titanium on a silicidation reaction.

TITANIUM SILICIDATION WITH HALOGEN LAMP ANNEALING

Titanium is deposited onto (100) silicon substrates by dc magnetron sputtering. Silicidation is performed with the halogen lamp annealing system which has an argon-purged quartz annealing chamber. Annealing temperature is controlled within ±1% to a preset value with a closed loop feedback system.

Figure 1 shows RBS spectra from (a) titanium silicided by a

conventional furnace with a extended tube at 700°C for 30 min in nitrogen and (b) by a lamp annealing system for 60 sec in argon, respectively. Titanium is well-known to oxidize easily. In case of the furnace annealing, titanium oxide is observed at surface, which is caused by a redisual oxygen gas during heat treatment. It is found that oxide-free and homogenious titanium silicide can be easily formed by the halogen lamp annealing for short time. This is due to the following two reasons; it is possible to purge an annealing chamber more perfectly, and a sample can be loaded or unloaded at sufficiently low temperature not to oxidize.

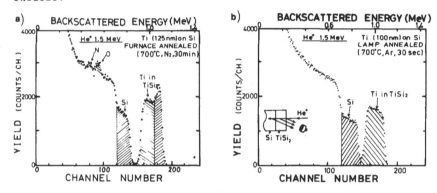

Fig.1 RBS spectra of titanium silicide formed by (a) furnace annealing, and (b) halogen lamp annealing.

In order to investigate the dependence of silicidation reaction on annealing temperature, titanium films on silicon are silicided by the halogen lamp annealing at 550-700°C in argon for 90 sec, as is shown in RBS spectra of Fig.2. Intermixing of titanium and silicon is clearly observed. Silicon atoms appear

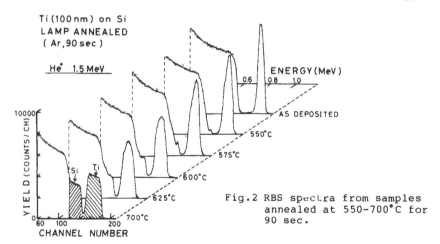

Fig.2 RBS spectra from samples annealed at 550-700°C for 90 sec.

at titanium surface even by the lowest temperature (550°C) annealing for only 90 sec. At 700 °C, homogeneous titanium silicide is formed. In order to identify the crystal phase of silicide formed at 625 °C and 700 °C, X-ray diffraction measurements are carried out, as is shown in Fig.3. In case of annealing at 625°C for 90 sec, only weak peaks associated with titanium monosilicide are observed, and annealing at 700°C for 90 sec results in a titanium disilicide layer which is the final phase of titanium silicide. The above result is consistent with the silicide composition obtained from the RBS spectrum height. Figure 4 shows a plot of sheet resistance of 60 nm titanium silicided at various annealing temperature. There is an increase up to about 30 ohm/□ at 550°C followed by a rapid decrease to 1.1-1.2 ohm /□ (resistivity is 15-17 μohm·cm) at around 650 °C. The increase of resistance seems to be caused by silicon diffusion into a titanium layer as is observed in RBS spectra in Fig.2.

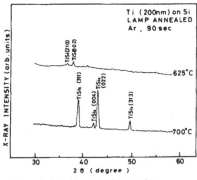

Fig.3 X-ray diffraction analysis of titanium silicide formed at 625°C and 700°C.

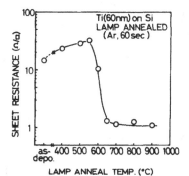

Fig.4 Sheet resistance as a function of annealing temperature.

At elevated annealing temperature, lateral growth of titanium silicide is observed. In Fig.5, length of lateral growth (L) as a function of annealing time and temperature (700°C , 800°C) is shown. L is found to be proportional to square root of annealing time, which indicates that silicidation reaction is a process limited by silicon diffusion between 700°C- 800°C [7]. L reaches more than 5 μm by annealing at 800°C for only 60 sec.

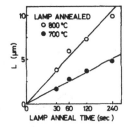

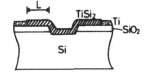

Fig.5 Length of lateral growth as a function of annealing time.

IMPURITY REDISTRIBUTION DURING SILICIDATION AND INFLUENCE ON SILICIDATION REACTION.

The redistribution of impurity atoms (arsenic or boron) at the silicon surface during the silicidation reaction is investigated. Arsenic or boron is implanted to silicon substrates; 1×10^{16} As$^+$/cm^2 at 35 keV or 4×10^{15} B$^+$/cm^2 at 10 keV. Annealing is performed at 900°C for 10 min in nitrogen to recover crystal damage. Then, titanium is sputter-deposited onto the silicon substrates, and the silicidation is carried out with the halogen lamp annealing in an argon ambient.

Redistribution of arsenic atoms during silicidation is studied by RBS, and is shown in Fig.6. In the as-deposited sample, a sharp RBS peak from arsenic atoms at silicon surface is observed. With the halogen lamp annealing at 700°C for 60 sec, arsenic atoms are incorporated into a formed silicide layer.

Unreacted titanium or formed titanium silicide are removed by HF, and depth profiles of boron in silicon are measured by SIMS, as is shown in Fig.7. With the silicidation, a boron profile moves in parallel toward the surface and the total concentration of boron decreases. This indicates that boron atoms in silicon which converted to silicide are incorporated into a formed silicide layer as well as the case of arsenic.

It is concluded that "snow plow effect" of arsenic and boron does not occur in titanium silicidation reaction.

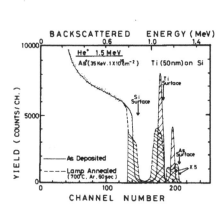

Fig.6 Redistribution of arsenic atoms at silicon surface: as-deposited and annealed at 700°C for 60 sec.

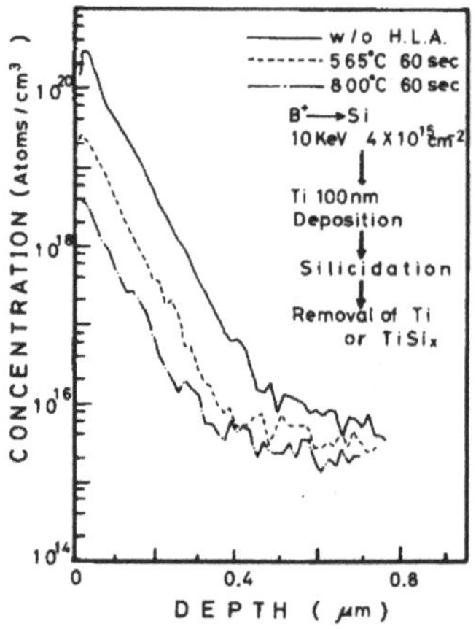

Fig.7 Depth profiles of boron atoms in silicon substrates measured by SIMS.

The influence of arsenic atoms implanted into titanium films on the silicidation reaction is also investigated. Titanium is deposited to a thickness of 95 nm onto silicon substrates and arsenic is implanted to the titanium film with a dose of 1×10^{16} cm^{-2} at 50 keV. Then, the silicidation reactions are performed by the halogen lamp annealing. Figure 8(a) shows RBS spectra from samples annealed at 650°C for 15-240 sec, and Fig.8(b) shows enlarged RBS spectra of arsenic atoms. It is found that at the initial stage of silicidation, arsenic atoms at titanium surface diffuse in the titanium layer, as shown in the spectra for 15 sec and 40 sec. While by annealing for 240 sec, arsenic forms a surface peak in the spectrum again. It indicates that arsenic atoms, which is initially spreading into titanium , are swept toward the surface as silicidation reaction proceeds.

Figure 9 shows RBS spectra from (a) an un-implanted sample and (b) arsenic implanted sample. In case of the titanium film

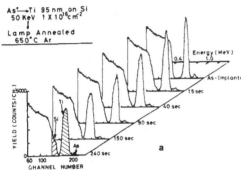

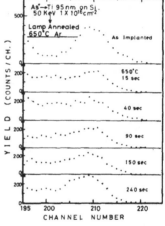

Fig.8 (a) RBS spectra showing redistribution of arsenic atoms in titanium by annealing at 650°C for 15-240 sec, and (b) enlarged spectra from arsenic atoms.

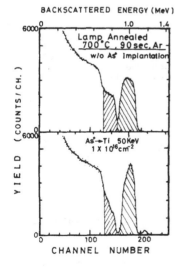

Fig.9 Influence of arsenic atoms implanted into titanium on silicidation.

without implantation, the halogen lamp annealing at 700°C for 90 sec results in a homogeneous titanium disilicide with sheet resistance of 0.6-0.7 ohm/□. On the other hand in case of arsenic implanted titanium film, silicidation reaction has not completed and sheet resistance of formed silicide does not reduce sufficiently (~4 ohm/□). It is concluded that arsenic atoms in titanium have an effect to retard silicidation reaction.

CONCLUSIONS

Sputter-deposited titanium films on silicon are silicided with the halogen lamp annealing. This method is found to be quite effective in forming an oxide-free and homogeneous titanium silicide layer. By RBS and X-ray diffraction measurements, the following results are obtained; (1) by annealing at 550 °C for only 90 sec, silicon atoms appear at titanium surface, (2) up to 625°C intermixing of titanium and silicon occurs, and crystal phase of the formed silicide is identified as titanium monosilicide, (3) the halogen lamp annealing at 700°C for only 60 sec results in a homogeneous titanium disilicide layer with, resistivity of 15-17 μohm·cm. Length of lateral growth over silicon oxide is proportional to square root of annealing time at 700°C and 800°C, which indicates that the silicidation reaction is a process limited by silicon diffusion. Length of lateral growth reaches more than 5 μm by annealing at 800°C for only 60 sec.
The redistribution of impurity atoms (arsenic or boron) at silicon or titanium surface during silicidation reaction is investigated. Both arsenic and boron atoms in silicon which is converted to silicide, are incorporated into a formed silicide layer and "snow plow effect" does not occur. Arsenic atoms at titanium surface diffuse in a titanium film uniformly at the initial stage of silicidation, rapidly. As silicidation reaction proceeds, arsenic atoms are swept toward the surface again. Arsenic atoms in titanium have an effect to retard silicidation reaction.

ACKNOWLEDGEMENTS

The authers would like to express their thanks to M.Sato of the Government Industrial Research Institute, Osaka, for offering the facilities of the van de Graaff accelerator, and also to Dr.H.Oka and Dr.H.Nakata for their encouragements.

REFERENCES

[1] S.P.Murarka, J.Vac.Technol.,17,775 (1980).
[2] T.P.Chow and A.J.Steckel, IEEE Trans. ED.,30,1480 (1983).
[3] Silicides for VLSI Applications, edited by S.P.Murarka
    (ACADEMIC PRESS, New York, 1983).
[4] S.P.Murarka and D.B.Fraser, J.Appl.Phys.,51,342 (1980).
[5] S.P.Murarka and D.B.Fraser, J.Appl.Phys.,51,350 (1980).
[6] S.P.Murarka, D.B.Fraser,A.K.Sinha and H.J.Levinstein, IEEE
    Trans.ED, 27,1409 (1980).
[7] R.W.Bower and J.W.Mayer, Appl.Phys.Lett.,20,359 (1972).

LUMINESCENCE OF Si-IMPLANTED InP
AFTER RAPID THERMAL ANNEALING

D. KIRILLOV[*] AND J. L. MERZ[**]
[*]Varian Research Center, Palo Alto, CA 94303
[**]University of California - Santa Barbara, Department of Electrical
and Computer Engineering, Santa Barbara, CA 93106

ABSTRACT

Changes in the luminescence spectra of InP caused by Si^+ implantation
and subsequent rapid lamp annealing were studied. It was found that the
best activation of dopants was obtained in the case of hot implantation
and lamp annealing in regimes close to melting of InP.

INTRODUCTION

InP is a direct gap semiconductor with strong luminescence which is
very sensitive to defects and dopants. It is convenient to use luminescence spectroscopic techniques for the study of annealing of ion implanted
InP. The main advantages of this technique are its nondestructive character and the possibility of obtaining detailed information on electronic
and lattice properties of the material.

EXPERIMENTAL RESULTS AND DISCUSSION

The undoped samples of InP used for the present studies were highly
polished (100) wafers, ~300-μm thick, and were n-type with a free carrier
concentration at room temperature of $1.22 \times 10^{16} cm^{-3}$. The samples were
implanted with Si^+ ions at a dose of $3.3 \times 10^{14} cm^{-2}$ and energy of 180 keV.
The samples were held at room temperature or 175°C during implantation.
After implantation, the samples were capped at room temperature with a
cathode-sputtered film of SiO_2, ~1000 Å thick, and were annealed by a lamp
pulse system. The samples were placed face down on a Si wafer and both
sides of the sample were heated by lamps in an inert Ar atmosphere. The
annealing temperature and time profiles were measured using a thermocouple
cemented onto the Si wafer. Maximum annealing temperatures and times were
limited by melting of the sample.

The implanted samples were also annealed in a furnace at 775°C for 15
min in order to compare furnace and lamp pulse annealing. Low power (~20
mW), weakly focused argon laser radiation with a wavelength of 5145 or 4880
Å was used for the excitation of the luminescence. The exciting radiation
was absorbed in a surface layer less than 2000 Å thick; the luminescence
therefore probed only the implanted part of the sample, since the projected
range of Si^+ ions in InP at 180 keV is ~2000 Å. The Raman spectra were also
measured in order to obtain additional information on crystalline structure
and doping.

Low Temperature Luminescence

The low temperature luminescence spectrum of virgin InP is shown in
Fig. 1(b). The maxima in the spectrum may be interpreted as follows [1].
The line A at ~1.417 eV corresponds to transitions involving excitons bound
to donors. The shoulder at 1.377 eV and the maximum D at 1.369 eV correspond to donor-acceptor pair recombination, and D' (1.327 eV) is the first

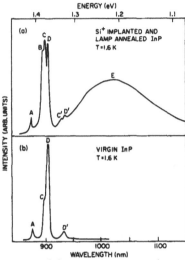

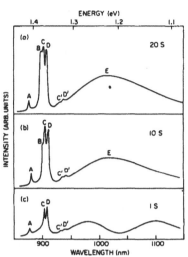

Fig. 1 (a) Low temperature (T = 1.8K) luminescence spectrum of InP hot implanted and lamp annealed at 900°C, 5s. (b) The spectrum of virgin InP.

Fig. 2 Low temperature spectra of hot implanted InP annealed by lamp at T = 850°C during (a) 20s, (b) 10s, (c) 1s.

LO-phonon replica of D. All these peaks are usually seen in undoped InP and are associated with commonly present unidentified defects.

Ion implantation produces a large number of defects which are effective centers for nonradiative recombination of photo-excited carriers. Only after thorough annealing could the band edge or impurity luminescence be detected. The luminescence spectrum of an InP sample implanted at 175°C and lamp annealed under optimum conditions [2,3] for 5s at 900°C is shown in Fig. 1a. New features appear in the spectrum 1(a) in comparison with 1(b). A broad maximum E at ~1.22 eV appears, which has already been observed earlier in Si-doped InP [4]. This band may correspond to transitions between free or weakly bound electrons and deep Si acceptors formed from substitutional Si at P sites. The maximum C and its LO phonon replica C' are stronger in intensity than in a virgin sample, and a new maximum B at 1.381 eV, which has not been observed before, appears in the spectrum 1(a). Because the B maximum appears only in well annealed Si-doped material, it may be associated with a Si impurity; a possible mechanism could involve the transitions between the Si donor and unidentified acceptor states. The peak C, coincident in energy with the shoulder C in the virgin sample, increases after Si implantation and annealing; it is likely that this results from another Si donor-acceptor pair recombination. These interpretations are supported by the measured temperature dependence of the luminescence intensity. The behavior of all lines corresponds to that of donor-acceptor recombination lines.

The transformation of the low temperature spectrum of hot implanted InP with increasing time of lamp annealing at 850°C is illustrated in Fig. 2. As the annealing time increases, the intensity of lines A, C, and D increases, showing significant improvement of the sample quality and the elimination of defects. The new features B and E, characteristic of Si doping, appear only in better annealed samples.

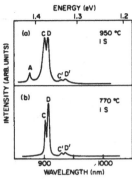

Fig. 3 Low temperature spectra of room temperature implanted InP lamp annealed during 1s at (a) T = 950°C and (b) T = 770°C.

The low temperature spectra of lamp annealed samples that have been implanted at room temperature are shown in Fig. 3. In this case, as Raman data show [5], the surface layer is made amorphous by implantation, unlike the hot implantation case, for which it remains single crystalline with a high density of defects. The spectrum 3(a) corresponds to the optimum annealing case. Annealing for 20s at 850°C gave a similar spectrum. It is important to note that the line B and the broad band E, which are present in the hot-implanted sample, are absent in this case, suggesting that the majority of Si atoms are nonsubstitutional due to incomplete reordering of the amorphous layer. Let us note also that the bound exciton maximum A is a sensitive probe of the crystalline quality. It is absent in the partially annealed sample, Fig. 3(b), and grows in intensity with improved annealing.

The low-temperature spectra of implanted samples annealed in the furnace at 775°C for 15 min. are given in Fig. 4. Band E is absent in the spectrum of the room temperature implanted sample, Fig. 4(b), showing that the activation of Si impurities is far from complete. However, this band has strong intensity in the spectrum of the hot implanted, oven-annealed sample Fig. 4(a). The line B, which we attribute to transitions between Si donors and shallow acceptors, is absent in the oven annealed sample. This suggests that the activation of Si impurities is worse in the case of oven annealing.

Room Temperature Spectra

In the case of room temperature spectra, all shallow donors are ionized and the annealed material becomes degenerate. This leads to an increase of the effective bandgap due to the Burstein-Moss effect. The bandgap shift of strongly doped n-type InP has been observed before [6,7]. Room temperature luminescence spectra of the hot implanted samples lamp annealed at 850°C for different time intervals are shown in Fig. 5. The room temperature spectrum of a virgin sample is also shown for comparison. Two maxima are present in the spectra of annealed samples. The maximum at ~1.15 eV corresponds to band E at ~1.22 eV in low temperature spectra and is believed to be due to transitions between free electrons and deep Si acceptors produced by Si atoms at P sites [4]. The peak at ~1.35 eV corresponds to band-to-band transitions. It broadens and shifts towards higher energy with increasing anneal time.

The shift of the luminescence maximum to higher energies is a manifestation of the Burstein-Moss effect, caused by the filling of band states under conditions of high carrier density. This effect is much stronger in InP for n-type doping because of the low density of states in the conduction band. The dominating transitions in this case of heavy doping occur between the electrons near the Fermi level in the conduction band and holes in the impurity tail of the valence band. The transitions are mainly indirect,

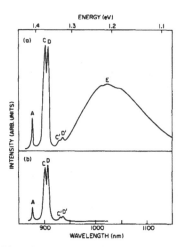

Fig. 4 Low temperature spectra of (a) hot implanted and (b) room temperature implanted InP after annealing in an oven.

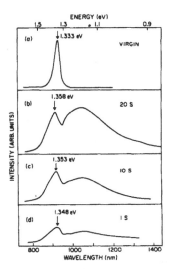

Fig. 5 Room temperature spectra of (a) virgin and (b)-(d) hot implanted and lamp annealed at 850°C InP with anneal time of (b) 20s, (c) 10s, and (d) 1s. Note the increase of energy of the band-to-band luminescence maximum due to increase of free electron concentration in a conduction band.

without conservation of wavenumber, due to strong scattering by impurities [8]. This increased concentration of electrons in the conduction band contributes to the Fermi level shift to higher energy and the corresponding increase of the effective bandgap. At the same time, bandgap shrinkage occurs due to the growth of the impurity induced tail of the valence band. The actual energy shift of the luminescence peak may be expressed as the difference in these two effects:

$$\Delta E = E_F - E_T ,$$

where E_F is the Fermi level for electrons (measured from the bottom of the conduction band) and E_T is the effective shift of the valence band induced by impurities (measured from the of the valence band of the intrinsic material). The concentration of free carriers may be evaluated from the experimentally measured luminescence energy shift [8,9]. Thus, for the highest energy shift observed, ~25 meV as seen in Fig. 5(b), the corresponding concentration of activated carriers is ~3 x 10^{18}cm^{-3} [9].

The room temperature spectra of hot implanted (a) and room temperature implanted (b) samples, oven annealed at 775°C for 15 min., are compared in Fig. 6. As can be seen, the quality of the hot implanted sample is much better than that of the room temperature implanted sample. The latter shows only some unidentified deep level luminescence without traces of band-to-band luminescence. The energy of the band-to-band luminescence maximum of Fig. 6(a), 1.348 eV, is shifted to higher energies in comparison with a virgin sample, 1.333 eV, but the shift is smaller than in the case of lamp annealing under optimal conditions, Fig. 5(b).

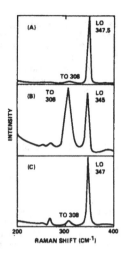

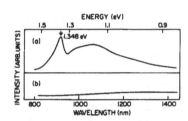

Fig. 6   Room temperature spectra of (a) hot implanted and (b) room temperature implanted InP after annealing in an oven.

Fig. 7   Room temperature Raman spectrum of InP; (A) virgin sample; (B) hot implanted InP after lamp annealing at 850°C during 20s; (C) unimplanted InP after lamp anneal- int at 900°C during 5s.

Raman Scattering

Room temperature Raman spectra are shown in Fig. 7. Only the LO-phonon line at 347.5 cm^{-1} allowed by selection rules is present in the spectrum of a virgin sample, Fig. 7(a). The prohibited TO-line at 308 cm^{-1} is hardly seen. The spectrum in the same scattering configuration of the hot implanted sample after lamp annealing at 850°C for 20s is shown in Fig. 7(b). All well annealed (by a lamp or in a furnace) samples implanted at room tempera- ture or at 175-200°C have similar Raman spectra. A new line coincident in energy with TO phonons appears in the spectrum. This line is expected to appear in Raman spectra of highly doped InP, due to formation of phonon- plasmon coupled modes (L$^-$ mode) [10]. We consider the free carrier contri- bution to this line to be dominant. As our data have shown, the hot im- planted sample remains single crystal after the implantation dose used in the present work, and the intensity of the prohibited TO mode remains small after implantation. Though low temperature annealing produces noticeable improvement of crystalline structure [5], only high temperature annealing produces activation of carriers and growth of the plasmon-phonon line. We were not able to detect unambiguously plasmon excitations of the higher energy branch, which allows a determination of the carrier concentration and mobility [10], most probably due to strong damping of plasmons due to defects. The expected position of this plasmon line for the spectrum in Fig. 7(b) was in the range of 950 cm^{-1}.

The spectrum of the unimplanted sample capped and annealed at 900°C for 5s is shown in Fig. 7(c). As can be seen from the figure, the spectrum is quite similar to that of the virgin sample. Therefore, it may be concluded that annealing at conditions close to melting conserves the single-crystal structure of the sample surface. But it is necessary to note that capping, annealing and removing of a cap cause strongly increased elastic Tyndall scattering of light which shows that defects are introduced into the surface layer.

TO mode may become allowed in the spectrum if the symmetry of the sample is changed, i.e., it becomes polycrystalline. Apparently, the contribution of scattering by TO phonons due to the polycrystallinity of the sample might be substantial in the case of room temperature implanted samples, which become amorphous after implantation [5]. Discrimination between contributions of polycrystallinity and free carriers to the TO line in the case of annealing of an amorphous material requires special studies.

CONCLUSION

We may conclude that photoluminescence is a good nondestructive diagnostic method to study ion implantation and lamp annealing in InP. Luminescence lines and bands characteristic of activated Si impurities introduced by ion implantation were found in low temperature and room temperature spectra.

We found that room temperature luminescence is an especially convenient diagnostic method. The intensity of the band-to-band room temperature luminescence grew with increasing quality of annealing, showing the elimination of deep traps which quenched the luminescence. The energy of the luminescence maximum increased at the same time, reflecting the electrical activation of shallow donor impurities introduced by implantation.

ACKNOWLEDGEMENTS

The authors are grateful to R. Kalish and S. Shatas for ion implantation and lamp annealing of samples and fruitful discussions.

REFERENCES

[1] C. Pickering, P. R. Tapster, P. J. Dean, and D. J. Ashen, GaAs and Related Compounds, 1982, Inst. Phys. Conf. Ser. No. 65 (Institute of Physics, London, England, 1982), p. 469.
[2] J. P. Lorenzo, D. E. Davis, K. J. Soda, T. G. Ryan and P. J. McNally, Laser-Solid Interactions and Transient Thermal Processing of Materials, eds. J. Narayan, W. L. Brown and R. A. Lenanas (North-Holland Publishing, New York, 1983), p. 683.
[3] A. N. M. Masum Choudhury, K. Tabatabaie-Alivi and C. G. Fonstaf, Appl. Phys. Lett. 43, 381 (1983).
[4] V. V. Negreskul, E. V. Russu, S. I. Radautsan and A. G. Cheban, Sov. Phys. Semicond. 9, 587 (1985).
[5] D. Kirillov and J. L. Merz, Energy Beam-Solid Interactions and Transient Thermal Processing, eds. J. C. C. Fan and N. M. Johnson, Materials Research Society Symposium Proceedings (Elsevier-North Holland, New York, 1984).
[6] G. G. Baumann, K. W. Benz and M. H. Pilkuhn, J. Electrochem. Soc. 123, 1232 (1976).
[7] F. Z. Hawrilo, Appl. Phys. Lett. 37, 1038 (1980).
[8] J. De-Sheng, Y. Makita, K. Ploog and H. J. Queisser, J. Appl. Phys. 53, 999 (1982).
[9] S. Bendapudi and D. N. Bose, Appl. Phys. Lett. 42, 287 (1983).
[10] A. Mooradian and G. Wright, Phys. Rev. Lett. 16, 999 (1966).

RAPID THERMAL-PULSED DIFFUSION OF Zn INTO GaAs

S.K. TIKU, J.B. DELANEY, N.S. GABRIEL AND H.T. YUAN
Texas Instruments Inc., P.O. Box 225936, M/S 134, Dallas, TX 75265

ABSTRACT

A rapid thermal process for the diffusion of Zn into GaAs has been developed to fulfill the need for highly doped p type layers in GaAs technology. The process uses a solid Zn:Si:O source layer and a quartz-halogen lamp system for the thermal drive-in. Surface concentrations of the order of $10^{20}/cm^3$ have been achieved with good depth reproducibility and low lateral diffusion. Specific contact resistance of Au:Zn/Au alloyed contacts fabricated using this process was in the 10^{-7} ohm-cm^2 range.

INTRODUCTION

There is a need for highly doped p+ layers in GaAs device technology for good ohmic contacts to p type layers. Beryllium implantation is generally used to fabricate p+ regions, and Au:Zn/Au metallization is deposited and alloyed to make ohmic contacts. This process gives highly irreproducible results. Heavily doped p+ layers (conc.$\geq 10^{20}/cm^3$) would ensure reproducible contact and a process less dependent on the details of the metallization scheme. Precisely controlled p-type layers are also needed for making junction type gates in JFET technology.

Zinc is an important acceptor type dopant in III-V compound semiconductors. Its diffusion is very fast and follows the interstitial-substitutional model. This model assumes that Zn moves much faster interstitially (than substitutionally) as a donor Zn_i, which is subsequently trapped by a Ga vacancy, V_{Ga}, thus producing a substitutional Zn acceptor. Since the Zn acceptor is mostly singly ionized, this reaction can be expressed as

$$Zn_i^+ + V_{Ga} \rightleftharpoons Zn_{Ga}^- + 2\,(h^+) \tag{1}$$

This charge state change of 2 gives rise to a concentration square (N^2) dependence of the diffusion coefficient. This dependence has been experimentally confirmed [1], and gives rise to an abrupt doping profile and a junction depth dependent on the surface concentration, as given by [2]:

$$x_j = 1.092\,(D_{sur} \cdot t)^{1/2} \tag{2}$$

where D_{sur} is the diffusion coefficient at the surface. Thus the junction depth is a function of the surface concentration and can be approximated to $(D_{sur} \cdot t)^{1/2}$.

Sealed tube and open tube processes are commonly employed for performing Zn diffusions into III-V compounds. These techniques use elemental zinc, zinc-arsenic compounds, diethyl zinc etc. as zinc sources [2]. Most of the traditional methods require the control of arsenic vapor pressure and surface concentration of Zn. The results are generally variable, consequently there is a need for a more reproducible diffusion process. A solid diffusion source (SiO :ZnO) has been used in the open tube method [3]. The source was deposited by the oxidation of a mixture of silane and diethyl zinc and covered by phospho-silicate glass. The open tube process took 20 minutes at 600°C for a typical diffusion of 3000 Å. Pure zinc or zinc oxide can not be used in such a process as extensive surface damage of the substrate results. Zinc ion-implantation is not a favored process as it needs an activation anneal, during which zinc redistribution and loss take place. Surface layers are damaged, activation is incomplete and mobility is low in implanted and annealed layers. The damage is not annealed out even at 800°C [4]. Electron beam or laser pulse diffusions of various dopants into GaAs have been attempted, however, these techniques suffer from defect generation problem [5], due to high thermal gradients in the wafer. In view of these facts, a diffusion process utilizing a solid source layer and rapid thermal annealing system based on optical heating that has a thermal time constant of seconds and no local temperature gradients, appears to be appropriate for Zn diffusion into GaAs. A similar process has been reported for diffusion of Si into GaAs [6], which unlike the process being reported, is critically dependent upon the nature of the cap layer.

PROCESS DESCRIPTION

A thin ($\sim$200 Å) layer of $ZnSi_x O_y$ was deposited on semi-insulating and n-type LEC GaAs using reactive sputtering. The concentration of Si in the films was about 50% that of Zn. The thickness of the source layer was more than sufficient as an infinite source of Zn and yet small enough to allow subsequent 500 Å silicon nitride cap layer deposition without peeling problems. The wafers were annealed in a quartz-halogen lamp system (HEATPULSE, A.G. Associates Inc.) under different temperature-time cycles. During the anneal the wafers were placed face up on a 4 inch Si slice. The temperature was measured using a thermocouple attached to a Si wafer placed near the substrate and calibrated using minute smears of glasses of known melting point. The cap and source layers were stripped off using dilute HF and HCl acids. The resulting diffusion depth was determined using step etching and sheet conductivity measurements; and SIMS profiling for a few selected samples. Fig 1 shows an almost linear drop in conductivity with etch depth confirming an almost flat diffusion profile. SIMS analysis also confirms a steep diffusion front and also shows a very high surface concentration of Zn, with an average concentration of $\sim$10^{20}/cm^3 in the diffused layer (Fig. 2). Diffusion depths from 500 Å to 5000 Å were achieved using different time-temperature pulses as shown in Fig.3. Table I lists the junction depth and the effective diffusion coefficient (at surface concentration) D_{sur} for a few temperatures. The values are in agreement with the published results (1.6 x10^{-12}cm^2/sec. at 600°C [6]) if an activation energy of about 1.5 eV is assumed.

Masked Diffusions

Selective doping using masked diffusions is usually required in integrated circuit processing. Both SiO_2 and Si_3N_4 have been used as mask materials in the past. Plasma CVD silicon nitride has proven to be more effective as a diffusion mask [7]. Lateral diffusion must be

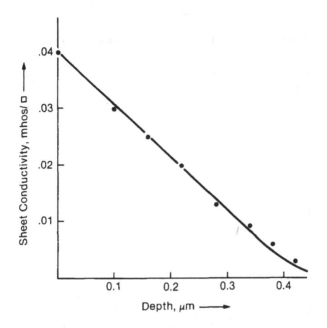

Figure 1.   Sheet conductivity vs. etch depth for a thermal pulse of 825°C for 10 seconds.

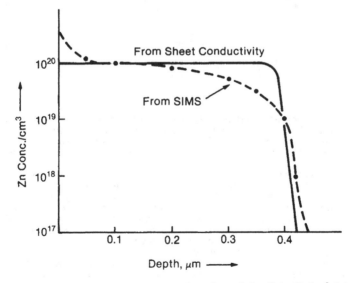

Figure 2.   Zn concentration as a function of depth in GaAs for a pulse of 825°C for 10 seconds.

Table I.

Diffusion Coefficient Under Different Thermal Pulse Conditions

Thermal Temp.°C	Pulse Time, sec.	Junction Depth μm	D_{sur} cm²/sec.
750	10	0.1	1×10^{-11}
775	10	0.2	4×10^{-11}
825	10	0.4	2×10^{-10}

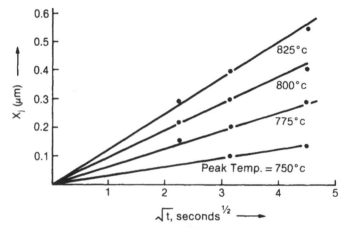

Figure 3.   Junction depth as a function of thermal pulse duration for Zn diffusion into GaAs.

avoided or minimized in a selective doping process. It has been shown that the lateral diffusion is a function of the interfacial stress between the mask layer and GaAs [8]. A small crack between the mask layer and the substrate surface must be absent. The lateral diffusion is worse when there is a large reservoir of dopant on the surface, since this provides the driving force for the diffusion. Plasma CVD silicon nitride was used in the present work as the mask. Use of a very thin layer (500-1000 Å) reduces stress. Fig.4 shows schematic drawings for two types of geometries used. Both were defined by lift-off and showed very little lateral diffusion. The extent of the lateral diffusion was below the detection limit of EBIC (Electron Beam Induced Current) technique; and transistors using these diffused layers for base contact, did not show any shorts for spacings of 1.5 μm. With high temperature pulses a thin p + skin (50-100 Å) may extend laterally as much as the depth or more depending upon the mask integrity. However, this can be easily etched off before subsequent processing.

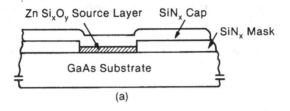

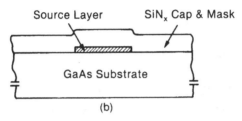

Figure 4.   Two geometries used for masked diffusions.

Ohmic Contact Results

Table II summarizes the results of the ohmic contact experiments. The transmission line method was used for the measurements. The high surface carrier concentration makes it possible to make contact with any metal (eg. Al), and also allows dry probing. The Au:Ge/Ni

Table II

Summary of Results

Surface Concentration	$\simeq$	$10^{20}/cm^3$
Average Mobility	$=$	60 cm/V-sec

Metallization System	Specific Contact Resistance ohm-cm^2
Au:Zn/Au System Furnace Alloyed	5×10^{-7}
Au:Ge/Ni System Pulse Alloyed	5×10^{-6}
Al (not sintered)	1×10^{-5}

488

system, normally used for n-type ohmic contacts, gives reasonably good contacts and thus eliminates the need for two metallization systems. Of course, the best results, (specific contact resistance in the middle 10^{-7} range) are achieved if a Au:Zn/Au alloy system is used.

In conclusion, the process gives reproducible p + layers in GaAs without the use of dangerous gases, causes no surface damage and has a fast turnaround time.

REFERENCES

1. D.J. Colliver, COMPOUND SEMICONDUCTOR TECHNOLOGY, Artech House Inc., (1976) p. 57

2. S.K. Ghandhi, VLSI FABRICATION PRINCIPLES, John Wiley and Sons, N.Y., (1983) p. 185

3. S.K. Ghandhi and R.J. Field, Applied Physics Letters, **38**, 267, (1981)

4. G.J. van Gurp, A.H. van Ommen, P.R. Boudewijn, D.P. Oosthoek and M.F.C. Willemsen, J. Appl. Phys., **55**, 338 (1984)

5. D.E. Davies, T.G. Ryan and J.P. Lorenzo, Appl. Phys. Lett., **37**, 443 (1980)

6. M.E. Greiner and J.F. Gibbons, Appl. Phys. Lett., **44**, 750 (1984)

7. C. Blaauw, A.J. SpringThorpe, S. Dzioba and B. Emmerstorfer, J. Electronic Mater., **13**, 251 (9184)

8. p. 195, Ref. 2

THE MICROSTRUCTURE OF TRANSIENTLY ANNEALED DONOR IMPLANTS IN GaAs

M.A. SHAHID, R. BENSALEM, and B.J. SEALY
Department of Electronic and Electrical Engineering
University of Surrey, Guildford, Surrey GU2 5XH, U.K.

ABSTRACT

Undoped SI (100) GaAs has been implanted with selenium and tin ions at room temperature at an ion energy of 300 keV and using ion dose in the range 1 x 10^{14} to 1 x 10^{15} ions cm^{-2}. Transient annealing at 1000°C and above has been studied using electrical measurements and transmission electron microscopy. The results show that tin implanted samples have comparatively higher values of electrical activity and mobility than those implanted with selenium ions. A difference in the microstructure of these two implants was observed. Selenium implanted samples show dislocation lines and loops possessing 1/2<110> Burgers vectors while tin implanted GaAs contains dislocation loops of 1/2<110> and 1/3<111> types and also dislocation lines having 1/2<110> Burgers vectors. Both types of defect in tin implanted samples are decorated with precipitates.

1. INTRODUCTION

Annealing of GaAs is a problem because of its tendency to dissociate at a high temperature. During the past few years rapid thermal annealing, has become an attractive alternative to conventional furnace annealing in order to reduce the decomposition of GaAs. This method employs a graphite strip heater [1,2], a multiply scanned electron beam [3,4] or more recently an incoherent light furnace [5,6]. Moreover, it is well known that p-type implants into GaAs can be activated relatively easily compared with the n-type implants which do not always produce high electrical activity even after annealing at a temperature of 900°C. It is generally felt that by reducing the dwell time at still higher temperatures, the electrical activity in a given n-type implant might be improved. In this paper we present results of such a study using selenium and tin ions and discuss the problems associated with such high temperature anneals along with the differences in the microstructure between these two n-type implants as seen by transmission electron microscopy.

2. EXPERIMENTAL METHOD

Undoped (100) semi-insulating GaAs was implanted at room temperature with selenium and tin ions in a non-channelling direction at an ion energy of 300 keV using doses of 1 x 10^{14}, 5 x 10^{14}, and 1 x 10^{15} ions cm^{-2}. Prior to annealing in an incoherent light furnace or on a graphite strip heater a two layer encapsulant was deposited in order to inhibit surface decomposition during annealing. This consisted of growing about 300Å of CVD Si_3N_4 on top of which about 600Å of reactively grown AlN was deposited. Annealing was carried out at a range of temperatures and times, the highest temperature being 1100°C for a few seconds. The electrical measurements were performed on clover leaf shaped samples using the Van der Pauw method. The depth profiles of electron mobility and concentration were measured using differential Hall effect and chemical stripping. The residual microstructure at various stages of ion implantation, encapsulation and annealing was investigated by transmission electron microscopy (TEM). Samples for TEM were prepared by mechanical polishing and chemical thinning from the back side of the samples. A JEOL 200CX scanning transmission electron microscope was used in order to record a two beam bright field image, a (g, 3g) dark field image and a selected area diffraction pattern from the same area of a sample. Stereo micrographs were also recorded in order to examine the in depth distribution of various types of defects.

3   RESULTS

3.1 Electrical Measurement

Dose dependence of the average values of sheet electron concentration for tin and selenium implants is shown in Figure 1. A comparison shows that the tin implanted samples have comparatively higher electrical activity throughout this dose range for otherwise similar conditions of implantation and annealing. Typical electron concentration and mobility profiles for samples implanted with tin and selenium for a dose of 1×10^{14} ions cm^{-2}

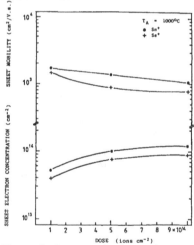

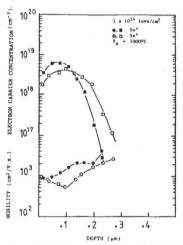

Figure 1. Average sheet carrier concentration as a function of ion dose.

Figure 2. Depth profiles of electron concentration and mobility for Sn^+ and Se^+ implanted GaAs.

are shown in Figure 2. Both of these samples have been annealed at a temperature of 1000°C for 20 seconds. The peak electron concentrations were about 7×10^{18} cm^{-3} and 6×10^{18} cm^{-3} for the tin and selenium implants respectively. The tin implanted samples have a higher electron mobility than those implanted with selenium ions. Hence, the measured sheet resistivity values for tin implanted samples are lower than those for selenium implanted material. The lowest sheet value of 27Ω/□ was recorded for a sample implanted with $1 \times 10^{15} sn^+cm^{-2}$ and annealed at 1090°C for 5 seconds. On the other hand, a value of 38Ω/□ was measured on a sample implanted with a similar dose of selenium and annealed at 1050°C for 9 seconds.

3.2 Transmission Electron Microscopy, TEM

The extent of damage in the ion implanted GaAs was investigated using transmission electron microscopy. Samples implanted at room with a high dose (> 10^{14} ions cm^{-2}) of a heavy ion at an energy of 300 keV show typically, a finely crystalline surface layer (Figure 3). Since the

Figure 3. A transmission electron micrograph of GaAs implanted with 1×10^{14} Sn^+cm^{-2} at an ion energy of 300 keV at room temperature.

implanted layer is finely crystalline, it produces broad rings in the diffraction pattern shown in the inset of this figure. In this micrograph, the single crystal diffraction spots arise from the substrate because the electron beam passed through this material before leaving the implanted layer.

In order to encapsulate the implanted GaAs, CVD Si_3N_4 was grown at a temperature of 635°C. This has resulted in partial annealing of the material (Figure 4). This micrograph shows relatively large islands of

Figure 4. Typical microstructure of partially annealed ion implanted GaAs during encapsulation by CVD Si_3N_4 at a temperature of 635°C. Ion dose = 1×10^{15} Sn^+ cm^{-2}, ion energy = 300 keV.

damage containing patches of contrast with some visible structural details within and also broad fringes indicating the presence of localized strain in these regions which are bounded by dislocation tangles. Moreover, a host of dislocation loops is present in relatively better parts of the material which shows regions of uniform background contrast. This effect is more clearly seen in the dark field micrograph of Figure 4b. The diffraction pattern (shown in the inset of Figure 4b) from such a sample, however, is a single crystal spot pattern.

When such a sample is annealed at a high temperature of 900°C for 30 seconds, almost all the strain centres of large dimensions disappear. The only defects present are dislocation lines and loops of varying shapes and sizes. The dark field micrographs of Figures 5a and 5b show this

 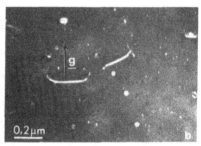

Figure 5. Se[+] implanted GaAs after annealing at 900°C. for 30 seconds. ion dose = 1 x 10^{14} Se[+] cm[-2], ion energy = 300 keV (a) 022 reflection. (b) 022 reflection.

situation for a sample implanted with 1 x 10^{14}Se[+]cm[-2] for the two orgothognal 022 reflections that lie in the (100) surface planes of GaAs. A comparison of these two micrographs shows that the dislocation lines have Burgers vectors of 1/2<110> which are also the displacement vectors for the perfect interstitial loops that lie on or near {110} planes. This conclusion has been arrived at by using a method of analysis described in detail elsewhere [6]. Figure 6 shows the effect of annealing at a still

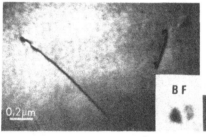

Figure 6. Se[+] implanted GaAs after annealing at 1040°C for 9 seconds. ion dose = 1 x 10^{14} Se[+]cm[-2]. The insets show enlarged diamond shaped defects in bright field (BF) and dark field (DF)

higher temperature of 1040°C for 9 seconds. The background is much clearer and the dislocation lines, which in fact are rather large half loops intersected by the free surface of the sample, are the major defects present at a relatively low density.

The stereo micrographs of Figure 7 show rather nicely the effect of annealing two different doses of tin implanted samples annealed at 1000°C for 5 seconds. The doses for Figures 7a and 7b respectively are 1 x 10^{14} and 1 x 10^{15}Sn[+]cm[-2]. A comparison shows that the higher dose still has a large concentration of very fine damage. Besides, unlike selenium implanted samples, the tin implanted GaAs has defects in the form of dislocation lines and loops, both of these defects being decorated with precipitates. Analysis [6] carried out on these samples reveals perfect extrinsic dislocation loops possessing 1/2<110> Burgers vectors lying on or near {110} planes along with an equal number of imperfect loops possessing 1/3<111> displacement vectors that lie on or near {111} planes.

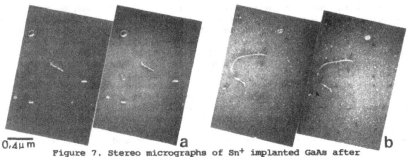

0,4 μm

a

b

Figure 7. Stereo micrographs of Sn$^+$ implanted GaAs after
annealing at 1000°C for 5 seconds. (a) 1 x 10^{14} Sn+ cm^{-2}
(b) 1 x 10^{15} Sn$^+$ cm^{-2}. Ion energy 300 keV.

Besides these defects a common feature of such a high temperature anneal at
1000°C and above is the presence of diamond shaped defects with an
associated fine structure (figure 7). Dark field micrographs show clearly
that they are volume defects. In bright field micrographs they appear as
two dark triangles separated by a line of no contrast always perpendicular
to the operative g-vector (figure 6). This effect is seen to take place
both in selenium and tin implanted samples following an anneal above
1000°C. There is, however, a problem with the high temperature annealing of
GaAs, that is, an occasional failure of the encapsulant. If the failure is
not too severe, the encapsulant can be removed easily by dissolution in HF
which is not the case if an acute failure of the cap takes place. In the
regions of partial failure, gross damage is produced which has typical
characteristics and is independent of implanted ion type. This damage
appears in the form of complicated patches of contrast associated with
tangles of dislocations resulting in regions of localized strain. The
micrograph of Figure 8 shows the effect of such a failure.

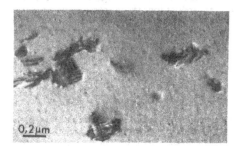

0,2 μm

Figure 8. A TEM micrograph
showing surface damage in
ion implanted GaAs due to
failure of the encapsulant
during annealing at 1000°C
for 5 seconds.

4. DISCUSSION

The results show that an electron concentration approaching the value
of about 10^{19}cm^{-3} can be achieved by implanting selenium or tin into GaAs
using a dose in excess of 1 x 10^{14} ions cm^{-2}. Such high electron
concentrations are measured only for samples annealed at a temperature of
1000°C and above. Moreover, tin implanted samples produce relatively higher
values of electron mobility and hence lower values of sheet resistivity

compared with those implanted with selenium ions. This can be useful in producing non-alloyed ohmic contacts to GaAs. The difference in the annealing behaviour as a function of ion-type is also important and it seems that this aspect has not been addressed so far. Some important differences in the microstructure of selenium and tin implanted GaAs are quite evident from the present experiments. For instance in the case of selenium implanted samples, the residual damage consists mainly of dislocation lines and perfect {110} loops possessing 1/2 <110> displacement vectors. There are no decorations observed in either the dislocation lines or loops. On the other hand, tin implanted samples have dislocation lines, perfect dislocation loops of 1/2<110> type and also an equal density of 1/3<111> faulted dislocation loops. Moreover, the dislocation lines and both types of loops have decorations which presumably are precipitates containing tin.

One aim of these experiments was to establish how high a temperature one can use safely in order to raise the level of electrical activity for a given high dose implant. It was noted that the electrical activity increases with temperature above 1000°C but there are related problems. The first problem is associated with the stability and reliability of the encapsulant. Our experiments have shown that the two layer encapsulant works successfully at temperatures of up to 1100°C for a few seconds. However, occasionally it tends to fail. The severity of its failure is reflected in the complexity and size of the patches of contrast due to damage, observed in TEM micrographs which also show residual strain and complex dislocation tangles. The second problem seems to be associated with the instability of GaAs at such high temperatures. Moreover, in the case of both types of ions, we have observed diamond shaped features 50 nm or less across and having a fine structure in them. These features are distributed throughout the depth of view as seen in a set of TEM stereomicrographs. They seem to retain their orientational relationship to the matrix and show well defined edges which are either visible or not, depending on the operative diffraction conditions. It seems that these defects are introduced in the bulk GaAs as a consequence of the high temperature anneal. To avoid their introduction it is necessary to set an upper temperature limit for annealing tin or selenium implanted GaAs. This limit seems to lie at a temperature of about 1000°C. Further work in this direction is in progress .

5. CONCLUSIONS

High temperature annealing of tin and selenium implanted GaAs has produced good electrical results. Tin implanted samples produce relatively higher values of electrical activity and mobility than those implanted with selenium ions. After annealing at 1000°C and above, a difference in the microstructure of residual damage in the tin and selenium implanted GaAs is observed. Defects in the selenium implanted samples are mainly dislocation lines and loops possessing 1/2<110> Burgers vectors. Tin implanted samples have both 1/2<110> and 1/3<111> type dislocation loops and 1/2<110> lines all of these defects being decorated with small precipitates. Moreover, there are small diamond shaped defects common to both types of ions. These may or may not be associated with the failure of the encapsulant which in general works successfully.

ACKNOWLEDGEMENTS:

The authors would like to thank the SERC for financial support, J. Mynard and staff of the Accelerator Laboratory for their help in ion implantation and staff of the Microstructural Studies Unit for their assistance with electron microscopy.

REFERENCES

1. R.K.Surridge, B.J.Sealy, A.D.E. D'Cruz and K.G.Stephens,
    Gallium Arsenide and Related Compounds
    Inst.Phys.Conf.Ser. No.33a, p.161, 1977.

2. B.J.Sealy, Microelec.J. 13, 21 (1982).

3. R.A.McMahon, H.Ahmed, R.M.Dobson and J.D.Speight,
    Elec.Lett. 16, 295 (1980).

4. M.A.Shahid, S.Moffatt, N.J.Barrett, B.J.Sealy and K.E.Puttick,
    Rad.Eff. 70, 291 (1983).

5. M.Arai, K.Nishiyama and N.Watanabe, Jap.J.Appl.Phys. 20,L214 (1981).

6. D.E.Davies, P.J.McNally, J.P.Lorenzo and M.Julian
    IEEE Elec.Dev.Lett. EDL-3, 102 (1982).

THE APPLICATION OF LASER ANNEALING TO THE FABRICATION
OF IMPATT DIODE STRUCTURES

ANTHONY E ADAMS AND L A HING
GEC Research Laboratories, Hirst Research Centre, Wembley, Middlesex UK

ABSTRACT

The conventional method for fabricating silicon IMPATT diode structures involves the epitaxial growth of successive n- and p-type layers onto a n$^+$ substrate followed by a boron diffusion to form the final p$^+$ layer. The high temperature time cycles experienced by the structure during these processes cause junction interfaces to become degraded through dopant diffusion. In this paper we examine the application of laser processing techniques to the epitaxial regrowth of low temperature deposited layers and report on the nature of the recrystallised material.

INTRODUCTION

The conventional method for the production of double drift IMPATT diodes involves the epitaxial growth of thin silicon layers onto highly doped n$^+$ (or p$^+$) substrates by the process of chemical vapour deposition (CVD). To ensure that high quality layers are produced, the technique requires that the substrate be heated to temperatures of at least 1000°C for several minutes, while an n-type and then a p-type layer is grown. To complete the active device structure, a boron drift is performed, again at a high temperature, to form the p$^+$ region. This temperature time cycle allows dopants to diffuse over significant distances causing the interfaces between the individual layers to become graded. For instance, arsenic can diffuse up from the substrate with the result that the n/n$^+$ interface becomes spread over several tenths of a micron. The spreading of junctions in this manner introduces additional series impedances which can have a deleterious effect on the diodes performance, such as a loss in operating efficiency, lower output power, and reduced operating frequency. Further, these problems become more acute as diodes which operate at higher frequencies are required.

As an alternative, we have been considering the application of laser annealing techniques [1-3] to these structures. We have previously applied ion-implantation and multiple pulse annealing techniques [4] to form abrupt p$^+$ layers [5] and shown that good device performance is obtained, as compared to boron drifted counterparts. However, to obtain an ideal device structure it is also necessary to consider methods of obtaining abrupt interfaces at the p/n junction and at the n/n$^+$ interface. To this end we have been examining the following : first a layer of silicon, up to 0.7 μm thick, is deposited onto an n$^+$ substrate in a conventional CVD reactor, but with the substrate temperature reduced to between 600/700°C. This is sufficient to prevent any upward diffusion of dopant from the substrate into the growing layer, but insufficient to promote epitaxial growth. The next stage is to use a scanning argon ion laser to epitaxially regrow the silicon film in the solid phase. It is well known that the diffusion lengths, associated with this process are no greater than a few lattice parameters [6]. Finally, two BF$_2$ ion-implantation and multiple pulse anneal schedules are used to form the p and p$^+$ regions respectively.

In this paper, we shall examine in detail, the use of cw laser annealing to epitaxially regrow the low temperature deposited silicon layers and discuss the resultant material properties. Also we compare the results of recrystallisation achieved by pulsed laser annealing.

EXPERIMENTAL

The cw laser annealing system used for this work has been described in detail elsewhere [7]. Essentially the system consists of an 18W argon ion laser operated on 'all lines'. After passing through an expanding telescope the beam is brought to a focus at the substrate surface, with a spot size of 35 μm. Microprocessor controlled galvanometers raster the beam across the sample surface with a scan velocity of 5 cm/s. The samples to be processed are mounted on a stainless steel chuck which is preheated to 350°C and the process schedule is controlled by a desk top micro-computer. The pulsed laser annealing system has also been previously described [5] and is based on a 1.5 J Q-switched ruby laser. A beam homogeniser [9] is used to ensure that the incident beam has a spatial uniformity of better than 10%, and an accurate optically-in-line energy meter is used to monitor the energy of each illuminating pulse. All the pulsed annealing was performed at room temperature. The silicon layers were grown onto both <111> and <100> silicon substrates in a standard CVD reactor by the decomposition of silane, with a substrate temperature less than 700°C. The substrates were cleaned 'in-situ' using a vapour phase HF etch and then thermally annealed at 1100°C. Some layers were doped during the deposit process.

To analyse the laser processed material several techniques have been used including transmission electron microscopy, electron diffraction, secco etching followed by scanning electron microscopy, and carrier concentration profiling by bevelling and two point spreading resistance measurements [10].

RESULTS

The as-deposited crystal structure of the layers was fine grain poly, with crystallite sizes of 100 Å or less. A range of beam powers from 1 to 9 Watts was used to process 0.7 μm thick layers. Below 3W no recrystallisation occurred. Above 5W the samples showed clear evidence of recrystallisation but it was also clear that this had occurred by surface melting. In the power range 3.5 - 5 Watts there was no evidence of melting and the layers were epitaxially regrown, as indicated by low energy electron channelling patterns. To ensure that no upward diffusion of dopant from the substrate occurred, samples with undoped layers were bevelled and examined using spreading resistance, Figure 1. If the film had melted, a significant

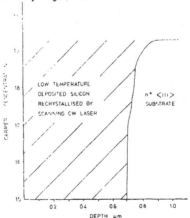

Figure 1 : Spreading resistance profile of low temperature deposited cw recrystallised undoped silicon.

amount of dopant would be expected to have entered the layer. However, these measurements, detected no evidence of arsenic diffusion. Some perturbation of the interface was visible, and this was attributed to a high density of defects present at the interface region. To confirm this, the following experiment was performed. The cw laser annealed layers 0.7 μm thick were irradiated with single pulses using the ruby annealer with energy densities ranging from 0.5 J/cm² to 2.5 J/cm². Before annealing, the surface of the regrown layers were largely featureless. This also remained the case for energy density values up to 2 J/cm², although some defects in the form of pits were visible Figure 2a. At 2.34 J/cm² the films became saturated with defects and <111> facets, Figure 2b, indicating that the recrystallisation front had regrown from a heavily defective region. Heat flow calculations, using a model described elsewhere [11], correlating melt depth with energy density as a function of film thickness Figure 3, indicated that this was the interface region.

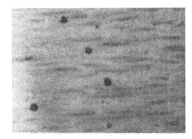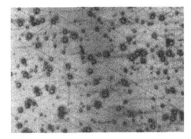

Figure 2a and 2b : Optical micrographs of surface of recrystallised layers before and after pulse annealing, respectively.

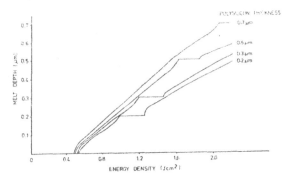

Figure 3 : Computed melt depth as a function of energy density and polysilicon film thickness.

To examine the nature of the crystal structure of the cw annealed films in greater detail, the films were etched using secco etch and examined in a scanning electron microscope Figure 4. The salient feature of the etched material is the presence of a high density of pits and defects. The resulting surface texture is similar to a sponge. This clearly shows that the as-deposited layers were not completely densified during the deposition process. Although the cw annealing process caused recrystallisation to occur, the lack of diffusion not only protected the original film substrate interface but also prevented movement of the silicon to anneal out extended defects. Further, because of the poor quality recrystallisation, layers

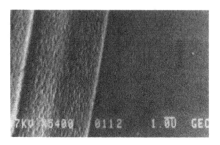

Figure 4 : Scanning electron micrograph of cw recrystallised film after secco etching.

doped during the deposition exhibited a concomitantly poor activation. A second feature of the micrographs is that the regrowth process is beam intensity dependent with the recrystallisation dropping off well within the dimensions of the scanning spot. In this case a step size of 8 μm was used between each successive scan. Where smaller steps sizes have been employed the region between each scan is much less visible, and better uniformity is obtained.

The application of pulsed laser annealing to the recrystallisation of the deposited silicon layers has also been investigated. For each layer thickness there was an associated energy density threshold for recrystallisation. This threshold was in good agreement to the predicted melt depths, Figure 3. These results were confirmed using secco etching. Films of different thickness were irradiated with a range of energy densities, and then etched, and the lowest energy density required for recrystallisation was correlated with Figure 3, to give Figure 5. Above the recrystallisation threshold high quality epitaxial regrowth occurred as shown by TEM investigation. Figure 6a is a typical result, this is for a 0.3 μm layer deposited onto a <100> substrate, irradiated with an energy density of 1.4 J/cm^2. Figure 6b shows the associated electron diffraction pattern. Despite the high quality regrowth, four point probe measurements showed that there was a significant upward diffusion of dopant from the underlying substrate. The surface sheet resistance dropped from being semi-insulating to 500 - 1000 Ω/square.

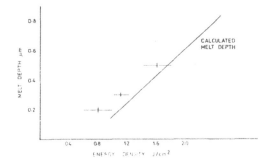

Figure 5 : Melt depth as a function of energy density for polysilicon layers on silicon.

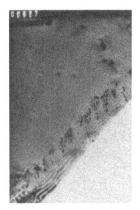 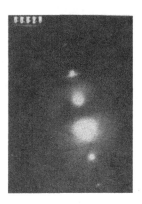

Figure 6a and 6b : Transmission electron micrograph for pulse recrystallised layer and associated diffraction pattern respectively.

In conclusion, we have shown that it is possible to epitaxially regrow low temperature deposited silicon thin films in the solid or liquid phase using a scanned cw or pulsed laser respectively. The pulsed laser process produces high quality regrowth, but the molten regime associated with this process allows underlying dopants to diffuse upwards into the layers degrading the original layer substrate interface. Solid phase recrystallisation using a cw laser inhibits this diffusion and maintains an abrubt interface however, the epitaxial regrowth is very dependent on the original quality of the deposited layers.

ACKNOWLEDGEMENT

I would like to acknowledge the useful contribution made to this work by Dr D J Godfrey, for the computer modelling, and Dr D B Holt of the Imperial College of Science and technology for the transmission electron micrographs.

REFERENCES

1   Laser and electron beam processing of materials, Ed. (C.W. White, P.S. Percy), MRS Proceedings, Academic Press 1980
2   Laser and electron beams solid interactions and materials processing, Ed. (J.F. Gibbons, L.D. Hess, T.W. Sigman), MRS Proceedings, North Holland 1981
3   Laser and electron beams interactions with solids, Ed. (B.R. Appleton, G.K. Celler), MRS Proceedings, North Holland 1982
4   C. Hill. Reference 2, 361
5   A.E. Adams and S.L. Morgan., Journal de Physique, Vol. C5, 1983, 433
6   J.F. Gibbons, Proceedings of Electrochem. Soc. Vol. 80-1, 980, 1
8   A.E. Adams and S.L. Morgan, Chemical Physics Vol.   , 1984,
9   A.G. Cullis, D.C. Webber and P. Baily, J. Phys. E. Sci. Instrum. Vol. 12, 1970,
10   M. Pawlik, GEC Jrnl. of Sci. Tech. Vol. 48(2), 1982, 119
11   D.J. Godfrey, A.C. Hill and C. Hill, J. Electrochem. Soc. 129(8), 1981, 1798

INCOHERENT LIGHT ANNEALING OF SELECTIVELY IMPLANTED
GaAs FOR MESFET APPLICATIONS

S.A. KITCHING, M.H. BADAWI, S.W. BLAND and J. MUN
Standard Telecommunication Laboratories Limited,
London Road, Harlow, Essex, England

ABSTRACT

    Capped and capless incoherent light annealing of high and low dose
silicon implants into GaAs have been compared with conventional capless
furnace annealing of the same implants. The yield and uniformity of DC
characteristics of selectively implanted depletion mode MESFET's
fabricated on 2-inch wafers annealed by the above three methods have also
been compared. Capped incoherent light annealing was found to give
results comparable to and in some cases better than conventional furnace
annealing both in terms of activation of the implants and also device
performance.

INTRODUCTION

    The use of incoherent light sources to anneal ion implanted GaAs is
becoming increasingly widespread [1-8]. The advantages of the method are
its relative simplicity and speed. Conventional furnace annealing
involves the use of sophisticated furnaces and wafer capsules and the
requirement to handle AsH_3 or As vapour. Thermal pulse anneals take of
the order of 15 seconds whereas the duration of a conventional furnace
anneal is usually 15 minutes or longer. The purpose of this paper is to
compare capped and capless incoherent light annealing with furnace
annealing.
    The comparison has been carried out on two levels:

(a) assessment of electrical characteristics following uniform implants
     of Si^+ at high and low doses, as a function of annealing schedule,
     and
(b) assessment of uniformity of the DC parameters measured for
     depletion-mode MESFETs on 2-inch wafers, as a function of annealing
     schedule

EXPERIMENTAL

Sheet Implant Anneals

    The parameters used in the first part of the experiment were chosen
to simulate as closely as possible the processes used to produce the
selectively implanted depletion mode MESFETs studied in the second part of
the experiment. All the wafers used in the experiment were 2 inch (100)
undoped semi-insulating GaAs wafers.
    The wafers were initially coated with 300 Å of vacuum evaporated
titanium. (This is the first step in the selective implantation
process). They were then given sheet implants of either 100 keV Si^{29}
ions at a dose of $7x10^{12}cm^{-2}$ to simulate the channel implant or
140 keV Si^{28} ions at a dose of $1x10^{14}cm^{-2}$ to simulate the source and
drain implant. The titanium layer was removed after implantation. The
wafers to be used for capped annealing in the incoherent light furnace
were coated on both sides with 400 Å of silicon nitride deposited in a
plasma enhanced chemical vapour deposition (PECVD) system.

The furnace annealed wafers were annealed uncapped face to face in a conventional furnace in an arsenic overpressure from a solid arsenic source at 850°C for thirty minutes. The incoherent light furnace used in these experiments was a Heatpulse 210T system manufactured by A.G. Associates. This consists of a gastight quartz chamber mounted between upper and lower banks of halogen lamps. The wafers to be annealed are placed on top of a 4-inch silicon wafer inside the chamber which is continuously flushed with nitrogen. Temperature monitoring is via a chromel-alumel thermocouple attached to a separate piece of silicon placed about 1 cm away from the silicon wafer. The displayed temperature is calibrated to take into account the temperature differential between the thermocouple and a silicon wafer placed on top of the 4-inch silicon support wafer. Thus the displayed and recorded temperature is not expected to accurately reflect the temperature of a GaAs wafer during the anneal.

The Heatpulse was operated in the "intensity mode" which causes the intensity of the lamps to be ramped up and down at a preset rate and also to be held at a given level for a preset time. The ramp rate chosen for these experiments was 50°C/s and the temperature of the anneals was 1050°C for two seconds. The capped wafers were annealed by simply placing them on the silicon support wafer however it was necessary to cover the uncapped wafers with a second 3-inch silicon wafer to prevent gross arsenic outdiffusion during the anneal. The capped wafers were heated by radiation from the upper bank of halogen lamps and also by conduction via the silicon support wafer which was heated by the lower bank of halogen lamps. The temperature experienced by the capped wafers during their anneal was probably fairly close to 1050°C. The uncapped wafers were heated by conduction alone via the silicon support wafer and the covering 3-inch silicon wafer and were therefore unlikely to have experienced a temperature as high as 1050°C during their anneals.

Carrier concentration profiles were obtained using the capacitance-voltage technique for the low dose implanted wafers and the differential Hall (Hall and strip) technique for the high does implants. Hall effect measurements were obtained using the van der Pauw technique.

Electrical Measurements

Capacitance-voltage (C-V) carrier concentration profiles obtained for the low dose (7×10^{12}cm^{-2}, 100 keV) wafers annealed by the three different methods are shown in figure 1. Capped thermal pulse annealing (TPA) produces the highest peak carrier concentration and the sharpest profile. This is presumably because the high temperature, short time anneal has succeeded in activating the implant but has not caused a large amount of diffusion of the silicon ions into the GaAs. Furnace annealed (FA) wafers exhibit a fairly high peak carrier concentration but a much broader profile, this is attributed to a significant amount of diffusion of the silicon ions during the long time furnace anneal. Capless thermal pulse annealing produces the lowest peak carrier concentration and a profile slightly narrower than that obtained for the furnace annealed samples. The low peak carrier concentration is probably due to the fact that the samples were subjected to a temperature much lower than 1050°C as described earlier. The relatively broad profile might also reflect an effective lengthening of the anneal time due to the increased thermal mass associated with the enclosing silicon wafers.

The Hall and Strip profiles obtained for the high dose implanted wafers are shown in figure 2. Capped thermal pulse annealing produces a very shallow profile with a high peak carrier concentration. Furnace annealing results in a much deeper profile with a lower peak carrier concentration. The capless thermal pulse annealed profile closely resembles the furnace

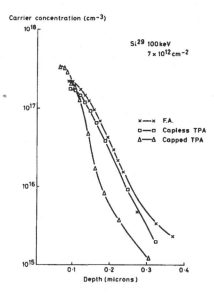

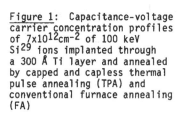

Figure 1: Capacitance-voltage carrier concentration profiles of $7 \times 10^{12} \text{cm}^{-2}$ of 100 keV Si^{29} ions implanted through a 300 Å Ti layer and annealed by capped and capless thermal pulse annealing (TPA) and conventional furnace annealing (FA)

annealed profile but is shallower. The corresponding mobility profiles are all very similar except that the thermal pulse annealed samples exhibit lower mobilities close to the surface and higher mobilities further into the GaAs compared with the furnace annealed samples.

Hall-Effect measurements obtained for the low dose implanted wafers are shown in table 1. The best results (i.e. lowest sheet resistance, highest mobility and highest sheet carrier concentration) were obtained for the furnace annealed samples. Capped thermal pulse annealing produced the next highest sheet carrier concentration with capless thermal pulse annealing producing the lowest. The lowest mobility was recorded for the capped thermally pulse annealed samples, which is indicative of incomplete removal of the implant damage during the short time anneal. Hall effect measurements obtained for the high dose ($1 \times 10^{14} \text{cm}^{-2}$, 140 keV) silicon implants are also shown in table I. Capped thermal pulse annealing has produced the highest sheet carrier concentration and furnace annealing has produced the lowest sheet carrier concentration. Once again the lowest mobility was recorded for the capped thermal pulse annealed samples.

It may be concluded from Table I that thermal pulse annealing of capped wafers appears to be more effective at activating the high dose implants than the low dose implants for the temperature/time regime studied. The opposite is true of furnace and capless thermal pulse annealing. This implies that different temperature/time regimes are required to optimally activate different energy/dose implants.

Unimplanted nitride capped wafers were pulse annealed under identical conditions to the implanted capped wafers. These wafers failed to show any n-type characteristics proving that silicon from the silicon nitride cap was not diffusing into the GaAs to produce the observed high levels of activation.

Table I

Van der Pauw - Hall effect measurements for low and high dose
silicon implants into GaAs annealed by capped and capless thermal
pulse annealing (TPA) and conventional furnace annealing (FA).

Dose (cm^{-2})	energy (keV)	anneal	Sheet Resistance (Ω/sq)	Mobility (cm^2/Vs)	Carrier conc. (cm^{-2})
7x10^{12}	100	FA	551	3613	3.2 E12
7x10^{12}	100	capless TPA	787	3200	2.5 E12
7x10^{12}	100	capped TPA	769	3050	2.7 E12
1x10^{14}	140	FA	182	2337	1.5 E13
1x10^{14}	140	capless TPA	162	2405	1.6 E13
1x10^{14}	140	capped TPA	64	2044	4.8 E13

Figure 2: Differential Hall
and Strip carrier concentration
and mobility profiles of
1x10^{14}cm^{-2} of 140 keV Si^{28}
ions implanted through
300 Å of Ti and annealed by
capped and capless thermal
pulse annealing (TPA) and
conventional furnace
annealing (FA)

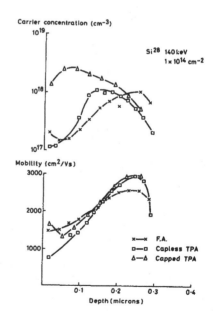

Selectively Implanted Devices

Depletion mode MESFETs with 1 μm gate lengths and 150 μm gate widths were fabricated on 2-inch GaAs wafers using a previously reported selective ion implantation technique [9]. The process consists of coating the GaAs wafer with a thin (300 Å) layer of titanium over which photoresist can be applied and delineated several times for a number of selective implantation steps. The use of a titanium barrier layer avoids any contamination of the GaAs surface by ion-hardened photoresist. The channel regions of the devices were selectively implanted with $7\times10^{12}cm^{-2}$ of Si^{29} ions at 100 keV and the source and drain regions were selectively implanted with $1\times10^{14}cm^{-2}$ of Si^{28} ions at 140 keV. Following the two-step implantation process, the photoresist and Ti layers were removed and the wafers were annealed by the three different methods described previously. The wafers for capped annealing in the Heatpulse system were coated with PECVD silicon nitride prior to annealing and this was removed after the anneal. AuGe/Ni was used as the metallisation for the ohmic contacts and CrAu was used for the Schottky barrier gates.

The DC parameters of the devices were measured using an automated system comprising an HP4145A semiconductor parameter analyser, an interfaceable probe station and an HP9826 computer for controlling the meaurements. Approximately 260 devices were measured over each 2-inch wafer and the yield of working devices on each wafer was greater than 90 percent. No degradation in yield was experienced for the thermally pulse annealed wafers compared with the furnace annealed wafers. The measured DC parameters, namely I_{DSS}, g_m and V_p of the tested FET's are shown in table II. The highest transconductance was obtained for the wafers that were coated with silicon nitride prior to their Heatpulse anneal. The best uniformity of DC parameters over a 2-inch wafer was obtained for the wafers annealed uncapped in the conventional furnace with an arsenic overpressure. The poorest uniformity was obtained for the wafers annealed uncapped between two silicon wafers in the Heatpulse. This was probably due to the non-uniform conduction of heat to the GaAs wafer by the two silicon wafers.

Table II

Mean values of DC parameters I_{DSS}, g_m and V_p and percentage standard deviations σI_{DSS}, σg_m and σV_p from their mean values for selectively implanted depletion mode MESFETs annealed by capped and capless thermal pulse annealing (TPA) and conventional furnace annealing (FA).

Anneal	I_{DSS} (mA)	σI_{DSS} percent	g_m (mS/mm)	σg_m percent	V_p (V)	σV_p percent
FA	50	7	125	6	3.6	5
Capless TPA	23	23	115	10	1.8	27
Capped TPA	45	13	144	8	2.8	11

CONCLUSIONS

The choice of annealing conditions for selectively implanted wafers is a compromise between the best conditions for annealing each of the individual implants. It has been demonstrated that capped thermal pulse annealing of silicon implants into GaAs can compete with conventional furnace annealing both in terms of activation of the implant and also the ensuing device parameters. Uncapped thermal pulse annealing has proved to be inferior to both capped pulse annealing and furnace annealing particularly in terms of uniformity of DC device parameters.

REFERENCES

[1] Arai, M., Nishiyama, K., and Watanabe, N., 'Radiation annealing of GaAs implanted with Si, Jpn. J. Appl. Phys., 1981, 20, pp L124-L126

[2] Davies, D.E., McNally, P.J., Lorenzo, J.P., and Julian, M., 'Incoherent light annealing of implanted layers in GaAs', IEEE Electron Device Lett., 1982, EDL-3, pp 102-103

[3] Ito, K., Yoshida, M., Otsubo, M., and Murotani, T, 'Radiation annealing of Si and S-implanted GaAs', Jpn. J. Appl. Phys., 1983, 22, pp L299-L300

[4] Kohzu, H., Kuzuhara, M., and Takayama, Y., 'Infrared rapid thermal annealing for GaAs device fabrication', J. Appl. Phys., 1983, 54, pp 4998-5003

[5] Kuzuhara, M., Kohzu, H., Takayama, Y., 'Electrical properties of S implants in GaAs activated by infrared rapid thermal annealing', J. Appl. Phys., 1983, 54, pp 3121-3124

[6] Tabatabaie-Alavi, K., Masum Choudhury, A.N.M., Fonstad, C.G., and Gelpey, J.C., 'Rapid thermal annealing of Be, Si and Zn implanted GaAs using an ultrahigh power argon arc lamp', Appl. Phys. Lett., 1983, 43, pp 505-507

[7] Rosenblatt, D.H., Hitchens, W.R., Shatas, S., Gat, A., Betts, D.A., 'Rapid thermal annealing of Si-implanted GaAs', Mat. Res. Soc. Symp., 1983, Boston, USA

[8] Badawi, M.H., Mun, J., 'Halogen lamp annealing of GaAs for MESFET fabrication', Electronics Letters, 1984, 20, pp 125-126

[9] Badawi, M.H., Dunbobbin, D.R., Mun, J., 'Selective implantation of GaAs for MESFET applications', Electronic Letters, 1983, 19, pp 598-600

RAPID THERMAL ANNEALING OF Si-IMPLANTED
GaAs FOR POWER FETs

H. KANBER,* R. J. CIPOLLI,* AND J. M. WHELAN**
*Torrance Research Center, Hughes Aircraft Company, Torrance, CA 90509
**Materials Science Dept., University of Southern California, Los Angeles,
CA 90089

ABSTRACT

Optimization and the advantages of rapid thermal annealing (RTA) for
the electrical activation of deep 300 keV Si^+ implants into GaAs are investigated and established for doses of 6 to 8×10^{12} cm^{-2}. These implant conditions are appropriate for power FETs. Results are compared with those based
on conventional controlled atmosphere capless furnace annealing (CAT).

The RTA yielded higher peak electron concentrations, high mobilities and
greater uniformities in the gateless FET saturation currents. The deep
implant results contrast with those for shallower implants for low noise
FETs. These differences are explained using a well-known implant damage
model.

INTRODUCTION

In this paper we compare the electrical characteristics of 300 keV Si^+
implants in GaAs after post-implant annealing using rapid thermal annealing
(RTA) and conventional furnace annealing using the controlled atmosphere
techniques (CAT) [1, 2]. The Si^+ energy at doses of $6-8 \times 10^{12}$ cm^{-2} are typical of those needed to fabricate power MESFETs in GaAs [3]. As will be shown,
there is a preference for the RTA approach to the annealing process. This
differs from our earlier study of shallower 100 keV Si implants [4], for
which there was no advantage with regard to the overall implant activation
providing one used the optimum RTA conditions.
        Potential advantages of rapid thermal annealing over conventional fur-
nace CAT annealing are that its rise time for heating to the desired temperature is very short as well as the briefness of the overall annealing period.
Advantage of the short rise time may be that damage reducing and implant
lattice site equilibration processes with their differing activation energies are more likely to occur simultaneously rather than sequentially when
the heating rates are approximately two orders of magnitude larger than is
typical for conventional furnace annealing. An inherent difficulty in RTA
annealing as currently practiced is the limited means available to prevent
surface degradation. This is not a problem if the CAT approach is rigorously
practiced [2], i.e. providing an As_4 overpressure over the GaAs surface which
is several orders of magnitude higher than the As dissociation pressure at
the elevated anneal temperature.
        The experimental section below compares the implanted GaAs characteris-
tics following various RTA schedules with those using optimized conventional
CAT furnace annealing. In the discussion section the observed differences
in the preference of one anneal mode over another for 100 keV and 300 keV
Si^+ implants in GaAs are interpreted in terms of the relative concentrations
and spatial distributions of implantation induced defects.

EXPERIMENTAL

The GaAs wafers were 2 inch diameter semi-insulating, unintentionally
doped and grown by the B_2O_3 liquid encapsulated Czochralski (LEC) method in
pyrolytic BN crucibles. Dislocation densities vary between the high 10^3 to
mid 10^4 cm^{-2} range and have a spatial W distribution across the diameter of
the wafer. The 300 keV ^{28}Si implants were made at room temperature with

does of 6, 7 and 8×10^{12} cm^{-2}. Fluences are estimated to be uniform over the wafer to within 1%. Overall electrical activation of the implanted Si requires annealing at temperatures between 850 and 950°C. At these temperatures and in a nominally inert atmosphere the GaAs surfaces tend to either lose As selectively and/or react with gaseous impurities such as O_2, H_2O and CO_2. Such surface deterioration is effectively minimized in the CAT furnace annealing in which a Pd diffused H_2 atmosphere is used which contains As$_4$ derived from a pure AsH$_3$ source.

The rapid thermal annealing system used was a commercial Heatpulse system [5]. The wafer is contained in a quartz isolation tube which positions it between upper and lower banks of halogen lamps, enclosed by water-cooled reflective walls. The GaAs wafer to be annealed was placed between a bottom four-inch Si wafer and a top GaAs wafer with the polished surfaces facing each other, thus called the face-to-face configuration. For SiO$_2$ encapsulated wafers, the top GaAs wafer was not used. A non-reactive gas is passed through the isolation tube at a rate of 5 l min^{-1}. N_2 from a liquid N_2 source and high purity Ar with less than 10 ppm of combined impurities have been used. As presently configured it is not feasible to use the higher purity Pd diffused H_2 containing As$_4$. The use of impermeable (to a degree) encapsulants such as SiO$_2$ and the face to face configuration with a second polished GaAs surface are limited means for reducing surface reactions. In addition the encapsulant produces additional dislocations because of the differences in thermal expansion coefficients. The face to face configuration does limit the gas flow to which the implanted surface is exposed.

Previous experience [1] with 100 keV Si$^+$ implants showed that RTA was satisfactory with respect to surface erosion for N$_2$ at 850°C, sometimes so for 890°C and unsatisfactory at 920°C. Preliminary experiments established that 10 second RTA times were superior to the 5 second for 300 keV Si$^+$ implants whereas 5 second 930°C schedule was preferred for 100 keV Si implants at comparable doses. SiO$_2$ (silox process) encapsulation reduces this extreme surface erosion but yields electrical results inferior to those for face to face annealing in high purity Ar ambients.

Secondary ion mass spectrometry (SIMS) was used to verify that there was no detectable change in the Si chemical profiles after annealing by either capless furnace CAT or RTA. An example of this illustrated in Figure 1 for a 300 keV Si implant at a dose of 8×10^{12} cm^{-2}. The SIMS (Cs beam) determined Si profiles at mass 28 after annealing by furnace and rapid thermal annealing coincide with the implanted profile. The depth scale was established to within six percent by measurements of the SIMS crater depths and absolute concentrations were calibrated by integration of the data points under the curve. Thus, any net differences in the electrical properties are associated with differences in the lattice site distribution of the Si, the migration of uncontrolled impurities and the relative extent of damage removal.

Three electrical evaluations were made: (a) net shallow donor concentration (taken as being the electron concentration) measurements from capacitance-voltage data on Au Schottky barrier diodes and summarized as the peak value, n_{peak}; (b) Hall mobilities, μ_n, based on the assumption of uniform current density in the active layer and (c) the source to drain saturation currents, I_{sat}, for a gateless MESFET whose channel dimensions are 15x200 μm. Particular attention is directed towards the I_{sat} data which reflects the product of the average thickness times the doping density in the implanted active area. Its unifomity over the implanted/annealed sample is a keystone to uniform device characteristics providing subsequent FET process steps are non-limiting.

Prior work [6] demonstrated that capless CAT furnace anneals were optimized at 890°C for 30 minutes. The preliminary RTA's were done with both face to face and SiO$_2$ encapsulation at 860 and 890°C for a Si$^+$ dose of 8×10^{12} cm^{-2} at 300 keV. For these experiments a two-inch wafer was quartered after implantation so that the radial distribution of as-grown imperfections would be reflected in each sample. These quartered wafer samples were then individually annealed. Three quarter sections were annealed for 10 seconds using RTA in a N_2 stream with the face to face configuration, one each at 860°C, 890°C and 920°C. The remaining fourth section was similarly annealed at 890°C after being encapsulated with SiO_2. Resultant peak carrier concentrations, room temperature mobilities and I_{sat} values are shown in Fig. 2. These results argued that the 10 second RTA at 890°C was somewhat superior to that at 860°C for the face to face configuration which is preferred to silox encapsulation as shown by the data in Fig. 2.

The next series of experiments was made using three consecutive wafers to the above wafer from the same ingot. These wafers were implanted with 300 keV $^{28}Si^+$ does of 6, 7 and 8×10^{12} cm^{-2}, quartered and three sections of each wafer were annealed for 10 seconds using RTA at 890, 920 and 950°C in Ar which contained less than 10 ppm of combined impurities. The remaining quarter sections from each wafer were furnace annealed using CAT at 890°C for 30 minutes. The resultant electrical properties are shown in Fig. 3. The most dominant characteristic favoring 10 second 890°C RTAs is the consistently higher I_{sat} values and the greater uniformity. Standard deviations in I_{sat} values for the RTA quarter sections were less than 4% whereas the CAT ones varied between 4 and 6%. It is difficult to find a generalized preferred RTA temperature from these data. Considerations leading to a choice of 890°C for 10 seconds are covered in the following section.

The final experimental point to make is that a significantly greater electrical activation is achieved by 10 second 890°C RTAs than by 5 to

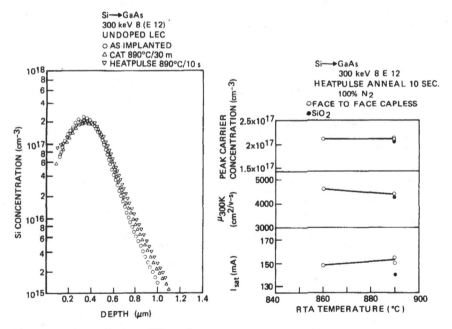

Fig. 1 SIMS chemical profiles of the Si concentration in undoped LEC GaAs after implantation and after capless furnace and Heatpulse annealing.

Fig. 2 Comparisons of peak carrier concentration, 300°K mobility and I_{sat} as a function of RTA temperature using capless face to face and SiO_2 encapsulants in N_2.

10 minute periods using conventional furnace heating at the same temperature even though samples in the latter may reach the desired temperature within 5 minutes.

DISCUSSION

For activation of 300 keV Si implants in semi-insulating GaAs at the relatively low doses described in this paper, we prefer 10 second RTA anneals at 890°C as a reasonable choice considering the presently available gas purities and compositions. Surface reaction rates for dissociation and etching by impurities in the gas phase increase non-linearly with temperature so that the lowest temperature for electrical activation is desireable. This is more so for the deeper 300 keV than the 100 keV Si implants because the anneal time is twice that of the 100 keV Si implants.

The irregular trend in n_{peak} and I_{sat} values for RTA temperatures between 890 and 950°C is troublesome but likely to be real because the extreme variations in I_{sat} for a given dose are 1.5 to 6 times the sum of the standard deviations of the quarters being compared. Variations reflect all the separate handling steps associated with one wafer at a time annealing in the RTA system. We tend to discount this as masking real changes occurring during the RTAs because clear temperature dependent monotonic trends were evident for the 100 keV Si implant RTAs [1]. Although these points need further study, it is clearly evident that 890°C, 10 second RTA is superior to the optimized 890°C, 30 min. furnace anneals at the same temperature for 300 keV Si implants at doses of 6, 7 and $8x10^{12}$ cm^{-2} in GaAs.

At the doses and energies used in this work the lattice after implantation is basically intact, i.e., non-amorphous. According to the stoichiometric disturbance model of Christel and Gibbons [7], the near-surface region contains excess concentrations of Ga and As vacancies whose maximum concentrations are at the surface. The transition to a deeper region in which the host atom concentrations exceed those of the lattice sites occurs at ~ 0.7 to 0.9 of the projected range, R_p, and the excess concentrations peak between R_p and R_p plus the straggle, ΔR_p. Relative concentrations of net vacancies increase nonlinearly with increasing energy and mass of the implant. The computed vacancy and excess concentrations for 50 and 100 keV B^{+} implants at a dose of $1x10^{13}$ cm^{-2} are shown in Fig. 4 to illustrate this point. This figure is adopted from Magee et al [8]. For simplicity we show the Ga and As vacancy and excess concentrations as being equal because of the relatively similar masses of Ga and As. Note that the excess vacancy concentrations have a strong dependence on energy. The actual values are higher because of the greater mass of Si. Transmission electron microscopy (TEM) [9, 10] and Rutherford BackScattering (RBS) [11] clearly indicate that much of the host atom damage is removed in relatively short times with furnace annealing at temperatures well below those necessary to achieve significant electrical activation of the Si implants. The excess host atom region is characterized after relatively low temperature furnace annealing by well-developed dislocation loops whereas the near-surface vacancy region is much less encumbered with dislocations. Diffusion of the vacancies towards the surface is likely to account for this observation. Furnace annealing studies at temperatures necessary for electrical activation of Si (890°C for 300 keV) implants agrees with this damage model and the dislocation generation relief within the peak excess host atom region as indicated by the Mn and Cr accumulation profiles [12-15]. Electrical activation of the implant by its diffusion to proper lattice sites is greatly enhanced by high vacancy concentrations. Thus, lower temperatures are required for optimal RTA with 300 keV than 100 keV Si implants. We propose that in the case of furnace annealing, the heating rate is sufficiently low so that these vacancy concentrations are seriously depleted by the time a suitable temperature is reached for the Si atoms to diffuse. For RTA this is not so pronounced because the heating rise time is about 100 times faster and the Si distribution more readily achieves its

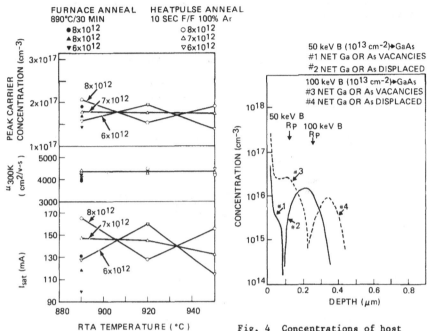

Fig. 3 Peak carrier concentration, $300^\circ K$ mobility and I_{sat} after capless face to face RTA as a function of RTA temperature in Ae for Si doses of 6, 7 and 8×10^{12} cm^{-2}. Corresponding values following the $890^\circ C$ 30 minute CAT furnace anneal are also shown.

Fig. 4 Concentrations of host atom net vacancies and net excesses as a function of depth for 50 and 100 keV B implanted GaAs. Adapted from Ref. 8.

equilibrium value by its enhanced diffusion constant because of the greater host atom vacancy concentrations.

CONCLUSION

For 300 keV Si$^+$ implants into undoped LEC GaAs at doses appropriate for power MESFETs, 6 to 8×10^{12} cm^{-2}, there is a distinct advantage of RTA over capless CAT furnace annealing provided both annealing techniques are optimized. This contrasts with the conclusions previously reached for 100 keV Si implants. An explanation for these differences is given. It is important to control the quality of the gaseous ambient in the RTA system and minimizing As losses by face-to-face annealing is preferable to the use of SiO$_2$ encapsulants.

ACKNOWLEDGEMENTS

The authors would like to thank G. Olson and R. Scholl of Hughes Research Laboratories for the use of their Heatpulse system and R. C. Rush

514

and P. Short of the Torrance Research Center for their technical help. The SIMS measurements were obtained from C. Evans and Associates, San Mateo, California. We also wish to thank Dr. H. Yamasaki and Dr. H. J. Kuno for their comments and support.

REFERENCES

1. R.M. Malbon, D.H. Lee and J.M. Whelan, J. Electrochem. Soc. 123, 1413 (1976).
2. H.B. Kim, J.M. Whelan, V.K. Eu and W.B. Henderson in : Proceedings of the Seventh Biennial Cornell Electrical Engineering Conference (Cornell University, Ithaca, NY, 1979) p. 121.
3. M. Feng, H. Kanber, V.K. Eu and M. Siracusa, Electronics Lett. 18, 1097 (1982).
4. H. Kanber, R.J. Cipolli, W.B. Henderson, and J.M. Whelan, 26th Electronic Materials Conference, Santa Barbara, CA. June 1984, paper L-3.
5. Heatpulse system manufactured by AG Associates, Mountain View, California 94043.
6. H. Kanber, unpublished.
7. L.A. Christel and J.F. Gibbons, J. Appl. Phys, 52, 5050 (1981).
8. T.G. Magee, H. Kawayoshi, R.D. Ormond, L.A. Christel, J.F. Gibbons, C.G. Hopkins, C.A. Evans, Jr. and D.S. Day, Appl. Phys. Lett. 39, 906 (1981).
9. D.K. Sadana, G.R. Booker, B.J. Sealy, K.G. Stephens and M.H. Badawi, Radiation Eff. 49, 183 (1980).
10. D.K. Sadana, T. Sands, and J. Washburn, Appl. Phys. Lett. 44, 623 (1984).
11. H. Kanber, M. Feng and J.M. Whelan, MRS Symposium on Ion Implantation and Ion Beam Processing of Materials, Boston, Mass., Nov. 83, Vol. 27 (North-Holland, New York, 1984), p. 365.
12. V. Eu, M. Feng, W.B. Henderson, H.B. Kim, and J.M. Whelan, Appl. Phys. Lett. 37, 473 (1980).
13. M. Feng, S.P. Kwok, V. Eu and B.W. Henderson, J. Appl. Phys. 52, 2990 (1981).
14. H. Kanber, M. Feng and J.M. Whelan, Appl. Phys. Lett. 40, 960 (1982).
15. H. Kanber, M. Feng and J.M. Whelan, J. Appl. Phys. 55, 347 (1984).

RAPID THERMAL ANNEALING OF
DEPOSITED SiO_2 FILMS*

J. WONG AND T.-M. LU, Center for Integrated Electronics and Physics
Dept., Rensselaer Polytechnic Institute, Troy, NY 12181

S.S. Cohen, General Electric Corporate Research and Development,
Schenectady, NY 12305

and

S. MEHTA, Ion Materials Systems, Eaton Corporation, Beverly, MA 01915

ABSTRACT

Silicon dioxide films have been deposited at an oxygen pressure and
substrate temperature as low as 10^{-2} Pa and 300°C, respectively, using a
partially ionized nozzle-beam technique. The refractive index of the
film is equal to that of the thermally grown silicon dioxide film, but a
frequency shift was observed in the infrared absorption peak at 1045 cm^{-1}
as compared to 1080 cm^{-1} of the thermally grown silicon dioxide film.
This frequency shift disappeared after the deposited film was annealed in
a rapid thermal annealing system utilizing an incoherent light source.
We emphasize that from the standpoint of device fabrication, although the
annealing temperature is high ($\sim$ 1100°C), the time is very short
(seconds), so that this process is still compatible to the requirements
of low temperature processing. Effects of rapid thermal annealing on
LPCVD and PECVD oxides are also studied and the results are compared to
that of the oxide deposited by the partially ionized nozzle-beam technique.

INTRODUCTION

Recently, a new annealing technique known as Rapid Thermal Annealing
(RTA) has been shown to have advantages over the conventional furnace
annealing (FA) method in many aspects [1-4]. In this method high temperature annealing (700°C to 1250°C), of short duration (in seconds) can be
achieved easily. For example, complete activation of the implanted
species can be carried out in a short duration with RTA [5]. This is
highly significant and is very useful for device fabrications which cannot withstand long time annealing that could lead to deeper diffusions.

Ionized Nozzle-Beam Deposition [6,7] (INBD) is a low temperature
process. The substrate temperature can be as low as 200°C. In this process, the source material is heated in a crucible to high temperature.
As the material is vaporized, it expands and ejects through a nozzle.
This nozzle beam is further ionized, accelerated by an applied electric
potential, and finally deposited on the substrate.

In this paper we report a comparison of RTA and FA treatments on
INBD, LPCVD and PECVD oxides on Si substrates. Ellipsometry, infrared
absorption spectroscopy, and etch rate analyses were used to characterize
the oxides before and after annealings. The RTA was found to have the
same effect as FA on the infrared absorption peak shift and the lowering
of etch rate on the annealed oxide samples.

*This work is supported in part by Semiconductor Research Corporation
and Eaton Corporation.

EXPERIMENTAL

Oxide Preparation

A detailed description of INBD oxide is described elsewhere [7]. The deposition parameters for INBD, CVD, and thermal oxides are summarized in Table I.

The INBD and thermal oxide samples were prepared on CZ 2" diameter N(100) 0.1-0.5Ωcm Si wafers, while the CVD oxides were prepared with CZ 3" diameter N(100) 3-5Ωcm Si wafers. Except for the PECVD oxides which have a thickness of 5 KÅ, all the other oxides were about 1 KÅ in thickness.

Rapid Thermal Annealing

The RTA system used in this experiment is a HEATPULSE 201M system [8]. The heating sources are high intensity quartz tungsten halogen lamps which have a power of 20 KW at full intensity. The lamp intensity is controlled by a microcomputer which provides trigger pulses to the lamp control triacs and is specified in percent of a full electrical intensity (100% = 20 KW). The parameters used to specify the heating cycle and their valid ranges are:

$$Heating\ Rate = 0.1 - 15\%/sec$$
$$Cooling\ Rate = 0.1 - 15\%/sec$$
$$Annealing\ Time\ (sec) = 1 - 1000$$
$$Annealing\ Intensity = 0 - 100\%$$

Intensity vs. substrate temperature calibration curve is available for this equipment [8].

Conventional Furnace Annealing

The annealing was carried out at about 1000°C for half an hour in a nitrogen ambient.

RESULTS AND DISCUSSION

In the present experiment, in order to simplify the study of the effect of RTA on INBD oxides and other deposited oxides, only one set of annealing parameters was chosen for the annealing cycle. They are: Heating Rate = 3%/sec, Annealing Intensity = 78% (∿ 1150°C), Annealing Time = 10 sec, and Cooling Rate = 3%/sec. Both RTA and FA experiments were carried out in nitrogen ambient.

Refractive Index Measurements

The sample thickness and refractive index were measured by the ellipsometry technique (Rudolph Auto ELII ellipsometer, λ = 632.8nm). As shown in Table II, the INBD oxides have refractive index very close to 1.46 (one obtained for thermal oxide) both before and after the annealing treatments. Similar results were observed for LPCVD and PECVD oxides. It is known that refractive index [9] is a good indicator of the stoichiometry of the silicon dioxide film. For example, silicon rich silicon dioxide [10] film was observed to have refractive index higher than 1.46. The results in Table II would appear to indicate that the three different deposited oxides have the same stoichiometry ratio as the thermal oxide both before and after annealings.

TABLE I. PROCESS SUMMARY

Oxide	Process Parameters
1. INBD	Source - Pure SiO_2 grain; ambient oxygen pressure at 10^{-2} Pa; Ionization current at 300ma; Acceleration voltage at 1 KV; and Substrate temperature at 300°C.
2. LPCVD	$SiH_4 + N_2O + O_2$ at 925°C.
3. PECVD	$SiH_2Cl_2 + N_2O$ at 350 to 400°C.
4. Thermal Oxidation	$H_2 + O_2$ at 850°C.

TABLE II. SUMMARY OF EXPERIMENTAL RESULTS

Oxide	Treatment	Thickness (Å)	Refractive Index-n	IR Absorption Peak (cm^{-1})	IR FWHM (cm^{-1})	Etch Rate* Å/min
INBD**	As Deposited	948	1.466	1045	110	500
	FA	874	1.458	1079	80	331
	RTA	910	1.462	1077	75	308
LPCVD	As Deposited	1072	1.459	1070	90	562
	FA	1048	1.454	1073	80	554
	RTA	1089	1.450	1070	85	582
PECVD**	As Deposited	5531	1.469	1050	171	410
	FA	5535	1.485	1077	140	208
	RTA	5397	1.449	1077	140	257
Thermal Oxidation		1153	1.460	1080	75	380

*Etching Solution: $HF/NH_4F = 1/15$
** Spectra of INBD and PECVD oxides are shown in Figures Ia and Ib, respectively.

518

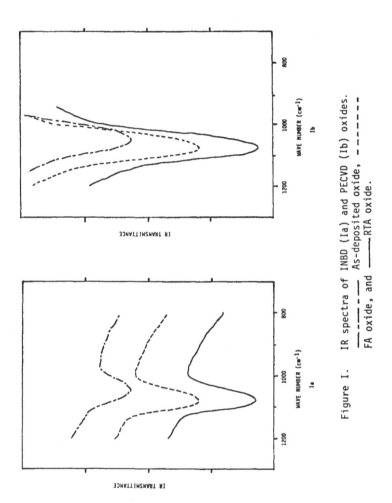

Figure I. IR spectra of INBD (Ia) and PECVD (Ib) oxides.
———— As-deposited oxide, — — — —
FA oxide, and ———— RTA oxide.

The thickness given in Table II are different because different samples were used for annealings and they do not necessarily reflect any possible changes of the thickness as a result of annealing.

Infrared Absorption

The infrared absorption spectra were taken with a double beam IR spectrometer (Perkin-Elmer Model 983G). The deposited INBD oxide has a broad absorption peak (wide FWHM) at 1045 cm^{-1} as compared to 1080 cm^{-1} for the thermal oxide. Similar shift of IR peak and broadening of absorption peak was also observed in the as-deposited LPCVD and PECVD oxides as shown in Table II and Figure I (INBD and PECVD oxides spectra only). The IR peak shift to lower wave number has also been observed in oxides deposited by other techniques such as, reactive sputtering deposition and evaporation [10]. The observed IR peak shifts might be due to the existence of defects, such as Si-O bond strain and porosity in the films [9-12].

After RTA or FA the INBD and PECVD oxides show a shift in IR peak towards 1080 cm^{-1}, the thermal oxide IR absorption peak. In addition, the IR absorption peak FWHM becomes narrower as shown in Figure I. This effect was also observed in deposited glass films [10,11]. Annealing densifies the deposited film. The total absorption intensity is proportional to the number of Si-O bonds only. The annealing process densifies the oxide and this increases the number of Si-O bonds per unit area. As the denser film absorps more, the FWHM decreases. In Figure Ib, the absorption is larger in the PECVD oxide, it is because its oxide is five times thicker than the others.

Comparatively small changes in the IR absorption peak of the LPCVD oxides might be due to the originally higher temperature of deposition, i.e., the LPCVD oxides were partially annealed during the deposition.

Etching

An etching solution of 1 part hydrofluoric acid to 15 part amonia fluoride was used to study the etch rates of the as-deposited oxide, FA and RTA oxides. From Table II, both INBD and PECVD oxides show a decrease of etch rate after annealing. This agrees with the IR peak shift results in which annealed oxides have a lower etch rate [9]. The almost constant etch rate of the LPCVD oxides again suggest that they were partially annealed during the deposition at a higher temperature. Note that annealed PECVD oxides have etch rates lower than thermal oxides. On the other hand, LPCVD oxides have higher etch rate. At this time, we do not have information to explain these results.

CONCLUSIONS

The results of the infrared absorption peak shift and etch rate measurements on the deposited INBD and PECVD oxides after RTA and FA suggest that the two techniques are comparable and RTA could be used as a substitute for long time FA. There is also an advantage in short processing time, only of the order of seconds and is a much more efficient process than FA. RTA is known to be able to control shallow junction diffusion in VLSI fabrications. The results in this report reassure the use of RTA in future VLSI fabrications.

ACKNOWLEDGEMENT

We thank Professor Murarka for invaluable discussions.

REFERENCES

1. Aron Gat, IEEE Electronic Device Letters, Vol. EDL-2, No. 4, April
    1981.

2. Kazuo Nishiyama, Michio Arai and Naozo Watanabe, Japanese J. Appl.
    Phys. 19(10), October 1980.

3. D. E. Davies, P. J. McNally, J. P. Lorenzo, and M. Julian, IEEE
    Electronic Device Letters, Vol. EDL-3, No. 4, April 1982.

4. Michio Arai, Kayuo Nishiyama and Naozo Watanabe, Japanese J. Appl.
    Phys., Vol. 20, No. 2, February 1981.

5. J. B. Lasky, J. Appl. Phys., 54, 6009 (1983).

6. I. Yamada and T. Takagi, Thin Solid Films 80, 105 (1981); I. Yamada,
    Proc. Int'l Ion Engineering Congress, ISIAT'83 and IPAT'83, Kyoto
    (1983). Conventionally, this is called the Ionized Cluster Beam
    Deposition. However, in the case of the deposition of silicon
    dioxide, the existence of oxide clusters has not been demonstrated;
    therefore, we prefer at this point to call this method the Ionized
    Nozzle-Beam Deposition.

7. Y. Minowa and K. Yamanishi, J. Vac. Sci. Technol., B1(4), 1148
    (1983); J. Wong, T.-M. Lu and S. Mehta, to be published in J. Vac.
    Sci. Technol., B, January/February 1985.

8. A trade mark of AG Associates, Inc.

9. W. A. Pliskin and H. S. Lehman, J. Electrochem. Soc., 112, 1013
    (1965).

10. W. A. Pliskin and S. J. Zanin, in Handbook of Thin Films, edited by
    Maissel and Glang, McGraw-Hill (1970).

11. Joe Wong, J. Electronic Material, Vol. 5, No. 2, 1976.

12. W. A. Pliskin and P. P. Castrucci, Electrochem. Technol., 6, 85
    (1968); J. Electrochem. Soc., 112, 148c (1965).

Silicon Crystal Growth from the Melt Over Insulators

ANALYSIS OF HIGH DOSE IMPLANTED SILICON BY HIGH DEPTH
RESOLUTION RBS AND SPECTROSCOPIC ELLIPSOMETRY AND TEM

T. LOHNER*, G. MEZEY*, M. FRIED*, L. GHIȚĂ**, C. GHIȚĂ**,
A. MERTENS***, H. KERKOW***, E. KÓTAI*, F. PÁSZTI*, F. BÁNYAI*,
GY. VÍZKELETHY*, E. JÁROLI*, J. GYULAI*, M. SOMOGYI****
*Central Research Institute for Physics, Budapest, P.O.B. 49,
Hungary
**Central Institute for Physics, Bucharest, MG 7, Roumania
***Humboldt University, 1040 Berlin, Invalidenstr. 42, GDR
****Institute for Technical Physics, Budapest, P.O.B. 76, Hungary

ABSTRACT

One of the applications of high dose ion implantation is to
form surface alloys or compound layers. The detailed characterization of such composite structures is of great importance.
This paper tries to answer the question: how can we outline, at
least, a qualitative picture from the optical properties measured by ellipsometry of high dose Al and Sb implanted silicon.
Attempts are done to separate the effect of implanted impurities
from the dominant disorder contribution to the measured optical
properties. As the ellipsometry does not provide information
enough to decide the applicability of optical models therefore
methods sensitive to the structure (channeling and TEM) were
applied too.

INTRODUCTION

Ellipsometric investigations of ion implanted silicon till
now, has mainly dealt with the characterization of the radiation
damage induced by energetic ions both on fully and partially
amorphous or on buried disordered layers [1-8]. Little effort
has been done to clarify the contribution of foreign atoms on
the optical parameters, perhaps, as for low and medium dose
ranges it is negligible. At high dose implantation,
($\phi > 5 - 7 \times 10^{15}$ atoms cm^{-2}), however, the ellipsometric parameters will seriously be affected by the presence of impurities
and their compounds which are formed with the host lattice. This
paper outlines the impurity effect for high dose Sb$^+$ and Al$^+$
implantation. Such low energy ion implantation can be used to
perform metal-semiconductor contacts [9]. The implanted layers
were additionally characterized with Rutherford Backscattering
(RBS) combined with channeling as a structure sensitive method
(disorder thickness, the quantity and distribution of antimony).

EXPERIMENTAL

Single crystal wafers cut to <100> direction were used in
the experiments. 10 keV Al$^+$ ions and 9 keV Sb$^+$ were implanted
at RT with doses of 0.1, 0.2, 0.5, 1, 2, 5×10^{16} atoms·cm^{-2}.
The thickness of the amorphous layer and the antimony distribution were measured by the high depth resolution channeling
applying glancing detection technique [10].

524

Ellipsometric measurements were made with a fixed wavelength ellipsometer operating at 632.8 nm. Selected samples
were measured by the spectroscopic ellipsometer [11] for 11
wavelength values in the 345-633 nm range.

RESULTS AND DISCUSSION

As it was pointed out earlier, in the medium dose range
and for fully amorphous layers, the effect of radiation damage
is the dominant and shields the effect of impurities [1-6].
Figure 1 shows the change of ellipsometric parameters for the
Al and Sb implants together with the theoretical spiral curve
which corresponds only to a single amorphous silicon layer of
increasing thickness and a native oxide layer. The experimental
verification of the later was done earlier [1,5,6,8].
Figure 2 shows channeling spectra of 9 keV Sb implants. The
thickness values of the disordered silicon layers measured by
means of channeling together with measured amount of antimony
are given in Table I.
For smaller doses, the ellipsometric parameters follow the
spiral curve within experimental accuracy of channeling
(Figure 1). For higher doses an alteration can be observed.
This is not surprising, first because RBS results give a considerable amount of Sb (Figure 2) second, an impurity-enhanced
oxidation occured during implantation [12]. For the highest
dose implantation, for example, approximately 50 atomic percent
antimony was found together with a compound of silicon oxide.
The complex refractive indices for the implanted samples
were determined by spectroscopic ellipsometry. As the thickness
of the implanted layers were learned from backscattering spectroscopy independently we were able to calculate the dielectric

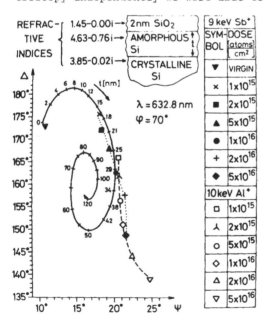

Figure 1.
The symbols show the
ellipsometric angles
(psi and delta)
measured on silicon
samples implanted
with different doses
of Sb and Al. The
continuous curve is
based on calculations
using a four-phase
model. The numbers
at the crossing bars
indicate the thick-
ness of the amorphous
silicon layer in
nanometers.

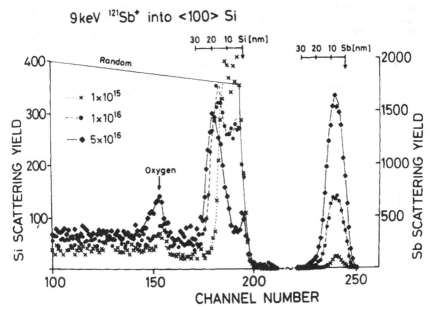

Figure 2. Channeling spectra of Sb⁺ implanted silicon.

Table I.
RBS and channeling data for implanted samples.

9 keV Sb⁺ into silicon						
Implanted dose ($\times 10^{16}$ atoms cm^{-2})	0.1	0.2	0.5	1	2	5
Measured dose ($\times 10^{16}$ atoms cm^{-2})	0.1	0.19	0.5	0.9	1.85	2.3
Measured thickness (nm)	16.5	18	20.5	23	23	24.5
10 keV Al⁺ into silicon						
Implanted dose ($\times 10^{16}$ atoms cm^{-2})	0.1	0.2	0.5	1	2	5
Measured thickness (nm)	26.5	30	32.5	35.5	27	42

function applying a three-phase model (air, heavily doped and damaged layer, crystalline silicon substrate). The calculations were performed using a computer program published by McCrackin [13]. The evaluations involve the dielectric function of single crystal silicon as it was determined by Aspnes and Studna [14]. Figure 3 shows the ε_2 functions for virgin and three different doses of Sb implantation. Figure 4 presents the ε_2 functions for virgin and three different doses of Al implantation.

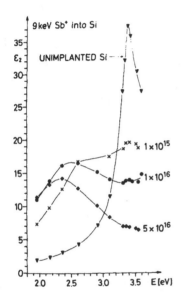 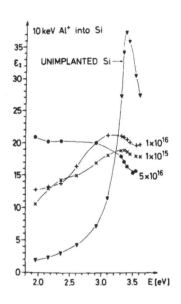

Figure 3. ε_2 spectra in the case of Sb$^+$ implants.

Figure 4. ε_2 spectra in the case of Al$^+$ implants.

In the case of Sb$^+$ implants the curve that corresponds to the lowest dose is similar to that obtained for medium dose As and P implantation by Luo et al. [15]. Besides, with increasing dose a shift was experienced in the ε_2 spectra [15]. In our case a similar trend was observed but our explanation rather concerns the reduction of the high energy part of the spectra which suggests that a metal-like behaviour is manifested in the implanted layer. In the case of Al$^+$ implantation the ε_2 spectrum of the highest dose shows the above mentioned trend.

TEM micrographs of the highest dose Sb and Al implants are displayed in Figure 5 and Figure 6, respectively. They show the non-homogeneous structure of the near-surface region. On basis of the work by Namavar et al. [16] these formations might be metallic agglomerates.

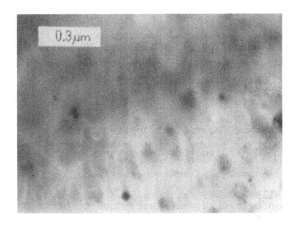

Figure 5. TEM image
of silicon implanted
with 5×10^{16} Sb^+/cm^2

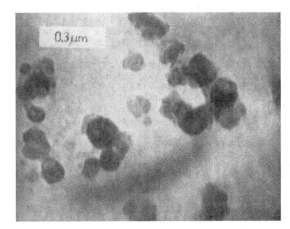

Figure 6. TEM image
of silicon implanted
with 5×10^{16} Al^+/cm^2

CONCLUSION

Present experiments show a strong impurity effect in the
ellipsometric parameters for high dose implantation of Sb and
Al into Si. Up to a dose value of 2×10^{15} atoms cm^{-2} the fixed
wavelength ellipsometry can still be used for fast disorder
thickness determination.

Investigation by spectroscopic ellipsometry suggested that
in the highest dose case a metal-like behaviour manifested at
present experimental conditions.

528

ACKNOWLEDGEMENT

The authors are grateful for helpful discussions to
Dr. Á. Barna (Inst. for Technical Physics, Budapest, Hungary)
and to Dr. Rose Takács (Iron and Steel Research and Development
Enterprise, Budapest) for TEM measurements.

REFERENCES

[1] T. Motooka and K. Watanabe, Journal of Applied Physics,
    51, 4125 (1980).
[2] J.P. Cortot and Ph. Ged, Applied Physics Letters, 41, 93
    (1982).
[3] M. Delfino and R.R. Razouk, J. Electrochemical Soc. 129,
    606 (1982).
[4] W.M. Paulson, S.R. Wilson, C.W. White and B.R. Appleton,
    in Defects in Semiconductors II (Mat. Res. Soc. Symp.
    Vol. 14, ed. S. Mahajan and J.W. Corbett, North-Holland,
    1983), 523.
[5] T. Lohner, G. Mezey, E. Kótai, F. Pászti, A. Manuaba and
    J. Gyulai, Nucl. Instr. and Meth. 209-210, 615 (1983).
[6] M. Fried, T. Lohner, E. Jároli, Gy. Vízkelethy, G. Mezey,
    J. Gyulai, M. Somogyi and H. Kerkow, Thin Solid Films,
    116, 191 (1984).
[7] Zh. Kusainov, B.N. Mukashev, K.H. Nusupov, V.V. Smirnov,
    Proc. Int. Working Meeting on Ion Implantation in Semi-
    conductors and Other Materials, Prague (1981) 19.
[8] T. Lohner, G. Mezey, E. Kótai, A. Manuaba, F. Pászti,
    A. Dévényi and J. Gyulai, Nucl. Instr. Meth. 199, 405 (1982).
[9] H. Klose, A. Mertens, R. Reetz and Ng. Tang, phys. stat.
    sol. (a) 77, 233 (1983).
[10] G. Mezey, E. Kótai, T. Lohner, T. Nagy, J. Gyulai and
    A. Manuaba, Nucl. Instr. and Meth. 149, 235 (1978).
[11] C. Ghiţă, L. Ghiţă, I. Baltog and M. Constantinescu,
    phys. stat. sol. (b) 102, 111 (1980).
[12] G. Mezey, T. Nagy, J. Gyulai, E. Kótai, A. Manuaba and
    T. Lohner, in Ion Implantation in Semiconductors, 1976
    Ed. F. Chernow, J.A. Borders and D.K. Brice (Plenum, 1977)
    49.
[13] F.L. McCrackin, NBS Technical Notes 479 (1969).
[14] D.E. Aspnes and A.A. Studna, Phys. Rev. B 27, 985 (1983).
[15] J. Luo, P.J. McMarr and K. Vedam, in Defects in Semi-
    conductors II (Mat. Res. Soc. Symp. Proc. Vol. 14.,
    Ed. S. Mahajan and J.W. Corbett, North-Holland, 1983), 529.
[16] F. Namavar, J.I. Budnick, A. Fasihuddin, H.C. Hayden,
    D.A. Pease, F.A. Otter and V. Patarini, in Ion Implanta-
    tion and Ion Beam Processing of Materials (Mat. Res. Symp.
    Proc. Vol. 27, Ed. G.K. Hubler, O.W. Holland, C.R. Clayton
    and C.W. White, North-Holland, 1984) 347.

CHANNELING EFFECT IN BORON AND BORON MOLECULAR ION-IMPLANTATION

M. DELFINO,* and B.M. PAINE**
* Philips Research Laboratories Sunnyvale, Signetics
  Corporation, Sunnyvale, CA 94088
**California Institute of Technology, Pasadena, CA 91125

ABSTRACT

The extent of boron channeled into interstitial and substitutional sites is measured for singly charged ions of B, BF, BCl, and BF_2 implanted into <100> silicon. We find that the most deeply penetrating boron atoms come to rest in interstitial sites with the consequence that electrical junctions are always more shallow than metallurgical junctions. This result is essentially independent of ion mass, energy, and fluence.

INTRODUCTION

At low energy and high fluence, boron ions implanted into crystalline silicon tend to scatter into <100> axial channels resulting in a tail in the implantation profile that deviates from the Lindhard, Scharff, and Schiott (LSS) prediction [1]. The dependence of this channeling tail on implantation parameters has been the subject of numerous investigations [2,3] because of its role in determining junction depths after activation annealing. Implanting into a random equivalent direction, by rotating and tilting the substrate away from the ion-beam, is most frequently used to minimize channeling [4]. However, complete supression of channeling and consequently the most shallow junctions occur only when the substrate surface is made amorphous to a depth that is greater than the deepest penetration of the implanted ions [5].

In this paper we examine the effect of axial channeling on both the atomic and carrier concentration profiles of boron and boron molecular ions implanted into <100> silicon. The effect of the fluence and the mass of the implanted ion on the extent of boron channeled into both substitutional and interstitial sites is evaluated.

EXPERIMENTAL METHODS

Boron was implanted as $^{11}B^+$, $^{30}BF^+$, $^{46}BCl^+$, and $^{49}BF_2^+$ into n-type <100> silicon at equivalent energies of 10- and 16-keV per boron atom to ion-doses of 6.0×10^{14}, and 1.2×10^{15}. Similarly, $^{11}B^+$ and $^{49}BF_2^+$ were implanted into substrates that were pre-amphorized by implanting with 1.5×10^{15} $^{28}Si^+$ cm^{-2} at 40-, 100-, and 200-keV. This schedule of implantations forms a 0.35-μm thick amorphous layer that is continuous to the surface. A 7 degree tilt angle with zero degree rotation was used for all implantations. The beam current was maintained at 1.0-μA cm^{-2} or less and the substrates were held at room temperature. After implantation, atomic boron concentration profiles were determined from secondary ion-mass spectrometry

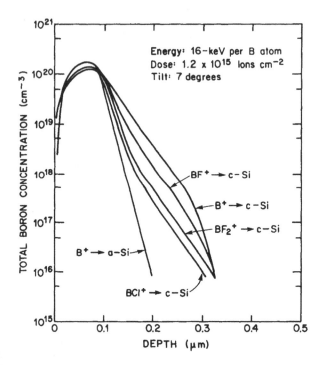

Figure 1. Atomic boron distributions as measured by SIMS. A constant sputtering rate of approximately 6.4×10^{-4} μm s^{-1} was assumed in deriving the depth scale.

(SIMS) depth profiling using an O_2^+ primary beam [6]. Carrier concentration depth profiles, assuming bulk mobilities, were obtained from spreading resistance measurements made on beveled samples after a two-step annealing treatment. First, the amorphized substrates were solid-phase epitaxially regrown by furnace annealing at 500 °C for 1h. in dry nitrogen. Then, cw laser annealing was used to electrically activate the implanted boron without causing it to measurably diffuse. The laser annealing was done, in room ambient, with an Ar-ion laser operating with 7.0 W multiline emission at a wavelength of 0.5-μm. A raster scan velocity of 8-cm s^{-1} with 50% lateral overlap was used. The back-surface wafer temperature was maintained at 420 °C during annealing.

RESULTS AND DISCUSSION

Figure 1 shows SIMS depth profiles for 16-keV boron ions implanted to a dose of 1.2×10^{15} cm^{-2} with the mass of the accelerated ion as a parameter. The implantation of $^{11}B^+$ into this amorphous substrate corresponds to when there is no measurable channeling. This assertion is supported by noting that the same concentration profile is obtained when $^{49}BF_2^+$ is

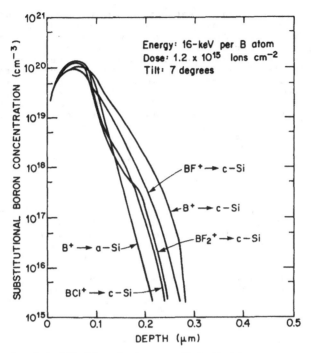

Figure 2. Substitutional boron distributions as measured by spreading resistance measurements.

implanted into amorphous silicon. The other boron molecular ions are assumed to behave similarily since the thickness of the amorphous layer is greater than the depth of their penetration. In view of this, several conclusions can be drawn from this data. First, in agreement with previous observation [3], less boron channels when it is implanted as a molecular ion, specifically a boron halide, than as a monatomic ion. Second, channeling is apparently more sensitive to the largest mass of the co-implanted atoms than to the mass of the accelerated ion itself. Consequently, $^{46}BCl^+$ is more effective in suppressing channeling than is $^{49}BF_2^+$. Third, a further reduction in channeling occurs because the number of co-implanted ions increases. This results in less channeling with $^{49}BF_2^+$ than with $^{30}BF^+$.

Figure 2 shows carrier concentration depth profiles for these same implanted samples after the low-temperature furnace and laser annealings. SIMS measurements show that the annealings do not cause the boron to diffuse within the instrumental resolution estimated at 0.02-μm. Consequently, we assume that these concentration profiles correspond to boron that is as-implanted into or near enough to substitutional sites to be electrically activated during the laser annealing. It is further assumed that after annealing all boron atoms are locked during any further heat treatment up to the melting point of silicon. Therefore, subtraction of the substitutional boron concentration (fig. 2) from the total boron concentration (fig.1) is considered to be a measure of the interstitial boron concentration. With these constraints in mind, we conclude that as a consequence of axial channeling, the most deeply penetrating boron ions come to rest in almost exclusively interstitial sites. This results in a large distinction between the electrical and metallurgical junction. A junction depth differential of approximately 0.07-μm is measured at a concentration level of 1×10^{16} cm^{-2}, and this is apparently not sensitive to the ion species. The electrical and metallurgical junctions coincide only when the ions have been implanted into the amorphous substrate. Furthermore, since interstitial diffusion is much more rapid than substitutional diffusion [7], this difference in junction depths will increase with additional heat treatments.

The results and conclusions obtained for the 6×10^{14} ions cm^{-2} dose samples are essentially the same as discussed here for the 1.2×10^{15} ions cm^{-2} dose. However, the depth differential between the electrical and metallurgical junctions is smaller with the lower dose because there is less channeling at lower fluences [3]. The effect appears even less pronounced at 10-keV, although the depth accuracy of the data becomes questionable below about 0.2-μm. This is especially true of the spreading resistance measurements, which are probably not accurate to any better than ± 0.02-μm.

ACKNOWLEDGEMENTS

The excellent technical assistance of P. Contreras (Philips Research Labs.) is sincerely appreciated.

REFERNCES

1. J. Lindhard, M. Scharff, and H. E. Schiott, "Range Concepts and Heavy Ion Ranges," Mat. Fys. Medd. Dan. Vid. Selsk, 33, 1-44, (1963).
2. T. M. Liu, and W. G. Oldham, "Channeling Effect of Low Energy Boron Implant in (100) Silicon," IEEE Electron. Dev. Lett., EDL-4, 59-62 (1983).
3. R. G. Wilson, "Boron, Fluorine, and Carrier Profiles for B and BF$_2$ Implants Into Crystalline and Amorphous Si," J. Appl. Phys., 54, 6879-6889 (1983).
4. A. E. Michel, R. H. Kastl, S. R. Mader, B. J. Masters, and J. A. Gardener, "Channeling in Low Energy Boron Ion Implantation," Appl. Phys. Lett., 44, 404-406 (1984).
5. B. L. Crowder, J. F. Ziegler, and G. W. Cole,"The Influence of the Amorphous Phase on Boron Atom Distributions in Ion Implanted Silicon," Ion Implantation in Semiconductors and Other Materials (Plenum, New York, 1973), pp. 257-266.
6. Charles Evans and Associates, San Mateo, CA 94402.
7. V. M. Zelevinskaya, G. A. Kachurin, N. P. Bogomyakov, B. S. Azikov, and L. L. Shirkov, "Capture of Impurity Atoms by Vacancies in Implanted Layers," Sov. Phys. Semi. 8, 164-166 (1974).

ION-PROJECTION LITHOGRAPHY FOR SUBMICRON
MODIFICATION OF MATERIALS

G. STENGL, ·H. LÖSCHNER, W. MAURER AND P. WOLF
FETEC , A 1070 Vienna, Austria

ABSTRACT

Ion-Projection Lithography (IPL) is a future production technique for
submicron electronic devices, wich combines the advantages of e-beam and
X-ray lithography without having their disadvantages. Like electrons, ions
can be accelerated, focused and deflected, and as is the case with X-rays,
scattering in resist layers is less pronounced as for e-beams. Experimental
results of IPL obtained with an Ion-Projection-Lithography-Machine IPLM-01
are presented: Ion images of self supporting masks ten times demagnified with
a geometrical resolution <0.25 µm printed into organic and inorganic resist
layers in high volume production oriented times.

INTRODUCTION

Ion beam processes gain increasing latitude in microelectronic device
fabrication. Ion implantation and dry etching (plasma and reactive ion
etching) are already common in the production of integrated circuits, re-
placing conventional furnace deposition and wet chemical processing. Until
now, ion beam processes have only been used in a non-lithographic way, i.e.,
patterns had to be defined by a (typically) photolithographic process
wich generates a mask for the broad ion beam.
Focused ion beam technologies offer new and exciting prospects for
submicron device fabrication. Compared to electron beams, ion beams suffer a
very small scattering in target materials. This characteristic offers a high
resolution lithographic technique. There is strong evidence that ion beam
lithography will be the leading pattern delineation technique in the
sub-0.5 µm and nanometer range [1,2,3].
Ion beam lithography has developed into three directions: Focused Ion
Beam Lithography (FIBL), Masked Ion Beam Lithography (MIBL) and
Ion Projection Lithography (IPL).
FIBL is a serial technique, whre a beam of ions from an ion source is
focused on a target and is scanned just as in a scanning e-beam system. Even
with the high current densities achived (1 A/ cm^2) FIBL writing speeds are
too slow for high volume production. The statistical nature of an ion beam
exposure, wich requires at least 1.000 ions to reach the smallest pixel,
prohibits the use of more sensitive resist layers to improve writing speeds
[4,5].
MIBL is 1:1 shadow printing with ions. For submicron resolution submicron
masks are required, wich are subjected to the same power density as the
target. Mask heating problems restrict this technique to the exposure of
sensitive organic resist layers.

ION PROJECTION LITHOGRAPHY

Using ions as information carrier in a demagnifying step and repeat
exposure system, submicrometer and even nanometer resolution can be obtained
combined with a very high depth of focus: mor than 100 micrometer.
IPL as a parallel exposure system permits significant lower exposure times
per chip than scanning systems by using masks that contain and store the
complex design patterns. Step and repeat exposure is mandatory for the

534

necessary chip-by-chip alignment to cope with wafer distortions induced by technological production steps. Demagnifying projection avoids the problems connected to the generation and maintenance of submicron masks.

The high power densities possible with IPL permit not only pattern transfer in conventional organic resists but extend lithography to new processes using Resistless Ion Beam Modification Techniques of Materials. Because of statistical limitations insensitive resist layers are mandatory for obtaining nanometer resolution.

Ion Projection Lithography meets the future requirements of high volume sub-0.5 μm IC production: Ion-projection exposure and modification of insensitive resist layers with sub 0.5 μm resolution and large depth of focus combined with an attractive throughput of wafers per hour.

ION PROJECTION LITHOGRAPHY MACHINE   IPLM-01

An Ion Projection Lithography Machine (IPLM) is an ion-optical imaging system designed to project ion images of open stencil masks at mx (10x or 5x) demagnification onto workpiece substrates (ion reduction printing).

Fig. 1 shows the building blocks of the IPLM-01: The ion source consists of a reliable duoplasmatron ion source with specially modified extraction geometry wich generates high-brilliance beams of hydrogen-, nitrogen-, or rare gas-ions (He^+, Ne^+, Ar^+, Xe^+), and a magnetic analyzer. A divergent ion beam illuminates an open-stencil mask with an ion energy between 5 keV and 10 keV. The immersion lens accelerates the ions to final energy of 60keV to 90 keV. The projective lens forms an image of the mask near its focal

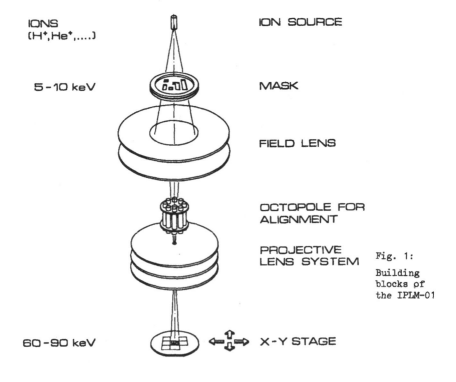

IONS
(H^+, He^+,....)   ION SOURCE

5-10 keV   MASK

                           FIELD LENS

                           OCTOPOLE FOR
                           ALIGNMENT

                           PROJECTIVE   Fig. 1:
                           LENS SYSTEM   Building
                                             blocks of
                                             the IPLM-01

60-90 keV   X-Y STAGE

plane at a nominal scale of 1:m (1:10 or 1:5). This scale is electrically
adjustable within ± 3%. The ion-optical system is operated with very small
numerical apertures, wich results in a very large depth of focus without
limiting resolution, as is the case in light optical systems.

Layer-to-layer alignment is performed by coarse mechanical adjustment
(to ±2 µm) followed by electrostatic deflection of the ion image (at an
accuracy of ± 0.05 µm and by rotation of the image with an axial magnetic
field. Electronic fine adjustment of x,y,θ and scale is possible without
moving parts. Different sensors will be incorporated into the alingment
system to measure the position of wafer marks with respect to ion test beams
generated by correspokdign mask openings.

At 10x demagnification ion beams shooting through the mask openings are
concentrated onto the wafer with an increase of the current density by two
orders of magnitude. Dependign on the ion species, an ion current density at
the wafer target up to 1 mA/cm^2 is possible. Fig. 2 summarises ion beam
processes an the typical exposure times to perform them in an IPLM.

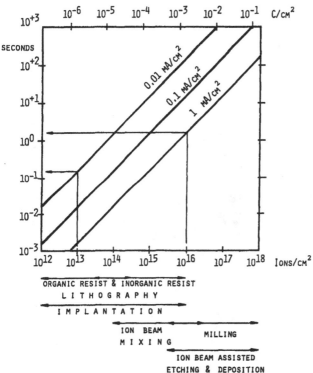

Fig. 2:

Exposure time for
ion beam processes
by IPLM-01.

IPLM-01 is built in a modular way to offer several developments incorporated into a R&D machine: Integration of an intensive single ion beam and
ion projection, integration of inspection (SIM) mode and focused ion mask
repair, and target chamber modifications for ion assisted etching and
deposition.

In order to fullfill the requirements of high volume production of
submicron ICs by Ion Projection Lithography, wich is exspected to start in
the ninties, an IPLM for pilot production will be developed about 1987.

IPLM-01: EXPERIMENTAL RESULTS

The experiments were performed with the IPLM-01 using a duoplasmatron ion source without mass analyzing unit. Therefor, different ion species were delivered to the mask and to the wafer target (according to the feeding gas predominant He^+). The earth magnetic field splits the ion trajectories into a set of images of different intensities.

The test mask used for the experiments was an electroplated metal foil generated on a silicon wafer with photoresist structures defining the mask openings. Mask heating is no serious problem due to the power density on the mask is three orders of magnitude lower than on the target, and a linear expansion of the mask is corrected by the electronic demagnification control. All results presented here were obtained with the same mask foil of 1.5 µm thickness resulting in a continous ion beam load of more than 250 hours.

A small part of the mask was structured by using a resolution test target. This resolution structures (property of Heidenhain, FRG) ten times demagnified shown in the following figures allow simple scale calculation by adding to every number of lines / mm the demagnification factor of ten.

Resolution tests in inorganic resist: SiO_2

Thermally grown silicon dioxide is used as an inorganic resist by ion bombardment enhanced etching (IBEE) of the ion exposed sites. Fig. 3 shows the basic mechanism: According to the dose of ions, ion bombarded oxide is etched in HF by a factor of 3 faster than undamaged oxide. Simple calculation yields a slope of the etched walls of the structures of about 70°. Fig. 4 is an overall view of structures in 0.2 µm thick SiO_2 generated by implantation of 1.2×10^{16} He^+ into former 0.3 µm thick oxide.

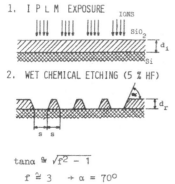

1. I P L M EXPOSURE

2. WET CHEMICAL ETCHING (5 % HF)

$$\tan\alpha \simeq \sqrt{f^2 - 1}$$
$$f \cong 3 \rightarrow \alpha = 70°$$

Fig. 3:

Pattern transfer into SiO_2 with IPL and wet etching

Fig. 4:

Resolution test pattern etched into SiO_2 by ion bombardment enhanced etching

Fig. 5 presents a close-up view of Fig. 4 showing lines and spaces of 0.2 µm width etched into silicon dioxide of 0.2 µm thickness. These SEM-micrographs demonstrate that wet chemical processing can be performed even below 0.5 µm in fairly thick layers by using ion projection lithography.

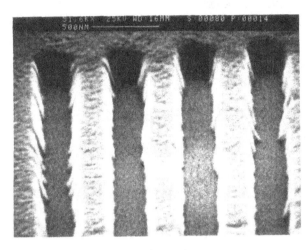

Fig. 5:

Lines and spaces of
0.2 μm width etched
in 0.2 μm after
ion projection of an
2 μm mask structure

Resolution test in organic resist layers

Several well-known organic resist layers have been exposed by the
IPLM-01 until now:

Nitrocellulose films of 0.4 μm thickness on silicon substrates have
shown to be very well suited for rapid adjustment and resolution evaluation
of the IPLM-01. A dose of 3×10^{14} He^+/cm^2 is sufficient to remove all the
nitrocellulose in the exposed areas without further development.

PMMA of various thicknesses can be patterned using the IPLM-01. Colour
changes of 0.4 μm thick PMMA on silicon wafers acieved by a wide range of
doses serves the same purpose as Nitrocellulose. Development in MIBK after
IPLM-01 exposure of 0.4 μm thick PMMA gives positive or negative tone
images (image reversal) for about 3×10^{13} and 5×10^{14} He^+/cm^2. Fig. 6 shows
IPLM-01 exposure of 1.5 μm thick PMMA on silicon with a dose of 5×10^{13} He/cm^2
developed in MIBK and isopropanol. The SEM micrograph shows the steep profile
of 0.45 μm lines.

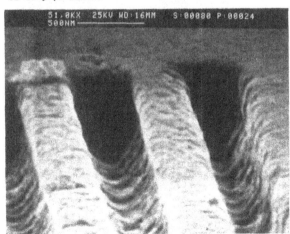

Fig. 6:

IPLM-01 exposure and
MIBK developed 0.45μm
lines and spaces in
1.5 μm PMMA

538

Electrostatic image deflection experiments

Applying a potential difference to the pre-projective lens octopole system the ion image projected onto the wafer target is deflected in x- or y-direction. The ion image of a 1250 lines/mm pattern is shown in the SEM-micrograph of fig. 7. Two exposures have been made, one with voltage applied to the octopole, one without. It is clearly seen, that the large deflection of about 9 microns does not interfere with image quality. Measurements show the possibility of deflecting the ion image up tu 50 microns in x- or y-direction.

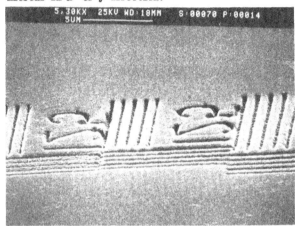

Fig. 7:

Electrostatic image deflection by the IPLM-01

Multibeam experiments have been performed using mask pinholes to generate submicron spots wich were staggered in x- and y-direction. Applications of this possibility of the IPLM-01 include the fabrication of linear grids with sub 0.5 μm periodicity.

Together with the image rotation possibility by the action of a weak magnetic field, wich is shown to rotate the whole ion image up to 0.3 degrees, the main requirements for alignment of subsequent lithographical layers are fullfilled.

SUMMARY

First experiments with the Ion Projection Lithography Machine IPLM-01 show geometrical resolution below 0.5 μm in organic and inorganic resist layers combined with a very high depth of focus. Electronic adjusment of the projected ion image on the wafer in x-,y-direction, in θ and in scale demonstrate the possibility to develop an efficient future production technique using resistless ion beam modification of materials.

REFERENCES

1. J.J. Muray, Semicond. International 7/4, 130 (1984)
2. W.L. Brown, 1984 Internat. Symposion Electron, Ion, Photon Beams,Torrytown
3. S. Namba, Ion Beam Microfabrication, Int. Ion Eng. Congr., Kyoto, 1983
4. W.L. Brown, T. Venkatesan, A. Wagner, Solid State Techn. 24/8, 60 (1981)
5. A. Wagner, Solid State Techn. 26/5, 97 (1083)

DISORDER PRODUCTION AT METAL-SILICON INTERFACES BY
MeV/amu ION IRRADIATION

R. L. HEADRICK* AND L. E. SEIBERLING**
*Department of Materials Science and Engineering, University of Pennsylvania,
Philadelphia, PA 19104
**Department of Physics, University of Pennsylvania, Philadelphia, PA 19104

ABSTRACT

We have shown that irradiation of Ag-Si and Au-Si interfaces by 14 MeV
0^{16} ions can produce non-registered silicon at the metal-silicon interface.
Evidence that this effect is due to electronic energy loss of the bombarding
ion is presented. The possible relationship of this effect to MeV-ion
enhanced adhesion is discussed.

INTRODUCTION

Several years ago, Griffith, et al. [1] showed that MeV heavy-ion
irradiation can cause significantly improved adhesion of evaporated thin
metal films to insulating substrates such as SiO_2 or Teflon. They suggested that the enhanced adhesion was associated with electronic sputtering
in insulators, perhaps through small scale mixing of atoms at the interface.
Subsequently, Mendenhall [2] found that adhesion of metal films to metals
and semiconductors could also be improved with MeV-ion bombardment. He
tried to measure mixing associated with adhesion at the Ag-Si interface
using high resolution Rutherford backscattering. However, despite a
reported depth resolution of 20A, no mixing was observed [2].

The lack of evidence for mixing and the fact that adhesion is easily
achieved with low doses in metal-metal systems may be construed as
evidence that electronic sputtering is not generally responsible for adhesion
[3]. However, Jacobson, et al. have found that 12 MeV oxygen ions will
improve adhesion of a copper film on aluminum, but only if the native
oxide is present [4]. If the oxide layer is removed by sputter etching in
vacuum prior to copper deposition, then no enhancement in adhesion is
observed. They suggest that the native oxide layer provides the necessary
insulator to trigger electronic sputtering.

The fact that interfacial mixing of more than 20Å has not been observed
in ion-adhered samples does not rule out sputter-induced mixing as the
cause of adhesion for two reasons. First, the average energy of electronically sputtered particles is less than a few eV [5,6], so mixing would not
be expected to exceed a few monolayers. Second, if atoms that are mixed
form new bonds across the interface, then one monolayer of mixing would be
adequate to substantially improve adhesion. We have recently reported the
results of an experiment designed to look for direct evidence of sputter-
induced mixing at the interface, on a monolayer level [7]. Our approach
was to use channeling, which is sensitive to disorder production in one
monolayer in a single crystal. If sputter-induced mixing, due to an interfacial oxide layer, is responsible for adhesion, then mixing should cause
disorder in a single crystal substrate adjacent to the oxide.

We observed disorder at a metal-silicon interface that increased
linearly with dose, with roughly one monolayer of disorder corresponding
to the (Scotch tape) threshold dose for adhesion. In this paper, we present

calculations of the expected disorder due to the nuclear stopping power of the bombarding ion. We also present evidence that the disorder we observed may be consistent with electronic sputtering in a native oxide at the interface.

EXPERIMENTAL RESULTS

The experiments were done in a non-standard geometry, that is transmission channeling through a thin (~5000Å) single crystal silicon window [8]. This geometry has several advantages, arising from the fact that the metal-silicon interface can be put on the beam-exit side of the crystal. There is strong evidence that Mev-ion enhanced adhesion depends on electronic rather than nuclear collisions [9,10]. Because nuclear collisions can cause much damage in a crystal, our goal was to reduce the nuclear collision probability and investigate the damage due to electronic effects alone. When an ion is well-channeled, the probability that it will displace a lattice atom due to a nuclear collision is reduced to zero [11]. By placing the interface at the beam-exit side, roughly 95% of the ions are well-channeled upon reaching the interface. This geometry also eliminates the possibility that primary nuclear recoils in the (randomly oriented) metal or oxide layer could contribute to interfacial damage.

Details of the experiment have been published elsewhere, and will not be repeated here [7]. The data are summarized in the figure. The top set of data points shows the amount of disordered silicon produced at the interface of a 40Å gold layer evaporated on a silicon <110> thin crystal as a function of beam dose. The middle set is for a 100Å silver layer, and the lower set for no metal layer. The data are well fit by a straight line. The uncertainty in the beam spot size is estimated to be 30 percent. Thus, the measured rates of disorder production, which are 0.11 Si atoms per incident ion for the Au-Si and 0.078 for the Ag on Si, are the same to within the estimated error. The zero dose intercept of each of these lines corresponds to the amount of interfacial oxide (see discussion).

In an attempt to check for thermal effects, the Au-Si data in figure 1 were repeated on a new spot of the sample that was 0.08 cm from the first. The open circles are the data taken from this spot. The slight increase in disorder (which is well within the estimated error) could be explained by a slight overlap in the irradiated spots. Thus, the effects of the beam occur only in the area that the beam actually hits. The experiment was also repeated with double the beam current and no change in the damage production rate was observed.

CALCULATIONS

The theory of radiation damage in solids by nuclear collisions can be used to estimate the expected nuclear contribution to interfacial damage. The theory of Kinchin and Pease [12] has been used successfully by Iwami et al. [13] to estimate the damage production rates for thin Ag and Pd over-layers on silicon substrates. These calculations were for a more ideal system than ours since the silicon was atomically clean and the metal layer was only a few monolayers thick. Here, we must take into account the effect of the interfacial oxide layer that the metal recoils must penetrate in order to cause damage in the silicon. The calculation is complicated further, since only a few monolayers of the metal near the interface contribute significantly to the damage.

According to the theory of Kinchin and Pease, damage cross sections can be calculated in a two step procedure. In a close impact parameter collision

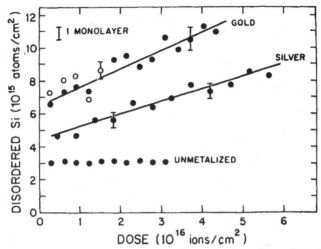

FIGURE 1 Disordered silicon produced at the interface of Au-Si,
Ag-Si, and unmetalized samples as a function of dose.

of an ion with a target atom, enough energy may be transferred to displace
the atom from its original position. The first step is to calculate the
total cross section, σ, for displacing a target atom:

$$\sigma = 4 \, M_1 Z_1^2 Z_2^2 E_R^2 \pi a_0^2 \, / \, (M_2 E_d E) \qquad (1)$$

In equation (1), M_1 and Z_1 are the projectile mass and atomic number,
respectively. M_2 and Z_2 are the target mass and atomic number. E_R is the
Rydberg energy (13.6 eV). E_d is the minimum energy to displace a target
atom, a_0 is the Bohr radius (0.529Å), and E is the energy of the incident ion.

The primary recoil atoms may have sufficient energy to displace other
nearby atoms in a collision cascade. The primary recoil has an average
energy of only several hundred eV and, therefore, displaces other atoms by
hard sphere collisions in which (on average) half of the recoil's energy
is transferred to another target atom. The total number of atoms displaced
per primary recoil after the collision cascade is N_d:

$$N_d = (1/2) \, (1 + \ln (\Lambda E / 2E_d)) \qquad (2)$$

where: $\qquad \Lambda = 4 \, M_1 M_2 \, / \, (M_1 + M_2)^2$

The total damage cross section in a bulk material is the product $\sigma \, N_d$.
As discussed above, the situation of a crystalline-amorphous interface is
more complicated than that of a bulk material. First, we are only inter-
ested in calculating the number of displaced silicon atoms from the crystal.
The primary recoil is assumed to be from the overlayer and since it must
recoil in a forward direction (away from the interface) the first atom
that it collides with will also be of the overlayer. The number of displaced silicon atoms per primary recoil then becomes $N_d' = (1/2)[N_d-2]$.

Second, only recoils from the first few monolayers of the overlayer
near the interface contribute to the displaced silicon. The recoils
undergo a collision once every lattice spacing and lose (on average) half
of their energy in each collision. Thus, if there are ℓ monolayers of

interfacial oxide, then a recoil from the mth monolayer of overlayer must have an energy of at least $2^{\ell+m-1}E_d$ to cross the interface. For a thick metal layer, a sum of $\sigma N_d'$ over all m converges. The expression for the damage production rate of silicon at the interface becomes:

$$R = N_d'' \, \sigma \, (Nt)_{mono} \tag{3}$$

Where:

$$N_d'' = 2^{-(\ell+1)} \, (\ln[\Lambda E / (2^{\ell+2} E_d)] - 3) \tag{4}$$

The factor $(Nt)_{mono}$ is the areal density of one monolayer of the overlayer.

Equations (3) and (4) have been used to calculate the damage production rates for the overlayers and interfacial SiO_2 layers. E_d has been taken to be 25 eV. The results of these calculations, along with the experimental results are tabulated in table 1. The oxide layers contribute, since they are amorphous and, therefore, also produce recoils. To calculate the contribution of the oxide layer, we have taken $\ell = 0$, since it lies directly on the Si surface. It can be seen from table 1 that the calculated damage rates for the samples with Ag and Au overlayers, $R_{overlayer}$, are far smaller than those observed experimentally. This is due to the very short range of the recoils. Since the oxide is directly in contact with the silicon, the total contribution of the interfacial oxide, R_{oxide}, is greater than that of the Ag and Au layers. Even if a significant amount of metal were to penetrate through the oxide layer by diffusion, the damage rates would still be several times smaller than those we have observed. In light of these results, it is difficult to explain the observed effects as a result of nuclear collisions alone.

TABLE 1 Calculated damage production contributions due to recoiling atoms in the oxide layer (R_{oxide}), and in the metal layer ($R_{overlayer}$). The experimentally measured damage rate (R_{exp}) is also tabulated.

Overlayer	SiO_2 thickness (monolayers)	R_{oxide}	$R_{overlayer}$	R_{exp}
none	4	9.3×10^{-3}	--	$< 2.4 \times 10^{-2}$
100Å Ag	5	9.6×10^{-3}	4.4×10^{-4}	7.8×10^{-2}
40Å Au	8	9.9×10^{-3}	3.7×10^{-5}	1.1×10^{-1}

DISCUSSION

Because of the discrepancy between the calculated and observed values in table 1, we must consider the possibility that another mechanism is responsible for the observed damage. As mentioned above, the observed effects are local to the irradiated area and insensitive to the beam current. These observations, along with the fact that neither Au nor Ag can readily react chemically with Si, lead us to believe that the observed effects cannot be explained as due to beam heating.

An interesting possibility is that the buildup of disordered silicon is due to oxidation of silicon at the interface. The oxygen would have to be supplied from the bulk or vacuum (1×10^{-8} Torr) interface by radiation enhanced diffusion. We have recently studied the stability of SiO_2 under 58 MeV Cu^{11+} ion bombardment [14]. No change in the amount of oxygen was observed (to a sensitivity of 10^{13} oxygen/cm^2) either at an oxidized surface or at a Ag-Si interface. These results indicate that no large scale (>100Å) diffusion of oxygen to or away from the interface occurs during MeV/amu ion bombardment. However, these experiments were performed on thick silicon targets and it is possible that 5000Å thick targets behave differently.

If we assume that the damage produced by our ion beam is caused by electronic sputtering in the interfacial oxide, then we must understand why no damage occured in the unmetalized case. Each of our samples was aged for several days in air after the metal layer was evaporated over a native oxide. Ponpon, et al. have shown that an oxide on silicon under a metal layer will continue to grow during exposure to air [15]. They have verified that the excess oxygen accumulates primarily at the interface rather than at the metal surface using SIMS, nuclear reaction analysis, and RBS. The growth of the oxide is, in fact, faster in the presence of the metal than it is on bare silicon. This is why the oxide is thicker on our gold and silver samples than on our unmetalized sample. Oui et al. [16] have measured the electronic sputtering yield of thin layers of SiO$_2$. They find that a thickness of greater than 3×10^{15} Si atoms/cm^2 in the oxide is necessary to trigger sputtering. These authors suggest that this may indicate that the erosion mechanism requires the presence of a coherent material. Our unmetalized sample, containing 3×10^{15} Si atoms/cm^2 in the oxide, showed no damage production with dose, but would not be expected to sputter.

The measured sputtering yield of Si for 20 MeV chlorine on thick SiO$_2$ is 2.74 Si atoms/ion [16]. The yield for 14 MeV oxygen, calculated assuming that the sputtering yield scales as $(dE/dx)^4$ [5], is 0.073 Si atoms/ion. Assuming stoichiometric sputtering, the yield of oxygen is 0.146 0 atoms/ion, slightly higher than our observed rate of damage production. The average energy of these sputtered atoms is expected to be less than a few eV [5]. Such low energy atoms could not produce damage in the silicon crystal through direct momentum transfer. We suggest, instead, that the sputtered oxygen atoms come to rest at interstitial positions in the silicon lattice. The proximity of oxygen atoms to the silicon, combined with the effects of a high density of electronic excitation, may allow silicon-silicon bonds to break and silicon-oxygen bonds to form. It should be pointed out that the Scotch tape threshold for adhesion for the Au-Si and Ag-Si systems corresponds to roughly one monolayer of damage in figure 1 [2].

This scenario would predict that damage production should cease when the oxide becomes sufficiently silicon-rich that it is no longer a good insulator. A saturation in damage production has not been observed, but the data thus far have not been carried beyond about 4×10^{15} disordered Si atoms/cm^2. Measurements of damage production in samples containing thicker oxides, both with and without metal layers should be able to settle the question of whether electronic sputtering is responsible for the damage. Initial results indicate that a thicker oxide without a metal layer does give rise to damage production. Experiments to look for a saturation effect are also planned.

It is tempting to ask if mixing, due to electronic sputtering in an interfacial oxide, is responsible for Mev-ion enhanced adhesion. Tombrello [3] has suggested that this is not consistent with their adhesion data, because metal-metal systems generally require a lower ion dose to pass the Scotch tape test for adhesion than do insulator-metal systems. Caution must be used, however, in interpreting the results of a threshold test for adhesion, such as the Scotch tape test. The Scotch tape test measures the difference in adhesion between the unirradiated film-substrate system and the Scotch tape-film system. This is because the test is passed just when the film-substrate system adheres as well as the tape-film system. The dose required, then, is larger when the initial film-substrate adhesion is poor, as it is in the case of metals on most insulators. What one desires to measure is the enhancement in adhesion for a given beam dose for the different systems. This requires a quantitative measurement of adhesion, as opposed to a threshold measurement. Using a scratch test, Jacobson et al. have found that adhesion of copper to aluminum can only be enhanced if an interfacial oxide layer is present [4]. Because all metals oxidize to some extent, and most samples tested for adhesion enhancement have not been sputter cleaned under vacuum before film deposition, adhesion enhancements quoted for metal-metal systems could be a result of electronic sputtering in the interfacial oxide.

CONCLUSIONS

We have shown that Mev/amu ion irradiation of metal-silicon interfaces causes a rapid buildup of disordered silicon at the interface. The damage production rate cannot be explained by nuclear collision effects alone within the framework of the Kinchin and Pease theory. Further, the assumption that the disorder is caused by an electronic interaction is consistent with the data. However, since little is known about the interaction of high energy ions with very thin insulating films, we cannot be certain that this is an electronic effect without further data.

We gratefully acknowledge the contributions of J. Klein to the experiments and the helpful comments of T. Tombrello. This work was supported by the NSF [PHY-8213598 and DMR-8216718] and the IBM Corp.

REFERENCES

1. J. E. Griffith, Y. Qui and T. A. Tombrello, Nucl. Instr. Meth. 198, 607 (1982).
2. M. H. Mendenhall, Ph.D. thesis, Caltech (1983).
3. T. A. Tombrello, Int. J. Mass. Spec. and Ion Phys., to be published (1984).
4. S. Jacobson, B. Jonsson and B. Sundqvist, Thin Solid Films 107, 89 (1983).
5. L. E. Seiberling, J. E. Griffith and T. A. Tombrello, Rad. Eff. 52, 201 (1980).
6. J. W. Boring, R. E. Johnson, C. T. Reimann, J. W. Garret, W. L. Brown and K. J. Marcantonio, Nucl. Instr. Meth. 218, 707 (1983).
7. R. L. Headrick and L. E. Seiberling, Appl. Phys. Lett. 45, 388 (1984).
8. L. C. Feldman, P. J. Silverman, J. S. Williams, T. E. Jackman and I. Stensgaard, Phys. Rev. Lett. 41, 1396 (1978).
9. T. A. Tombrello, Mater. Res. Soc. Symp. Proc. 27, 173 (1984).
10. I. V. Mitchell, J. S. Williams, P. Smith and R. G. Elliman, Appl. Phys. Lett. 44, 193 (1984).
11. L. C. Feldman, J. W. Mayer and S. T. Picraux, Materials Analysis by Ion Channeling, Academic Press, New York (1982).
12. G. H. Kinchin and R. S. Pease, Rep. Prog. Phys. 18, 1 (1955).
13. H. Iwami, R. M. Tromp, E. J. Van Loenen and F. W. Saris, Physica 116B, 328 (1983).
14. A. M. Behrooz, D. M. Shadovitz, L. E. Seiberling, D. P. Balamuth, R. W. Zurmuhle and R. L. Headrick, to be published.
15. J. P. Ponpon, J. J. Grob, A. Grob, R. Stuck and P. Siffert, Nucl. Instr. Meth. 149, 647 (1978).
16. Yuanxun Qiu, J. E. Griffith, Wen Jin Meng and T. A. Tombrello, Rad. Eff. 70, 231 (1983).
17. T. A. Tombrello, Mater. Res. Soc. Symp. Proc. 25, 173 (1984).

ION-INDUCED MIXING IN Ni-SiO$_2$ BILAYERS

T. C. BANWELL*, M-A. NICOLET*, P. J. GRUNTHANER**, AND T. SANDS[+]
*California Institute of Technology, Pasadena, California 91125
**Jet Propulsion Laboratory, Pasadena, California 91109
[+]Lawrence Berkeley Laboratory, Berkeley, California 94720

ABSTRACT

We report on our studies of Ni transport induced by 300 keV Xe irradiation of 25 nm Ni films evaporated on thermally grown SiO$_2$ at Xe fluences of 10^{13}-10^{16} cm^{-2} and at temperatures of 300-750 K during irradiation. Cross-sectional TEM, and selective etching combined with 2 MeV He backscattering spectrometry and ESCA were used to profile the Xe and Ni within the SiO$_2$. At 300 K, backscattering shows cascade mixing dominates, although only ~ 1/35 that predicted by cascade theory, with most of the Ni in the SiO$_2$ contained in a resolution limited peak adjacent to the SiO$_2$ interface. TEM shows that this Ni is contained in a 5 nm band, 5 nm below the interface as Ni oxide clusters. Examination of the satellite structure of the Ni 2p line by XPS also shows this band is predominantly Ni^{2+}. At 750 K, the near-surface peak vanishes and only recoil implantation is evident. Ni^0 is evident by XPS in samples irradiated at 300 K, though not at higher temperatures. We explain our results in terms of phase separation during cooling of the collision cascade.

INTRODUCTION

It is well established that atomic transport in solids can be induced by energetic ion irradiation [1]. The transport processes are initiated by energy deposited in the solid by the incident ion. The early stages involve high energy collisions (>> 1 eV) which are dominated by screened coulomb scattering [2]. An important transport process occurring in this regime is recoil implantation [3,4]. As the energy dissipates, a second regime is entered wherein electronic effects (chemical bonding) are important. This regime may be dominated by heats of mixing and cohesive energies of the solids constituents, as well as defect chemistry [2,5,6]. Atoms undergo diffusional mixing during both regimes, possibly biased by chemical effects in the later regime. These processes occur in a volume of order 10^1-10^2 Å diameter and 10^2-10^3 Å depth about the incident ion track. Specific details, however, are still mostly unknown. We report on our investigations of the mechanisms in the mixing of Ni/SiO$_2$ bilayers by 290 keV Xe irradiation. Previous work demonstrated that effects of chemical origin dominate interfacial mixing in this system [7].

EXPERIMENTAL

Substrates with a $\sim$ 6500 Å SiO$_2$ layer were prepared by 1100°C steam oxidation of polished <111> Si wafers. 250 Å Ni layers were deposited on the SiO$_2$ by e-beam evaporation. Xenon implantations were made at sample temperatures of 300, 570 and 750-770 K with 290 keV Xe (R_p = 390 Å, ΔR_p = 140 Å in Ni) [8] to a fluence ϕ of 10^{13}-10^{16} cm^{-2}. The free Ni layer was subsequently removed using a Nichrome etch, or with boiling 12 M HCl as described elsewhere [9]. Both etchants produced identical results.

2 MeV He backscattering spectrometry was used to measure the amount and distribution of the Ni and Xe remaining in the SiO$_2$ after irradiation and removal of free Ni. Special care was taken to minimize background interference from pulse pile-up in the spectrometry electronics, and from slit-scatter of the incident He beam.

Selected samples were examined using x-ray photoelectron spectroscopy (XPS). A gold metal surface was used to calibrate the spectrometer (84.00 eV for Au (4f 7/2) peak); however, charging of the SiO$_2$ sample surface produced substantial peak shifts (4-5 eV). Consequently, the Si(2p 3/2) and background carbon C(1s) core lines served as internal references. Depth profiles were also obtained by sequential etching and XPS analysis as described elsewhere [10].

Selected samples were also examined by high-resolution cross-sectional TEM. Sections of a sample, epoxied face-to-face, were thinned to $\sim$ 200 Å using Ar ion-milling. The microscope magnification was accurately measured by imaging the adjacent SiO$_2$/Si interface under the same imaging conditions as used to image the ion-mixed region. The {111} Si lattice fringes were then used to calibrate the negatures from the area of interest.

RESULTS

Figure 1 shows typical backscatter profiles (log plot) of Ni in SiO$_2$, these after 6-6.6x10^{15} cm^{-2} Xe irradiation at the various temperatures. At 300-570 K, most of the Ni is contained in a resolution limited peak just below the SiO$_2$ surface. The surface peak, suppressed slightly at 570 K, is greatly diminished at 770 K. A deep exponential tail characteristic of recoil implantation is evident beyond 400 Å [3,4]. For a given temperature, the amount of Ni in the recoil tail exhibited a linear ϕ dependence. The increase from 300 K to 570 K produces no discernible change in the recoil tail. This tail is modified at 770 K.

Both recoil implantation and diffusional mixing may contribute to Ni transport. This suggests that the areal density of Ni in SiO$_2$; [Ni]$_s$, may follow

$$[Ni]_s = R\phi + A\phi^{1/2} \qquad (1)$$

over a limited range of Xe fluence ϕ [7].

Figure 2 shows a graph of $[Ni]_s\phi^{-1/2}$ versus $\phi^{1/2}$ for various irradiation temperatures. A linear relationship suggested by eqn. (1) is apparent for $\phi > 10^{15}$ cm^{-2}. The apparent net recoil yield (slope R) decreases from 300 K to 570 K, even though the deep tail is unchanged, and recovers some at 770 K. The net isotropic diffusional mixing (intercept A) decreases monotonically with increasing temperature. For comparison, the transport in Ni/Si bilayers attributed to collisional mixing is indicated by the dashed line on top [2,11]. The net mixing in Ni/SiO$_2$ is considerably smaller! The linear ϕ behavior at 300 K for $\phi^{1/2} < 0.47(10^{15}$ cm$^{-2})^{1/2}$ has been attributed to Poisson statistics and the fact that cascade volumes do not overlap at very low fluences; the number of (disjoint) cascade volumes increases as $\phi\cdot$ (effective cascade area) [12]. The linear ϕ dependence at 570 K for $\phi^{1/2} < 1.3(10^{15}$ cm$^{-2})^{1/2}$ is not attributed to Poisson statistics since it would correspond to an unreasonably smaller cascade area.

Figure 3(a) is a cross-sectional TEM bright-field image of a sample, before etching, implanted at 300 K with 6E15 Xe/cm^2. A sharp interface is evident. The Ni layer again overshadows the Ni in the SiO$_2$. Figure 3(b) is a high-resolution image of the sample in Fig. 3(a) after removal of the free Ni. The position of the SiO$_2$ surface is indicated by arrows. This image shows that the Ni is contained in crystalline clusters with $\sim$ 21 Å diameter. The majority of the clusters lie in a 50 Å band approximately 50 Å below the SiO$_2$ surface. The Ni in the vicinity of the recoil tail ($\sim$ 400 Å) is also in the form of clusters. There is a relatively small variation in the apparent size of the clusters in Fig. 3(b). Figure 4(a) is a high-resolution TEM image of the crystalline clusters near the surface of the sample imaged in Fig. 3(b). Figure 4(b) is a higher magnification image of the region boxed in Fig. 4(a). The most common plane spacing observed in the crystalline clusters is 2.41 ± 0.04 Å. Occasionally, clusters were observed in a low index orientation (crossed fringes) suggesting a cubic (probably f.c.c.) structure. The orientation of the particle in Fig. 4(b) is near <110>. Crossed {111} planes are visible. We conclude that the clusters probably have f.c.c. structure with a_0 = 4.17 ± 0.07 Å. This could correspond to either NiO or Ni$_2$O$_3$ [13]. The presence of Ni in additional forms was not perceptible in this sample by TEM.

Figure 5 shows the Ni 2p photoelectron spectra of a sample implanted at 300 K with 6×10^{15} Xe cm^{-2}. The dominant peak at 857 eV binding energy, with its satellite $\sim$ 6 eV higher in binding energy, is identified as the Ni (2p 3/2) core lines of a Ni oxide [14]. The corresponding Ni(2p 1/2) core

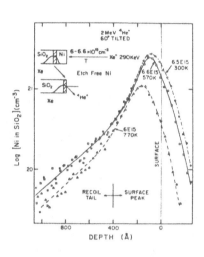

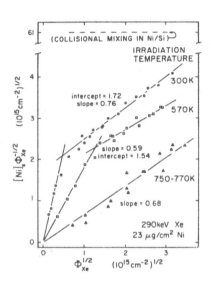

Fig. 1 He backscatter profiles of
Ni in SiO$_2$ for samples irradiated
with 6-6.6x10^{15} cm^{-2} Xe at temperatures of 300, 570 and 770 K.

Fig. 2 Graph of $[Ni]_s \Phi^{-1/2}$ versus
$\phi^{1/2}$ for samples at 300, 570 and
750-770 K during Xe irradiation.

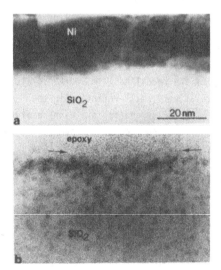

Fig. 3 (a) Cross-sectional
TEM bright-field image of Ni/
SiO$_2$ after room temperature
6x10^{15} cm^{-2} Xe irradiation.
(b) High-resolution image of
sample in (a) after removal of
free Ni. Arrows indicate SiO$_2$
surface.

line is evident at 874 eV. There is a slight shift to higher binding energies,
presumably due to size effects [15], which precludes specific identification
of the oxide. Not seen by TEM, however, is the presence of unoxidized Ni

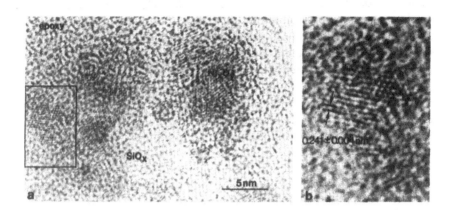

Fig. 4 (a) High-resolution TEM image of Ni oxide clusters near the surface of sample imaged in Fig. 3(b). (b) Higher magnification image of boxed region in (a). Orientation of Ni oxide particle is near <110>. Crossed {111} planes are visible.

metal indicated by the Ni (2p 3/2) core line at 853 eV binding energy. The intensity of this line approximately doubled after ∿ 50 Å of SiO_2 was removed.

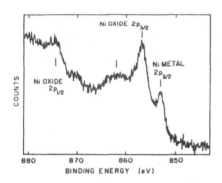

Fig. 5 Ni 2p electron spectra of room temperature 6×10^{15} cm^{-2} Xe irradiated sample after etching. Binding energy is corrected for charging.

Profiles of samples irradiated at 570-770 K showed the same Ni oxide peaks, though located deeper than for 300 K irradiation, but did indicate the presence of Ni in any other state. Examination of the Si 2p core line did not reveal the presence of Ni silicide at any temperature.

DISCUSSION

Chemical effects are not expected to influence mixing in the collisional regime [2]. The differences between Ni/Si and Ni/SiO$_2$ indicated in Fig. 2, as well as the discontinuity of Ni across the Ni-SiO$_2$ interface shown in Figs. 1,4 leads us to adopt the following model for mixing of Ni/SiO$_2$ bilayers: During the initial collisional regime, the cascade is "hot" enough that all the constituents are miscible and Ni diffuses into the SiO$_2$. This phase is characterized by an effective diffusion parameter $(Dt)_{col}$ which is similar to that for Ni/Si bilayers. As the cascade cools, the Ni concentration exceeds the instantaneous equilibrium level in the SiO$_2$. The Ni-SiO$_2$ interface becomes a sink for Ni and a substantial fraction of the Ni leaves the SiO$_2$. This segregation phase is characterized by a temperature dependent effective diffusion parameter $(Dt)_{seg} \gg (Dt)_{col}$. The formation mechanism of Ni oxide clusters during this time is not clear. At 570-770 K, $(Dt)_{seg}$ becomes large enough that the shallow recoils are also involved in the segregation phase.

The significant $\phi^{1/2}$ dependence at 300-570 K and the uniform size of the NiO clusters indicates that the clusters dissociate when enveloped in a cascade. This model can explain why the peak Ni concentration in the SiO$_2$ saturates at 4-6x10^{21} cm^{-3} [8], substantially below that available from the Ni layer, without invoking interfacial impediments. However, the role of equilibra between Ni^0 and oxidized Ni and the precipitation processes remains unclear. Reaction products have not been explicitly accounted for either.

CONCLUSIONS

A speculative model for ion mixing in Ni/SiO$_2$ bilayers has been developed supported by strong evidence obtained from the combined utilization of backscatter spectrometry, high resolution cross-sectional TEM and XPS analysis. This system joins a growing number for which consideration of thermodynamic processes play an essential role in understanding their behavior in ion mixing.

ACKNOWLEDGMENTS

The authors gratefully acknowledge the technical assistance provided by R. Gorris and M. Parks (Caltech), and useful discussions with Drs. B. M. Paine, W. L. Johnson (Caltech), and F. Grunthaner (JPL, Pasadena). The project benefitted from the interest and encouragement of Dr. T. M. Reith (IBM, Tucson). The research was financially supported in part by IBM/GPD (Tucson).

REFERENCES

1. J. W. Mayer, B. Y. Tsaur, S. S. Lau, and L. S. Hung, Nucl. Instr. and Meth. 182/183, 1 (1981).

2. B. M. Paine and R. S. Averback, to be published in Nucl. Instr. and Meth. B, March 1985.
3. S. Dzioba and R. Kelly, J. Nucl. Mat. 76, 175 (1978).
4. L. A. Christel, J. F. Gibbons, and S. Mylroie, Nucl. Instr. and Meth. 182/183, 187 (1981).
5. Y. T. Cheng, M. Van Rossum, M-A. Nicolet, and W. L. Johnson, Appl. Phys. Lett. 45, 185 (1984).
6. W. L. Johnson, Y. T. Cheng, M. Van Rossum, and M-A. Nicolet, (unpublished).
7. T. C. Banwell and M-A. Nicolet in, Ion Implantation and Ion Beam Processing of Materials, edited by G. K. Hubler, O. W. Holland, C. R. Clayton, and C. W. White, (Elsevier Science Pub. Co., New York, 1984), Mat. Res. Soc. Symp. Proc. Vol. 27, p. 109.
8. J. P. Biersack and J. F. Ziegler in, Ion Implantation Techniques, (Springer-Verlag, New York, 1982), p. 157.
9. T. Banwell, B. X. Liu, I. Golecki, and M-A. Nicolet, Nucl. Instr. and Meth. 209/210, 125 (1983).
10. P. J. Grunthaner, R. P. Vasquez, and F. J. Grunthaner, J. Vac. Sci. Technol. 17, 1045 (1980).
11. U. Shreter, F. C. T. So, B. M. Paine, and M-A. Nicolet in, Ion Implantation and Ion Beam Processing of Materials, edited by G. K. Hubler, O. W. Holland, C. R. Clayton, and C. W. White, (Elsevier Science Pub. Co., New York, 1984), Mat. Res. Soc. Symp. Proc. Vol. 27, 109.
12. T. Banwell, N. R. Corngold, and M-A. Nicolet, (unpublished).
13. R. P. Elliot, Constitution of Binary Alloys, First Supplement, (McGraw-Hill, New York, 1965), p. 661.
14. K. S. Kim and R. E. Davis, J. Electron. Spectrosc. 1, 251 (1972/73).
15. P. J. Grunthaner, Ph.D. Thesis, California Institute of Technology, (1980).

Pulsed Ion-Beam Induced Reactions of Ni and Co
with Amorphous and Single Crystal Silicon

J. O. Olowolafe[*] and R. Fastow
Department of Materials Science
Cornell University
Ithaca, New York 14853

Abstract

Thin layers (~1,000 A) of Ni and Co have
been reacted with both (100) and amorphous
silicon (a-Si) using a pulsed ion beam. Samples
were analyzed using Rutherford backscattering,
x-ray diffraction, and transmission electron
microscopy. Rutherford backscattering showed
that the metal/a-Si and metal/(100)-Si reaction
rates were comparable. Both reactions began at
the composition of the lowest eutectic. For
comparison, furnace annealing of the same
structures showed that the reaction rate of Ni
with amorphous silicon was greater than with
(100) Si; Co reacted nearly identically with
both substrates. Diffraction data suggest that
pulsed ion beam annealing crystallizes the
amorphous silicon before the metal/a-Si reaction
begins.

INTRODUCTION

Metal/amorphous silicon reaction rates are, in many
cases, greater than metal/single crystal silicon reaction
rates. The mechanisms which are responsible depend on the
system. For Ni/Si and Pd/Si (forming Ni_2Si and Pd_2Si),
this dependence arises from the diffusion limited nature of
these reactions, coupled with the influence of substrate
orientation on silicide grain size [1,2]. The formation of
$NiSi_2$ and $CoSi_2$ similarly show a dependence on the
crystalline nature of the substrate. A laterally uniform
disilicide was produced at lower temperatures when NiSi and
CoSi were reacted with amorphous silicon. It was argued
that these effects may have been due to the positive
formation energy of the amorphous Si [3,4]. For the
disilicides formed from Er, Ti, and Cr, lower reaction
temperatures and laterally uniform growth was also observed
for reactions with amorphous silicon. These effects were
attributed to a lower concentration of impurities at the
metal/a-Si interface [5-7]. This paper will investigate

[*] Department of Physics, University of Ife, Ile-Ife, Nigeria

whether there are differences between metal/a-Si and metal/(100)-Si reactions using pulsed ion beams.

Pulsed ion beams have been used to react Ni, Co, Pt, Pd, and Ir with single crystal silicon [8-12]. In each case, the initially reacted layer had the composition of the lowest eutectic. Marker experiments have shown that the reaction starts at the interface and occurs in the liquid state, before either the silicon substrate or the metal overlayer melts [8,11]. In this work low energy density irradiation of Ni on (100) Si resulted in an interfacial layer having the composition $Ni_{50}Si_{50}$. Transmission electron microscopy showed that this layer was polycrystalline NiSi. Irradiation of Co on (100) Si also resulted in an interfacial layer. This polycrystalline layer, however, was non-uniform in composition and consisted of Co_2Si, $CoSi_2$, and possibly CoSi.

EXPERIMENT

Two different types of samples were fabricated using e-beam deposition. The first type was a single, 800 A thick layer of Ni on a (100) silicon substrate. The second type was a bilayer having the structure (100)-Si/metal/a-Si. The metal was either Ni or Co ,and the thicknesses of the evaporated layers were 1,400 A (metal) and 3,900 A (silicon). The base pressure during deposition was $5x10^{-7}$ torr. Immediately before loading, the silicon wafers were rinsed in a 50 % HF solution. After deposition the samples were cut into 1cm x 1cm squares to be pulsed ion beam annealed or furnace annealed.

Pulsed ion beam annealing was carried out under a vacuum of $1 x 10^{-5}$ torr. The average ion energy was 280 keV and the total incident energy density was varied from 0.4 J/cm^2 to 0.8 J/cm^2. The pulse length (full width at half maximum) was 70 ns. The beam was composed of 40 % protons and 60 % heavy impurities. A more complete description of the beam is given elsewhere [13]. The total incident energy density was measured by the temperature rise of a nickel platelet of known thermal mass.

Thermal annealing of the samples having the sandwich structure was done under a vacuum of $5 x 10^{-8}$ torr. The samples were annealed for 1/2 hr at temperatures between 250 and 430 C. Rutherford backscattering spectroscopy was then used to determine the silicide thicknesses at the metal/(100)-Si and the metal/a-Si interfaces. For the single layer Ni/Si samples, layer thicknesses were measured as a function of annealing time at a single temperature (275 C). This was done to check for time delays in the silicide formation at the metal/(100)-Si interface.

Analysis of the reacted samples was carried out using Rutherford backscattering and x-ray diffraction (Guinier camera and densitometer). Computer simulations were used to estimate the surface temperature of the samples during pulsed ion beam annealing.

RESULTS

The effect of pulsed ion beam annealing on single layer Ni/Si samples is shown in figure 1. The incident energy densities were 0.5 J/cm^2, and 0.6 J/cm^2 for the sample with the larger step. The thicknesses of the silicide layers were 420 A and 1000 A, and the composition was $Ni_{50}Si_{50}$ (the lowest eutectic compositions for this system occur at $Ni_{44}Si_{56}$, and $Ni_{54}Si_{46}$). X-ray and TEM diffraction showed that only the phases Ni and NiSi were present.

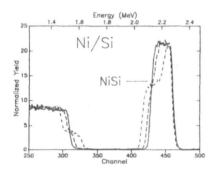

Fig. 1
Rutherford backscattering spectra of pulsed ion beam annealed Ni/Si. Incident energy densities were 0,0.5, and 0.6 J/cm^2.

A planar TEM image of a fully reacted Ni/Si sample is shown in figure 2. The film was composed of polycrystalline NiSi with an average grain size of 1,700 A.

Fig. 2.
Transmission electron micrograph of pulsed ion beam annealed Ni/Si (fully reacted).

Rutherford backscattering spectra of the Ni sandwich structures are shown in figure 3. Pulsed ion beam annealing at 0.5 J/cm^2 caused the Ni to react equally with both the (100) Si and the a-Si. The thickness of the reacted layer was 520 A, and the composition was $Ni_{50}Si_{50}$. X-ray diffraction showed that polycrystalline Si, NiSi, and Ni were present. The same type of sample, after furnace annealing at 275 C for 1/2 hr is shown in figure 3c. In this case, the reaction with amorphous silicon was greater than with (100) Si.

A plot of Ni_2Si thickness vs (time)$^{1/2}$, made with the single layer Ni/Si samples, showed that there was no time delay for the silicide formation . The composition of both silicide layers was $Ni_{66}Si_{33}$; and x-ray diffraction showed only the presence of Ni_2Si. This is consitent with published data for anneals at 275 C [14].

Computer simulations of one dimensional heat flow in silicon were used to calculate the surface temperature during pulsed ion beam annealing (fig 4). The simulation took into account the experimentally measured current and voltage traces as well as the measured impurity concentration of the beam. The energy dependent stopping power and the temperature dependent thermal conductivity of silicon were used to model the substrate. This simulation is described in detail elsewhere [13]. For a 0.5 J/cm^2 pulse, the surface temperature reached a maximum of 1,100 C.

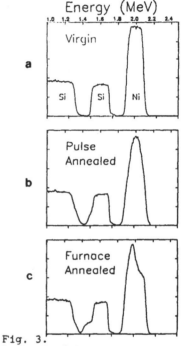

Fig. 3.
Rutherford backscattering spectra of a-Si/Ni/(100)-Si a) unannealed, b) pulse annealed, c) furnace annealed.

This was below the melting points of Ni (1453 C) and Si (1412 C), but above the lowest Ni-Si eutectic (964 C). The average cooling rate between 1,100 C and 900 C was 1.2 x 10^{10} K/sec. Discontinuities in the first derivative of the temperature vs time curve were artifacts of the programming (values of the beam voltage and current were taken at 20 ns intervals).

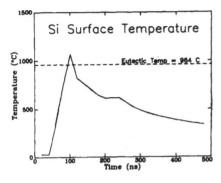

← Figure 4.
Computer simulation of surface temperature during pulse ion beam annealing (0.5 J/cm^2).

Rutherford backscattering spectra of the Co sandwich structures are shown in figure 5. Pulsed ion beam annealing caused the Co to react equally with both the crystalline and the amorphous silicon. Below an energy density of 0.6 J/cm^2, no reaction was observed. The reacted layers had a graded composition, in contrast to the uniform step in the Ni spectrum. X-ray diffraction of this sample revealed the presence of polycrystalline Co, Si, and Co$_2$Si (figure 6). TEM diffraction of single layer Co/Si samples showed, in addition, the presence of CoSi$_2$, and possibly CoSi^2 [8].The backscattering spectra of·an (100)-Si/Co/a-Si sample thermally annealed at 430 C for 1/2 hr (figure 4c) showed no difference in reaction rates at the two interfaces. The step in the RBS spectrum corresponded to the composition Co$_{66}$Si$_{33}$.

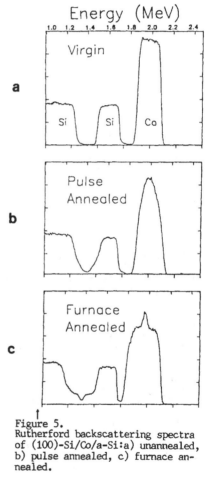

Figure 5.
Rutherford backscattering spectra of (100)-Si/Co/a-Si:a) unannealed, b) pulse annealed, c) furnace annealed.

X-ray diffraction confirmed that the phase Co$_2$Si was present, which is consistent with published data for anneals at this temperature [14].

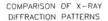

COMPARISON OF X-RAY
DIFFRACTION PATTERNS

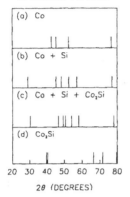

(a) Co

(b) Co + Si

(c) Co + Si + Co,Si

(d) Co,Si

20 30 40 50 60 70 80

2θ (DEGREES)

Fiq. 6.
X-ray (CuKα) diffraction
spectra of the a-Si/Co/(100)-Si
structure a) unannealed,
b) pulse annealed at 0.57 J/cm².
c) pulse annealed at 0.7 J/cm².
d) furnace annealed at 430 C for
1/2 hr.

DISCUSSION

At low incident energy densities, pulsed ion beam induced reactions are well behaved. In general, a low energy pulse will react only the interface without melting the silicon substrate or the metal overlayer [8]. For the case of nickel on silicon, compositionally uniform layers consisting of polycrystalline NiSi were produced at energy densities of 0.5 and 0.6 J/cm². Only at energy densities greater than 0.75 J/cm² was cellular structure observed. For the case of cobalt on silicon, at 0.7 J/cm², again only the interface reacted. However the reacted layer was non-uniform in composition. This behavior has been attributed to the two eutectics in the Co-Si phase diagram having nearly the same temperature (1195 C and 1259 C), but widely differring compositions ($Co_{77}Si_{23}$ and $Co_{23}Si_{77}$) [8].

Pulsed ion beam, and thermally (furnace) induced metal/silicon reactions differ in the reaction mechanisms (liquid vs solid state), and in the final composition and crystal structure of the reacted layers. When the metal is reacted with amorphous · silicon, there are additional differences. The first difference is that furnace annealing causes a larger reaction of Ni with amorphous silicon. Pulsed ion beam annealing, however, produces comparable reactions. The second difference is that the unreacted silicon remains amorphous after furnace annealing (at the annealing temperatures used in this experiment). Pulsed ion beam annealing, however, crystallizes the unreacted amorphous silicon. Furthermore, when Co was reacted with amorphous silicon, the silicon crystallized at an energy density lower than that needed to form the silicide layer.

It is likely that the amorphous silicon underwent explosive crystallization. There have been reports of explosive crystalization for both CW and pulsed laser irradiated amorphous silicon [15,16]. This mechanism is the only one consistent with the present experimental results. The amount of silicon which could have undergone solid phase crystallization is insignificant (less than 1.0 A) given the short times involved [17]. Melting and regrowth of the amorphous silicon was also impossible at these low energy densities. The silicon was observed to crystallize at the beginning of the Ni-Si reaction (eutectic temperature of 964 C); however, the melting temperature of amorphous silicon is 1200 ±50 C [16]. The occurance of explosive crystallization would also explain why there was no difference between the metal/a-Si and the metal/(100)-Si reactions. If the silicon crystalized at or below the eutectic temperature then there would be no thermodynamic difference between the two Si layers. If this is the case, then only in those systems where the deepest eutectic is below the temperature required for explosive crystallization can any difference in metal/a-Si vs metal/(100)-Si reactions be observed.

CONCLUSION

Pulsed ion beam induced Ni/Si and Co/Si reactions begin at the eutectic compositions (for both amorphous and (100) Si). For the Ni-Si reactions, the silicide layer had a uniform composition of $Ni_{50}Si_{50}$. For the Co-Si reactions, the silicide layer had a non-uniform composition. The thickness of the silicide layers after pulsed ion beam annealing was the same for both amorphous and (100) silicon. This may have been due to the crystallization of the amorphous silicon by the pulse. Thermal (furnace) reactions of Ni with amorphous and (100) silicon showed that the reaction rates with the amorphous silicon were faster. Thermal reactions of Co with amorphous and (100) silicon showed no measurable differences in the reaction rates.

Acknowledgements

The authors would like to thank L. Doolittle for use of the RBS simulation program (RUMP); and J.W. Mayer and J. Gyulai for helpful discussions. We also thank the Cornell Plasma Physics Laboratory (D. Hammer) for use of the pulsed ion beam. R.Fastow was supported in part by MSC, and J.O. Olowolafe by an IAEA fellowship.

REFERENCES

1. J.O. Olowolafe, M-A. Nicolet, and J.W. Mayer, Thin Solid Films, 38(1976).

2. N. Cheung, S.S. Lau, M-A. Nicolet, and J.W. Mayer, Proc. Symp. on Thin Film Interfaces and Interactions , edited by J.E.E. Baglin, J.M. Poate (Electrochemical Society, Princeton, N.J. 1980)vol80-2, p.494.

3. C.-D. Lien, M-A. Nicolet, and S.S. Lau, Phys. Stat. Sol.(a) 123(1984).

4. C.-D. Lien, M -A. Nicolet, and S.S. Lau, Appl. Phys. A 34, 249-251(1984).

5. L.S. Hung, J. Gyulai, J.W. Mayer, S.S. Lau, M-A. Nicolet, J. Appl. Phys. $\underline{54}$, 5076(1983).

6. C.S. Wu, S.S. Lau, T.F. Kuech, B.X. Liu, Thin Solid Films $\underline{104}$, 175(1983).

7. C.-D. Lien, M-A. Nicolet, S.S. Lau, Appl. Phys. A (in press, 1984).

8. R. Fastow, J.W. Mayer, T. Brat, M. Eizenberg, J.O. Olowolafe, (submitted for publication in Appl. Phys. Lett.).

9. T. Brat, M. Eizenberg, R. Fastow, C.J. Palmstrom, J.W. Mayer, Journal of Applied Physics (accepted for publication).

10. J.E.E. Baglin, R.T. Hodgson, W.K. Chu, J.M. Neri, D.A. Hammer, J.J. Chen, I.B.M. Research Report (3/31/81).

11. C.J. Palmstrom and R. Fastow, Laser- Interactions and Thermal Processing of Materials, edited by J. Narayan, W.L. Brown, R.A. Lemmons, Materials Research Society, North Holland Publishing Co. (1982). p.715.

12. R. Fastow (to be published).

13. R. Fastow. Y. Maron, J.W. Mayer, Phys. Rev B. $\underline{31}$ (1985).

14. K.N. Tu and J.W. Mayer, Thin Films- Interdiffusion and Reactions, edited by J.M. Poate, K.N. Tu, and J.W. Mayer, Electrochemical Society, John Wiley and Sons, Inc. (1978).

15. G. Auvert, D. Bensahel, A. Perio, V.T. Nguyen, G.A. Rozgonyi, Appl. Phys. Lett. $\underline{39}$(9), 724(1981).

16. M.O. Thompson, G.J. Galvin, J.W. Mayer, P.S. Peercy, J.M. Poate, D.C. Jacobson, A.G. Cullis, N.G. Chew, Phys. Rev. Lett. $\underline{52}$(26), 2360(1984).

17. L. Csepregi, J.W. Mayer, and T.W. Sigmon, Phys. Lett. A $\underline{54}$, 157(1975).

Silicon Crystal Growth from the Melt Over Insulators

SOLID-LIQUID INTERFACE INSTABILITY IN THE ENERGY-BEAM RECRYSTALLIZATION OF SILICON ON INSULATOR

EL-HANG LEE
Monsanto Electronic Materials Company, St. Peters, MO 63376
Present address: AT&T Technologies, Engineering Research Center,
P. O. Box 900, Princeton, NJ 08540

ABSTRACT

An attempt has been made to systematically sort out the characteristic modes of morphological transition in the energy beam recrystallized thin film silicon on insulating substrates, and to relate them to the mechanisms of solid-liquid interface stability breakdown. Stable to unstable breakdown modes include faceted, cellular, and dendritic configurations as well as transient and composite configurations thereof. These primary modes of breakdown then lead to the secondary modes of breakdown which constitute the sub-boundary formation. The mechanics of the primary (interface) breakdown and that of the secondary (sub-boundary) breakdown must be clearly differentiated in understanding the breakdown process. Constitutional supercooling and absolute supercooling models have been used to explain the various interface instabilities.

INTRODUCTION

One of the most intriguing phenomena in the current practices of molten-zone recrystallization of amorphous, thin-film silicon on insulating substrates (SOI hereafter) is the breakdown transition of a defect-free seeded growth into a spurious network of lineage defects called sub-boundaries [1-14]. The breakdown transition has been universally observed irrespective of the energy-beams employed, whether graphite strip-heater [1-4], high-power lamp [5-6], electron-beam [8-9], or laser-beam [10]. As a molten-zone travels from seed region (where the silicon film is in contact with the silicon crystal substrate) to the SOI region (where the silicon film lies on an insulating layer), the crystallization proceeds free of defects for up to 100 microns and then abruptly breaks down into sub-boundaries, as illustrated in Fig. 1.

Several studies have now shown that the sub-boundaries are by-products of roughened solid-liquid interface (interface hereafter) at the trailing edge of a molten-zone. So far, in the historical perspective, the interface breakdown was first observed in the form of sawtooth-faceted configurations [1,3]. Later, curved, cellular breakdown interfaces were also observed [2]. In the midst of these seemingly unrelated observations, Lee [12-14] reported systematic development of faceted, cellular and dendritic configurations with decreasing thermal gradients. Recently, Limanov and Givargizov [15] also reported the three types of breakdown configurations. In more recent studies the breakdown was observed not only in these elemental modes but also in transient [15,16] or composite modes [16]. Transient modes include faceted-to-cellular or cellular-to-dendritic configurations, and composite modes include faceted/dendritic or cellular/dendritic configurations. Also, a study [17] reported sub-boundaries speculated to be due to the mixture of faceted breakdown and cellular breakdown.

All these studies collectively suggest that the interface (primary) breakdown is a precursor to the sub-boundary (secondary) breakdown. The disposition of the secondary breakdown, then, must be understood on the

564

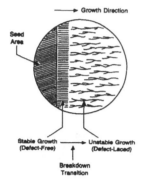

Growth Direction

Seed
Area

Stable Growth ⟶ Unstable Growth
(Defect-Free)   (Defect-Laced)

Breakdown
Transition

FIGURE 1. Schematic illustration
of the growth stability breakdown
transitions that are observed in
the current practices of beam-
induced, lateral seeded-
recrystallization of SOI.

basis of the primary breakdown characteristics. The characteristics of
the facet-induced sub-boundaries, for example, may be different from
those of cellular-induced sub-boundaries. The periodicity of subboundaries might be another good example of such a dependence [18]. Once
the primary breakdown is initially locked into one of the above configurations, indications are that it remains largely invariant throughout the
course of recrystallization, maintaining consistent sub-boundary characteristics, unless a perturbation is introduced.

We thus have focused our attention on the behavior of interface
breakdown transition and have attempted to systematically sort out the
various characteristic modes. We then attempted to relate the associated
mechanisms with the traditional understanding of interface
instabilities. Insofar as the traditional understanding is still incomplete, so is the understanding of SOI interface instabilities. Nevertheless, the attempt is considered important, and even essential, because
such an understanding should provide a conceptual basis to prevent the
sub-boundary formation. It may be worth pointing out that the utilization of such an understanding has indeed demonstrated extending the
defect-free seeded growth far beyond the currently observed limits
[12,13].

RECRYSTALLIZATION SCHEME

Figure 2 illustrates a sample and energy-beam configuration that has
been typically used for the recrystallization study by the author (and
many others as well). The samples were 1.0 um thick, amorphous thin-film
silicon deposited on thermally oxidized (100) silicon wafers and capped
with a 2.0 um thick oxide layer. The thermal oxide layer was 0.2-0.5 um
thick and was patterned in 500 um wide stripes separated by 3 um wide
grooves. The 3 um wide grooves, where the thin film silicon is in contact with the underlying silicon substrate, served as a seed as well as a
heat-sink. Such patterned samples have been found to be quite effective
for the seeded-crystallization study largely due to dimensional flexibility, repeatability, and, above all, well-regulated thermal geometry.
Samples being heated to 1100-1300°C on a base heater were scanned under a
2 mm wide graphite strip-heater (glowing between 1700-2000°C at 1 mm from
the sample), or a line-shaped Ar⁺ laser beam at a speed between 1 and
5 mm/sec. in an argon ambient.

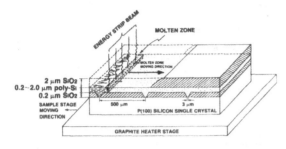

<figure>FIGURE 2. Schematic diagram of the energy beam-induced recrystallization technique.</figure>

PRIMARY MODES OF BREAKDOWN (INTERFACE BREAKDOWN)

Faceted Breakdown

Figure 3(a) shows a sawtooth-faceted breakdown that has grown out of a stable growth from the seed. Thermodynamically, silicon is known to favor faceted growth [19]. In SOI, the faceted growth has been observed mainly in sawtooth configurations [1,3,12,13,15], where the edges are two-dimensional projections of (111) planes. In these cases, the sub-boundaries grow out from the corner of each sawtooth pattern in one-to-one correspondence. Along the [100] direction the sawtooth configuration has been observed in isosceles triangle shapes [18]. In other growth directions between [100] and [110], the sawtooth assumes modified shapes, depending on the local growth direction and thermal conditions [18]. In contrast to these, non-sawtooth-facets also have been observed when the growth is along the [110] direction. Figure 4 shows a portion of stripe-patterned SOI that has been recrystallized with a strip-heater crossing

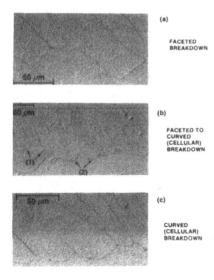

(a)

FACETED
BREAKDOWN

(b)

FACETED TO
CURVED
(CELLULAR)
BREAKDOWN

(c)

CURVED
(CELLULAR)
BREAKDOWN

FIGURE 3. Secco-etched morphologies of (a) the sawtooth-faceted configuration, (b) the faceted-to-cellular transient configuration, and (c) the curved (cellular) configuration. Note that the sub-boundaries originate at the corner of each breakdown in a manner of one-to-one correspondence.

the [110]-oriented seed groove. The long, stretched line of defects running parallel to the seed grooves in the middle of the stripe is a
faceted front that has frozen in collision with the seeded growth from
both seed grooves. Similar phenomena have been observed in electron-beam
recrystallized SOI, too [9]. In these cases, no sub-boundaries have been
observed. It may be assumed that the sawtooth breakdown is a necessary
condition for sub-boundary breakdown. It is not clear, however, whether
it is a sufficient condition.

SCAN DIRECTION ↑

500 μm

FIGURE 4. Non-sawtooth breakdown of the seed crystallization that has been obtained as
a result of growth collision
from both sides of the seed
grooves. The growth direction
is in [110].

Curved (Cellular) Breakdown

This type of breakdown is known to occur as a result of impurity segregation, and is typically characterized by non-faceted, curved interfaces [20]. Traditionally, cellular growth is observed extensively in
metallic substances which do not favor faceted growth. Crystallography
does not play an important role in cellular growth [21(b)]. Figure 3(c)
shows a set of curved configurations in contrast to the sawtooth-faceted
configurations. Again, the sub-boundaries originate from each corner of
the curved interfaces, and form similar patterns to that of the facet-
induced ones. Cellular breakdown in silicon is quite intriguing in that
silicon favors faceted growth as mentioned above [19].

Dendritic Breakdown

Traditionally, dendritic configurations are known to develop in a
highly supercooled melt and form well-defined branches in preferred
directions [21]. In SOI, a good example of dendritic breakdown has been
reported in Ref. [15]. Similarly, uniquely distinct, needle-like dendrites have been also observed, consistent with the traditionally known
behavior [12,13,16]. In dendritic growth, crystallographic factors control the shape and growth direction, as the growth occurs under sufficiently low thermal-gradient conditions [21(b)]. Our observations have
been found to be consistent with such trends in that the needle-like dendrites developed mainly in low-temperature gradient regions immediately
following the faceted or curved (cellular) breakdown [12,13]. The growth
aligns consistently with the [110] direction. The traditional mode of
dendritic breakdown has been found to contribute to the formation of
sub-boundaries [15], but the specific relation between the dendritic
needles and sub-boundaries has not become clear at this point. Certainly
they have been observed to form growth striations in composite configurations with other modes of breakdown.

TRANSIENT/COMPOSITE BREAKDOWN

Faceted-to-Cellular Transient Breakdown

Occasionally, configurations have been observed that are neither fully sawtooth-shaped [Fig. 3(a)] nor fully rounded [Fig. 3(c)], but somewhat in between, as shown in Fig. 3(b). Here the edges are still straight and the inside corners are sharp, but the outside corners start curving. It is believed that these have developed as an intermediate, transient configuration, as a faceted breakdown transforms into a curved breakdown. The transition from a faceted to a nonfaceted interface has been traditionally recognized by several researchers [22,23] in the study of interface breakdown. Miller, et al. [24], for example, reported a nonfaceted transition in O-terphenyl or Salol materials under lowered thermal gradient conditions. It has been attributed to the anisotropy of growth, possibly involving impurities, under the conditions of large supercooling. Similar transition has also been reported recently by Limanov and Givargizov in the recrystallization of thin-film silicon [15], and has been related to the variation of film thickness. In our study, the transition is attributed to the interface thermal gradients and to the impurities under various supercooling conditions. It should be pointed out that the interface thermal gradient could be related to the variation of film thickness [12].

Cellular/Dendritic Composite Breakdown

Fig. 5(a) shows an array of curved breakdown that is combined with dentritic needles to form what may be referred to here as cellular/-dendritic composite configurations. The slightly bent, short tails, emanating from each interior corner, are cellular boundaries which would lead to regular sub-boundaries. These tails can be clearly distinguished from the needle-like segments described above. Fig. 5(b) hypothetically

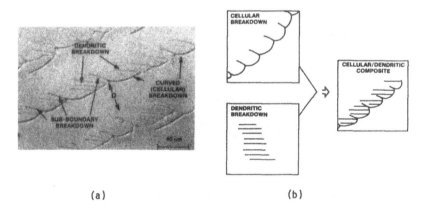

(a)   (b)

FIGURE 5.   (a) Secco-etched cellular/dendritic composite configuration. The breakdown occurs after a certain range of stable growth; (b) hypothetical illustration of the cellular/dendritic composite formation observed in (a).

illustrates how the two distinctly different modes could have merged to form the Fig. 5(a) composite. (Since this composite is an after-the-fact observation, the order of development is not discernible at this time.) Occasionally, a more regular array has been observed where the dendritic segments maintained periodic spacings at the corner of each curved interface [16]. Traditionally, composite configurations, recognized as "cellular-dendrites," have been observed under specific conditions of constitutional supercooling between the cellular and dendritic growth regimes [25]. In such cases, both modes are known to coexist [25]. While the mechanism is not well understood at this time, the composite formation is quite intriguing.

Breakdown Sequence

Fig. 6 shows a series of interfaces that have frozen at various stages of breakdown. Initially, the seeded growth along the seed groove progresses free of defects up to a distance of 50 um. The growth then breaks down into sawtooth-faceted or curved interfaces, with sub-boundaries breaking out at each corner. After another defect-free progression, the growth develops into a next set of defect arrays. Fig. 6(a), for instance, shows an interface that has frozen immediately before a breakdown. It is basically smooth and devoid of any perturbation across the entire growth front. Fig. 6(b) shows an interface that has started developing cusps as a result of interface destabilization and is about to break down into cellular morphology with curved interfaces. In Fig. 6(c) is shown an interface that has broken down into an array of cellular configuration. Finally, Fig. 6(d) shows an array of advanced cellular growth that is coupled with dendritic needles to form cellular-/dendritic composite configurations. Fig. 7 hypothetically illustrates the sequence of these breakdown processes in a systematic way. Traditionally, systematic transitions of smooth interfaces into cellular and dendritic interfaces have been well recognized under various conditions of constitutional supercooling [20(b)].

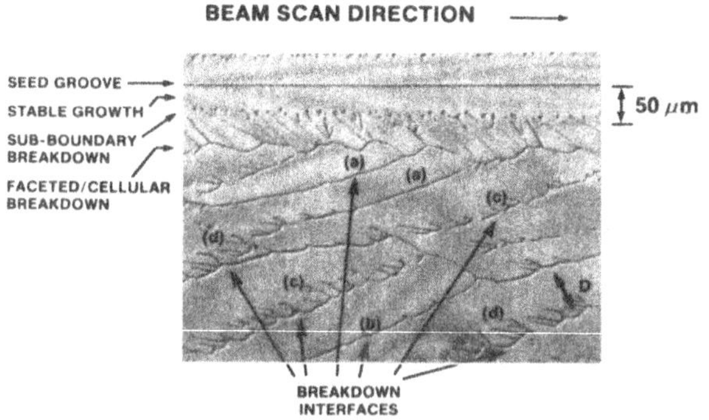

FIGURE 6. Morphological variations near a seed groove which display an array of transitional breakdown sequence. The solidification progresses in the lower-right direction as a result of convolution between two thermal gradients: one determined by cooling through the seed groove and the other by thermal beam profile. Note again that the breakdown occurs after a distance of stable growth.

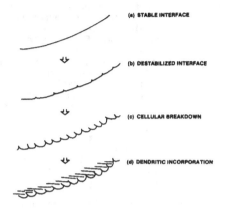

FIGURE 7. Hypothetical illustration of the systematic breakdown transition sequence observed in Fig. 6.

SECONDARY MODES OF BREAKDOWN (SUB-BOUNDARY FORMATIONS)

The above study suggests that the primary (interface) breakdown is necessary to promote the secondary breakdown constituting the sub-boundaries. In faceted or cellular breakdown the sub-boundaries grow out at each interior corner in one-to-one correspondence. Dendritic breakdown is also responsible for the sub-boundary breakdown, as observed here as well as in other studies [12-16]. In general, the characteristics of sub-boundaries from each primary mode of breakdown seem to differ from each other [18]. More study is needed to fully understand the correlation between the primary modes of breakdown and the secondary modes of breakdown for each category of breakdown discussed above. It must be made clear at this point that the fundamental mechanisms of secondary (sub-boundary) breakdown might be different from those of primary (interface) breakdown. While the latter may include excessive supercooling, the former may include oxide precipitate effect [26], liquid-pocket formation [27], and/or nucleation-limited process [28], all of which may generate dislocation networks to be punched out continuously behind an advancing breakdown interface to form subboundaries.

DISCUSSION

The above results may be summarized with a flowchart shown in Fig. 8. It collectively suggests that in order to prevent sub-boundaries, one must prevent the primary interface breakdown. It is therefore important to understand the cause of solid-liquid interface instabilities which might be responsible for the interface breakdown.

It is generally known that the stability of an interface is perturbed under the conditions of excessive supercooling. [Under the conditions of a stable supercooling, as shown in Fig 9(a), the interface morphology is known to be smooth and to cause no breakdown.] There are two known cases of excessive supercooling: absolute supercooling [Fig. 9(b)] and constitutional supercooling [Fig. 9(c)]. Absolute supercooling is caused by

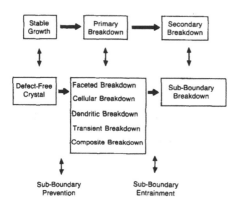

FIGURE 8. An overview of the breakdown process. In order to cause sub-boundaries, a stable, defect-free solid-liquid interface should first make a breakdown transition into faceted, cellular, dendritic, transient and/or composite configuration as primary modes of breakdown. The interface breakdown then causes the sub-boundaries to develop as a secondary effect. In order to prevent sub-boundaries, the interface breakdown must be prevented. Once formed, the sub-boundaries may be entrained.

the evolution of heat of solidification in the pure melt, which induces the interface temperature higher than that in the melt, so that a negative temperature gradient incurs at the interface. This interface instability then allows uncontrolled breakdown into the highly supercooled melt, mainly in the form of dendritic configuration [21,29]. In impure melts, the impurities are constantly segregated into the melt and depress the melting point at the interface. If the actual temperature in the melt is lower than the equilibrium temperature, constitutional supercooling can occur, under which conditions a stable interface can break down into a faceted growth, a cellular growth, and in the limit, a dendritic growth, as the temperature gradient decreases [30].

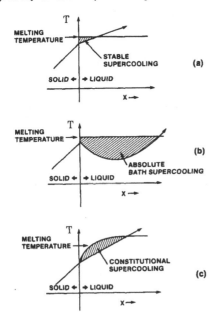

FIGURE 9. Three cases of supercooling: (a) Supercooling in this case allows stable solidification, preventing any breakdown; (b) Supercooling in this case becomes excessive to induce negative thermal gradient at the solid-liquid interface, causing uncontrollable dendritic shooting; (c) Supercooling in this case is caused by the impurity segregation in the liquid and can induce stability breakdown when the thermal gradient in the liquid falls below the equilibrium value determined by the impurity concentration. The morphological breakdown in this case may include faceted, cellular, dendritic and the combination thereof, depending on the degree of supercooling or the level of thermal gradient.

From a morphological point of view, the observed breakdown (and the cellular breakdown in particular) suggests that constitutional supercooling could be largely responsible for the interface breakdown, although absolute supercooling might also be intricately mixed therein. In either case, the interface temperature gradient is an important parameter to determine the fate of interface stability. The interface temperature gradient, in turn, could be a function of many parameters, such as molten-zone temperature, zone width, zone speed and the heat dissipation (or cooling).

The temperature gradient effect upon the onset of interface instability in SOI has been observed in many examples, some of which have been already discussed elsewhere [12-14]. Briefly to summarize, the stable, sub-boundary-free growth region from the seed is attributed to a highly efficient cooling through the seed opening, which effectively maintains the interface temperature gradient high enough to prevent the onset of excessive supercooling. At a point away from the seed, where the seed cooling becomes less effective, the interface becomes destabilized and develops one or a combination of the above-mentioned breakdown modes. It has been also experimentally observed that the stable, defect-free region from the seed can be extended to longer distances by varying the temperature gradient in this region.

The variation of the stable growth region has been effectively explained in terms of the "incubation distance" concept used in the context of constitutional supercooling [20(c),31]. This concept seems to explain also the phenomena that occurred in Figs. 5(a) and 6, where a set of breakdown invariably occurs after a short distance of stable growth in an alternating pattern. As the interface progresses, it may take some distance for the segregated impurities to accumulate enough concentration to induce constitutional supercooling. The destabilized interface may then break down into a variety of morphology, depending on the degree of supercooling. A hypothetical calculation of incubation distances has shown a reasonable agreement with the observed values [14]. In some of the studied samples, carbon [14,32] and oxygen [14,33-35] have been observed. Trace amounts of metallic impurities have also been observed [2,18]. The evidences of impurity, incubation distance, and the diversified breakdown modes in association with the temperature gradient variation, all seem to suggest (at least in a macroscopic sense) the constitutional supercooling as the most probable cause of interface instability in the initial phase of breakdown transition. Despite these and other evidences [11,12] the cause of stability break- down still seems to be a controversial issue that requires further understanding. At least at this point the microscopic role of impurities and temperature gradients seems to be a key to the fundamental understanding of these phenomena. These aspects will be discussed in future communications.

SUMMARY

An attempt has been made to systematically sort out the characteristic modes of solid-liquid interface breakdown transition from a stable interface to an unstable interface. The observed modes include faceted, cellular and dentritic, along with the transient and composite configurations. Constitutional supercooling and absolute supercooling models have been used to explain the cause of observed breakdown transition. The primary (interface stability) breakdown phenomenologically leads to the secondary (sub-boundary) breakdown. The mechanics associated with the sub-boundary breakdown are by no means the same as those of the interface stability breakdown, and so are to be understood independently. There are indications that once the interface breakdown is locked into one of the discussed configurations, it tends to retain the equilibrium configuration throughout the course of recrystallization. Then the sub-boundary generation pattern also remains locked-in, maintaining consistent characteristics, unless a perturbation is introduced to alter the conditions of primary mode of breakdown.

ACKNOWLEDGMENTS

The author expresses sincere appreciation to P. Doerhoff, R. Crepin, R. Craven, R. Hockett, R. Sandfort, and J. Moody, all of Mansanto Electronic Materials Company. He also sincerely acknowledges G. A. Rozgonyi of North Carolina State University, G. Celler, V. Zaleckas, and M. Edsall of AT&T.

REFERENCES

1. M. W. Geis, H. I. Smith, B.-Y. Y. Tsaur, John C. C. Fan, D. J. Silversmith, and R. W. Mountain, J. Electrochem. Soc., 129, 2812 (1982).
2. H. J. Leamy, C. C. Chang, H. Baumgart, R. A. Lemons and J. Cheng, Mater. Lett., 1, 33 (1982).
3. M. W. Geis, H. I. Smith, D. J. Silversmith, R. W. Mountain, J. Electrochem. Soc., 130, 1178, (1983).
4. R. F. Pinizzotto, H. W. Lam, and B. L. Vaandrager, Appl. Phys. Lett., 40, 388 (1982).
5. A. Kamgar and E. Labate, Mater. Lett., 1, 91 (1982).
6. T. J. Stultz and J. F. Gibbons, Appl. Phys. Lett., 41, 824 (1982).
7. D. P. Vu, M. Haond, D. Bensahel and M. Dupuy, J. Appl. Phys., 54, 437 (1983).
8. J. A. Knapp and S. T. Picraux, Mater. Res. Soc. Symp. Proc. 13, 557 (North-Holland, New York, 1983).
9. Y. Hayafuji, T. Yanada, S. Usui, S. Kawado, A. Shibata, N. Watanabe, M. Kikuchi and K. E. Williams, Appl. Phys. Lett., 43, 473 (1983).
10. D. Herbst, M. A. Bosch, and S. R. Tewksbury, IEEE Electron Dev. Lett., EDL-4, 280 (1983).
11. R. A. Lemons and M. A. Bosch and D. Herbst, Mater. Res. Soc. Symp. Proc. 13, 581 (North-Holland, New York, 1983).
12. E. H. Lee, Mater. Res. Soc. Symp. Proc., 23, 471 (North-Holland, New York, 1984).
13. E. H. Lee, Appl. Phys. Lett., 44, 959, (1984).
14. E. H. Lee, in VLSI Science and Technology/1984, K. E. Bean and G. A. Rozgonyi, Editors, Proceedings of the Second International Symposium on VLSI Science and Technology, Vol. 84-7, The Electrochemical Society, pp 250-266.
15. A. B. Limanov and Givargizov, Mater. Lett. 2, 93, (1983).
16. E. H. Lee and G. A. Rozgonyi, paper presented at the International Crystal Growth Conf. (ACCG-6/IC&GE-6), Atlantic City, July 16-20, 1984 (to be published in the J. Cryst. Growth).
17. T. Stultz, paper presented in Mater. Res. Soc. Symp., Albuquerque, NM, Feb. 27-29, 1984.
18. E. H. Lee (to be published).·
19. See for example (a) D. P. Woodruff, The Solid-Liquid Interface, Cambridge University Press, Cambridge, (1973), pp 45; (b) D. T. J. Hurle, in Crystal Growth: An Introduction, P. Hartman, Ed., North-Holland, New York, (1983), pp 233-239.

20. See for example (a) D. P. Woodruff, The Solid-Liquid Interface, Cambridge University Press, Cambridge (1973), p 84; (b) R. A. Laudise, The Growth of Single Crystals, Prentice-Hall, Englewood Cliffs, (1970) pp 104-109; or (c) B. Chalmers, Principles of Solidification, John Wiley and Sons, New York, (1964) pp 150-168.

21. See for example (a) B. Chalmers, Principles of Solidification, John Wiley and Sons, New York, (1964) pp 63-95; (b) W. C. Winegard, Metallurgical Reviews, 6, 57-99 (1961); or (c) R. A. Laudise, The Growth of Single Crystals, Prentice-Hall, Englewood Cliffs, (1970), p 209.

22. D. T. J. Hurle, Mechanisms of Growth of Metal Single Crystals from the Melt, Pergamon Press, New York, 1962, pp 79-147.

23. W. A. Tiller and J. W. Rutter, Canad, J. Phys., 34, 96 (1956).

24. C. E. Miller, J. Cryst. Growth, 42, 357 (1977).

25. (a) B. Chalmers, Principles of Solidification, John Wiley and Sons, New York, (1964) p 164; (b) G. A. Chadwick, in Fractional Solidification, Vol. 1, M. Zief and W. R. Wilson, Eds., Marcel Dekker, New York, 1967, pp 113-135.

26. A. Kamgar, G. A. Rozgonyi, and R. Knoell, Mater. Res. Soc. Symp. Proc., 13 (North-Holland, New York, 1983) p 569.

27. R. Pinizzotto, J. Cryst. Growth, 63, 559 (1983).

28. L. Pfeiffer, this conference.

29. (a) J. C. Brice, The Growth of Crystals from the Melt, North-Holland, New York, 1965; also, (b) The Growth of Crystals from the Liquids, North-Holland, New York, 1973.

30. (a) D. W. Jones in Crystal Growth, Vol. 1, C. H. L. Goodman, Ed., Plenum Press, London, 1981, p 236, (b) also, p 190 in Ref. [21(c)].

31. W. A. Tiller, K. A. Jackson, J. W. Rutter, and B. Chalmer, Acta. Metallurgica, 1, 428 (1953).

32. R. F. Pinizzotto, H. W. Lam, B. L. Vaandrager, Appl. Phys. Lett., 40, 388 (1982).

33. J. C. C. Fan, B. Y. Tsaur, C. K. Chen, J. R. Dick, and L. L. Kazmerski, Appl. Phys. Lett., 44, 1086 (1984).

34. C. I. Drowley and T. I. Kamins, Mater. Res. Soc. Symp. Proc., 13, (North-Holland, New York, 1983) p 511.

35. D. K. Biegelson, N. M. Johnson, D. J. Bastelink and M. D. Moyer, Mater. Res. Soc. Proc., 1 (North-Holland, New York, 1981) p 487.

CHARACTERIZATION, CONTROL, AND REDUCTION OF SUBBOUNDARIES IN SILICON ON INSULATORS

M. W. GEIS, C. K. CHEN, HENRY I. SMITH, R. W. MOUNTAIN, AND C. L. DOHERTY
Lincoln Laboratory, Massachusetts Institute of Technology, Lexington, Massachusetts 02173

ABSTRACT

Subboundaries are the major crystalline defects in thin semiconductor films produced by zone-melting recrystallization (ZMR). Using transmission electron microscopy (TEM) and chemical etching we have analyzed the angular discontinuity and defect structure of subboundaries in ZMR Si films. Annealing in oxygen has resulted in the elimination of dislocation bands from sizable regions of some films. Calculations suggest that cellular growth due to constitutional supercooling may not occur in some Si ZMR.

Subboundaries have been consistently observed in films of AgCl,[1] Ge,[2] salol,[3] InSb[4,5] and Si[6-9] produced by zone-melting recrystallization (ZMR). The present study of subboundaries in Si was concentrated on the two basic structures shown in Fig. 1. When subjected to ZMR in a graphite-strip-heater oven with improved thermal characteristics,[10] these structures gave different subboundary morphologies, as shown in Fig. 2 by optical micrographs of the etched surface of two films. The thinner Si films (Fig. 1b) contained branched subboundaries (Fig. 2a), while the thicker Si films with a thicker lower oxide (Fig. 1a) contained a lower density of subboundaries, without branching[11,12], as shown in Fig. 2b. Figure 2b also shows several locations where a series of tangled dislocation clusters forms a linear array, with defect-free material between the successive clusters. The thicker Si films required chemical thinning for transmission electron microscopy (TEM) evaluation. During thinning, the crystallographic defects in the film etch faster than the defect-free areas. Defects observed by TEM in etched films often had a "sperm" appearance, as shown by the micrograph and diagram of Fig. 3. Whether a diluted Secco etch[13] or a HF-HNO$_3$ etch was used to thin the 2-μm Si film, etching revealed all the defects subsequently observed in the film by TEM. Some thicker films that had not been chemically thinned were also examined by TEM. These films had the same defect characteristics as the thinned films, although it was difficult to obtain high resolution micrographs without thinning.

The angular discontinuities across the subboundaries present in ZMR Si films associated with the two structures of Fig. 1 were markedly different. The angular discontinuity across a subboundary can be characterized by three angles, α, β, and γ, as shown in Fig. 4. Figure 5 shows plots of the subboundary angular discontinuities over a region of a 0.5-μm-thick ZMR Si film. The two angles β and γ have average changes of approximately one degree, while α is consistently zero to within the experimental accuracy of ± 0.2 deg. This is to be compared with Fig. 6, which shows similar plots for a 2-μm-thick Si film with a 2-μm-thick lower oxide. Only β discontinuities were consistently non-zero. Figure 7 shows a comparison of the total discontinuity determined from β and γ for the thin and thick film cases. The structure with thicker Si and SiO$_2$ exhibits substantially smaller angular discontinuities than the thinner structure.

The thicker ZMR films often have a variety of defects as shown in Fig 8. Often there are linear arrays of regularly spaced dislocation clusters, between which the material is nearly defect-free, as shown in the TEM micrograph of Fig. 9. In other locations there are narrow bands of random dislocations, as shown in the TEM micrograph of Fig. 10. Annealing films

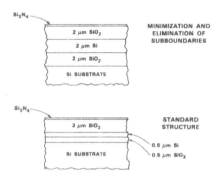

Figure 1. Two basic structures used for this study of zone-melted-recrystallization (ZMR) of Si films.

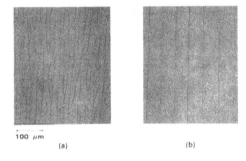

(a)   (b)

Figure 2. Optical micrographs of the etched surfaces of two ZMR Si films. (a) The Si and SiO$_2$ films were 0.5 μm thick, as depicted in Fig. 1b. (b) The Si and SiO$_2$ films were 2 μm thick, as depicted in Fig. 1a.

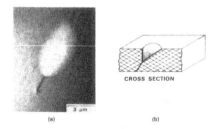

(a)   (b)

Figure 3(a). Transmission electron micrograph of a dislocation in a 2-μm-thick Si film that has been chemically thinned. (b) Cross-sectional diagram of the etched dislocation.

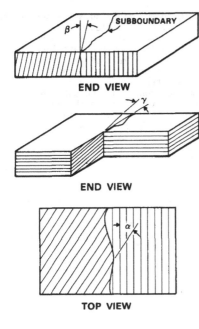

END VIEW

END VIEW

α

TOP VIEW

Figure 4. Schematic drawing showing the three angles α, β and γ used to define the angular discontinuity across a subboundary.

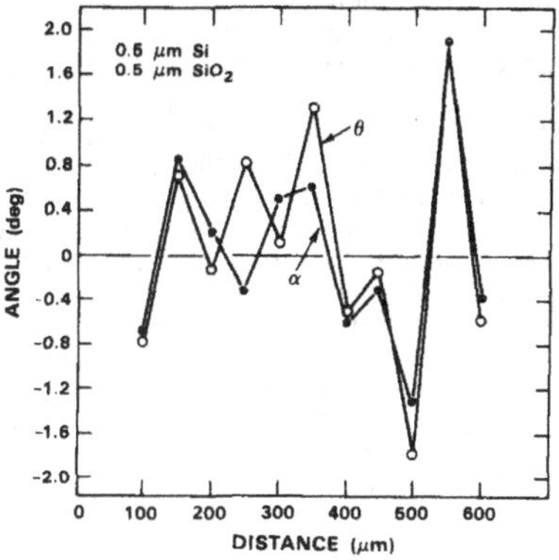

Figure 5. Plot of the angles β and γ for a 0.5-μm-thick recrystallized Si film.

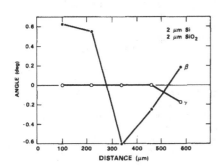

Figure 6. Plot of the angles β and γ for a 2-μm-thick recrystallized Si film.

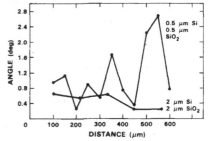

Figure 7. Comparison of the total angular discontinuity across the subboundaries, calculated as the square root of the sum of the squares of β and γ, for 0.5- and 2.0-μm-thick Si films.

containing such bands for 8 hours at 1250°C in an oxidizing atmosphere often produces nearly defect-free material with only a few isolated dislocations.[14] Optical micrographs of such a film, which was chemically etched to reveal defects, are shown in Figs. 11 a and b. Figure 11b, a micrograph taken with white light, shows that a large area of the film is free of any visible defects. Figure 11a shows the same film viewed under monochromatic light, so that variations in the Si thickness produce intensity differences due to interference between beams reflected from the front and back surfaces of the film. The defect-free region shows a thickness variation on the order of 100 nm. The thickness variation indicates that the liquid-solid interface was faceted in this region during ZMR, but stable subboundaries were not formed at the interior corners of the interface. Figure 11 also shows several defect arrays that were not removed by annealing. Figure 12 shows TEM micrographs of a 0.5-μm-thick ZMR Si film and a 2-μm-thick recrystallized and annealed film. The thinner film contains many obvious subboundaries while the thicker film contains only an occasional dislocation over an area of a few square millimeters.

Several authors[15-18] have discussed cellular growth due to constitutional supercooling as a cause of subboundaries in ZMR films, so only a brief discussion will be given here. Table 1 lists the impurities commonly present in graphite-strip-heater recrystallized Si films, their concentrations estimated from secondary ion mass spectrometry,[19] and their segregation coefficients.[20] The thermal gradient, G, necessary to stabilize the liquid-solid interface against cellular growth is given by the equation[21]

$$G \approx C R m (1-k) / (D k), \qquad (1)$$

where C is the atomic fraction of the impurity, R is the speed of solidification, m is the liquidus slope, k is the segregation coefficient, and D is the diffusion coefficient of the dopant in liquid Si[20] (assumed to be 3×10^{-4} cm^2 sec^{-1}). The values of G calculated from this expression for the impurities in Si and ZMR speeds of 1 and 0.1 mm sec^{-1} are listed in Table 1. For the more commonly used ZMR speed of 1 mm sec^{-1}, G is of the order of 100 K cm^{-1}, which is within an order of magnitude of the estimated thermal gradient.[22] For ZMR speeds in excess of 1 mm sec^{-1}, or for higher impurity concentrations than those listed in Table 1, constitutional supercooling may be a factor in subboundary formation. The ZMR speeds used in the work reported here varied from 0.1 to 0.2 mm sec^{-1}. For these speeds, the calculated value of G is about 10 K cm^{-1}, which is much less than the estimated gradient. Hence, it is unlikely that constitutional supercooling is a factor in the present work.

Stress induced by thermal gradients in the Si film during ZMR and by expansion upon freezing may give rise to subboundaries and dislocations that follow the tracks of the interior corners. In some earlier work on the crystallization of copper, thermal gradients were believed responsible for subboundaries.[23] When copper was zone melted to form single crystals, subboundaries were formed even though the copper was of high purity. When the crystals were subsequently annealed at temperatures a few degrees below the melting point of copper, the subboundaries annealed out, provided the density of dislocations associated with the subboundaries was below a critical value. The presence of subboundaries after the zone melting process was attributed to the thermal gradients during zone melting, which were sufficient to cause plastic deformation. Once the thermal gradients were removed the defects could be annealed out.

The stress present during ZMR of Si films is sufficient to plastically deform the substrate Si wafer. This deformation results from slip in the

100 μm

Figure 8. Optical micrograph of the etched surface of a 1-μm-thick ZMR Si film showing the different types of defects.

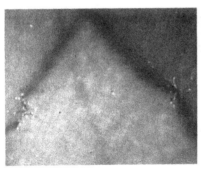

├─ 5 μm ─┤

├─ 5 μm ─┤

Figure 9. Transmission electron micrograph of a chemically thinned 2-μm-thick ZMR Si film showing dislocation clusters.

Figure 10. Transmission electron micrograph of a chemically thinned 2-μm-thick ZMR Si film showing a band of random dislocations. Note that the triangular shaped defect, which is believed to be a microtwin, is parallel to <110> in the film.

wafer along (111) planes. Clearly, the Si film should also be subject to plastic deformation. We speculate that such deformation can induce subboundaries.

In summary, we have measured the dependence on film thickness of the angular discontinuities across subboundaries in ZMR Si films. In 2-μm-thick Si films on 2-μm-thick SiO$_2$, periodic clusters of dislocations and bands of random dislocations that follow the tracks of the interior corners are frequently observed. The latter dislocations often can be eliminated from large areas by annealing in O$_2$ for several hours. At the scanning rates used in these experiments cellular growth due to constitutional supercooling is unlikely. Deformation due to stress induced by thermal gradients and expansion on freezing may give rise to subboundaries.

ACKNOWLEDGEMENTS

The authors would like to acknowledge the expert technical assistance of P. M. Nitishin and fruitful discussions with A. J. Strauss.

This work was sponsored by the Department of the Air Force and the Defense Advanced Research Projects Agency.

H. I. Smith, one of the authors, is with the Department of Electrical Engineering and Computer Science, Massachusetts Institute of Technology, Cambridge, MA, and is a consultant at Lincoln Laboratory.

TABLE 1

Properties of Impurities in Molten Si

Impurity	C	k	m (K per atomic fraction)	G (K cm^{-1})	
				for R = 1 mm sec^{-1}	for R = 0.1 mm sec^{-1}
O	2 to 4 x 10^{-5}	0.3-1.20	420	6.7	0.67
Cu	2 x 10^{-7}	4 x 10^{-4}	600	96	9.6
C	2 x 10^{-5}	6 x 10^{-2}	520	62	6.2
N	2 x 10^{-8}	7 x 10^{-4}	600	56	5.6

  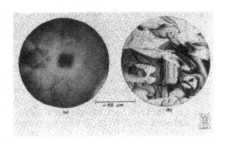

200 μm

Figure 11. Optical micrographs of the etched surface of a 2-μm-thick recrystallized Si film after annealing in O_2. (a) Micrograph obtained with monochromatic illumination. (b) Micrograph obtained with white light.

Figure 12. Transmission electron micrographs of two ZMR Si films. (a) Micrograph obtained from an annealed and chemically thinned 2-μm-thick film. (b) Micrograph obtained from a 0.5-μm-thick film.

REFERENCES

1. J. M. Hedges and J. W. Mitchell, Phil. Mag. A 44, 223 (1953), and J. W. Mitchell, Proc. R. Soc. London A 371, 149 (1980).

2. J. Maserjian, Solid-State Electron. 6, 477 (1963).

3. K. A. Jackson and C. E. Miller, J. Crystal Growth 42, 364 (1977).

4. A. R. Billings, J. Vac. Sci. Technol. 4, 757 (1969).

5. C. C. Wong, C. J. Keavney, H. A. Atwater, C. V. Thompson, and H. I. Smith, Energy Beam-Solid Interactions and Transient Thermal Processing, J. C. C. Fan and N. M. Johnson, eds. (Elsevier, North Holland, New York, 1984), p. 627.

6. T. O. Sedgwick, R. H. Geiss, S. W. Depp, V. E. Hanchett, B. G. Huth, V. Graf, and V. J. Silvestri, J. Electrochem. Soc. 129, 2802 (1982).

7. M. W. Geis, H. I. Smith, B-Y. Tsaur, J. C. C. Fan, D. J. Silversmith, and R. W. Mountain, J. Electrochem. Soc. 129, 2812 (1982).

8. M. W. Geis, H. I. Smith, D. J. Silversmith, R. W. Mountain, and C. V. Thompson, J. Electrochem. Soc. 130, 1178 (1983).

9. J. R. Davis, R. A. McMahon, and H. Ahmed, in Laser-Solid Interactions and Transient Thermal Processing of Materials, J. Narayan, W. L. Brown, and R. A. Lemons, eds. (Elsevier North Holland, New York, 1983), p. 563.

10. C. K. Chen, M. W. Geis, H. K. Choi, B-Y. Tsaur, and J. C. C. Fan, in Energy Beam-Solid Interactions and Transient Thermal Processing, D. K. Biegelsen, G. A. Rozgonyi, and C. V. Shank, eds. (Materials Research Society, 1985), this volume.

582

11. L. Pfeiffer, T. Kovacs, and K. W. West, in Energy Beam-Solid Interactions and Transient Thermal Processing, D. K. Biegelsen, G. A. Rozgonyi and C. V. Shank, eds. (Materials Research Society, 1985), this volume.

12. H. A. Atwater, H. I. Smith, C. V. Thompson, and M. W. Geis, Mater. Lett. 2, 269 (1984).

13. F. Secco D'Aragona, J. Electrochem. Soc. 119, 948 (1972).

14. L. Pfeiffer and co-workers have also demonstrated the removal of isolated defects in ZMR Si films by annealing (personal communication).

15. H. J. Leamy, C. C. Chang, H. Baumgart, R. A. Lemons, and J. Cheng, Mater. Lett. 1, 33 (1982).

16. R. A. Lemons, M. A. Bosh, and D. Herbst, in Laser-Solid Interactions and Transient Thermal Processing of Materials, J. Narayan, W. L. Brown, and R. A. Lemons, eds. (Elsevier North Holland, New York, 1983), p. 581.

17. E-H. Lee, in Energy Beams-Solid Interactions and Transient Thermal Processing, D. K. Biegelsen, G. A. Rozgonyi, and C. V. Shank, eds. (Materials Research Society, 1985), this volume.

18. J. C. C. Fan, B-Y. Tsaur, C. K. Chen, J. R. Dick, and L. L. Kazmerski, Appl. Phys. Lett. 44, 1086 (1984).

19. R. F. Pinizzotto, F. Y. Clark, S. D. S. Malhi, and R. R. Shah, in Comparison of Thin Film Transistors and SOI Technology, H. W. Lam, ed. (Elsevier, North Holland, New York, 1984).

20. W. Zulehner and D. Huber, "Czochralski-Grown Silicon" in Crystals: Growth, Properties and Applications Vol. 8, H. C. Freyhardt, ed. (Springer-Verlag, New York 1982), pp. 1-143.

21. J. D. Verhoeven, Fundamentals of Physical Metallurgy (John Wiley and Sons, New York, 1975).

22. H. E. Cline, J. Appl. Phys. 54, 2683 (1983).

23. F. W. Young and J. R. Savage, J. Appl. Phys. 35, 1917 (1964).

IMPROVED CRYSTAL PERFECTION IN ZONE-RECRYSTALLIZED Si FILMS ON SiO$_2$

Loren Pfeiffer, K. W. West, Scott Paine*, and D. C. Joy, AT&T Bell Laboratories, Murray Hill, NJ 07974

ABSTRACT

We review recent results of our Graphite Strip Heater Si-on-Insulator (SOI) effort: (i) recrystallization of SOI films on 100 mm wafers, (ii) model of subboundary pattern formation in SOI films, (iii) low defect density SOI films by ultra slow scanning of the melt zone, (iv) low defect density SOI films by patterned openings in the cap oxide overlayer.

SOI Recrystallization of 100 mm Wafers:

We have converted our scanned zone melting graphite strip heater equipment[1] to accommodate 100 mm dia. wafers. Figure 1 shows a 100 mm SOI wafer processed in this equipment. The Si film is 0.5 μm thick over 1 μm of thermal SiO$_2$ and capped with 2 μm of LPCVD SiO$_2$. The crystallization conditions were as follows: Ar gas ambient 300 torr, base temperature 1185°C, wire 1 mm by 1 mm at 2250°C, and scan velocity 1.4 mm/sec. After scanning the wafer surface was mirror smooth and featureless. The cap oxide was then removed and the film was defect etched[2] to bring out the grain boundaries and subboundaries. The recrystallization was not seeded and one can see about 16 or so major grain boundaries across the film. The single crystal areas within these regions contain the usual subboundary networks reported by us and others.[3,4] The wafer shown deviates from ideal flatness by a maximum of 26 μm without using a vacuum chuck. If held in a chuck its flatness is equivalent to a virgin unprocessed

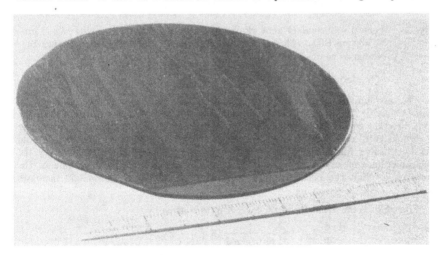

Fig. 1.   Strip heater recrystallized SOI film on 100 mm wafer. The wafer was defect etched and high contrast photography was used to emphasize the grain boundaries.

wafer. The scanning was from top to bottom as shown in the photo. The upper heater was turned off near the end of the scan leaving the small area of polysilicon which shows in the photo as a lighter shade of grey due to the action of the Schimmel[2] defect etch. Strip Heater melt recrystallization is thus shown to be able to produce large wafers having a flatness, surface morphology, and SOI crystal structure suitable for subsequent VLSI processing.

Model of Subboundary Pattern Formation:

Networks of low angle grain boundaries always form spontaneously whenever a submicron Si film on a non-epitaxial substrate is scan melt recrystallized by any method. A typical subboundary network obtained with a strip heater is shown in Fig. 2. Most of the effort in the SOI field is directed at the elimination of these Si subboundary networks or at the mitigation of their effects. We have developed a model of subboundary pattern formation[5] that accounts for their characteristic branched structure, and that explains the apparent long-range mutual attraction that pairs of neighboring subboundaries appear to have for one another before merging.

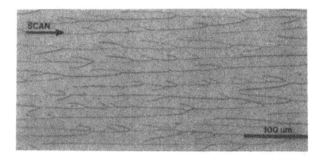

Fig. 2. Typical subboundary network obtained by a strip heater scanning of a .5 μm Si film on SiO$_2$. The sample was Schimmel defect etched to enhance the Nomarski contrast.

The assumptions required for the model (see Fig. 3) are well grounded in experiment: (i) The Si films tend to recrystallize with <010> approximately along the scan direction and (100) texture. (ii) The melt-solid interface consists of alternating (111) and (11$\bar{1}$) crystal facets which may extend up to the melting isotherm but not into the superheated region. (iii) The growth rate of each facet is limited only by the nucleation rate of new monolayers and is independent of the spreading rate of the monolayers. (iv) The nucleation rate is zero at the melting isotherm and increases linearly with the undercooling.

From Fig. 3, the dynamics of the facet lengths is given by

$$\frac{d}{dt}\ell_{2i} = \frac{1}{\sin(\alpha_1+\alpha_2)}(R_{2i-1} - R_{2i+1})$$

$$\frac{d}{dt}\ell_{2i+1} = \frac{1}{\sin(\alpha_1+\alpha_2)}(R_{2i+2} - R_{2i}) ,$$

(1)

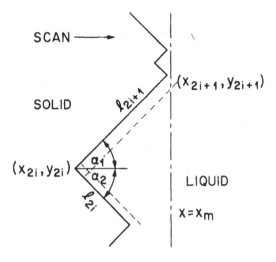

Fig. 3.   Melt-solid interface geometry assumed for the model. The dashed lines are a portion of this interface a short time later.

where, R_i is the normal growth rate of the ith facet. If the heat source is scanned at a velocity v, and x_m is the position of the melting isotherm, then $x_m = vt$. By our assumptions if follows that

$$R_{2i} = \beta_2 \int_{x_{2i}}^{x_{2i-1}} \frac{(x_m - x)\,dx}{\cos \alpha_2} = \beta_2 \ell_{2i} \left(x_m - \frac{x_{2i-1} + x_{2i}}{2} \right)$$

and similarly,   (2)

$$R_{2i+1} = \beta_1 \ell_{2i+1} \left(x_m - \frac{x_{2i+1} + x_{2i}}{2} \right) ,$$

where β_1 and β_2 are kinetic factors determined by the tilt between the plane of the Si film and the growing (111) and (11$\bar{1}$) facets. For films of (100) texture $\beta_1 = \beta_2$.

Plots of the dynamics of the facet growth were made by integrating Eqs. (1) and (2) numerically using a Runge-Kutta technique. The rule for generation of new facets was based on the mechanism described by Geis et al.[4] If a peak formed by two adjacent facets approaches the melting point, it flattens and indents forming new (111) and (11$\bar{1}$) facets. Initial indentations of only $10^{-5}\,\ell_{2i}$ are sufficient for stable growth.

Figure 4 is a plot generated by the model showing the time development of the facets and the loci of the interior corners of the faceted growth front. The growth is chosen to be exactly along <010> with no tilt, thus $\alpha_1 = \alpha_2 = \pi/4$, and $\beta_1 = \beta_2$. Initially all facets were chosen to have equal length except for a 5% perturbation of one. The perturbation propagates rapidly, and the system quickly settles into a quasi-stable pattern of facet generation and annihilation.

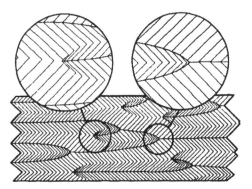

Fig. 4.   Computer plot generated by Eqs. (1) and (2) showing the time development of the facets and the loci of the interior corners of the faceted growth front, moving from left to right.

The insets show interior corner nucleation and annihilation in more detail. Interior corners newly formed at x_m recede towards lower temperature isotherms, because they are bounded by short facets which grow more slowly than the longer facets of the second neighbors on both sides. An interior corner is always drawn toward the side with the shorter facet. See also dashed line in Fig. 3. This is why neighboring interior corners come together more and more rapidly as their separation decreases. While the model does not explicitly require that adjacent interior corners annihilate in pairs, they nearly always do so, because the dependence of the growth rate on undercooling tends to drive interior corners to a common temperature isotherm.

Figure 5 shows plots of the loci of interior corners of the faceted interface under several simulated growth conditions. Note the striking similarities to the experimental subboundary networks of Fig. 2. This harmony between simulation and experiment provides strong evidence for the hypothesis that subboundaries originate at the interior corners of a faceted growth front.

Faceted growth along axes rotated away from <010> and tilted away from (100) texture are shown in the two lower simulations. The top simulation shows the effect of reducing the scan velocity from v to v/4 half way through the scan. Geis et al[4] have reported experimental evidence that for (100) Si films, the average distance between subboundaries varies approximately as the square root of scan speed. We have confirmed this result for experimental films grown in our laboratory, and find that the average distance between facet tracks in our simulations also shows this dependence.

We conclude there is a close correspondence between the loci of interior corners of the (111), (11$\bar{1}$) faceted growth front, and the patterns formed by experimental low angle grain boundaries in melt recrystallized thin Si films. It is remarkable how well the patterns generated by the model and experiment agree. We point out that these calculations do *not* specify the details of the subboundary formation at the interior corners. Indeed, we note that nothing in this work demands that subboundaries *must* form on the interior corners of the faceted front. It is quite possible to envision a dislocation-free single crystal film growing from a melt of perfect purity with a faceted growth front but without subboundary formation.

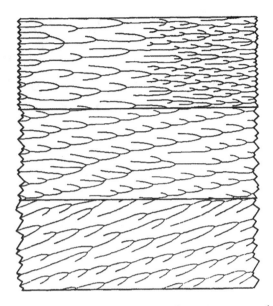

Fig. 5.　Computer generated plots of the loci of interior corners (referred to the solid) of the faceted melt front during growth.

Ultra Slow Scan Melting to Reduce Defects:

We find that the defect density in recrystallized SOI thin films is substantially reduced if the scanning velocity of the melt zone is lower than ~100 μm/sec.[6] This is believed to be below range of strict applicability of the facet model just discussed, because there is now enough time for the crystal to efficiently reject most melt impurities during growth, and moreover the higher thermal gradients now favor nucleation primarily at the highly undercooled regions very near the interior corners of the faceted front. Figure 6 shows a 1.0 μm thick Si film recrystallized by scanning the melt zone at 50 μm/sec. The Schimmel Defect etch[2] has brought out a single clear subboundary near the center of the photo and several others defined only by an intermittent succession of disconnected etch pits running approximately parallel to it at about 50 μm lateral spacing. These disconnected etch pits indicate that the Si film has recrystallized either without a continuous subboundary network[7] or at least with subboundaries of very much less misalignment than the usual 1° or 2°.

Figure 7 is a Scanning Electron Microscope (SCM) microphoto obtained in electron channeling backscatter contrast of a larger area of another 1.3 μm thick Si film melt-recrystallized at a scan rate of only 30 μm/sec in our graphite strip heater. The subboundaries are now seen as approximately parallel strips about 60 μm apart. The weak contrast change across each subboundary is indicative of very little misalignment. This sample contained many cm^2 of material recrystallized in this way. In order to directly measure the misalignment across each subboundary, the SEM was set in electron diffraction channeling mode[8] by rocking the incident electron beam through a cone of about 7° half angle. This produced the diffraction image shown in Fig. 8. From the known electron diffraction conditions one can calculate that the misalignment

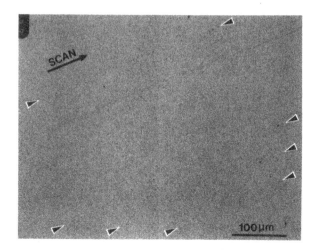

Fig. 6. Nomarski micrograph of SOI film melt scanned at 50 μm/sec. After melting the cap oxide was removed and the film was defect etched using Schimmel's etch. Marker arrows indicate rows of disconnected etch pits.

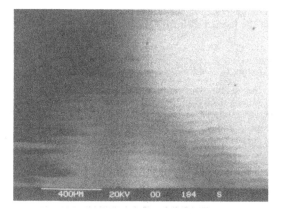

Fig. 7. Electron channeling contrast SEM microphoto of another SOI film scanned at 30 μm/sec.

across each subboundary is a diving twist of 0.1° or less. Note the real space image of the subboundaries super-imposed on the diffraction image.

If we have indeed produced large areas several cm across with subboundary misalignments of only about 0.1°, we might expect to get an ideal [4]He channeling spectrum, because [4]He channeling at 2.0 MeV has little sensitivity to lattice misalignments of this size. Figure 9 confirms this expectation in a quite dramatic way. It is the first report of unseeded thin SOI melt-recrystallized film with a RBS χ_{min} value at the ideal 3%.

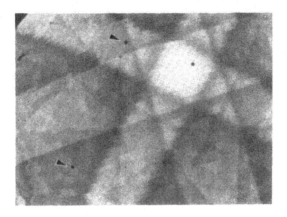

Fig. 8.   Electron diffraction channeling imaged over the same film as Fig. 7. The diffraction conditions show that the misalignment across each subboundary is about 0.1° or less.

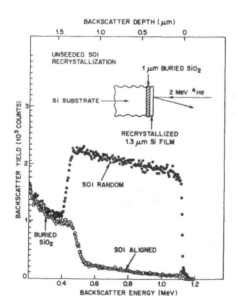

Fig. 9.   Rutherford backscattering channeling of 2.0 MeV ^{4}He from the same slow scanned SOI film as Figs. 7 and 8. This is the first report of an ideal 3% χ_{min} for an unseeded melt-recrystallized SOI thin film.

In experiments where all conditions were held constant except for scan speed we obtained similar results for slowly scanned 0.5 μm Si films on 1 μm buried SiO_2, and have established that it is the ultra slow scanning that is responsible for these improved results. What then is the mechanism of the effect? We speculate that ultra slow advancement of the solidification front improves the Si recrystallization by allowing more time for the growing crystal to reject impurities from the melt. Further we believe that the dominant impurity in these Si melts is probably dissolved SiO_2 from the upper and lower encapsulating layers.

Openings in The Cap Oxide to Reduce Defects:

For Si films 20 μm or thicker we have previously reported[1] that areas larger than 1 mm by 1 mm may be recrystallized over buried SiO_2 in our strip heater without forming subboundaries. This Si although free of subboundaries does, however, contain dislocations and defects in other configurations at a level of about 10^8 cm^{-2}. We will now discuss a new effect involving the etching of openings in the 2 μm thick SiO_2 cap that greatly reduces the number density of these residual defects.[9]

The effect is rather dramatically illustrated in Figs. 10 and 11. Shown there are Nomarski microphotos of angle lapped 10:1 sections of 30 μm Si films recrystallized by scanning the melt zone along the <110> direction above 1.6 mm by 1.6 mm buried oxide islands. Figure 10 shows a defect etched section under a continuous 2 μm thick deposited SiO_2 cap. We see the grain boundaries near each edge of the buried oxide where the melt fronts from the side-seeding met the main bottom to top melt scan. Between these collision-front boundaries there are no subboundaries, but there are many isolated dislocations and other defects. The section on the right side of Fig. 11, on the other hand, recrystallized with no visible defects. The difference is that in this case the

Fig. 10.   Defect etched Nomarski microphoto of an angle lapped (10:1) section of a 30 μm Si film recrystallized over a 1.6 mm by 1.6 mm buried oxide island and under a continuous SiO_2 cap.

2 μm SiO₂ cap contained long open grooves laying bare the underlying Si film in 8 μm wide strips on 100 μm centers.

These openings that are patterned in the cap oxide before recrystallization have at least two potentially beneficial effects: (i) They provide some stress relief to accommodate the volume expansion of the recrystallizing Si liquid. (ii) They provide a vent path for impurities that might otherwise exceed the solid solubility limit in the Si film. SIMS measurements using a Cs ion beam show that the Si under the cap openings contains less than one third the oxygen measured under the continuous cap. This and other evidence leads us to favor venting as the mechanism of the effect, and to propose that the impurity vented is oxygen. The oxygen is most probably introduced during the time the Si film is molten, and arises because the molten Si partially dissolves the upper and lower SiO₂ encapsulating layers.

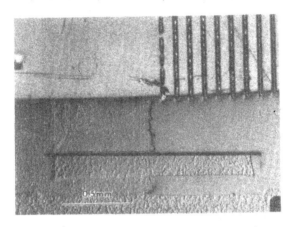

Fig. 11. Defect etched Nomarski microphoto of an angle lapped (10:1) section of a 30 μm Si film recrystallized over a 1.6×1.6 mm buried oxide island. The capping SiO₂ was continuous on the left, but contained vent and relief openings on the right.

Acknowledgements

The authors are most pleased to acknowledge George Gilmer, Wim vanSaarloos, Terry Kovacs, and Fred Stevie for there indispensible contributions to portions of this work.

* Permanent address, MIT, Department of Physics, Cambridge, Mass.

1. L. Pfeiffer, G. K. Celler, T. Kovacs, and McD. Robinson, Appl. Phys. Letts. *43*, 1048 (1983).

2. D. G. Schimmel, J. Electrochem. Soc. *126*, 479 (1979).

3. L. Pfeiffer, J. M. Gibson, and T. Kovacs, Mat. Res. Soc. Symp. Proc. *25*, 505 (1984).

4. M. W. Geis, H. I. Smith, D. J. Silversmith, R. W. Mountain, and C. V. Thompson, J. Electrochem. Soc. *130*, 1178 (1983).

5. L. Pfeiffer, S. Paine, G. H. Gilmer, W. vanSaarloos and K. W. West, Phys. Rev. Letts., (to be published).

6. L. Pfeiffer, K. W. West and D. C. Joy, Appl. Phys. Letts., (to be published).

7. The group at MIT Lincoln Labs appears to see similar etching behavior in certain of their SOI films. See the paper by M. W. Geis et al., in this symposium volume.

8. D. C. Joy, D. E. Newbury and D. L. Davidson, J. Appl. Phys. *53*, R81 (1982).

9. L. Pfeiffer, T. Kovacs and K. W. West, Appl. Phys. Letts., (to be published).

HIGH-VOLTAGE ELECTRON MICROSCOPY INVESTIGATION OF SUBGRAIN BOUNDARIES IN RECRYSTALLIZED SILICON-ON-INSULATOR STRUCTURES

H. BAUMGART[*] and F. PHILLIPP[**]
* Philips Laboratories, A Division of North American Philips Corporation, 345 Scarborough Road, Briarcliff Manor, New York 10510
** Max Planck Institute for Metals Research, Heisenbergstr. 1, D-7000 Stuttgart 80, West Germany

ABSTRACT

The microstructure of high-quality recrystallized Si films on SiO_2 substrates produced by CO_2 laser induced zone-melting has been investigated by high voltage electron microscopy (HVEM). Subgrain boundaries represent the major defects in these recrystallized films. The origin of the subboundaries has been traced to periodic internal stress concentrations occurring at the faceted growth interface. These highly localized stresses cause plastic deformation of the growing single crystal film by nucleation of an array of slip dislocations. The mechanism responsible for the formation of subgrain boundaries has been revealed to be polygonization, where thermally activated dislocations rearrange themselves into the lower energy configuration of the low angle grain boundary.

INTRODUCTION

In the quest for novel silicon-on-insulator (SOI) technologies zone-melting recrystallization (ZMR) has been one of the more successful techniques. The renewed interest in thin semiconductor films has led to extensive research of zone-melting recrystallization induced by laser [1], electron-beam [2], high power halogen lamp [3], and graphite strip heater [4]. This modified Bridgeman crystal growth technique has proved to be very effective in producing large areas of single crystal silicon films on amorphous insulators. The most interesting feature of zone-melting recrystallized Si films is the fact that they are single crystalline films containing subgrain boundaries as the dominant lattice defect. Subboundaries in ZMR Si films on SiO_2 have been shown to have no significant effect on the performance of MOSFET's but they do affect minority carrier transport. For a better understanding of the influence of subgrain boundaries on carrier transport it is highly desirable to have accurate knowledge about the microscopic core structure of subboundaries. Presently the origin of the subgrain boundaries has not been established. A key element for a deeper understanding of the basic lateral melt growth mechanism in thin Si films on SiO_2 is detailed information about the nucleation process of subgrain boundaries in addition to knowledge about the kinetics of the growth interface. The aim of this paper is an investigation of the sources and precursor stages of the characteristic subgrain boundaries in recrystallized Si films. During this study we performed high voltage transmission electron microscopy (HVEM) combined with optical Nomarski microscopy and scanning electron microscopy (SEM) and correlated the information with in-situ observation of the kinetics of the solid-liquid interface front.

SAMPLE PREPARATION

For the substrate samples we used both 4" fused quartz wafers and 4" Si wafers; the latter were prepared by thermally oxidizing the Si wafers to a thickness of 1.0 μm. The polycrystalline Si films were then deposited by low pressure chemical vapor deposition (LPCVD) from SiH_4 at 620°C to a thickness of 0.5 μm or 2.0 μm. After that the poly-Si films were capped by a 2.0 μm SiO_2 layer. The zone-melting recrystallization of the poly-crystalline Si film was performed with a CO_2 laser. In our experimental arrangement the laser beam was focussed with a cylindrical lens into a

594

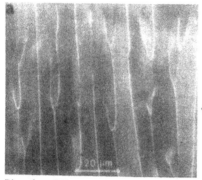

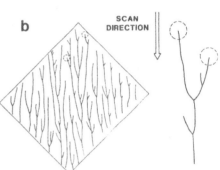

Fig. 1a. SEM micrograph of recrystallized Si film on insulator showing typical surface mophology with subgrain boundaries.

Fig. 1b. Schematic illustrating the abrupt emergence of subboundaries and their subsequent coalescence with other boundaries.

strip which formed a slightly convex ≈ 4 mm wide molten zone in the Si film. The wafer was first heated to a background temperature of 1100°C by a bottom graphite heater and was then moved under a stationary laser beam at typical scan rates of ≈ 2.0 mm/s. After completion of non-seeded zone-melting crystal growth the surface morphology of the recrystallized Si film was studied by optical microscopy. To do this the oxide capping layer had to be stripped by buffered HF and the subgrain boundaries were delineated by a quick dip in Schimmel etch. We chose (HVEM) for the investigation of the interior structure of the Si films. In order to prepare the films for TEM analysis we cut small discs from the laser processed wafers by ultrasonic machining and immersed them for several hours in concentrated HF to dissolve the isolation oxide under the Si film. The Si films eventually lift off the substrate and float to the surface.

RESULTS AND DISCUSSION

Zone-melting grown Si films on SiO$_2$ always display characteristic lineage structures, which are referred to as subgrain boundaries in the literature. Figure 1a shows a typical example of a Schimmel etched surface of a recrystallized Si film in an SEM micrograph. The etched surface reveals a lineage structure of subboundaries roughly parallel to the growth direction and of nearly constant spacing. It is apparent that all sub-boundaries start abruptly somewhere in the crystal. Generally the subboundaries follow the scan direction for a while until, sooner or later, they coalesce with a second subboundary in a characteristic hook-like node. Indeed, superposition with other subboundaries in nodes occurs very frequently along the path of a subboundary. Our aim is to clarify the origin and nature of this peculiar defect structure. In order to determine the underlying mechanism of subboundary formation it is necessary to search for the precursor of this defect and to analyze the details of its nucleation process. In the schematic of Fig. 1b we have encircled the area of interest for materials analysis just before the first appearance of individual subboundaries in the surface etch pattern. What is happening in the crystal before the point where each boundary lineage structure appears first, is best investigated by high voltage electron microscopy (HVEM). The advantage of the HVEM stems from the fact, that the entire recrystallized SOI film can be analyzed (up to several um in thickness) without pre-thinning the sample, so that the original grown-in lattice defect structure can be studied. Employing a 1.2 MV AEI EM-7 HVEM microscope we were able to study the inner structure of a subboundary from its source

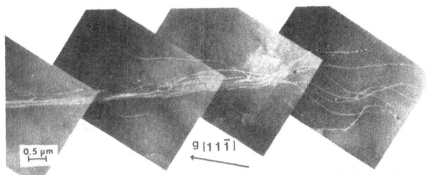

Fig. 2. High voltage transmission electron micrograph displaying the source of a subboundary followed by rearrangement of the initial random dislocation configuration into a low angle grain boundary in a polygonization process.

to its end in a node point of convergence with another boundary. A representative result is displayed in the darkfield micrograph of Fig. 2. It was necessary to rotate the specimen 45° with the (11$\bar{1}$) diffraction vector as rotation axis in order to observe the individual dislocations. The major microstructural defects, seen within the boundary, are long arrays of slip dislocations. This is direct proof that the observed subboundaries really consist of low angle grain boundaries [3,5]. On the average ≈ 20 dislocations were found in the boundaries of 0.5 µm thick recrystallized Si films. Dislocation separation in the boundary was ≈ 25 nm and the tilt angles introduced by the 20-25 dislocations in a tilt boundary was determined to be in the range of 0.8 - 1.2°. What is new in Fig. 2 is that we were able to analyze the source of an individual low angle grain boundary. Unlike conventional 100 - 200 KV microscopes, where typically 80% of the defects to be studied are etched away during the specimen thinning procedure, an HVEM microscope allows the analysis of large area specimens of considerable thickness and provides therefore the best chance of actually finding a subboundary source. In ZMR Si films the subboundary sources consist of randomly dispersed tangles of slip dislocations with their nucleation points pinned to the free surfaces of the thin film . Figure 3 shows a close-up of the slip dislocation network forming the precursor of a low-angle grain boundary. For a better 3-dimensional visualization of the grown-in defect configuration in the ZMR

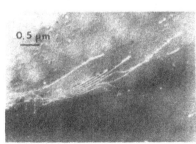

Fig. 3. HVEM close-up of the origin of a typical low angle grain boundary.

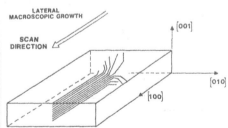

Fig. 4. Schematic diagram of dislocation arrays illustrating the nature and origin of a low angle grain boundary during ZMR of Si films on SiO$_2$ as deduced from our HVEM analysis.

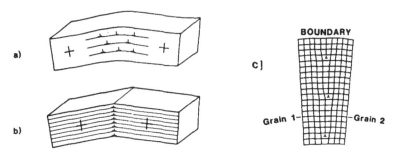

Fig. 5. Schematic representation of the formation of a low angle grain boundary by polygonization. (a) Initial random dislocation distribution at internal stress concentration. (b) Rearrangement of dislocations to form low grain angle boundary. (c) Geometrical model of vertical array of dislocations in a low angle grain boundary.

film Fig. 4 presents a schematic model of the dislocation sources at the free surfaces depicting the sub-boundary precursor and the vertical array of slip dislocations that form the low angle grain boundary. Within the framework of dislocation theory [6] our observations lead to the conclusion, that the nucleation of slip dislocations at the subboundary source is the result of high local internal stresses occurring at periodic stress concentrations in the growing Si film which cause plastic deformation. Whenever stress concentrations exceed a critical value of about $G/30$, where G denotes the shear modulus of the material, dislocations are nucleated [7].

Basically this accounts for the initial nucleation of the subboundary precursor form, while the subsequent transformation of the uncorrelated tangle of slip dislocations into a low angle grain boundary is caused by the process of polygonization. This effect is explained in its most basic form in Fig. 5. Figure 5a depicts the initial random distribution of slip dislocations in the locally plastically deformed crystal, which corresponds to the subboundary precursor stage. According to Hull and Bacon [7] the energy of a crystal containing slip dislocations can be reduced, if the dislocations line up one above the other, producing a more stable vertical dislocation wall as seen in Fig. 5b. Such a dislocation wall is actually a low angle grain boundary which is also known as a tilt boundary. The geometry of such a boundary is depicted in the schematic of Fig. 5c. This rearrangement of dislocations in configurations of lower energy involves

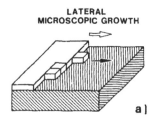

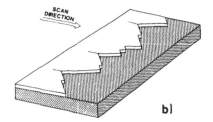

Fig. 6a. Simplified model of lateral microscopic growth by attachments of atoms at the edges of steps.

Fig. 6b. Representation of (111) faceted instantaneous growth front where small growth ledges generate new facets.

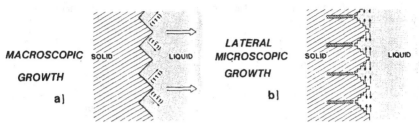

Fig. 7. Model of the solid-liquid interface in top view. (a) Macroscopic growth occurs by superposition of laterally advancing microscopic growth steps. (b) Location of internal stress concentration at inner corners of (111) facets where impurities are trapped.

climb and glide. Such a process can only occur when a plastically deformed crystal is thermally activated, which is provided by the 1100°C substrate heating during ZMR. This mechanism accounts for the peculiar defect configuration in Fig. 2 which presents a particularly striking example of polygonization.

To answer the question what causes these high local internal stress concentrations we had to correlate the HVEM results with in-situ observations of the crystal growth phenomena during ZMR of Si films. We and several other authors [8,9,1] have found that the liquid-solid interface front exhibits classical faceted or cellular morphology. E.H. Lee studied the non-planar interface in a systematic way and obtained with decreasing temperature gradients morphologies from sawtooth faceted to cellular and dendridic [10]. The faceted interface was attributed to absolute supercooling and the cellular and dendridic forms to constitutional supercooling. These non-planar solidification fronts can be understood by considering the lateral microscopic growth mechanism according to E. Bauser et al. [11,12]. It is well known that on an atomistic level crystal growth occurs by the addition of new atoms at the edges of steps as' schematically shown in Fig. 6a. Macroscopic growth arises from a superposition of lateral microscopic growth steps. When Si is grown from the melt, the instanteneous growth facets are bound by the close packed (111) lattice planes. Figure 6b shows the case in zone-melting recrystallization when the growth front was observed to be faceted with (111) planes. These growth front facets are not atomically flat but do contain enough steps that enable lateral microscopic growth to proceed. These steps can lead to new facets as indicated in Fig. 6b by fluctuations in the lateral growth speed. Microscopic facet growth proceeds then by the lateral motion of monomolecular and equidistant growth steps. The basic aspects of lateral microscopic film growth are schematically represented in Fig. 7. As a consequence of this growth mode impurity inhomogeneity and microsegregation may develop. At the inner corners of the (111) oriented facets, local

DEVELOPMENT OF GROWTH INTERFACE

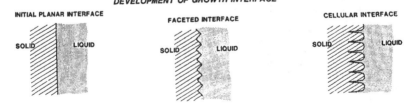

Fig. 8. Schematic representation of the major morphological interface instabilities. High local internal stresses arise at the inner corners of the facets or cells causing heterogeneous nucleation of slip dislocations.

coalescence of monomolecular steps occurs and leads to the preferential incorporation of impurity atoms into the lattice which get trapped there. The impurities in ZMR Si films are most likely oxygen atoms that were introduced into the molten zone by partial dissolution of the SiO_2 layers bordering the Si film. It has indeed been found by J.C. Fan et al., [13] using secondary ion mass spectroscopy that oxygen is strongly concentrated at the subboundaries and their location has been demonstrated to coincide with the coalescence point at the inner corners of the growth facets. Therefore we suggest that the inner corner of newly formed facets is the location where the high internal stress concentrations are produced due to impurity incorporation and impingement of growth steps. The impurities cause lattice dilation and the coalescence of growth steps causes constraints on the anomalous volume expansion of solidifying Si. Both effects sum up and are responsible for the local internal stresses that lead to dislocation generation. As shown in Fig. 8 faceting is not the only non-planar morphology of the solid-liquid interface. In the case of a cellular interface impurities rejected by the solidifying Si form a boundary layer at the growth interface wherein the freezing point is depressed. This last-to-freeze liquid in the cells exerts the compressive stress by its volume expansion during solidification.

CONCLUSION

Using the fact that semiconductor crystals grow exclusively layer by layer via a lateral microscopic growth mechanism we explained the periodic occurrence of high internal stress concentrations during zone-melting of Si films on SiO_2. This in turn leads to local plastic deformation by nucleation of slip dislocations in order to reduce the stresses. The mechanism responsible for the formation of the low angle grain boundaries was found to be the process of polygonization, which results in a more stable dislocation configuration of lower energy. Upon thermal activation, low angle grain boundaries form to accommodate those dislocations which are not annihilated by other dislocations or at free surfaces. The relative stability of low angle grain boundaries results from the absence of stress on the slip planes of the individual dilocations and the cancellation of long range stress fields.

REFERENCES

[1] A.B. Limanov and E.I. Givargizov, Mat. Letters 2, 93 (1983).
[2] J.A. Knapp and S.T. Picraux, Mat. Res. Soc. Symp. Proc., Vol. 23, p. 533, Elsevier Science Publishing Co., New York (1984).
[3] M. Haond, D.P. Vu and D. Bensahel, J. Appl. Phys., 54 (7) 3892 (1983).
[4] M.W. Geis, H.I. Smith, B.Y. Tsaur, J.C. Fan, D.J. Silversmith and R.W. Mountain, J. Electrochem. Soc., 129, 2813 (1982).
[5] Y. Komem and Z.A. Weinberg, J. Appl. Phys., 56 (8) p. 2213 (1984).
[6] J.P. Hirth and J. Lothe, Theory of Dislocations, McGraw-Hill, New York (1968).
[7] D. Hull and D.J. Bacon, Introduction to Dislocations, p. 175-196, Pergamon Press (1984).
[8] H.J. Leamy, C.C. Chang, H. Baumgart, R.A. Lemons and J. Chen, Mat. Letters 1, 33 (1982).
[9] R.A. Lemons, M.A. Bosch and D. Herbst, Mat. Res. Soc. Symp. Proc., Vol. 13, p. 581, Elsevier Science Publishing Co., New York (1983).
[10] El-Hang Lee, Mat. Res. Soc. Symp. Proc., Vol 23, p. 471 (1984) and this volume (1985).
[11] E. Bauser and G.A. Rozgonyi, J. Electrochem. Soc. Vol. 129, No. 8, 1782 (1982).
[12] E. Bauser in Festkörperprobleme (Advances in Solid State Physics) Volume XXIII, p. 141, P. Grosse, ed., Vieweg, Braunschweig (1983).
[13] J.C.C. Fan, B-Y Tsaur, and C.K. Chen, Appl. Phys. Lett. 44, (11) 1086 (1984).

SUBBOUNDARY SPACING AND APPEARANCE IN LASER ZONE-MELTING

RECRYSTALLIZATION OF SILICON ON AMORPHOUS SUBSTRATE

J.P. JOLY, J.M. HODE, J.C. CASTAGNA
L.E.T.I. - C.E.A. - I.R.D.I. - Commissariat à l'Energie Atomique
LETI - CENG - 85 X - 38041 Grenoble Cedex - FRANCE

ABSTRACT

Recrystallized silicon films on amorphous substrates are mainly characterized by subgrain boundaries separated by a few microns. Using seeding
from the silicon substrate (lateral epitaxy), subboundary free areas adjacent to the seed are achieved in the direction of the beam scanning. We have
demonstrated the large influence of the growth direction and of the thickness of the silicon layer together on the incubation distance and on the
spacing of the subboundaries. Surprisingly, a variation of the growth (scan)
velocity of two orders of magnitude (from 1 to 70 cm/s) has no noticeable
influence on these parameters. A 1 x 10 elliptical Ar+ laser beam has been
used in these experiments. An interpretation of these results in terms of
lateral (step) growth and stress will be given.

INTRODUCTION

The major defects present in molten zone recrystallized silicon on
insulator films are grain boundaries. Even the large angle boundaries can be
avoided easily, the small angle ones are often encountered. An exact interpretation of their formation has never been given. If they have a poor influence on the large size (channel length) MOS transistors electrical
characteristics, they lead to very large leakage currents for the smallest
ones (channel length less than 3 μm) [1]. This is a severe drawback if the
material is to be applied in advanced technology. Accordingly, a big effort
has been concentrated both on the technological ways to suppress the defects
from the active area of the transistor (lateral epitaxy, selective heating)
and on the understanding of their formation. To go farther in this understanding, one way is to look at the experimental factors influencing the appearance, density and shape of these defects. This work has been done in the
case of strip-heater heating where the molten zone is very wide [2]. We wanted to make a similar study in the laser case (small molten zone). The analysis in the laser case is more difficult : no natural texture of the film
is obtained and the standard (round) molten zone shape does not allow a uniform growth direction along the solid-liquid interface. Consequently, we decided to perform the experiments with special conditions : the laser beam
was focused into an elliptical elongated spot in order to obtain a roughly
constant orientation of the interface ; we have used the lateral epitaxy
technique and the defects have been investigated in the zone adjacent to the
seeding areas. This allows us to check the following parameters : the distance of appearance of the subgrainboundaries (SGB), their density and shape
versus the crystalline orientation of the substrate, the crystalline orientation of the growth direction (molten zone orientation) and other parameters like the growth velocity and the thicknesses of the layers.
The purpose of this paper was also to analyse these results with respect to the fundamental mechanisms which can explain the subboundary formation : stresses and the liquid-solid interface. Some experimental features
obtained on the strain in the film and on the fine interface shape are used
together with these results to confirm the hypothesis of Pinizzotto making
the stress induced by the local heating the major cause of the subboundary
formation.

EXPERIMENT

A conventional LOCOS process was used first on [100] silicon wafers with a 200 nm thick final oxide layer and patterns of different sizes and shapes. All the edges of these patterns were oriented parallel to a [110] direction. After a dilute HF solution etching in order to remove the native oxide layer, the wafers were rapidly loaded into a low pressure CVD reactor where silicon films 250, 500 and 1000 nm thick were deposited. A 20 nm thick thermal oxide was grown and a 20 nm thick nitride deposited as encapsulating layers for the recrystallization. This bi-layer encapsulating structure was preferred to the standard nitride monolayer since cracking of the nitride was often observed after recrystallization. The oxide layer prevents this difficulty.

The wafers were recrystallized using a coherent CR 18 Ar+ laser. The beam was transformed by an elliptical afocal telescope into a parallel 1 x 10 elliptical beam. A 200 mm focal length flat field lens focused this beam, so that a spot of about 15 x 150 μm was obtained on the sample. It was scanned by a galvanometric mirror system. The preheating of the wafers was varied between 200° to 600° C. The beam power was maintained in a narrow range to obtain melting on the seeding areas but not under the oxide of the isolated areas. The orientation of the major axis of the beam with respect to the crystal orientation has been changed either by a rotation of the telescope or of the sample. The crystalline quality of the film has been characterized using standard Schimmel etching.

RESULTS

Three stages of crystal growth of the film in the direction of the scanning can be distinguished, starting with the seeding edge (figure 1). A first region adjacent to this edge is defect free. At a distance D from the edge, extended accumulation of defects at least one micron wide appear (second stage). They correspond to the feather-like figures described by Lam and co-workers [3]. These figures have a well defined period, d_F, and length, D_F. They then transform into thin lines identified as subgrain boundaries. These latter are separated by a mean distance d. In fact, these boundaries can join together to form a single boundary. New boundaries can also appear farther into the film.

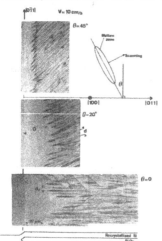

Fig. 1 : Micrographs of the silicon film near the seeding edge after defect etching (Schimmel) for three different growth directions

We systematically checked the influence of several experimental parameters on the average values of these characteristic distances. These parameters were the angle θ of the major axis of the beam with the [110] direction corresponding to the seeding edge, the thickness, e, of the silicon film, the scan velocity, and the substrate temperature. For a given thickness of the silicon layer, the angle θ has a major influence on these distances and on the shape of the defects (figure 1). If the major axis is close to a [010] direction (θ = 45°), the appearance distance D is small, the feather-like figures are close together and short. The subboundaries are also close together, and their period d is well defined. Their direction is [001] and stable. This means that appearance of new boundaries is infrequent. Twins often occur instead of subboundaries in this case. If the beam is [011] oriented (θ = 0), the D values are much larger, spreaded and sensitive to a small misorientation of the liquid-solid interface. The subboundaries appear after very long and wide feather-like figures. They are farther apart and are either roughly [011] oriented (perpendicular to the interface) or [001] oriented. Accordingly, their spacing is not well established. New feather-like figures and subboundaries appear frequently, even far from the seed.

The variations of the characteristic average distances with θ between 0 and 45° are given on figure 2 for a 500 nm thick silicon layer. A factor of 2 to 3 exists between the [010] and the [011] case.

If the angle θ is now fixed, the grain boundary appearance and density is greatly influenced by the silicon layer thickness, as can be seen on figure 3. The variation of the characteristic distances (D, d_F, d) is approximatively linear. Even the size of the feather-like figures increases as the layer thickness increases. The boundary structure, appearance and density for all thicknesses are similar to each other with a similarity ratio proportional to the film thickness.

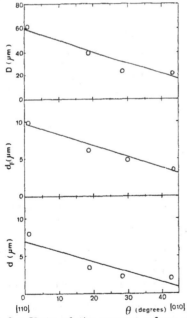

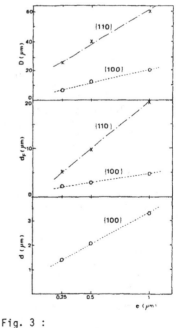

Fig. 2 : Plots of the average values of D, dF and d versus the growth direction θ . The parameters are defined on fig. 1

Fig. 3 :

D, d_F , and d versus the silicon film thickness, e

The scan (growth) velocity has a very weak influence on the boundary structure and on the characteristic distances regardless of the film thickness and the angle θ (figure 4). This can be seen for a very large range of velocity (from 0.3 to 700 mm/s).

The influence of the preheating is not obvious in the range we studied (200° C to 600° C). Nevertheless, we can see in figure 5, that if we add the results of other authors [4], it is clear that the subboundary density is roughly proportional to the difference (Tm - Tp) between the melting temperature of silicon and the preheating temperature. This dependance is obtained for a 500 nm thick layer and a [100] growth direction.

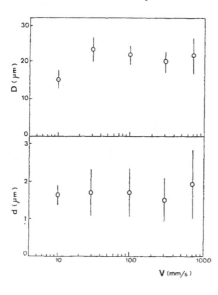

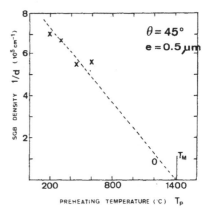

Fig. 5 :

Plot of the average value of d (SGB spacing) versus the preheating temperature Tp (x our results, 0 from ref. [4])

Fig. 4 : Plots of the average values of D and d versus the growth (scan) velocity V

DISCUSSION

The lack of influence of the growth velocity on the subboundaries spacing is somewhat surprising. In the experiments of metals or alloys solidification, this spacing is theoretically a decreasing function of the velocity ($V^{-1/2}$) [5]. The mechanism is related to the constitution supercooling phenomenon due to the presence of impurities in the material or to its binary nature. Geis and coworkers [2] in strip heater silicon layer recrystallization have found an increasing variation of the cell size with the velocity ($V^{1/2}$). In our case, the deposited film is very pure (cracking of silane). Even if impurities (oxygene, nitride) can be incorporated in the film during the melting from adjacent layers, the thermal gradient is so large at the interface that the constitution supercooling criterion cannot be satisfied. Accordingly, the cause of the boundary formation is probably not in the thermodynamic instability of the interface. The lack of influence of the velocity strongly confirms this point.

But the thermodynamic stability does not imply a perfectly flat interface, particularly in the silicon case. The covalent crystals like silicon or germanium are characterized by very strong differences in the kinetic growth behaviour of the different crystallographic faces. The (111) faces are the slow growth faces. The mechanism is known as lateral growth [6]. Steps one or several interplanar distances high nucleate and propagate to ensure the overall growth rate. The experimental (111) faceting of the interface confirms this growth mechanism in our case. This experimental evidence has been obtained previously in the strip heater case [2]. We have been able to obtain this evidence in the laser case by a rapid quenching of the molten zone near a seeding edge. The collapse of two opposite solidification fronts results in a defect which can be delineated by Schimmel etching (figure 6a). The reconstruction of the initial interface shape of each front confirms the (111) nature of the faces.   From these quenching experiments, several important features have been deduced :

. The faceting appears just after the seeding edge (figure 6b), before the subboundaries appear. (This can be seen when both collapsing interfaces have propagated a small distance from two opposite seeding edges).

. When the subboundaries appear, they are often correlated with the reentrant corners of the faceting (figure 6a), but it is not systematic. In the [110] case, even if they appear farther, subboundaries appear all the same and without interface faceting (figure 6c).

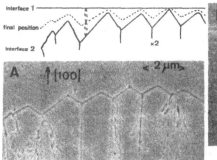

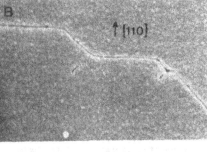

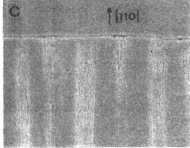

Fig. 6 : SEM micrographs of a recrystallized film after a rapid quenching of the molten zone

a : with grain boundaries on one side and a [100] growth direction

b : Without any grain boundaries on each side

c : With grain boundaries or feather-like figures on one side and a [110] growth direction

The following conclusions can be made from these experiments : the faceting alone does not explain the subboundary formation, since subboundaries and facets can exist separately. Nevertheless, a correlation exists between both of them. It is not surprising that defects like subboundaries are correlated with the (111) regular faces. It is well known, particularly from the dendritic growth of germanium and silicon [7], that defects like twins are more efficient as a source of growth front steps than two dimensional nucleation.

To go farther, we still need a hypothesis on the cause of subboundary formation. At the present time, the most probable explanation has been proposed by Pinizzotto [8]. The boundaries can be the results of the large stresses in the solid film near the interface due to local heating, and to the supported nature of the film (caping layers, underlying structure). An experimental confirmation of this large stress increasing from the seeding edge has been given by Zorabedian and Alter [9]. We have been able to confirm this point by looking carefully at the crystalline orientation of the film after the seeding by electron selective area diffraction.

Between the seeding edge and the subboundaries appearance, the crystalline orientation is surprisingly not constant. A rocking of a few degrees can be seen in about 20 μm. This demonstrates a very high strain in the film. This large strain which already exists in the solid film at the high temperatures near the solid-liquid interface during the recrystallization, is the consequence of a very large stress. This stress, increasing from the seeding edge can become so large after a certain distance D to be responsible for dislocation and subboundary formation. The induced SGBs release a part of the stress. The subboundary density would, therefore, be an increasing function of the overall stress. This assumption is confirmed by the behaviour of the subboundary structure with all the experimental factors : the overall stress is probably a decreasing function of the film thickness and of the preheating temperature, but does not depend on the scan velocity.

The behaviour of the subboundary structure with θ is more complicated. If the overall stress does not depend on θ , the development of the faceting probably leads to a strong modulation of the local stress which explains a variation of the appearance and of the density of the defects with θ .

CONCLUSION

We have demonstrated the strong influence of the silicon film thickness and of the growth direction on the appearance and repartition of the subgrain boundaries in the lateral epitaxy experiments. These defects appear farther from the seed and are more distant from each other, the closer the growth direction from a [110] orientation is and the thicker the film is. At the opposite, variation of the beam scan velocity (growth velocity) has a weak influence on the results. A high strain in the recrystallized film have been demonstrated. Experimental evidence of (111) faceting has been obtained from quenching experiments in both cases : with subboundaries and without subboundaries. These experiments confirm that the overall stress in the film is the major cause of the subboundaries formation.

From a practical point of view (microelectronics applications), the goal is to obtain the largest possible defect-free areas. From our studies with a standard film thickness (0,5 μm) and a standard preheating (500° C) 20 to 30 μm wide defect-free isolated islands or stripes can be obtained with the lateral epitaxy technique. But an elongated beam with the major axis parallel to a [110] orientation must be used instead of the standard round beam for which it is very difficult to obain constant growth direction.

Furthermore, the use of large scan velocity (30 cm/s without morphological problems) and of an elongated beam (100 to 200 μm long) leads to high processing rate (a four inch wafer in five minutes) with a low preheating temperature. This processing rate can be still improved with a higher preheating which will allow a longer molten zone with the available laser power. This demonstrates that the Ar+ laser can be an industrial tool for the SOI processing.

REFERENCES

[1] E.W. Maby, H.A. Atwater, A.L. Keigler and N.M. Johnson, Appl. Phys. Lett. 43, 482 (1983)

[2] M.W. Geis, H.I. Smith, B.Y. Tsaur, J.C.C. Fan, D.J. Silversmith and R.W. Mountain, J. Electrochem. Soc. 129, 2812 (1983)

[3] H.W. Lan, R.F. Pinizzotto and A.F. Tash, J. Electrochem. Soc. 128, 1981 (1981)

[4] D.B. Rensch and J.Y. Chen, IEEE Electron Device Letters EDL 5 (2), 38 (1984)

[5] G.F. Bolling and W.A. Tiller, J. Appl. Phys. 31, 2040 (1960)

[6] J.W. Cahn, W.B. Hillig and G.W. Sears, Acta Metallurgica 12, 1421 (1964)

[7] D.A. Davydov, V.N. Maslov, Sov. Phys. Cryst. 9, 393 (1965)

[8] R.F. Pinizzotto, J. Cryst. Growth, 63, 559 (1983)

[9] P. Zorabedian, F. Adar, Appl. Phys. Lett. 43, 177 (1983)

SUBBOUNDARY FREE SUBMICRONIC DEVICES

ON LASER-RECRYSTALLIZED SILICON ON INSULATOR

A.J. AUBERTON-HERVE, J.P. JOLY, J.M. HODE, and J.C. CASTAGNA
L.E.T.I. - C.E.A. - I.R.D.I. - Commissariat a l'Energie Atomique
LETI-CEN.G-85X-38041 GRENOBLE CEDEX-FRANCE

ABSTRACT

Seeding from bulk silicon (lateral epitaxy) has been used in Ar+ laser recrystallization to achieve subboundary free silicon on insulator areas. On these areas C.MOS devices have been performed using almost entirely the standard processing steps of a bulk micronic C-MOS technology. n - MOS transistors with channel length as small as 0.3 um have shown very small leakage currents. This is attributed especially to the lack of subboundaries. A 40 % increase in the dynamic performances in comparison with equivalent size C-MOS bulk devices has been obtained (93 ps of delay time per stage for a 101 stages ring oscillator with 0.8 μm of channel length). This is the best result presented so far on recrystallized SOI. No special requirements are needed in the lay out of the circuit with the chosen seed structure. Furthermore an industrial processing rate for the laser recrystallization processing has been achieved using an elliptical laser beam, a high scan velocity (30 cm/s) and a 100 μm line to line scan step (a 4' wafer in 4 minutes).

INTRODUCTION

The silicon on insulator (SOI) technology has several theoretical advantages on the conventional bulk approach for the VLSI MOS applications and especially in the C-MOS ones. The complete dielectric isolation between each individual transistor allows in principle :
. A closer spacing between the devices and hence a higher integration,
. A simplification of the process (no well formation),
. A definitive solution to the latch up problem.
Other advantages are coming from the reduction of the source and drain junction surfaces : radiation hardening, reduction of the junction capacitances leading to an improvement of the circuit speed.
But, to take full advantage fo these characteristics, several conditions have to be satisfied : the crystalline quality of the silicon layer must be as good as possible and close to the quality of the bulk wafers. The number of fixed and fast electrical states at the underlying interface and in the insulator must be as low as possible. If these conditions are not satisfied, the electrical characteristics of the MOS transistors can be greatly affected. Especially large leakage currents can be seen for the n-channel transistors.
Amongst the different SOI technologies, the zone melting recrystallization (ZMR) by lasers, strip-heaters, lamps ... has several promising features : the starting wafer is a silicon one (low price, no limitation in size), the underlying dielectric is thermal silicon dioxide. This latter is well known for its high quality in term of interface electrical states.
The crystalline quality of the recrystallized silicon film can be rather low on a macroscopic scale due to the presence of grain- or subgrainboundaries'(GB, SGB) but it is in fact very good on a microscopic scale apart from the boundaries as can be seen on the TEM micrograph of figure 1.

It has been demonstrated [1] that these extended defects can act as preferential diffusion path for the source and drain implants, so that tremen-
dous leakage currents appear if the channel length of the transistors is
smaller than about 3 microns.

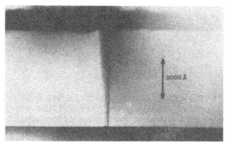

Figure 1 : TEM cross section of the recrystallized film with a
subgrainboundary

Accordingly, special efforts have been made to obtain GB free active
areas. Two techniques have been proposed : lateral epitaxy [2] and selective
heating [3].

In the lateral epitaxy technique, the silicon substrate is used as a
seed through openings in the underlying oxide. A SOI stripe adjacent to the
seeding edge is free from any SGBs. We have performed a systematic study of
the influence of different parameters on this phenomenon. The results of
this study are published in the same session. From a practical point of
view, the following conclusions can be deduced from this work : the SGBs appear further from the seeding edge if the growth direction is close to a
[110] direction and if the silicon film is thick. For a 0.5 μm thick layer,
a maximum 30 μm width of defect free area can be obtained.

The aim of this paper was to test if, using the lateral epitaxy technique, the ZMR can be suitable for an advanced technology. More precisely, we
wanted to verify if low leakage can be achieved even on small channel length
transistors. We wanted also to demonstrate that the defect free active SOI
areas can be obtained without special design rules induced by the seeds and
furthermore, that this laser technique is compatible with an industrial processing rate.

EXPERIMENTAL

We used a C-MOS device vehicle test designed to study an advanced micronic bulk technology. Individual p and n channel MOS transistors (with channel length running from 0.3 to 3 μm and channel width running from 5 to 20
μm), 101 stages, C-MOS ring oscillators (with 0.8 μm channel length p and n
channel transistors) can be processed with this vehicle. An additional level
of masking has been obtained from the standard field oxide level by a simple
inversion of polarity and a small oversize. This mask has been used to perform a first localized 200 nm thick oxydation on the (100) p-type 6-8 Ω.cm
silicon wafers. This oxide is used as the underlying isolation dielectric
(figure 2). After etching of the LOCOS masking layers, a 500 nm thick polysilicon layer has been deposited by LPCVD. A double layer of polysilicon
oxide (20 nm thick) and deposited LPCVD nitride (20 nm) have been used as
capping layers for the recrystallization. This operation was achieved using
an elliptical Ar+ C-W laser beam of about 20 x 100 μm2 in size. The beam was
scanned perpendicularly to the major axis of the elliptical spot, with a velocity of about 30 cm/s. The wafers were preheated at 500° C. The molten zo-

ne was approximatively 90 μm wide on the bulk areas. Accordingly, a 70 μm step has been applied on the raster scanning. The major axis was [110] oriented. This orientation corresponds to the first oxide edge patterns (seeding edge) orientation. The power of the laser was adjusted in order to melt the silicon layer and a fine top layer of the substrate on the non-isolated (seeding) areas but not to melt the substrate under the oxide. This allows a suitable lateral epitaxy without detrimental effects.

The crystalline quality of a typical SOI active area after this treatment can be checked on figure 3 after defect (Schimmel) etching. It can be seen that subgrainboundaries have been effectively avoided. This is the case if the SOI area does not exceed 30 μm in size in the direction of the scanning. Almost all the active transistors areas are in this case in an advanced technology. Residual defects (dislocations) can be seen along the edge of the SOI area. They probably come from the difference in the thermal behaviour between the SOI and the bulk areas. Individual dislocations (slip lines) can also be seen in the bulk areas.

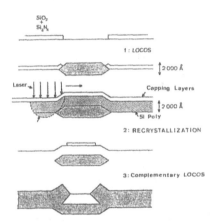

Figure 2 : The specific processing steps of the SOI technology

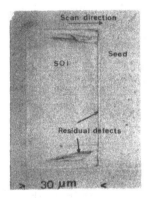

Figure 3 : The crystalline quality of (and near) an SOI island (after Schimmel etching)

After removal of the caping layer and thinning of the silicon layer from 500 nm to 180 nm by dry-etching, a second complementary LOCOS (600 nm thick) corresponding to the standard field oxide has been performed. It allows a complete dielectric isolation of the silicon island from each other (figure 2). Thereafter, all the steps of a C-MOS technology have been performed except those corresponding to the well formation and to the field implantation (gate oxide thickness : 25 nm, poly Si gate : 400 nm).

The following channel doping conditions have been applied :

. A Boron implantation with a peak at the back interface for the N-MOS (figure 4A),

. No implantation (natural p type doping after recrystallization) or a double implantation of Boron and Phosphorus leading to a buried junction (figure 4b). Accordingly, two kinds of P-MOS transistors have been obtained : deep depletion and weakly depletion mode transistors.

RESULTS

All types of transistors exhibit very good characteristics. The leakage currents even for the very small channel length n-MOS transistors are very low (0.2 pA/µm of channel width) (see figure 5). This important and new result for the transistors fabricated on recrystallized films can be interpreted in the following way :

. Since the subgrainboundaries have been avoided, no preferential doping paths are present in the film. Furthermore, the temperature of the MOS processing has been kept low (850° C).
. There is almost no parasitic channel at the back interface. This is due to a moderate fixed charge density at this interface and to the deep Boron implant.

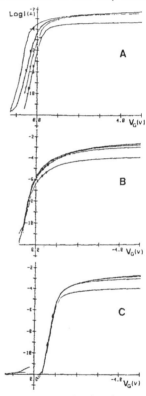

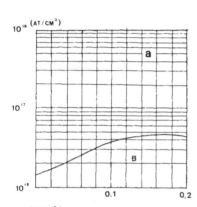

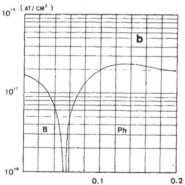

Figure 4 : Dopant profiles deduced from SUPREM simulation for

a) N-MOS transistors

b) Weakly depleted P-MOS transistors

Figure 5 : Characteristics Log ID(VG)

a) N-MOS transistor (L = 0.3 µm)

b) Weakly depleted P-MOS transistors (L = 1 µm)

c) Deep depleted P-MOS transistors (L = 1 µm)

To obtain more precisely this charge density, we have measured the variation of the upper transistor VT$_u$ versus back gate polarization and the variation of the back transistor VT$_b$ versus upper gate polarization in the deep depleted P-MOS transistors (see figure 6).

If we assume that the saturation value of the threshold voltage VTs in the curve corresponds to the complete depletion of the layers and that the fixed charges in the upper oxide are negligible with respect to the back oxide ones, one obtains :

$$(V_T)_{b.s} = (\Phi_{MS})_b - \frac{Q_F}{(C_{ox})_b} - \frac{qWN_A}{(C_{ox})_b}$$

$$(V_T)_{u.s} = (\Phi_{MS})_u - \frac{qWN_A}{(C_{ox})_u}$$

Where the superscript b and u correspond respectively to the back and upper channel, Φ_{ms} is the work function difference, Cox the oxide capacitance, W the thickness of the silicon layer. According to these equations, N$_A$ (the natural doping) and Q$_F$ (the back fixed charge density) can be calculated and we obtain N$_A$=810^{15} /cm3 and Q$_F$ = 10^{11} /cm2.

From the subthreshold characteristics of the back transistor, an interface fast state density of about 610^{11} /cm2/eV is obtained. These values agree quite well with other results obtained on recrystallized films structures [4].

Typical threshold voltages of 0.4 V for N-MOS, - 0.4 V for weakly depleted PMOS and - 0.7 V for deep depleted PMOS have been measured for the upper transistors (with no back bias).

The 101 stage C-MOS ring oscillators with the weakly depleted P-MOS transistors exhibit very interesting characteristics. Delay per stage of 93 ps under 5 V have been measured. Both types of transistors have an effective channel length of 0.8 μm and an effective channel width of 5 μm. Such velocity device performances are the best published so far on recrystallized films.

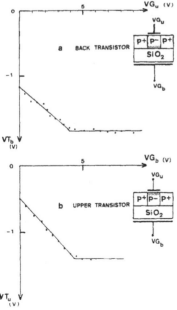

Figure 6 : Variation of the back transistor threshold voltage versus the upper gate bias (A), variation of the upper transistor threshold voltage versus the substrate (back gate) bias (B) in the case of the deep depleted transistors.

CONCLUSION

From these results, the feasability of submicronic devices using the technique of zone melting recrystallization has been established. Very low leakage currents have been obtained on these devices. This is particularly due to the lack of subgrainboundaries in the active areas. We have demonstrated that the seed openings can be adjusted to the active areas so that no special design rules are required. Furthermore, an industrial processing rate can be obtained using an elliptical LASER beam.

Of course, these results have to be obtained on a larger scale to confirm the validity of the technique. An effort must be made to avoid the dislocation formation in the seeding areas. Even if these defects do not affect the transistors characteristics, they can induce wafer deformation, creating adversely problems in a complete MOS process.

REFERENCES

[1] E.W. Maby, H.A. Atwater, A.L. Keigler and N.H. Johnson, Appl. Phys. Lett. 43, 482 (1983)

[2] H.W. Lam, R.F. Pinizzotto and A.F. Tasch, Jr, J. Electrochem. Soc, 128, 1981 (1981)

[3] J.P. Colinge, E. Demoulin, D. Bensahel and G. Auvert Appl. Phys. Lett. 41, 346 (1982)

[4] T.I. Kamins, Technical Digest of the 1982 IEDM p 420

RECENT ADVANCES IN Si AND Ge ZONE-MELTING RECRYSTALLIZATION

C. K. CHEN, M. W. GEIS, H. K. CHOI, B-Y. TSAUR, AND JOHN C. C. FAN
Lincoln Laboratory, Massachusetts Institute of Technology
Lexington, Massachusetts 02173-0073

ABSTRACT

By improving the thermal uniformity and stability of our graphite-strip-heater oven, we have been able to significantly improve the overall quality of ZMR Si films. We have observed unbranched subboundaries and new types of defects that are less extended than the usual sub-boundaries. The overall wafer flatness has been improved so that total warp is less than 4 μm for 3-inch wafers. We have also utilized the ZMR technique for producing thin Ge-on-insulator films.

INTRODUCTION

In recent years, substantial efforts have been directed toward the development of a technology for producing high-quality single-crystal semiconductor films on insulating substrates [1-4]. These efforts have been motivated by the potential of thin-film devices for achieving higher packing density, speed, and radiation resistance than bulk devices, and by the potential of Si-on-insulator (SOI) structures for accomplishing the three-dimensional integration of electronic circuits. Of the several SOI approaches currently under investigation [5], we have developed one of these, zone-melting recrystallization (ZMR) using graphite strip heaters, to the point where 2- and 3-inch-diameter wafers can be recrystallized over their entire surface [6-8], and we have used the recrystallized SOI material to fabricate majority carrier devices and integrated circuit chips with properties comparable to those of analogous bulk devices [9,10]. However, before the full potential of SOI technology using graphite strip heaters can be realized, techniques must be developed to eliminate or minimize the remaining material problems. In this paper we will discuss subboundaries and other less extended defects that are present in recrystallized SOI films, and report recent results on the flatness of recrystallized wafers. We will then discuss the ZMR of Ge-on-insulator films by the graphite-strip-heater technique.

THERMAL UNIFORMITY

Several features have been incorporated in the graphite-strip-heater oven to permit uniform heating of wafers to 1150°C with a center-to-edge variation of only 5°C, and to achieve improved thermal stability of the entire system and the liquid-solid interface in particular. A graphite heat sink is used between the wafer and lower strip heater to provide a large thermal mass for minimizing fluctuations in the base temperature of the wafer and to provide more uniform heating of the wafer via radiative coupling. Both the upper and lower strip heaters are mounted in a manner permitting free thermal expansion without bowing, since bowing would prevent uniform recrystallization across the entire surface of 3-inch wafers.

DEFECTS

The most important imperfections remaining in recrystallized Si films are low-angle subgrain boundaries or subboundaries. As discussed in detail by Geis, et al. [11], the improvements in the overall thermal stability and uniformity of the graphite-strip-heater system have resulted in major advances in our effort to eliminate subboundaries. We will summarize these developments briefly here. Figure 1 shows Nomarski micrographs of two recrystallized SOI films that have been defect etched to reveal the subboundaries. The subboundaries in Fig. 1(a), which was obtained for an SOI wafer in which the Si film and underlying SiO₂ layer are each ~0.5 μm thick, are branched. A dramatically different morphology is shown in Fig. 1(b), which was obtained for a wafer in which the Si film and underlying SiO₂ layer are ~1 μm thick. Here the subboundaries are straight and unbranched, as a result of the extreme thermal stability of the liquid-solid interface during recrystallization. These subboundaries are also more widely spaced, indicating that the thermal gradient was lower [12]. Until recently

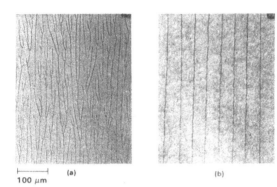

Fig. 1. Optical micrographs of (a) branched and (b) unbranched subboundaries in recrystallized SOI films.

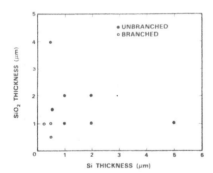

Fig. 2. Regions of branched and unbranched subboundaries as a function of Si and SiO₂ thickness.

such unbranched subboundaries were rarely observed in films less than 5 μm thick. Now, however, with the improvements in the overall thermal uniformity and stability of the graphite-strip-heater oven, we routinely obtain this morphology over most of the area of films as thin as 1 μm and have even observed it in some 0.5-μm films. Recent results on the dependence of subboundary morphology on the thickness of the Si film and the SiO₂ layer are summarized in Fig. 2. Increasing the Si thickness increases the heat of fusion released and the resistance to lateral heat flow, while increasing the SiO₂ thickness reduces flow to the substrate. These effects tend to reduce the thermal gradients and temperature fluctuations during recrystallization, favoring the formation of unbranched subboundaries in thicker films.

An even more significant development described by Geis, et al. [11] is illustrated by Fig. 3, which is a Nomarski micrograph of a recrystallized film 2 µm thick that was defect etched. As shown by TEM, the subboundaries have been replaced by less extended defects, either dislocation clusters arranged in linear arrays or individual dislocations forming diffuse bands, and the remaining area is nearly defect-free. Furthermore, the individual dislocations can be removed by annealing, although the clusters appear to be thermally stable. So far we have observed these new defects only in limited areas of 2-µm films, but we are now hopeful that further improvements in the strip-heater technique will make it possible to entirely eliminate the subboundaries even from 0.5-µm films.

Fig. 3. Optical micrograph of a recrystallized film showing un-branched subboundaries and other less extended defects.

BARE Si   RECRYSTALLIZED SOI

Fig. 4. Interference patterns, with a 4 µm/fringe sensitivity, for 3-inch (a) bare Si wafer and (b) recrystallized SOI wafer.

WAFER FLATNESS

Having discussed the crystallographic defects in ZMR SOI films, we shall next consider the flatness of SOI wafers. Stringent flatness requirements are imposed by the photolithographic process used in VLSI fabrication. The ZMR wafers can now meet the requirements for fabricating circuits with ~2 µm geometries. Figure 4 shows interference patterns obtained with a high-angle-of-incidence laser flatness monitor for a bare Si wafer, on the left, and an SOI wafer after recrystallization, on the right. These patterns, which were measured with the wafers mounted on a vacuum chuck, provide a topographic map of the wafer surface in which each interference fringe corresponds to a height of ~4 µm. The recrystallized wafer, like the bare Si wafer, is found to have a total warp of less than 4 µm, peak to valley. Without the vacuum chuck, the free standing bow of a recrystallized 3-inch SOI wafer ranges up to 80 µm. A significant reduction in warp has been achieved over the past year by means of a number of changes in ZMR technique. These include the improvement discussed above in the thermal uniformity of the graphite-strip-heater system, a reduction in the heating and cooling rates before and after ZMR to ~4 °C/sec, the use of 0.020-inch thick Si substrates to provide increased mechanical strength, and an increase in the scan speed, which reduces the time that the wafer is exposed to radiation from the upper heater.

Although the problem of warp has been essentially solved, after ZMR the substrates show a small amount of slip. However, the slip planes, which result in 1000 Å ripples in the substrate, do not produce any corresponding thickness variation or defects in the Si films and do not pose a problem for device fabrication.

616

As a further step in our utilization of the graphite-strip-heater technique for zone-melting recrystallization, we have recently initiated an effort to determine the usefulness of the technique for growing high-qualtiy single-crystal Ge films on SiO₂-coated Si wafers [13]. By providing

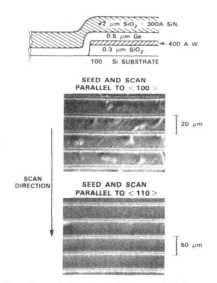

substrates for epitaxial growth of GaAs, the growth of such films could make a major contribution to monolithic GaAs/Si (MGS) device integration. Fig. 5(a) shows the sample structure that has yielded the best overall surface morphology to date. A thin W film is deposited by e-beam evaporation on an SiO₂-coated <100> Si substrate. Stripe openings for seeding are then lithographically defined and formed by selectively etching away the W and SiO₂ to expose the Si substrate. A polycrystalline layer of Ge is deposited by e-beam evaporation, followed by capping layers of sputtered SiO₂ and Si₃N₄. The W buffer layer and the capping SiO₂ and Si₃N₄ layers are necessary to insure wetting by the Ge film and to obtain smooth surface morphology.

The effect of seed and scan orientation on the surface morphology of the ZMR Ge films is shown in Fig. 5 by Nomarski micrographs of two recrystallized samples that were defect etched. The upper sample, with seed and scan direction parallel to <100> directions of the Si substrate, contains many microtwins, which are symptomatic of a growth front with {111} facets, as has been observed for Si. This sample does not contain a tungsten buffer layer and hence is more susceptible to cap failure and agglomeration of the Ge film. The lower sample, with seed and scan direction parallel to <110> and with a tungsten buffer layer, shows a very smooth surface morphology with very few defects. Our best results to date have been achieved for 50 μm square Ge/W/SiO₂ islands with edges parallel to <110> directions. Figure 6 shows low- and high-magnification Nomarski micrographs of such a sample after ZMR and defect etching. A thickness variation of ~200 Å was observed within the islands. The surface showed a slight texturing, and careful inspection revealed a few low-angle grain boundaries. The individual grains are more clearly delineated by scanning electron microscopy, as shown in Fig. 7. We have verified by x-ray and electron diffraction that the Ge films have the <100> orientation of the substrate, and no W is detected in the Ge films by Auger analysis, which has a detection limit of ~1%.

Fig. 5. (a) Seeded Ge-on-insulator sample structure. (b) Optical micrographs of Ge-on-insulator wafers after ZMR with seed and scan parallel to <100> and <110> directions.

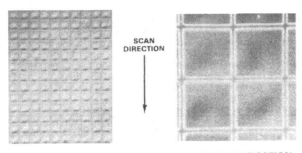

SCAN DIRECTION

LOW MAGNIFICATION   **HIGH MAGNIFICATION**

Fig. 6. Nomarski micrographs of seeded 50-μm Ge-on-insulator islands after ZMR and defect etching.

Fig. 7. Scanning electron micrograph of seeded 50-μm Ge-on-insulator islands after ZMR.

SUMMARY

As a result of improvements in the thermal uniformity and stability of the graphite-strip-heater system, unbranched subboundaries have been obtained in recrystallized Si films as thin as 0.5 μm, and less extended defects have been formed in 2-μm films. The unbranched subboundaries are indicative of extreme thermal stability at the liquid-solid interface during ZMR, and the new defects signal a major advance in the effort to eliminate subboundaries altogether. We have prepared recrystallized 3-inch SOI wafers with a total peak-to-valley warp of less than 4 μm, the same as before ZMR. Finally, we have described the structure and seed orientation used to obtain smooth surface morphology and good crystal alignment for ZMR Ge-on-insulator films.

ACKNOWLEDGEMENTS

The authors would like to acknowledge the expert technical assistance of M. J. Button, M. C. Finn, M. N. Gilson, R. W. Mountain, and P. M. Nitishin, and fruitful discussions with A. J. Strauss. This work was sponsored by the Department of the Air Force and the Defense Advanced Research Projects Agency.

618

REFERENCES

1. <u>Laser and Electron Beam Processing of Materials,</u> edited by C. W. White and P. S. Peercy (Academic Press, New York, 1980).
2. <u>Laser and Electron Beam-Solid Interactions and Material Processing,</u> edited by J. F. Gibbons, L. D. Hess, and T. W. Sigmon, Materials Research Society Symposia, vol. 1 (North Holland, New York, 1981).
3. <u>Laser and Electron Beam Interactions with Solids,</u> edited by B. R. Appleton and G. K. Celler, Materials Research Society Symposia, vol. 4 (North Holland, New York, 1982).
4. <u>Laser-Solid Interactions and Transient Thermal Processing of Materials,</u> edited by J. Narayan, W. L. Brown, and R. A. Lemons, Materials Research Scoiety Symposia Proceedings, vol. 13 (North Holland, New York, 1983).
5. See J. Cryst. Growth <u>63</u> (3), (1983), an issue devoted to SOI technology.
6. J. C. C. Fan, M. W. Geis, B-Y. Tsaur, IEDM Technical Digest (Washington, DC, 1980), p. 845; M. W. Geis, H. I. Smith, D. A. Antoniadis, D. J. Silversmith, J. C. C. Fan, and B-Y. Tsaur, Electronic Materials Conference, Santa Barbara, CA, 1981; and E. W. Maby, M. W. Geis, Y. L. LeCoz, D. J. Silversmith, R. W. Mountain, and D. A. Antoniadis, Electron Dev. Lett. EDL-2, 241 (1981).
7. M. W. Geis, H. I. Smith, B-Y. Tsaur, J. C. C. Fan, D. J. Silversmith, and R. W. Mountain, J. Electrochem. Soc. <u>129</u>, 2812 (1982).
8. J. C. C. Fan, B-Y. Tsaur, R. L. Chapman, and M. W. Geis, Appl. Phys. Lett. <u>41</u>, 186 (1982), and references therein.
9. B-Y. Tsaur, M. W. Geis, J. C. C. Fan, D. J. Silversmith, and R. W. Mountain, Appl. Phys. Lett. <u>39</u>, 909 (1981).
10. B-Y. Tsaur, J. C. C. Fan, R. L. Chapman, M. W. Geis, D. J. Silversmith, and R. W. Mountain, IEEE Electron Dev. Lett. <u>EDL-3</u>, 398 (1982).
11. M. W. Geis, C. K. Chen, H. I. Smith, R. W. Mountain, and C. L. Doherty, paper A7.2, Materials Research Society, 1984 Fall Meeting, Boston, MA.
12. M. W. Geis, H. I. Smith, D. J. Silversmith, and R. W. Mountain, J. Electrochem. Soc. <u>130</u> 1178 (1983).
13. For results of similar approaches, see M. Takai, T. Tanigawa, K. Gamo, and S. Namba, Japn. J. Appl. Phys. <u>22</u>, L624 (1983) and T. Nishioka, Y. Shinoda, and Y. Ohmachi, J. Appl. Phys. <u>56</u>, 336 (1984).

SI-ON-INSULATOR FORMATION USING A LINE-SOURCE ELECTRON BEAM

J. A. Knapp, Sandia National Laboratories, Albuquerque, NM 87185, USA

ABSTRACT

A line-source electron beam has been used to melt and recrystallize isolated Si layers to form Si-on-Insulator structures, and the process is simulated by heat flow calculations. Using sample sweep speeds of 100-600 cm/s and peak power densities up to 75 kW/cm^2 in the 1 x 20 mm beam, we have obtained single-crystal areas as large as 50 x 350 μm. Seed openings to the substrate are used to control the orientation of the regrowth and the heat flow in the recrystallizing film. A finite-element heat flow code has been developed which correctly simulates the experimental results and which allows the calculation of untried sample configurations.

INTRODUCTION

Several laboratories have been exploring the use of directed energy sources, both line and spot shaped, to melt and recrystallize isolated Si layers to form large-grain or single-crystal Si-on-Insulator structures (SOI).[1] There are several possible advantages to a melt/recrystallization approach using e-beams at high power densities and sweep speeds, and we and other groups have been exploring their use.[2-6] Our work has been with a line-source electron beam, using sample sweep speeds of 100-600 cm/s and peak power densities up to 75 kW/cm^2.[2,3] The sample spends a relatively short time at high temperature, minimizing thermal effects such as wafer contamination or distortion. Thermal gradients near the surface are steeper at high sweep speeds, which may allow a wider window of processing parameters between melting of the isolated Si layer and melting of the substrate. Of course, the use of a line shaped beam also eliminates the problems associated with overlapping of rastered spot beams.

In forming SOI at high sweep speeds, it has been necessary to use a seeded configuration to control the film crystallographic orientation through openings in the isolation layer.[3] For our experiments, seed openings are arranged as parallel stripes along a <110> direction, spaced from 5 to 200 μm apart. Since thermal conduction through the seed openings is much higher than through the isolation oxide, the seed opening geometry also serves to control the local heat flow and hence the pattern of recrystallization. By sweeping the beam either along a <110> direction, parallel to the seed openings, or at an angle of up to 45°, smooth regrowth is obtained, with no protrusions and only one low-angle grain boundary midway between seed stripes. A finite-element heat flow code has been developed in order to better understand the experiment and to simulate other sample or beam configurations. Computations using this code correctly predict the observed lateral regrowth when sweeping the beam along the seed openings. We are presently using the code to optimize the sample structures for larger areas of seeded growth.

An example of the experimental results, together with the corresponding numerical simulation, is presented here. A more extensive paper, summarizing both the experiments and the calculations, will be published later.[7]

This work performed and Sandia National Laboratories supported by the U.S. Department of Energy under Contract # DE-AC04-76DP00789.

RESULTS

Samples consist of 4" Si<100> wafers, on which are deposited isolation layers of either 1 or 2 μm SiO_2. Polycrystalline Si is deposited in layers of 0.5, 1.0, or 2.0 μm thickness, followed by a capping layer of 1 or 2 μm of SiO_2. Some samples have an additional cap of 50 nm Si_3N_4. The electron beam system is described in detail elsewhere.[2,7-8] The beam is ~ 2 cm long, with a measured width of 1.0-1.2 mm full width at half maximum. For the SOI studies, beam energies of 30-38 keV are used, with peak power densities of up to 75 kW/cm^2. Samples are heated in situ to 400-600°C, while being swept under the beam at speeds up to 600 cm/s.

Figure 1(a) shows an optical micrograph of a sample treated with a 62 kW/cm^2 beam swept at 396 cm/s along the direction of the seed openings. The long stripes in the micrograph are the seed openings to the substrate. The isolation oxide is 1 μm thick, with a 1 μm thick poly Si layer and a 2 μm thick SiO_2 capping layer. In the micrograph, the seed openings are 50 μm apart, and the resulting lateral regrowth is smooth, since single-crystal regrowth has occurred. This effect is observed for seed stripe spacings of 50 μm or less, but not for spacings of 100 μm or more.

Panel (b) of the figure shows another area of the same sample after removal of the cap and treatment with a Secco etch.[9] to delineate the defects and grain boundaries. In the lower part of the micrograph, the seed openings are again 50 μm apart, and a low-angle grain boundary is seen midway between stripes. In the upper part of micrograph (b), away from the seed stripes, protrusions and grain boundaries appear which are indicative that the regrowth was randomly oriented. The protrusions are due to the trapping of small pockets of liquid Si at the intersection of grains which are solidifying: as these pockets recrystallize, the only avenue for their expansion is upward through the capping layer.

The heat flow during experiments such as shown in Fig. 1(a) and (b) is calculated using the classical explicit method to solve a finite-element approximation to the heat flow equation. A cross-section of the sample is divided into small sub-volumes (usually 1x1 μm near the surface), each of which can be either Si or SiO_2, with the appropriate temperature dependent heat capacities and thermal conductivities. The phase change in Si is accounted for, as well as the depth distribution of the e-beam energy deposition. Recrystallization is assigned in the calculation as the point at which a molten element has lost the heat of fusion. No attempt is made to model nucleation effects or the undercooling of the recrystallizing Si.

By doing the calculation for the two-dimensional YZ plane perpendicular to the seed stripes and parallel to the long axis of the beam, the effect of scanning a gaussian-shaped line beam along the X direction can be approximated by a spatially uniform beam with a gaussian time dependence. Since even at 1 mm the beam is much wider than the typical thermal diffusion length on these time scales, thermal diffusion along the direction of the beam sweep can be neglected. The heat flow is strongly dominated by flow into the substrate (the Z direction) and, for short distances near seed openings, perpendicular to the seeds (the Y direction). By assigning the YZ plane of temperature for each time step to an X position equal to velocity x time, the calculation serves as a very good approximation to a full three-dimensional heat flow solution of a steady state scanning beam.

Figure 1(c) shows the results of such a calculation for the experimental results in Fig.1(b). The figure plots the position of the recrystallizing Si front in the isolated layer as a function of position behind a steady-state beam scanning at 396 cm/s. Examination of the figure clearly shows that the layer freezes first at the seed openings, with the front proceeding out from the seed for 20-25 μm. At distances sufficiently far from the seeds, the heat loss through the underlying oxide is enough for the layer to begin freezing before the front reaches that region. These

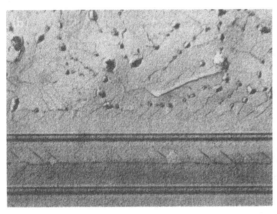

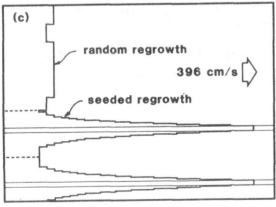

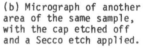

(c)

random regrowth

396 cm/s

seeded regrowth

100 um

Fig.1 (a) Nomarski contrast optical micrograph of an SOI sample treated with a 62 kW/cm^2 beam swept at 396 cm/s. The sample structure was 1 μm isolation/1 μm poly-Si/2 μm cap. The seed openings are 5 μm wide. Substrate temperature was 400°C.

(b) Micrograph of another area of the same sample, with the cap etched off and a Secco etch applied.

(c) Calculation of the recrystallization front for the experiment of (b).

areas would show randomly oriented regrowth while the areas near the seeds would exhibit single-crystal regrowth. The dashed lines mark the last elements to recrystallize, and would be an approximate boundary between random and seeded regrowth. Where two fronts meet between seeds, the dashed line would indicate where a single low-angle grain boundary is expected. The correspondence between the calculated features of Fig. 1(c) and the experiment of Fig.1(b) is quite good and gives confidence that the code correctly models the heat flow.

SUMMARY

A finite-element solution to the heat flow in layered, seeded SOI structures has been developed and applied to our line-source e-beam treatment of such structures. The predictions of seeded lateral regrowth of the isolated Si layer match the experimental observations. More extensive details of the calculation and experimental results will be presented elsewhere.[7]

ACKNOWLEDGEMENTS

Discussions with S. T. Picraux were invaluable. The help of B. L. Draper in wafer and mask preparation was essential. Technical assistance by R. E. Asbill is gratefully acknowledged.

REFERENCES

[1] See, for example, the papers in Energy Beam-Solid Interactions and Transient Thermal Processing, J. C. C. Fan and N. M. Johnson, eds. (North Holland, NY 1984).

[2] J. A. Knapp and S. T. Picraux, Laser-Solid Interactions and Thermal Processing of Materials, J. Narayan, W. L. Brown, and R. A. Lemons, eds. (North Holland, NY 1983), p.557.

[3] J. A. Knapp and S. T. Picraux, ref. 1, p.533.

[4] J. R. Davis, R. A. McMahon and H. Ahmed, ref. 2, p.563.

[5] T. Inoue, K. Shibata, K. Kato, T. Yoshii, I. Higashinakagawa, K. Taniguchi, and M. Kashiwagi, ref. 1, p. 523.

[6] Y. Hayafuji, T. Yanada, H. Hayashi, K. E. Williams, S. Usui, S. Kawado, A. Shibata, N. Watanabe, and M. Kikuchi, ref. 1, p. 491.

[7] J. A. Knapp and S. T. Picraux, to be published.

[8] J. A. Knapp and S. T. Picraux, J. Appl. Phys. 53, 1492 (1982).

[9] F. Secco d'Aragona, J. Electrochem. Soc. 119, 948 (1972).

COMPUTER MODELING OF LATERAL EPITAXIAL
GROWTH OVER OXIDE

LYNN O. WILSON AND G. K. CELLER
AT&T Bell Laboratories, 600 Mountain Ave., Murray Hill, NJ 07974

ABSTRACT

Radiative melting of Si with an extended stationary heater and its crystallization are modeled numerically. The two-dimensional model provides the velocity and shape of the solid-liquid interface, above and below a buried oxide structure with slit-shaped openings. The results show qualitative agreement with the experimental data. Superheating and undercooling are included in the calculation and the amount of superheating is confirmed experimentally.

INTRODUCTION

Controlled crystallization of radiatively melted silicon films has been used extensively to form single crystalline layers on insulating substrates [1]. In parallel with these experiments, there have been some attempts to model temperature profiles induced in silicon with pulsed or cw lasers and with scanned line sources. Most published calculations are for a one-dimensional case, and the two-dimensional solutions do not include motion of the solid-liquid interface [2-4].

Here we model temperature distribution and motion of the melt front in a two-dimensional domain. Our solution is valid for the case of uniform high intensity irradiation of a thick Si film deposited over an oxidized Si wafer [5]. Narrow slits in the buried oxide are required for epitaxial growth, affecting strongly the heat flow and necessitating the two-dimensional solution. A second SiO_2 layer covers uniformly the silicon surface. Its presence is included in the value of the surface emissivity.

FORMULATION

The computational domain in Figure 1 represents a cross-section of a portion of the wafer. The domain can be reflected about its right edge and the structure repeated, to give a periodic array with oxide layers of half width $w1$ and seeding windows of half width $w2$.

We assume that heat transport within the wafer is primarily due to conduction, that heat input from the lamps is via coupling at the wafer top surface, and that heat loss is via radiation at the top and bottom surfaces. The motion of the melt/solidification front $\mathbf{u}$ is governed by the difference in the heat fluxes in the liquid and solid silicon at the interface.

The governing non-dimensional equations are [6]

$$Pe \ \frac{\partial T}{\partial t} = \nabla \cdot (K \nabla T) , \tag{1}$$

$$\frac{d(\mathbf{n}_f \cdot \mathbf{u})}{dt} = \frac{R}{Pe \ S} \left[\mathbf{n}_f \cdot (K \nabla T) \Big|_{\mathbf{u}^+} - \mathbf{n}_f \cdot (K \nabla T) \Big|_{\mathbf{u}^-} \right] , \tag{2}$$

with boundary conditions

$$T(\mathbf{u}(\mathbf{x},t),t) = 0 \qquad \text{at the melt/solidification front } \mathbf{u}(\mathbf{x},t), \tag{3}$$

$$\mathbf{n} \cdot (K \nabla T) \quad \begin{aligned} &= 0 & &\text{at the sides} \quad x = 0, \ x = w1 + w2, \\ &= \epsilon[P(t) - (1+RT)^4] & &\text{at the top} \quad y = d1 + d2 + d3, \\ &= -\epsilon(1+RT)^4 & &\text{at the bottom} \quad y = 0, \end{aligned} \tag{4}$$

and the initial condition

$$\mathbf{u}(\mathbf{x}, 0) = \mathbf{u}_0(\mathbf{x}) . \tag{5}$$

$K(\mathbf{x}, t)$ is the dimensionless thermal conductivity, ϵ the emissivity, and $P(t)$ the power input from the oven. $Pe = \rho C d^2/(\kappa \tau)$ is the Péclet number, $R = \sigma d T_m^3/\kappa$ is the radiation number, and $S = L/(CT_m)$ is the Stefan number. Also, $\mathbf{n}$ represents the unit outward normal to a surface and $\mathbf{n}_f$ the unit normal at the front, pointing from the solid to the liquid.

Using values typical of silicon for the density $\rho = 2.41$ gm/cm^3, heat capacity $C = 1.40$ J/(gm·K), melting temperature $T_m = 1685°$K, and latent heat $L = 1810$ J/gm, plus the Stefan-Boltzmann constant $\sigma = 5.672\times10^{-12}$ J/(sec cm^2(·K)^4), and a characteristic length $d = 10\mu$m, time $\tau = 1$ sec, and thermal conductivity $\kappa = 0.1$ W/(cm·K), we obtained $Pe = 2.5\times10^{-5}$, $R \doteq 2.7\times10^{-4}$, and $S \doteq 1.03$. Accordingly, we simplified the non-linear two-dimensional Stefan problem (1)-(5) somewhat by neglecting the left hand side of (1) (quasi-static approximation) and ignoring the T^4 dependence in (4).

The resulting equations were solved numerically using a Cray-1. All results will be presented in a dimensional form. The dimensional temperature T is given by $T = T_m(1+RT)$.

ISOTHERMS

The temperature distribution in the structure depends upon the position of the melt or solidification front and the power input, among other things. Some characteristics of the thermal profile can be discussed in general, though. We do so here, but exclude cases in which the front intersects the oxide layer, as they may involve superheating or undercooling.

In these computations, we used the following dimensions for the domain: $d1 = 160\mu$m, $d2 = 4\mu$m, $d3 = 60\mu$m, $w1 = 96\mu$m, $w2 = 48\mu$m. The thermal conductivities for solid Si, liquid Si, and SiO_2 respectively were 0.216, 0.64, 0.02 W/(cm·K). The emissivities of the bottom and top (liquid) surfaces were $\epsilon_B = 0.62$ and $\epsilon_L = 0.58$. The latter value has been derived for a rough Si surface covered with a glassy layer that traps most of the incident light through total internal reflection [7,8]. The power was $P = 1.05P_2$, where $P_2 = (1+\epsilon_B/\epsilon_L)\sigma T_m^4$ is the threshold needed to maintain equilibrium with one surface fully molten [7].

Figure 2 shows an isotherm plot when the solid/liquid interface is above the oxide layer; Figure 3 shows a similar plot when the interface is assumed to be well below the oxide layer.

Notice that there is a steep temperature gradient across the oxide. The gradient is less steep in the solid silicon and is least steep in the liquid silicon. The isotherms are compressed in the oxide and expand within the window region. At the wafer top, the temperature is slightly greater over the oxide layer than it is over the window region. The wafer bottom has a nearly uniform temperature. This uniformity is even more pronounced if a thicker substrate ($d1$) is used in the calculations.

The length characterizing the distance over which the influence of the oxide layer can be felt significantly is slightly in excess of 50μm. The temperature changes by $\sim 1.2°$ over 100μm of solid silicon, $\sim 0.4°$ over 100μm of liquid silicon, and $\sim 0.6°$ across the 4μm of isolation oxide.

MOTION OF THE FRONT

We present calculations for the time dependent motion of the front. We assume, somewhat arbitrarily, that it is initially planar, 4μm below the top surface of the wafer. The power $P(t)$ is assumed to be held constant at a value $1.05 P_2$ for 17 seconds, then ramped down to zero linearly with time over the next 60 seconds.

Figures 4 and 5 form composite calculated pictures of the positions of the front at various instants of time during the melting and freezing processes, respectively.

Melting

Even if the melt starts as planar and parallel to the surface, it quickly readjusts itself so that it is deeper over the oxide. It then propagates downward toward the oxide with only slight additional steepening. The melt front contacts the top of the oxide strip at the left hand edge first (i.e., in the center of the full strip). The contact point quickly propagates to the right, toward the window, and the remaining portion of the front steepens. The front flips rapidly as it goes through the seeding window, changing shape from concave downward to concave upward. Accurate computations for this regime are exceedingly difficult to perform. The melt front then contacts the bottom of the oxide strip at the window. The contact point moves swiftly to the left, along the bottom of the strip. Below the oxide, the front is steeper than it was when it was above it. As the melt front continues to travel downward, it tends to flatten out.

Freezing

When freezing begins, the solidification front moves upward toward the oxide, steepening as it does so. After contacting the oxide, it does a quick flip through the window, essentially mimicking the behavior of

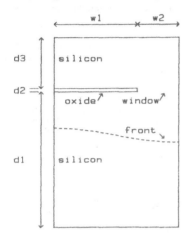

Fig. 1a: The computational domain.

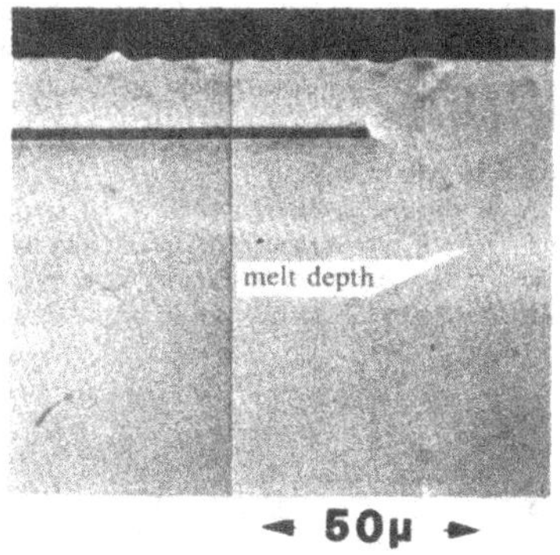

Fig. 1b: Optical micrograph of polished cross-section of recrystallized sample after 10 second Schimmel etch.

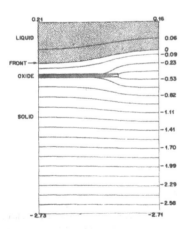

Fig. 2: Typical isotherm distribution when front is above the oxide. Numbers refer to deviation from T_m.

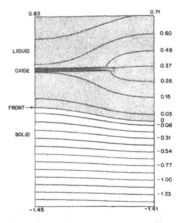

Fig. 3: Typical isotherm distribution when front is below the oxide. Numbers refer to deviation from T_m.

the melt front in reverse. The shape of the solidification front just above the oxide is similar to that of the melt front at a comparable position. The solidification front moves upward with very little change of shape.

Surface freezing

When the top surface starts freezing, the emissivity of the solidified portion jumps to a value $\epsilon_S > \epsilon_L$ (we used $\epsilon_S = 0.84$), and more power is coupled into the wafer. This tends to slow the motion of the front in the vicinity of the contact point with the surface. Mathematically, there can be an instability, the physical consequences of which have yet to be determined.

Rate of movement of front

A one-dimensional model of the front motion indicates that the front will move linearly downward with time when the power input is constant, and parabolically upward when the power is ramped down linearly. Although the behavior is far more complicated in this two-dimensional model, it conforms semi-quantitatively to the basic concepts of the simpler model.

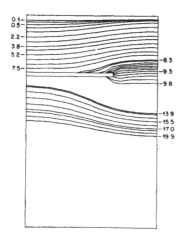

Fig. 4: Calculated melt front positions. Numbers refer to time in seconds.

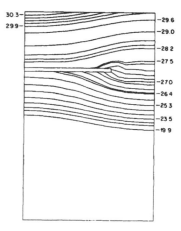

Fig. 5: Calculated solidification front positions. Numbers refer to time in seconds.

SUPERHEATING AND UNDERCOOLING

When the front contacts the oxide strip, the temperature profile in the structure may change radically, particularly as the front passes through the seeding window. A critical parameter is P/P_2.

Suppose that $P/P_2 > 1$, so that melting is taking place, and that the melt front is going through the seeding window. Power in excess of P_2 applied to the surface in the vicinity of the seeding window affects the advancement of the melt front. However, power applied more than a few tens of microns away from the surface above the window will have an insignificant effect on the front. If we suppose that melting below the oxide layer cannot be initiated at a position unattached to the present melt front, the power in excess of P_2 may go into superheating of a portion of the device below the oxide layer. The wider the oxide strip is, the greater the amount of superheating may be.

Conversely, if $P/P_2 < 1$, some of the silicon above the oxide layer may become undercooled during the freezing process.

Figures 6 and 7 are isotherm plots for a domain with $w1 = 600\mu m$, all other dimensions as before. In Figure 6, $P/P_2 = 1.05$, and the melt front is assumed to be planar, at the bottom of the seeding window. Note that a region under the oxide strip is superheated and that the amount of superheating increases with distance from the window. Conversely, in Figure 7, $P/P_2 = 0.95$, and the solidification front is assumed planar, at the top of the seeding window. A region over the oxide strip is undercooled. Since the thermal gradient in liquid silicon is approximately a factor of three smaller than that in solid silicon, the region affected by undercooling is larger.

We show in Figure 8 the magnitude of the superheating at the left edge of the structure, under the oxide, as a function of the half width $w1$ of the oxide strip. The superheating increases almost exponentially with the width of the strip. Results for undercooling are nearly symmetric.

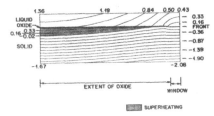

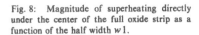

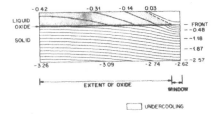

Fig. 6: Isotherms during superheating. The half width $w1$ of the oxide is $600\mu m$.

Fig. 7: Isotherms during undercooling. The half width $w1$ of the oxide is $600\mu m$.

Fig. 8: Magnitude of superheating directly under the center of the full oxide strip as a function of the half width $w1$.

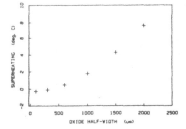

COMPARISON WITH EXPERIMENTS

It is not possible to monitor directly the interface motion. The maximum extent of the melt can be detected, however, by cross-sectioning and chemical staining of a recrystallized sample. One example is shown in Fig. 1(b). In Fig. 9 a sequence of maximum melt penetrations is shown for a series of irradiation energies. This is approximately equivalent to a series of positions during a single melt, and indeed the shapes of the experimental profiles are similar to the calculated ones in Fig. 4.

Superheating and undercooling were included in the model since we observed superheating experimentally [9] (we cannot directly detect undercooling). The most direct evidence came from Si wafers with the buried oxide half-width $w1 = 920\ \mu m$ over part of the area and $w1 = 1840\ \mu m$ over the rest of each wafer. After recrystallization, traces of melting were detected on the sand-blasted back sides of the samples as shown in Fig. 10, but only directly under the centers of the wide oxide stripes. This is in excellent agreement with calculated superheating values of $\sim2°$ and $\sim7°C$ for 920 and $1840\mu m$ half-width, (see Fig. 7), since the temperature drop across $600\mu m$ distance between the buried oxide and the back surface is also $7°C$.

628

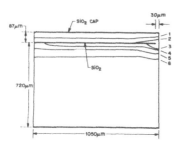

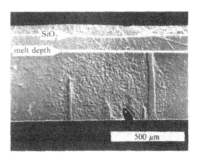

Fig. 9: Composite picture of maximum melt penetrations found experimentally for a series of irradiation energies.

Fig. 10: Etched cross section of a recrystallized sample, showing the depth of the planar melt and the localized melt spikes originating from the superheated back surface.

SUMMARY

Radiative melting and recrystallization of silicon with a buried oxide layer has been modelled numerically in two dimensions. The influence of slit-shaped openings in the oxide on temperature distribution and on propagation of the solid-liquid interface has been determined and shown to agree with the experimental data. Superheating and undercooling of silicon away from the openings in oxide have been included in the model. The calculated value of superheating agrees with the measurement.

ACKNOWLEDGMENT

Contributions of Lee Trimble to the experimental part of the paper are gratefully acknowledged.

REFERENCES

[1] "Single-Crystal Silicon on Non-Single-Crystal Insulators," G. W. Cullen, editor, J. Cryst. Growth **63**, No. 3 (1983).

[2] C. Y. Chang, Y. K. Fang, B. S. Wu, and R. M. Chen, Mat. Res. Soc. Symp. Proc. **23** (1984), 497.

[3] I. D. Calder, ibid, p. 507.

[4] J. M. Hode, J. P. Joly, and P. Jeuch, ibid, p. 513.

[5] G. K. Celler, McD. Robinson, L. E. Trimble, and D. J. Lischner, J. Electrochem. Soc. **132** (Jan. 1985).

[6] L. O. Wilson and G. K. Celler, in preparation.

[7] G. K. Celler, L. E. Trimble, and L. O. Wilson, these Proceedings.

[8] E. Yablonovitch, "Statistical Ray Optics," J. Opt. Soc. Am. **72** (1982), 899.

[9] G. K. Celler, K. A. Jackson, L. E. Trimble, McD. Robinson, and D. J. Lischner, Mat. Res. Soc. Symp. Proc. **23** (1984), 409.

THERMAL PROFILES IN SILICON-ON-INSULATOR (SOI) MATERIAL RECRYSTALLIZED WITH SCANNING LIGHT LINE SOURCES

KATSUHIKO KUBOTA *, CHARLES E. HUNT, and JEFFREY FREY
Cornell University School of Electrical Engineering, Ithaca, NY 14853
*) on leave from Hitachi, LTD., Musashi Works, Tokyo, JAPAN

ABSTRACT

A two dimensional solution of the classical heat equation is obtained and used to predict thermal profiles during line source zone melting recrystallization of silicon on insulators. A macroscopic solidification model is used to find the extent of the molten zone in multilayered structures. The problems of convergence associated with moving phase boundaries are reduced by using transformed temperature and the enthalpy model. The resultant isotherms, obtained at varying zone scan speeds, indicate optimum experimental conditions.

INTRODUCTION

We describe here a two-dimensional calculation to obtain thermal profiles during recrystallization of multilayered and SOI structures using a scanning line energy source, such as white light. The model used is shown in Figure 1. Effects of source scanning, temperature dependence of thermal conductivities and film reflectivities, melting of Si, and stacking of thin films of differing materials are all considered. The five degrees of freedom in the model (power density, film thicknesses, initial substrate temperature, zone width, and scanning velocity) allow the calculated solutions to be used to indicate optimum processing parameters. Comparisons with one-dimensional calculations indicate that an empirical correlation can be used to reduce the problem to three degrees of freedom in certain cases, making process optimization possible using simpler and cheaper one-dimensional solutions.

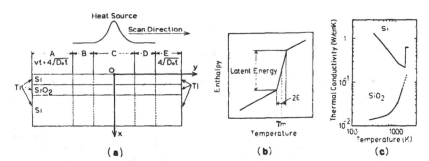

FIGURE 1: SCHEMATIC ILLUSTRATION FOR THERMAL CALCULATION
(a) System model: SOI regrowth with a moving line source.
(b) Approximated temperature-enthalpy relation for Si.
(c) Thermal conductivities for Si and SiO₂,
used in the calculations

CALCULATION METHODS

The heat equation is rewritten with transformed temperatures and then solved for enthalpy using finite-difference numerical methods. The enthalpy model [1,2] is used to calculate the shape and thermal profile of the interface of the Si molten region without involving the complicated boundary conditions associated with a moving phase boundary. Since the thermal conductivities are temperature dependent, with discontinuities at the melting points, the Kirchhoff transformation [3] is performed on all temperatures in the calculations to linearize their values. A moving coordinate system is employed to cover the wide scan region with concentration on the area of interest, i.e., the region near the heat source. The model considers strictly macroscopic thermal conduction, with both solidification kinetics, which may be orientation dependent, and microscopic phenomena such as constitutional supercooling associated with impurities, ignored.

The classical heat equation relates the change in internal energy with time, to heat flow by diffusion. Neglecting volume changes, the internal energy change equals that of the enthalpy, leading to the following two-dimensional equation in the coordinate system moving with the heat source:

$$\frac{\partial H(\Theta(T))}{\partial t} = \frac{\partial^2\Theta}{\partial x^2} + \frac{\partial^2\Theta}{\partial y^2} + v\,\frac{\partial H(\Theta(T))}{\partial y} \qquad \text{for Si} \qquad (1)$$

where:

$$\Theta(T) = \int k_{Si}(T')dT'$$

and:

H	:	Enthalpy (Si)
Θ	:	Transformed Temperature (Si)
k_{Si}	:	Thermal Conductivity (Si)
t	:	Time
x	:	Depth Direction
y	:	Scan Direction
v	:	Scan Velocity

Equation (1) and an analogous relation for SiO_2 are approximated by discretization to form matrix equations which are solved by the Newton-Raphson method [4]. The boundary conditions are:

$$-\frac{\partial\Theta}{\partial x} = (1-R)P_{max}exp(-\frac{y^2}{a^2}) \qquad \text{at top surface} \qquad (2)$$

$$T = T_i \qquad at \ y = -(vt + 4\sqrt{D_o t}\,) \text{ and } y = 4\sqrt{D_o t} \qquad (3)$$

Also, the temperatures and their gradients are continuous at each film interface, and the temperature gradient at the backside surface is zero. The incident power spatial distribution is taken as gaussian with amplitude P_{max} and characteristic length a. R is the temperature dependent reflectivity, and T_i is the initial substrate temperature. Energy absorption at elevated temperatures takes place at the top surface; heat loss by radiation and convection at the surfaces, calculated to be less than $100\ W/cm^2$ even at the Si melting point, is neglected. Equation (3) approximates the boundary conditions to infinity, since the temperature converges to the initial temperature T_i in a distance several times longer than the thermal diffusion length in one dimensional calculations. Thermal diffusivity is taken as constant in this model.

PHYSICAL CONSTANTS

A piecewise-linear approximation is used for the temperature-enthalpy relation for Si as shown in Figure 1b. The melting point is taken as all temperatures within ϵ (= 0.5K) of 1685K to speed convergence. The enthalpy slope within this region totals $1660\ J/cm^3K$; this rise in enthalpy corresponds to the latent heat of fusion for Si. The solid and liquid phase heat capacities are both taken to be constant at $2.21\ J/cm^3K$ outside of the narrow melting region in Figure 1b.

The thermal conductivity of Si is modeled as in Figure 1c using an empirical expression [5]. As with the enthalpy, the melting point singularity is avoided by using an average intermediate value when a nodal temperature lies within the range +/- 2ϵ. The solid phase thermal conductivity of SiO_2 is also modeled using an empirical expression [6]. Figure 2 shows reflectivities calculated using optical constants available in the literature [7,8,9]. A cap (to prevent agglomeration during recrystallization) is required in all practical systems. Above 1200K for a sample without a Si_3N_4 cap, the reflectivity is seen to be nearly constant at 40%, with a rapid increase to 70% at the Si melting point. Adding a 50 nm Si_3N_4 cap reduces the reflectivity to about 15%, below the melting point, and to 35% above. Other than changing the reflectivity, the effect of this very thin film on the thermal profile of the overall system is negligible; as a result, the values of reflectivity with a cap are used in all calculations, but the film system used in the calculations includes no cap.

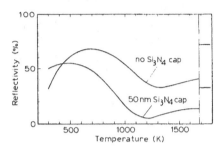

FIGURE 2: CALCULATED REFLECTIVITY IN
MULTILAYERED SYSTEM

CALCULATED RESULTS

Figure 3 shows the molten regions and isotherms at various power densities and initial temperatures. In each case, the steady state condition is shown in the vicinity of the heat source. This range of power densities is of interest when using high power incoherent light line sources for the melt zone. The temperature difference across the SiO_2 is seen to be about 10K, increasing with higher power density and/or lower initial temperatures. This temperature difference also increases with inter-layer oxide thickness, which thus should be considered a crucial parameter for process margin in multilayered systems where redistribution of underlying impurities must be avoided. The molten region, which lags behind the center of the heat zone, enlarges with increases in initial temperature or power density, as expected. Simple solidification theory indicates that the crystal growth rate is directly proportional to the thermal gradient in the solid phase at the moving phase boundary (under the assumption that the thermal gradient in the melt is very small). Therefore, the optimum regrowth condition is obtained when the temperature gradient is steepest and most uniform along the melting point isotherm. Of the five profiles shown in Figure 3, the combination of 20 kW/cm^2 and 1273K corresponds to this optimum condition, since the steep temperature gradient is parallel to and adjacent with, the solid-liquid interface.

With scanning line sources, the minimum energy transfer to the underlying layers occurs when the Si melting point isotherm extends only to the insulator/top-film interface. The scan velocity should be fast enough to just meet this requirement. Using the optimum conditions of initial temperature and power density found from the profiles of Figure 3, the Si melting point isotherms are shown at various scan velocities in Figure 4. For these conditions, the optimum velocity is found to be 32.3 cm/sec. It is also evident that melting of the substrate material occurs for scan speeds less than 30 cm/sec, indicating that the process margin for optimum scan velocity is less than 10% at these high power conditions.

Figure 5 shows the relationship between optimum scan velocity and power density for three initial temperatures. The proper scan velocities increase with power density as P^n, where n, as shown in the figure, is near 2. This dependence is explained with the help of a series of one-dimensional calculations over a wide range of power densities, as shown in Figure 6. The time needed to melt the top Si layer, which is inversely proportional to the scan velocity, decreases with increasing power density with a slope of about $-\frac{1}{2}$. Qualitatively, this indicates that heat diffusion dominates over energy absorption by the latent heat of fusion in the power range considered. Since the diffusion length of heat varies as $t^{\frac{1}{2}}$, energy spread has a $t^{-\frac{1}{2}}$ dependence on the power density; hence the scan velocity dependence on power density is as P^2.

632

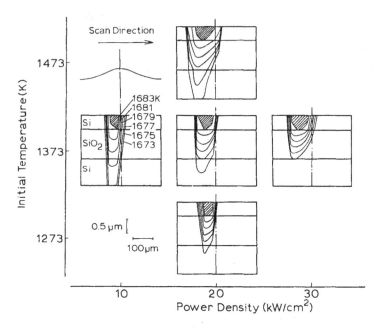

FIGURE 3: THERMAL PROFILES AT SCAN SPEEDS JUST MELTING Si FILM

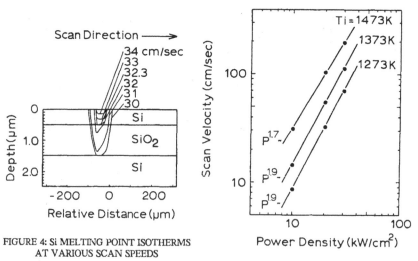

FIGURE 4: Si MELTING POINT ISOTHERMS
AT VARIOUS SCAN SPEEDS

FIGURE 5: RELATIONSHIP BETWEEN POWER
DENSITY AND OPTIMUM SCAN SPEED

USE OF ONE-DIMENSIONAL SIMULATIONS

The near-constant value of exponent n over a wide range, indicating the velocity dependence on power density, as suggested by Figure 6, shows that the use of one-dimensional thermal simulations may be suitable for many recrystallization process optimization calculations. Use of the three degrees of freedom (film thickness, power density, and initial sample temperature), which are all that is needed for one-dimensional calculations, can lower computational expense. Figure 7 shows the general correlation between one and two-dimensional calculations in terms of optimum scan velocity, assuming the scan speed in the one-dimensional case to be the width of the heat source divided by the time needed to melt the top film (as shown in Figure 6). This correlation factor is nearly .7, varying slightly with power density and initial temperature. This particular numerical value results corresponds to the fact that in the gaussian profile of the heat zone (used in the two dimensional model), the maximum temperature lies within the region $(-a < y < a)$ (ie. within the e^{-1} region), because of heat diffusion. Energy source distributions other than gaussian will have correspondingly different correlation factor values.

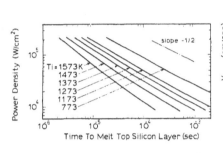

FIGURE 6: 1-D CALCULATIONS OF Si TOP LAYER MELT TIMES

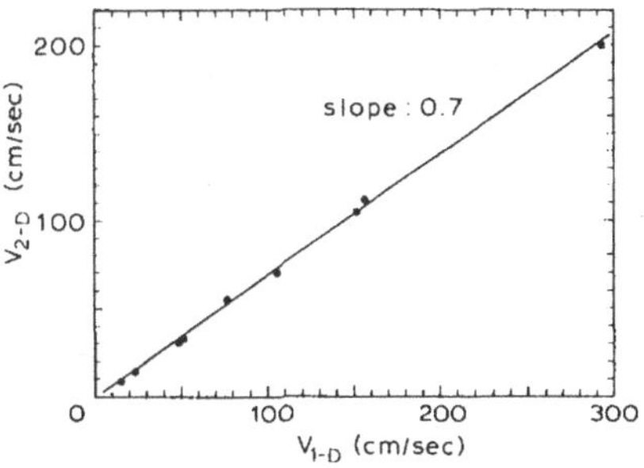

FIGURE 7: CORRELATION BETWEEN ONE AND TWO DIMENSIONAL CALCULATIONS

SUMMARY

A thermal calculation method useful for study of multi-layer SOI line source zone melting has been presented. The resulting profiles, calculated with particular reference to a high power incoherent light line source, are applicable to all systems with comparable absorbed power densities, and indicate the shape of the molten zone and the isotherms. Calculated results show that the moving solidification front thermal profile is a strong function of the five processing control variables, power density, film thickness, initial sample temperature, heat zone width, and scan velocity. Knowledge of the temperature profile makes process optimization of the solidification process possible. In some cases these five degrees of freedom can be optimized through the use of simpler one-dimensional calculations in which empirical approximations are incorporated.

The authors would like to thank R. Byrnes for his help with computational techniques. This work was supported by the Semiconductor Research Corporation Center of Excellence for Microscience and Technology.

REFERENCES

[1] G.H. Meyer, SIAM J. Num. Anal. 10 (522) 1973

[2] N. Shamsundar and E.M. Sparrow, ASME J. Heat Trans. 97 (333) 1975

[3] M. Lax, Appl. Phys. Lett. 33 (786) 1978

[4] B. Carnaham, H.A. Luther, and J.O. Wilkes, "Applied Numerical Methods", John-Wiley & Son, Inc. 1969

[5] A.E. Bell, RCA Review 40 (295) 1979

[6] TPRC Data Series, "Thermophysical Properties of Matter", IFI/PLENUM 1970

[7] K.M. Shuarev, B.A. Baum, and P.V. Gel'd, Sov. Phys. S. S. 16 (2111) 1975

[8] Y.J. van der Meulen and N.C. Hien, J. Opt. Soc. Am., 64 (804) 1974

[9] G.E. Jellison, Jr. and F.A. Modine, Appl. Phys. Lett., 41 (180) 1982

[10] H.E. Cline, J. Appl. Phys., 55 (2910) 1984

KINETICS OF RADIATIVE MELTING OF Si.

G. K. CELLER, L. E. TRIMBLE, AND LYNN O. WILSON,
AT&T Bell Laboratories, Murray Hill, NJ 07974.

ABSTRACT

During radiative melting, a silicon surface breaks up into coexisting solid and liquid regions with spacing dependent on incident flux, thermal parameters, and crystalline properties of the sample. The space-averaged reflectivity becomes a function of the incident photon flux, profoundly affecting the transfer of energy and the rate of melting.

We explain time evolution of the molten surface morphology and present data relating depth of melting to the incident photon flux for bulk Si and Si with buried oxide. The data prove the existence of a steady state transition region in which melting is only superficial and time-independent.

IN SITU DETECTION OF SURFACE BREAKUP DURING MELTING

Crystalline Si, irradiated with a stationary radiative heater from one side and radiatively cooled from all sides, melts on the irradiated surface by breaking up into faceted molten segments of typical dimensions from 100 to 1000 μm that are bounded by the (111) planes [1,2]. The recrystallized surface shows traces of faceted melting because the planarity of the surface in the melted regions is lost. Moreover, the recrystallized surface is covered with surface ripples that are quite regular over the crystalline surface and follow the grain boundaries if material is polycrystalline. We interpreted these ripples as frozen imprints of solid lamellae in the liquid, similar to those observed in laser melting of thin films [3-5].

By modifying the lamp furnace [6] to view directly the molten surface with a long focal length microscope, we were able to record the evolution of the surface morphology during melting. The incident radiant flux was increased at a constant rate over 60 sec. When the melt temperature T_m = 1685 K was reached, small square-shaped molten regions appeared at random on the irradiated polished surface and grew in size. At a critical size of ~100 μm there was sufficient undercooling in the center of each square molten region to nucleate a microscopic solid inclusion. It is not possible to determine directly whether nucleation occurred at the surface oxide or whether localized regrowth from the solid below was present instead. The latter is more likely in view of the perfect crystallographic alignment of the recrystallized silicon with the substrate. Once formed, each solid inclusion expanded into a network of thin solid stripes that spread over the molten regions with a regular spacing between stripes. The growth of each network followed the expansion of the faceted melt region in which it was located. This is illustrated in Fig. 1, where optical micrographs of the partially molten surface are shown. Black stripes inside the rectangular molten regions are the solid lamellae since reflectivity of the solid is less than that of the liquid. As the large faceted regions grow, they start approaching other such regions and occasionally merge into larger regions. More often, they "repel" each other. i.e. their growth stops so that there is an approximately uniform random network of solid areas separating the molten zones. An example of such repulsion can be seen in Fig. 1(b), where an elongated molten region avoided a smaller melt segment, leaving a gap between the two.

The size and number of molten segments is a function of ramping and maximum input power. Rapid increase of power causes larger superheating of the surface and simultaneous melt nucleation at more points than would be caused by slow heating. On the other hand, slow heating causes formation of large rectangular melt areas starting from just a few nuclei. In that case, the bottom of each molten pool is less faceted than during rapid melting. When power is reduced, freezing starts on the solid lamellae, and the last to freeze regions form solid ripples at the surface where previously the liquid stripes were located.

Fig. 1. Two views of a partially molten silicon surface inside the lamp furnace. The micrograph (b) was taken a few seconds after (a) to show evolution of faceted melting with the increase in incident power. The molten segments are elongated because of a lateral temperature gradient caused by removal of one lamp for better visibility.

KINETICS OF DEEP MELTING

Melting of single crystalline Si with a low defect density is very inhomogeneous, as was discussed in the preceding section. On the other hand, when a high density of defects is present in the surface region, silicon melts uniformly on a macroscopic scale. The depth of melting is then well defined and is controlled by the incident energy flux and the emissivities of both front and back surfaces.

Emissivities

The emissivity of a polished Si surface near the melt temperature T_m is ~0.6. Surface roughness increases the emissivity only slightly [7], to ~0.62. Low reflectivity (high emissivity) is achieved by encapsulating a rough surface with a transparent overlayer, since it traps most of the diffusely reflected light by total internal reflection. Following the calculation of Yablonovitch [7], we obtain $\epsilon = 0.84$ for a rough solid Si surface near T_m encapsulated with a SiO_2 layer. The emissivity of such a surface after melting decreases relatively slightly, to 0.58, because the roughness of silicon is preserved by the solid SiO_2. In contrast, for smooth Si without a cap, the emissivity decreases by more than a factor of two, from 0.60 to 0.28, and for smooth and encapsulated Si it changes from 0.72 to 0.37.

Transition range in incident power.

When faceted melting is suppressed by crystalline defects, the silicon surface breaks up into microscopic solid and liquid regions upon reaching T_m at a threshold incident power P_1. The presence of the solid inclusions or lamellae is necessitated by the increase in reflectivity of the melting Si surface. As the incident power is slowly raised further, the fraction f of the surface area which is liquid adjusts until the absorbed power and the heat loss from both surfaces balance. At a higher power P_2, the surface is completely molten (f=1) and any further increase in the input radiation leads to a molten volume proportional to the incident energy. The calculated relationship between the incident and absorbed power is plotted in Fig. 2, for three different samples. The curve (c) is for the simplest case of a smooth Si surface, the curve (b) for a smooth surface coated with a SiO_2 cap, and the curve (a) for a rough encapsulated surface. In all three cases the back surface is assumed to be the same, with the emissivity $\epsilon_B = 0.62$. Since we want to examine the effects near T_m, we also assume that the solid and liquid Si emissivities are temperature independent. In the transition region between P_1 and P_2, the emissivity ϵ of the irradiated front surface depends on the ratio of solid to liquid at the surface and is given by $\epsilon = (1-f)\epsilon_S + f\epsilon_L$. The melt fraction f is not measured directly. However, ϵ can be expressed in terms of the incident power P_I and the back surface emissivity ϵ_B only:

$$\epsilon = \epsilon_B[(P_I/\sigma T_m^4) - 1]^{-1}. \tag{1}$$

The above dependence has been derived from the equality of the absorbed flux and the total loss term:

$$\epsilon P_I = (\epsilon_B + \epsilon)\sigma T_m^4, \tag{2}$$

where σ is the Stefan-Boltzmann coefficient. Notice that the emissivity in the transition region is not directly dependent on ϵ_S or ϵ_L but that the two threshold values P_1 and P_2 are.

$$P_1 = (\epsilon_B/\epsilon_S + 1)\sigma T_m^4, \tag{3}$$

$$P_2 = (\epsilon_B/\epsilon_L + 1)\sigma T_m^4, \tag{4}$$

The threshold for deep melting P_2 depends strongly on surface conditions and is lowest for a rough encapsulated surface, as shown in Fig. 2.

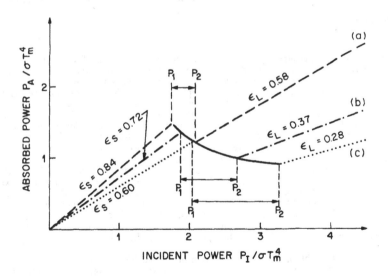

Fig. 2. Calculated coupling of radiative flux to Si for three different conditions of the surface exposed to light: (a) a rough Si surface covered with a thick oxide film, (b) a smooth Si surface covered with a thick oxide film, and (c) a smooth Si surface without oxide. P_1 and P_2 thresholds shown for all three cases.

Experimental confirmation of the behavior predicted by the curve (a) in Fig. 2 is shown in the following three figures. In Fig. 3, the depth of maximum melting is plotted as a function of the maximum incident power for crystalline Si wafers with a sandblasted top surface covered by a 2 μm layer of SiO_2. The most striking feature is that the melt depth curve consists of two straight segments, the first of which has a slope close to zero. This first segment identifies the transition region, where the surface molten fraction changes to compensate for increased power input. The difference between the continuous and dashed lines is interpreted as being caused by a deviation from steady state conditions when the input power is changed rapidly. For such a rapid change, some net energy gain by the wafer (and hence some deep melting) is possible before the average surface reflectivity readjusts to a higher value. Therefore, the dashed curve provides a more accurate value of P_2.

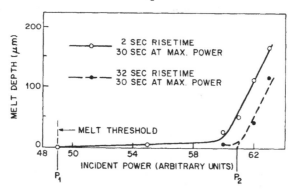

Fig. 3. Maximum melt depth vs. incident power for crystalline Si with a sandblasted surface covered by a 2 μm SiO_2 layer. The risetime was measured from a power setting ~5% below P_1.

In Fig. 4, the measured melt depth is plotted as a function of time at maximum incident power, nominally at two different power settings. In fact, there was a gradient of incident power across the 10 cm wafer diameters. One side of each 10 cm wafer consistently melted more deeply than the opposite side, with the difference across the wafer being equivalent to about 2% change in flux. Therefore, the three curves labeled 62 are approximately equivalent to power settings of 61.5, 62, and 62.5, and a similar spread holds for the lower power of 60. The main feature of the plot is the constant melt depth for $P_I < P_2$ and the dependence on time that tends towards linearity as the power is increased. The two powers selected were just below and above P_2. The difference in the time dependence is quite striking.

From a technological point of view, more interesting is the behavior of Si wafers covered with patterned oxide films and further coated with thick polysilicon and an oxide cap. The melt depth vs. maximum incident power for such wafers is plotted in Fig. 5. The curve is quite similar to the one for bulk Si, with two linear segments defining the P_1 and P_2 thresholds. The slope of the first segment is higher than in Fig. 3, possibly because the melt temperature is slightly depressed at the grain boundaries. Also the melt-solid interface is perturbed by the openings in the SiO_2 [8,9], so the average data values were plotted.

It is difficult to obtain absolute calibration of the incident powers, but for the sake of comparing our data with the calculated curve (a) shown in Fig. 2 it is sufficient to compare the ratios of P_2 to P_1. They are 1.24 and 1.25 for the bulk and Si-on-insulator, respectively, and agree quite well with the value of 1.19 obtained theoretically. This indicates that the emissivity calculations based on light trapping are correct and confirms that microscopic solid lamellae are universally present in the transition range of input powers, regardless of crystalline properties, surface texture or film thickness.

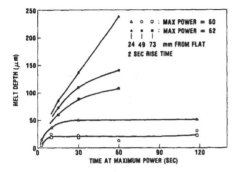

Fig. 4. Maximum melt depth in crystalline Si from Fig. 3, vs. time at the maximum power.

Fig. 5. Maximum melt depth vs. incident power for Si with a patterned SiO_2 layer buried 80 μm below a rough encapsulated surface.

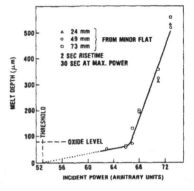

Theory of deep melting

If the incident power $P_I(t)$ is ramped linearly from P_2 to P_{max} over Δ_1 seconds, held constant at P_{max} for Δ_2 seconds, and then ramped down to zero linearly with time over Δ_3 seconds, it can be shown that the depth of melting $\delta_m(t)$ is given by [10]

$$
\begin{aligned}
\delta_m(t) &= \delta_o + \frac{\beta}{2\Delta_1}(\frac{P_{max}}{P_2} - 1)t^2 & 0 \leqslant t \leqslant \Delta_1, \\
&= \delta_o + \beta(\frac{P_{max}}{P_2} - 1)(t - \Delta_1) & \Delta_1 \leqslant t \leqslant \Delta_1 + \Delta_2, \\
&= \delta_o + \beta(\frac{P_{max}}{P_2} - 1)(t - 0.5\Delta_1) - \frac{\beta P_{max}}{2\Delta_3 P_2}[t - (\Delta_1 + \Delta_2)]^2 & \Delta_1 + \Delta_2 \leqslant t, \quad (5)
\end{aligned}
$$

when δ_o is the initial melt depth at $t = 0$, $\beta = (\epsilon_L + \epsilon_B)\sigma T_m^4/\rho L$, and ρ is an average density of Si and L is the latent heat. Thus, the melt depth increases parabolically during the initial ramping up stage and linearly during the constant power stage, then decreases parabolically during the ramping down stage. This is illustrated in Fig. 6 for hypothetical, but representative, parameter values.

Consequently, the maximum melt depth varies linearly with the time at maximum incident power and approximately linearly with the excess power $P_{max} - P_2$. Although it is difficult to measure the absolute power accurately, theory and experiment are in semi-quantitative agreement.

640

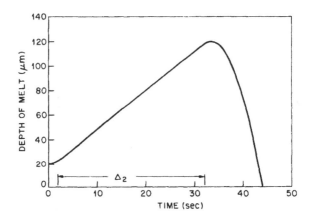

Fig. 6. An example of the calculated melt depth versus time for $\delta_0 = 20 \ \mu m$, $\Delta_1 = 2$ sec, $\Delta_2 = 30$ sec, $\Delta_3 = 60$ sec, $P_{max} = 1.025 \ P_2$, $\epsilon_L = 0.58$, $\epsilon_B = 0.62$, $\rho = 2.41$ g/cm^3, and L $= 1810$ J/g.

SUMMARY

Melting of silicon with a radiant flux has been characterized theoretically and experimentally. We have shown that there is a well defined transition region of input powers in which the surface is broken up into liquid and solid regions so that the space averaged emissivity is intermediate to the emissivities of the solid and liquid and is a function of the incident power. In this transition region there is no net absorption of energy. Above the threshold P_2 for complete surface melting, the melt depth is proportional to the incident flux in excess of P_2. The value of P_2 depends strongly on the surface conditions and is particularly low for rough encapsulated surfaces because of light trapping in the transparent overlayer.

REFERENCES

[1] G. K. Celler, McD. Robinson, L. E. Trimble and D. J. Lischner, Appl. Phys. Lett. **43**, 868 (1984).

[2] G. K. Celler, K. A. Jackson, L. E. Trimble, McD. Robinson, and D. J. Lischner, Mat. Res. Soc. Symp. Proc. **23**, 409 (1984).

[3] W. G. Hawkins and D. K. Biegelsen, Appl. Phys. Lett. **42**, 358 (1983).

[4] M. A. Bosch and R. A. Lemons, Phys. Rev. Lett. **47**, 1151 (1981).

[5] K. A. Jackson and D. A. Kurtze, J. Cryst. Growth (1985).

[6] D. J. Lischner and G. K. Celler, Mat. Res. Soc. Symp. Proc. **4**, 759 (1982).

[7] E. Yablonovitch, J. Opt. Soc. Am. **72**, 899 (1982).

[8] G. K. Celler, McD. Robinson, L. E. Trimble, and D. J. Lischner, J. Electrochem. Soc. **132**, 211 (1985).

[9] Lynn O. Wilson and G. K. Celler, these Proceedings.

[10] Lynn O. Wilson, G. K. Celler, and L. E. Trimble, in preparation.

SOI TECHNOLOGIES: DEVICE APPLICATIONS AND FUTURE PROSPECTS

B-Y. TSAUR
Lincoln Laboratory, Massachusetts Institute of Technology,
Lexington, Massachusetts 02173-0073

ABSTRACT

Silicon-on-insulator (SOI) technologies have four major applications: very-large-scale integrated circuits (ICs), high-voltage ICs, large-area ICs, and vertical ICs. This paper will review the recent progress made in these areas and discuss the prospects of various SOI technologies for achieving commercialization.

INTRODUCTION

Extensive efforts have recently been directed toward the development of Si-on-insulator (SOI) technology for producing high-quality, device-worthy Si films on insulating substrates. These efforts have been motivated by the potential of thin-film devices for a variety of device applications: very-large-scale integrated circuits (ICs), high-voltage ICs, large-area ICs, and vertical ICs. Each application has its unique requirements on device structure and performance and therefore imposes different requirements on SOI material properties. In this paper, we will consider the rationale of SOI technology for these applications, discuss the SOI device structure and material requirements for each, and review the state-of-the-art development in both materials and devices for the various SOI technologies. We will examine the status of competing non-SOI technologies and address some of the crucial issues such as cost and manufacturing feasibility that are pertinent to the commercialization of SOI technology.

VERY-LARGE-SCALE INTEGRATED CIRCUITS

The continuing reduction in feature size has been a basic trend in the evolution of Si integrated circuits. However, as device dimensions and device separation shrink below 1 μm, many problems emerge for bulk Si technology. For example, parasitic coupling between devices or adjacent circuits through the Si substrate can severely limit VLSI circuit performance. The SOI technology offers many advantages over conventional bulk technology. (1) It achieves complete device isolation, which can minimize capacitive coupling between various circuit elements over the entire IC chip and eliminate latchup in CMOS circuits. (2) Because of the simplicity of device isolation, minimum device separation is determined only by the limitations of lithography and therefore higher packing density is possible. (3) Reduction in parasitic capacitances and chip size should lead to an increase in circuit speed. (4) Improved radiation hardness is expected because SOI circuits have much less charge collection volume than bulk Si circuits. (5) The SOI technology should permit simpler processing and offer greater flexibility in circuit design and layout, both of which should lead to reductions in chip cost.

Most efforts to utilize SOI technology for VLSI applications have concentrated on CMOS, which is rapidly becoming the mainstream VLSI technology. However, there is also increasing interest in the development of SOI bipolar devices and merged CMOS-bipolar circuits, which combine high-density, low-power CMOS digital capability with high-speed bipolar analog capability on the same chip. For CMOS applications, the SOI structure usually consists of a Si film 0.2-0.5 μm thick on an oxidized bulk Si wafer. The thickness of the buried oxide layer should be selected to minimize parasitic capacitance (including junction, line and coupling capacitances) and to allow suitable heat dissipation. For bipolar device applications, the SOI film should be thicker (> 1 μm) and preferably incorporate a heavily doped buried layer to reduce collector resistance. High majority-carrier mobility and long minority-carrier lifetime are both essential, imposing stringent requirements on material quality. The SOI wafers should be flat within a few micrometers to satisfy the lithography requirement in VLSI processing. Finally, the technique for producing SOI wafers should be scalable and have the capability of volume production at a reasonable cost.

Three SOI technologies have been extensively investigated: beam recrystallization, formation of buried oxide by oxygen implantation, and oxidation of porous silicon. The processing details of each technology have been reported previously in several reviews [1] and will not be discussed here. Summarized below are major advances accomplished recently in both material development and device applications.

Beam Recrystallization

Improvement in temperature uniformity and reduction of thermal stress have permitted zone-melting recrystallization of full-size SOI wafers (up to 3 inches in diameter) by the graphite-strip-heater technique. The wafers are almost free of slip lines and warpage, and they are comparable in flatness to bulk wafers. The SOI films still contain subboundaries and have slight surface roughness of the order of a few hundred angstroms. Ring oscillators with good speed performance and medium-scale ICs such as 240-gate CMOS gate arrays have been fabricated in these films. In addition, two merged CMOS-bipolar technologies utilizing zone-recrystallized SOI films have been demonstrated [2]. In each case a single sequence of processing steps was used to fabricate fully isolated CMOS devices and vertical bipolar transistors on the same Si wafer. As shown schematically in Fig. 1, the

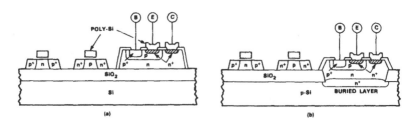

Fig. 1. Schematic diagrams of two configurations with SOI/CMOS and bipolar devices. The bipolar device is fabricated in (a) the SOI film or (b) epitaxial layers grown selectively on the Si substrate.

CMOS devices were fabricated in a SOI film, while the bipolar devices were fabricated either in the SOI film [Fig. 1(a)] or in epitaxial Si layers grown selectively on the Si substrate [Fig. 1(b)]. In the latter case the bipolar transistors incorporate a buried collector layer. Good electrical characteristics with current gain of more than 100 were obtained for both SOI and epitaxial bipolar devices, as shown in Fig. 2.

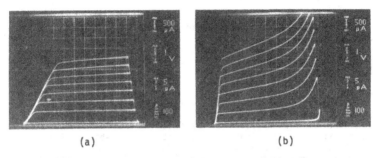

(a)   (b)

Fig. 2. Common emitter I-V characteristics for bipolar transistors: (a) SOI, (b) epitaxial.

To minimize or eliminate subboundaries, various patterning and seeding techniques have been employed. Most of the experiments were performed with scanning CW laser or electron beams. Significant progress has been made in obtaining defect-free Si films in selected areas. So far, the throughput obtained with laser and electron beam techniques has been low because the beam size is small and multiple overlapping scans are required to process large-area samples. Scaled-up systems using these techniques are currently under development. A high-power argon-laser pattern generator is being developed by research groups at Mitsubishi and Tokyo Electron in Japan for processing up to 6-inch wafers. This system will have the capability of pattern alignment and selective chemical vapor deposition. In Great Britain, a high throughput e-beam recrystallization system is being developed at Cambridge University for processing 5-inch wafers in 30 s with carousel-to-carousel loading capability.

Formation of Buried Oxide

Full-size SOI wafers (up to 4 inches in diameter) with excellent uniformity and flatness have been obtained by oxygen implantation and subsequent solid-state epitaxial growth. The films contain threading dislocations with densities of $10^6 - 10^8$ cm^{-2} that are usually aligned along [110] directions. Very impressive LSI circuit results have been achieved with this technology. A 1K-bit CMOS static random access memory (SRAM) circuit with 1.5 µm channel length was reported [3] by NTT Research Laboratory in Japan. This circuit exhibits an access time of 12 ns. More recently, Texas Instruments has reported [4] the fabrication of a 4K SRAM with a 2.5-µm design rule, which has an access time of 55 ns.

The key requirement for successful implementation of buried oxide technology is the development of a high-beam-current ion implantation machine. A machine that can deliver at least 100 mA at an energy of more than 150 keV is needed for production (~ ten 4-inch wafers per hour). The development of such a machine presents a challenging engineering problem. Because of increasing interest expressed by the IC community, a U.S.

manufacturer (Eaton) and a British government institute (Harwell) have committed themselves to this development, giving some encouragement concerning the future of this SOI technology.

Oxidation of Porous Silicon

In this process, selected surface regions of a Si wafer are isolated by oxidation of porous Si that is formed by a selective anodization process. The major difficulties experienced with this technology have been wafer warpage and defect generation in the isolated Si regions due to the large thermal stress induced by prolonged oxidation. Recently, by employing high-pressure oxidation at reduced temperatures for shorter times, significant improvement in wafer flatness (warp < 25 μm for 4-inch wafers) has been obtained [5] and the SOI wafers are free of defects. Several LSI circuits have been fabricated using this process. A 2K-gate CMOS gate-array circuit has been fabricated [5] by OKI in Japan with good yield and 50% higher speed than equivalent bulk circuits. A 16K-bit CMOS SRAM fabricated with a 2-μm design rule by NTT in Japan has a 35 ns access time [6].

The advantages of this SOI technology are the high material quality and the potential for high throughput. However, because anodization is a wet-chemical process, reproducible formation of uniform porous Si layers and precise control of SOI island size and separation may be difficult. If these difficulties can be overcome and warpage can be further reduced, the oxidation of porous Si should be an attractive SOI technology.

Important Issues

Despite the promising results that have been demonstrated for all three technologies, the question remains as to whether commercialization of these technologies will take place in the near future. We will examine some of the critical issues such as material quality, cost and manufacturing feasibility in this section.

The oxidation-of-porous-Si technology currently produces the highest quality SOI material. In addition, the process has high throughput and is potentially low in cost. However, since the isolation process is an integral part of the entire device fabrication procedure, this technology is useful only for customized integrated circuits. Therefore, successful development of this technology will benefit the in-house fabrication of specific circuits but will probably have limited impact on the IC community in general.

From the manufacturing point of view SOI materials produced by beam recrystallization and oxygen implantation are equivalent to bulk Si wafers and would be useful for many general-purpose applications. However, both beam-recrystallized and buried-oxide SOI materials currently contain significant concentrations of defects, namely subboundaries and dislocations, respectively. Subboundaries are believed to be more detrimental to device performance than dislocations despite the higher density of the latter. This difference can be explained by reference to Fig. 3, which shows schematically an MOS device with 1-μm channel length that has either a dislocation [Fig. 3(a)] or a subboundary [Fig. 3(b)] in the channel region. The presence of defects in the channel region can affect the device performance in three ways: (1) reduction in carrier mobility, (2) enhanced dopant diffusion along defects, giving rise to high leakage current, and (3) degradation of the gate oxide, affecting threshold voltage uniformity and gate oxide reliability. Neither subboundaries nor dislocations should significantly degrade carrier mobility as long as the trapping-state density near these defects is reasonably low. Increased leakage due to enhanced dopant diffusion is likely to occur for subboundaries but not for dislocations because of the threading nature of the latter [see Fig. 3(a)]. Finally, subboundaries should have a more adverse effect on the gate oxide because they are planar defects, whereas

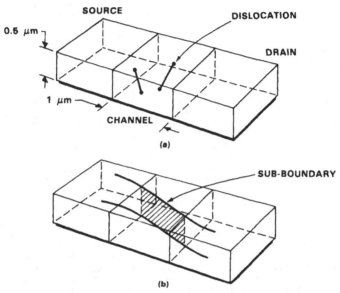

Fig. 3. Schematic structures of SOI/MOSFETs that have (a) threading dislocations or (b) subboundary in the channel region.

dislocations are line defects. We believe that subboundaries are unacceptable for VLSI applications, and their elimination is essential for the success of beam-recrystallization SOI technology. Some recent progress toward this goal is being reported at this meeting by workers at Lincoln Laboratory [7]. In spite of the high dislocation densities in buried-oxide SOI materials, devices fabricated in these materials exhibit excellent uniformity in threshold voltage, mobility, and leakage current. In this respect they are indistinguishable from bulk devices.

A question frequently asked is what wafer cost would make SOI technology competitive with conventional bulk Si technology. It should be realized, however, that it is the cost and performance of the chips fabricated from the wafers that directly determine whether or not a new technology is worth pursuing. The approximate cost of a chip can be estimated from the following expression:

$$\text{chip cost} \simeq \frac{\text{wafer cost + manufacturing cost}}{\text{chips per wafer} \times \text{yield}} \qquad (1)$$

We shall first use this expression to estimate the cost of a VLSI chip fabricated on a bulk Si wafer by conventional technology. The cost of a 5-inch Si wafer is ~$20. The manufacturing cost is a complex figure that includes the costs of circuit design, processing, and testing. For a state-of-the-art CMOS process, this cost is about $300. The total number of chips on a 5-inch wafer is estimated to be ~2000 for typical LSI chips a few millimeters on a side. If we assume the yield is 30%, according to Eq. (1) the chip cost is $0.53. If we assume the same cost for an SOI chip, we can work backwards to calculate the permissible cost of an SOI wafer. As mentioned earlier, the SOI structure permits simpler processing (saving one or two photolithographic steps) and offers greater layout flexibility. Therefore, the manufacturing cost should be lower than that for bulk processing, and we assume that it ranges from $250 to $280. As shown by SOS circuits, the SOI structure allows higher packing density than bulk (especially for CMOS circuits), resulting in smaller chip size. We assume an area savings of 20% and therefore an increase in number of chips to ~2400 per wafer. For high-quality SOI materials it is not unrealistic to assume the fabrication yield is as high for SOI circuits as for bulk circuits and perhaps even higher because of the simpler processing. Under these assumptions the permissible cost of an SOI wafer would range from $100 to $130, from five to seven times the cost of a bulk Si wafer.

This simple analysis suggests that if high-quality SOI wafers were commercially available now, they would be readily adopted by the IC community even though their cost were a few times that of bulk Si wafers. Unfortunately, commercial SOI wafers are not expected to be available in the near future. Meanwhile, bulk Si technology is progressing rapidly. For example, development of advanced device isolation techniques such as trench isolation [8] is expected to result in packing densities as high as for SOI, elimination of latchup, and simplification in circuit layout and fabrication procedure. If this is accomplished, SOI technology will no longer be superior to bulk technology except for some marginal improvement in speed performance. Therefore SOI wafers would have to be comparable in cost to bulk Si wafers in order to be competitive. How soon SOI technology becomes readily available will probably determine its future for VLSI applications.

To complete this discussion of VLSI circuits, we should point out that SOI is a potential technology for radiation-hardened circuits, which are needed for space and military applications. For these applications, high performance and good reliability are of paramount importance and cost is of secondary concern. We believe that SOI will play an important role in space and military markets if high-quality SOI wafers become available.

HIGH-VOLTAGE INTEGRATED CIRCUITS

As in the development of digital electronics, the evolution of power electronics is following the trend of monolithic integration in order to achieve better performance, greater functional versatility and lower cost. Although the level of integration for power ICs is not as high as for digital ICs, material preparation and device fabrication are in many cases more complex and difficult. A typical power IC consists of two basic units: a low-voltage control circuit and high-voltage output power devices. Integration of these two types of devices on bulk Si requires special isolation techniques that usually involve complicated fabrication procedures, and the isolation regions often occupy large areas, resulting in poor circuit density. Conventional dielectric isolation [9] produces wafers with SiO_2-isolated Si islands supported by a thick poly-Si substrate. This technique is not only costly but introduces defects in the Si islands. The SOI technologies, which provide simple and complete device isolation, lend themselves naturally to high-voltage IC applications.

There are many ways to implement high-voltage ICs utilizing SOI structures. Figure 4 illustrates several approaches. The control circuits can be either MOS (mostly CMOS) or bipolar devices, which are usually operated at a supply voltage of 5-10 V. The output power devices can be either offset-gate high-voltage MOS transistors that operate at voltages up to 1000 V but at relatively low current (< 1 A), or high-power bipolar transistors that operate at medium voltages (50-200 V) but have high current (up to 50 A) capability. Depending on the specific circuit application, the control circuit and power devices can be fabricated in the SOI region or in the Si substrate. Complete device isolation can be achieved for all the cases shown in Fig. 4.

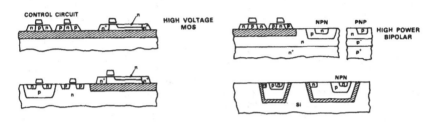

Fig. 4. High-voltage ICs utilizing SOI structures.

As shown schematically in Fig. 4, the drain (n^+ region) of the offset-gate MOS transistor is separated from the gate (or channel region) by a lightly doped n-region to reduce the electric field near the channel and therefore achieve high breakdown voltage. Fabrication of offset-gate high-voltage MOS transistors has been demonstrated in SOI materials prepared by laser recrystallization [10] or oxygen implantation [11] . Usually, the high-voltage MOS devices have relatively large dimensions, and the low-voltage control circuits are much less complex than digital VLSI circuits. Therefore, the material requirement is not as stringent as that for VLSI circuits. Beam-recrystallized SOI materials, in which the thickness of buried oxide underneath the Si film can be conveniently increased for high-voltage operation, are particularly suitable. The presence of subboundaries will probably not significantly affect device performance, so that high-voltage ICs with reasonable yield can be produced. Buried-oxide SOI formed by oxygen implantation can also be used for high-voltage ICs. However, the thickness of the buried oxide layer is fixed at about ~0.5 μm so that the voltage is limited to ~ 300 V.

For high-power bipolar transistors, a relatively thick (25-50 μm) SOI film is needed to provide high current capability. The SOI material should be of high crystalline quality with long minority-carrier lifetime. By using seeded growth, subboundary-free SOI films 25-50 μm thick have been produced by the graphite-strip-heater technique. The initial films prepared in this manner contained many point defects, which were generated by thermal stress and the precipitation of impurities such as oxygen. By using a "cap-relief" structure to allow escape of impurities and expansion of the Si film for stress reduction during the recrystallization process, point defects have been almost completely eliminated [12].

So far, only a limited research effort has been devoted to SOI development for high-voltage ICs, which have enormous potential for commercial applications in the areas of automobile electronics, power regulators, motor controls, consumer audio or video electronics, fluorescent light ballasts, etc. Since the existing bulk and dielectric isolation technologies for high-voltage ICs have many drawbacks, there is a good opportunity for SOI technology to play an important role in this area. Several SOI materials already produced are of sufficient quality for a number of high-voltage applications. The SOI community should pay more attention to this market segment, which may provide the best opportunity for the commercial use of SOI technology in the near future.

LARGE-AREA INTEGRATED CIRCUITS

It has been proposed that SOI structures consisting of a device-quality Si film on an insulating transparent substrate are potentially useful for the fabrication of active-matrix-addressed flat-panel displays. Such structures would allow monolithic integration of matrix switching elements and peripheral control circuits on the same substrate to achieve high system performance. Several laboratories have fabricated high-performance thin-film transistors (TFTs) and small-scale ICs such as ring oscillators and dynamic shift registers [13] in Si films prepared by beam recrystallization on bulk fused-silica substrates. Small functional liquid-crystal display panels [14,15] have also been successfully demonstrated using beam-recrystallized Si-on-quartz materials and conventional IC processing technology.

The key issue confronting SOI technology for display applications is cost. At present the cost of quartz substrates is too high for many applications, especially large-area panels. In addition, conventional IC processing may be too costly, and processing of large-area panels (for example, 10 x 10 inch size) would require dedicated equipment that is not available. The SOI technology is competing with several alternative TFT technologies based on materials such as amorphous Si, small-grain polycrystalline Si, and CdSe that promise lower cost because they can be deposited by low-temperature processes on inexpensive glass substrates. The TFTs fabricated in these materials are inferior in performance to those fabricated in beam-recrystallized SOI films. However, they are adequate for matrix switching elements, although marginal at best for peripheral control circuitry. Therefore it may be advantageous to use a system that monolithically combines low-cost TFTs deposited on a glass panel as switching elements with high-performance control circuits fabricated in a beam-recrystallized SOI film on the same panel. If a beam-recrystallization SOI technique is to be used, it would be necessary to develop a low-temperature processing procedure that is compatible with the fabrication of the matrix switching TFTs.

Another type of large-area IC offering a possible application for SOI technology is the monolithic microwave integrated circuit (MMIC). As shown schematically in Fig. 5, a typical MMIC consists of a few field-effect

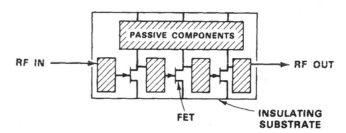

Fig. 5. Schematic structure of a monolithic microwave IC.

transistors (FETs) used as amplifiers or switches together with many passive components such as resistors, capacitors and inductors, which occupy most of the IC chip area. The chip size can range from a few millimeters to a few centimeters on a side. The substrate is normally an insulator to reduce dielectric loss and parasitic capacitance. Such MMICs have been widely produced in GaAs using semi-insulating GaAs substrates. Some work has been done on SOS structures but the results have generally been unsatisfactory because of the poor quality of the Si films. Silicon-on-quartz (SOQ) structures produced by beam recrystallization should be very attractive for this application because of the low dielectric constant and low dielectric loss of quartz.

Low-noise microwave MESFETs have been fabricated [16] in SOQ films prepared by the graphite-strip-heater recrystallization technique. Figure 6 shows a schematic cross section and photomicrograph of one such device,

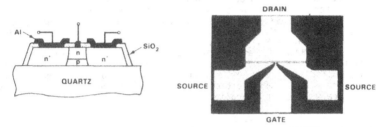

Fig. 6. Si-on-quartz low-noise microwave MESFET.

which has a gate length of ~1.4 μm, gate width of 200 μm, and source-drain spacing of ~3 μm. The maximum DC transconductance is ~45 mS/mm. Several similar devices have been tested for microwave performance. The maximum available power gain calculated from the measured S parameters for a typical device is shown in Fig. 7. The maximum frequency of oscillation,

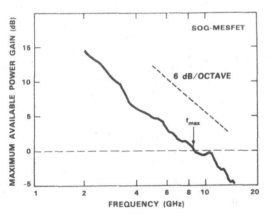

Fig. 7. Maximum available gain as a function of frequency for SOQ MESFET.

f_{max}, is ~9 GHz. The best f_{max} we have measured is ~14 GHz. The device of Fig. 7 has been operated as an amplifier. The minimum noise figure was 2.5 dB at 1.2 GHz, with an associated gain of 10.4 dB. This performance is superior to that of state-of-the-art devices fabricated in SOS [17]. Further improvement in microwave performance should be achieved by reducing the gate length and optimizing other device paramaters.

In MMICs, the active devices occupy only a small portion of the chip, and the passive components are fabricated directly on the insulating substrate. Selective beam recrystallization of a patterned SOI film in localized regions should be the most desirable approach. Considerable research is needed to explore the potential of SOI techniques for this new application area.

VERTICAL (3-D) INTEGRATED CIRCUITS

The trends in VLSI circuitry are toward greater complexity, higher speed, lower power, and greater functional versatility. These trends have been supported primarily by a continuing reduction in feature size, which results in higher device switching speed but lower signal propagation speed along the interconnects. This problem is further amplified as the chip size is increased to achieve higher levels of integration, with the result that circuit performance is often limited by interconnect delay. The 3-D integration of multiple IC layers provides a new possibility for achieving not only multiple functions on a single chip but also high-speed performance due to short wiring. In addition, 3-D integration would allow novel system architecture such as parallel signal processing and new software development.

The key material issue that will determine the usefulness of SOI technology in 3-D integration is whether a device-quality SOI film can be formed on a substrate without affecting the structure and performance of devices already fabricated on the substrate. Laser and e-beam recrystallization have been shown to produce SOI layers with negligible effects on underlying device performance. Significant progress has been made recently in optimizing the recrystallization process to achieve nearly defect-free Si islands over large areas. Several simple 3-D ICs consisting of two layers of active devices have been demonstrated. For example, CMOS ring oscillators [18] exhibiting reasonably good yield and promising speed performance have been fabricated with devices of one type (for example, p-channel transistors) in the Si substrate and the other type (n-channel transistors) in the recrystallized SOI film. In another experiment, an SOI structure was formed on a Si substrate in which a dynamic shift register circuit had been fabricated, and another shift register circuit was then fabricated in the recrystallized SOI film [19]. A divide-by-two circuit using two-layer 3-D integration, with the devices in the two layers electrically connected, has been operated up to 88 MHz [20].

Development of 3-D ICs is an ambitious and difficult task. Many technical problems must be addressed.

(1) All the 3-D circuits fabricated to date employ poly-Si as the interconnect material for the bottom-layer devices because conventional Al metallization is incompatible with the high processing temperature used in fabricating the top-layer devices. Because of their high sheet resistance, however, poly-Si interconnects degrade circuit speed. Therefore advanced interconnect materials, such as refractory metals or their silicides, will have to be used for satisfactory speed. In addition, interconnection between circuits in two different layers requires development of new via hole technology because the separation between the layers is usually at least 1 µm.

(2) Surface planarization techniques have to be developed for the intermediate insulating layers so that the SOI films will have the smooth surface morphology needed for device fabrication. Planarization becomes increasingly important as the number of layers increases, since otherwise the surface topography will become progressively more irregular.

(3) The characteristics of SOI devices are sensitive to the bias voltages applied below the insulating layer. Large bias voltages can lead to an increase in leakage current and a shift in threshold voltage. This

effect must be taken into consideration in the design and fabrication of 3D ICs. The problem can usually be solved by increasing the thickness of the insulating layer or inserting a layer that provides electric field shielding between adjacent layers of active devices.

(4) Heat dissipation becomes more difficult for layers above the Si substrate. Circuit design should therefore be oriented toward low-power technology such as CMOS and partitioning the system so that circuits requiring the most power are located in or close to the Si substrate.

(5) In conventional 2-D technology, the chip size is increased to achieve higher levels of integration, whereas the area of 3-D ICs can remain about the same as the number of layers is increased. The yield model for conventional 2-D ICs, according to which yield decreases exponentially with chip area, will have to be replaced by a new model for estimating the yield of 3-D ICs in order to assess the production feasibility.

Although the research and development effort required to implement 3-D integration as a commercial technology will undoubtedly require many years, this effort is well justified by the potential advantages of the technology. Since the effort is not under immediate time pressure, its scope can be broad enough to include new concepts and applications. For example, 3-D integration could encompass the integration of different semiconductor materials, such as Si and GaAs/GaAlAs, in order to achieve the monolithic integration of optoelectronic and electronic devices. Research on 3-D integration should have great benefits for future generations of integrated circuits.

SUMMARY

Significant progess has been made in the development of SOI technologies for four different IC applications. For very-large-scale ICs, the buried-oxide and oxidation-of-porous-Si technologies have achieved satisfactory material quality for circuit fabrication. However, successful commercialization of these technologies would require the development of high-throughput production systems for SOI wafers in the immediate future, before the rapid advances in bulk Si technology overcomes the present advantages of SOI.

For high-voltage ICs, several beam recrystallization techniques have produced suitable SOI structures, since the material requirements are not as stringent as those for VLSI applications. In this area there is a good opportunity for SOI technology to compete with conventional technologies.

For large-area ICs such as flat-panel displays, it is difficult for beam-recrystallization SOI technology to compete with potentially low-cost thin-film technologies that employ low-temperature processing of materials such as amorphous Si on inexpensive glass substrates. In order to combine SOI control circuits monolithically with low-cost switching elements for display panels, it will be necessary to develop high-throughput beam recrystallization systems and low-temperature processing techniques. Monolithic microwave ICs are a promising new area of application for SOI, with initial results that strongly encourage further research.

Finally, although realization of 3-D ICs for practical use cannot be expected in the near future, their development is a worthwhile long-range project. Continuing research in this area should lead to new device concepts, processing technologies, circuit design methodologies, and system architectures that could benefit IC development for years to come.

ACKNOWLEDGMENTS

The author acknowledges A. J. Strauss for many helpful discussions. This work was sponsored by the Department of the Air Force and the Defense Advanced Research Projects Agency.

REFERENCES

1. See, for example, "Single-Crystal Silicon on Non-Single-Crystal Insulators," edited by G. W. Cullen, J. Cryst. Growth 63, (1983).
2. B-Y. Tsaur, R. W. Mountain, C. K. Chen, and J. C. C. Fan, IEEE Electron Device Lett. EDL-5, 461 (1984).
3. K. Izumi, Y. Omura, M. Ishikawa, and E. Sano, in Digest of Technical Papers, 1982 Symposium on VLSI Technology, Kanagawa, Japan.
4. C. E. Chen, T. G. W. Blake, L. R. Hite, S. D. S. Malhi, B. Y. Mao, and H. W. Lam, IEEE SOS/SOI Technology Workshop, Hilton Head Island, South Carolina, October 1984.
5. F. Otoi, K. Anzai, H. Kitabayashi, and K. Uchiho, IEEE SOS/SOI Technology Workshop, Hilton Head Island, South Carolina, October 1984.
6. T. Mano, T. Baba, H. Sawada, and K. Imai, in Digest of Technical Papers, 1982 Symposium on VLSI Technology, Kanagawa, Japan.
7. C. K. Chen, M. W. Geis, H. K. Choi, B-Y. Tsaur, and J. C. C. Fan, Paper A7.6 presented in this symposium.
8. T. Yamaguchi, S. Morimoto, G. H. Kawamoto, H. K. Park, G. C. Eiden, in Digest of Technical Papers, p. 522, IEDM, Washington, DC, 1983.
9. K. E. Bean and W. R. Runyan, J. Electrochem. Soc. 124, 5C (1977).
10. C. I. Drowley and T. I. Kamins, MRS Spring Meeting, Albuquerque, New Mexico, February 1984.
11. M. Akiya, K. Ohwada, and S. Nakashima, Electron. Lett. 17, 640 (1981).
12. L. Pfeiffer, T. Kovacs, and K. W. West, Paper A7.3 presented in this symposium.
13. A. Chiang, M. H. Zarzycki, W. P. Menli, and N. M. Johnson, in Energy Beam-Solid Interactions and Transient Thermal Processing, edited by J. C. C. Fan and N. M. Johnson (Elsevier, New York, 1984), p. 551.
14. T. Nishimura, A. Ishizu, T. Matsumoto, and Y. Akasaka, MRS Spring Meeting, Albuquerque, New Mexico, February 1984.
15. W. G. Hawkins, D. J. Drake, N. B. Goodman, P. J. Hartman, MRS Spring Meeting, Albuquerque, New Mexico, February 1984.
16. B-Y. Tsaur, H. K. Choi, C. K. Chen, C. L. Chen, R. W. Mountain, and J. C. C. Fan, IEDM, San Francisco, December 1984.
17. R. J. Naster, Y. C. Hwang, and S. Zaidel, in Digest of Technical Papers, p. 72, ISSCC, 1981.
18. S. Kawamura, N. Sasaki, T. Iwai, R. Mukai, M. Nakano, and M. Takagi, in Digest of Technical Papers, p. 364, IEDM, Washington, DC, 1983.
19. S. Akiyama, S. Ogawa, M. Yoneda, N. Yoshii and Y. Terui, in Digest of Technical Papers, p. 352, IEDM, Washington, DC, 1983.
20. T. Nishimura, K. Sugahara, Y. Akasaka, and H. Nakata, Ext. Abs., 16th Conference on Solid State Devices and Materials, p. 527, Kobe, Japan, 1984.

LASER RECRYSTALLIZATION AND 3D INTEGRATION

J.P. COLINGE
CNET, BP98, 38243 Meylan, France

ABSTRACT

There are various methods for producing device-worthy Silicon-on-Insulator films, most, however, are unsuitable for fabrication of 3D integrated structures. The laser recrystallization technique is currently the only one which has produced single-crystal devices for 3D ICs. Improvements on this technique have been such that defects such as grain boundaries can be localized and even eliminated. High speed CMOS circuits with VLSI features have been realized as well as new devices which take advantage of the 3D arrangement of vertically integrated structures. Although 3D integration is still in the early stages of development, it has already opened up new perspectives for applications such as high speed circuits, dense memories, and sensors.

INTRODUCTION

Current integrated circuits are essentially 2D devices made according to planar technologies. Within the last four years, however, increasing research efforts have aimed at the realization of vertically integrated structures , more commonly referred to as "3D" IC technologies. In these techniques, active silicon layers are stacked upon one another, and devices are made in them. 3D integrated circuits may be considered as the simple stacking of "independent" 2D planar layers of devices. Advantage, however, can be taken of the 3D arrangement of the interconnections and active areas to create new devices such as high density memory cells and logic gates. Furthermore, the advantages of Silicon-on-Insulator (SOI) over bulk technologies, such as their latch-up-free nature, are also provided by 3D structures. All the 3D devices realized up to now have been made via cw laser recrystallization of polysilicon, since this technique can be used to deposit energy within a very thin superficial layer. Control of the thermal profile in the silicon film during quenching is the key for obtaining device-worthy single-crystal islands. Techniques such as sophisticated beam shaping and both selective or indirect laser annealing have therefore been developed to induce thermal profiles suitable for the growth of large single-crystals.

This paper reviews the state-of-the-art techniques in the fields of laser recrystallization of silicon films and 3D integrated devices. The prospects of these technologies will be discussed from both the material quality and device points of view.

3D INTEGRATION: STATE-OF-THE-ART

Many different approaches to 3D integrated structures have been reported in the literature, all based on two key issues: growth of device-worthy silicon films on an amorphous insulator (typically SiO2) and realization of devices in upper layers that does not jeopardize the underlying structures already fabricated.

The different approaches published up to now can be classified in the following way:

a: Multilayer SOI, in which SOI layers are stacked upon one another, the O-th layer being the substrate. The different layers are interconnected using a classical metallization process and there is no field-effect interaction between devices belonging to different layers (Figure 1)/1,2,3/.

b: Stacked CMOS, in which a single gate is shared by the n- and the p-channel transistor /4,5,6/. Still greater advantage can even be taken from the 3D arrangement of this type of structure when the drains as well as the gates of the complementary transistors are shared /7/ (Figure 2). This merging capability of stacked CMOS structures offers a large increase in packing density with respect to conventional bulk CMOS /8/. 64K CMOS SRAMS have been made using this stacking principle /9/. · In the stacked transistor CMOS approach (STCMOS), all n-channel transistors are placed in the substrate and the p-channel devices are made in the recrystallized layer, which is of poorer quality than the substrate. STCMOS devices are therefore best adapted for nMOS-oriented CMOS technologies.

c: Staggered CMOS /10/, which is devoted to the realization of specific devices, such as SRAM cells. This approach provides very dense structures (figure 3) in which the drain of one transistor serves as gate for another, and so on. This process is unfortunately incompatible with standard processes.

d: Double-gate devices, like the Cross-MOS inverter /11/ (Figure 4), where a single active layer is controlled by two gates, one over it (top gate), and the other below it (bottom gate).

e: "Mezzanine" devices, in which part of the structure is made in the bulk (storage capacitor), and part of it is made on the field oxide (access transistor). Here advantage is taken from the 3D arrangement of different silicon layers to create dense, high-performance DRAM cells /12,13,14/ (Figure 5).

It is worthwhile noting that although there are many techniques for producing device-worthy SOI films (ELO, SIMOX, FIPOS, various beam recrystallization techniques), with the exception of passivated polysilicon devices /9/, all the existing 3D devices have been realized using a laser recrystallization step. Indeed, laser recrystallization is the only technique providing device-worthy silicon films on an already processed substrate that neither induces overheating nor damages devices fabricated in the underlying layers, owing to the unique features of superficial heating and short time involved. (E-beam recrystallization is quite similar to laser annealing, but it induces damage in the underlying layers /15/).

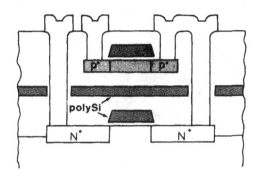

Figure 1: Example of Multilayer SOI CMOS inverter /1-3/. The upper transistor is made in laser-recrystallized silicon.

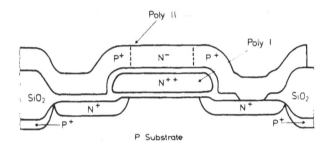

Figure 2: Stacked CMOS inverter, after /7/.

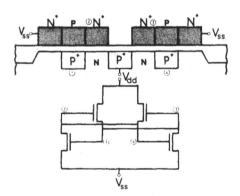

Figure 3: Staggered CMOS and equivalent circuit, after /10/.

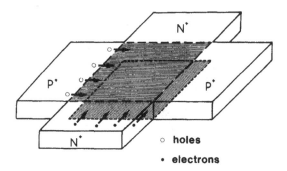

Figure 4: Cross MOS Inverter, after /11/. Gates are not represented, for clarity.

○ holes

• electrons

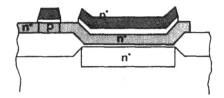

Figure 5: Dynamic RAM cell made in laser-recrystallized silicon /12/.

LASER RECRYSTALLIZATION TECHNIQUES

Pulsed lasers, the first lasers used for polysilicon recrystallization, can only bring about an increase in grain size of up to a few microns. This kind of laser is no longer used in film growth experiments. CW CO2 lasers are used for recrystallization of silicon films on quartz substrates /16/, but their long wavelength precludes superficial heating required for 3D circuit fabrication. The most widely used laser for SOI and 3D processing is the CW argon laser. Owing to its 500 nm wavelength, energy is deposited only within a thin superficial layer. Hence, melting and recrystallization of a polysilicon film can be obtained without raising temperature in the underlying active layers to over 900°C (a 1 um-thick oxide layer is assumed between each silicon layer). Furthermore, the short dwelling time (< 1 ms) involved implies virtually no dopant diffusion within the underlying devices. Within the last four years, many improvements have been made in laser recrystallization, all of which aim at producing defect-free silicon films or islands.

The key for growing large (single-)crystals during laser recrystallization is controlling the shape of the trailing edge. (The trailing edge is the liquid-solid interface at the rear of the molten zone.)

The beam profile of a CW laser being gaussian, chevron-like recrystallization patterns are observed after scanning a polysilicon sample with a conventional focused circular laser beam (Figure 6a). This pattern is caused by the convex shape of the trailing edge in the wake of the advancing spot. Crystal growth proceeds along the thermal gradient (perpendicular to the trailing edge), i.e. crystallites grow from the edges of the scan line towards its center. On the other hand, a concave trailing edge can be obtained by using modified beam profiles such as the doughnut-shaped beam /17/ or partially masked beams /18,19/. In this last case, however, part of the laser power is wasted (Figure 6b). When a concave trailing edge is produced, the center of the scanned line quenches before its edge, and crystal growth proceeds outwards from the already-grown crystal towards the edge of the scanned line. Random nucleation inwards from the edges is thus ruled out.

Tailoring of the trailing edge can be managed by many different ways. One can keep the laser spot circular and tailor the trailing edge through patterning the films instead. For example, prior to recrystallization one can pattern into islands the polysilicon film in order to obtain a concave trailing edge within these islands (moated island structure) /20,21/. One can also tailor the trailing edge by patterning the oxide layer beneath the polysilicon film. In this case, variations in thermal conductivity between the substrate and the silicon film create the desired shape in trailing edge /22/.

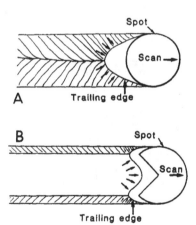

Figure 6: a) Chevron recrystallization pattern.
           b) large grain growth using a partly masqued beam.

Another method, which provides excellent and reproducible results, is the patterning of an antireflection cap on top of the silicon film to be recrystallized /23,24/. Figure 7 shows how tailoring of the trailing edge is achieved using stripes of an antireflecting material (either SiO2, Si3N4, or Si3N4 on SiO2). In this case, the microfloating zone is composed of a succession of concave interfaces. All defects, among them grain boundaries, are then swept underneath the antireflecting stripes by the thermal gradient, and defect-free crystals are obtained, with localized defects between them. It is worth noting that the location of the defects is controlled by lithography, so that defects can be placed in the field area of the circuits. This technique has already been used to recrystallize the second active layer of 3D integrated circuits /3/.

Further improvements on the method have been obtained by using a raster scan of the laser perpendicular to the antireflection stripes in order to induce a pseudo-linear advancing molten zone /25,26,27/. By opening a seeding window at the beginning of the SOI structure <100> crystal areas 200 um long and 5 mm wide can be grown in which the location of the remaining defects (subboundaries) is fully controlled (Figures 8 and 9). These defects can be removed by etching or selective oxidation during further processing steps /26/.

A concave trailing edge can also be obtained using an "indirect annealing" technique in which polysilicon islands are capped by an oxide and a dummy polysilicon layer. The islands melt owing to the heat flow from the uppermost silicon layer down towards the substrate, such that a concave temperature profile is obtained in the islands during solidification /28/.

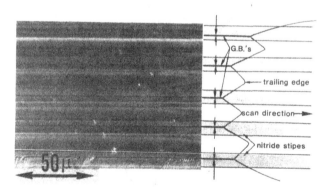

Figure 7: Tailoring of the trailing edge and large grain growth obtained using antireflection (AR) stripes. Grain boundaries are straight and located beneath the AR stripes.

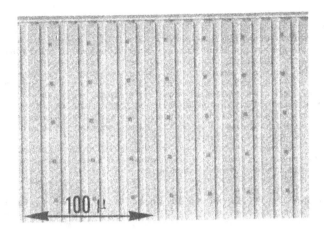

Figure 8: Single-crystal area grown using AR stripes, horizontal oscillatory scan and slow downward motion. The seeding window is at the top of the photograph. The sample is decorated using Secco etch, and etch pits are used to reveal the <100> orientation.

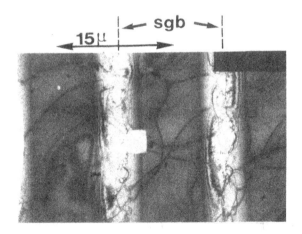

Figure 9: TEM photograph of the crystal of Figure 8. The 15 um-wide grains are separated by subgrain boundaries.

660

DEFECTS IN THE RECRYSTALLIZED FILMS

The most dramatic defects found in laser-recrystallized silicon films are grain boundaries. When located in a transistor, a grain boundary parallel to the current flow can induce a short-circuit between source and drain because of enhanced dopant diffusion along the grain boundary /29/. When the boundary is perpendicular to the carrier flow, it brings about an uncontrolled shift of the threshold voltage and it decreases the mobility of the carriers in the channel /30/. Subgrain boundaries. which are simply dislocation networks between slightly misoriented crystals /31/, seem to have a much weaker electrical activity.

Since the most recent laser recrystallization techniques now provide films and islands free of (sub)grain boundaries, defects the presence of which was formerly masked by the activity of the grain boundaries have now become apparent. Impurities, such as oxygen and nitrogen in the cap layers can dissolve into the silicon during laser processing and create residual film doping (oxygen) or surface states (nitrogen). For example, surface state densities of 4 E 12 cm-2 are observed in samples recrystallized using an Si3N4 encapsulant. This density is reduced to 1 E 11 cm-2 when an oxide cap is used. Reduction of surface states is observed simultaneously at both the upper and lower Si-SiO2 interfaces.

Figure 10 shows typical Id(Vgl,Vg2) curves of a SOI transistor which has been recrystallized using SiO2 antireflection stripes. Threshold voltages are 0.6 and 3 volts at the front and the back interfaces, respectively. If nitride stripes are used, threshold voltages around -2 and -25 volts are observed, giving rise to large leakage currents. Leakage currents can also occur along the transistor edges. Transistor isolation using RIE patterning of mesa islands are therefore be more suitable than LOCOS isolation since the <100> edges of mesas offer minimum surface density features.

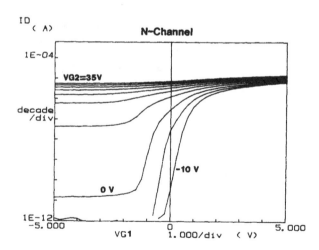

Figure 10: Id(Vgl,Vg2) characteristics of a SOI transistor. L=6 um, W=20 um. Upper gate oxide is 100 nm thick and lower gate oxide is 1 um thick. Nd=5E15 cm-3.

CONCLUSIONS

        Laser recrystallization seems to be one of the most
appropriate techniques for producing device-worthy silicon films in
3D integrated circuits, as only a localized superficial heating of
samples is produced. Recent laser annealing techniques provide grain
boundary-free films and islands in which channel mobility attains
values of 600 cm2/V.s (electrons) and 230 cm2/V.s (holes). Leakage
currents on the order of 1 pA/micron can be obtained. 3D CMOS test
circuits that comply with VLSI design rules (L=3 microns) have been
produced /3/ and operate at clock frequencies of 78 MHz at Vdd=5
volts. Furthermore, owing to the 3D arrangement, new high density
memory cells as well as new types of devices based on 3D field-effect
interactions between stacked active silicon layers have now become
feasible. After only a few years of research in this field,
outstanding results have been obtained and suggest that 3D
integration will be a dramatic breakthrough for the integrated
cicuits during the coming decade.

REFERENCES

1.   S.Kawamura, N.Sasaki, T.Iwai, R.Mukai, M.Nakano, M.Takagi
      Proceedings of IEDM, 364 (1983)
2.   S.Akiyama, S.Ogawa, M.Yoneda, N.Yoshii, Y.Terui
      Proceedings of IEDM, 352 (1983)
3.   T.Nishimura, K.Sugahara, Y.Akasaka, H.Nakata
      Ext. Abstr. of the Conf. on Solid State Devices and Materials
      Kobe, 527 (1984)
4.   J.F.Gibbons, K.F.Lee
      IEEE Electron Dev. Lett. $\underline{1}$, 117 (1980)
5.   G.T.Goeloe,E.W.Maby,D.J.Silversmith,R.W.Mountain,D.A.Antoniadis
      Proceedings of IEDM, 554 (1981)
6.   A.L.Robinson, D.A.Antoniadis, E.W.Maby
      Proceedings of IEDM, 530 (1983)
7.   J.P.Colinge, E.Demoulin, M.Lobet
      IEEE Trans. on Electron Devices $\underline{29}$, 585 (1982)
8.   B.Hoefflinger, S.T.Liu, B.Vajdic
      IEEE J. of Solid-State Circuits $\underline{19}$, 37 (1984)
9.   S.D.S.Malhi, R.Karnaugh, A.H.Shah, L.Hite, P.K.Chatterjee,
      H.E.Davis, S.S.Mahant-Shetti, C.D.Gosmeyer, R.S.Sundaresan,
      C.E.Chen, H.W.Lam, R,A.Haken, R.F.Pinizzotto, R.K.Hester
      Presented at the Device Research Conference
      Santa Barbara, Paper VB-1 (1984)
10.   E.W.Maby, D.A.Antoniadis
      Paper presented at the MRS Spring Meeting
      Albuquerque (1984)
11.   J.F.Gibbons, K.F.Lee, F.C.Wu, G.E.J.Eggermont
      IEEE Electron Dev. Lett. $\underline{3}$, 191 (1982)
12.   R.D.Jolly, T.I.Kamins, R.H.McCharles
      IEEE Electron Dev. Lett. $\underline{4}$,8 (1983)
13.   J.C.Sturm, M.D.Giles, J.F.Gibbons
      IEEE Electron Dev. Lett. $\underline{5}$, 151, (1984)
14.   J.F.Gibbons, K.F.Lee
      Proceedings of IEDM, 111 (1982)
15.   S.Saitoh, K.Higuchi, H.Okabayashi
      Jpn. J. Appl. Phys. $\underline{22}$, Suppl. 22-1, 197 (1983)
16.   N.M.Johnson, D.K.Biegelsen, H.C.Tuan,M.D.Moyer, L.E.Fennell
      IEEE Electron Device Letters $\underline{3}$, 369, (1982)
17.   S.Kawamura, J.Sakurai, M.Nakano, M.Takagi
      Appl. Phys. Lett. $\underline{40}$, 232 (1982)

18. T.J.Stultz, J.F.Gibbons
    Appl. Phys. Lett. 39, 498 (1981)
19. J.M.Hode, J.P.Joly, P.Jeuch
    Techn. Dig. of ECS Spring Meet.,Montréal, 232 (1982)
20. D.K.Biegelsen, N.M.Johnson, D.J.Bartelink, M.D.Moyer
    Appl. Phys. Lett. 38, 150 (1981)
21. G.E.Possin, H.G.Parks, S.W.Chiang, Y.S.Liu
    Proceedings of IEDM, 424 (1982)
22. S.Kawamura, N.Sasaki, M.Nakano, M.Takagi
    J. Appl. Phys. 55, 1607 (1984)
23. J.P.Colinge, E.Demoulin, D.Bensahel, G.Auvert
    Appl. Phys. Lett. 41, 346 (1982)
24. J.P.Colinge, E.Demoulin, D.Bensahel, G.Auvert
    Jpn. J. Appl. Phys., Suppl 22-1, 205, (1982)
25. J.P.Colinge, D.Bensahel, M.Alamome, M.Haond, J.C.Pfister
    Electronics Letters 19, 985 (1983)
26. J.P.Colinge, D.Bensahel, M.Alamome, M.Haond, C.Leguet
    Energy-beam Interactions and Transient
    Thermal Processing, Ed. by J.C.C.Fan and N.M.Johnson
    North-Holland, New York, 597 (1984)
27. C.I.Drowley,, P.Zorabedian, T.I.Kamins
    Energy-beam Interactions and Transient
    Thermal Processing, Ed. by J.C.C.Fan and N.M.Johnson
    North-Holland, New York, 465 (1984)
28. R.Mukai, N.Sasaki, T.Iwai, S.Kawamura, M.Nakano
    Proceedings of IEDM, 360 (1983)
29. K.K.Ng, G.K.Celler, E.I.Povilonis, R.C.Frye, H.J.Leamy
    and S.M.Sze, IEEE Electr. Dev. Lett. 2, 316 (1981)
30. J.P.Colinge, H.Morel, J.P.Chante
    IEEE Trans. on Electron Devices 30, 197(1983)
31. M.Haond, D.P.Vu, D.Bensahel, M.Dupuy
    J. Appl. Phys. 54, 3899 (1983)

SINGLE CRYSTALLINE SOI SQUARE ISLANDS FABRICATED BY LASER
RECRYSTALLIZATION USING A SURROUNDING ANTIREFLECTION CAP AND
SUCCESSIVE SELF-ALIGNED ISOLATION UTILIZING THE SAME CAP

R. MUKAI, N. SASAKI, T. IWAI, S. KAWAMURA, and M. NAKANO
Fujitsu Limited, IC Development Division, Kawasaki 211, Japan

ABSTRACT

A new laser recrystallizing technique has been developed
for high density SOI-LSI's. This technique produces single
crystalline silicon islands on an amorphous insulating layer
without seed. Square windows are opened at arbitrary places in
an antireflection cap over a polycrystalline film on an amorphous insulating layer. Grain boundaries of the polycrystalline Si
in the window are removed completely at the subsequent laser-recrystallization step. Single crystalline silicon islands are
formed by self-aligned etching of silicon film which was covered
by the antireflection cap. This technique is an effective
method for fabricating high density SOI-LSI's, since the single
crystalline islands can be fabricated at arbitrarily selected
places. Yield of the grain-boundary-free islands was 95%; the
size of the island is $10 \times 20\mu m$, and the irradiation overlap of
laser-beam traces is 70%.

INTRODUCTION

Recently, three-dimensional integration(1,2,3) has been realized by
silicon on insulator(SOI) technology utilizing laser-recrystallization technique. Doughnut-shaped beam laser-recrystallization(4), twin-beam laser-re-
crystallization(5), and laser-recrystallization using stripe patterns of
antireflection cap(6) have been developed for producing a grain-boundary
-free stripe region on an amorphous insulating layer. However, these recrystallization techniques have such a restriction for circuit desiners
that active area of transistors should be aligned into rows. Further development of an effective recrystallizing method is needed to fabricate high
density SOI-LSI's.
This paper describes a laser recrystallization of an area defined by a
surrounding antireflection cap over deposited polycrystalline silicon film
on an amorphous insulating layer without seed and successive self-aligned
isolation utilizing the same cap. Different kinds of square shape of single
crystalline silicon island on an amorphous insulating layer are fabricated
by this process. These islands can be fabricated at arbitrarily selected
places, which suggests that this technique is a more effective method for
fabricating high density SOI-LSI's.

LASER RECRYSTALLIZATION OF AN AREA DEFINED BY THE SURROUNDING ANTI-
REFLECTION CAP

We formerly showed that single crystalline silicon square islands on
an amorphous insulating layer were obtained by keeping the interior of the
island cooler than the periphery during recrystallization by the indirect
laser heating technique(7). The new laser recrystallizing technique in
this paper simply and stably produces such a desired temperature profile.

664

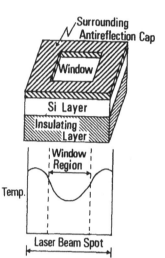

Fig. 1. (a) Schematic view of the sample for the laser recrystallization of an area defined by the surrounding antireflection cap. (b) Temperature profile induced in silicon layer during the recrystallization. The temperature profile realizes that grain boundaries are confined in the surrounding region under antireflection cap.

Table 1. Structure for the present technique.

Surrounding Antireflection Cap	Si$_3$N$_4$ 300Å/SiO$_2$ 300Å
Si Layer	Poly-Si 4000Å
Insulating Layer	Si$_3$N$_4$ 1000Å/SiO$_2$ 1 μm
Substrate	Si Wafer

Table 2. Laser irradiation conditions.

Laser	CW Ar Ion
Power	8-14 W
Scan Speed	1-5 cm/sec
Sub.Temp.	450 °C in air

Figure 1(a) shows the schematic view of the sample for this technique. A window is opened in the antireflection cap where a single crystalline island should be fabricated, as shown in fig.1(a). The laser power absorption in the window area is lower than that in the surrounding region covered by the antireflection cap. Therefore, the interior of the area is kept cooler than the periphery during recrystallization, as shown in Fig.1 (b). Table 1 shows sample structure for the laser recrystallization technique in this work. A 0.4μm thick polycrystalline silicon film was deposited on an amorphous insulating layer by low pressure chemical vapor deposition(LPCVD). The amorphous insulating layer consists of a 0.1μm thick Si$_3$N$_4$ film deposited by LPCVD and a 1μm thick thermal oxide. The cap consists of Si$_3$N$_4$(300Å) and SiO$_2$(300Å). After laser irradiation for the recrystallization, the cap can be used for a self-aligned isolation process. Table 2 shows the laser irradiation conditions in this work.

Figre 2 shows two silicon films recrystallized by this technique after chemical delination of grain boundaries. The cap was removed, after laser irradiation in this case. Square patterns in Fig.2 were induced by mass transport of silicon during the laser recrystallization. The silicon film thickness of the surrounding region covered by the cap was decreased. Figure 2(a) shows a SEM micrograph of a completely recrystallized silicon film. The area defined by the window of the cap was included in the melt width of the laser beam trace; the width was 80μm in this case. In the square regions, no grain boundaries are observed. Grain boundaries are confined to the surrounding region covered by the cap. The size of the

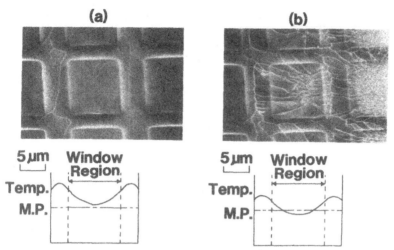

Fig. 2. SEM micrographs of two recrystallized silicon film after removing the cap and delineating grain boundaries chemically, and the temperature profiles of the silicon films during laser irradiation. (a) The silicon film is completely melted and no grain boundary is observed in square regions surrounded by the antireflection cap. These square patterns are made by mass transport of silicon during the laser recrystallization. (b) Central part of the square region did not melt, which suggests that the desired temperature profile is obtained in the square region during laser irradiation.

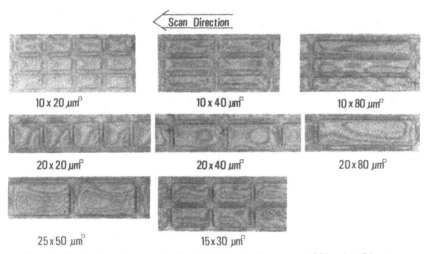

Fig. 3. Optical micrograph of eight kinds of recrystallized silicon film with grain boundary free regions after removing the cap and delineating grain boundaries chemically.

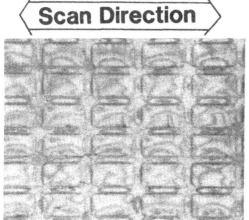

Fig. 4. Optical micrograph of a recrystallized silicon film after removing the cap and delineating grain boundaries chemically. The recrystallization was performed without special care to arrange laser beam trace to square patterns but neighbouring traced were overlapped for each other. No grain boundaries are observed in 95% of square regions. These square regions were defined by the cap during the laser recrystallization.

10 μm   Beam Diameter:80 μm
         Overlap:70%

square region is 10 x 20μm. Figure 2(b) shows a SEM micrograph of a recrystallized silicon film for the irradiation at lower power; silicon film in the square region was incompletely melted but the silicon film region was completely melted by the laser irradiation. This SEM micrograph confirms that the desired temperature profile was obtained in the region during laser recrystallization; the interior of the area was kept cooler than the periphery. The square region was placed inside of the melt width of laser beam trace; the width was 50μm in this case. The grain boundary patterns in the recrystallized silicon film show that nucleation began from the interior of the square region to the surrounding region.

Figure 3 shows another example of the recrystallized silicon films after removing the surrounding antireflection cap and delineating of grain boundaries chemically. Those square regions were placed inside of the melt width of the laser beam trace of 80μm wide. In these recrystallized silicon film, no grain boundaries are observed in all square regions included in the melt width of the laser beam trace.

Figure 4 shows a silicon film recrystallized without special care to arrange laser beam trace to the square patterns but neighbouring traces are overlapped for each other. The laser beam diameter was 80μm and the overlap was 70%. No grain boundaries are observed in 95% of square regions in the laser annealed area. The square regions were defined by the cap during laser irradiation for the recrystallization. The size of the area defined by the cap was 10 x 20μm. The residual 5% square regions show some grain boundaries.

SELF-ALIGNED ISOLATION PROCESS

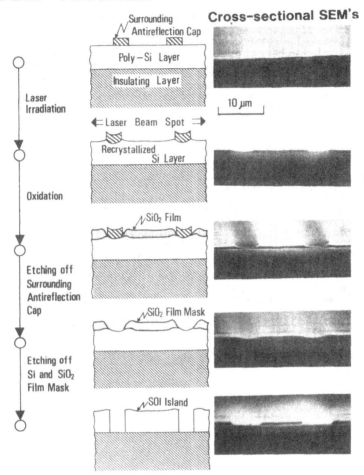

Fig. 5. Schematic representation of the self-aligned isolation process.

SELF-ALIGNED ISOLATION OF GRAIN BOUNDARY FREE REGIONS

The grain boundary free regions produced by laser recrystallization utilizing the surrounding antireflection cap can be isolated by self-aligned process with the same cap. The self-aligned isolation process was realized, as shown in Fig.5. A 2000Å thick thermal oxide was grown in the area by using the surrounding antireflection cap as a mask. After etching off the cap, made up the 300Å thick Si_3N_4 and the 300Å thick SiO_2 films, those regions defined by the antireflection cap were isolated by etching off silicon between them using the 2000Å thick oxide as a mask. This etching was performed in CF_4 plasma. Finally the 2000Å thick oxide was etched off.

SUMMARY

Single crystalline SOI square islands are fabricated by the laser re-crystallization of an area defined by the surrounding antireflection cap and successive self-aligned isolation utilizing the same cap. This laser recrystallization technique is an effective method for producing grain boundary free region in silicon film on an amorphous insulating layer. This laser recrystallization produces different kinds of square shape of grain-boundary-free region at arbitrarily selected place in the silicon film without seed, which suggests that this technique is an effective method for fabricating high density SOI-LSI's.

REFERENCE

1. N.Sasaki, S.Kawamura, T.Iwai, R.Mukai, M.Nakano, and M.Takagi, The 15th Conf. Solid State Devices and Materials, Tokyo, 1983, Late News A-3-7LN. Suppl. Extended Abstract p.24.
2. N.Sasaki, S.Kawamura, T.Iwai, M.Nakano, K.Wada and M.Takagi, The 16th Conf. Solid State Devices and Materials, Kobe, 1984, Late News LC-12-6. Final Program and Late News Abstracts p.72.
3. S.Kawamura, N.Sasaki, T.Iwai, R.Mukai, M.Nakano and M.Takagi, 1984 Symp. VLSI Tech. Diges. Tech. Papers, p.44(1984).
4. S.Kawamura, J.Sakurai, M.Nakano, and M.Takagi, Appl. Phys. Lett. 40, 394 (1982).
5. N.Sasaki, R.Mukai, T.Izawa, M.Nakano, and M.Takagi, Published in Appl. Phys. Lett. Nov. 15 Issue, (1984).
6. J.P.Colinge, E.Demoulin, D.Bensahel, and G.Auvert, Appl. Phys. Lett. 41, 346(1982).
7. R.Mukai, N.Sasaki, T.Iwai, S.Kawamura, and M.Nakano, Appl. Phys. Lett. 44, 994(1984).

TWO-STEP LASER RECRYSTALLIZATION OF SILICON STRIPES IN SiO$_2$ GROOVES FOR CRYSTALLOGRAPHIC ORIENTATION CONTROL

K. EGAMI, M. KIMURA, T. HAMAGUCHI*, and N. ENDO**
Fundamental Research Laboratories, *R&D Planning and Technical Service Division, and
**Microelectronics Research Laboratories, NEC Corporation, Miyazaki, Miyamae-ku,
Kawasaki 213, JAPAN

ABSTRACT

We demonstrate a new two step laser recrystallization for crystallographic orientation control. In the cw Ar ion laser recrystallization of silicon stripes in the structure consisting of SiO$_2$ grooves/polycrystalline Si sublayer/backing substrates, first, one edge of poly-Si stripes is intentionally recrystallized under relatively low laser power and a long dwell time in order to form a strong <100> texture with lamellar grains, second, poly-Si stripes are fully recrystallized using the above <100> texture as seed crystals by scanning a laser beam along the stripes. We discuss a strong <100> texture formation related to partially molten state in the first process of secondary seed formation, and use of a grooved structure with poly-Si sublayer suppressing edge nucleation during lateral epitaxy.

INTRODUCTION

In laser recrystallization, the principal objectives are to enlarge the grain size and to control the crystallographic orientation with respect to device worthy fabrication. To minimize competitive nucleation and thereby enhance the grain size a concave solid-liquid interface is desired. This can be achieved through thermal profile control, e.g. using patterning of silicon, anti-reflection layer, etc. [1,2]. Lateral epitaxy by seeded solidification (LESS) is a typical method that uses a Si wafer as a seed crystal [3,4]. However, for realizing a multilayer Si on insulator (SOI) strucutre, such as for three dimensional devices, it is desirable to control the orientation without the use of the Si substrate as a seed crystal, because the reservation of a seed area adds to process complexity and uses an additional area.

As a first approach, a double laser recrystallization using a cw Nd:YAG laser and a cw Ar ion laser was developed to overcome the above problems [5]. This technique consists of two processes; one is a secondary seed formation process and the other is a grain growth process. Since a strong <100> texture can be obtained by cw Nd:YAG laser recrystallization [6], this laser was applied for the secondary seed formation. And for the grain growth process, a SOI structure having Si stripes embedded in SiO$_2$ grooves/polycrystalline silicon sublayer and SiO$_2$ insulator was developed for suppressing edge nucleation by the edge heating effect [7]. However, when using a transparent substrate, such as a quartz glass with small thermal expansion coefficient, microcracks are generated. Furthermore, a cw Nd:YAG laser having small absorption coefficient for Si [8] is destructive for underlying Si layers of multilayer SOI. Until now, it was difficult to form a strong <100> texture by a cw Ar ion laser recrystallization of Si on a SiO$_2$ layer. Recently we have shown that a cw Ar laser at very long dwell times is also effective in producing highly textured films [9]. We therefore, here, introduce a two step process using only an Ar laser which has the beneficial aspect of creating negligible damage in underlying layers.

EXPERIMENTS

A schematic view of our sample is shown in Fig. 1. We prepared two kinds of substrates; (1) sapphire and (2) surface oxidized silicon wafers. First, a chemical vapor deposited (CVD) polycrystalline silicon film was deposited on both substrates with 2 in. diameter and 330µm thickness. Next, either a CVD SiO$_2$ film was deposited on the poly-Si surface or a thick poly-Si film was partly oxidized to form SiO$_2$/poly-Si/substrate structure for a single layer SOI. Grooves were then formed in the surface of the SiO$_2$ layer with a conventional UV lithography and dry etching process. Typical values of the groove width (w) was 10-26µm. The groove isolation width (x) was 2-10µm. The groove depth (t) was 0.6μm. As a final active layer, a low pressure chemical vapor deposited(LPCVD) polycrystalline silicon film was deposited on the grooved SiO_2 surface at 700 °C [6]. The fabrication of poly-Si stripes in grooves was carried out by a chemical and mechanical polishing. In this process of planarization, only the undesirable poly-Si, which was not in SiO_2 grooves, was removed. The surface step height obtained was below 0.06μm.

In laser recrystallization, as mentioned above, Si stripes were recrystallized by two processes, one is a secondary seed formation with a strong <100> texture and the other is a subsequent grain growth. In the secondary seed formation, one edge of Si stripes was intentionally recrystallized under relatively low power and slow scan speed giving rise to a long dwell time (30-40 msec). The spot size of cw Ar ion laser was 20-90μm in diameter and the scanning speed was 1-3mm/s. The power range was 6-13W. As shown in Fig. 1, the laser beam was scanned along the X direction. In the subsequent grain growth, Si stripes were recrystallized by scanning a laser beam along the Y direction using the above <100> texture as a seed crystal. The spot size of the annealing beam was 50-150μm in diameter with scanning speed of 10-20mm/s. The power range was 8-15W. The substrate temperature was kept at 300 °C in air to reduce the laser power by increasing the optical absorption in the silicon.

After recrystallization, the samples were chemically etched with Secco etchant to reveal grain boundaries. These were examined by optical microscopy and an electron channeling pattern(ECP) method [10]. The electron channeling pattern, which shows the crystallographic orientation in a microscopic area, was observed by rocking electron beam, 3-6μm in diameter, with an angular width of 30 ° and an accelerating voltage of 25-40 kV.

RESULTS AND DISCUSSIONS

Previously the authors have presented the grooved SOI/poly-Si sublayer structure which causes edge heating and suppression of edge nucleations for the grain enlargement [7]. Here, the pattern size effect was investigated for the stable subsequent grain growth. Results of using sapphire substrates are shown in Fig. 2. The samples, given in Fig. 2 (a) and 2 (b), were recrystallized under the same annealing conditions. For the small isolation width (2 μm), given in Fig. 2(a), small grains were formed in the center of the stripes and large grains in the side area. This indicates insufficient laser power density and clearly shows the edge heating effect. In Fig. 2(b), the single crystal film without grain boundaries was formed under an optimum laser power density. The wider isolation (5-10μm) led to a more stable concave solid-liquid interface resulting in long single crystal Si stripes (max. 3.8 mm). A similar result was obtained in the case of Si substrates. Of course here, much more energy must be absorbed in the poly-Si sublayer under the wide isolation region due to the larger thermal conductivity of the substrates. On the other hand, in the case of quartz glass substrate with a small thermal conductivity, a narrow isolation width (1.5μm) can result in the stable concave solid-liquid interface [7].

Previously a strong <100> texture formation by cw Ar ion laser has been investigated [9]. We believe that the initial poly-Si film structure plays an important role in achieving a strong <100> texture formation through the partially molten state [6], in addition to other contributing factors [11,12]. So, the 700 °C LPCVD poly-Si film with a preferred <100> orientation is preferable for a strong <100> texture formation by laser recrystallization [6]. Consequently, it has been found that the large dwell time annealing (30-40msec) for surviving <100> oriented crystallites in the partially molten state enables one to form a strong <100> texture with lamellar Si grains in spite of using high thermal conductivity substrates [9].

Therefore using relatively low laser power and long dwell time annealing, we carried out the secondary seed formation. In lateral epitaxy by seeded solidification, a part of the Si substrates is the primary seed crystal. Here we replace the intrinsic seed with an intentionally grown secondary seed crystal. Figure 3 shows the surface photographs on the stripe patterned films after using a 35-40msec dwell time. The Si lamellae were seen to align along the groove on the sapphire substrate as shown in Fig. 3(a). Using Si substrates, although the grain size was small (1 x 5 μm), Si lamellae were resolidified, as shown in Fig. 3 (b). As previously reported [12], it was difficult for Si grains with lamellar features to be resolidified on the continuous Si film/SiO_2/Si

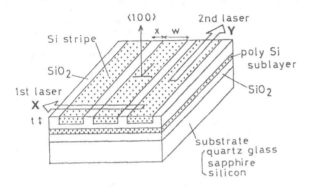

Fig.1 Schematic view of a present SOI structure with a poly-Si sublayer under the SiO₂ grooves. The thickness of poly-Si sublayer is 0.3-0.4μm.

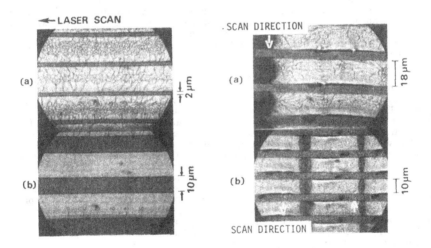

Fig.2 Surface photographs showing the pattern size effect of laser recrystallized Si stripes in grooves after Secco etching.

Fig.3 Surface photographs of Si lamellae at the stripe patterned film.
(a):on the sapphire substrate annealed with a beam spot of 80μm, a scanning speed of 2mm/s and a power of 7W.
(b):on the Si substrate annealed with a beam spot of 35μm, a scanning speed of 1mm/s and a power of 8W.

substrate. Thus, in the patterned Si films, a rapid thermal dissipation along both lateral and vertical direction of the substrate could be suppressed, especially along the lateral direction.

Extensive ECP observations gave us a new result that the elongated direction of Si lamellae is a $<110>$ direction in plane orientation, as shown in Fig. 4. Intentionally, a Si lamella having a misorientation 24 deg. relative to the stripe, was chosen for study. The ECP corresponding to this grain indicated by an arrow shows that the elongated direction of the lamella has a $<110>$ orientation in the plane in addition to a $<100>$ direction parallel to the substrate normal.

Now, we discuss the stable formation of Si lamellae and the alignment parallel to the stripe direction. The Si lamellae, given in Fig. 5, appear to be resolidified due to the thermal profile produced by the present SOI structure. Figure 5(b) shows the estimated thermal profile in the stripe on scanning a laser beam across the stripes. Thus, it seems that Si lamellae stably align along the isotherms parallel to the AA' direction. The fact that the elongated direction of Si lamellae resolidified from the partially molten state is parallel to a $<110>$ direction is supported by recent work of Celler et al. [16, 17]. They have shown that on a (100)Si wafer surface directly heated by a radiative source the molten areas are rectangular (often approximately square) with edges exactly parallel to the $<110>$ directions and the molten region is fully faceted, bounded by the (111)planes. Figure 6 shows an illustrative configuration of Si lamellae. It is believed that the interface between underlying SiO_2 and the base of Si lamellae is exactly parallel to the (100)plane to minimize the interfacial energy [14], however, that the side of Si lamella is parallel to (100)planes is not necessary. Thus, assuming two (111)planes one can satisfy the crystallographic geometry, given in Fig. 6.

In the subsequent grain growth, since the Si lamellae once annealed have smaller optical absorption coefficients than that of an initial polysilicon film, the grain growth from seed area smoothly occurred. The laser beam spot covering three or four stripes could bring about one or two single crystal Si stripes without defects. Figure 7(a) shows a surface photograph after Secco etching to reveal grain boundaries, including the secondary seed area and the grain growth area. A Si stripe in the center part of the photo seems to be grown from a lamellar grain being used as a seed crystal. As shown here, using the large thermal conductivity substrates, such as sapphire (0.3 W/deg. cm) and silicon (1.5 W/deg. cm), single crystal Si stripes (0.5-1 mm in length) could be grown without microcracks induced by thermal stress. On the other hand, the sapphire and silicon wafers are also large thermal expansion substrates in comparison with quartz glass. During a laser recrystallization, only a surface layer was heated to high temperature, however, the whole substrate was held at low temperature (300 °C). Then, the total mismatch of thermal dilation between the very near surface layer and the backing substrate can be reduced by using substrates with larger thermal expansion coefficient than a quartz glass. Figure 7(b) shows ECP indicating a $<100>$ orientation along the substrate normal obtained in a single crystal Si stripe. At the present time, many ECP observations revealed that the in-plane orientation of the recrystallized Si stripes by this two step technique is random. As shown in Fig. 5, although the probability of the Si lamellae aligned parallel to the stripe direction is not low, the in-plane orientation control by this two step technique is not perfect now. A stable formation of Si lamellae with both the $<110>$ orientation parallel to the stripe direction and the $<100>$ orientation normal to the substrate surface will offer the possibility for controlling in plane orientation by this method.

As previously mentioned, Si stripes in the grooves which were simply recrystallized by scanning cw Ar ion laser along the Y direction were randomly oriented with respect to the substrate normal direction [7]. Here, we have found an effect for the orientation control by this two step laser recrystallization. In strip heater zone melting technique, Smith et al. indicated that the transition region which is seen at the beginning of zone melt edge acts as seed crystals [12, 14]. Probably, the $<100>$ oriented crystallites normal to the substrate surface can survive and grow in this transition region. Furthermore, because a low scanning speed (several mm/s) and a wide melting zone in strip heater technique work well for the large grain growth and the orientation control, such a single step recrystallization can overcome the above two problems. However, in laser recrystallization, it is difficult to form both the transition region and the grain growth area by a single laser scanning. As Hawkins et al. reported, the partially molten state can be realized by in a specific power range P_1-P_2 [15], the partially molten state

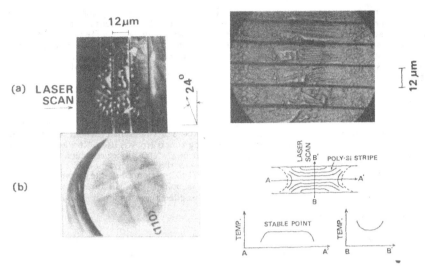

Fig.4 (a)SEM photograph after lamellar melting corresponding to electron channeling pattern(b) showing a ⟨100⟩ direction parallel to the substrate normal and a ⟨110⟩ direction parallel to the Si lamella in plane axis.

Fig.5 Surface photograph(upper) of the secondary seed area and the estimated thermal profile(lower) on the aligned Si lamellae parallel to the stripe direction.

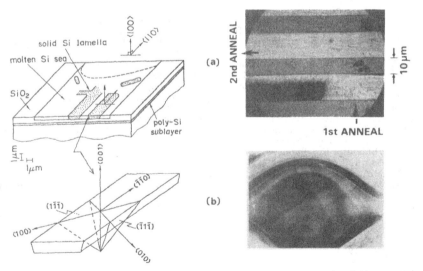

Fig.6 Schematic Si lamellae resolidified from the partially molten state and the crystallographic configuration with two (111) planes at the side.

Fig.7 Surface photograph of the secondary seed area and the grain growth area after Secco etching(a) and the electron channeling pattern(b) of the grain growth area.

zone at the beam edge is very narrow in rapid laser scanning for a full melting. Furthermore, in a short dwell time, the surviving of $<100>$ oriented crystallites that act as seed crystals for the subsequent grain growth may be imperfect. Therefore, it was necessary to separate the recrystallization into two processes: one is a secondary seed formation process under a low laser power density and a long dwell time, and the other is a subsequent grain growth process under a high power density on the grooved SOI structure with a poly-Si sublayer.

SUMMARY

We have demonstrated a two step laser recrystallization for crystallographic orientation control. For a stable grain growth, the SOI structure in which Si stripes are put in SiO_2 grooves/poly-Si sublayer/SiO_2/backing substrate was used. It was pointed out that a long dwell time annealing brings about a strong $<100>$ texture which can act as a secondary seed crystal using lamellar features in the partially molten state. In the subsequent grain growth, long (0.5-1mm) and $<100>$ oriented single crystal Si stripes were formed with no microcracking on both sapphire and silicon substrates. This technique can potentially be used to fabricate unseeded three dimensional devices.

ACKNOWLEDGEMENT

The authors acknowledge Dr. H Tsuya and Dr. H. Watanabe for their encouragement and helpful discussions. We also express our thanks to M. Eguchi for his technical assistance. This work was performed under the management of the R&D Association for Future Electron Devices as a part of the R&D Project of Basic Technology for Future Industries sponsored by the Agency of Industrial Science and Technology, MITI.

REFERENCES

[1] D.K.Biegelsen, N.M.Johnson, D.J.Bartelink, and M.D.Moyer, Mat. Res. Soc. Symp. Proc. 1, 487 (1981).
[2] T.J.Stultz and J.F.Gibbons, Appl. Phys. Lett. 39, 498 (1981).
[3] M.Tamura, H.Tamura, and T.Tokuyama, Jpn. J. Appl. Lett. 44, L23 (1980).
[4] H.W.Lam, R.F.Pinizzotto, and A.F.Tasch, Jr., J. Electrochem. Soc. 128, 1981 (1981).
[5] K.Egami, M.Kimura, and T.Hamaguchi, Appl. Phys. Lett. 44, 962 (1984).
[6] M.Kimura and K.Egami, Appl. Phys. Lett. 44, 420 (1984).
[7] K.Egami, M.Kimura, and T.Hamaguchi, Appl. Phys. Lett. 43, 1023 (1983).
[8] G.E.Jellison, Jr., and F.A.Modine, Appl. Phys. Lett. 41, 180 (1982).
[9] K.Egami, M. Kimura, Appl. Phys. Lett. 45, 854 (1984).
[10] C.G.Van Essen, E.M.Schulson, and R.H.Donaghey, Nature 225, 847 (1970).
[11] M.W.Geis, D.A. Antoniadis, D.J. Silversmith, R.W. Mountain, and H.I. Smith, Appl. Phys. Lett. 37, 454 (1980).
[12] H.I.Smith, C.V.Thompson, M.W.Geis, R.A.Lemons, and M.A. Bösch, J.Electrochem. Soc. 130, 2050 (1983).
[13] M.A.Bösch and R.A.Lemons, Phys. Rev. Lett. 47, 1151 (1981).
[14] M.W.Geis, H.I.Smith, B-Y, Tsaur, J.C.C.Fan, D.J.Silversmith, and R.W.Mountain, J. Electrochem. Soc. 129, 2812 (1982).
[15] W.G.Hawkins and D.K.Biegelsen, Appl. Phys. Lett. 42, 358 (1983).
[16] G.K.Celler, K.A.Jackson, L.E.Trimble, McD.Robinson, and D.J.Lischner, Mat. Res. Soc. Symp. Proc. 23, 409 (1984).
[17] G.K. Celler, McD. Robinson, L.E. Trimble, and D.J. Lischner, Appl. Phys Lett. 43, 868 (1983).

PHOTODETECTOR ARRAYS IN LASER-RECRYSTALLIZED SILICON INTEGRATED WITH AN OPTICAL WAVEGUIDE

R.W. WU*, H.A. TIMLIN*, H.E. JACKSON**, AND J.T. BOYD*
* Solid State Electronics Laboratory, Department of Electrical and Computer Engineering, University of Cincinnati, Cincinnati, Ohio 45221-0030
** Department of Physics, University of Cincinnati, Cincinnati, Ohio 45221-0011

ABSTRACT

Integrated detection of light propagating in an optical waveguide by a photodetector array fabricated directly on the waveguide surface is demonstrated. Laser recrystallization of LPCVD polysilicon patterned with periodically-spaced anti-reflection stripes is utilized. Lateral p-i-n photodiode elements formed by ion implantation are characterized by reverse leakage currents of $< 10^{-11}$ amp and a typical breakdown voltage of 50 volts. Optical response is found to be linear over a dynamic range of greater than 55 dB.

INTRODUCTION

Detection of light propagating in a low loss planar optical waveguide in an integrated way is recognized as an important step in forming fully integrated optical devices. Recently the technique of laser annealing has successfully reduced loss in several thin-film optical waveguides to very low values [1-3]. Very recently the integration of a photodetector with the waveguide has been attempted. Photodetectors have been formed in amorphous silicon deposited on a LiNbO3 substrate [4] and in laser recrystallized silicon on a LiTaO3 substrate [5]. In the first case poor device performance resulted, while device performance in the second was limited by small crystalline grain size. In the latter case the photodetectors were not integrated with a waveguide and integrated waveguide detection was not demonstrated. We report here the formation of a high quality photodetector array in laser recrystallized silicon deposited on a silicon nitride planar waveguide and integrated waveguide detection. In the experiments reported here, we utilize a Si3N4 planar waveguide, but such photodetector arrays could be formed on planar optical waveguides utilizing other materials as well.

LASER RECRYSTALLIZATION

Laser recrystallization of polycrystalline silicon has been successfully achieved using a technique first utilized by Colinge et al. [6]. Essentially the growth of grain boundaries is confined to specific locations by using a pattern of anti-reflection coatings. In our case a 650Å layer of Si3N4 was deposited on the oxide-covered polysilicon and lithographically patterned to form a periodic array (28 micron period) of 8 micron wide anti-reflection stripes. An argon-ion laser with a output power of 9-11 watts illuminates a layer of 0.45 micron thick LPCVD polysilicon capped with a 150Å layer of SiO2 and is scanned at a velocity of 20 cm/sec uni-directionally parallel to the anti-reflection stripes. The focused spot of the laser is 55 microns with a scan-to-scan spacing of 17 microns. Because more laser power is transmitted through the anti-reflection stripes, those regions are heated to a higher temperature so that recrystallization begins midway between the anti-reflection stripes.

Grain boundaries thus become localized along the anti-reflection stripes where two adjacent regions undergoing recrystallization meet.

A photograph of the results of laser recrystallization is displayed after Seeco etching in Fig. 1. Four adjacent regions of single crystal silicon are seen, each 20 microns wide and greater than 250 microns long.

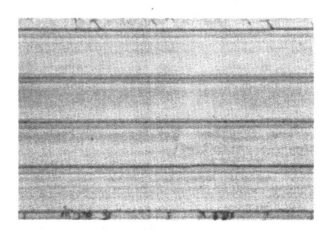

Fig. 1. Photomicrograph of laser recrystallized polysilicon after Seeco etching. Grain boundaries are seen to be confined under anti-reflection stripes. The periodic spacing is 28 microns.

By optimizing exposure parameters, multiple adjacent single crystal regions each 20 microns wide and several millimeters long have been formed with grain boundaries in between. Because these single crystal regions are periodically spaced, they are ideal for fabrication of a photodetector array where one photodetector is positioned in the center of each single crystal region. The periodic spacing of the photodector array is thus the same as that of the anti-reflection stripes. The grain boundaries then lie in the isolation region between adjacent photodetector elements; these regions are removed by plasma etching after junction formation to eliminate any adverse effect the presence of the grain boundary might have on electrical crosstalk.

INTEGRATED WAVEGUIDE STRUCTURE

Figure 2 shows a cross-sectional view in a plane normal to the array axis of one element of the complete integrated waveguide structure. The 0.14 micron Si_3N_4 waveguide is deposited by LPCVD on a 1.6 micron SiO_2 layer thermally grown on a silicon substrate. The photodiode sturcture was fabricated by ion implantation with boron, boron, and phosphorus doses of 3×10^{14} cm^{-2}, 6×10^{11} cm^{-2}, and 1×10^{15} cm^{-2}, and energies of 40 keV, 60 keV, and 150 keV for the p+, p- and n+ implants, respectively. The photodetectors here are edge-illuminated so that the quantum efficiency is determined by the length of the photodetector normal to the array axis. As this dimension does not effect array resolution, i.e. corresponds to the long dimension of the crystalline grains, it can be sufficiently long to yield high values of quantum efficiency [7].

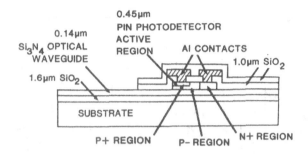

Fig. 2. A cross-sectional view of one element of the integrated optical photodetector array fabricated on a planar Si3N4 optical waveguide.

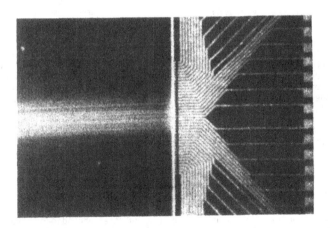

Fig. 3. A photomicrograph showing the laser streak propagating in the waveguide from the left to the right in the figure. The metal pattern of the photodetector array in the right half of the figure is illuminated by scattered light and has a period of 28 microns.

In Fig. 3 light propagating in the waveguide (streak in left half of photomicrograph) illuminates the metal pattern of a 40 element photodetector array. The photodetector array is one part of an extensive test pattern including isolated pn junctions and MOS devices formed to evaluate the performance of these devices formed on the laser recrystallized silicon.

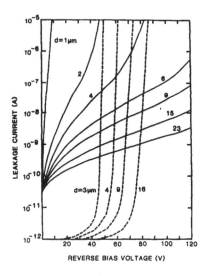

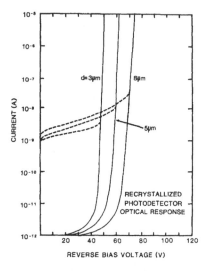

Fig. 4. Comparison of current-voltage charateristics of photodiodes formed in non-recrystal-polysilicon (solid curves) and photodiodes formed in laser recrystallized silicon (dashed curves). Cuves are given for several values of d, the spacing between the p+ and n+ regions.

Fig. 5. Current-voltage character-istics of an element of the photo-detector array formed on recrystal-lized silicon with light propagating in the waveguide (dashed curves) and without light propagating in the waveguide (solid curves). Detection of light from the waveguide provides a photocurrent response of 3-4 orders of magnitude for a input He-Ne laser power of 4 mW.

PHOTODIODE CHARACTERIZATION AND INTEGRATED WAVEGUIDE DETECTION

The key question concerning the use of the photodiode array is the quality of the individual photodiodes. In Fig. 4 we compare the current-voltage characteristics of photodiodes formed in non-recrystallized polysilicon (solid curves) with those formed in laser recrystallized silicon (dashed curves). Several curves are displayed for different values of d, the separation between the p+ and n+ regions. Note that a high reverse leakage current and a soft breakdown are observed for the photodiodes formed on polysilicon. In contrast, for the photodiodes formed on laser recrystallized silicon much lower leakage currents (in the 10^{-12} amp range) and a sharper breakdown characteristic with breakdown voltages of about 50 volts are observed. The low leakage currents observed here are about 3 orders of magnitude smaller than those previously reported [8] for a lateral diode formed in laser recrystallized silicon, and correspond to leakage current densities somewhat less than the ~ 10^{-5} amp/cm^2 previously reported [8].

In order to demonstrate integrated waveguide detection by the photodetectors, light from a 4 mW He-Ne laser operating at 632.8 nm was prism-coupled into the Si$_3$N$_4$ waveguide and thus to a photodetector element. The I-V curves in Fig. 5 show the detector response in the presence of light in the waveguide (dashed curves) and well as the response in the absence of light in the waveguide (solid curves). Integrated waveguide detection is demonstrated by these curves which show a three order of magnitude increase in current when light is present in the waveguide. Figure 6 displays the optical dynamic range observed for several photodetectors. The dynamic range was measured to be 55 dB for d=5 microns, increasing to 60 dB for a device with d=27 microns.

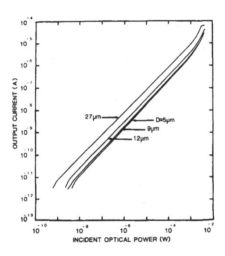

Fig. 6. Plot of detector output current as a function of incident laser power for several values of d.

SUMMARY

In summary, we have demonstrated integrated detection of light propagating in an optical waveguide by forming a photodetector array directly on the waveguide surface. A high quality photodetector array was formed in laser recrystallized silicon and provided an optical dynamic range of greater than 55 dB for the integrated waveguide detection.

ACKNOWLEDGEMENTS

We acknowledge the contributions of John L. Janning of NCR Microelectronics Division, Miamisburg, Ohio, to the successful fabrication of the devices described above. This research was supported in part by the Air Force Office of Scientific Research and the National Science Foundation.

REFERENCES

1. S. Dutta, H.E. Jackson, and J.T. Boyd, J. Appl. Phys. 52, 3873 (1981).

2. S. Dutta, H.E. Jackson, J.T. Boyd, F.S. Hickernell and R.L. Davis, Appl. Phys. Lett. 39, 206 (1981).

3. S. Dutta, H.E. Jackson, J.T. Boyd, R.L. Davis, and F.S. Hickernell, IEEE J. Quant. Electron. QE-18, 800 (1982).

4. J. Yumoto, H. Yajima, Y. Seki, and J. Schimada, Appl. Phys. Lett. 40, 632 (1982).

5. R.E. Reedy and S.H. Lee, Appl. Phys. Lett. 44, 19 (1984).

6. J.P. Colinge, E. Demoulin, D. Bensahel, and G. Auvert, Appl. Phys. Lett. 41, 346 (1982).

7. C.L. Fan and J.T. Boyd, Appl. Opt. 22, 3297 (1983).

8. G.E.J. Eggermont and J.G. de Groot, IEEE Electron. Device Lett. EDL-3, 156 (1982).

X - RAY STUDIES OF STRAIN IN
LASER-CRYSTALLIZED SOI FILMS

MICHEL RIVIER* AND F. REIDINGER**, G. GOETZ***, J. McKITTERICK***,
A. CSERHATI***
  *Michel River, IBM France, BP 58, 91120 Corbeil-Essonnes, France
  **F. Reidinger, Allied Corporate Laboratories, CRL-206, PO Box 1021 R,
  Morristown, NJ 07960
***G. Goetz, Bendix Aerospace Technology Center, 9140 Old Annapolis Road,
  Columbia, MD 21045
***J. McKitterick, Bendix Aerospace Technology Center, 9140 Old Annapolis
  Road, Columbia, MD 21045
***A. Cserhati, Bendix Aerospace Technology Center, 9140 Old Annapolis Road,
  Columbia, MD 21045

ABSTRACT

We have used x-ray diffraction to measure the strain perpendicular to
the substrate surface in laser crystallized silicon films on oxidized
silicon and fused quartz substrates. The dependence of the strain on grain
orientation was determined and the influence of the scan speed, the insulating oxide thickness, and subsequent high temperature exposure was
examined. Maximum strain was obtained for grains oriented with the (100)
plane parallel to the substrate surface. The strain decreased with
increasing angle between the surface plane and the (100) plane of the grains.
The stress parallel to the surface in the variously oriented grains was
calculated from the stiffness tensor, assuming an isotropic, in-plane stress,
and a variation similar to the strain was found. The strain found on
oxidized wafers was about half that on fused quartz. Its dependence upon the
oxide thickness (0.2 um to 1.0 um) was not significant for scan speeds under
10 cm/sec. Similarly, the variation in strain with scan speed was very small
for speeds below 10 cm/sec. Scan speeds above 50 cm/sec caused significant
increases in the strain.

The measured strain was reduced by high temperature anneals. A 1100°C
anneal reduced the average strain by 60% and caused a clear reduction in
grain imperfections (as determined by diffracted beam width). However, a
900°C anneal increased the diffracted beam widths even though the average
strain was reduced by about 30%.

APPROACH

Polysilicon films of 0.5 um thickness were deposited using LPCVD, on
fused quartz substrates and 100 silicon substrates with thermally grown
oxides of thicknesses 0.2 um, 0.5 um, and 1.0 um. A 2.0 um thick LPCVD
oxide cap was deposited on the polysilicon films and was kept in place
throughout the processing and measurement steps. The polysilicon was
crystallized by a CW Ar ion laser with a 100 um spot size. Scan speeds were
varied from 0.1 cm/sec to 400 cm/sec with the power adjusted to obtain the
same molten line width at each speed. A 60% overlap was used for all test
areas. All samples were examined "as crystallized" except for a set with
oxide thickness 1.0 um and scan speed 100 cm/sec which were annealed at
900°C, 1000°C, and 1100°C for one hour to simulate typical processing
temperatures.

The x-ray measurements were made with a diffractometer operating with
Cu K_1 radiation. The lattice constant perpendicular to the substrate surface
was derived directly from the Bragg angle measurement for a number of crystal
planes (Figure 1). It is estimated that the measured values of the Bragg
angles have a maximum error of +0.01°

682

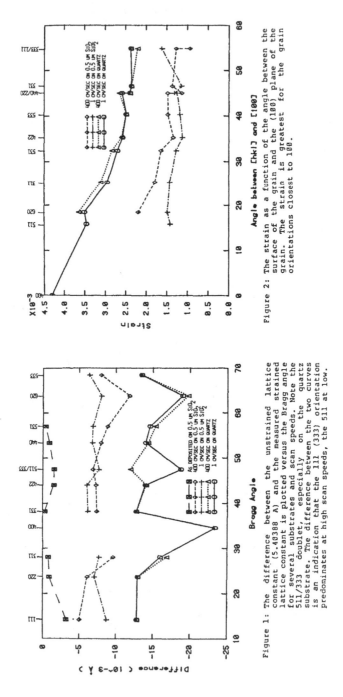

Figure 1: The difference between the unstrained lattice constant (5.40388 Å) and the measured strained lattice constant is plotted versus the Bragg angle for several substrates and scan speeds. Note the 511/333 doublet, especially on the quartz substrate. The difference between the two curves is an indication that the 111 (333) orientation predominates at high scan speeds, the 511 at low.

Figure 2: The strain as a function of the angle between the surface of the grain and the (100) plane of the grain. The strain is greatest for the grain orientations closest to 100.

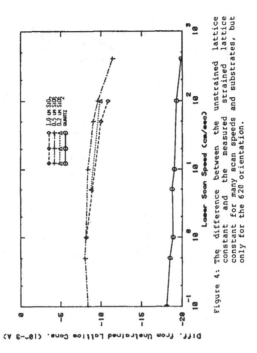

Figure 4: The difference between the unstrained lattice constant and the measured strained lattice constant for many scan speeds and substrates, but only for the 620 orientation.

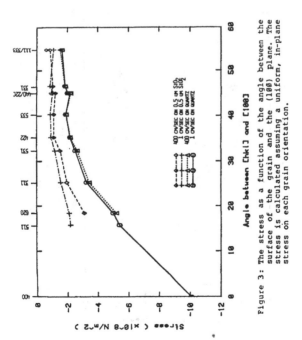

Figure 3: The stress as a function of the angle between the surface of the grain and the (100) plane. The stress is calculated assuming a uniform, in-plane stress on each grain orientation.

684

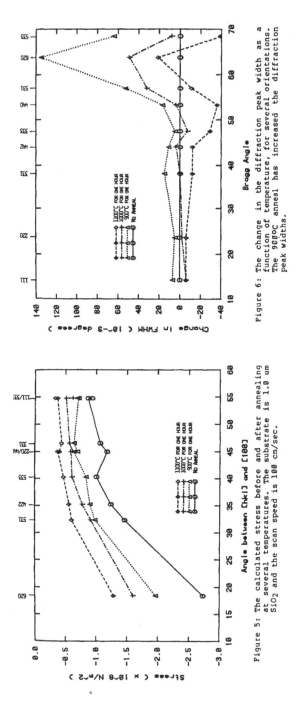

Figure 6: The change in the diffraction peak width as a function of temperature, for several orientations. The 900°C anneal has increased the diffraction peak widths.

Figure 5: The calculated stress before and after annealing at several temperature. The substrate is 1.0 um SiO2 and the scan speed is 100 cm/sec.

RESULTS

A. Strain vs. Grain Orientation

From the lattice constant, the strain is easily calculated. The strain was found to vary significantly with grain orientation for all crystallized films, and is shown in Figure 2 as a function of the angle between the (100) plane of the grain and the substrate surface. We have split the 511/333 doublet (see Figure 1) by assuming the high scan speed result is due to a 333 reflection and the low scan speed due to the 511, since the (111) orientation predominates at high scan speeds. Figure 3 shows the calculated stress that results in the measured strains, assuming the stress is in-plane and uniform within the grains. These results indicate a sizable variation in stress between the differently oriented grains.

B. Strain vs. Scan Speed

Some indication of the weak dependence of strain on scan speed can be seen in Figures 2 and 3 from the small difference between the results for 1 cm/sec and 400 cm/sec. This is more clearly shown in Figure 4 where the strained lattice constant is shown for several substrates and many scan speeds. Measurements of the (620) grains are shown, since these have higher accuracy. It can be seen that there is little dependence on scan speed or oxide thickness for speeds of 10 cm/sec and less. For higher scan speeds there is a significant increase in strain.

C. Influence of High Temperature Anneal

The reduction of in-plane stress during processing at elevated temperatures is shown in Figure 5. The relative change is similar for all grain orientations and even a 900°C anneal causes significant reduction in stress. Changes in the width of the diffraction peaks produced by the elevated temperature anneals are shown in Figure 6. Noteworthy is fact that the 900°C anneal increased the width of the peaks, while the 1100°C anneal reduced the widths.

CONCLUSIONS

1. Both strain and stress are a function of grain orientation. They are maximum for grains with (100) planes parallel to the surface of the film (Figures 2 and 3) and decrease with increasing angle between the (100) and surface planes.

2. Stress in films crystallized on quartz substrates is about twice that for films crystallized on oxidized silicon wafers for all grain orientations and has little dependence on scan speed.

3. Stress in films crystallized on oxidized wafers has little dependence on scan speed or oxide thickness for scan speeds below 10 cm/sec. Increasing scan speeds above 50 cm/sec causes significant increases in the stress, particularly for oxide thicknesses greater than 0.5 um.

4. Normal processing temperatures significantly reduce the "as crystallized" stress and can reduce grain imperfections. However, lower processing temperatures (i.e., below the stress relieving temperature for SiO_2 films) can increase grain imperfections, as seen by the increase in diffraction beam widths.

ADHESION EFFECTS ON THE RECRYSTALLIZATION OF SILICON FILMS

C. E. BLEIL AND J. R. TROXELL Electronics Department, General Motors
Research Laboratories, Warren, Michigan   48090-9055

ABSTRACT

Laser processing of thin films of amorphous or polycrystalline silicon on
insulator substrates, such as the glass normally used for liquid crystal
displays, frequently leads to film thickness variations which are unacceptable for device fabrication. Some thickness variations are caused by the
high surface tension of molten silicon and the poor adhesion of the silicon
to the substrate. Techniques to reduce this problem by increasing the adhesion of the film to silicon dioxide coated Corning 7059 glass substrates
have been investigated. Two different approaches were used. First, silicon
ions were implanted into the silicon-glass interface to increase the direct
bonding of the silicon to the silicon dioxide. Second, layers of material
known to exhibit better adhesion to both silicon and silicon dioxide were
introduced between the silicon film and the glass substrate. Both techniques produced films which, after subsequent laser processing, showed significantly reduced thickness variations. These procedures make it possible
to laser process thin films of silicon on Corning 7059 glass substrates
under conditions which produce large grain polysilicon films without producing unacceptably large thickness variations or film cracking.

INTRODUCTION

Large grain polycrystalline films of silicon (polysilicon) have been prepared on high temperature substrates such as silicon dioxide coated silicon
wafers, quartz, or sapphire.[1-2] However, considerable interest has been
expressed in the preparation of device quality semiconductor films on low
temperature substrates such as glass. Some success in this direction has
been achieved with amorphous[3] or fine grain polysilicon[4] but, the low
mobilities which usually result limit the application of such films.
Attempts to improve the carrier mobilities by increasing the grain size
through laser processing have been effective. However, difficulties are
frequently encountered resulting from the narrow range of laser power
(energy window) required to achieve reproducible results. The narrowness of
this energy window for a fixed film thickness is in large measure the result
of the thermal gradient profile, the high surface tension of molten silicon,
and the relatively low adhesion of the silicon to typical substrate materials. Two other properties of the recrystallization process are important to
the attainment of good quality crystalline silicon films. The first of
these is related to the volume change which occurs between amorphous (a-Si)
and crystalline (c-Si) silicon. This volume change can be as large as 20%
based on the extremes of density for the two states. The second is related
to the expansion of silicon during freezing. In the latter case, when one
establishes the required thermal gradient and appropriate dwell times for
large grain growth, significant material transport can occur leading to a
rough surface texture. It appears, therefore, that there are four process
objectives to be met to assure minimum thickness variations during recrystallization. They are:

1. Reduce random surface material transport,
2. Minimize the volume change during recrystallization,
3. Control the thermal gradients, and

4. Maximize the adhesion to the silicon substrate.

The first two of the above have received considerable attention[5-9] and will not be discussed further. The third item[10,11], while important to this work, is not the major thrust of this study. The control of thermal gradients will be addressed in a subsequent publication. This paper will deal primarily with the enhancement of adhesion between the recrystallized silicon film and the underlying substrate.

EXPERIMENTAL PROCEDURE

Several experiments were proposed to study the role of adhesion. First, we tried to isolate the effects of the temperature profile and second, to test the adhesive compatibility of selected films interposed between the silicon and the glass. These experiments fell into two categories:

1. Ion induced modification of the silicon-glass interface, and
2. Introduction of thin layers between the glass and the silicon.

All the films have been prepared on a standard substrate consisting of Corning 7059 borosilicate glass covered with a 1 μm deposited film of silicon dioxide. The silicon dioxide was selected to provide a chemically uniform surface. Most of the silicon depositions were made by electron beam evaporation. Some of the samples were implanted with boron at 1×10^{11} /cm² for resistance measurements after processing. Samples prepared for the evaluation of adhesion were not overcoated with silicon dioxide or silicon nitride. Although ample evidence is available in the literature to illustrate the benefits of overcoating,[8] the overcoat was omitted to provide a uniform response independent of the overcoat layer. The lack of a surface restraining film permits mass transport, and results in a sharper delineation of the ratio of cohesive to adhesive forces relative to the substrate. Thus, the experimental comparisons became more sensitive to variations of adhesion. In addition, we thus avoid the introduction of a film (silicon dioxide) which has a thermal expansion coefficient which is much smaller than that of either the silicon or the substrate glass. In this way, we hope to minimize the film cracking problems which are relatively common when silicon dioxide layers are present.

Initially, an a-Si film was compared to a polysilicon film to assess the effects of volume changes during the crystallization process. This first experiment consisted of a silicon wafer covered with a 1 μm film of silicon

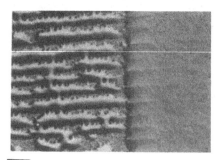

100 μm

Fig. 1: Severe dewetting of silicon film from a silicon dioxide coated wafer. The smooth processed area, showing large grains, was first scanned at low power to produce fine grain polysilicon, followed by high power scanning of the entire wafer. The dewetted area coincided with the area which did not receive the initial low power scan.

dioxide to isolate the wafer from the subsequent deposit of 0.5 μm a-Si. The wafer was then heated to 570 K and a portion of it was subjected to a very low power laser scan. A 50 by 75 mm area in the center of the 75 mm wafer was accordingly converted to a very fine grain polysilicon. (i.e., the laser exposure time was long enough to produce fine grain polysilicon, but short enough to avoid significant silicon mass transport.) In this case, the incident laser power was 8 W and the scan rate was 10 cm/s. The sample was then scanned in a direction perpendicular to the first and at several different power levels, ranging from 8 to 15 watts, and a scan rate of 10 cm/s. The result showed smooth processing over most of the preprocessed region and a very strong furrowing outside that area (see Fig. 1). While this figure dramatizes the problems which can occur when adhesion is inadequate, we cannot offer a simple explanation for these results. This experiment was intended to show the stress effects of volume changes during processing which could lead to dewetting in the furrows. However, interpretation of the results shown in Fig.3 is complicated by several additional features. It is known that a-Si has a low temperature absorption coefficient which is an order of magnitude greater than c-Si near the argon ion wavelength (~500nm).[12] Since the scan rate was relatively slow (10 cm/s), if one assumes that explosive crystallization occurred ahead of the scan and below the melting temperature, then there should have been no difference between the two zones. Similarly, if the a-Si melted directly and required the same energy input, it should have shown little difference from the c-Si layer even though it may have reached its melting point a few nanoseconds earlier because of the higher initial absorption. The explanation which seems most probable is that the thermal dissipation of the heated film by the 350 μm thick silicon substrate provided a substantial heat sink for the heat from the laser beam. This would have the effect of increasing the thermal gradient laterally and suppressing any delay between spontaneous crystallization of the a-Si and its melting. In addition, if the effective melting temperature of the a-Si is significantly lower than that of c-Si[13], then for the processing conditions shown, we may have produced an increased stress on the substrate interface in the region which was not pre-processed. Alternatively, it is possible that silicon dioxide which grew on the surface of the pre-processed region during the low power scan could have aided in preventing furrowing, or that the low power processing resulted in improved adhesion of the silicon to the substrate.

With these difficulties in mind a new set of experiments was devised to establish the role of adhesion independent of the effects of crystallization. Only a-Si films were used to ensure this independence. This procedure provided a uniform response and maximized the stress effects related to crystallization. Moreover, when the results are applied to polysilicon films one should expect an even greater latitude in the processing parameters provided that the silicon film does not dewet the substrate. The results of these experiments are shown in Figs. 2 through 7.

All films were deposited on Corning 7059 glass substrates (0.1 cm thick, which were initally coated with 1.0 μm of chemical vapor deposited silicon dioxide. The amorphous silicon films were electron beam deposited, while the polysilicon film in Fig. 6 was deposited by atmospheric pressure chemical vapor deposition. All of these samples were laser processed at a substrate temperature of 570 K under vacuum conditions of 10^{-3} Pa. In each case, several different scan rates and laser power levels were used to establish the range for desirable processing including the melting of the silicon film. For the regions shown in these figures, scan rates were approximately 10 cm/s, except for the sample which incorporated the ion implanted polymethyl methacrylate (which exhibits characteristics of silicon carbide) layer. That sample was scanned at 60 cm/s. The samples shown in Figs. 2 and 3 were processed with an incident laser power of about 2.5 W.

690

Fig. 2: Low power laser processed silicon film on borosilicate glass showing the onset of furrowing in a 0.16 μm thick film.

Fig. 3: Laser induced crystallization of a 0.16 μm amorphous silicon film deposited on a borosilicate glass slide and implanted with 1×10^{16} Si+ ions/cm^{-2} at 200 keV. Note that the furrowing observed in Fig. 4 is absent.

The thicker sample shown in Fig. 4 was processed at 3.5 W, while the ion implanted polymethyl methacrylate sample was processed with an incident laser power of 5.3 W. The remaining samples, shown in Figs. 6 and 7, incorporated thicker heat absorbing layers, and required laser powers of about 12 W for the processing shown.

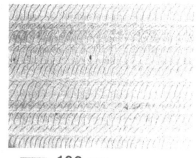

Fig. 4: Furrowing produced by laser crystallization of a 0.25 μm amorphous silicon film deposited on a borosilicate glass slide.

Fig. 5: Laser induced crystallization of a 0.25 μm amorphous silicon film deposited on a 0.1 μm thick film of ion implanted polymethyl methacrylate.

—— 100 μm

—— 100 μm

Fig. 6: Laser induced crystallization of a 1.0 μm thick
chemical vapor deposited polysilicon film. The top 0.15 μm
of the film was implanted with
1x10¹⁶ Si+/cm⁻² at 200 keV in
order to make it amorphous prior
to recrystallization.

Fig. 7: Laser induced crystallization of a 0.5 μm thick
amorphous silicon film deposited
on 0.1 μm of silicon nitride and
1.0 μm of chromium.

RESULTS

The processing shown in Figs. 2 and 4 for a-Si films of two different thicknesses deposited directly on glass is typical for our standard processing
conditions. At lower power levels one may avoid the furrowing but only very
small grain polysilicon is produced. Long residence times for the laser
beam, which are required to achieve large grain polysilicon, invariably
result in the destruction of the film. For comparable film thicknesses
(Figs. 3 and 5, respectively) good processing was achievable, using the
adhesion enhancement techniques shown, at even higher laser powers, and
without furrowing or film cracking.

The four techniques illustrated in Figs. 3,5,6 and 7 all show uniform processing without furrowing. In each case a significant increase in the
energy window for laser processing is observed. In some cases at the highest power levels used, considerable cracking of the film occurred but, even
under these conditions of extreme temperature and stress, furrowing did not
occur. Based on the published literature [14] of related problems, the
cracking is most likely due to differential thermal expansion. By eliminating silicon dioxide from the film system, we have minimized the cracking
problem for most processing conditions. For the processing conditions
shown, neither furrowing nor cracking was observed over areas in excess of 1
cm².

The results of the experiments reported here clearly show the importance of
adhesion of the silicon films to the substrate and the importance of the
thermal profile during laser processing. Moreover, the clean uniform processing associated with each of the methods for adhesion enhancement suggests that a variety of techniques will suffice to improve the adhesion.

However, higher power levels required for some of the interposed layers may increase the heat dissipation problems for thin silicon films. The result shown in Fig. 7 is for a 0.5 μm silicon on a 1 μm chromium interlayer and shows that the chromium serves as an effective heat sink. Further investigations of the use of such heat sinking layers is underway and will be reported elsewhere.

ACKNOWLEDGEMENTS

The authors gratefully acknowledge the very helpful assistance of Art Fritz and John Biafora of the Electronics Department for the preparation of the sample materials.

REFERENCES

1. R.A. Laff and G.L. Hutchins, IEEE Trans. Electron Devices Ed-21,743 (1974).

2. M.A. Bosch and R.A. Lemons, Appl. Phys. Lett. 40, 166 (1982).

3. M. Matsumura and H. Hayama, Proc. IEEE, 68, 1349 (1980).

4. M. Matsui,Y. Shiraki, Y. Katayama, K.L.I. Kobayashi, A. Shintani and E. Maruyama, Appl. Phys. Lett., 37, 936 (1980).

5. J. Stephen, B.J. Smith and N.G. Blamires, Laser and Electron Beam Processing of Materials, Ed. C.W. White and P.S. Peercy Academic Press, N.Y. 1980, p. 639 et sequel.

6. T. Stultz, J. Gibbons, Mat. Res. Soc. Proc. 13, 463 (1983).

7. M.W. Geis, Henry I. Smith, B-Y. Tsaur, John C.C. Fan. P.J. Silversmith, R.W. Mountain and R.L. Chapman, Ibid 477.

8. E. Yablonovitch and T. Gmitter, 1983 MRS Ann. Meeting, Boston, Extended Abs., p. 36.

9. Z.A. Weinberg, V.R. Deline, T.O. Sedgwick, S.A. Cohen, C.F. Aliotta, G.J. Clark and W.A. Langford, Appl. Phys. Lett. 43, 1105 (1983).

10. Y.I. Nissim, A. Lietoila, R.B. Gold, and J.F. Gibbons, J. Appl. Phys., 51, 274 (1980).

11. M.L. Burgener and R.E. Reedy, J. Appl. Phys., 53, 4357 (1982).

12. J.S. Williams, W.L. Brown, H.J. Leamy, J.M. Poate, J.W. Rodgers, D. Rousseau, G.A. Rozgonyi, J.A. Shelnutt and T.T. Sheng, Appl. Phys. Lett. 33, 542 (1978).

13. E.P. Donovan, F. Spaepen, D. Turnbull, J.M. Poate,and D.C, Jacobson, Appl. Phys. Lett., 42, 698 (1983)

14. W.M. Paulson and S.R. Wilson, J. of Electronic Materials, 12, 107 (1983).

PART VIII

Silicon Solid Phase
Crystal Growth on Insulators

SOLID PHASE EPITAXY OF UHV-DEPOSITED AMORPHOUS Si
OVER RECESSED SiO$_2$ LAYER.

M. TABE and Y. KUNII
Atsugi Electrical Communication Laboratory, NTT,
1839-Ono, Atsugi-shi, Kanagawa 243-01, Japan.

ABSTRACT

Lateral solid phase epitaxy (L-SPE) of ultra-high-vacuum
(UHV) deposited amorphous Si (a-Si) over patterned SiO$_2$ has
been studied to produce monocrystalline silicon-on-insulator
(SOI) films. When employing UHV-deposited a-Si, it is
essential for L-SPE to reduce step height at the pattern
boundary. This is because low density a-Si including columnar
voids is formed at the step wall by the self-shadowing effect
and SPE region does not extend across the low density a-Si
area. L-SPE growth distance of 7 μm was achieved by low
temperature annealing (575°C, 20 hr) on a planar substrate with
recessed SiO$_2$ patterns. Another deposition technique of a-Si
for SPE, i.e., chemical vapor deposition is reviewed for
comparison.

INTRODUCTION

Recently, a new approach to realize Si crystalline films on
insulating substrates (SOI's), i.e., lateral solid phase
epitaxy (L-SPE) of deposited amorphous Si (a-Si) has been
intensively studied by several authors [1-4]. In this technique, a-Si first grows vertically on a seed region and then
grows laterally over a patterned SiO$_2$ by heating at low temperatures (550-650°C). L-SPE distance is usually determined by
competition between SPE growth and random grain growth in a-Si.
The film surface is satisfactorily smooth after SPE and the
impurity profile in the underlying layer is not deformed because of the low temperature treatment, in comarison with the
film surface and impurity profile after liquid phase epitaxy by
laser, electron beam, or carbon-heater annealing. It may be
difficult for SPE to form a large SOI area. However, the large
area is not always required for LSI's, because the active area
has become smaller with an increase in packing density.
Ohmura et al. reported that 1-μm-growth of L-SPE occurred
in vacuum-deposited a-Si films when high-dose Si$^+$ implantation
was used to improve the film quality [1]. Also, Ishiwara
et al. reported that 6-μm-growth occurred in L-SPE over an SiO$_2$
stripe in a dense a-Si film [3]. The dense film was formed by
deposition of polycrystalline Si on a heated substrate in
ultra-high-vacuum (UHV) and subsequent amorphization by
high-dose Si$^+$ implantation. In addition, Roth indicated that
the rate of random nucleation was enhanced near the seed/oxide
boundary [4].
We have studied two different methods for a-Si deposition,
i.e., chemical vapor deposition (CVD) by decomposition of SiH$_4$
and deposition from an e-beam heated Si source in UHV. In
both methods, following conditions are commonly required for
SPE without help of implantation; (i) a-Si/substrate interface
in the seed region should be clean enough to achieve vertical
SPE (V-SPE) and (ii) as-deposited a-Si should exclude
crystalline nuclei and impurities such as Ar, O or C [5].

The CVD method is of advantage to surface coverage at steps on the substrate. However, it is rather difficult in CVD to control a-Si thickness because the deposition rate sensitively depends on the wafer temperature. In addition, high temperature heating at 1100°C is necessary for substrate surface cleaning. On the other hand, UHV-deposited a-Si can be precisely controlled in thickness by in-situ measurement and the substrate cleaning temperature is as low as 800°C, although the step coverage is inferior to CVD. It is demonstrated that the step coverage problem can be solved by making the SiO$_2$ patterns recessed.

In this paper, first we review the authors' previous results of CVD a-Si. Second, we describe the recent results of UHV-deposited a-Si and the effect of recessed SiO$_2$ structure on L-SPE. Finally, it is shown that facet formation at the SPE front, which commonly occurs in L-SPE, is closely related to the microscopic SPE mechanism.

SPE OF CVD a-Si

Deposition Conditions of a-Si

SPE of CVD a-Si had not been reported until when the authors succeeded in establishing substrate surface cleaning steps for the advance of SPE across the a-Si/substrate interface [6]. As the first cleaning step, heating at 1100°C in H$_2$ ambient is necessary for removing native oxide on the substrate surface (in the seed region). The subsequent cleaning step during lowering temperature is etching of Si surface by dilute HCl for preventing foreign atom adsorption onto the surface. It is known that, for vapor phase epitaxial (VPE) growth, the substrate surface is cleaned enough only by the first cleaning step. However, both steps are necessary for SPE, because deposition temperature of a-Si (500-650°C) is much lower than that of VPE (1000-1200°C) so that foreign atom adsorption is enhanced. Immediately after the substrate surface cleaning, Si films were deposited by decomposition of SiH$_4$ with Ar (or N$_2$) carrier gas at 500-650°C. Ar (or N$_2$) carrier gas was used because the Si deposition rate with H$_2$ carrier gas is one order of magnitude lower than that with Ar (or N$_2$) [7].

The deposition rate increases with increasing temperature (an activation energy of 2.2 eV below 550°C). This temperature dependence indicates that even small temperature deviation deteriorates thickness controllability. UHV-deposition is superior to CVD in this point, because the deposition rate is independent of the substrate temperature.

The characteristics of deposited Si are also essential to SPE [6]. Transmission electron microscopy (TEM) and transmission electron diffraction (TED) observation showed that the film deposited at 700°C was polycrystalline, whereas the films below 650°C were "amorphous". However, even in "amorphous phase", many micro-crystallites were observed in the film deposited at 650°C. Since micro-crystallites grow in random orientation during heat treatment, a low deposition temperature is preferable for SPE.

V- and L-SPE of CVD a-Si

Figure 1 shows Rutherford backscattering (RBS) channeling spectra during V-SPE growth for samples deposited at 550 and 620°C. Here, annealing temperature was 600°C. For the sample deposited at 550°C, SPE front advanced with annealing time, although slight retardation of SPE due to residual contamination on the substrate surface was observed at the initial stage. After 60 min annealing, the SPE front reached the topmost surface and the reflection electron diffraction (RED) showed an identical pattern with the (100) substrate. On the other hand, for the sample deposited at 620°C, SPE growth was obstructed because of random grain growth of micro-crystallites. The inclined spectrum implies inclusion of random grains in the SPE layer.

V-SPE depends on deposition rate as well as deposition temperature. It was observed that V-SPE growth rate increased with an increase of a-Si deposition rate [6]. This result is explained by foreign atom incorporation during deposition; incorporated amount of SPE-obstructive impurities (O, N or Ar) would be determined by the ratio of foreign atom to SiH_4 partial pressure. Therefore, low deposition temperature and high deposition rate are preferable for SPE growth. However, as the deposition temperature decreases, it becomes difficult to obtain sufficiently clean substrate surface in addition to the rapid decrease in deposition rate. Therefore, from the view points of substrate surface cleaning and deposition rate, 550-580°C was employed as a deposition temperature within the present experimental conditions.

For L-SPE experiments, SiO_2 stripe patterns on (100) substrates were formed parallel to [001] and [011] directions using conventional photolithography process. After substrate surface cleaning in a manner as described before, a-Si was deposited with SiH_4 ($2-6 \times 10^{-2}$ atm)/Ar (or N_2) at 550-580°C, which satisfy the requirements of low temperature and high deposition rate (typically 400 A/min) for SPE.

Cross-sectional SEM images during L-SPE at 575°C are shown in Fig. 2 for the [001] SiO_2 stripe. It was observed that L-SPE advanced with annealing time and specific atomic planes (facets) were formed at the SPE front. The facet orientation

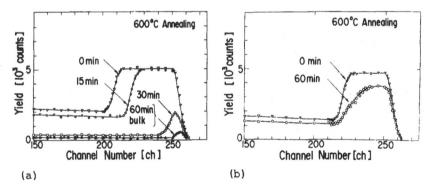

(a)   (b)

Fig. 1. RBS channeling spectra during V-SPE for samples deposited at (a) 550°C and (b) 620°C. Annealing temperature was 600°C.

at the initial stage was approximately (110) because the observed angle between the facet and (100) was 42-50° which agrees with 45° of the angle between (110) and (100). As the SPE advanced, another facet, i.e. ($\bar{1}$10) facet, appeared. Eventually SPE stopped because of random grains.

As an example, SOI with a 4-μm-wide [001] SiO_2 stripe was demonstrated; a-Si layer crystallized over the stripe after annealing at 600°C for 1 day (Fig. 3). The epitaxial layer did not include grains with different orientations. In a case of

On SiO_2 On Si

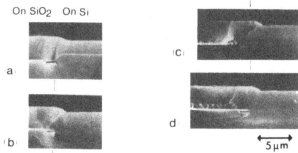

Fig. 2. Cross-sectional SEM photographs during L-SPE for [001] stripe. 575°C annealing for (a) 3 hr, (b) 6 hr, (c) 9 hr, and (d) 12 hr.

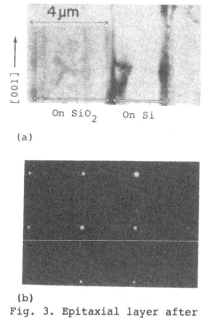

(a)

(b)

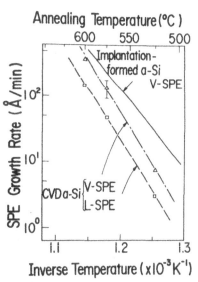

Fig. 3. Epitaxial layer after annealing at 600°C for 1 day.
(a) TEM micrograph.
(b) TED pattern on SiO_2 stripe.

Fig. 4. Dependence of L-SPE growth rate on annealing temperature. V-SPE growth rates of CVD a-Si and implantation-formed a-Si [9] are also plotted.

10-µm-wide [001] stripe, the whole surface layer crystallized epitaxially but a deep region just above the SiO_2 layer became polycrystalline at a distant place from the stripe edge.

For the [011] stripe, a (111) facet was formed at the SPE front and L-SPE was extremely suppressed. In addition, {111} twins were formed at the stripe edge.

The facet formation suggests that L-SPE over the SiO_2 stripe is equivalent to V-SPE in the direction normal to the facet plane in an atomic scale. Annealing temperature dependence of L-SPE growth rate is plotted in Fig. 4. V-SPE growth rates of CVD and implantation-formed a-Si [9] are also plotted. The ratio of L-SPE to V-SPE is consistent with the previous results of orientation dependence of V-SPE growth rate [9], considering the secant factor between facet orientation and L-SPE growth direction.

L-SPE distance is determined through competition between L-SPE and random grain growth. It was found that L-SPE distance gradually increased with decreasing annealing temperature. However, it should be noted that temperatures below 550°C are inappropriate for practical use, because annealing period extremely increases with decreasing temperature.

SPE OF UHV-DEPOSITED a-Si

Deposition Conditions of a-Si

In this experiment, a Si MBE machine for 4-inch wafers was used for deposition of a-Si. The substrate was uniformly heated by a graphite heater and Si atoms were evaporated toward the substrate by a sweepable e-gun. The working pressure during deposition was less than 5×10^{-8} Torr.

Substrate heating at 800°C in UHV was employed as substrate surface cleaning process before deposition. During the heating, native oxide was removed through a chemical reaction with substrate Si [10-12]. Although small amount of carbon contamination might be remained on the surface even after the cleaning step [13,14], it did not affect SPE growth and appreciable delay of V-SPE was not observed. Thus, it was confirmed that the cleaning temperature, which was much lower than CVD, was satisfactory for SPE of UHV-deposited a-Si.

a-Si was deposited at the rate of 10 A/s after cooling the substrate below 200°C. Temperature below 400°C is required for a-Si deposition. Above 400°C, single- and poly-crystalline Si is deposited on the seed and SiO_2 regions, respectively. Deposition rate or film thickness was easily controlled by a quartz-oscillator thickness monitor because the sticking coefficient of Si on the substrate is unity and is independent of substrate temperature.

Post-Heating Effect on SPE

It is known that oxygen in air penetrates into a-Si layer and V-SPE is obstructed owing to the oxygen inclusion when the as-deposited a-Si is exposed to air without post-heating (densification) in the UHV chamber [15]. Reduction of a-Si thickness during V-SPE at 568°C is plotted in Fig. 5 for samples with and without post-heating (500°C for 30 min). The post-heated film grew epitaxially with a flat a-Si/SPE-Si interface up to the topmost surface after 60 min annealing.

700

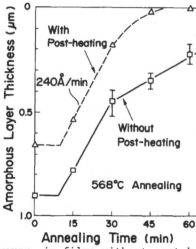

Fig. 5. Dependence of a-Si thickness on annealing time during V-SPE growth at 568°C.

However, in films without post-heating, the interface undulated during V-SPE and did not advance epitaxially up to the surface. The initial 10 min delay of V-SPE was due to the time for samples to reach the furnace temperature and did not originated from contamination at the a-Si/substrate interface. V-SPE growth rate was 240 Å/min at 568°C which agreed with that for implantation-formed a-Si (160±100 Å/min) [16], and was larger than that of CVD a-Si (80Å/min). This difference in growth rate is due to the difference in SPE-obstructive impurity incorporation or film structure.

Post-heating functions as densification of a-Si films. However, we could not find appreciable difference in density for the films with and without post-heating probably because of an inevitable measurement error; both films had the same value, 2.29 g/cm^3, within accuracy of ±0.11 g/cm^3. On the other hand, spin density due to dangling bonds (g-value of 2.006 [8]) decreases from 2.1×10^{19} to 1.4×10^{19} spins/cm^3 by the post-heating. The latter value, 1.4×10^{19} spins/cm^3, is approximately the same as that of CVD a-Si [6].

L-SPE of UHV-deposited a-Si

In spite of the good results of V-SPE, there is a problem peculiar to UHV-deposition that L-SPE stops at the SiO_2 pattern edge step [3]. This phenomenon reflects that the density of a-Si deposited on the side wall of the SiO_2 stripe is much lower than that on a planar region because of the self-shadowing effect [17,18]. The self-shadowing effect, which becomes remarkable as the incident atomic beam is inclined to the surface, causes columnar growth surrounded by voids. In addition to the self-shadowing effect in the microscopic scale, there is another problem that the step coverage is inferior to CVD; a smaller amount of Si atoms are deposited on the side wall of SiO_2 than on a planar region because the normal component of the incident Si flux to the side wall is very small. Therefore, reduction of the step using a recessed SiO_2 stripe is necessary for SPE of UHV-deposited a-Si.

Figure 6 shows the experimental procedure for L-SPE on recessed [001] SiO_2 stripes (0.2 µm thick). The recessed structure was formed using "self-aligned planer oxidation technology (SPOT)" and the resultant step height at the pattern edge was 150 A. In this case, the seed region was higher than the SiO_2 stripe. For reference, unrecessed SiO_2 stripes (0.1-0.3 µm thick) were also formed on the wafers.

As shown in Fig. 7, in the wafer with 0.1-µm-steps, L-SPE occurred over the SiO_2 stripe from the step of Si beam side, where Si atoms were deposited on the side wall of SiO_2 with the incident angle (the angle between the incident beam and the perpendicular line to the wall-surface) of 10-20°. On the other hand, polycrystalline Si grew from the step of opposite side, where the incident angle was 40-50°. In the wafer with 0.3-µm-steps, L-SPE occurred from neither side and polycrystalline layer covered over SiO_2 pattern after annealing at 575°C for 20 hr.

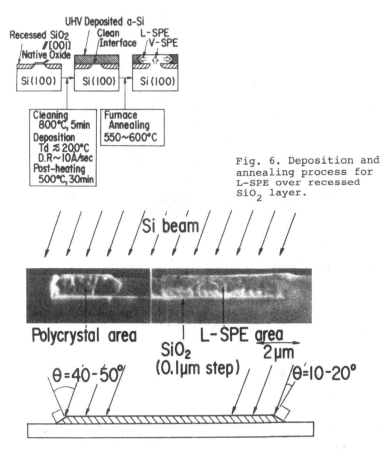

Fig. 6. Deposition and annealing process for L-SPE over recessed SiO_2 layer.

Fig. 7. Incident angle effect on crystallization for 0.1-µm-step unrecessed SiO_2 stripe (550°C annealing for 2 days).

According to SEM observation, the SPE front advanced with annealing time over recessed SiO_2 stripes at 575°C (Fig. 8). The corresponding V- and L-SPE growth is plotted in Fig. 9. The change in SPE growth rate is consistent with the facet formation at the single/amorphous interface. In an early stage, the (110) facet was formed at the pattern edge and then moved in the lateral direction. While the (110) facet advanced, another facet, i.e. ($\bar{1}$10) facet, appeared as in the case of CVD a-Si. After the lateral growth of 3 μm at the growth rate of 120 Å/min, the two {110} facets became disordered and the corresponding L-SPE growth rate decreased (55 Å/min). In this stage, {111} micro facets were probably formed. The normalized growth rates of "Lateral (1)" and "Lateral (2)" can be calculated from the dependence of the normalized V-SPE growth rate on the substrate orientation [9], putting the V-SPE growth rate on the (100) substrate at unity.

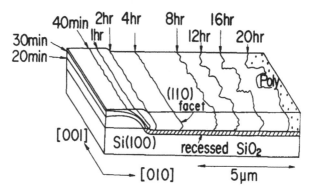

Fig. 8. SPE growth on planar substrate with recessed [001] SiO_2 stripe (575°C annealing).

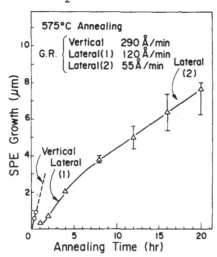

Fig. 9. Dependence of SPE growth on annealing time (575°C annealing).

The calculated values (Lateral (1): 0.47, Lateral (2): 0.26) agree well with the experimental values (Lateral (1): 0.41, Lateral (2): 0.19). Finally, 7-μm L-SPE distance was obtained after annealing at 575°C for 20 hr. The dislocation density was 10^6 cm^{-2} on the seed distant from the SiO$_2$ stripe, and 10^8 cm^{-2} at the SiO$_2$ stripe edge. Thus, it was demonstrated that the planar substrate with recessed SiO$_2$ stripes was effective and suitable for L-SPE of UHV-deposited a-Si.

TWO-BOND FORMATION MODEL

Drosd and Washburn have reported on the SPE mechanism (D-W model) [19] assuming that an a-Si atom must form at least two undistorted bonds for SPE. The facet formation is explained by the D-W model and the boundary conditions that an a-Si atom cannot form undistorted bonds sticking to the SiO$_2$ region [20].

For the [001] stripe, an a-Si atom forms undistorted bonds with seed crystalline atoms, breaking and/or distorting the bonds to the other a-Si atoms, if it is distant from the boundary (SiO$_2$). However, if an a-Si atom is at the boundary ("A" in Fig. 10(a)), it cannot form two undistorted bonds with seed crystalline atoms and remains in amorphous phase even after completion of SPE of the first layer. Successively, the second layer grows in SPE on the first SPE layer, but recedes from the boundary by √2 of (110) plane spacing. In this case, atom "B" remains in amorphous phase being affected by amorphous atom "A". Consequently, a (110) facet grows at the interface. After or during the facet formation, the (110) SPE front advances at a smaller growth rate according to the D-W model. For the [011] stripe, a (111) facet is formed through a similar process to the (110) facet (Fig. 10(b)). During [111] SPE, twin nuclei can be created and micro-twin formation occurs, as predicted by the model. Thus, facet formation is phenomenologically interpreted by the two-bond formation model.

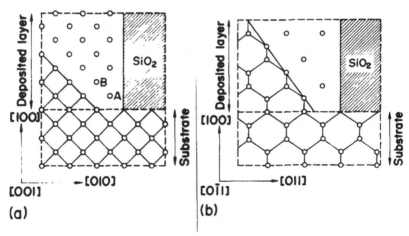

Fig. 10. Facet formation based on two-bond formation model for (a) [001] and (b) [0$\bar{1}$1] SiO$_2$ stripes.

CONCLUSION

We have studied SPE of CVD and UHV-deposited a-Si for fabrication of an SOI structure. UHV-deposition is superior to the CVD from the viewpoints of thickness control and low cleaning temperature. However, a recessed SiO_2 structure is required to achieve L-SPE in UHV-deposited a-Si. When using the recessed structure, 7-μm overgrowth is obtained at an annealing temperature as low as 575°C. In addition, we have found that specific atomic planes (facets) are formed at SPE front. It is shown that the facet formation is based on microscopic SPE mechanism.

ACKNOWLEDGMENT

The authors wish to thank K.Kajiyama for valuable discussions, and T.Sakai and M.Kondo for their encouragement throughout this work.

REFERENCES

1. Y.Ohmura, Y.Matsushita and M.Kashiwagi, Jpn.J.Appl.Phys., 21, L152(1982).
2. Y.Kunii, M.Tabe and K.Kajiyama, J.Appl.Phys., 54, 2847(1983).
3. H.Ishiwara, H.Yamamoto, S.Furukawa, M.Tamura and T.Tokuyama, Appl.Phys.Lett., 43, 1028(1983).
4. J.A.Roth, Mat. Res. Soc. Symp., ABSTRACTS p.48 (1983).
5. E.F.Kennedy, L.Csepregi, J.W.Mayer and T.W.Sigmon, J.Appl.Phys., 48, 4241(1977).
6. Y.Kunii, M.Tabe and K.Kajiyama, Jpn.J.Appl.Phys., 21, 1431(1982).
7. Y.Ban, H.Tsuchikawa and K.Maeda, Semiconductor Silicon (Abstract of Electrochem. Soc. Meeting, 1973), p.292.
8. M.H.Brodsky, R.S.Title, K.Weiser and G.D.Pettit, Phys.Rev., B1, 2632(1970).
9. L.Csepregi, E.F.Kennedy, T.W.Mayer and T.W.Sigmon, J.Appl.Phys., 49, 3906(1978).
10. B.A.Joyce, R.R.Bradley and G.R.Booker, Phil.Mag., 15, 1167(1967).
11. M.Tabe, K.Arai and H.Nakamura, Jpn.J.Appl.Phys., 20, 703(1981).
12. M.Tabe, Jpn.J.Appl.Phys., 21, 534(1982).
13. R.C.Henderson, R.B.Marcus and W.J.Polito, J.Appl.Phys., 42, 1208(1971).
14. R.C.Henderson, J.Electrochem.Soc., 119, 772(1972).
15. S.Saitoh, T.Sugii, H.Ishiwara and S.Furukawa, Jpn.J.Appl.Phys., 20, L130(1981).
16. G.L.Olson, J.A.Roth and L.D.Hess, Proceedings of the U.S.-Japan Seminar on Solid Phase Epitaxy and Interface Kinetics, Oiso, Japan (1983).
17. A.G.Dirks and H.J.Leamy, Thin Solid Films, 47, 219(1977).
18. D.K.Pandya, A.C.Rastogi and K.L.Chopra, J.Appl.Phys., 46, 2966(1975).
19. R.Drosd and J.Washburn, J.Appl.Phys., 53, 397(1982).
20. Y.Kunii, M.Tabe and K.Kajiyama, J.Appl.Phys., 56, 279(1984).

FORMATION OF Si-ON-INSULATOR STRUCTURE UNDER SOLID PHASE GROWTH

M.MIYAO, M.MONIWA, T.WARABISAKO, H.SUNAMI AND T.TOKUYAMA
Central Research Lab., Hitachi Ltd., Kokubunji, Tokyo 185, Japan

ABSTRACT

Two possible solutions to the problem of nucleus growth encountered in lateral solid-phase epitaxial growth over insulating films are discussed. A stress field originating from the thermal expansion coefficient for Si and SiO_2 acts as the driving force behind preferential nucleation. Utilization of underlying Si_3N_4 films successfully eliminated nucleii growth at topographically irregular portions. In addition, single crystallization of poly-Si nucleii was achieved on SOI structures for the first time. Lateral growth speed (v [cm/s]) of $1.6 \times 10^6 \exp(-3.9/kT[eV])$ was obtained during high-temperature annealing (≥ 1000 C).

INTRODUCTION

Laser and electron-beam induced recrystallization of deposited poly-Si films on insulating layers is raising expectations of achieving new device structures using a Si-on-insulator (SOI) material system [1]. The bridging epitaxy, i.e., seeded lateral epitaxy, is regarded as the most realistic approach for VLSI application, because precise control of the crystal orientation is only possible through the seeding process. However, there remain some problems mainly concerned with crystal growth by the liquid phase scheme , when 3-dimensional LSIs are considered. Surface ripple of the recrystallized layer and redistribution of impurity atoms in the underlying Si substrate are the problems to be solved.

An alternative SOI technique in which only low temperature annealing (≤ 600 C) is necessary has been proposed by Ohmura et al [2], Kunii et al [3] and Yamamoto et al [4]. The basic idea is lateral solid phase epitaxy (L-SPE) in which deposited amorphous Si films are initially grown vertically at the seeding region and then laterally over a SiO_2 film. However, random crystallization occurrs in the amorphous Si layer at the same time. This restrains propagation of L-SPE, and also degrades crystal quality of the grown layer.

In line with this, the present paper describes characteristics of L-SPE layers. Problems concerning random nucleation and nucleii growth were encountered. Possible solutions for this problem are proposed.

EXPERIMENTS

Si substrates ((100) orientation) were covered by SiO_2 stripes and square islands. Thickness of the SiO_2 films were 10 - 100 nm and pattern-edges were made parallel to the [011] axes. After cleaning of the sample surfaces, Si films were deposited (deposition rate: 0.1 nm/sec, deposition thickness: 1 - 1.5 μm) under ultra-high vacuum (UHV) conditions (10^{-10} mbar). During deposition, some samples were kept at room temperature (sample(I)) and others were kept at 750 C (sample(II)). After annealing (400 - 1200 C, 1 - 20 hrs) in a dry N_2 atmosphere, both sample types were studied by Nomarski optical microscopy and micro-probe reflection high energy electron diffraction (μ-RHEED) [5].

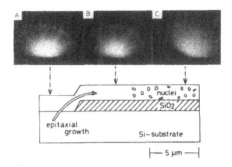

Fig.1   μ-RHEED observation of lateral solid
phase epitaxial growth layer

LATERAL SOLID PHASE EPITAXIAL GROWTH OVER SiO_2

Annealing characteristics of an amorphous Si layer deposited on SOI
structure, i.e, sample (I), were initially investigated. Crystal quality
after furnace annealing (600 C, 30 min) is shown as a function of distance
from the seeding area.   Results indicated that amorphous Si turned into
crystalline state at both the seeding area (A) and the SiO_2 region (B). In
addition, the crystal orientation of both regions was exactly identical to
that of the Si substrate. Consequently, this lateral crystal growth over SiO_2
was confirmed to have originated from solid-phase epitaxial growth.
    However, RHEED pattern of (C) indicated that a polycrystalline state was
formed at the region far from the seeeding area during annealing. This is
because crystallization occurred randomly until L-SPE growth propagated to
this region. As a result, lateral crystal growth was limited to a few microns
in the present experiments.
    Consequently, in order to enlarge L-SPE growth area, development of
certain techniques, i.e, (1) enhancement of L-SPE velocity, (2)
single-crystallization of poly-Si nuclei and (3) retardation of nucleation,
becomes essential.   Yamamoto and one of the present authors successfully
enhanced L-SPE velocity by doping P^+ ions in an amorphous Si very recently
[6].   Therefore, present efforts are being focused to on investigating the
other two improvements, i.e., (2) and (3).

SINGLE-CRYSTALLIZATION OF POLY-Si NUCLEII

The basic idea of single-crystallization of poly-Si nucleii in SOI
structures comes from a recently reported epitaxial alignment method of
poly-Si directly deposited on a Si substrate [7,8]. To examine this idea, we
investigated the annealing characteristics of a molecular-beam-epitaxy sample
on a SOI structure, i.e., sample (II) (poly-Si: 1.4 μm thick, SiO_2: 0.5 μm
thick).   The crystal structure of the as-deposited sample obtained by u-RHEED
is shown in Fig.2-(A). It indicates not only that crystal Si was epitaxially
grown on the Si substrate region during deposition ((A)) , (i.e., molecular
beam epitaxial growth), but also that only poly-Si was deposited on the SiO_2
pattern regions ((B),(C)). This crystal structure change after annealing was

carefully investigated. It was found that structures did not change significantly during low-temperature annealing ($\leq$ 800 C). However, when the annealing temperature exceeded 1000 C, the poly-Si turned into a crystalline state. An example (1150 C, 1 hr) where all poly-Si on the SiO_2 pattern was completely recrystallized is shown in Fig. 2-(b). In addition, crystal orientations were exactly identical to that of the Si substrate. In this way, single-crystallization of poly-Si in an SOI structure was achieved for the first time. This newly developed method of lateral epitaxial alignment is a useful means of reducing poly-Si nucleii randomly generated during L-SPE growth.

Growth speed under lateral epitaxial alignment was estimated from the various samples. It is compared in Fig.3 with SPE speed [9] (crystallization velocity of amorphous Si). Alignment velocity (v [cm/sec]) was estimated to be $1.6 \times 10^6 \exp(-3.9/kT[eV])$. This velocity is almost the same as that obtained for vertical epitaxial alignment (single-crystallization of poly-Si on Si) [7,8]. Activation energy of the epitaxial alignment (3.9 eV) corresponds to the self-diffusion energy of Si atoms [10]. This value is twice that of SPE [9]. Consequently, epitaxial alignment is only effective at high temperature annealing ($\geq$ 1000 C). When only low-temperature processing (600 - 800 C) is required for SOI fabrication, control of nucleation in the amorphous Si becomes more important. This triggered additional research, i.e., investigation of the nucleation phenomena.

NUCLEATION CONTROL IN AMORPHOUS-Si

Random nucleation in an amorphous Si layer on SiO_2 films is investigated using sample (I), where amorphous Si (1.0 μm) was deposited on SiO_2 films of different thicknesses (0.1 μm and 0.01 μm). One example of nucleation of poly-Si obtained by a Nomarsky optical micrograph after furnace annealing (600 C, 3 or 5 hrs) and Wright etching is shown in Fig.4. In this figure, two different types of nucleii are clearly visible. One type is generated ramdomly above the flat SiO_2 region (type(A)). No distinctive phenomena were observed in either thick or thin film regions. The other type of nucleii was found only along the SiO_2 steps (type(B)).

Growth speed of nucleii for types (A) and (B) was measured as a function of annealing temperature. The results are summarized in Fig.5. For type (A) nucleii, activation energy of 2.1 eV was obtained, which corresponds to the dissociation energy of the Si - Si bond. However, for type (B) nucleii, activation energy decreased to 1.7 eV. Therefore, only type (B) nucleii were found after the shorter annealing time (Fig.4-(a)). Such nucleation at the SiO_2 steps is a serious problem in device fabrication. This is because, in actual device processing, amorphous Si is deposited on SiO_2 films with complicated structures.

In order to clarify the preferential nucleation mechanism for type (B), crystal orientation of the sample in Fig.4-(a) was examined as a function of depth by a combination of μ-RHEED and etching techniques. The RHEED pattern at the sample surface indicated that the surface region is an amorphous state. A polycrystalline signal was obtained only from the region deep in the deposited Si layer, i.e., deeper than 0.5 μm from the sample surface. Stress distribution in Si calculated by Blech et al. [11] and Moniwa et al. [12], which are based in the thermal expansion difference between Si and SiO_2, indicates that a large stress field exists in the Si layer near the SiO_2 films. In particular, they are concentrated at the topographically irregular portions of the SiO_2 films , i.e., at the SiO_2 steps.

These results suggest that a stress field is the driving force behind type (B) nucleii growth. In order to confirm this speculation, nucleation in amorphous Si deposited on Si_3N_4 films having the same thermal expansion coefficient was investigated. No preferential nucleation at the Si_3N_4 steps was found. This indicates that elimination of stress fields is essential for retarding nucleii growth.

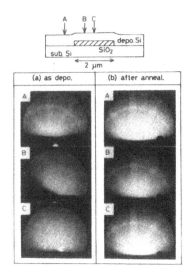

Fig.2 μ-RHEED observation of epitaxial alignment in poly-Si: (a) as-deposited, (b) after annealing (1150C, 1hr)

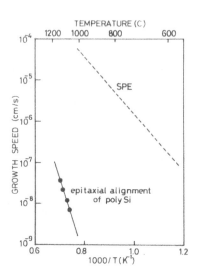

Fig.3 Growth speed of lateral epitaxial alignment in poly-Si and solid phase epitaxy in amorphous Si

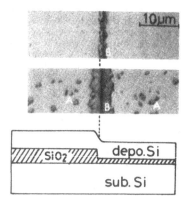

Fig.4 Nucleation of poly-Si in a-Si deposited on SiO$_2$ films: (a) 600C, 3hr annealing, (b) 600C, 5hr annealing

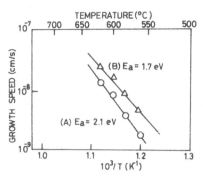

Fig.5 Growth speed of nucleii as a function of annealing temperature: (A) nucleii in the flat SiO$_2$ region, (B) nucleii at SiO$_2$ steps

SUMMARY

Annealing characteristics of UHV-deposited Si layers were inveatigated to control nucleation and nucleii growth encountered in L-SPE. Nucleation in amorphous Si was found to occur preferentially at SiO_2 steps. Elimination of the stress field, i.e., utilization of Si_3N_4 film, successfully retarded nucleii growth. In addition, lateral epitaxial alignment of poly-Si was obtained during high temperature annealing ($\geq$ 1000 C) for the first time. This successfully diminished the number of poly-Si nucleii. Both improvements will be of great help in establishing SOI formation under the solid-phase process.

ACKNOWLEDGEMENTS

The authors would like to express their thanks to Prof.S.Furukawa, Dr.H.Ishiwara and Mr.H.Yamamoto of the Tokyo Institute of Technology for their valuable suggestions and discussions. They also wish to thank Drs. M.Tamura, Y.Shiraki , N.Natsuaki, A.Ishizaka, M.Ichikawa and T.Doi of C.R.L Hitachi Ltd. for their helpful comments during the course of this study.

REFERENCES

(1) Laser-Solid Interactions and Transient Thermal Processing of Materials, edt: J.Narayan, W.L.Brown and R.A.Lemons (North-Holland, New-York, 1983)
(2) Y.Ohmura, Y.Matsushita and K.Kashiwagi; Jpn. J. Appl. Phys. 21, 1152, (1982)
(3) Y.Kunii, M.Tabe amd K.Kajiyama; J. Appl. Phys. 54, 2847 (1982)
(4) H.Yamamoto, H.Ishiwara, S.Furukawa, M.Tamura and T.Tokuyama; Proc. 15th Conf. Solid State Devices and Materials p.89 (Tokyo, 1983)
(5) M.Ichikawa and K.Hayakawa; Jpn. J. Appl. Phys. 21, 145 (1982)
(6) H.Yamamoto, H.Ishiwara, S.Furukawa, M.Tamura and T.Tokuyama; 15th Symp. "Ion Implantation in Semicond. and Sub-micron Fabrication" held at Inst. Phys. Chem. Res. (Wako-shi, Japan 1984) p.5
(7) B.Y.Tsaur and L.S.Hung; Appl. Phys. Lett. 37, 648, (1982)
(8) N.Natsuaki, M.Tamura, T.Miyazaki and Y.Yanagi; ibid (4) p.47
(9) L.Csepregi, E.F.Kennedy, T.J.Gallagher, J.W.Mayer and T.W.Sigmon; J. Appl. Phys. 48, 4234, (1977)
(10) R.M.Barger and R.P.Donovan; Fundamantals of Silicon Integrated Device Technology (Prentice-Hall, Englewood Cliffs, 1976), vol.I, p.205
(11) J.A.Blech and E.S.Meieran; J. Appl. Phys. 38, 2913 (1967)
(12) M.Moniwa, R.Tsuchiyama, M.Miyao, H.Sunami and T.Tokuyama; to be published Appl. Phys. Lett.

SOLID PHASE PROCESSES FOR SEMICONDUCTOR-ON-INSULATOR

C.V. THOMPSON
Department of Materials Science and Engineering, Massachusetts Institute of
Technology, Cambridge, MA. 02139

ABSTRACT

A wide variety of techniques for producing device-quality semiconductor
films on insulating substrates (SOI) are being studied. Processes which
provide low defect density films at low temperatures and which do not require seeding from a single crystal substrate would offer the greatest
flexibility. While such processes do not currently exist, approaches based
on crystallization of amorphous silicon or grain growth in polycrystalline
silicon are being investigated. Development of either approach requires
careful control of film properties and improved understanding of the fundamental materials processes involved. Theory and experiments on surface-
energy-driven secondary grain growth (SEDSGG) are briefly reviewed.
Controlled SEDSGG may provide a low temperature means of obtaining low
defect density films of a variety of materials on a common substrate.

I. INTRODUCTION

In recent years, research on new processes for obtaining electronic-
device-quality semiconductor films on insulating substrates (SOI) has generated widespread interest. This interest is due, in part, to the ease with
which device isolation can be accomplished in single layer SOI structures.[1]
There is also interest in application of SOI structures in new multilayer or
3-dimensional devices[2] and circuits. While the determination of what are
acceptable crystalline defect densities (and types) depends on the specific
application, a successful SOI process should, in general, lead to films with
few defects. Also, for applications in multilayer devices, processing of
subsequent layers cannot be allowed to cause dopant redistribution in
initial layers. This latter requirement has motivated research on low
temperature techniques for obtaining silicon, or more generally, semiconductor films on insulators. In this paper, two low temperature approaches
to SOI will be briefly discussed. In both approaches, control of solid
phase processes is sought. In Section II, a brief description of techniques
involving controlled crystallization of amorphous silicon will be given. In
these techniques, crystal growth must occur while nucleation of new crystals
is avoided. A brief survey of factors which influence crystal growth and
nucleation in thin silicon films will be given. In Section III, theory and
experiments on secondary grain growth phenomena in silicon films will be
reviewed.

II. SOI BY CRYSTALLIZATION OF AMORPHOUS SILICON

Condensed phase processes for SOI are summarized and categorized in
Table I. The approach that has been the subject of the greatest amount of
research has involved __melting__ and directional resolidification, or "recrystallization", of silicon films that are confined between layers of SiO_2.[3]
This process has been called zone melting recrystallization[4] (ZMR) or, when
seeded from the substrate, lateral epitaxy by seeded solidification[5] (LESS).
It can be accomplished using resistively heated graphite strips, lasers,
electron beams and high intensity lamps. Single crystal Si films can be
obtained when solidification is seeded by the single crystal substrate
through lithographically defined holes in the thermal oxide (Fig. 1a).

Alternatively, single crystals with controlled orientations can be obtained without seeding (Fig. 1b) via controlled crystallization using lithographically defined solidification barriers.[6-8] Liquid phase processes need not involve scanned heat sources but can also be accomplished through rapid blanket or uniform anneals[9]. If multiple seed holes are created in the thermal oxide, resolidification of melted films can proceed from a controlled number of locations. The success of this technique requires that growth proceeds only from seed windows, or in other words, that nucleation of new crystals does not occur. This approach can also be applied in crystallization of __solid__ amorphous films of silicon. This process is known as lateral solid phase epitaxy (LSPE).

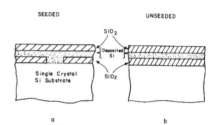

Figure 1. Cross-sectional views of films used for (a) seeded and (b) unseeded crystal growth in Si films.

In LSPE,[10-13] amorphous silicon is deposited onto a thermally oxidized silicon wafer. The deposited silicon film contacts the single crystal substrate through lithographically defined holes in the oxide. Under proper conditions, heating will result in seeded crystallization of the amorphous film. Crystallization fronts will emerge from each hole and progress through the film until they impinge. As in the liquid processes, nucleation of new crystals must be avoided. Detailed descriptions of LSPE have been

TABLE I

	SEEDED	UNSEEDED	
Liquid to Crystal ($l \rightarrow x$)	— Lateral Epitaxy by Seeded Solidification (LESS)	— Zone - Melting Recrystallization (ZMR) with Film Patterning	High Temperature ($\geq 1412°C$)
Amorphous to Crystal ($a \rightarrow x$)	— Lateral Solid Phase Epitaxy (LSPE) — Separation by Implanted Oxygen (SIMOX)	— Seed Selection by Ion Channeling (SSIC)	Low ($550°-625°C, a \rightarrow X$) to Intermediate ($\geq 1000°C$, anneal)
Crystal to Crystal ($x \rightarrow x$)		— Secondary Grain Growth (SGG) (Uniform and Scanned Heating)	Intermediate ($\geq 1000°C$)

Constraints of
Lattice Matching and
Substrate Access

given in papers immediately preceeding this one. It should be noted that films obtained by LSPE often have high defect densities which can be reduced through the use of an annealing step at 1150°C.[14]

An alternative solid phase approach to SOI involves the separation of silicon layers through implantation of oxygen[15,16] (SIMOX). In this process a high dose (~2x10[18] cm[-2]) of oxygen is implanted into a single crystal silicon wafer to ranges of 0.3 to 0.5μm. Little ion induced damage occurs in the top surface layer of the wafer. However, the silicon near the peak oxygen concentration is completely amorphized. Subsequent annealing leads to crystallization seeded from the surface and to the formation of a stoichiometric silicon dioxide layer. It should be noted that a polycrystalline silicon region can occur adjacent to the oxide layer and that annealing above 1000°C improves electronic properties.[16]

Both LSPE and SIMOX require seeding from single crystal substrates. While schemes for fabrication of multilayer structures can incorporate these techniques, processes which do not involve substrate seeding would offer greater flexibility. One such technique involves the use of ion bombardment to amorphize all but a few grains in a polycrystalline film of Si. Grains which have orientations which provide for ion channeling should remain crystalline. Through use of implantation from two directions, the 3-dimensional orientations of surviving grains can be specified. Subsequent heating of partially amorphized films can, under the right conditions, lead to growth of the selected grains without nucleation of new grains. This technique, called seed selection by ion channeling (SSIC), has been proposed and tested by Reif and coworkers.[17,18] Recent results confirm that the direction of ion bombardment is critical in determining the mean crystallographic texture of the resulting films.

In all processes involving crystallization of amorphous silicon (a → x processes) it is necessary that crystal growth occur without nucleation of new crystals. The theory of crystal nucleation and growth with specific reference to elemental semiconductors has been reviewed by Spaepen and Turnbull.[19] Ideally, in a → x processes, conditions should be sought which lead to increases in crystal growth rates without increases, and preferably with decreases in the crystal nucleation rate. In contrast to crystal nucleation, crystal growth in amorphous silicon (especially as created by ion implantation) has been extensively characterized experimentally.[5] It is known, for example, that growth rates depend strongly on composition. It is especially interesting to note that while O, N and C reduce[20] growth rates, P, As and B enhance[21] growth rates. These results suggest that film purity, and therefore the means of film formation, could strongly affect crystal growth rates. It has also been clearly demonstrated that the rate of crystal growth in amorphous silicon is strongly dependent on crystal orientation.[19,22,23] One would expect that crystallization rates will be influenced by the density of a film and the stress in film, both of which vary not only with deposition technique but with deposition conditions (eg, substrate temperature). Crystallization rates should also be affected by film geometry. For example, capillarity arguments would suggest that in LSPE, aperture widths and film thicknesses should influence crystallization rates.

Less is known about crystal _nucleation_ in amorphous silicon. It is difficult to speculate as to which, if any, of the factors listed above will lead to increases in crystal growth rates without leading to increases in crystal nucleation rates. However, because critical radii in crystal nucleation[24] are typically less than 10Å, it seems likely that most changes in film geometry will not directly effect crystal nucleation rates. Fundamental studies of crystal nucleation in amorphous semiconductors would presumably augment research on a → x processes for SOI.

III. GRAIN-GROWTH-BASED PROCESSES

An alternative solid phase approach to fabrication of SOI structures is through control of grain growth in polycrystalline semiconductor films. Recent research[25-28] has shown that in ultrathin films (<1000Å) of semiconductors, secondary grain growth can lead to grain sizes many times greater than the film thickness. More importantly, the orientations of the grains that result from this process are nonrandom and can be controlled. In this section the conditions required for secondary grain growth will be reviewed.

Uniform heating of polycrystalline films can result in an increase in the average grain size through grain growth. This process is usually driven by the energy reduction associated with the decrease in grain boundary area. In **normal** grain growth (Fig. 2a-c), the distribution in grain sizes remains monomodal while the average grain size increases through growth and elimination of grains. In thin films of silicon, normal grain growth leads to the formation of columnar grains with radii, r_n, roughly equal to the film thickness, h. Because normal grain growth is driven by the reduction in grain boundary energy, the rate of normal grain growth decreases when grain sizes on the order of the film thickness are attained.

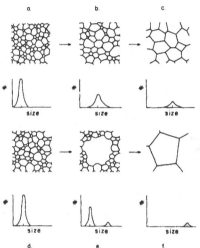

Figure 2. Grain size distribution for normal (a-c) and secondary (d-f) grain growth.

A second grain growth phenomenon often occurs in thin films which have reached the condition $r_n \cong h$. In this process, known as **secondary** or abnormal **grain growth**, a small fraction of the normal grains (those resulting from normal grain growth) continue to grow at high rates, and a bimodal distribution in grain sizes (Fig. 2e) develops. Ultimately, secondary grains grow until they impinge on other secondary grains, resulting in a monomodal grain size distribution (Fig. 2f) and average grain sizes much greater than the film thickness.

Figure 3 illustrates a secondary grain in a thin film of silicon (note that the dark curved lines in the grain are bend contours, an artifact of TEM sample preparation and imaging, and are not crystalline defects). Figure 4 shows a schematic diagram of a secondary grain. Defined in Fig. 3 are the film thickness, h, the secondary grain diameter d_s ($r_s = d_s/2$), and the average normal grain diameter d_n ($r_n \approx d_n/2$). The top and bottom surfaces of the film have associated with them a surface energy per unit area, γ, which is strongly dependent on crystallographic orientation. It is postulated that those grains that become secondary grains have orientations which correspond to the minimum surface energy, γ_{min}. Given an average surface energy for normal grains, $\bar{\gamma}$, a measure of the surface energy anisotropy is given by $\Delta\gamma = \bar{\gamma} - \gamma_{min}$. When surface energy anisotropy supplies a large part of the driving force for grain growth, all secondary grains should have the same crystallographic planes parallel to the substrate

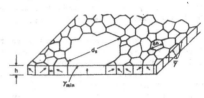

Figure 3. Secondary grain in a 750Å thick film doped to a calculated level of 1×10^{21} cm^{-3} P and annealed at 1100°C.

Figure 4. Schematic diagram of normal and secondary grains.

(i.e., they should have uniform or near-uniform texture).

If we consider a cylindrical region of film which has undergone secondary grain growth, the energy change that results is given by[30]

$$\Delta F = \frac{2\Delta\gamma}{h} - \frac{\beta\gamma_{gb}}{r_n} + \frac{2\gamma_{gb}}{r_s} \qquad [1]$$

where it has been assumed that the top and bottom surfaces of the film are identical (hence the factor of 2 in the first term) and where β is a constant[30] approximately equal to 1 (assuming hexagonal columnar normal grains). The first term in Eq. 1 represents the energy decrease associated with the decrease in surface energy, the second term represents the decrease in energy associated with the elimination of grain boundaries in the normal grain matrix, and the third term represents the energy cost associated with the creation of the secondary grain. In order for growth to occur, ΔF must be less than zero. Although the accuracy of Eq. 1 is limited when $r_s \cong r_n$[30], it suggests that surface energy anisotropy can play an especially important role in the initial stages of secondary grain growth. If we assume that $r_s \gg r_n$ and that r_n remains approximately equal to h, Eq. 1 simplifies to[26]

$$\Delta F \cong \frac{-(2\Delta\gamma + \gamma_{gb})}{h} \qquad [2]$$

which can be considered the driving force for secondary grain growth.

If secondary grain growth is treated as the growth of a cylindrical secondary grain into a __homogeneous matrix__, the rate of secondary grain growth should be proportional to ΔF[30,31],

$$\dot{r} = -M \, \Delta F = M_o \exp\left(\frac{-Q}{kT}\right) \Delta F \qquad [3]$$

where M is a mobility term and M_o is a weakly temperature dependent kinetic constant. [A more detailed discussion of M is given elsewhere[30].] Q in Eq. 4 is an activation energy which characterizes the rate limiting activated atomic process in the motion of secondary grain boundaries. Equations [1]–[3] provide a description of the _pre-impingement_ secondary grain growth rate, r_s. They suggest that r_s should be strongly temperature dependent, should increase with decreasing film thickness and should increase with increasing grain boundary mobility. They also suggest that r_s should increase with γ_{gb} and $\Delta\gamma$, and that secondary grains should have orientations which minimize the surface energy.

Figure 5 shows experimental results[29] for 750Å thick films of Si deposited by low pressure chemical vapor deposition (LPCVD) onto a thermally oxidized Si wafer and encapsulated by films of CVD SiO_2. Detailed experimental conditions will be reported elsewhere.[29] Note, however, that the silicon films were doped by ion implantation to the calculated compositions shown (all other conditions were kept constant). Figure 5 shows differences in the pre-impingement secondary grain sizes and the average normal grain sizes versus the reciprocal of the annealing temperature. The results were obtained after 20 minute anneals. Note first that, as expected, r_s–r_n in isochronal anneals is Arrhenian in temperature, showing an activation energy of about 3.8 eV. Note also the striking effect of phosphorous content on the secondary grain growth rate and, presumably, on the secondary grain boundary mobility. [It is interesting that the activation energy for P diffusion is approximately equal to 3.7 eV.[32]]

In order to obtain kinetic information from films in which secondary grain growth has led to partial impingement, it is useful to measure the areal fraction, x, of a film that has undergone transformation to secondary grains. Figure 6 shows x plotted versus annealing temperatures after isochronal 20 minute anneals.[29] The calculated phosphorous content has been kept constant while film thickness was varied. These results suggest that again, as expected, the secondary grain growth rate increases with decreasing film thickness. [The dashed lines represent "Johnson-Mehl"[24,30] fits.]

The dominant texture of the secondary grains obtained in this research is (111). In pure Si films it might be expected that (100) orientations minimize the Si/SiO_2 surface energy.[33,34] In these films, however, it is likely that P segregates to the Si/SiO_2 interface. Phosphorous segregation may lead to changes in the magnitudes of the surface energy or even to a change in the orientation corresponding to the minimum surface energy. Note that if one assumes that surface energy scales with broken bond densities,

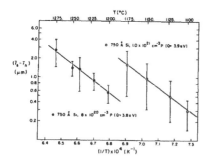

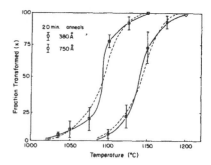

Figure 5. Secondary grain radius, r_s, minus normal grain radius, r_n, after 20 minute anneals at temperatures T. Also shown are calculated doping levels.

Figure 6. Areal fraction of a film transformed, x, after 20 min. anneals at temperatures T. Both films were doped with P to calculated levels of 1×10^{21} cm^{-3}.

one would expect that γ_{min} occurs at (111) orientations.[35,36] Also, Jaccodine has measured free surface energies for (100), (110), and (111) Si and finds that $\gamma_{(111)}$ is lowest.[35]

Three factors which affect secondary grain growth rates have been briefly discussed; temperature, film thickness, and grain boundary mobility (specifically as effected by P doping). As in amorphous to crystal transitions, there are many factors which should affect secondary grain growth. It is known, for example, that O and Cl reduce the rate of normal grain growth in Si.[37] Also, film density and the state of stress of a film should affect the kinetics of grain growth. Initial grain sizes and orientations will not only affect the growth rates but also the final secondary grain sizes. All of these factors are likely to change with film deposition techniques and conditions. Secondary grain growth should also be strongly effected by the nature of the substrate and the encapsulating film (e.g., surface composition, structure, and cleanliness). The influence of many of these parameters have not yet been fully explored and/or exploited.

Because secondary grain growth can be driven by surface energy anisotropy, changes in the topography of a surface can be used in controlling the kinetic path of this solid phase process. For example, in recent experiments[28] 300Å thick germanium films were deposited onto SiO_2 surfaces which had been lithographically patterned with a surface relief structure having a square-wave cross-section, a period of 2000Å, and a depth of 100Å. This patterning of the substrate increases the driving force for surface-energy-driven secondary grain growth and leads to growth of grains with specific in-plane orientations, as well as specific textures. It was found that secondary grain growth in these germanium films led to the formation of grains with (100) texture and with a strong tendency for alignment of ⟨100⟩ in-plane directions along the groove directions. Were this affect to work perfectly, impingement of secondary grains would result in a single crystal film. This process is a solid phase form of graphoepitaxy.[38]

Single crystal films or films composed of grains which meet at low angle grain boundaries might result directly from secondary grain growth (SGG) in very thin semiconductor films with artificial, anisotropic surface topography. SGG may also be usable as a process for producing single crystal islands which could serve as seeds for subsequent epitaxial crystal growth. For example, islands which have undergone secondary grain growth could be covered by amorphous deposited films. Subsequent controlled crystallization might yield films with very large grains. Cline[39,40] has recently proposed that secondary grain boundaries can be propogated through a polycrystalline film using a scanned heat source in a process similar to those described in refs. 4 and 5. In combination with film patterning, a scanning technique might also provide single crystal films with controlled orientations.

In experiments reported so far, secondary grain growth in silicon has required annealing at temperatures in excess of 1000°C (1050°C to 1300°C, depending on conditions discussed above). Secondary grain growth leads directly to grains or crystals with low defect densities (no dislocations and for fewer than one twin per grain[29]). It should be noted that while crystallization of amorphous silicon can occur at less than 600°C, the resulting crystals often have high defect densities. Removal of these defects, if possible, requires annealing at temperatures greater than 1000°C. While there is currently no demonstrated process for obtaining device quality SOI through grain growth, this approach appears to offer the possibility of obtaining low defect density films at low to intermediate temperatures and without the requirement of substrate seeding.

IV. SUMMARY

Two low temperature solid phase approaches to SOI have been described, controlled crystallization of amorphous silicon and controlled grain growth in polycrystalline silicon. While processes which involve melting of silicon are well developed and may be adequate for some applications, it is felt that in other applications, especially in construction of multilayer devices, lower temperature processes will be preferred. Processes which avoid the use of single crystal substrates for seeding offer greater flexibility in that they bypass the constraints of lattice matching and seed access. For these reasons, unseeded solid phase processes are viewed as offering, in the long term, the greatest flexibility.

A brief discussion of the materials properties which influence crystallization of amorphous silicon has been given. Crystal growth must occur without nucleation of new crystals. Under some circumstances, this can be accomplished in silicon at temperatures less than 600°C. Higher temperature anneals (>1000°C) are often required in order to reduce defect densities.

The theory and some experimental results for secondary grain growth in silicon were reviewed. Surface-energy-driven secondary grain growth (SEDSGG) can result in films with grain sizes much larger than the film thickness and in grains with uniform or near uniform crystallographic texture. The temperature required for significant SEDSGG can be lowered through changes in the film composition and the film thickness. The impact of other factors were also briefly discussed. Secondary grain growth in silicon films has been observed at temperatures as low as 1050°C. The grains resulting from secondary grain growth have low defect densities. Surface-energy-driven secondary grain growth may provide the basis for a flexible low-temperature process for SOI. SEDSGG occurs in both Si and Ge and in a variety of metals[30,41,42], and is presumably a near universal phenomenon. SEDSGG can occur in films on both crystalline and amorphous substrates and need not require lattice matching. Grain-growth-based processes may ultimately provide the means of producing device quality films of a variety of materials on a common substrate.

ACKNOWLEDGEMENTS

The author would like to thank Henry I. Smith for many helpful discussions and for a critical review of the manuscript. Most of the experimental work on grain growth in Si was performed by Hyoung-June Kim and will be described in detail elsewhere. This work was supported by the Semiconductor Research Corporation under contract number 83-01-033.

REFERENCES

1. H.W. Lam, A.F. Tasch, Jr. and R.P. Pinizzotto, Chap. 1 in "VLSI Electronics Microstructure Science", Vol. 4, Ed., Norman G. Einspruch, Academic Press, New York (1982).
2. D.A. Antoniadis, Mat. Res. Soc. Symp., 23, 587 (1984).
3. "Materials Research Society Symposia Proceedings, Vol. 4, (1982), Vol. 13, (1983), and Vol. 23, (1984).
4. M.W. Geis, H.I. Smith, B.-Y. Tsaur, J.C.C. Fan, D.J. Silversmith and R.W. Mountain, J. Electrochem. Soc. 129, 2812 (1982).
5. J.C.C. Fan, M.W. Geis and B.-Y. Tsaur, Appl. Phys. Lett. 38 365 (1981).
6. H.A. Atwater, H.I. Smith and M.W. Geis, Appl. Phys. Lett. 41, 747 (1982).
7. H.A. Atwater, C.V. Thompson, H.I. Smith and M.W. Geis, Appl. Phys. Lett. 43, 1126 (1983).

8. D.K. Biegelsen, N.M. Johnson, D.J. Bartelink and M.D. Moyer, Appl. Phys. Lett. 38, 150 (1981).

9. G.K. Cellar, McD. Robinson and D.J. Lischner, Appl. Phys. Lett. 42, 99 (1983).

10. Y. Ohmura, Y. Matsushita and M. Kashiwagi, Jpn. J. Appl. Phys. 21 L152-154 (1982).

11. Y. Kunii, M. Tabe and K. Kajiyama, Jpn. J. Appl. Phys. 22, Supplement 22-1, 605 (1983).

12. J.A. Roth, G.L. Olson and L.D. Hess, Mat. Res. Soc. Symp. Proc. 23, 431 (1984).

13. M. Tabe and Y. Kunii, this proceedings.

14. Y. Kunii, M. Tabe and K. Kajiyama, J. Appl. Phys. 54, 2847 (1983).

15. K. Izumi, M. Doken and H. Ariyoshi, Electronics Letts. 14 593 (1978).

16. K. Izumi, Y. Omura and S. Nakashima, Mat. Res. Soc. Symp. Proc., 23 443 (1984).

17. R. Reif and J.E. Knott, Electronics Letts. 17, 586 (1981).

18. K.T-Y. Kung, R.B. Iverson and R. Reif, to be published in Materials Letters.

19. F. Spaepen and D. Turnbull, Chap. 2 in "Laser Annealing of Semiconductors", ed. J.M. Poate and J.W. Mayer, Academic Press, New York (1982).

20. E.F. Kennedy, L. Csepregi, J.W. Mayer and T.W. Sigmon, J. Appl. Phys. 48, 4241 (1977).

21. L. Csepregi, E.F. Kennedy, T.J. Gallagher, J.W. Mayer and T.W. Sigmon, J. Appl. Phys. 48, 4234 (1977).

22. L. Csepregi, E.F. Kennedy, J.W. Mayer and T.W. Sigmon, J. Appl. Phys. 49, 3906 (1978).

23. H. Yamamoto, H. Ishiwara, S. Furukawa, M. Tamura and T. Tokuyama, Mat. Res. Soc. Symp. Proc. 25, 511 (1984).

24. J.W. Christian, "The Theory of Transformations in Metals and Alloys - Part I", second edition, Pergamon Press, New York (1975).

25. C.V. Thompson and H.I. Smith, Electronic Materials Conference, Burlington, VT (1983).

26. C.V. Thompson and H.I. Smith, Appl. Phys. Lett. 44, 603 (1984).

27. T. Yonehara, C.V. Thompson and H.I. Smith, Mat. Res. Soc. Symp. Proc. 23, 627 (1984).

28. T. Yonehara, H.I. Smith, C.V. Thompson and J.E. Palmer, Appl. Phys. Lett. 45, 631 (1984).

29. H.J. Kim and C.V. Thompson, to be published.

30. C.V. Thompson, submitted to J. Appl. Phys.

31. D. Turnbull, Trans AIME 191, 661 (1951).

32. J.S. Makris and B.J. Masters, J. Electrochem. Soc. 120 1253 (1973).

33. H.I. Smith, C.V. Thompson, M.W. Geis, R.A. Lemons and M.A. Bosch, J. Electrochem. Soc. 130, 2050 (1983).

34. D.K. Biegelsen, L.E. Fennell and J.C. Zesch, Appl. Phys. Lett. 45, 546 (1984).

35. R.J. Jacodine, J. Electrochem. Soc. 110, 524 (1963).

36. A. Szilagyi, Ph.D. Thesis, Dept. of Physics, Massachusetts Institute of Technology (1984).

37. R. Angelucci, M. Severi and S. Solmi, Mat. Chem. and Phys. 9, 235 (1983).

38. M.W. Geis, D.C. Flanders and H.I. Smith, Appl. Phys. Lett. 35, 71 (1979).

39. H.E. Cline, J. Appl. Phys. 55, 2910 (1984).

40. H.E. Cline, J. Appl. Phys. 55, 4392 (1984).

41. J.L. Walter and C.G. Dunn, Trans. AIME 227, 185 (1963).

42. R.W. Cahn, Chap. 19 in "Physical Metallurgy", ed., R.W. Cahn, North-Holland, Amsterdam (1970).

GRAIN GROWTH PROCESSES DURING TRANSIENT ANNEALING
OF As-IMPLANTED, POLYCRYSTALLINE-SILICON FILMS

S.J. KRAUSE*, S.R. WILSON**, W.M. PAULSON**, and R.B. GREGORY**
*Dept. of Mechancial and Aerospace Engineering, Arizona State University,
Tempe, AZ 85287
**Semiconductor Research and Development Laboratory, Motorola, Inc., 5005
E. McDowell Road, Phoenix, AZ 85008

ABSTRACT

    Polycrystalline silicon films of 300 nm thickness were deposited on
oxidized wafer surfaces, implanted with As, and annealed on a Varian IA 200
rapid thermal annealer. Transmission electron microscopy was used to study
through-thickness and cross sectional views of grain size and morphology of
as-deposited and of transient annealed films. A bimodal distribution of
grain sizes was present in as-deposited polycrystalline silicon films. The
first population was due to columnar growth of some grains to a final
average diameter of 20 nm. The second population of small equiaxed grains
of 5 nm average diameter were formed early in the deposition process.
During transient annealing grains in the first population grew rapidly up
to 280 nm equiaxed grains. After this the growth rate decreased due to the
grain size reaching the thickness of the film. Grains in the second
population grew rapidly up to a size of 150 nm, after which the growth rate
was lowered due to grains impinging upon one another. The grain growth
processes for both populations have been described with a modified model
for interfacially driven grain growth. This model accounts for diffusion
and grain growth which occur with rapidly rising and falling temperatures
during short annealing times characteristic of transient annealing
processes.

INTRODUCTION

    Polycrystalline silicon is commonly used as an interconnect and as a
gate material for most metal-oxide-semiconductor (MOS) devices. To reduce
resistivity of polycrystalline silicon, it is ion implanted and annealed
during the same steps as are the source and drain of the MOS device.
Depending on the temperature of deposition, the silicon may be completely
amorphous, partially amorphous and partially crystalline, or entirely
crystalline [1]. The polycrystalline silicon itself may have a single or
bimodal distribution of grain sizes [2] which may or may not be heavily
textured [3]. During furnace annealing of single crystal silicon the
diffusion of As and Si may be enhanced by the presence of As dopant [4].
In phosphorous doped polycrystalline silicon, the enhanced diffusion rate
of Si (due to P dopant) increases the grain growth rate [5]. This process
has been modeled by Wada and Nishimatsu [6] by an interfacial energy driven
grain growth model. In an earlier paper we described the grain growth of
polycrystalline silicon during transient annealing with a modified model
for interfacially driven grain growth [7]. In this paper we have
characterized dopant enhanced grain growth of a bimodal grain size
distribution of As doped polysilicon during the short times and high
temperatures characteristic of transient annealing.

EXPERIMENTAL

    The substrates used in these experiments were 3 inch (100) Si wafers
on which a 0.1 micron SiO$_2$ film was grown. Undoped polysilicon, about 300

nm thick, was deposited on the oxidized wafer surface at 625°C by low pressure (about 50 Pa) chemical vapor deposition of silane. The films were then implanted with [75]As (60 keV, 5.0 x 10[15] / cm²) and capped with sputtered SiO_2. Wafers were transient annealed for 10 to 25 seconds using a Varian IA 200 rapid isothermal annealer with a resistively heated graphite sheet. The annealer has been described elsewhere [8]. The nominal heater set point temperatures were 1150°C and 1200°C as measured by a thermocouple located 2 mm behind the heater. The true wafer temperature, which differs from that of the set point temperature, is measured by an optical pyrometer located directly behind the wafer. The pyrometer was calibrated by correlating pyrometer readings to temperature measurements from a thermocouple located on a silicon wafer which had reached thermal equilibrium at specific elevated temperatures in the rapid thermal annealer. A typical time – temperature profile for an 1150° set point temperature is shown in Figure 1. The temperature measured at 0 seconds is not indicative of the wafer, but is due to radiation from the heater element passing directly through the wafer, since silicon is essentially transparent to infrared radiation at room temperture. As the wafer temperature rises the transmitted radiation decreases. The reading of 875°C between about 3 and 7 seconds is the minimum reading on the pyrometer and not the wafer temperature. After 7-10 seconds the wafer temperature begins to exceed 875°C and the temperature in the figure is indicative of the wafer temperature. The wafer temperature rises rapidly between 10 and 20 seconds and then remains essentially constant between 20 and 30 seconds. The wafer temperature exceeds the heater set point temperature because the thermocouple measuring and controlling the heater temperature is not in direct contact with the heater, but is about 2 mm away. Therefore, the real heater temperature is higher than the set point temperature.

The grain size in the polysilicon films was determined by transmission electron microscopy (TEM). The films were bright and dark field imaged at 200 kV in a JEOL 200CX by standard techniques. Grain size distributions were determined by measuring between 30 and 100 grains from a micrograph and plotting number of grains versus grain size. A typical distribution for an as-deposited film is shown in Figure 2 where the two maxima for the grain sizes have been specified as 20 nm and 5 nm.

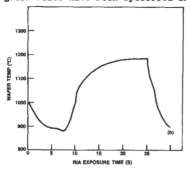

Figure 1. Temperature vs. time for heater set point of 1150°C and exposure time of 25 seconds.

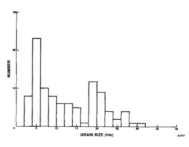

Figure 2. Grain size distribution for as-deposited polycrystalline film.

RESULTS AND DISCUSSION

A tabulation of grain sizes as a function of annealing conditions is presented in Table I. The nominal annealing conditions are presented as exposure time and as set point temperature. We have analyzed our results, as discussed later, in terms of an effective time at peak temperature.

TABLE I Grain size as a function of annealing conditions

Set Point Temp.(^{o}C)	Peak Wafer Temp.(^{o}C)	Exposure Time(sec.)	Effective Time(sec.)	$\sqrt{Dt^*}/T$ (cm/K$^{1/2}$x10^{-6})	Grain size(nm) 1st	2nd
25	–	–	–	–	20	5
1150	910	10	1.0	0.2	47	5
1150	1075	12.5	1.65	1.1	103	31
1150	1177	15	2.0	2.5	178	89
1150	1210	17.5	2.82	3.7	275	156
1150	1271	20	6.75	8.0	295	168
1150	1276	25	11.5	10.2	310	195
1200	1145	10	1.0	1.5	260	120
1200	1297	15	3.25	6.3	300	160
1200	1305	20	8.0	10.5	315	175
1200	1343	25	11.5	15.3	300	150

The as-deposited films, prior to annealing, displayed a bimodal distribution of grain sizes. The two populations of grain sizes are shown in Figures 3 and 4 for through-section and cross section views, respectively. The first population is composed of elongated grains of a columnar morphology with an average diameter of 20 nm. These evidently arise from prefential growth of some grains during deposition. The second population is composed of small equiaxed grains about 5 nm in diameter. These are located chiefly at the bottom of the film and must form early in the deposition process. The cross section view shows smaller grains at the bottom of the film. Duffy et al [3] have also reported observing a bimodal distribution of grains for polysilicon films deposited at 622°C. They stated that large, highly oriented grains at least 70 to 100 nm in size were located in a matrix of smaller randomly oriented grains with crystallite size of about 10 nm.

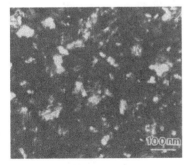

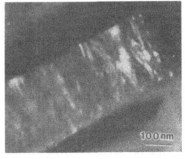

Figure 3. Through-section view of as-deposited film.

Figure 4. Cross-section view of as-deposited film

In this study, the bimodal distribution of crystallite sizes was preserved in the grain growth process during transient annealing. However, both populations assumed equiaxed morphologies as grain growth progressed. Figures 5 and 6 show through-section and cross section views after 12.5

seconds of annealing at a heater set point temperature of 1150°C. Two different sets of grain sizes with averages of about 31 nm and of about 103 nm are clearly evident in these views. At the longest times up to 25 seconds and the highest annealing temperatures of up to 1305°C the two grain size populations grow to maximum diameters of about 160 nm and 330 nm. Our earlier work showed that arsenic implanted films, at the concentration level used here, showed grain growth at a rate of about 1.4 times that of unimplanted films due to the dopant enhanced diffusion of silicon [7]. These results were consistent with earlier work of Wada and Nishimatsu [6] on grain growth in furnace annealed polycrystalline silicon. In examining cross section views of the sample annealed 12.5 seconds and of samples with longer annealing times and/or higher annealing temperatures, it was observed that both populations of the distribution grew into an equiaxed morphology. The progression of grain growth during transient annealing is shown in Figure 7 where grain size is plotted as a function of annealing time for films subjected to treatment with an 1150°C set point heater temperature. Significant grain growth does not begin until annealing time has reached 10 seconds. Grain growth occurs rapidly between 10 and 20 seconds, but then slows down. This is consistent with our electrical measurements which show significant mobility increases from 10 to 20 seconds, but little increase after 20 seconds [7]. This is also consistent with other studies which also show increases in mobility with decreasing grain boundary area which is due to a reduction in depletion zones associated with grain boundaries [9].

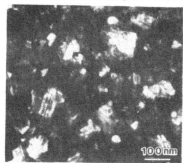
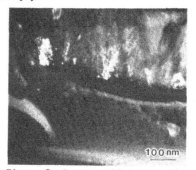

Figure 5. Through-section view of film annealed at 1150°C set point for 12.5 seconds.

Figure 6. Cross section view of film annealed at 1150°C set point for 12.5 seconds.

The curves for the two populations in Figure 7 appear similar to S-shaped curves which are observed for recrystallization of metals during the process of grain nucleation, growth, and impingement [10]. However, most studies of grain growth of metals and other materials have been done at constant temperature. In this study significant diffusion and grain growth occur during the rapid temperature rise and fall and must be taken into account. In order to describe grain growth during transient annealing we have developed a modified model for interfacial energy driven grain growth. This model successfully describes grain growth of both populations in this study and seems to account for the appearance of the S-shaped curves representing the grain growth.

Wada and Nishimatsu [6] have shown that grain growth in doped polysilicon is driven by interfacial energy, and grain size ($\bar{r}$) is therefore proportional to the square root of time at constant temperature. However, we have modified the interfacial energy model in two ways to compensate for heating and cooling times during transient annealing. First, the effect of changing temperature has been compensated for by calculating the diffusion coefficient (D) of Si in polycrystalline Si based on the peak temperature of the wafer. Second, an effective time (t*), as described by Shewmon [11], is calculated based on the time-temperature profile, the peak temperature (T), and the appropriate activation energy. In this study we have used a value of 2.4 eV for activation energy of diffusion of Si in polycrystalline Si as specified by Wada and Nishimatsu [6] in their study of grain growth of doped polysilicon. Based upon a classical model for grain growth, as discussed by Wada and Nishimatsu [6], by McLean [10], and others, a new, modified model for interfacially driven grain growth can be given by the relationship: $\bar{r} \propto \sqrt{Dt^*}/T$. D may be calculated from $D = D_0 \exp(E_a/kT)$ where D_0 is the proportionality constant, (and is assumed here to be 1 cm²/sec. for simplicity), E_a is the activation energy, and k is Boltzman's constant.

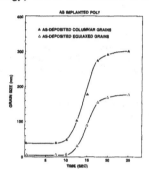

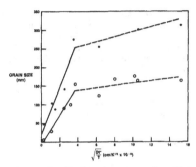

Figure 7. Grain size as a function of annealing time for an 1150°C set point temperature.

Figure 8. Grain size as a function of $\sqrt{Dt^*}/T$ for 1150°C and 1200°C set point temperatures.

In Figure 8 grain size is plotted as a function of Dt^*/T for samples annealed at various time-temperature conditions. Both populations of grain sizes from the bimodal distributions have been plotted. The closed circles (●) represent sizes of grains of the first population which have grown from the larger as-deposited grains with the columnar morphology. The open circles (o) represent sizes of grains which have grown from the second population of smaller as-deposited grains with an equiaxed morphology.

A linear least squares fit was applied to the data for each population up to a value of $\sqrt{Dt^*}/T$ less than 4 x 10^{-6} cm/K$^{1/2}$, as shown by the solid lines (—). This resulted in a correlation coefficient of approximately 0.98 for both grain size populations. At values of $\sqrt{Dt^*}/T$ higher than 4 x 10^{-6} cm/K$^{1/2}$ there is a break in the least squares fits due to geometrical constraints affecting grain growth. For the first population, the one with larger grains, the growth rate is lowered when the grain size reaches the thickness of the film. A similar effect has been observed for furnace annealing of polysilicon films by Jain and Overstraeten [5] and in transient annealing of polysilicon films by Pinizzotto [12]. For the second population, the one with the smaller grains, the growth rate is reduced when the grains impinge upon one another. Since only a fraction of the smaller grains actually participate in grain growth, they do not impinge upon one another until they reach a diameter of about 150 nm. Similar effects for metals have been documented by McLean [11] and others. The regions for reduced growth rates are represented by dashed lines (---) on the plot because the scatter of the data is high. The reduced growth rates also explain the flattening of the S-shaped curves at longer annealing times as shown in Figure 7. Thus, we find that, prior to geometrical interference, grain growth processes during transient annealing of doped polysilicon can be described by a modified model for interfacially driven grain growth.

SUMMARY AND CONCLUSIONS

1. For polysilicon films deposited at 625°C a bimodal distribution of grains is present which is composed of large columnar grains and smaller equiaxed grains.

2. The bimodal distribution of grain sizes is generally preserved in grain growth during transient annealing in As doped polysilicon films, but the original grains from both populations grow into larger, equiaxed grains during annealing.

3. The grain growth rate of both populations of the bimodal distribution is lowered by geometrical constraints at later stages of annealing. The growth rate of the larger grains in the first population decreases when grain diameters approach the film thickness. The growth rate of smaller grains in the second population decreases when the grains impinge upon one another.

4. A modified model for interfacial energy driven grain growth has been applied to describe grain growth of a bimodal distribution of grain sizes during transient annealing.

REFERENCES

1.   T.I.Kamins, M.M.Mandurah, and K.C.Saraswat, J. Electrochem. Soc., 125, 927 (1978).

2.   M.T.Duffy, J.T.McGinn, J.M.Shaw, R.T.Smith, and R.A.Soltis, RCA Review, 44, 287 (1983).

3.   R.R.Anderson, J. Electrochem. Soc., 120, 1540 (1973).

4.   J.M.Fairfield and B.J.Masters, J. Appl. Phys., 38, 3148 (1967).

5.   E.T.Tannenbaum, Solid-State Electronics, 2, 123 (1961).

6.   Y. Wada and S. Nishimatsu, J. Electrochem. Soc., 125, 927 (1978).

7.   S.J.Krause, S.R.Wilson, W.M.Paulson, and R.B.Gregory, Appl. Phys. Lett., 45, 778 (1984).

8.   S.R.Wilson, R.B.Gregory, W.M.Paulson, A.H.Hamdi, and F.D.McDaniel, Appl. Phys. Lett., 41, 978 (1982).

9.   J.Y.W.Seto, J. Appl. Phys., 47, 5168 (1976).

10.   D. McLean, Grain Boundaries in Metals, Oxford University Press, London (1957).

11.   P.G.Shewmon, Diffusion in Solids, McGraw-Hill, New York (1963).

12.   R.F. Pinizotto, Proc. Mat. Res. Soc., February (1984), in press.

MODIFYING CRYSTALLOGRAPHIC ORIENTATIONS OF POLYCRYSTALLINE Si FILMS USING ION CHANNELING

K. T-Y. KUNG, R. B. IVERSON, AND R. REIF
Department of Electrical Engineering and Computer Science
Massachusetts Institute of Technology, Cambridge, MA 02139

ABSTRACT

   Polycrystalline silicon films 4800 Å thick deposited via low pressure chemical vapor deposition on oxidized silicon wafers have been amorphized by silicon ion implantation and subsequently recrystallized at 700°C. Due to channeling of the ions through grains whose <110> axes were sufficiently parallel to the beam, these grains survived the implantation step and acted as seed crystals for the solid-phase epitaxial regrowth of the film. This work suggests the feasibility of combining ion implantation and furnace annealing to generate large-grain, uniformly oriented polycrystalline films on amorphous substrates. It is a potential low-temperature silicon-on-insulator technology.

INTRODUCTION

   In recent years there has been considerable interest in growing device-quality Silicon layers On Amorphous insulators (SOA). One possible application of such a process is in the fabrication of novel device structures and three-dimensional integrated circuits. As reported in the literature, high-quality silicon layers have been formed on amorphous substrates using a number of approaches including: Lateral Epitaxy by Seeded Solidification (LESS) using a laser [1], an electron beam [2], or a graphite strip heater [3] as heat source; graphoepitaxy [4]; and Lateral Solid-Phase Epitaxy (LSPE) [5]. Unfortunately, the high operating temperatures associated with LESS and graphoepitaxy render them unsuitable for three-dimensional integration. LSPE suffers from the requirement of lateral seeding, which restricts structure topology and imposes a penalty on packing density.
   In an earlier communication [6], a novel low-temperature process was proposed for the formation of a large-grain, uniformly oriented polycrystalline film on an arbitrary non-seeding substrate. In this process, a polycrystalline film of the desired material is first deposited. An ion implantation step is then employed to selectively amorphize the film. (The implant schedule should be chosen to amorphize all except a few similarly oriented grains where significant ion channeling occurs.) In the final step, a low-temperature anneal is carried out to recrystallize the film via a solid-phase epitaxial process using the surviving grains as seeds. It was suggested that once perfected this process may be used to generate active layers of silicon on amorphous insulators for device and integrated circuit fabrications. This approach is attractive for SOA because the process temperatures are low and it does not require lateral seeding. Hereafter, this process is referred to as Seed Selection through Ion Channeling (SSIC).
   Experiments done in the past have demonstrated the capability of SSIC to produce large-grain polycrystalline silicon films on SiO_2 [7,8]. In the works of Kwizera and Reif [7], polysilicon films obtained by Low Pressure Chemical Vapor Deposition (LPCVD) were amorphized by silicon ion implantation and recrystallized at 525°C to achieve an average grain size of about 1 µm. The results of Iverson and Reif [8] are more encouraging, where 7 µm grains were observed in an LPCVD polysilicon film after multiple-angle phosphorus

implants and annealing at 700°C. One question that was raised concerned the origins of these large grains: were they regrown from seeds selected via SSIC, or were they actually regrown from crystallites generated by spontaneous nucleation? To answer this question, the two groups have each proposed an indirect argument to show that the recrystallizations must have been seeded by crystallites that survived due to ion channeling [7,8]. Nevertheless, the most convincing way to identify the origins of these large grains is to check their orientations. Seeding by the selected grains should lead to a uniformly oriented polycrystalline film, while seeding by spontaneously nucleated crystallites will result in a random distribution of orientations. In this article, we present an experiment that demonstrates the orientation effect of SSIC for polysilicon films deposited on SiO_2.

EXPERIMENTAL

Starting with <100> and <111> silicon wafers, a 1000 Å SiO_2 layer was thermally grown in a dry oxidizing ambient. The two wafer orientations were needed to allow x-ray investigation of several crystallite directions in the film without interference from the substrate. Undoped polysilicon films 4800 Å thick were then deposited via LPCVD at 620°C. Based on an earlier report by Kamins et al. [9], films deposited at this temperature contain a predominant {110} texture. Five sets of samples were implanted, each with 210 keV Si^{28} ions to a dose of 1×10^{15} cm^{-2}, but at directions of incidence tilted by 0°, 1°, 3°, 5°, or 7° relative to the substrate normal. Each implant direction was expected to preserve only grains whose <110> axes were parallel to the beam. While the ion current was low (below 0.3 $\mu A/cm^2$), the samples were cooled with liquid nitrogen to below -170°C during the entire implant cycle. A low implant temperature was preferred in order to enhance ion channeling in properly oriented grains by reducing acoustic vibrations of the crystalline lattice, and to avoid self-annealing of damage in misoriented grains due to beam-heating. Annealing of the implanted samples was carried out at 700°C for 4 h in a nitrogen ambient.

At the end of each processing step, the film structure was examined by both Transmission Electron Microscopy (TEM) and x-ray diffraction. The x-ray work was done using an x-ray monochromater and sample rotation apparatus schematically illustrated in Fig. 1. This apparatus allows measurement of the x-ray intensity diffracted from any particular set {hkl} of crystalline planes, designated by the Bragg angle θ, as a function of the angle ϕ between the substrate plane and the plane formed by the incident and reflected beams. This is equivalent to measuring the angular distribution of the crystallographic orientation <hkl> within the film. For each sample we have checked for x-ray diffractions from the {400}, {220}, {111}, {311}, and {331} planes. It was found that only diffractions from the {220} planes, if any, could be discerned above the noise.

RESULTS AND DISCUSSION

For the as-deposited sample, TEM analysis shows that it is polycrystalline with an average grain size of about 700 Å. The x-ray intensity diffracted from the {220} planes of this sample recorded as a function of ϕ is shown in Fig. 2. A finite diffracted intensity exists for 70° < ϕ < 90°, while only scattered noise is present for ϕ < 70°. Also, this plot is symmetric with respect to the substrate normal (ϕ = 90°). This figure shows that most of the <110> directions in the as-deposited film must be confined within ±20° of the substrate normal. These observations are consistent with those reported by Kamins et al. [9].

For the five sets of as-implanted samples, TEM indicates that they are completely amorphous; that is, there is no trace of surviving crystallites.

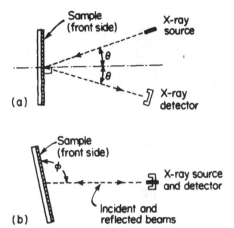

Fig. 1 Illustrations of the x-ray monochromater and sample rotation apparatus used in this experiment:
(a) top view with $\phi = 90°$;
(b) side view with $\phi > 90°$.

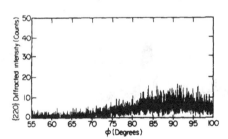

Fig. 2 {220} diffracted intensity as a function of ϕ for the as-deposited sample.

Also, no x-ray diffraction from any set of crystalline planes was observed, as can be expected from completely amorphous films. These results imply that the implantation step has completely amorphized the polysilicon films and are counter to our anticipation that some grains should have survived due to ion channeling. Nevertheless, it is possible that only a very small volume of each selected grain has survived, and the total surviving volume is too small to be detectable by either technique.

After post-implantation annealing, all five sets of samples have recrystallized. From TEM, the average grain diameter in each of the films is 2500-3500 A. The {220} diffracted intensities (recorded as functions of

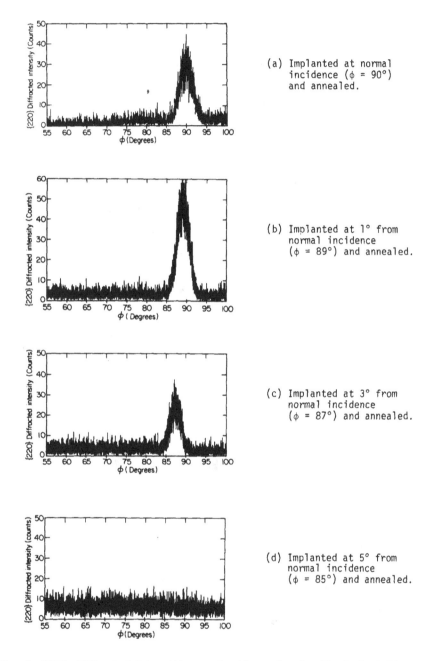

(a) Implanted at normal incidence ($\phi = 90°$) and annealed.

(b) Implanted at 1° from normal incidence ($\phi = 89°$) and annealed.

(c) Implanted at 3° from normal incidence ($\phi = 87°$) and annealed.

(d) Implanted at 5° from normal incidence ($\phi = 85°$) and annealed.

Fig. 3 {220} diffracted intensities as functions of ϕ for the implanted/annealed samples.

φ) for the first four samples are shown in Figs. 3(a)-(d). In Figs. 3(a), (b), and (c) for the 0°, 1°, and 3° implanted/annealed films, most of the <110> directions in each sample have been confined within ±4° of the corresponding implant direction (φ = 90°, 89°, and 87°, respectively). However, Fig. 3(d) is quite different for the 5° implanted/annealed sample, where the <110> directions have become more randomly oriented than in the as-deposited film. The x-ray result for the 7° implanted/annealed sample is similar to that shown in Fig. 3(d), indicating that the implant/anneal treatment has also randomized the <110> directions in this film.

For the first three samples, the x-ray results can be explained by seed selection through ion channeling. The ion implantation step had amorphized all except those grains whose <110> axes were sufficiently parallel to the beam, and these grains acted as seed crystals in a subsequent solid-phase epitaxial process to recrystallize the film.

For the last two cases, the results imply that the ion bombardment process must have destroyed practically all the crystallites, so that during regrowth spontaneous nucleation dominated, leading to randomly oriented films. However, in Fig. 2 for the as-deposited film, the {220} diffracted intensity decreases only very slightly as φ varies from 90° to 83°. Through a simple argument, this means that the initial number of <110> grains available for channeling decreases only by a small fraction as the ion beam is tilted away to 7° from normal incidence. Therefore, it is not clear why, when the tilt angle was switched from 0°, 1°, or 3° to 5° or 7° with all other conditions remaining unchanged, the implantation step completely amorphized the film. Further studies of implant damage and regrowth mechanisms are necessary before a convincing argument may be constructed.

In order to ensure that the ion implantation step is the source of grain size enhancement and orientation effects, some as-deposited films were annealed with the implanted ones. These unimplanted but annealed films have the same structure as the as-deposited film, as indicated by TEM and x-ray. This is not surprising since a 700°C anneal should not alter the structure of the polysilicon film [9].

This experiment was aimed at demonstrating the orientation capability of SSIC. A different implant direction was selected for each of five samples. In the three successful cases (0°, 1°, and 3° from normal incidence), <110>-axially oriented films were obtained; that is, the <110> directions are parallel to the implant direction within a small angular spread. In theory, this angular spread is limited by the channeling half angle for the <110> direction (around ±2° [10]) and the raster scan angle of the ion implanter (±0.5° for the machine used), if spontaneous nucleation is avoided. We believe that the observed angular spread of ±4° can be further minimized. Work is currently underway to determine the effect of varying the implant dose and regrowth temperature for different implant directions.

Finally, for fabricating devices and integrated circuits three-dimensionally oriented films are required -- films within which every set of crystalline directions <hkl> are lined up. To achieve this, it is necessary to implant in at least two different directions [8]. Work is also being carried out to experimentally determine the feasibility of obtaining three-dimensionally oriented films using two-angle implants.

CONCLUSION

Polysilicon films 4800 Å thick deposited via low pressure chemical vapor deposition at 620°C on oxidized wafers have been amorphized by silicon ion implantation and subsequently recrystallized at 700°C. Channeling of the ions through grains whose <110> axes were sufficiently parallel to the beam allowed these grains to survive the implantation step and act as seed crystals for the solid-phase epitaxial regrowth of the film. In each of three successful films, most of the <110> directions are confined within ±4°

of the implant direction. This experiment suggests the feasibility of combining ion implantation and furnace annealing to generate large-grain, uniformly oriented polycrystalline films on amorphous substrates while maintaining low processing temperatures.

ACKNOWLEDGMENTS

This project was funded by the National Science Foundation, Grant No. 83-03450ECS, and was carried out using the facilities of the M.I.T. Center for Materials Science and Engineering. One of the authors (KTYK) wishes to express his gratitude for the support provided by a Digital Equipment Corporation Graduate Fellowship. Part of the implantation work was done at the California Institute of Technology, where access granted by Prof. M-A. Nicolet and assistance from F. C. T. So are highly acknowledged. The x-ray work was done at the M.I.T. Lincoln Laboratories, with access granted by Drs. A. J. Strauss, M. W. Geis, and S. H. Groves and assistance from E. L. Mastromattei. We are grateful to Prof. H. I. Smith for helpful criticisms.

REFERENCES

1. H. W. Lam, R. F. Pinizzotto, and A. F. Tasch, Jr., J. Electrochem. Soc. 128 1981 (1981); G. K. Celler, L. E. Trimble, K. K. Ng, H. J. Leamy, and H. Baumgart, Appl. Phys. Lett. 40 1043 (1982).
2. T. Inoue, K. Shibata, K. Kato, T. Yoshii, I. Higashinakagawa, K. Taniguchi, and M. Kashiwagi, Mat. Res. Soc. Symp. Proc. 23 523 (1984); J. A. Knapp and S. T. Picraux, Mat. Res. Soc. Symp. Proc. 23 533 (1984).
3. R. F. Pinizzotto, H. W. Lam, and B. L. Vaandrager, Appl. Phys. Lett. 40 388 (1982); J. C. C. Fan, B-Y. Tsaur, R. L. Chapman, and M. W. Geis, Appl. Phys. Lett. 41 186 (1982).
4. M. W. Geis, D. C. Flanders, and H. I. Smith, Appl. Phys. Lett. 35 71 (1979); M. W. Geis, D. A. Antoniadis, D. J. Silversmith, R. W. Mountain, and H. I. Smith, Appl. Phys. Lett. 37 454 (1980).
5. J. A. Roth, G. L. Olson, and L. D. Hess, Mat. Res. Soc. Symp. Proc. 23 431 (1984); H. Yamamoto, H. Ishawara, S. Furukawa, M. Tamura, and T. Tokuyama, Mat. Res. Soc. Symp. Proc. 25 511 (1984).
6. R. Reif and J. E. Knott, Electron. Lett. 17 586 (1981).
7. P. Kwizera and R. Reif, Appl. Phys. Lett. 41 379 (1982); P. Kwizera and R. Reif, Thin Solid Films 100 227 (1983).
8. R. B. Iverson and R. Reif, Mat. Res. Soc. Symp. Proc. 27 543 (1984).
9. T. I. Kamins, M. M. Mandurah, and K. C. Saraswat, J. Electrochem. Soc. 125 927 (1978).
10. Estimated using: (a) the channeling half angle $\psi_{1/2} = 6°$ for 103 keV Ra ions in <110> Si measured by C. Jech, Phys. Lett. 39A 417 (1972); and (b) the approximate relation $\psi_{1/2} \propto (Z/E)^{1/2}$ discussed in H. Grahmann, A. Feuerstein, and S. Kalbitzer, Rad. Effects 29 117 (1976).

Author Index

Subject Index